GW01164025

THE 16th CAMBRIDGE WORKSHOP ON COOL STARS, STELLAR SYSTEMS AND THE SUN

COVER ILLUSTRATION:

Logo from Cool Stars 16 Poster
L. Walkowicz, U. C. Berkeley

ASTRONOMICAL SOCIETY OF THE PACIFIC
CONFERENCE SERIES

A SERIES OF BOOKS ON RECENT DEVELOPMENTS IN ASTRONOMY AND ASTROPHYSICS

Volume 448

EDITORIAL STAFF

Managing Editor: Joseph Jensen
Associate Managing Editor: Jonathan Barnes
Publication Manager: Pepita Ridgeway
Editorial Assistant: Cindy Moody
LaTeX Consultant: T. J. Mahoney

MS 179, Utah Valley University, 800 W. University Parkway, Orem, Utah 84058-5999
Phone: 801-863-8804   E-mail: aspcs@aspbooks.org
E-book site: http://www.aspbooks.org

PUBLICATION COMMITTEE

Don McCarthy, Chair
The University of Arizona

Jill Bechtold
University of Arizona

Marsha J. Bishop
National Radio Astronomy Observatory

Gary J. Ferland
University of Kentucky

Scott J. Kenyon
Smithsonian Astrophysical Observatory

Doug Leonard
San Diego State University

Lynne Hillenbrand
California Institute of Technology

René Racine
Université de Montréal

Travis Rector
University of Alaska Anchorage

Ata Sarajedini
University of Florida

ASPCS volumes may be found online with color images at http://www.aspbooks.org.
ASP monographs may be found online at http://www.aspmonographs.org.

For a complete list of ASPCS volumes, ASP monographs, and
other ASP publications see http://www.astrosociety.org/pubs.html.

All book order and subscription inquiries should be directed to the ASP at
800-335-2626 (toll-free within the USA) or 415-337-2126,
or email service@astrosociety.org

ASTRONOMICAL SOCIETY OF THE PACIFIC
CONFERENCE SERIES

Volume 448

THE 16th CAMBRIDGE WORKSHOP ON COOL STARS, STELLAR SYSTEMS AND THE SUN

Edited by

Christopher M. Johns–Krull
Dept. of Physics & Astronomy, Rice University, Houston, TX 77005, USA

Matthew K. Browning
CITA, University of Toronto, Toronto, ON M5S3H8, Canada

Andrew A. West
Dept. of Astronomy, Boston University, Boston, MA 02215, USA

SAN FRANCISCO

ASTRONOMICAL SOCIETY OF THE PACIFIC
390 Ashton Avenue
San Francisco, California, 94112-1722, USA

Phone: 415-337-1100
Fax: 415-337-5205
E-mail: service@astrosociety.org
Web site: www.astrosociety.org
E-books: www.aspbooks.org

First Edition
© 2011 by Astronomical Society of the Pacific
ASP Conference Series
All rights reserved.

No part of the material protected by this copyright notice may be reproduced or utilized in any form or by any means—graphic, electronic, or mechanical, including photocopying, taping, recording, or by any information storage and retrieval system—without written permission from the Astronomical Society of the Pacific.

ISBN: 978-1-58381-776-6
e-book ISBN: 978-1-58381-777-3

Library of Congress (LOC) Cataloging in Publication (CIP) Data:
Main entry under title
Library of Congress Control Number (LCCN): 2011938620

Printed in the United States of America by Sheridan Books, Ann Arbor, Michigan.
This book is printed on acid-free paper.

Contents

Foreword xi
    S. L. Hawley

Participants xiii

Part I. Formation and Evolution

Mid-Infrared Variation in Young Stars 5
    L. M. Rebull (Invited Speaker)

Innovative Imaging of Young Stars: First Light ExPo Observations 15
    S. V. Jeffers, H. Canovas, C. U. Keller, M. Min, and M. Rodenhuis

Accretion Makes a Splash on TW Hydrae 23
    N. Brickhouse

Constraining the Evolution of Brown Dwarf Binarity as a Function of Age: A Keck LGS AO Search for Brown Dwarf and Planetary Mass Companions to Upper Scorpius Brown Dwarfs 31
    B. Biller, K. Allers, M. Liu, L. M. Close, and T. Dupuy

Extreme Coronal Mass Ejections in Young Low-Mass Stars 43
    A. N. Aarnio, K. G. Stassun, S. P. Matt, W. J. Hughes, and S. L. McGregor

Finding the Youngest Planets (Poster Finalist) 53
    C. J. Crockett, N. Mahmud, L. Prato, C. M. Johns-Krull, P. Hartigan, D. T. Jaffe, and C. A. Beichman

Revealing the Chamaeleon: First Detection of a Low-mass Stellar Halo Around the Young Open Cluster η Chamaeleontis (Poster Winner) 61
    S. J. Murphy, W. A. Lawson, and M. S. Bessell

HST/COS Spectra of DF Tau and V4046 Sgr: First Detection of Molecular Hydrogen Absorption Against The Lyman α Emission Line (Poster Finalist) 69
    H. Yang, J. L. Linsky, and K. France

Part II. Fundamental Parameters

Lithium Depletion in Solar Type stars: Lithium and Planet Presence 81
    S. G. Sousa, N. C. Santos, G. Israelian, E. Delgado Mena, J. Fernandes, M. Mayor, S. Udry, C. Domínguez Cerdeña, R. Rebolo, and S. Randich

Model Atmospheres From Very Low Mass Stars to Brown Dwarfs 91
    F. Allard, D. Homeier, and B. Freytag

Low-Mass Eclipsing Binaries: Observations vs. Theory 99
    J. C. Morales, I. Ribas (Invited Speaker), and C. Jordi

Testing Theory with Dynamical Masses and Orbits of Ultracool Binaries 111
    T. J. Dupuy, M. C. Liu, and M. J. Ireland

New Low-Mass Eclipsing Binaries from Kepler 121
    J. L. Coughlin, M. López-Morales, T. E. Harrison, N. Ule, and D. I. Hoffman

Dust and Chromospheres in M Dwarfs (Poster Winner) 131
    M. S. Bessell

Characterization of High-Energy Emissions of GKM Stars using Wide Binaries
with White Dwarfs (Poster Finalist) 139
    S. Catalán, A. Garcés, and I. Ribas

Rotational Velocities of Very Low Mass Binaries (Poster Finalist) 147
    Q. M. Konopacky, A. M. Ghez, B. A. Macintosh, R. J. White, T. S. Barman,
    E. L. Rice, and G. Hallinan

Convective Core Overshoot and Mass Loss in Classical Cepheids: A Solution to
the Mass Discrepancy? (Poster Finalist) 155
    H. R. Neilson, M. Cantiello, and N. Langer

Part III. Time Domain

Asteroseismology of Cool Dwarfs and Giants with Kepler 167
    R. L. Gilliland (Invited Speaker)

Starspots and Stellar Rotation: Stellar Activity with *Kepler* 177
    L. M. Walkowicz (Invited Speaker) and G. S. Basri

Weather at the L/T Transition: A Large *J*-Band Survey for Variability of Cool
Brown Dwarfs 187
    J. Radigan, R. Jayawardhana, D. Lafrenière, and É. Artigau

The Galactic M Dwarf Flare Rate 197
    E. J. Hilton, S. L. Hawley, A. F. Kowalski, and J. Holtzman

Observations of Late-Type Stars with the Infrared Spatial Interferometer 207
    E. Wishnow, C. Townes, V. Ravi, S. Lockwood, H. Mistry, W. Fitelson,
    W. Mallard, and D. Werthimer

A Search for Periodic Optical Variability in Radio Detected Ultracool Dwarfs: A
Consequence of a Magnetically-Driven Auroral Process? (Poster Finalist) . 219
    L. K. Harding, G. Hallinan, R. P. Boyle, R. F. Butler, B. Sheehan, and A. Golden

Part IV. Magnetic Fields and Magnetic Activity

Solar Energetic Events, the Solar-Stellar Connection, and Statistics of Extreme
Space Weather 231
    C.J. Schrijver (Invited Speaker)

New Insights into Stellar Magnetism from the Spectropolarimetry in All Four
Stokes Parameters 245
    *O. Kochukhov, F. Snik, N. Piskunov, S.V. Jeffers, C.U. Keller, V. Makaganiuk,
J.A. Valenti, C.M. Johns-Krull, M. Rodenhuis, and H.C. Stempels*

Magnetic Fields on Cool Stars 255
    A. Reiners (Invited Speaker)

The Age-Rotation-Activity Relation: From Myrs to Gyrs 269
    *K. R. Covey, M. A. Agüeros, J. J. Lemonias, N. M. Law, A. L. Kraus, and the
PTF Collaboration*

Global-scale Magnetism (and Cycles) in Dynamo Simulations of Stellar
Convection Zones (Poster Winner) 277
    B. P. Brown, M. K. Browning, A. S. Brun, M. S. Miesch, and J. Toomre

Magnetic Field Measurements on the Classical T Tauri Star
BP Tauri (Poster Finalist) 285
    W. Chen and C. M. Johns-Krull

The Mouse that Roared: A SuperFlare from the dMe Flare Star EV Lac Detected
by Swift and Konus-Wind (Poster Finalist) 293
    *R. A. Osten, O. Godet, S. Drake, J. Tueller, J. Cummings, H. Krimm, J. Pye,
V. Pal'shin, S. Golenetskii, F. Reale, S. R. Oates, M. J. Page, and A. Melandri*

Observations of X-ray Flares in G-K Dwarfs by XMM-Newton (Poster Finalist) . 301
    J. C. Pandey and K. P. Singh

The Activity and Rotation Limit in the Hyades (Poster Finalist) 313
    U. Seemann, A. Reiners, A. Seifahrt, and M. Kürster

Part V. Galactic Context

New Surveys for Brown Dwarfs and Their Impact on the IMF 323
    J. D. Kirkpatrick (Invited Speaker)

L Dwarf Kinematics 333
    S. J. Schmidt, A. A. West, and S. L. Hawley

A Very Cool, Very Nearby Brown Dwarf Hiding in the Galactic Plane 339
    *P. W. Lucas, C. G. Tinney, B. Burningham, S. K. Leggett, D. J. Pinfield,
R. Smart, H. R. A. Jones, F. Marocco, R. J. Barber, S. N. Yurchenko,
J. Tennyson, M. Ishii, M. Tamura, A. C. Day-Jones, A. Adamson, F. Allard, and
D. Homeier*

Low–Mass Stars in the Sloan Digital Sky Survey: Galactic Structure,
Kinematics, and the Luminosity Function 347
    J J. Bochanski (Invited Speaker)

Contents

Looking for Systematic Variations in the Stellar Initial Mass Function 361
    N. Bastian, K. R. Covey, and M. R. Meyer

Implications of Radial Migration for Stellar Population Studies 371
    R. Roškar, V. P. Debattista, S. R. Loebman, Ž. Ivezić, and T. R. Quinn

Searching for Ultra-cool Objects at the Limits of Large-scale Surveys (Poster Finalist) 379
    D. J. Pinfield, K. Patel, Z. Zhang, J. Gomes, B. Burningham, A. C. Day-Jones, and J. Jenkins

Part VI. Splinter Sessions

Habitability of Planets Orbiting Cool Stars 391
    R. Barnes, V. S. Meadows, S. D. Domagal-Goldman, R. Heller, B. Jackson, M. López-Morales, A. Tanner, N. Gómez-Pérez, and T. Ruedas

Aspects of Multi-Dimensional Modelling of Substellar Atmospheres 403
    C. Helling, E. Pedretti, S. Berdyugina, A. A. Vidotto, B. Beeck, E. Baron, A. P. Showman, E. Agol, and D. Homeier

Young Stars in the Time Domain: A CS16 Splinter Summary 415
    K. R. Covey, P. Plavchan, F. Bastien, E. Flaccomio, K. Flaherty, S. Marsden, M. Morales-Calderón, J. Muzerolle, and N. J. Turner

Ultracool Dwarf Science from Widefield Multi-Epoch Surveys 429
    N. R. Deacon, D. J. Pinfield, P. W. Lucas, M. C. Liu, M. S. Bessell, B. Burningham, M. C. Cushing, A. C. Day-Jones, S. Dhital, N. M. Law, A. K. Mainzer, and Z. H. Zhang

Splinter Session "Solar and Stellar Flares" 441
    L. Fletcher, H. Hudson, G. Cauzzi, K. V. Getman, M. Giampapa, S. L. Hawley, P. Heinzel, C. Johnstone, A. F. Kowalski, R. A. Osten, and J. Pye

The Radio–X-ray Relation in Cool Stars: Are We Headed Toward a Divorce? . . 455
    J. Forbrich, S. J. Wolk, M. Güdel, A. Benz, R. Osten, J. L. Linsky, M. McLean, L. Loinard, and E. Berger

Planet Formation Around M-dwarf Stars: From Young Disks to Planets 469
    I. Pascucci, G. Laughlin, B. S. Gaudi, G. Kennedy, K. Luhman, S. Mohanty, J. Birkby, B. Ercolano, P. Plavchan, and A. Skemer

Juvenile Ultracool Dwarfs 481
    E. L. Rice, J. K. Faherty, K. Cruz, T. Barman, D. Looper, L. Malo, E. E. Mamajek, S. Metchev, and E. L. Shkolnik

Frontiers in X-ray Astronomy - CS16 Splinter Session 493
    J. Robrade, P. C. Schneider and K. Poppenhaeger

The M4 Transition: Toward a Comprehensive Understanding of the Transition into the Fully Convective Regime 505
    K. G. Stassun, L. Hebb, K. Covey, A. A. West, J. Irwin, R. Jackson, M. Jardine, J. Morin, D. Mullan, and I. N. Reid

Fundamental Stellar Properties from Optical Interferometry 517
    G. T. van Belle, J. Aufdenberg, T. Boyajian, G. Harper, C. Hummel, E. Pedretti,
    E. Baines, R. White, V. Ravi, and S. Ridgway

Determining the Metallicity of Low-Mass Stars and Brown Dwarfs: Tools for
    Probing Fundamental Stellar Astrophysics, Tracing Chemical Evolution
    of the Milky Way and Identifying the Hosts of Extrasolar Planets 531
    A. A. West, J. J. Bochanski, B. P. Bowler, A. Dotter, J. A. Johnson, S. Lépine,
    B. Rojas-Ayala, and A. Schweitzer

Author Index 545

Foreword

The 16th Cambridge Workshop on Cool Stars, Stellar Systems and the Sun (Cool Stars 16) was held on the University of Washington campus in Seattle, Washington during August 28 - September 2, 2010. By all accounts it was a great success, with nearly 350 attendees, a five day poster session, twelve splinter sessions, eighteen contributed talks and twelve invited talks. Excellent refreshments, local excursions, and a lively boat trip/banquet enriched the social interaction. Even the weather cooperated for the most part, with only the planned hike getting cancelled due to infamous Seattle rain.

The five major science themes of the meeting covered the usual topics: formation and evolution, fundamental parameters, magnetic fields and activity; together with two timely new topics: time domain (especially focussing on Kepler results) and galactic context (particularly contributions from the Sloan Digital Sky Survey). The splinter sessions filled in the details in many of the new, exciting areas of cool stars science, ranging from interferometry to planets around cool stars to flares and Xrays, and from detailed modelling to metallicities and radii to wide field surveys to young stars and brown dwarfs.

As usual, many people contributed countless hours to make the meeting a success. Chief among these were the chair of the Local Organizing Committee, Sarah Garner, and the rest of the LOC: John Bochanski, Kevin Covey, James Davenport, Suzanne Hawley, Eric Hilton, Adam Kowalski, Victoria Meadows, Sarah Schmidt, and John Wisniewski. We were also fortunate to have the able assistance of Trish Dobson and Ruth Paglierani from the Conference Connection. Various members of the Scientific Organizing Committee toiled over the science program, chaired the invited sessions, served as liaisons to the splinter sessions, and carried out the poster judging. The SOC was chaired by Suzanne Hawley and the members were: Tom Ayres, Isabelle Baraffe, Matt Browning, Adam Burgasser, Andrea Dupree, Fabio Favata, Mark Giampapa, Manuel Guedel, Moira Jardine, Hugh Jones, Elizabeth Lada, Mihalis Mathioudakis, Rachel Osten, Nikolai Piskunov, Chris Johns-Krull, Neill Reid, Jurgen Schmitt, John Stauffer, Beate Stelzer, Jeff Valenti, Andrew West, and Lucianne Walkowicz. Special thanks go to the SOC members who are editors of these proceedings: Chris Johns-Krull, Matt Browning and Andrew West.

The poster judging resulted in 15 poster finalists, whose papers appear in the print copy of the proceedings. From among the finalists, the three prize winners were awarded a free copy of the proceedings and an Amazon gift card. Congratulations to Ben Brown (first place), Simon Murphy (second place), and Mike Bessell (third place).

We thank the following organizations for financial assistance: The National Science Foundation, the NASA Astrobiology Institute, the European Space Agency, the Astrophysical Research Consortium/Apache Point Observatory, the University of Washington College of Arts and Sciences and the University of Washington Department of Astronomy.

And finally, the SOC discussed several worthy proposals for Cool Stars 17. We look forward to the next meeting in this series, to be hosted in Barcelona, Spain during June, 2012.

Suzanne Hawley, for the CS16 SOC
University of Washington, Seattle, Washington, USA, February, 2011

Cool Stars 16 Scientific Organizing Committee (Top) and Local Organizing Committee (Bottom)

Participants

E. Agol, University of Washington, ⟨agol@astro.washington.edu⟩
R. Akeson, NExScI/Caltech, ⟨rla@ipac.caltec.edu⟩
F. Allard, CRAL-ENS, ⟨fallard@ens-lyon.fr⟩
P. Allen, Franklin and Marshall College, ⟨peter.allen@fandm.edu⟩
M. Ammler-von Eiff, IfA, University of Goettingen, ⟨mammler@uni-goettingen.de⟩
J. Andersen, Boston University, ⟨janmaire@bu.edu⟩
R. Anderson, Geneva Observatory, ⟨richard.anderson@unige.ch⟩
K. Apps, Sussex University, ⟨appssol3@hotmail.com⟩
E. Artigau, Université de Montréal, ⟨artigau@Astro.umontreal.ca⟩
M. Asplund, Max Planck Institute for Astrophysics, ⟨asplund@mpa-garching.mpg.de⟩
J. Aufdenberg, Embry-Riddle Aeronautical University, ⟨aufded93@erau.edu⟩
T. Ayres, CASA, University of Colorado, ⟨Thomas.Ayres@Colorado.edu⟩
E. Baines, Naval Research Laboratory, ⟨ellyn.baines.ctr@nrl.navy.mil⟩
D. Baker, University of Leicester, ⟨deab1@star.le.ac.uk⟩
N. Baliber, Caltech, ⟨baliber@lcogt.net⟩
I. Baraffe, University of Exeter, ⟨burdett@astro.ex.ac.uk⟩
B. Barber, University College London, ⟨rjb@star.ucl.ac.uk⟩
T. Barman, Lowell Observatory, ⟨barman@lowell.edu⟩
R. Barnes, University of Washington, ⟨rory@astro.washington.edu⟩
E. Baron, University of Oaklahoma, ⟨baron@ou.edu⟩
M. Barsony, Space Science Institute, ⟨fun@alumni.caltech.edu⟩
N. Bastian, University of Exeter, ⟨bastian@astro.ex.ac.uk⟩
F. Bastien, Fisk University, ⟨fabienne.bastien@gmail.com⟩
A. Becker, University of Washington, ⟨acbecker@gmail.com⟩
B. Beeck, IfA, Goettingen University, ⟨beeck@astro.physik.uni-goettingen.de⟩
P. Beiersdorfer, Space Sciences Laboratory, UC, Berkeley, ⟨beiersdorfer@ssl.berkeley.edu⟩
C. Bell, University of Exeter, ⟨bell@astro.ex.ac.uk⟩
A. Benz, ETH, Zurich, ⟨benz@astro.phys.ethz.ch⟩
S. Berdyugina, Kiepenheuer-Institut fuer Sonnenphysik, ⟨secr@kis.uni-freiburg.de⟩
A. Berndt, AIU, Jena, ⟨alex@astro.uni-jena.de⟩
C. Bertout, Institut d'Astrophysique & Observatoire de Paris, ⟨claude.bertout@obspm.fr⟩
M. Bessell, RSAA, ⟨bessell@mso.anu.edu.au⟩
W. Bhatti, Johns Hopkins University, ⟨waqas@pha.jhu.edu⟩

Participants

B. Biller, IfA, ⟨bbiller@ifa.hawaii.edu⟩
J. Birkby, IOA, Cambridge, ⟨jlb@ast.cam.ac.uk⟩
C. Blake, Princeton, ⟨cblake@astro.princeton.edu⟩
J. Bochanski, MIT, ⟨jjb@mit.edu⟩
P. Boeshaar, UC, Davis, ⟨boeshaar@physics.ucdavis.edu⟩
E. Bohm-Vitense, University of Washington, ⟨office@astro.washington.edu⟩
R. Bonito, UNIPA-INAF-OAPA, ⟨sbonito@astropa.unipa.it⟩
S. Boudreault, Mullard Space Science Laboratory, ⟨szb@mssl.ucl.ac.uk⟩
B. Bowler, University of Hawaii, ⟨bpbowler@ifa.hawaii.edu⟩
T. Boyajian, Georgia State University, ⟨tabetha@chara.gsu.edu⟩
I. Brandão, CAUP, ⟨isa@astro.up.pt⟩
attendeeN. Brickhousenbrickhouse@cfa.harvard.eduSAO
S. Brittain, Clemson University, ⟨sbritt@clemson.edu⟩
G. Bromage, University of Central Lancanshire, ⟨jmconnell1@uclan.ac.uk⟩
B. Brown, University of Wisconsin, ⟨bpbrown@astro.wisc.edu⟩
A. Brown, CASA, University of Colorado, ⟨Alexander.Brown@colorado.edu⟩
J. Brown, MPE, ⟨jbrown@mpe.mpg.de⟩
M. Browning, CITA, ⟨mattorb@gmail.com⟩
E. Bubar, University of Rochester, ⟨eric.bubar@rochester.edu⟩
A. Burgasser, UC, San Diego, ⟨aburgasser@ucsd.edu⟩
B. Burningham, University of Hertfordshire, ⟨b.burningham@herts.ac.uk⟩
J. Caballero, Centro de Astrobiologia Madrid, ⟨caballero@astrax.fis.ucm.es⟩
P. Cargile, Vanderbilt University, ⟨p.cargile@vanderbilt.edu⟩
K. Carpenter, NASA GSFC, ⟨Kenneth.G.Carpenter@nasa.gov⟩
B. Carter, University of Southern Queensland, ⟨carterb@usq.edu.au⟩
L. Casagrande, Max Planck Institute for Astrophysics,
⟨luca@mpa-garching.mpg.de⟩
S. Catalan, University of Hertfordshire, ⟨s.catalan@herts.ac.uk⟩
G. Cauzzi, INAF, Osservatorio di Arcetri, ⟨gcauzzi@arcetri.astro.it⟩
W. Chen, Rice University, ⟨wc2@rice.edu⟩
A. Cody, Caltech, ⟨amc@astro.caltech.edu⟩
O. Cohen, SAO, ⟨ocohen@cfs.harvard.edu⟩
R. Collet, Max Planck Institute for Astrophysics, ⟨remo@mpa-garching.mpg.de⟩
J. Coughlin, New Mexico State University, ⟨jlcough@nmsu.edu⟩
K. Covey, Cornell University, ⟨kcovey@astro.cornell.edu⟩
C. Crockett, Lowell Observatory, ⟨crockett@lowell.edu⟩
K. Cruz, American Museum of Natural History, ⟨kellecruz@gmail.com⟩

M. Cushing, NASA/JPL, ⟨michael.cushing@gmail.com⟩
S. Daemgen, ESO, ⟨sdaemgen@eso.org⟩
T. Dall, ESO, ⟨tdall@eso.org⟩
J. Datson, Tuorla Observatory, University of Turku, ⟨juclda@utu.fi⟩
J. Davenport, University of Washington, ⟨jrad@astro.washington.edu⟩
C. Davison, Georgia State University, ⟨davison@chara.gsu.edu⟩
A. Day-Jones, Universidad de Chile, ⟨adjones@das.uchile.cl⟩
N. De Lee, University of Florida, ⟨ndelee@astro.ufl.edu⟩
N. Deacon, IfA, University of Hawaii, ⟨ndeacon@ifa.hawaii.edu⟩
C. Deen, The University of Texas at Austin, ⟨deen@astro.as.utexas.edu⟩
P. Delorme, University of St. Andrews, ⟨pd10@st-andrews.ac.uk⟩
P. DeStefano, , ⟨paul.destefano@willamettealumni.com⟩
S. Dhital, Vanderbilt University, ⟨saurav.dhital@vanderbilt.edu⟩
S. Dieterich, Georgia State University, ⟨dieterich@chara.gsu.edu⟩
S. Domagal-Goldman, University of Washington, ⟨shawn.goldman@gmail.com⟩
A. Dotter, STScI, ⟨hill@stsci.edu⟩
G. Doyle, Armagh Observatory, ⟨jgd@arm.ac.uk⟩
A. Dupree, CfA, ⟨dupree@cfa.harvard.edu⟩
T. Dupuy, University of Hawaii, ⟨tdupuy@ifa.hawaii.edu⟩
E. Dzifcakova, Astronomical Institute Ondrejov, ⟨elena@asu.cas.cz⟩
C. Epstein, The Ohio State University, ⟨epstein@astronomy.ohio-state.edu⟩
B. Ercolano, University of Exeter, ⟨barbara@astro.ex.ac.uk⟩
B. Espey, Trinity College, Dublin, ⟨Brian.Espey@tcd.ie⟩
J. Faherty, Stony Brook/American Museum, ⟨jfaherty@amnh.org⟩
M. Fanelli, NASA AMES Research Center, ⟨michael.n.fanelli@nasa.gov⟩
E. Flaccomio, INAF, Osservatorio di Palermo, ⟨ettoref@astropa.unipa.it⟩
K. Flaherty, University of Arizona, ⟨tinnin@as.arizona.edu⟩
S. Fleming, University of Florida, ⟨scfleming@astro.ufl.edu⟩
L. Fletcher, University of Glasgow, ⟨lyndsay@astro.gla.ac.uk⟩
J. Forbrich, CfA, ⟨jforbrich@cfa.harvard.edu⟩
B. Freytag, CRAL, ⟨Bernd.Freytag@ens-lyon.fr⟩
D. Fügner, AIP, ⟨dfuegner@aip.de⟩
J. Gagne, Université de Montréal, ⟨jonathan.gagne@umontreal.ca⟩
J. Gallardo, Universidad de Chile, ⟨gallardo@das.uchile.cl⟩
D. Garcia-Alvarez, IAC/GRANTECAN, ⟨david.garcia@gtc.iac.es⟩
S. Gaudi, The Ohio State University, ⟨gaudi@astronomy.ohio-state.edu⟩
D. Geisler, Universidad de Concepcion, ⟨dgeisler@astro-udec.cl⟩

K. GEISSLER, SUNY Stony Brook, ⟨geissler@astro.sunysb.edu⟩
C. GELINO, IPAC/Caltech, ⟨cgelino@ipac.caltech.edu⟩
K. GETMAN, Pennsylvania State University, ⟨gkosta@astro.psu.edu⟩
M. GIAMPAPA, NSO, ⟨giampapa@noao.edu⟩
R. GILLILAND, STScI, ⟨gillil@stsci.edu⟩
J. GOMES DA SILVA, Centro de Astrofisica da Universidade do Porto, ⟨jntgds@gmail.com⟩
T. GOMEZ, University of Washington, ⟨gomezt@u.washington.edu⟩
P. GONDOIN, ESA, ⟨pgondoin@rssd.esa.in¿⟩
C. GRADY, Eureka Scientific, ⟨Carol.A.Grady@nasa.gov⟩
E. GRIFFIN, Herzberg Institute of Astrophysics, ⟨Elizabeth.Griffin@nrc.gc.ca⟩
S. GU, Yunnan Observatory, ⟨shenghonggu@ynao.ac.cn⟩
M. GUEDEL, University of Vienna, ⟨manuel.guedel@univie.ac.at⟩
M. GULLY-SANTIAGO, The University of Texas at Austin, ⟨gully@astro.as.utexas.edu⟩
H. GÜNTHER, CfA, ⟨hguenther@cfa.harvard.edu⟩
K. HAISCH, Utah Valley University, ⟨Karl.Haisch@uvu.edu⟩
G. HALLINAN, NRAO & UC, Berkeley, ⟨gregg@astro.berkeley.edu⟩
C. HAMILTON, Dickinson College, ⟨hamiltoc@dickinson.edu⟩
R. HAMILTON, New Mexico State University, ⟨rthamilt@nmsu.edu⟩
L. HARDING, National University of Ireland, Galway, ⟨harding.leon@gmail.com⟩
G. HARPER, Trinity College, Dublin, ⟨graham.harper@tcd.ie⟩
J. HARTMAN, CfA, ⟨jhartman@cfa.harvard.edu⟩
P. HARVEY, The University of Texas at Austin, ⟨pmh@astro.as.utexas.edu⟩
M. HASTIE, MMT Observatory, ⟨mhastie@mmto.org⟩
S. HAWLEY, University of Washington, ⟨slh@astro.washington.edu⟩
S. HEAP, NASA GSFC, ⟨sara.heap@gmail.com⟩
L. HEBB, Vanderbilt University, ⟨leslie.hebb@vanderbilt.edu⟩
P. HEINZEL, Academy of Sciences, ⟨pheinzel@asu.cas.cz⟩
R. HELLER, Hamburger Sternwarte, ⟨rheller@hs.uni-hamburg.de⟩
C. HELLING, University of St. Andrews, ⟨ch80@st-and.ac.uk⟩
E. HILTON, University of Washington, ⟨hilton@astro.washington.edu⟩
D. HOMEIER, IfA, Goettingen University, ⟨derek@astro.physik.uni-goettingen.de⟩
J. HORNBECK, University of Louisville, ⟨jbhorn02@louisville.edu⟩
S. HOYER, Universidad de Chile, ⟨shoyer@das.uchile.cl⟩
W. HUANG, University of Washington, ⟨hwenjin@astro.washington.edu⟩
H. HUDSON, SSL, UC, Berkeley, ⟨hhudson@ssl.berkeley.edu⟩
J. HUGHES, Seattle University, ⟨jhughes@seattleu.edu⟩

C. HUMMEL, ESO, ⟨chummel@eso.org⟩
L. INGLEBY, University of Michigan, ⟨lingleby@umich.edu⟩
J. IRWIN, CfA, ⟨jirwin@cfa.harvard.edu⟩
R. JACKSON, Keele University, ⟨r.j.jackson@epsam.keele.ac.uk⟩
B. JACKSON, NASA GSFC, ⟨decaelus@gmail.com⟩
W.-C. JAO, Georgia State University, ⟨jao@chara.gsu.edu⟩
M. JARDINE, University of St. Andrews, ⟨mmj@st-andrews.ac.uk⟩
V. JATENCO-PEREIRA, Universidade de São Paulo, ⟨jatenco@astro.iag.usp.br⟩
S. JEFFERS, University of Utrecht, ⟨s.v.jeffers@uu.nl⟩
R. JEFFRIES, Keele University, ⟨rdj@astro.keele.ac.uk⟩
J. JENKINS, Universidad de Chile, ⟨jjenkins@das.uchile.cl⟩
E. JENSEN, Swarthmore College, ⟨ejensen1@swarthmore.edu⟩
C. JOHNS-KRULL, Rice University, ⟨cmj@rice.edu⟩
J. JOHNSON, Caltech, ⟨johnjohn@astro.caltech.edu⟩
C. JOHNSTONE, University of St. Andrews, ⟨cpj2@st-andrews.ac.uk⟩
H. JONES, University of Hertfordshire, ⟨h.r.a.jones@herts.ac.uk⟩
N. JOSHI, Goettingen University, ⟨nandan@astro.physik.uni-goettingen.de⟩
E. JOSSELIN, GRAAL - Universite Montpellier II, ⟨eric.josselin@univ-montp2.fr⟩
S. KAFKA, DTM/CIW, ⟨skafka@dtm.ciw.edu⟩
K. KARKARE, Caltech, ⟨kkarkare@caltech.edu⟩
J. KASTNER, Rochester Institute of Technology, ⟨jhk@cis.rit.edu⟩
G. KENNEDY, Institute of Astronomy, University of Cambridge, ⟨gkennedy@ast.cam.ac.uk⟩
K. KINEMUCHI, NASA AMES Research Center, ⟨kkinemuchi@gmail.com⟩
R. KING, University of Exeter, ⟨rob@astro.ex.ac.uk⟩
J. KIRKPATRICK, Caltech, ⟨davy@ipac.caltech.edu⟩
A. KLUTSCH, Universidad Complutense de Madrid, ⟨klutsch@astrax.fis.ucm.es⟩
O. KOCHUKHOV, Uppsala University, ⟨oleg.kochukhov@fysast.uu.se⟩
Q. KONOPACKY, LLNL, ⟨konopacky1@llnl.gov⟩
R. KONSTANTINOVA-ANTOVA, Institute of Astronomy BAS, ⟨renada@astro.bas.bg⟩
H. KORHONEN, ESO, ⟨hkorhone@eso.org⟩
A. KOWALSKI, University of Washington, ⟨adamfk@u.washington.edu⟩
A. KRAUS, IfA, University of Hawaii, ⟨alk@ifa.hawaii.edu⟩
D. LAFLAMME, Université de Montréal, ⟨denise.laflamme@umontreal.ca⟩
J. LAMBERT, GRAAL - Universite Montpellier II, ⟨julien.lambert@univ-montp2.fr⟩
N. LAW, University of Toronto, ⟨law@di.utoronto.ca⟩
A. LEBRE, GRAAL-University of Montpellier, ⟨agnes.lebre@univ-montp2.fr⟩

S. Leggett, Gemini Observatory, ⟨sleggett@gemini.edu⟩
J. Lemonias, Columbia University, ⟨jenna.lemonias@gmail.com⟩
L. Lenz, IfA, Goettingen University, ⟨lealenz@astro.physik.uni-goettingen.de⟩
S. Lepine, American Museum of Natural History, ⟨lepine@amnh.org⟩
J. Lepson, UC, Berkeley/SSL, ⟨lepson@ssl.berkeley.edu⟩
J. Lim, University of Hong Kong, ⟨jjlim@hku.hk⟩
J. Linsky, JILA/University of Colorado, ⟨jlinsky@jilau1.colorado.edu⟩
N. Lodieu, IAC Tenerife, ⟨nlodieu@iac.es⟩
S. Loebman, University of Washington, ⟨sloebman@yahoo.com⟩
L. Loinard, CRyA-UNAM, ⟨l.loinard@crya.unam.mx⟩
D. Looper, Institute for Astronomy, ⟨dagny@ifa.hawaii.edu⟩
M. Lopez Garcia, Universidad Complutense de Madrid, ⟨mal@astrax.fis.ucm.es⟩
M. Lopez-Morales, CIW-DTM / CSIC-IEEC, ⟨mercedes@dtm.ciw.edu⟩
P. Lucas, University of Hertfordshire, ⟨d.j.pinfield@herts.ac.uk⟩
K. Luhman, Penn State University, ⟨kluhman@astro.psu.edu⟩
J. Lutz, University of Washington, ⟨jlutz@astro.washington.edu⟩
G. Mace, UC, Los Angeles, ⟨gmace@astro.ucla.edu⟩
N. Mahmud, Rice University, ⟨naved@rice.edu⟩
L. Malo, University of Montreal, ⟨malo@astro.umontreal.ca⟩
E. Mamajek, University of Rochester, ⟨emamajek@pas.rochester.edu⟩
S. Marsden, James Cook University, ⟨marsden@usq.edu.au⟩
Y. Martinez Osorio, Uppsala University, ⟨yeisson.osorio@fysast.uu.se⟩
M. McLean, Harvard University, ⟨mmclean@cfa.harvard.edu⟩
V. Meadows, UW, ⟨youngk@uw.edu⟩
S. Metchev, Stony Brook University, ⟨metchev@astro.sunysb.edu⟩
M. Mittag, Hamburger Sternwarte, ⟨mmittag@hs.uni-hamburg.de⟩
S. Mohanty, Imperial College London, ⟨s.mohanty@imperial.ac.uk⟩
D. Montes, UCM. Universidad Complutense de Madrid, ⟨dmg@astrax.fis.ucm.es⟩
M. Morales-Calderon, Caltech, ⟨mariamc@ipac.caltech.edu⟩
D. Morgan, Boston University, ⟨dpmorg@bu.edu⟩
J. Morin, Dublin Institute for Advanced Studies, ⟨morin@cp.dias.ie⟩
P. Muirhead, Cornell University, ⟨muirhead@astro.cornell.edu⟩
S. Mukherjee, Jawaharial Nehru University, ⟨saumitramukherjee3@gmail.com⟩
D. Mullan, Univ of Delaware, ⟨mullan@udel.edu⟩
S. Murphy, Australian National Unversity, ⟨murphysj@mso.anu.edu.au⟩
D. Murray, University of Hertfordshire, ⟨j.h.hough@herts.ac.uk⟩
J. Muzerolle, Space Telescope Science Institute, ⟨muzerol@stsci.edu⟩

Participants xix

K. Muzic, University of Toronto DAA, ⟨muzic@astro.utoronto.ca⟩
J. Neff, College of Charleston, ⟨neffj@cofc.edu⟩
H. Neilson, University of Bonn, ⟨hneilson@astro.uni-bonn.de⟩
V. Neves, Centro de Astrofísica da Universidade do Porto, ⟨vasco.neves@astro.ua.pt⟩
D. Nguyen, University of Florida, ⟨dcnguyen@astro.ufl.edu⟩
E. O'Gorman, Trinity College Dublin Ireland, ⟨eamon.ogorman@gmail.com⟩
D. O'Neal, Keystone College, ⟨douglas.oneal@keystone.edu⟩
A. Önehag, Uppsala University, ⟨annao@fysast.uu.se⟩
M. Opher, George Mason University, ⟨mopher@gmu.edu⟩
R. Osten, STScI, ⟨osten@stsci.edu⟩
A. Palacios, GRAAL-CNRS/ Montpellier II University, ⟨palacios@graal.univ-montp2.fr⟩
J. Pandey, Aryabhatta Res. Inst. Of Obs. Sciences, Nainital, ⟨jeewan@aries.res.in⟩
S. Parker, University of New South Wales, ⟨sparker@phys.unsw.edu.au⟩
J. Parks, Georgia State University, ⟨parksj@chara.gsu.edu⟩
I. Pascucci, STScI, ⟨pascucci@stsci.edu⟩
J. Patience, University of Exeter, ⟨patience@astro.ex.ac.uk⟩
M. Pecaut, University of Rochester, ⟨mpecaut@pas.rochester.edu⟩
E. Pedretti, SUPA University of St. Andrews, ⟨ep41@st-andrews.ac.uk⟩
P. Petit, Université de Toulouse, ⟨petit@ast.obs-mip.fr⟩
M. Petr-Gotzens, ESO, ⟨mpetr@eso.org⟩
I. Pillitteri, SAO, ⟨ignazio.pillitteri@tiscali.it⟩
J. Pineda, MIT/Caltech, ⟨jspineda@alum.mit.edu⟩
D. Pinfield, University of Hertfordshire, ⟨D.J.Pinfield@herts.ac.uk⟩
M. Pinsonneault, The Ohio State University, ⟨pinsonneault.1@osu.edu⟩
N. Piskunov, Uppsala University, ⟨piskunov@fysast.uu.se⟩
P. Plavchan, NExScI/Caltech, ⟨plavchan@ipac.caltech.edu⟩
K. Poppenhaeger, Hamburg Observatory, ⟨katja.poppenhaeger@hs.uni-hamburg.de⟩
L. Porter, Max Planck for Astrophysics, ⟨laurap@mpa-garching.mpg.de⟩
J. Pye, University of Leicester, ⟨pye@star.le.ac.uk⟩
A. Quirrenbach, Landessternwarte Heidelberg, ⟨A.Quirrenbach@lsw.uni-heidelberg.de⟩
J. Radigan, University of Toronto, ⟨radigan@astro.utoronto.ca⟩
V. Ravi, UC, Berkeley, ⟨ehwnw@lmi.net⟩
L. Rebull, SSC/JPL/Caltech, ⟨rebull@ipac.caltech.edu⟩
N. Reid, STScI, ⟨inr@stsci.edu⟩
A. Reiners, IfA, Goettingen University, ⟨Ansgar.Reiners@phys.uni-goettingen.de⟩

Participants

T. Reinhold, Goettingen University, ⟨ reinhold@astro.physik.uni-goettingen.de ⟩
C. Reyle, Besancon Observatory, ⟨ celine@obs-besancon.fr ⟩
B. Riaz, STScI, ⟨ basmah@stsci.edu ⟩
E. Rice, American Museum of Natural History, ⟨ erice@amnh.org ⟩
S. Ridgeway, NOAO, ⟨ ridgway@noao.edu ⟩
A. Riedel, Georgia State University, ⟨ riedel@chara.gsu.edu ⟩
I. Ribas, CSIC-IEEC, Spain, ⟨ iribas@ice.csic.es ⟩
J. Robert, Université de Montréal, ⟨ jasmin@astro.umontreal.ca ⟩
J. Robrade, Hamburger Sternwarte, ⟨ jrobrade@hs.uni-hamburg.de ⟩
J. Roche, Trinity College, Dublin, ⟨ joroche@tcd.ie ⟩
M. Rodriguez-Ledesma, MPIA, ⟨ ledesma@mpia.de ⟩
B. Rojas-Ayala, Cornell University, ⟨ babs@astro.cornell.edu ⟩
R. Roskar, University of Washington, ⟨ roskar@astro.washington.edu ⟩
S. Saar, SAO, ⟨ saar@cfa.harvard.edu ⟩
J. Sanz Forcada, Centro de Astrobiologia, ⟨ jsanz@cab.inta-csic.es ⟩
J. Schlieder, Stony Brook University, ⟨ jschlied@ic.sunysb.edu ⟩
S. Schmidt, University of Washington, ⟨ sjschmidt@astro.washington.edu ⟩
T. Schmidt, Astrophysical Institute (AIU) Jena, ⟨ tobi@astro.uni-jena.de ⟩
C. Schneider, Hamburger Sternwarte, ⟨ christian.schneider@hs.uni-hamburg.de ⟩
K. Schrijver, Lockheed Martin Advanced Technology Center, ⟨ schrijver@lmsal.com ⟩
A. Schweitzer, Hamburger Sternwarte, ⟨ aschweitzer@hs.uni-hamburg.de ⟩
U. Seemann, ESO, ⟨ useemann@eso.org ⟩
A. Seifahrt, UC, Davis, ⟨ seifahrt@physics.ucdavis.edu ⟩
C. Sennhauser, ETH Zurich, ⟨ csennhau@astro.phys.ethz.ch ⟩
W. Sherry, NSO/NOAO, ⟨ wsherry@noao.edu ⟩
E. Shkolnik, Carnegie Institution of Washington, ⟨ shkolnik@dtm.ciw.edu ⟩
A. Showman, University of Arizona, ⟨ showman@lpl.arizona.edu ⟩
H. Singh, University of Delhi, ⟨ singh@iucaa.ernet.in ⟩
B. Sitarski, UC, San Diego, ⟨ bsitarski@ucla.edu ⟩
M. Skelly, Laboratoire Astrophysique de Toulouse Tarbes, ⟨ mskelly@ast.obs-mip.fr ⟩
A. Skemer, University of Arizona, ⟨ askemer@as.arizona.edu ⟩
S. Skinner, University of Colorado, ⟨ stephen.skinner@colorado.edu ⟩
D. Soderblom, STScI, ⟨ drs@stsci.edu ⟩
S. Sorahana, The University of Tokyo ISAS/JAXA, ⟨ sorahana@ir.isas.jaxa.jp ⟩
S. Sousa, CAUP Portugal, ⟨ sousasag@asteo.up.pt ⟩

K. Stassun, Vanderbilt University, ⟨ keivan.stassun@vanderbilt.edu ⟩
J. Stauffer, Caltech, ⟨ stauffer@ipac.caltech.edu ⟩
B. Stelzer, INAF – Osservatorio Astronomico di Palermo, ⟨ stelzer@astropa.unipa.it ⟩
E. Stempels, Uppsala University, ⟨ Eric.Stempels@fysast.uu.se ⟩
D. Stooksbury, University of Georgia, ⟨ stooks@engr.uga.edu ⟩
M. Stumpf, Max-Planck-Institute for Astronomy, ⟨ stumpf@mpia.de ⟩
P. Szkody, University of Washington, ⟨ szkody@astro.washington.edu ⟩
H. Tabernero, Universidad Complutense de Madrid, ⟨ htg@astrax.fis.ucm.es ⟩
A. Tanner, Georgia State University, ⟨ angelle.tanner@gmail.com ⟩
R. Tata, Instituto de Astrofisica de Canarias, ⟨ rrtata@iac.es ⟩
P. Teixeira, ESO, ⟨ Pteixeir@eso.org ⟩
S. Terebey, California State University, Los Angeles, ⟨ sterebe@calstatela.edu ⟩
P. Thorman, UC, Davis, ⟨ thorman@physics.ucdavis.edu ⟩
C. Tinney, University of New South Wales, ⟨ c.tinney@unsw.edu.au ⟩
B. Tofflemire, University of Washington, ⟨ tofflb@u.washington.edu ⟩
N. Turner, JPL/Caltech, ⟨ neal.turner@jpl.nasa.gov ⟩
N. Ule, New Mexico State University, ⟨ nmule@nmsu.edu ⟩
J. Valenti, STScI, ⟨ valenti@stsci.edu ⟩
G. van Belle, ESO, ⟨ gvanbell@eso.org ⟩
M. Varady, Purkyne University, ⟨ mvarady@physics.ujep.cz ⟩
A. Vidotto, University of St. Andrews, ⟨ Aline.Vidotto@st-andrews.ac.uk ⟩
M. Vieytes, Instituto de Astronomía y Física del Espacio, ⟨ mariela@iafe.uba.ar ⟩
A. Wachter, Universidad de Guanajuato DA, ⟨ astrid@astro.ugto.mx ⟩
L. Walkowicz, UC, Berkeley, ⟨ lucianne@astro.berkeley.edu ⟩
G. Wallerstein, University of washington, ⟨ wall@astro.washington.edu ⟩
F. Walter, Stony Brook University, ⟨ fwalter@mail.astro.sunysb.edu ⟩
W. Wang, Max-Planck-Institute for Astronomy, ⟨ wwang@mpia.de ⟩
S. Wende, IfA, Goettingen University, ⟨ sewende@astro.physik.uni-goettingen.de ⟩
A. West, Boston University, ⟨ aawest@bu.edu ⟩
E. Whelan, Dublin Institute for Advanced Studies, ⟨ whelane@ujf-grenoble.fr ⟩
R. White, Georgia State University, ⟨ white@chara.gsu.edu ⟩
E. Wishnow, UC, Berkeley, ⟨ ehwnw@lmi.net ⟩
S. Wolk, CfA, ⟨ swolk@cfa.harvard.edu ⟩
U. Wolter, Hamburger Sternwarte, ⟨ uwolter@hs.uni-hamburg.de ⟩
B. Wood, Naval Research Laboratory, ⟨ brian.wood@nrl.navy.mil ⟩
H. Worters, SAAO, ⟨ hannah@saao.ac.za ⟩
N. Wright, CfA, ⟨ nwright@cfa.harvard.edu ⟩

I. YAMAMURA, Institute of Space and Astronautical Science JAXA, ⟨yamamura@ir.isas.jaxa.jp⟩

H. YANG, JILA University of Colorado, ⟨haoyang@colorado.edu⟩

Z. ZHANG, University of Hertfordshire, ⟨z.zhang7@herts.ac.uk⟩

The University of Washington campus where CS16 was held. Meeting participants are waiting for the bus to the banquet.

Part I
Formation and Evolution

Mid-Infrared Variation in Young Stars

L. M. Rebull

Spitzer Science Center, MS 220-6, 1200 E. California Blvd., Pasadena, CA, 91125, USA

Abstract. Since 2003, the Spitzer Space Telescope has provided groundbreaking views of Galactic star formation in bands from 3.6 past 24 microns. During the cryogenic mission (the first 5.5 years), variability of young stars at these bands was noted, although typically with just a few epochs of observation. The cryogen ran out in 2009, and we are now in the warm mission era where the shortest two bands (3.6 and 4.5 microns) continue to function essentially as before. The phenomenal sensitivity and stability of Spitzer at these bands has enabled several dedicated monitoring programs studying the variability of young stars at timescales from minutes to years. The largest of these programs is YSOVAR (Stauffer et al.), but there are several smaller programs as well. With at least as many as 2200 young star light curves likely to come out of this, these programs as a whole enable more detailed study of the young star-disk interaction in the infrared for a wider range of ages and masses than has ever been accomplished before. Early results suggest a wide variety of sources of variability, including dust clouds in the disk, disk warps, star spots, and accretion. This contribution will review some of the most recent results from these programs.

1. Overview of Young Stars

The general outline of the formation of low-mass stars has been widely accepted for at least 20 years (see, e.g., Bertout 1989). An initial molecular cloud collapses onto itself, forming an envelope and then a disk around a central mass; jets help regulate angular momentum in the early phases and perhaps the interaction of the magnetic field with the circumstellar disk regulates the angular momentum in later stages (e.g., Königl 1991, Shu et al. 2000).

Figure 1 shows the basic "anatomy" of a young stellar object (YSO), at ages of ~1-5 Myr, when there is still a substantial circumstellar disk but no envelope or jets. In this Figure, the circumstellar disk is flared at the outer edges, and the inner edge is truncated by the protostellar magnetic field. The completely convective young star is rotating quickly, and as such has a strong magnetic field. Accreting matter follows the field lines, and crashes onto the protostar near the magnetic poles. The active young star produces flares in X-rays, ultraviolet from the accretion shocks, emission lines from the accretion columns, and infrared from the disk itself. Near-infrared (NIR) emission originates closer into the central object than mid-infrared (MIR). Note that even in this simple picture, very few of these properties are likely to be constant even over relatively short time intervals; rotation, accretion, flares, and even inhomogeneities forming and dispersing in the disk are all highly dynamic processes.

Figure 1 also shows (on the right) a schematic, simple approximation for the relationship between peak emission from the disk and distance from the central protostar. Some protostars have disk emission starting at wavelengths as short as the NIR JHK, 1-2 μm; these disks likely are quite close in to the central object, on the order of $\sim 20 R_*$. However, in the MIR (such as the Spitzer Space Telescope bands at 3.6-8 μm), we sample disk properties much further out, from $\sim 30 R_*$ to $\sim 200 R_*$. In reality, this is a vast simplification, and heated inner disk walls and/or rims, system inclination, disk-photosphere contrast, and many other properties in addition to the temperature of the central object affect what location in the disk a given wavelength samples. In the relatively extreme case of HH 30, a nearly edge-on disked young star studied with the Hubble Space Telescope in the optical and NIR, indications can be seen of a light beam (or shadow?) from the central source sweeping across the flared disk with a period of ~ 7.5d (Duran-Rojas et al. 2009, Watson & Stapelfeldt 2007). Reality is complicated.

In the rest of the contribution, I will attempt to address whether young stars really do vary in the MIR, and if so, on what timescale. An important next step in understanding any variability is determining whether the source of any MIR variation is really at ~ 10s of R_* or at some other significantly different distance.

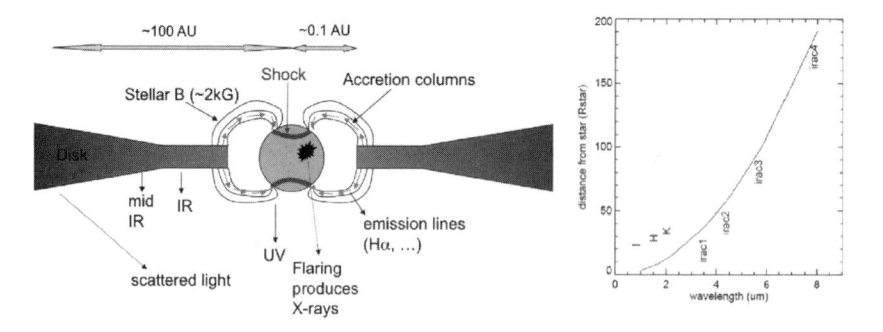

Figure 1. LEFT: Anatomy of a young star, after Hartmann (1998). Not to scale! Note that the mid-infrared emission comes from relatively far out in the disk. RIGHT: Simplified version of relationship between peak emission wavelength and distance from the protostar. For a typical low-mass protostar, the near-infrared bands I, H, and K sample within $\sim 10 R_*$; the four IRAC bands at 3.6, 4.5, 5.8, and 8 μm are sensitive to disk properties at ~ 30–$\sim 200 R_*$.

2. MIR YSO Variability in the Pre-Spitzer Era

Prior to the advent of the Spitzer Space Telescope (Werner et al. 2004), there were occasional published references to variability in YSOs in mid-infrared wavelengths, such as the following.

Prusti & Mitskevich (1994) originally set about looking for variations in all the repeated observations of Herbig AeBe (HAeBe) stars found at 12 and 25 μm in the Infrared Astronomy Satellite (IRAS) data taken in 1983. However, they found that source confusion was prohibitive, and focused their study on two HAeBe stars, AB Aur and WW Vul. They found significant variations on timescales (t) of months. They suggested that cometary clumps or a clumpy wind were plausible explanations for the variations observed.

Liu et al. (1996) reported that they found MIR variations in their ground-based data, ranging in amplitude from 30-300%, on timescales of days to years. They pointed out that the MIR variations most likely do not have the same origin as the optical/NIR variability. They suggest that since most of the MIR is from disk, then the cause of the variability must be there too. In order to achieve the variations that they observed, they postulate that the mass accretion rate ($\dot{M}$) varies by an order of magnitude. Small-amplitude changes in the MIR could be due to reprocessed accretion luminosity, whereas larger changes could be due to disk accretion rate, a disk instability, or outflow activity.

Abraham et al. (2004) took on the relatively difficult task of comparing Infrared Space Observatory (ISO) data taken in 1995-98 to MIR data taken at other bands with other facilities/instruments (such as MSX) at other epochs. They studied 7 FU Ori objects, and found weak MIR variability on timescales of years (over 1983–2001).

The next year, Barsony et al. (2005) reported on ground-based observations in the MIR of embedded objects in the ρ Ophiuchi cloud core. By comparison to ISO data, they found significant variability in 18 out of 85 objects detected, on timescales of years. They found such variability in all spectral energy distribution (SED) classes with optically thick disks, and suggest that this might be due to time-variable accretion.

Later that year, in a large paper covering the MIR properties of the Orion Nebula, Robberto et al. (2005) reported on MIR variability in Orion, almost as an afterthought. They found variations up to ~1 mag, on timescales of ~2 years. They invoke changes in $\dot{M}$, activity in the circumstellar disk, or changes in the foreground A_v to explain the variations they see.

Finally, Juhasz et al. (2007) report on the ISO variability of SV Cep. SV Cep is a UX Orionis-type variable, the generic properties of which include intermediate-mass YSOs with short (t ~days-weeks) eclipse-like events in the optical. These could be edge-on self-shadowed disks, for example. This study is the only one (at least, the only one of which we have knowledge) reporting on a monitoring campaign conducted with ISO itself (as opposed to comparison of ISO data to data taken with other instruments/facilities). They obtained contemporaneous optical monitoring data over ISO's lifetime (1995–1998) to aid in the interpretation of the MIR light curves. They found significant MIR variability on t ~25 months; the MIR variations were anti-correlated with the optical variations but the far-IR variations were correlated with optical. They suggest a self-shadowed disk with a puffed-up inner rim, but find that this model does not do well at reproducing the MIR variations; again, $\dot{M}$ variations are invoked to explain the MIR variations.

3. Results in the Spitzer Era

3.1. Introduction to Spitzer

The Spitzer Space Telescope (Werner et al. 2004) is an 85 cm, f/12 telescope. Before the on-board cryogen was exhausted, it operated at <~5.5 K, and was background-limited at 3-180 μm. It has two science cameras (Infrared Array Camera – IRAC – Fazio et al. 2004 and the Multiband Imaging Photometer for Spitzer – MIPS – Rieke et al. 2004), plus a low/moderate resolution spectrograph (Infrared Spectrograph – IRS – Houck et al. 2004). Launched August 2003 into an Earth-trailing orbit, it was 10-1000 times more sensitive than the 1983 IRAS mission.

The cryogen ran out in May 2009, and the telescope passively remains at ~30 K. At this temperature, the IRAC 3.6 and 4.5 μm channels still operate essentially as they did before cryogen exhaustion, which is still 120-1000 times faster than VLT or Keck. This portion of the mission is "Spitzer-Warm", and NASA has committed to fund ~3 years of warm operations. As part of the Warm Mission, large (>500 hours), coherent observing programs were solicited, called "Exploration Science" programs.

The cryogenic Spitzer legacy for star formation research is substantial. There are multi-band maps of ~300 square degrees of the Galactic plane, with >100 million sources. There are maps of ~70 square degrees in nearby (d <500 pc) star-forming regions, with ~8 million total sources in Taurus, Ophiuchus, Perseus, Chamaeleon, Serpens, Auriga, Cepheus, Lupus, Orion clouds, etc. Conservatively, we estimate that there are ~20,000 YSOs in this rich data set.

Spitzer is a superb telescope for photometric monitoring because it is stable (better than 1%) and sensitive, wide-field (a single IRAC field of view is 5′ on a side), Earth-trailing (so no orbital day/night aliasing), and it observes at bands sensitive to both photospheres and dust. In the Warm Mission era, we have the same amount of observing time as in the cryogenic era, and "just" 2 channels. There are several Exploration Science and smaller programs exploring the time domain with Spitzer.

3.2. Variability at Spitzer Bands

YSO variability at Spitzer bands is unambiguously apparent, and the torrent of papers on the subject is still ramping up. In the below, I discuss the papers in the order in which they appeared in the published literature.

The Legacy program "Cores to Disks" (c2d; Evans et al. 2003, 2009) took two epochs of observation (both IRAC and MIPS) separated by several hours to allow for asteroid removal. Several different papers (Alcala et al. 2008 and references therein) looked for variation between these two epochs (on timescales of ~3-6 hrs), and did not find anything believable (within ~25%) at wavelengths 3.6-24 μm.

Another Legacy program, "Surveying the Agents of a Galaxy's Evolution" (SAGE; Meixner et al. 2006) studied the Large Magellanic Cloud (LMC), again in two epochs (both IRAC and MIPS), but this time separated by ~3 months. Vijh et al. (2009) report on all of the variables found by comparing these two maps. They found mostly asymptotic giant branch (AGB) stars, which they point out is not entirely unexpected. However, we note here that optical variability is one of the defining characteristics of YSOs, and AGB stars are the most common "contaminant" in Spitzer selection of YSO candidates; having the right MIR colors plus MIR variability does not ensure that a given object is necessarily a YSO. Vijh et al. (2009) find 29 massive (=HAeBe) YSO candidates out of nearly 2000 variables, which they interpret to mean that at least 3% of all massive YSOs are variable. They also report that the amplitude of variability is often greatest at 24 μm, perhaps because most of their YSO SEDs peak at 24 μm (or longer).

The first high-cadence monitoring of young stars in IRAC bands was conducted by Morales-Calderón et al. (2009). The stars in IC 1396A were monitored twice a day for 14 d, plus every ~12 s for 7 hrs. More than half of the YSOs showed variations, from ~0.05 to ~0.2 mag, on a wide variety of timescales, which enables the first possible serious physical interpretations of the variations. About 30% of the YSOs had quasi-periodic variations, on timescales of ~5-12d periods, which they interpreted as 1 or 2 high-latitude spots illuminating inner wall of the circumstellar disk, plus a large inclination angle. Two objects have variations on timescales of ~hours, but no color in the variations, which is interpreted as flares, and/or possibly $\dot{M}$ flickering. Other light curves are more likely due to varying $\dot{M}$ or disk shadowing. About 20% of the IC 1396A YSOs vary on t ~days, without color changes, which could be due to $\dot{M}$ variations, and/or rapidly evolving spots. There are three objects that vary on timescales of days, with color variations, which the authors interpreted as radial differential heating of the inner disk, and possible inner disk obscurations. There were 46 variables not identified as YSOs (e.g., without a discernible IR excess); possibly they are YSOs or even AGBs, but more data are needed to interpret these. Larger amplitude variables tend to also be more embedded objects, but an order of magnitude change in $\dot{M}$ is needed to match the light curves, so this is probably not the dominant factor. A simple starspot is insufficient to explain the variability, but a hotspot combined with disk inhomogeneities does work. Also in the data was a young δ Scuti star, with a 3.5 hr periodicity on top of a ~9 d period.

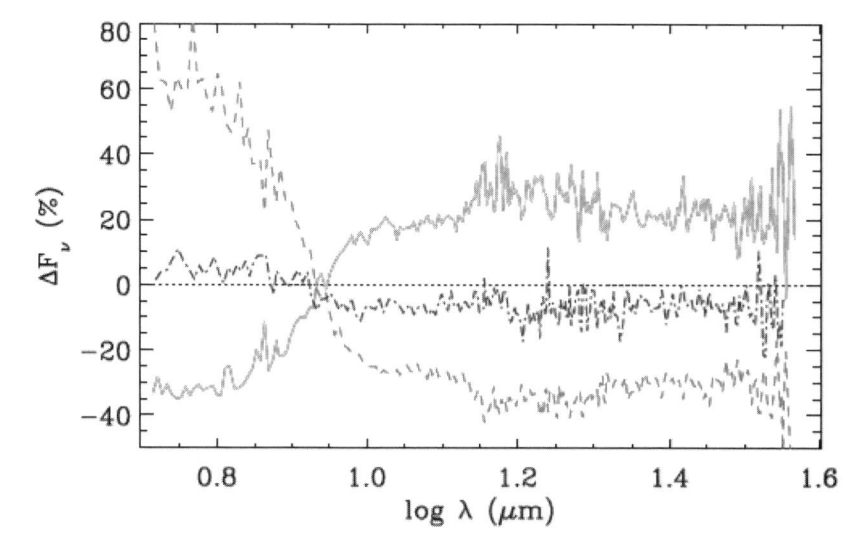

Figure 2. Figure 1b from Muzerolle et al. (2009). Difference spectra between the first and second epochs (solid green), second and third epochs (dashed magenta), and third and fourth epochs (dash-dot blue), as a percentage change in flux. First epoch was 2007 Oct 9, second epoch was 2007 Oct 16, third epoch was 2008 Feb 24, and fourth epoch was 2008 Mar 2. Variations "pivot" at ~ 8.5 μm, and are as large as 20-30% in a week.

Working in IC 348, Muzerolle et al. (2009) report specifically on the variations they observed in the T Tauri star LRLL 31. This object is identified specifically as a "transition disk", meaning that it falls in a category of object thought to be in transition between a primordial, thick disk and a disk actively forming planets with gaps and structure in the disk created by protoplanets. The SED for this object suggests a large inner hole or gap. Muzerolle et al. (2009) initially noticed variations in IRAC+MIPS (3.6-24 μm) observations taken over t ~5 months. They used both IRS and MIPS (5-40 μm) to further probe these variations on timescales of ~days to ~months. In the IRS spectra, reproduced in Figure 2, they found that the variations pivot at a point ~8.5 μm, and they found variations of 20-30% within a week. They also found variations at 24 μm on t ~1 day; recall Figure 1 – note that variations on those timescales are certainly not very far away from the star, even at that wavelength. Muzerolle *et al.* (2009) interpret these observations as vertical variations of an optically thick annulus located close to the star. Variations in $\dot{M}$ (up to a factor of 5) could be contributing here, or a companion causing gap, or even a warp.

Giannini *et al.* (2009) conducted observations of the Vela Molecular Ridge (VMR-D), with just IRAC. The two maps were taken ~6 months apart. They simply accept variability of YSOs at the MIR bands as a defining characteristic of YSOs, as a statement of fact, and do not attempt to further justify it. This suggests a change in culture in the community. Giannini *et al.* (2009) conclude that 19 (out of ~200) are likely variable young stars.

Bary *et al.* (2009) obtained IRS spectra of 11 actively accreting T Tauri stars in Taurus-Auriga; 2 of the 11 (DG Tau and XZ Tau) had significant variation in the 10 μm silicate feature on timescales pf ~months to years (not weeks). They point out that this timescale is consistent with the source of the variations being motions of dust in the disk at R <~1 AU, and not with a clumpy dust envelope. Disk shadowing could still be possible, especially at the longer timescales. The possibility remains that there are binary companions to these objects as well. They had difficulty in fitting the line profile with existing models, suggesting that similar problems encountered by other investigators fitting single-epoch observations of other sources may ultimately be due to similar time-dependencies in those other sources. In any case, vertical mixing and disk winds are likely to be significant components of the source of the variability.

4. YSOVAR

John Stauffer leads the Exploration Science program (from Spitzer's Cycle 6) entitled, "Young Stellar Object Variability: Mid Infrared Clues to Accretion Disk Physics and Protostar Rotational Evolution," or YSOVAR. We were allocated 550 hours to conduct the first sensitive MIR (3.6 and 4.5 μm) time series photometric monitoring of several star-forming regions on timescales of ~hours to years. Our fields include ~1 square degree of Orion (centered on the Orion Nebula Cluster) plus smaller ~25 square arcminute regions in 11 other well-known SFRs: AFGL 490, NGC 1333, Mon R2, NGC 2264, Serpens Main, Serpens South, GGD 12-15, L1688, IC1396A, Ceph C, and IRAS 20050+2070. Details of our fields, as well as a complete list of our collaborators, can be found at our website: http://ysovar.ipac.caltech.edu.

For our observations, we typically obtain ~100 epochs/region (sampled ~twice/day for 40d, less frequently at longer timescales). We started obtaining data in Sep. 2009 and will be obtaining data through June 2011. At the completion of our program, there should be good light curves for at least ~2200 YSOs! We are also obtaining simultaneous (or nearly simultaneous) ground-based monitoring at I_c, J, and K_s, which aid significantly in our ability to interpret the light curves. (NB: if anyone in the community is interested in helping obtain such data, please contact us at ysovar-at-ipac.caltech.edu.)

Note that we include under the YSOVAR umbrella some affiliated programs such as J. Stauffer's Cycle 7 Orion follow-up on some of our targets discussed below, P. Plavchan's Cycle 6 Rho Oph intensive monitoring, K. Covey's Chandra/Spitzer Ceph C monitoring, and J. Forbrich's GGD 12-15 Chandra/Spitzer monitoring. As of this writing, there are five clusters with at least some data: Orion, L1688, Ceph C, IC 1396A, IRAS 20050.

Morales-Calderón *et al.* (2010, 2011; see also this volume) report on the early results from the YSOVAR monitoring of Orion. We find variability in ~65% of the objects with infrared excess (Class I+II) and ~30% of the objects without infrared excess (Class III). It should not be surprising that there is tremendously diverse behavior exhibited by these variables. Figure 3 shows just a sample of some of the light curves from some of the objects in Orion. The shape of the light curves likely have origins in slow changes in $\dot{M}$, changes in the $\dot{M}$ geometry, flares, photospheric spots, disk warps, and some causes yet to be identified! The contemporaneous optical and NIR data sometimes have a similar shape and amplitude as the MIR light curves, sometimes the NIR has a much larger amplitude, and sometimes the NIR variations are much smaller or not variable at all. In some cases, the NIR variations are phase-shifted with respect to the MIR. (See Morales-Calderón *et al.* 2010 for example light curves and more discussion.)

Because the emission in the MIR is likely coming from the disk (thermal dust emission) as well as the photosphere, the variations we see are likely due to variability in the disk as well as the photosphere. Thus, it is in general harder to derive a period for the central YSO for our target objects than from light curves, say, in I_c, where most of the emission comes from the photosphere (and spots rotating into and out of view generate rotationally modulated light curves). For just 16% of the variable objects with infrared excess (Class I+IIs) can we derive a period, and most of those are the ones with smaller excesses (90% of those are Class IIs, 10% are Class Is). For members without an IR excess, 40% are variables, and most of those are periodic. We can report >100 new periods. Of the Orion members with period measurements in the literature, we recover about 45% of those. There are also 10 eclipsing binaries, 5 of which are new discoveries (Morales-Calderón 2011).

One significant class of variables that we have discovered have AA Tau-like variations (see Bouvier *et al.* 2007 and references therein for discussion of AA Tau). These "dipper" stars have narrow flux dips, on timescales of days, and typically more than one dip are seen over our 40 d window; see Figure 4. In order for us to categorize a given object as a dipper, we require that the dip is seen in more than one epoch unless there are corroborating data at another band. Any optical or J band corroborating data must have the dips be deeper by at least 50%. The "continuum" of the light curve must be flat enough that dip "stands out." We find 38 Class I or II objects (~3%) in our set that are dippers, and we interpret this variability as structure in the disk, such as clouds or warps.

Other upcoming results include the following. Plavchan *et al.* (priv. comm.) report that WL 4 is still eclipsing, 10 years after the 2MASS calibration data (Plavchan *et al.* 2008) were taken. This system is probably a quasi-stable disk eclipsing a binary system like KH-15D. Muzerolle, Flaherty *et al.* (priv. comm.; see also this volume) studied IC 348 and find IRAC variability (on timescales of days to years) in 56% of Class 0/I objects, 69% of Class II objects, and 58% of the transition disks. Moreover, even at 24 μm, 60% of Class 0/I, 40% of Class II, and 40% of transition disks vary! They also find dips in the light curves like the YSOVAR dippers.

The YSOVAR data set (as well as the associated programs) are certain to yield interesting results in the coming years. For lack of space, I have not addressed any possible monitoring results from Herschel or WISE, much less any recent non-MIR

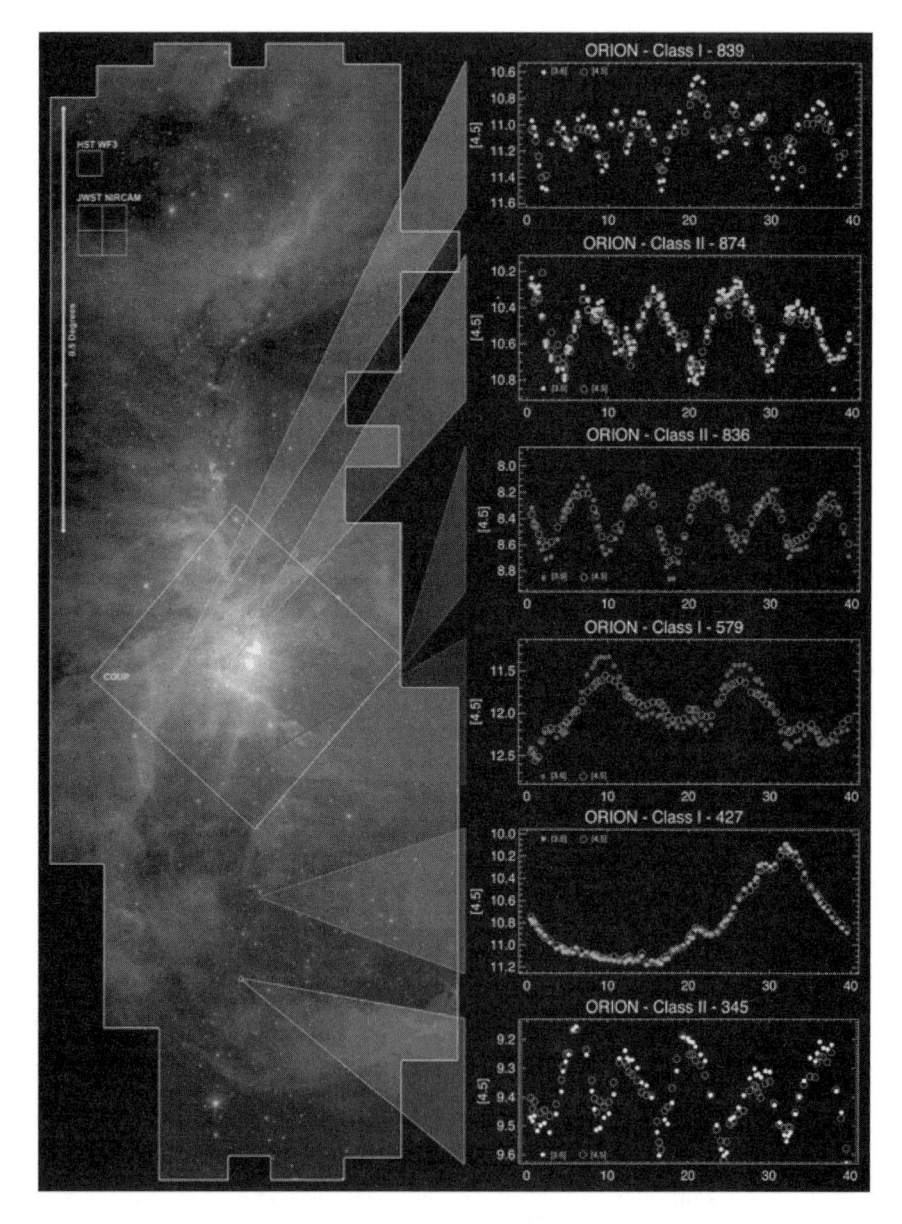

Figure 3. Image depicting some of the variability found in YSOVAR observations in Orion. The image on the left is the two-band (3.6 and 4.5 μm) Spitzer composite. Indications of the relative sizes of 0.5°, and the Hubble WF3 and the JWST NIR-CAM fields of view are in the upper left. The Chandra COUP field is indicated, centered on the Trapezium. For each of the light curves depicted, the solid point is 3.6 μm and the hollow point is 4.5 μm. Note the diversity of behavior exhibited by these variables.

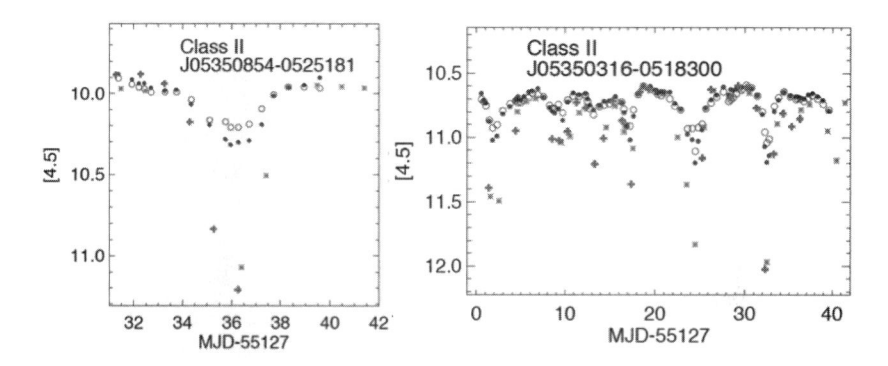

Figure 4. Two examples of AA Tau-like ("dipper") variability in our Orion data, from Morales-Calderón et al. (2010). Solid blue circles are 3.5 μm, open black circles are 4.5 μm, red or green ∗ are J, and magenta + are I_c. The light curves are shifted in y-axis to align to the [4.5] "continuum" level.

monitoring of young stars, such as CoRoT monitoring of NGC 2264 (see, e.g., Alencar et al. 2010 and references therein for more information).

5. Conclusions

While 15 years ago, we were as a community uncertain as to whether young stars vary in the mid-infrared, the literature suggested at least small variations on timescales of months to years, likely due to the disk. However, with the advent of the Spitzer Space Telescope and its stable, sensitive, wide-field platform for monitoring young stars, it has become unambiguous that yes, young stars vary in the mid-infrared, and they vary on pretty much any timescale that one cares to observe them (much as they do at many other bands). While definitive physical explanations for all of the tremendous diversity of variability types is still elusive, strong candidates for some types of variation are emerging. Some of the variability is clearly due to photospheric spots, much is due to structure in the disk, some is variation in mass accretion rate. Rotation, and the dynamic nature of the young star-disk system, are both clearly important. The answers are still forthcoming!

Acknowledgments. I wish to acknowledge many helpful conversations with J. Stauffer, M. Morales-Calderón, P. Plavchan, K. Covey, J. Carpenter, and the rest of the YSOVAR team. I also wish to thank J. Muzerolle for pre-publication access to his results. This work is based in part on observations made with the Spitzer Space Telescope, which is operated by the Jet Propulsion Laboratory, California Institute of Technology under a contract with NASA. Support for this work was provided by NASA through an award issued by JPL/Caltech.

References

Ábrahám, P., et al., 2004, A&A, 428, 89
Alcalá, J., et al., 2008, ApJ, 676, 427
Alencar, S., et al., 2010, A&A, 519, A88

Barsony, M., Ressler, M., & Marsh, K., 2005, ApJ, 630, 381
Bary, J., et al., 2009, ApJL, 706, 168
Bertout, C., 1989, ARA&A, 27, 351
Bouvier, J., et al., 2007, A&A, 463, 1017
Durán-Rojas, M. C., Watson, A., Stapelfeldt, K., & Hiriart, D., 2009, AJ, 137, 4330
Evans, N. J., et al., 2003, PASP, 115, 965
Evans, N. J., et al., 2009, ApJS, 181, 321
Fazio, G., et al., 2004, ApJS, 154, 10
Giannini, T., et al., 2009, ApJ, 704, 606
Hartmann, L., 1998, "Accretion processes in star formation", Cambridge, UK ; New York : Cambridge University Press, 1998. (Cambridge astrophysics series ; 32) ISBN 0521435072.
Houck, J., et al., 2004, ApJS, 154, 18
Juhász, A., et al., 2007, MNRAS, 374, 1242
Königl, A. 1991, ApJL, 370, 39
Liu, M., et al., 1996, ApJ, 461, 334
Meixner, M., et al., 2006, AJ, 132, 2268
Morales-Calderón, M., et al., 2009, ApJ, 702, 1507
Morales-Calderón, M., et al., 2010, ApJL, submitted
Morales-Calderón, M., et al., 2011, in preparation
Muzerolle, J., et al., 2009, ApJL, 704, 15
Plavchan, P., et al., 2008, ApJK, 684, 37
Prusti, T., & Mitskevich, A., 1994, in The nature and evolutionary status of Herbig Ae/Be stars. Astronomical Society of the Pacific Conference Series, Vol. 62. Proceedings of the First International Meeting held in Amsterdam, 26-29 October 1993, San Francisco: Astronomical Society of the Pacific (ASP), edited by Pik Sin The, Mario R. Perez, and Edward P. J. Van den Heuvel, p.257
Rieke, G., et al., 2004, ApJS, 154, 25
Robberto, M., et al., 2005, AJ, 129, 1534
Shu, F., et al. 2000, in Protostars and Planets IV, ed. V. Mannings, A. P. Boss & S. S. Russell (Tucson: University of Arizona Press), p. 789
Vijh, U., et al., 2009, AJ, 137, 3139
Watson, A., & Stapelfeldt, K., 2007, AJ, 133, 845
Werner, M., et al., 2004, ApJS, 154, 1

The 16th Cambridge Workshop on Cool Stars, Stellar Systems and the Sun
ASP Conference Series, Vol. 448
Christopher M. Johns-Krull, Matthew K. Browning, and Andrew A. West, eds.
© 2011 Astronomical Society of the Pacific

Innovative Imaging of Young Stars: First Light ExPo Observations

S. V. Jeffers, H. Canovas, C. U. Keller, M. Min, and M. Rodenhuis

Astronomy Institute, Utrecht University, P.O.Box 80000, 3508 TA, Utrecht, The Netherlands

Abstract. We have developed an innovative imaging polariemter, ExPo, that excels in the imaging of the circumstellar environments of young stars. The basic physics that ExPo exploits is that starlight reflected from a star's circumstellar environment becomes linearly polarised, making it easily separable from unpolarised starlight. Our preliminary results, from the William Herschel Telescope in La Palma, show that ExPo has successfully detected several known protoplanetary disks out to a much larger distance and at a finer resolution than previously observed. ExPo has also made a significant number of new detections of protoplanetary disks and stellar outflows. We use innovative data analysis techniques, related to speckle interferometry, to detect the innermost parts of the disk to much closer than any other techniques operating at visible wavelengths. In this paper I present highlights of ExPo's first light observations.

1. Why Polarimetry?

Imaging circumstellar environments is difficult because they are many orders of magnitude fainter than their central stars, yet it is the only way of fully understanding and characterising these disks. Methods that are commonly employed to remove unwanted stellar light involve highly sophisticated adaptive optics systems, coronagraphs and nulling interferometers (Perryman 2003). However, by directly measuring the linearly polarized light reflected from a circumstellar disk it is possible to clearly discriminate between light from circumstellar environments and unpolarized light from the central star. In particular using polarimetry is advantageous because of its ability to constrain the distribution of dust sizes, providing information on the sources, sinks and dynamics of dust grains in circumstellar environments Krivov et al. (2000). The basic physics of ExPo is illustrated in Figure 1.

1.1. Complications

Additionally challenges arise in the imaging of circumstellar environments, which necessitate innovative instrumental design and meticulous data analysis. These generally arise from a small angular separation of the disc from the central star, circumstellar environments having a high contrast ratio $\approx 10^{-4}$ for discs and $\approx 10^{-9}$ for exoplanets. The Earth's atmosphere also contributes by seeing and atmospheric polarisation effects and then there are unwanted instrumental effects.

16   Jeffers et al.

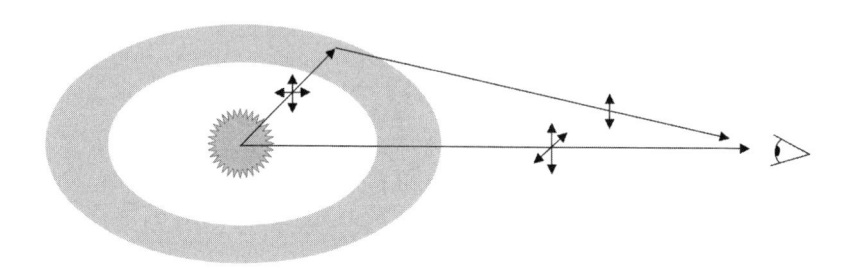

Figure 1. ExPo uses the basic physics that light reflected from a star's circumstellar environment is linearly polarised, while starlight remains unpolarised.

2. The ExPo Polarimeter

ExPo is a highly sensitive, visible-light imaging polarimeter that is optimized for detecting circumstellar material. It is currently a visitor instrument at the 4.2m William Herschel Telescope on La Palma. In its current implementation, ExPo essentially comprises a fast modulator (FLC - ferro-electric liquid crystal), which acts as a half waveplate, with the Dual Beam polarimetric technique.

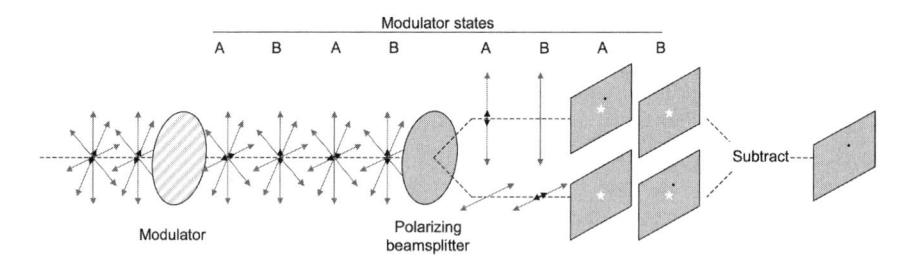

Figure 2. Schematic design of ExPo. As described in more detail in the text, light enters ExPo from the left-hand side. It is then separated into perpendicular polarisation states, labelled A and B states, using a modulator.

A schematic diagram of ExPo is shown in Figure 2 for a star-planet system, where instead the planet linearly polarised. Star light enters the instrument from telescope at the left-hand side of the image. It then passes through the modulator where it is separated into two orthogonal polarisation directions that alternate (labelled as A and B frames). The polarising beam splitter then separates these two orthogonal polarisation directions into two separate beams which are simultaneously imaged on the same camera. The resulting A image shows a star with a (linearly polarised) planet in the upper panel and only a star in the lower panel, while the B frame shows only a star in the upper panel and a star with a planet in the lower frame. Subtracting these frames results in only the linearly polarised planet. To demonstrate this concept, a planet at a separation of 1" from a star was simulated in the lab at a contrast ratio of 10^{-4}. The results are shown in Figure 3 for Stokes I, where it is impossible to locate the planet, and in linearly polarised light.

By rotating the FLC polarization modulator, we can measure the complete linear polarization signal (Stokes Q and U) from the circumstellar environment. ExPo works between 450 and 900 nm and can use the full wavelength range as well as filters to select specific wavelength ranges. A fast and sensitive imaging (EMCCD) camera is used to take short exposures (0.028s). The fast polarization modulation combined with the true dual-beam system minimizes the effect of seeing and flat-field errors and has been used successfully in solar physics (Keller 1996) and and allows us to reach polarization sensitivities at the 10^{-5} level. A more detailed description of the instrument is given by Rodenhuis et al. (2008).

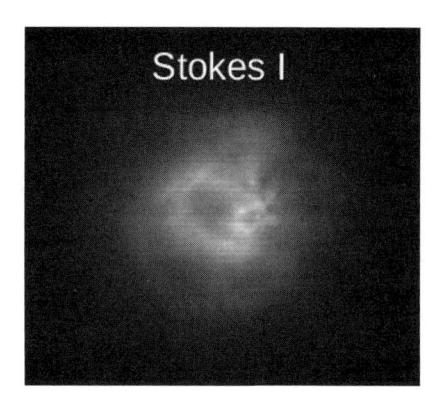

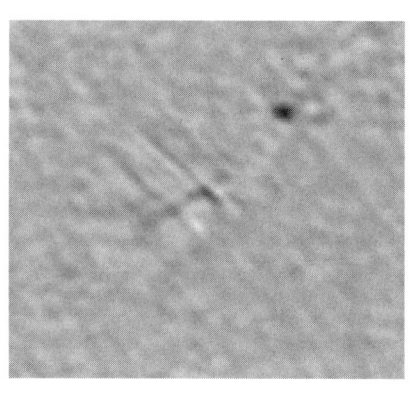

Figure 3. A concept demonstration of ExPo using laboratory simulations of a star with a planet at 1" Jeffers et al. (2008). The images show a Stokes I (left-hand side), where only the star is visible and on the right-hand side the subtracted images that results in only the linearly polarised planet. The planet is at a contrast ratio of 10^{-4}.

3. Data Analysis

In order to push the instrument to as close as possible to the diffraction limit of the telescope, it is necessary to employ careful data analysis techniques that minimise atmosphere effects and instrumental effects.

3.1. Image Combination

As ExPo is based on the dual beam method we have investigated methods of extracting the polarisation signal from the two images. There are two methods commonly used, one is the method based on the image ratios and the other based on image differences. After careful investigation (e.g. Canovas et al. (in preparation)) the image difference method was chosen to be most suitable for analysing ExPo data. To isolate the linearly polarised signal, essentially the A and B frames, as previously described, are combined in the following way:

$$P = 0.5[(A_{left} - A_{right}) - (B_{left} - B_{right})] \quad (1)$$

This approach has also been used for other dual beam systems such as Keller et al.(1996). Additionally, other effects that need to be accounted for are in the form of atmospheric effects and instrumental effects.

3.2. Atmospheric Effects

Atmospheric effects that impact the sensitivity of our observations are mainly in the form of seeing and atmospheric. By taking extremely short exposures i.e. 0.028s we significantly reduce the impact of seeing on our observations. Atmospheric dispersion is corrected for by the inclusion of an ADC (Atmospheric Dispersion Corrector) at the entrance to the instrument.

3.3. Instrumental Effects

To minimise any instrumental effects we use a master dark frame, computed from an average of at least 8000 darks and with cosmic rays removed, a master flat field which is an average of at least 1000 dome flats at each FLC position. Instrumental polarisation effects are minimised by using a polarisation compensator as described in Rodenhuis et al.(in preparation). A more in-depth description of the data analysis techniques are given by Canovas et al (in preparation).

4. First Light Results

In this section we present the results of our first observing runs at the 4.2m William Herschel Telescope, La Palma. The observing runs were in December/January 2009/10 and May 2010.

4.1. Targets

During our first observing runs our main science targets were the circumstellar environments of young stars, both Herbig Ae/Be and T Tauri stars. Additionally we also observed several debris disks and transitional disks. For calibration purposes we also observed non-polarised standard stars at regular altitude intervals, polarised stars, and diskless stars.

In this section a selection of first light ExPo images are presented that illustrate the potential ExPo has for imaging circumstellar environments. We show one image per class for T Tauri, Herbig Ae/Be, and Post AGB stars. The images shown are in polarised intensity. An in-depth analysis of these stars and additional images of other protoplanetary discs will be published shortly.

4.2. T Tauri star

An ExPo image of a classical T Tauri star is shown in Figure 4. The disk is indicated as a dark lane that curves through the centre of the image, which is indicated by the FWHM of the intensity profile. The polarisation intensity vectors of the disk are aligned and perpendicular to the central star. Also clearly present are two jets emanating from either side of the disk. Spectroscopic studies by Kwan et al. (2007) show that the jet emanating from this star originates in the disk. A more detailed analysis of this target, including colour images and radiative transfer disc modelling, will be published in Jeffers et al. (in preparation).

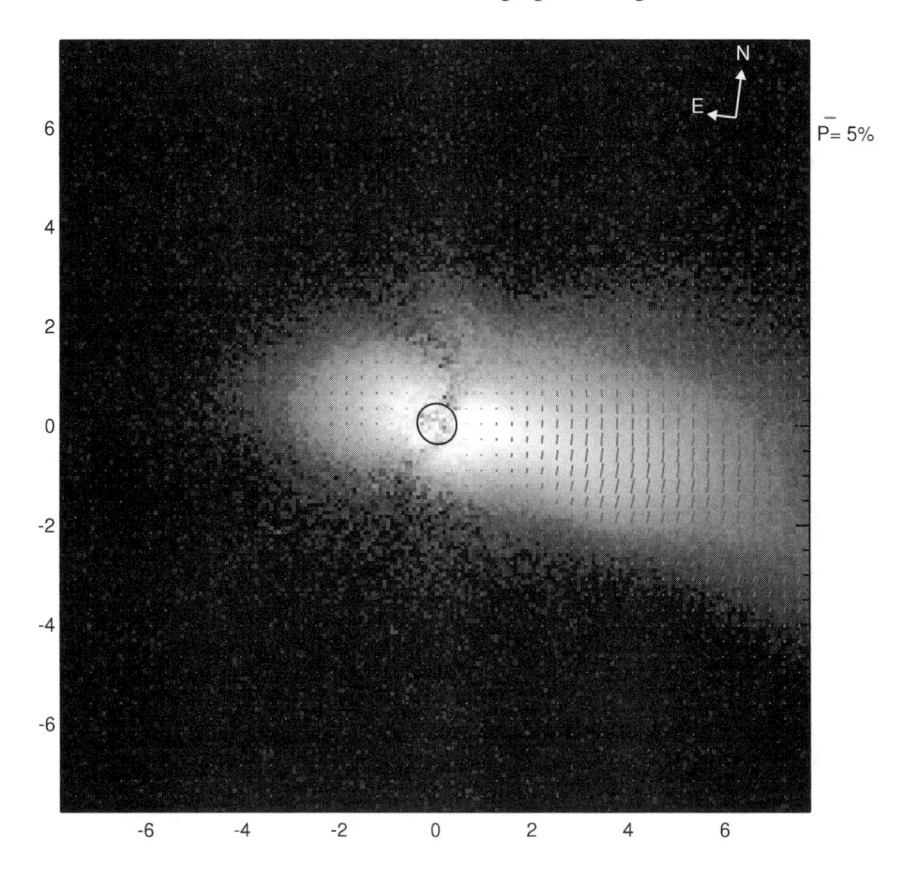

Figure 4. A polarised intensity image of a T Tauri star with a edge-on circumstellar disk and two large jets. The black circle in the centre of the image represents of the fwhm of the intensity image. The dark, and slightly curved dark lane also at the centre of the image indicates the location of the edge-on circumstellar disk. This image shows clearly how ExPo excels in imaging the full circumstellar environment of young stars.

4.3. Herbig Be star

For the Herbig Be disks we show an ExPo image of the young protoplanetary disk MWC 147. This image shows a large molecular cloud with P Cygni profiles that indicates a strong stellar wind. Comparison with images in the infrared show only the left-hand side of the image, trace the PAH emission (e.g. large carbonaceous grains), while the ExPo observations indicate the presence of dust.

4.4. Post AGB star

As part of our observing programme we observed the post-AGB star R CrB. R CrB stars are one of the most enigmatic types of variable stars that frequently show irregular decreases, of up to 9 magnitudes, in their light curves. These minima are considered to be caused by a regular 'dust puffs' that obscure the starlight. At the time of observation

20   Jeffers et al.

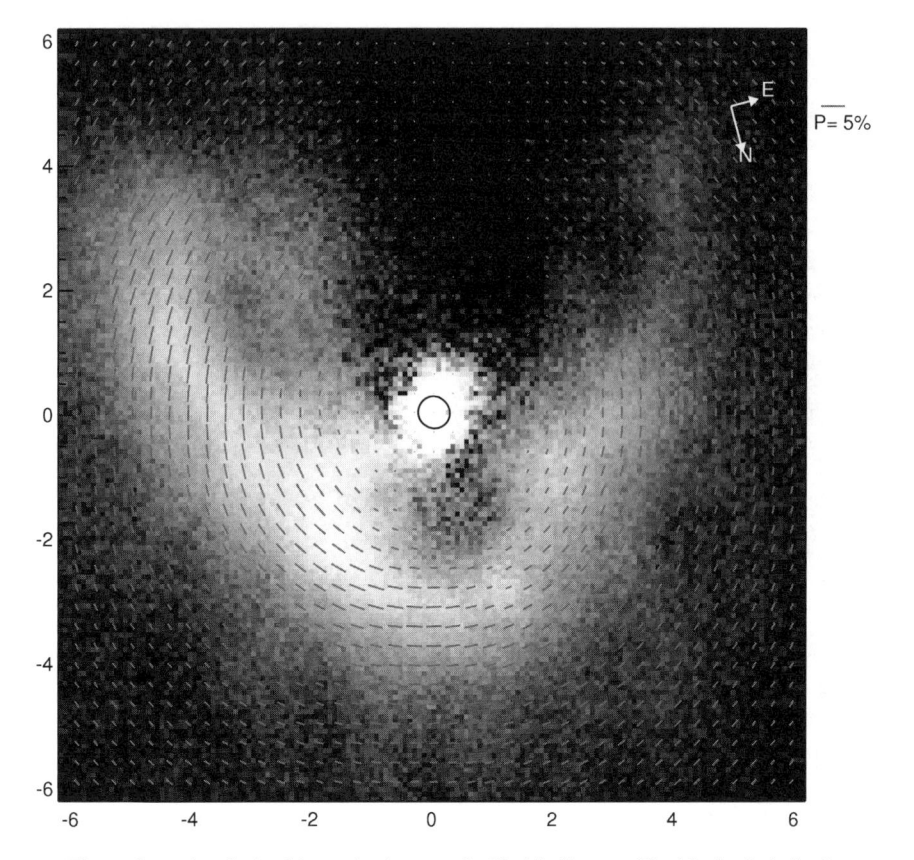

Figure 5. A polarised intensity image of a Herbig Be star. The black circle in the centre of the image represents of the fwhm of the intensity image. The disk of this star is concentrated in the centre of the image while the large V-shaped material is considered to be its molecular cloud.

(May 2010) R CrB was in a minimum state, meaning that the obscuring dust cloud was acting as a coronograph. This obscuration enabled us to see another dust cloud in polarised light. This image is shown in Figure 6, where the scattering dust cloud is located to the lower right of the central star, and is indicated by polarisation vectors that are perpendicular to the star. The star, located in the centre of the image, shows some high noise levels that are likely to result from the polarisation of the obscuring cloud. We have determined that the grain sizes in the 'Scattering Cloud' are Carbon grains with a minimum size of $0.005 \mu m$ and a maximum size of $0.25 \mu m$ and follow a size distribution with a slope of $n(r) \alpha\ r^{-3.5}$. The scattering is assumed to be backwards, i.e. the scatting cloud is located behind the central star.

5. Future Prospects

Our immediate plans are to complete a survey of protoplanetary discs comprising both Herbig Ae/Be and T Tauri stars. We will also use images taken at different filters,

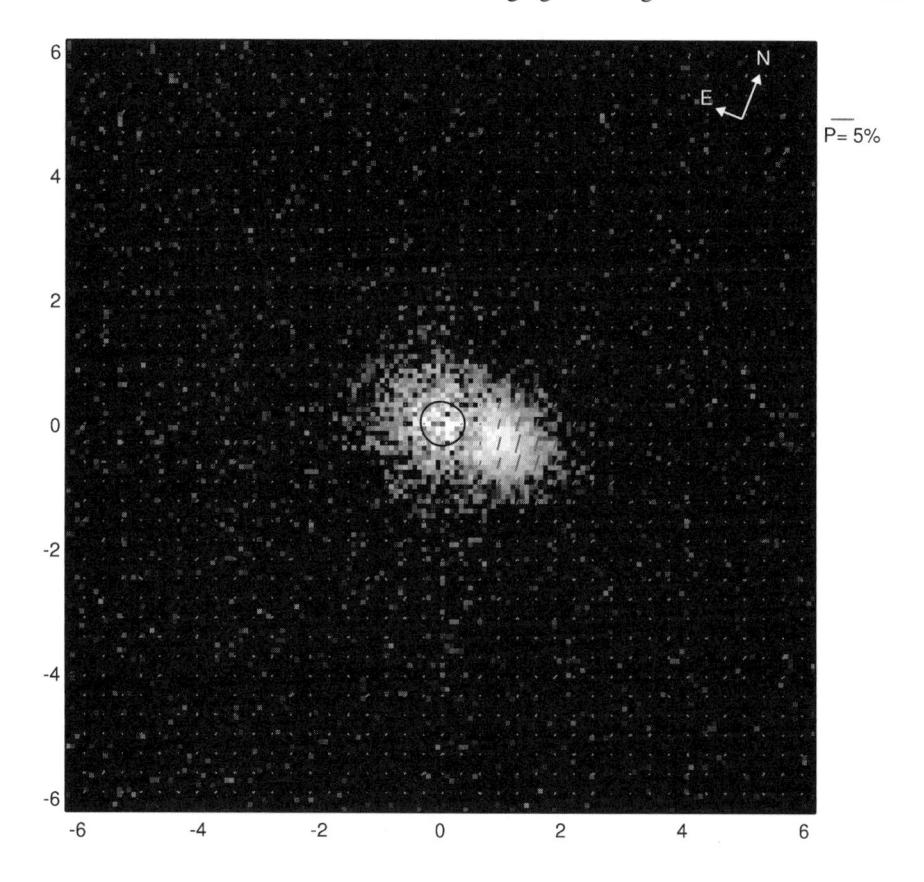

Figure 6. Polarised intensity image of the post-AGB star R CrB. The black circle in the centre of the image represents of the fwhm of the intensity image. At the time of observation R CrB was in a minimum phase. The cause of this obscuration is the ejection of a dust puff which mimics a coronograph by dimming the star by 8 magnitudes. Since we were fortunate to be awarded observing time during this minimum, it was possible to observe another dust cloud close to R CrB clearly in polarised images.

and the degree of polarisation, to further understand the disc composition and grain settling in the disk. This will probe the structure of the disc with greater sensitivity that previously achieved. We will also use ExPo to observe the circumstellar environments of post-AGB and post-RSG stars. As shown in this paper we have already achieved spectacular results for determining the dust properties of R CrB. The future plans for the data analysis developments include being able to resolve inner disks to within 0.2". We are already in the process of developing this technique which uses the polarised intensity images with a type of 'lucky imaging'.

Future instrumental developments include the upgrading of ExPo with an Adaptive Optics system, and an IFU (integral field unit) both due to come online in mid-2011. Further ahead instruments such as the ZIMPOL arm of SPHERE and the EPOL part of

EPICS are based on the same design concept as ExPo. SPHERE is currently expected to come online in 2013.

Acknowledgments. We acknowledge support from NWO in the form of a VICI grant to C.U.Keller, and SVJ thanks Alaska Airlines for never finding her luggage.

References

Jeffers, S. V., Miesen, N., Rodenhuis, M., & Keller, C. U. 2008, Proc. SPIE, 7014, 70147B
Keller, C. U. 1996, Solar Phys., 164, 243
Krivov, A. V., Mann, I., & Krivova, N. A. 2000, A&A, 362, 1127
Kwan, J., Edwards, S., & Fischer, W. 2007, ApJ, 657, 897. arXiv:astro-ph/0611585
Perryman, M. A. C. 2003, in Astronomy, Cosmology and Fundamental Physics, edited by P. A. Shaver, L. Dilella, & A. Giménez, 371
Rodenhuis, M., Canovas, H., Jeffers, S. V., & Keller, C. U. 2008, Proc. SPIE, 7014, 70146T

Accretion Makes a Splash on TW Hydrae

Nancy S. Brickhouse

Harvard-Smithsonian Center for Astrophysics, 60 Garden St., Cambridge, MA 02138 USA

Abstract. The *Chandra* Large Program on the Classical T Tauri star TW Hydrae (489 ksec, obtained over the course of one month) brings a wealth of spectral diagnostics to the study of X-ray emission from a young star. The emission measure distribution shows two components separated by a gap (i.e. no emission measure in between). Light curves for the two components can then be constructed from the summed light curves of the appropriate individual lines. The two light curves show uncorrelated variability, with one large flare occurring only in the hot component. We associate the hotter component with the corona, since its peak temperature is ~10 MK. Ne IX line ratio diagnostics for temperature and density indicate that the source of the cooler component is indeed the accretion shock, as originally reported by Kastner et al. (2002). The temperature and density of the accretion shock are in excellent agreement with models using mass accretion rates derived from the optical. We require a third component, which we call the "post-shock region," from line ratio diagnostics of O VII. The density derived from O VII is lower than the density derived from Ne IX, contrary to standard one-dimensional model expectations and from hydrodynamics simulations to date. The column densities derived from the two ions are also significantly different, with the column density from O VII lower than that from Ne IX. This post-shock region cannot be the settling flow expected from the cooling of the shock column, since its mass is 30 times the mass of material that passes through the shock. Instead this region is the splash of stellar atmosphere that has been hit by the accretion stream and heated by the accretion process (Brickhouse et al. 2010).

1. Introduction

Accreting astrophysical systems show common properties across a diversity of sources, e.g. many systems have a disk of gas and dust surrounding the central object. In particular, stars with strong magnetic fields, from low-mass pre-main sequence stars to white dwarfs and neutron stars in binary systems, can truncate the accretion disk and channel material to a "hot spot" near the pole (Konigl 1991). For a young star with a dipole field, the inner edge of the accretion disk is of order several stellar radii from the star. Thus the accreting material accelerates to approximately the free-fall velocity and shocks near the surface of the star (Calvet & Gullbring 1998).

In young stars, the presence of an accretion hot spot can be deduced from a number of observable features by comparison with similar non-accreting stars, including an excess of ultraviolet continuum, a dilution of the photospheric absorption spectrum by excess optical continuum (known as "veiling"), strong, broad emission lines such as Hα, and a large discontinuity at the Balmer jump (e.g., Bouvier et al. 2007). Measure-

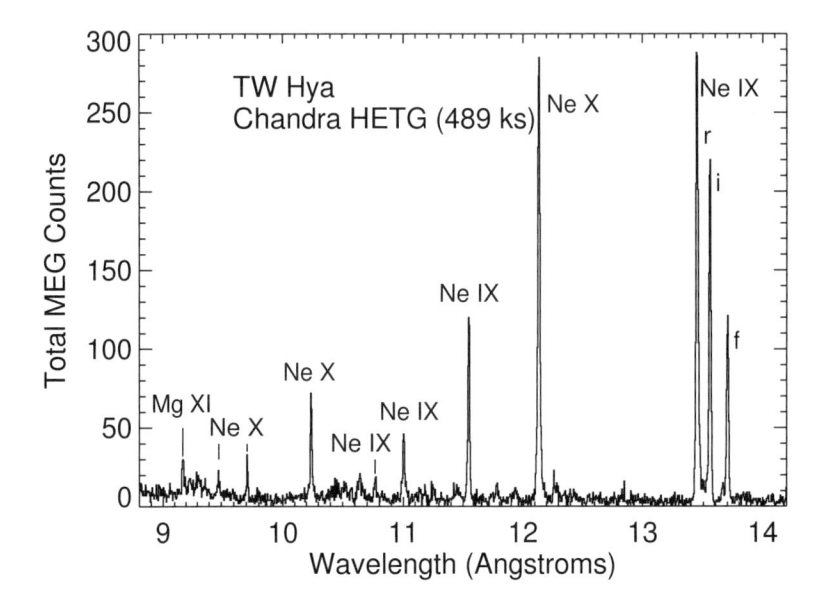

Figure 1. Section of the *Chandra* High Energy Transmission Grating spectrum of TW Hya, using the Medium Energy Grating (MEG), showing strong lines of Ne IX and Ne X, as well as the weak line of Mg XI. The resonance, intercombination, and forbidden lines of Ne IX are labeled r, i, and f, respectively.

ments of such features provide estimates of the star's mass accretion rate which agree with each to other to about an order of magnitude.

In principle, high resolution X-ray emission spectra from young stars can provide a more direct test of this accretion process, since the X-rays formed in the shock originate from the accreting material itself, rather than from shock-heated material in the underlying stellar atmosphere. X-ray line ratio diagnostics give measurements of the electron temperature and density of the gas for comparison with the shock model predictions.

For the most part, however, the X-ray emission from young stars seems to originate in a hot and rather violent solar-type corona, with temperatures far in excess of the few MK expected from the accretion shock (e.g., Stassun et al. 2007). Furthermore, many of the spectra show intrinsic absorption of the soft X-ray emission, possibly by the gas in the star-forming region, the disk (in edge-on systems), and/or the pre-shock accretion stream. The exact location of the shock relative to the stellar surface is not known, but estimated theoretically by balancing the ram pressure of the gas with the pressure of the stellar atmosphere. Thus it is also possible that the shock is buried in the photosphere and unobservable (Drake 2005).

A decade past the launches of the *Chandra* and *XMM-Newton* X-ray observatories, high resolution grating spectra of a dozen or so accreting stars have now been obtained (e.g., Guedel & Telleschi 2007). Of these, only five show evidence for the high density associated with an accretion shock, with TW Hydrae (TW Hya) being the first discovered (Kastner et al. 2002). Kastner et al. (2002) found $T_e \sim 3$ MK and $N_e \sim 10^{13}$ cm^{-3} for TW Hya. Magnetic field strengths are of order kGauss (Johns-Krull 2007; Yang et al. 2008a,b). Thus the plasma β (ratio of thermal to magnetic pressure) < 1.0, and the magnetic field controls the plasma flow. This paper summarizes new results from Brickhouse et al. (2010) on TW Hya, with an emphasis on new application of spectroscopic techniques.

2. The *Chandra* Large Program on TW Hya

We were awarded a *Chandra* Large Program on TW Hya to establish whether the X-ray emission was associated with accretion, and if so, to determine the physical properties of the accretion shock. During the period between 2007 Feb 15 and 2007 Mar 3, *Chandra* observed TW Hya four times using the High Energy Transmission Grating (HETG) in combination with the ACIS-S detector, for a total exposure time of 489.5 ksec. We also obtained spectroscopy and photometry from a ground-based campaign from four continents (Africa, Asia, Australia, and S. America), in order to search for any correlations between X-ray and optical emission (A. K. Dupree et al. 2010, in preparation). The summed high resolution X-ray spectrum exhibits strong emission lines from H- and He-like ions of O, Ne, Mg, and Si and from Fe L-shell ions. A region of the spectrum showing a preponderance of Ne lines is shown in Figure 1. The list of measured lines is given in Brickhouse et al. (2010).

3. X-ray Spectral Analysis

High signal-to-noise, high resolution X-ray spectroscopy tells us the physical conditions of an astrophysical plasma. Since X-ray observations of stars are "photon-starved," long exposure times with the high resolution gratings are required even for relatively nearby, bright objects. The data demand good spectroscopic models with accurate and complete atomic rate data. The use of diagnostic line ratios in the work presented here is a significant improvement over traditional analysis, made possible by long-term efforts in laboratory astrophysics. Specifically, the temperature derived using recent calculations of the He-like Ne IX line ratios is a better indicator of the shock temperature than that determined from a global fit to the spectrum (effectively in this case from the ratio of Ne IX to Ne X lines).

3.1. Emission Measure Distribution Analysis

The emission measure distribution is critical to our ability to isolate any shock emission from coronal emission. In this section we discuss the methods used to determine the emission measure distribution. Additional details are found in Brickhouse et al. (2010).

For an emission line the intensity I_λ at a given electron temperature T_e is given by $\frac{\varepsilon(T_e)}{4\pi R^2} \int_{T_e} N_e N_{hydrogen} dV$, where ε_λ is the atomic emissivity at T_e (including a factor for the elemental abundance relative to hydrogen), R is the distance to the source, N_e is the electron number density, $N_{hydrogen}$ is the hydrogen number density, and the integral is taken over the volume V of emitting gas. The emission measure at T_e is defined as $\int_{T_e} N_e N_{hydrogen} dV$, in units of cm^{-3}. In a collisionally ionized stellar corona, the X-ray emission arises from a broad range of plasma temperatures associated with the heating and cooling of magnetically confined loops of plasma. Convolving ε as a function of T_e with the emission measure distribution and summing the result give the predicted line intensity. In practice the emission measure distribution is iterated to fit either the global spectrum or a set of line fluxes.

With the exception of Ne IX, atomic data from ATOMDB V1.3[1] are used for the spectral analysis (Smith et al. 2001). For Ne IX we use new calculations of Chen et al. (2006), which include the effects of resonance excitation in the collision strengths. Experimental benchmarks of the temperature diagnostic using the Ne IX atomic rate data agree with this theory to within 7%. These calculations were made following indications from *Chandra* observations of stellar coronae that the He-like diagnostic ratios in ATOMDB were inconsistent with other diagnostics (Ness et al. 2003; Testa et al. 2004). The new calculations increase the sum of the forbidden and intercombination line prediction by ~25%, and thus bring temperatures up by a factor of 2. These calculations now appear to be reliable (Smith et al. 2009).

Brickhouse et al. (2010) describe the approach, which begins with fitting a two-temperature model to the global spectrum, allowing variable abundances and a single absorber. Line ratio diagnostics are then used to further constrain the model. For example, the lower temperature determined from Ne IX is fixed in the model because the Ne X emission comes from both components of the model. The higher temperature component is then broadened into a Gaussian shape to allow for non-isothermal plasma. Tests with different shapes and with a broadened lower temperature component did not find better fits to the data.

Figure 2 shows the high temperature emission measure distribution for TW Hya, compared with stellar coronae for three stars. In general for the coronal stars, emission is present at all temperatures up to some peak.[2] On the other hand, structure in the emission measure distribution, e.g. the sharp peak in Capella at 6 MK, is not well understood. Hydrostatic models of magnetic "loop" structures do not produce steep gradients and thus dynamic processes must be important in determining the struture of stellar coronae. Since there is no predicted shape of the coronal emission measure distribution, the Gaussian broadened function seems adequate for our purpose.

The emission measure distribution of TW Hya is notable for an extremely prominent, narrow feature at 2.5 MK and a gap between 2.5 MK and 6 MK. Some emission measure below 2.5 MK is allowed but not required. The gap is real, as lines produced in this temperature range (e.g. Mg XI) are observed to be weak in the spectrum, and would be overpredicted with more emission measure. To our knowledge, such a gap has never been observed in non-accreting systems. Thus we consider that the 2.5 MK peak and the hotter component of the emission measure distribution have distinct origins.

Using the model emission measure distribution convolved with the line emissivity functions, one can then determine which emission lines come from which component. It is interesting to note that lines of Ne X and to some extent Fe XVII are from both components while the rest of the lines are strongly dominated by only one component. Light curves constructed from the two groups of lines show uncorrelated variability, with one large flare occurring only in the hot component. We take the broad hot com-

---

[1] http://cxc.harvard.edu/atomdb

[2] Below about ~1 MK lines are outside the bandpass of the HETG, and measurements in the extreme ultraviolet are needed; interstellar absorption makes the measurement of these lines difficult. For HR 1099, the lack of emission measure below 1 MK reflects low signal-to-noise in existing data.

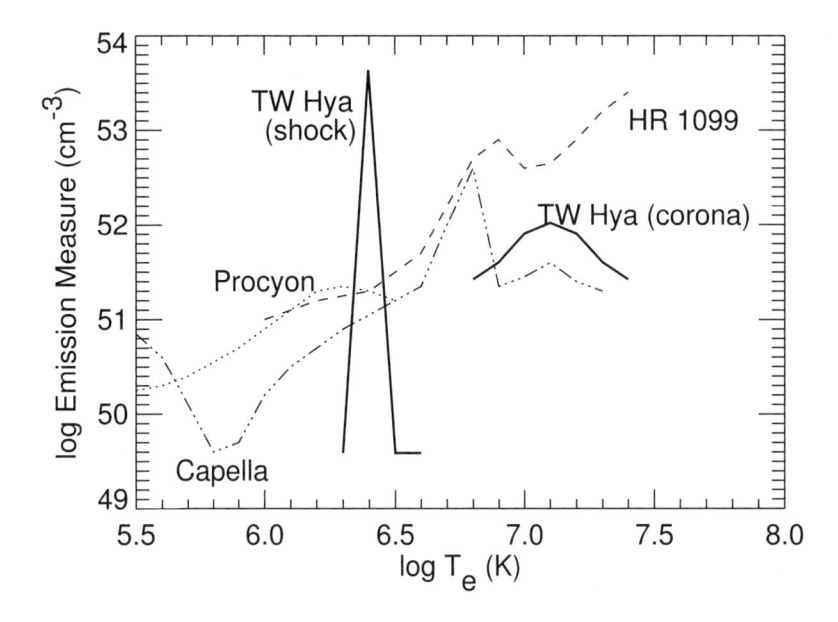

Figure 2. Emission measure distribution as a function of T_e for a sample of cool star coronae: Capella (dash-dotted), HR 1099 (dashed), and Procyon (dotted) from Brickhouse & Drake (2000) and the accreting star TW Hya (solid) from Brickhouse et al. (2010).

ponent to be coronal in nature. In the next section, we discuss the evidence that the 2.5 MK component results from a shock produced in the accretion flow.

3.2. Diagnostics of the Accretion Shock

Shock models for the X-ray emission predict electron temperature and density as a function of distance from the shock. He-like ions provide diagnostic lines for these quantities and allow us to test the shock model. For TW Hya, the accelerating material flowing from the disk is expected to reach a velocity close to the free-fall velocity ~500 km s^{-1}. The predicted temperature at the shock front depends on the gravitational field of the star. For given stellar parameters, the mass accretion rate is proportional to the density of the accretion stream times the filling factor (i.e. the area of the "hot spot" relative to the stellar surface area). With an observed density diagnostic line ratio from a specific ion, the emitting volume can be determined from the emission measure. In standard one-dimensional models, the density increases from the shock front down to the atmosphere of the star and the shocked gas slows down and cools radiatively. The path length for a given ion can be calculated from the model to give an estimate of the filling factor (Cranmer 2008).

He-like diagnostic lines in the *Chandra* data are available from Mg XI, Ne IX, and O VII. The Mg XI lines are weak and blended but appear consistent with the parameters derived from Ne IX. Brickhouse et al. (2010) find the values $T_e = 2.50 \pm 0.25$ MK, in excellent agreement with the model prediction, and $N_e = 3.0 \pm 0.2 \times 10^{12}$ cm^{-3}. The density from Ne IX is in good agreement with the low end of the range of mass accretion rates determined by other means (Muzerolle et al. 2001), for the estimated filling factor.

While the plasma parameters at the shock front support the simple theory, the post-shock flow is not observed. Instead, the cooler emission represented by O VII is less dense than expected by a factor of 7. The density is lower than that from Ne IX and thus cannot be the cooling column. With emission measure and electron density for both O VII and Ne IX we can also compare the volume and mass. The volume determined from O VII is 300 times larger, while the mass from O VII is 30 times larger, than those parameters found from Ne IX, respectively.

Brickhouse et al. (2010) suggest that the stellar atmosphere must be heated by the impact of the accretion stream, up to more than 1 MK. This ionized gas is thus available to be confined in closed magnetic loops or to be swept out in a stellar wind (e.g., Dupree et al. 2005). In such a dynamic, turbulent scenario, one-dimensional models are inadequate to describe the interaction between the accretion stream and the stellar atmosphere. Magnetohydrodynamic (MHD) simulations show that the hot spots formed on the surface are inhomogenous, with a central hot, dense region surrounded by a cooler, less dense region (Romanova et al. 2004; Orlando et al. 2010; Sacco et al. 2010). In the low plasma β regime applicable to TW Hya (Yang et al. 2008b), Sacco et al. (2010) find that high velocity (and thus temperature), low density streams are formed high in the chromosphere and are not absorbed, while low velocity (and thus temperature), high density streams are preferentially absorbed since the length for photons to travel through the chromosphere is larger.

While the issues of inhomogeneity and absorption in the chromosphere are likely to be important for systems with high mass accretion rates, the observations of TW Hya cannot be explained by these models. The models of Sacco et al. (2010) would predict the higher temperature Ne IX streams to have lower density and less absorption than the O VII streams, the opposite of what is observed (see also the next section). The observed 30 times larger mass from the lower temperature O VII would also imply that this component is the dominant accretion component, and therefore that most of the accreting material is coming from only a few stellar radii, which seems unlikely. Furthermore, the mass accretion rate would be higher, inconsistent with optical data. Finally we note that all systems observed at grating resolution show excess soft X-ray emission, associated with O VII (Guedel et al. 2007; Robrade & Schmitt 2007). Thus the heating of significant stellar atmosphere may be a ubiquitous process that needs to be taken into account when theoretical models confront X-ray observations.

3.3. Absorption by the Accretion Stream

Ratios of the resonance series lines of He-like lines are sensitive to both temperature and absorption. Temperatures determined from G-ratios break the degeneracy and allow us to measure the absorbing column density. Brickhouse et al. (2010) find that at least two absorbers with significantly different column densities are required, with less absorption of the cooler, less dense material that produces O VII than of the hotter, more dense material that produces Ne IX. Since interstellar absorption is negligible (Kastner et al. 2002), this absorption must be intrinsic to the system. Given the pole-on orientation with inclination angle of 7° (Qi et al. 2004), the near-neutral, pre-shock accretion stream is in the line of sight and almost certainly the X-ray absorber. In this case, the lower absorption and larger volume at O VII would be consistent with an expansion beyond the accretion column. Absorption of coronal emission below 6 MK may also explain the large gap in the emission measure distribution.

The Ne IX He-like ion resonance series lines rule out any significant line opacity. The optical depth for lines of a given ion scales as $gf_{osc}\lambda$. Since the oscillator strengths decrease rapidly with increasing n for the series, the Ne IX 2p→1s Heα line would be much more strongly affected by resonance scattering than the other lines. An optical depth of 0.4 would be required to match the Heα/Heβ ratio, but this would underpredict Heγ by 50% (Brickhouse et al. 2010). Absorption of all the Ne IX lines by a neutral or near-neutral plasma provides a good fit to the data. This finding supports the idea that absorption by the accretion stream suppresses the observed X-ray emission (Gregory et al. 2007), rather than the idea that the shock forms too deep in the stellar atmosphere to be observed. Again, if the depth of the shock depends on the mass accretion rate, it will be important to compare systems with high mass accretion rate to TW Hya with sufficient detail.

3.4. Non-thermal Line Broadening

Measuring the full width at half maximum (FWHM) widths of the strongest lines produced in the shock, Brickhouse et al. (2010) obtain a velocity of 183 ± 16 km s^{-1}. At 2.5 MK the Doppler FWHM is 75 km s^{-1} indicating a subsonic turbulent velocity of 165 ± 18 km s^{-1}. Line shifts are consistent with zero. These velocities support the idea that stellar atmosphere is heated to > 1 MK by the accretion impact.

4. Conclusions

High resolution X-ray spectroscopy of TW Hya with the *Chandra* HETG provides diagnostic tests of accretion models. The theoretical Ne IX line ratio diagnostics should be highly accurate, given the excellent agreement between theoretical and laboratory astrophysics studies of electron-impact excitation. Their application to the accretion shock in TW Hya is a major success of laboratory astrophysics. Considerable effort is needed to bring other important line diagnostics to comparable levels of high accuracy.

The one-dimensional shock model for TW Hya is in good agreement with the temperature and density measured using Ne IX line ratios. On the other hand, the observed O VII cannot arise in the post-shock column of radiatively cooling gas because its electron density and column density are too low and its mass is too high. Instead, the softest X-ray emission appears to result from material in the stellar atmosphere that is heated by the accretion process. This material is at sufficiently high density that confinement by coronal magnetic field loops is likely. Highly ionized gas is also available to feed nearby open field lines that may channel outflows in the form of winds or jets.

How broadly can the results from TW Hya be applied to other accreting stars, especially systems with higher mass accretion rates? TW Hya is certainly an extreme case. Its accretion rate is relatively low (given its relatively old age). The coronal emission is less extreme than found in many other young stars: in fact, the emission measure of the shock is comparable to the emission measure of the corona.

On the other hand, the new view of turbulent, churned-up stellar atmosphere — heated to soft X-ray emitting temperatures — must replace the vision of gently settling post-shock gas in all accreting stars. Furthermore, since systems with higher mass accretion rates dump even more energy on the stellar surface than is the case for TW Hya, we also need to consider what other physical processes might be in play. Magnetic processes such as wave heating and reconnection are likely to occur, and may contribute to enhanced coronal activity, winds and jets in such systems.

Acknowledgments. The author acknowledges collaborators S. R. Cranmer, A. K. Dupree, G. J. M. Luna, and S. J. Wolk. This work was supported by Chandra GO7-8018X from NASA to the Smithsonian Astrophysical Observatory (SAO) and by NASA contract NAS8-03060 to SAO for the Chandra X-ray Center.

References

Bouvier, J., Alencar, S. H. P., Harries, T. J., Johns-Krull, C. M., & Romanova, M. M. 2007, in Protostars and Planets V, edited by B. Reipurth, D. Jewitt, & K. Keil (Tucson: U. Arizona Press), vol. 951, 479
Brickhouse, N. S., Cranmer, S. R., Dupree, A. K., Luna, G. J. M., & Wolk, S. 2010, ApJ, 710, 1835
Brickhouse, N. S., & Drake, J. J. 2000, Rev. Mex. Astron. & Astrofis, 9, 24
Calvet, N., & Gullbring, E. 1998, ApJ, 509, 802
Chen, G.-X., Smith, R. K., Kirby, K. P., Brickhouse, N. S., & Wargelin, B. J. 2006, Phys.Rev.A, 74, 2709
Cranmer, S. R. 2008, ApJ, 689, 316
Drake, J. J. 2005, in Proceedings of the 13th Cambridge Workshop on Cool Stars, Stellar Systems and the Sun, edited by F. Favata, G. Hussain, & B. Battrick (ESA: European Space Agency), 519
Dupree, A. K., Brickhouse, N. S., Smith, G. H., & Strader, J. 2005, ApJ, 625, L131
Gregory, S. G., Wood, K., & Jardine, M. 2007, MNRAS, 379, 35
Guedel, M., Skinner, S. L., Mel'Nikov, S. Y., Audard, M., Telleschi, A., & Briggs, K. R. 2007, A&A, 468, 529
Guedel, M., & Telleschi, A. 2007, A&A, 474, L25
Johns-Krull, C. M. 2007, ApJ, 664, 975
Kastner, J. H., Huenemoerder, D. P., Schulz, N. S., Canizares, C. R., & Weintraub, D. A. 2002, ApJ, 567, 434
Konigl, A. 1991, ApJ, 370, L39
Muzerolle, J., Calvet, N., & Hartmann, L. 2001, ApJ, 550, 944
Ness, J.-U., Brickhouse, N. S., Drake, J. J., & Huenemoerder, D. P. 2003, ApJ, 598, 1277
Orlando, S., Sacco, G. G., Argiroffi, C., Reale, F., Peres, G., & Maggio, A. 2010, A&A, 510, 710
Qi, C., Ho, P. T. P., Wilner, D. J., Takakuwa, S., Hirano, N., Ohashi, N., Bourke, T. L., Zhang, Q., Blake, G. A., Hogerheijde, M., Saito, M., Choi, M., & Yang, J. 2004, ApJ, 616, L11
Robrade, J., & Schmitt, J. H. M. M. 2007, A&A, 473, 229
Romanova, M. M., Ustyugova, G. V., Koldoba, A. V., & Lovelace, R. V. E. 2004, ApJ, 610, 920
Sacco, G. G., Orlando, S., Argiroffi, C., Maggio, A., Peres, G., Reale, F., & Curran, R. L. 2010, A&A, 522, 55
Smith, R. K., Brickhouse, N. S., Liedahl, D. A., & Raymond, J. C. 2001, ApJ, 556, L91
Smith, R. K., Chen, G.-X., Kirby, K. P., & Brickhouse, N. S. 2009, ApJ, 700, 679
Stassun, K. G., van den Berg, M., & Feigelson, E. 2007, ApJ, 660, 704
Testa, P., Drake, J. J., & Peres, G. 2004, ApJ, 617, 508
Yang, H., Johns-Krull, C. M., & Valenti, J. A. 2008a, AJ, 133, 73
— 2008b, AJ, 136, 2286

Constraining the Evolution of Brown Dwarf Binarity as a Function of Age: A Keck LGS AO Search for Brown Dwarf and Planetary Mass Companions to Upper Scorpius Brown Dwarfs

Beth Biller,[1,2] Katelyn Allers,[3] Michael Liu,[1] Laird M. Close,[4] and Trent Dupuy[1,5]

[1] *Institute for Astronomy, University of Hawaii, 2680 Woodlawn Drive, Honolulu, HI, 96822 USA*

[2] *Max-Planck-Institut für Astronomie, Königstuhl 17, D-69117 Heidelberg, Germany*

[3] *Department of Physics and Astronomy, 153 Olin Science, Bucknell University, Lewisburg, PA 17837*

[4] *Steward Observatory, University of Arizona, 933 N. Cherry Ave., Tucson, AZ 85721, USA*

[5] *Harvard-Smithsonian Center for Astrophysics, 60 Garden St., Cambridge, MA 02138, USA*

Abstract. We searched for binary companions to 20 brown dwarfs in Upper Scorpius (145 pc, 5 Myr, nearest OB association) with the facility infrared camera NIRC2 and the Laser Guide Star adaptive optics system on the 10 m Keck II telescope. We discovered a close companion (0.14″, 20.9±0.4 AU) to the very low mass object SCH J16091837-20073523. From spectral deconvolution of integrated-light near-IR spectroscopy of SCH1609 using the SpeX spectrograph (Rayner et al. 2003), we estimate primary and secondary spectral types of M6±0.5 and M7±1.0, corresponding to masses of 79±17 M_{Jup} and 55±25 M_{Jup} at an age of 5 Myr and masses of 84±15 M_{Jup} and 60±25 M_{Jup} at an age of 10 Myr. For our survey objects with spectral types later than M8, we find an upper limit on the binary fraction of <9% (1-σ) at separations greater than 10 AU. We combine the results of our survey with previous surveys of Upper Sco and similar young regions to set the strongest constraints to date on binary fraction for young substellar objects and very low mass stars. The binary fraction for low mass (<40 M_{Jup}) brown dwarfs in Upper Sco is similar to that for T dwarfs in the field; for higher mass brown dwarfs and very low mass stars, there is an excess of medium-separation (10-50 AU projected separation) young binaries with respect to the field. These medium separation binaries will likely survive to late ages.

1. Introduction

Numerous brown dwarf binaries have been discovered in the the field (Close et al. 2003; Burgasser et al. 2003; Burgasser et al. 2006). Almost all of these have separations <15 AU, with the majority having separations within <7 AU. This tight binary distribution was initially viewed as evidence for the ejection scenario of brown dwarf formation. In the ejection scenario, brown dwarfs are stellar embryos which are expelled from their natal subclusters due to interaction with other subcluster members, therefore cutting off accretion. Only tight brown dwarf binaries can survive an ejection event (Reipurth & Clarke 2001).

In the last decade a population of wide (>15 AU separation) brown dwarf and "planetary mass" (<13 M_{Jup}) binaries have been discovered in young (<12 Myr) nearby clusters (Luhman 2004; Chauvin et al. 2005; Kraus et al. 2005, 2006; Allers 2006; Jayawardhana & Ivanov 2006; Close et al. 2007; Konopacky et al. 2007; Todorov et al. 2010). These recent results suggest that the multiplicity properties of young (~few Myr) substellar objects in star-forming regions may be substantially different from the old (~few Gyr) field population. If numerous, these young binaries also provide serious constraints for current theories of brown dwarf formation, since such wide binaries cannot be formed by a non-dissipative ejection model. However, most of these objects were either discovered serendipitously, are from surveys with unpublished statistics, or are from very small (≤13 object) surveys, so it is unknown how significant a population they form. Here, we conduct a systematic survey to search for such binaries in Upper Sco, the nearest OB association to the Earth.

2. Sample Selection and Observations

At an age of ~5 Myr and a distance of 145 pc (Preibisch et al. 2002), the Upper Scorpius OB association is one of the nearest sites of ongoing star-formation and is at an intermediate age point between very young clusters such as Taurus (< 1 Myr) and older field objects (~100 Myr). Additionally, Upper Sco is somewhat denser than nearby T associations such as Taurus and Chamaeleon. Binarity of young objects is expected to vary both as a function of age and environment. Thus, Upper Sco provides a key binarity data point, intermediate in both age and density.

Some low mass stars and high mass brown dwarfs in Upper Sco have already been studied for binarity (Kraus et al. 2005, Kraus et al. 2008) . Here, we extend binarity results to lower mass brown dwarfs and planetary mass objects in Upper Sco. We surveyed a sample of 20 substellar objects in Upper Sco with reported spectral types of M7.5 or later. These objects were selected from those with spectroscopic followup (Lodieu et al. 2008) from the near-IR photometric and proper motion surveys of Lodieu et al.(2007), Slesnick et al.(2008). Indeed, these are the least massive objects currently known in Upper Sco, with estimated masses of <40 M_{Jup}.

Objects were observed with integration times of 1-20 minutes with the facility infrared camera NIRC2 (narrow mode) and the Laser Guide Star adaptive optics system (Bouchez et al.2004; Wizinowich et al. 2004) on the 10 m Keck II telescope. Observations were conducted on the nights of 2007-07-17, 2008-07-27, 2009-05-29, 2009-05-30, and 2009-06-30. Search observations were conducted in the K_S filter ($\lambda_{central}$ = 2.146 μm). The observed object FWHM varied from 3 to 7 pixels, with most objects appearing slightly elongated in the direction of the tip-tilt reference star.

3. Discovery of a Brown Dwarf Companion to SCH 16091837-20073523

A close candidate companion (0.14″) was detected around SCH J16091837-20073523 (henceforth SCH 1609) with colors consistent with a young substellar object (Fig. 1).

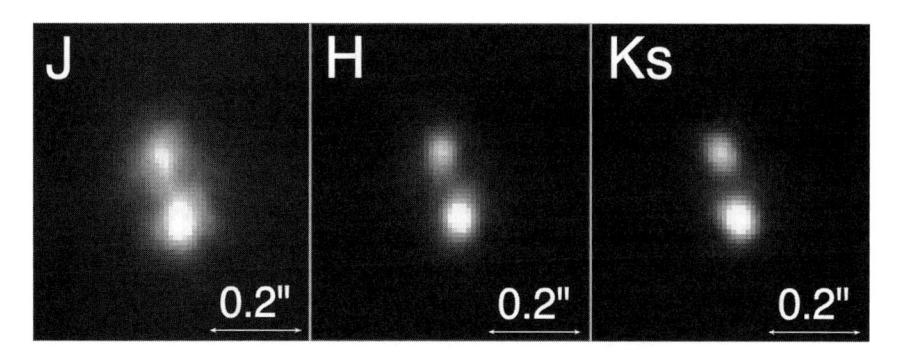

Figure 1. Left: J, H, and K_s-band images of SCH 16091837-20073523AB obtained with NIRC2 and the LGS AO system of the 10m Keck II telescope. North is up, east is left. Note that primary and companion both appear slightly elongated in the direction towards the tip-tilt star. The confirmed companion is at 0.144±0.002" separation and PA=15.87±0.13° with flux ratios of ΔJ = 0.51±0.09, ΔH = 0.51±0.03, and ΔK_s = 0.46±0.01 mag. We estimate masses of 47.4±11.7 M_{Jup} and 33.5±6.0 M_{Jup} for primary and companion respectively.

Photometry and astrometry for this object is presented in Table 1. For details on photometry and astrometry, see Biller et al. (submitted). The companion possesses very similar colors to its primary, suggesting that it is a true substellar companion. SCH 1609 was reobserved with NIRC2 at Keck II on 1 May 2010. The overall quality of the dataset was poor; however, we acquired sufficient good frames to demonstrate that the companion likely has common proper motion with the primary. Measuring centroid positions of the primary and companion (as the 2nd epoch data is not high enough quality for psf-fitting photometry), the companion moved by <0.7 pixels relative to the primary between epochs, consistent with the errors in our simple center-of-light centroiding. As no directly measured proper motion is available for SCH 1609 we adopt the mean value of (-11, -25) mas yr^{-1} for Upper Sco here (de Bruijne et al 1997; Preibisch et al. 1998). At a distance of 145 pc, parallax motion for Upper Sco is quite small – ~7 mas. As the parallax factor in the 2nd epoch observation was similar to that in the first, we neglect parallax here. With a pixel scale of 0.009942±0.00005" per pixel for the NIRC2 narrow camera, we would have expected the companion to move ~2.3 pixels relative to the primary between epochs if it was a background object at a much larger distance. Thus, this is likely a proper motion pair.

We obtained integrated-light near-IR spectroscopy of SCH1609 on 2010 September 14 (UT) using the SpeX spectrograph (Rayner et al. 2003) on the NASA Infrared Telescope Facility. A series of 12 exposures of 30 seconds each were taken, nodding along the slit, for a total integration time of six minutes. Our observations were taken at an airmass of 1.57, and the seeing recorded by the IRTF was 0.''9 The data were taken using the Low-Res prism with the 0.''5 slit aligned with the parallactic angle, producing a 0.8–2.5 μm spectrum with a resolution (R=$\lambda/\Delta\lambda$) of ~150. For telluric correction of our SCH1609 spectrum, we observed a nearby A0V star, HD 149827 and obtained calibration frames (flats and arcs). The spectra were reduced using the facility reduction pipeline, Spextool (Cushing et al. 2004), which includes a correction for telluric absorption following the method described in Vacca et al. (2003).

Table 1. Properties of the SCH1609AB System

	Primary	Secondary
Distance	145±2 pc[a]	
Age	5 Myr[b]	
Separation	0.144±0.002″ (20.9±0.4 AU)	
Position Angle	15.87°±0.13°	
ΔJ (mag)	—	0.51±0.09
ΔH (mag)	—	0.51±0.03
ΔK_S (mag)	—	0.46±0.01
J (mag)	13.53±0.09[c]	14.04±0.09
H (mag)	12.90±0.04[c]	13.41±0.04
K_S (mag)	12.56±0.03[c]	13.01±0.03
$J - K_S$ (mag)	0.97±0.09	1.03±0.09
$J - H$ (mag)	0.63±0.10	0.63±0.10
$H - K_S$ (mag)	0.34±0.05	0.40±0.05
$\log \frac{L}{L_\odot}$	-2.04±0.12	-2.23±0.12
Spectral Type	M7±0.5	M6±1.0
T_{eff}	2990±60 K	2850±170 K
Estimated Mass (5 Myr)	79±17 M_{Jup}	55±25 M_{Jup}
Estimated Mass (10 Myr)	84±15 M_{Jup}	60±25 M_{Jup}

[a]Preibisch et al.(2002)
[b]Preibisch & Zinnecker (1999)
[c]from 2MASS

SCH1609AB was assigned a composite optical spectral type of M7.5 by Slesnick et al. (2008). We determine spectral types for each component by comparing our integrated light near-IR spectrum of SCH1609AB to synthetic composites generated from template near-IR spectra of known members of Upper Scorpius (Brendan Bowler, priv comm). Our Upper Scorpius templates have optical spectral types ranging from M4 to M8.5. We verified that our templates have near-IR spectral types (calculated using the H_2O index of Allers et al. 2007) that agree to within 1 subtype with their optical types. The best fitting template is the composite spectrum of UScoCTIO 75 (M6, Ardila et al. 2000; Preibisch et al. 2002) and DENIS-P J155605.0-210646 (M7; Martin et al. 2004, Slesnick et al. 2008). We assign spectral types of M6.0 ± 0.5 to SCH1609A and M7.0 ±1.0 to SCH1609B, where uncertainties are determined from the spectral types of synthetic composite spectra where $\chi^2 \geq \chi^2_{min} + 1$, significantly earlier than the combined M7.5 spectral type from Slesnick et al. (2008).

We estimate the masses and effective temperatures of SCH 1609 AB based on the DUSTY models of Chabrier et al. (2000) and the temperature scale of Luhman (2004). The age of Upper Scorpius has been measured as 5 Myr, with a spread of up to 2-3 Myr (Preibisch & Zinnecker 1999; Slesnick et al. 2008), but more recent work suggests ages as old as 10 Myr (Eric Mamajek, private communication). Thus to account for age spread, we estimate masses at both 5 and 10 Myr ages, corresponding to masses of 79±17 M_{Jup} and 55±25 M_{Jup} at an age of 5 Myr and masses of 84±15 M_{Jup} and 60±25 M_{Jup} at an age of 10 Myr.

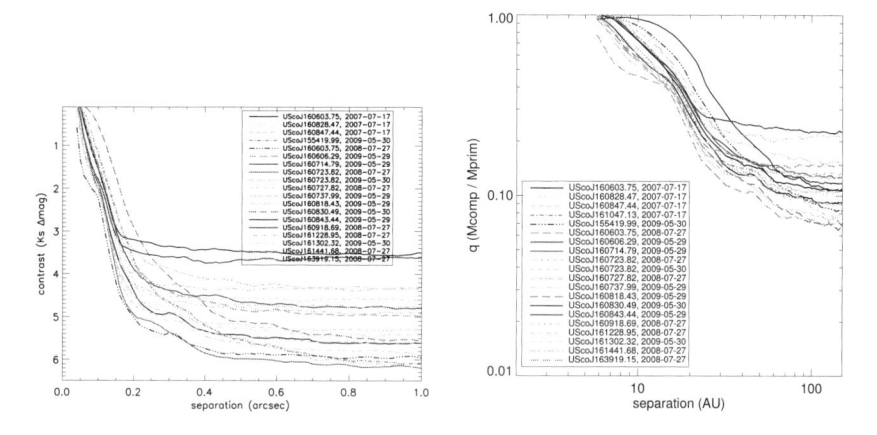

Figure 2. Left: 5σ contrast curves for 18 survey objects from Lodieu et al. (2008). Noise levels after data reduction were calculated as a function of radius by calculating the standard deviation in an annulus (with width equal to approximately the FWHM of the PSF) centered on that radius. Noise curves were then converted to contrast in Δ magnitudes by dividing by the measured peak pixel value of the object. In general, we achieve contrasts of >4 mag at separations of ≥0.4", sufficient to detect a 2MASS 1207 analogue at the distance of Upper Sco. Right: Mass ratio (q) vs. separation using the DUSTY models. Contrasts were converted to minimum detectable mass ratios using the models of Chabrier et al. (2000) at an adopted age of 5 Myr. For the best 75% of our data, we are complete to q~0.8 at 10 AU and complete to q~0.2 at ≥20 AU.

We estimate the semimajor axis of SCH 1609AB's orbit from its observed separation. Assuming a uniform eccentricity distribution between 0 < e < 1 and random viewing angles, Dupuy & Liu (2010) compute a median correction factor between projected separation and semimajor axis of $1.10^{+0.91}_{-0.36}$ (68.3% confidence limits). Using this, we derive a semimajor axis of $23.0^{+19.0}_{-7.5}$ AU for SCH 1609AB based on the observed separation in June 2009.

4. Achieved Contrasts and Limits on Minimum Detectable Companion Masses

$5\text{-}\sigma$ contrast curves are presented in Figure 2. Noise levels after data reduction were calculated as a function of radius by calculating the standard deviation in an annulus (with width equal to approximately the FWHM of the PSF) centered on that radius. Noise curves were then converted to contrast in Δ magnitudes by dividing by the measured peak pixel value of the object. Contrasts were converted into absolute magnitudes using photometry reported in Lodieu et al. (2008) and Slesnick et al. (2008), and adopting a distance of 145 pc for Upper Sco. A filter transform was calculated from K to K_s band using the spectra from Lodieu et al. (2008).

To test the fidelity of our contrast curves, we inserted and retrieved simulated objects in our data. Objects were simulated as 2 dimensional gaussians with FWHMs from fits to the primary and contrasts from our measured contrast curves. Objects simulated with contrasts from our measured curves were retrieved with S/N≥5 for all survey targets down to separations of 0.07" – thus, our measured contrast curves are a reliable estimate of the detectable contrasts for potential companions down to separations of 0.07".

Contrasts were converted to minimum detectable mass ratios using the models of Chabrier et al. (2000) at an adopted age of 5 Myr and assuming a similar bolometric correction between each target and any potential companion (Figure 2). We note that for the best 75% of our data we are complete for all binaries with q≥0.8 at separations >10 AU and all binaries with q≥0.2 at separations >50 AU.

5. Discussion

We find an upper limit on the binary fraction (>10 AU) of 9% (1-σ) for the 18 objects we surveyed from Lodieu et al. 2008 (calculated via the method of Burgasser et al. 2003). (We exclude the sources observed from the Slesnick et al. 2008 sample since they appear so much brighter than the Lodieu et al. sources and likely have masses > 40 M$_{Jup}$).

5.1. Brown Dwarf Binary Fraction as a Function of Mass

We compare our measured binary fraction to that of more massive brown dwarfs and very low mass stars in the Upper Sco embedded cluster. Kraus et al. (2005) surveyed 12 brown dwarfs and very low mass stars with ACS on HST. These objects have estimated masses of 0.04 – 0.1 M$_\odot$ and thus comprise a higher mass sample than our survey. Kraus et al. (2005) discovered three binaries in this sample, one of which (USco-109 AB) is below the sensitivity of our survey to detect, with a projected separation of only ~5 AU. Thus, for the purposes of comparison, we adopt a binary fraction of 2/12 = 17^{+15}_{-6} % for the Kraus et al. (2005) sample. To determine the likelihood that these two binomial distributions are the same, we adopted the Fischer exact test method used by Ahmic et al. (2007) and described in the appendix of Brandeker et al. (2006). The likelihood that the Kraus et al. (2005) sample is drawn from the same distribution as ours is 0.15. Thus, as noted by previous authors, the binary fraction in Upper Sco continues to decrease with decreasing primary mass.

This comparison is limited by the relatively small number of objects observed in Upper Sco. Thus, to improve statistics, we have compiled a larger list using objects from similar surveys of other young, nearby regions – specifically Taurus (<1 Myr, 145 pc, objects from Kraus et al. 2006; Konopacky et al. 2007) and Chamaeleon (<3 Myr, 160 pc, objects from Ahmic et al. 2007). We include only objects that would have been detected at the sensitivity level of our survey and limit this analysis to nearby clusters (<200 pc) since more distant clusters (e.g. NGC 1333, IC 348, Serpens) are more than 250 pc distant and do not reach comparable sensitivity levels at 10 AU. All selected surveys have similar sensitivity levels (complete to q~0.8 at 10 AU, complete to q~0.2 - 0.3 at ≥20 AU) so it is unlikely that our survey would have discovered a binary at a separation >10 AU missed by these other surveys, and vice versa. We adopt 3 mass bins for this analysis: (1) high mass (0.07 – 0.1 M$_\odot$), with 6 binaries detected out of 23 objects surveyed (6 objects from Ahmic et al. 2007, 5 from Kraus et al. 2006, 6 from Kraus et al. 2005, 4 from Konopacky et al. 2007, and the two objects from the Slesnick et al. 2008 sample from the current survey), (2) medium mass (0.04 – 0.07 M$_\odot$), with 0 binary detected out of 18 objects surveyed (4 objects from Ahmic et al. 2007, 6 from Kraus et al. 2005, and 8 from Kraus et al. 2006), and (3) low mass

(<0.04 M$_\odot$), with 0 binaries detected out of 25 objects surveyed (7 objects from Kraus et al. 2006 and the 18 objects from the Lodieu et al. 2008 sample surveyed herein). We note that while a number of additional binaries are known in this mass range, e.g. 2MASS 1207AB (Chauvin et al. 2005), 2M 1622 (Allers 2006; Allers et al. 2006; Jayawardhana & Ivanov 2006; Allers et al. 2007; Close et al. 2007), and 2MASS 0441 (Todorov et al. 2010), survey statistics are not available for these objects and thus we cannot include them in our sample. Binary fractions and likelihoods between bins are presented in Table 2. As expected, binary fraction decreases monotonically with primary mass. The likelihood that the low mass bin (<0.04 M$_\odot$) objects share the same binary fraction as the high mass bin (>0.07 M$_\odot$) is less than 0.02.

Table 2. Statistical Sample Comparison

Sample 1	Sample 2	Likelihood
Upper Sco vs. Field		
<0.04 M$_\odot$ BDs in Upper Sco 0 / 18, <9%	Field T Dwarfs 0 / 32, <5%	1.0
Upper Sco Mass Comparison		
<0.04 M$_\odot$ 0 / 18, <9%	0.04–0.1 M$_\odot$ 2 / 12, 17^{+15}_{-6}%	0.15
Multiple Regions Mass Comparison		
<0.04 M$_\odot$ 0 / 25, <7%	0.07–0.1 M$_\odot$ 6 / 23, 26^{+11}_{-7}%	0.01
<0.04 M$_\odot$ 0 / 25, <7%	0.04–0.07 M$_\odot$ 0 / 18, <9%	1.0
0.04–0.07 M$_\odot$ 0 / 18, <9%	0.07–0.1 M$_\odot$ 6 / 23, 26^{+11}_{-7}%	0.03

5.2. Brown Dwarf Binary Fraction as a Function of Age

By ages of 1 Gyr, most of our survey objects will have cooled to become T dwarfs. Thus, it is interesting to compare the primordial binary fraction of these objects to the binary fraction of similar objects in the field. Our survey is only sensitive to companions at projected separations of >10 AU, however, this is a particularly interesting separation space to probe, as older field T dwarf binaries rarely have separations this large (Burgasser et al. 2003, 2006). In fact, of the 32 T dwarfs surveyed in Burgasser et al. (2003, 2006), no companions were detected with separation >10 AU (down to q≥0.4, i.e. comparable sensitivity limits to our survey). This places an upper limit on the binary fraction >10 AU of 5%. Again using the Fischer exact test method, we found a likelihood of 1 – i.e., given the small sizes of both of these samples, they are very likely drawn from the same parent sample. Thus, the very low mass brown dwarf binary fraction appears to be similar for both young and field objects.

Do higher mass objects (>0.07 M$_{Sun}$) in young clusters also have a similar binary fraction (>10 AU) as their counterparts in the field? We compare the binary fraction (>10 AU separation) for 6 binaries discovered out of 23 young objects (the "high mass" bin from the previous section) drawn from binarity surveys of Upper Sco (Kraus et al. 2005), Taurus (Kraus et al. 2006; Konopacky et al. 2007), Chamaeleon (Ahmic et al. 2007) and this work with that of 1 binary (>10 AU separation) discovered from 39 field M8–L0.5 objects from Close et al. 2003. These two samples share a similar mass range (primary mass between 0.07 – 0.1 M$_\odot$), but very different wide binary fractions: 26^{+11}_{-7}% for the young sample vs. $2.6^{+5.4}_{-0.06}$% for the old field sample. Using the Fischer exact test, the likelihood that these two samples are drawn from the same binomial distribution is 0.01. Thus, for objects with mass > 0.07 M$_\odot$, there is an overabundance of 10–50 AU separation very low mass binaries in young clusters relative to the field.

Upper Sco is a somewhat older and higher density OB association, while Taurus and Chamaeleon are more diffuse, younger T clusters. Thus, we also compare binary fraction between these two different ages and environments. Combining the sample described in the previous section and separating by region, we find 3 binaries detected from 34 objects in Taurus and Chamaeleon and 3 binaries detected from 32 objects in Upper Sco. Binary fraction is nearly the same between the two, although it is important to note that the sample in Taurus and Chamaeleon is systematically higher mass than that in Upper Sco (dominated by the 18 very low mass brown dwarfs surveyed in this paper.) Thus, given the trend in binary fraction with mass, the binary fraction in Upper Sco may be considerably higher than in Taurus.

5.3. Stability of 10-50 AU Separation Binaries in Young Nearby Starforming Regions

Up to ~25% of very low mass (henceforth VLM) stars and substellar objects in young star forming regions may have companions at separations >10 AU. However, very low mass star / brown dwarf binaries with separations >15 AU are rare in the field. Of ~100 VLM binaries compiled at vlmbinaries.org, only ~10% have separations >15 AU. Assuming a binary fraction of ~10%, less than 1% of field VLMs have companions at separations >15 AU (Close et al. 2007).

Close et al. (2007) suggest that very wide (>50 AU) young brown dwarf binaries are disrupted within the first 10 Myr of their existence by interactions with stars in their natal cluster. To set limits on the survival time of a young wide binary in its natal cluster, they adopt the analytic solution of Fokker-Plank (FP) coefficients from Weinberg et al. (1987) which describes the advective diffusion of a binary due to stellar encounters. Namely, from this solution, the time t_* necessary to evaporate a binary with initial semimajor axis a_0 is:

$$t_* \sim 3.6 \times 10^5 (\frac{n_*}{0.05 pc^{-3}})(\frac{M_{tot}}{M_\odot})(\frac{M_*}{M_\odot})^{-2}(\frac{V_{rel}}{20 km s^{-1}})(\frac{a_0}{AU})^{-1} \qquad (1)$$

where n_* is the number density of stellar perturbers of mass M_* and relative velocity V_{rel}. Using this relationship, Close et al. (2007) determine that young wide VLM binaries such as 2M 1207AB will not survive 10 Myr of interactions with 0.7 M$_\odot$ stellar perturbers with a number density n_* of 1000 pc^{-3}. Thus, Close et al. (2007) show that most of these binaries will not survive to join the field if born in a dense starforming region. We determine here whether the same is true for 10 – 50 AU binaries. While Close et al. (2007) assume a number density of nearby stars of 1000 pc^{-3}, which is appropriate near dense core regions, it is probably too high for objects in diffuse areas of Taurus or Chamaeleon. Assuming a typical density of 100 pc^{-3} for for Taurus, Ophiuchus, and Upper Sco, we repeat this calculation for the six binaries that fall into our highest mass bin (specifically, CFHT-Tau 7, CFHT-Tau 17, and CFHT-Tau 18 from Konopacky et al. 2007, USco-55 and USco-66 from Kraus et al. 2005, and SCH 1609AB, the newly discovered binary presented herein). We find that all of these binaries are quite stable and will survive >10 Myr in either a 100 pc^{-3} environment or a 1000 pc^{-3} environment (i.e. long enough to join the field population). An environment with stellar densities >10^4 pc^{-3} (equivalent to the Trapezium cluster in Orion) is necessary to disrupt these binaries on <10 Myr timescales.

The existence of a significant population of these binaries presents a conundrum, since very low mass stars and brown dwarf binaries with separations >15 AU are rare in the field. However, the field brown dwarf population encompasses a mix of objects which formed in a variety of different star-forming regions. Close et al. (2007) suggest that brown dwarf binaries with separations >20 AU are found rarely in the field because they can only form in low-density star-forming regions, while the majority of field objects formed in denser initial regions where any such binary would be disrupted. However, other authors have suggested that most stars in the field likely form in OB associations like Upper Sco (Konopacky et al. 2007, Preibisch & Mamajek 2008), so this population of young, wide binaries with separations of 10-50 AU remains problematic.

6. Conclusions

We searched for binary companions to 20 brown dwarfs in Upper Scorpius (145 pc, 5 Myr, nearest OB association) with the facility infrared camera NIRC2 and the Laser Guide Star adaptive optics system on the 10 m Keck II telescope. This survey is the most extensive to date for companions to very young (5 Myr), very low mass (<40 M_{Jup}) cluster brown dwarfs. We discovered a close companion (0.14″, 20.9±0.4 AU) to the M7.5 (combined spectral type) brown dwarf SCH J16091837-20073523. From spectral deconvolution of integrated-light near-IR spectroscopy of SCH1609 using the SpeX spectrograph (Rayner et al. 2003), we estimate primary and secondary spectral types of M6±0.5 and M7±1.0, corresponding to masses of 79±17 M_{Jup} and 55±25 M_{Jup} at an age of 5 Myr and masses of 84±15 M_{Jup} and 60±25 M_{Jup} at an age of 10 Myr.

For our survey objects with spectral types later than M8, we find an upper limit on binary fraction of <9% (1-σ) at separations greater than 10 AU. As expected from similar mass binaries in the field, we find that the binary fraction (>10 AU separations) appears to decrease monotonically with mass for young brown dwarfs. However, while proto-T-dwarfs (M<40 M_{Jup}) have a similar wide (>10 AU) binary fraction as field T dwarfs, there exists an anomalous population of wide higher mass binaries (0.07 – 0.1 $M_{\odot}$ primaries, separations of 10–50 AU) at young ages relative to older ages. This may suggest that field objects predominately formed in higher stellar density regions than nearby cluster objects.

Acknowledgments. The data presented herein were obtained at the W.M. Keck Observatory, which is operated as a scientific partnership among the California Institute of Technology, the University of California and the National Aeronautics and Space Administration. The Observatory was made possible by the generous financial support of the W.M. Keck Foundation. The authors wish to recognize and acknowledge the very significant cultural role and reverence that the summit of Mauna Kea has always had within the indigenous Hawaiian community. We are most fortunate to have the opportunity to conduct observations from this mountain. B.B. was supported by Hubble Fellowship grant HST-HF-01204.01-A awarded by the Space Telescope Science Institute, which is operated by AURA for NASA, under contract NAS 5-26555.

References

Ahmic, M., Jayawardhana, R., Brandeker, A., Scholz, A., van Kerkwijk, M. H., Delgado-Donate, E., & Froebrich, D. 2007, ApJ, 671, 2074
Allers, K. N. 2006, Ph.D. Thesis,
Allers, K. N., Kessler-Silacci, J. E., Cieza, L. A., & Jaffe, D. T. 2006, ApJ, 644, 364
Allers, K. N., et al. 2007, ApJ, 657, 511
Ardila, D., Martín, E., & Basri, G. 2000, AJ, 120, 479
Bouchez, A. H., et al. 2004, proc. SPIE, 5490, 321
Brandeker, A., Jayawardhana, R., Khavari, P., Haisch, K. E., Jr., & Mardones, D. 2006, ApJ, 652, 1572
Burgasser, A. J., Kirkpatrick, J. D., Reid, I. N., Brown, M. E., Miskey, C. L., & Gizis, J. E. 2003, ApJ, 586, 512
Burgasser, A. J., Kirkpatrick, J. D., Cruz, K. L., Reid, I. N., Leggett, S. K., Liebert, J., Burrows, A., & Brown, M. E. 2006, ApJS, 166, 585
Chabrier, G., Baraffe, I., Allard, F., & Hauschildt, P. 2000, ApJ, 542, 464
Chauvin, G., Lagrange, A.-M., Dumas, C., Zuckerman, B., Mouillet, D., Song, I., Beuzit, J.-L., & Lowrance, P. 2005, A&A, 438, L25
Close, L. M., Siegler, N., Freed, M., & Biller, B. 2003, ApJ, 587, 407
Close, L. M., et al. 2007, ApJ, 660, 1492
Cushing, M. C., Vacca, W. D., & Rayner, J. T. 2004, PASP, 116, 362
de Bruijne, J. H. J., Hoogerwerf, R., Brown, A. G. A., Aguilar, L. A., & de Zeeuw, P. T. 1997, Hipparcos - Venice '97, 402, 575
Jayawardhana, R., & Ivanov, V. D. 2006, Science, 313, 1279
Konopacky, Q. M., Ghez, A. M., Rice, E. L., & Duchêne, G. 2007, ApJ, 663, 394
Kraus, A. L., White, R. J., & Hillenbrand, L. A. 2005, ApJ, 633, 452
Kraus, A. L., White, R. J., & Hillenbrand, L. A. 2006, ApJ, 649, 306
Kraus, A. L., Ireland, M. J., Martinache, F., & Lloyd, J. P. 2008, ApJ, 679, 762
Lodieu, N., Hambly, N. C., Jameson, R. F., Hodgkin, S. T., Carraro, G., & Kendall, T. R. 2007, MNRAS, 374, 372
Lodieu, N., Hambly, N. C., Jameson, R. F., & Hodgkin, S. T. 2008, MNRAS, 383, 1385
Luhman, K. L. 2004, ApJ, 614, 398
Martín, E. L., Delfosse, X., & Guieu, S. 2004, AJ, 127, 449
Preibisch, T., Guenther, E., Zinnecker, H., Sterzik, M., Frink, S., & Roeser, S. 1998, A&A, 333, 619
Preibisch, T., Brown, A. G. A., Bridges, T., Guenther, E., & Zinnecker, H. 2002, AJ, 124, 404
Preibisch, T., & Zinnecker, H. 1999, AJ, 117, 2381
Preibisch, T., & Mamajek, E. 2008, Handbook of Star Forming Regions, Volume II, 235
Rayner, J. T., Toomey, D. W., Onaka, P. M., Denault, A. J., Stahlberger, W. E., Vacca, W. D., Cushing, M. C., & Wang, S. 2003, PASP, 115, 362
Reipurth, B., & Clarke, C. 2001, AJ, 122, 432
Slesnick, C. L., Hillenbrand, L. A., & Carpenter, J. M. 2008, ApJ, 688, 377
Stamatellos, D., Hubber, D. A., & Whitworth, A. P. 2007, MNRAS, 382, L30
Stamatellos, D., & Whitworth, A. P. 2009, MNRAS, 392, 413
Todorov, K., Luhman, K. L., & McLeod, K. K. 2010, ApJ, 714, L84
Vacca, W. D., Cushing, M. C., & Rayner, J. T. 2003, PASP, 115, 389
Weinberg, M. D., Shapiro, S. L., & Wasserman, I. 1987, ApJ, 312, 367
Wizinowich, P. L., et al. 2004, proc. SPIE, 5490, 1

Plenary and Splinter Sessions at Cool Stars 16

The 16th Cambridge Workshop on Cool Stars, Stellar Systems and the Sun
ASP Conference Series, Vol. 448
Christopher M. Johns-Krull, Matthew K. Browning, and Andrew A. West, eds.
© 2011 Astronomical Society of the Pacific

Extreme Coronal Mass Ejections in Young Low-Mass Stars

A. N. Aarnio,[1] K. G. Stassun,[2,3] S. P. Matt,[4] W. J. Hughes,[5] and S. L. McGregor[5]

[1] *Department of Astronomy, University of Michigan, 830 Dennison Building, 500 Church Street, Ann Arbor, MI, 48109, USA*

[2] *Department of Physics and Astronomy, Vanderbilt University, VU Station B, 1807, Nashville, TN, 37235, USA*

[3] *Department of Physics, Fisk University, 1000 17th Avenue N., Nashville, TN, 37208, USA*

[4] *Laboratoire AIM Paris-Saclay, CEA/Irfu Université Paris-Diderot CNRS/INSU, 91191 Gif-sur-Yvette, France*

[5] *Department of Astronomy and Center for Integrated Space Weather Modeling, Boston University, 725 Commonwealth Avenue, Boston, MA, 02215, USA*

Abstract. Two long-standing questions in the study of young, low-mass stars are: (1) What are the mechanisms that govern the observed order-of-magnitude decrease of stellar angular momentum during pre-main-sequence evolution, and (2) What are the physical drivers of X-ray production in these stars at up to 10^4 times the solar value? Application of solar flare models to the most powerful X-ray flares observed among T Tauri stars in Orion suggests that the flares are produced by magnetic loop structures with lengths of up to tens of stellar radii. We present new results demonstrating that, for the majority of these stars, the extremely large flaring structures are not anchored to or stabilized by circumstellar disks. Given the energy and size scales involved, mass losses (e.g., via stellar coronal mass ejections—CMEs—associated with these flares) at such large effective lever arms could shed substantial angular momentum. To begin estimating the attendant angular momentum losses of such extreme CMEs in young stars, we have assembled from the solar literature a database of ~10,000 X-ray flares and CMEs, from which we determine for the first time the empirical relationship between solar X-ray flare energy and CME ejected mass. Finally, we demonstrate how our flare flux/CME mass relationship can be used to estimate stellar angular momentum loss via extreme CMEs in young stars.

1. Introduction

Being pre-main sequence and solar mass, T Tauri Stars (TTS) tell us, in essence, what our Sun was like when it was very young. TTS and their protoplanetary disks are ideal astrophysical laboratories for studying the solar system during the epoch of planet formation. At the same time, there is no astrophysical laboratory like the present-day Sun for studying in detail the microphysics needed to understand solar-type processes involved in early stellar evolution.

Presently, it is not well understood how stellar X-ray emission is produced. Observations indicate that T Tauri X-ray emission is similar to solar X-ray emission, but with total luminosities several orders of magnitude greater than the present-day Sun. Potential explanations are scaled up, solar-like coronae (e.g., Stassun et al. 2004, 2006, 2007) and accretion-driven X-ray emission (Kastner et al. 2002). It also remains unclear how solar-type stars shed up to an order of magnitude of angular momentum as they contract onto the main sequence. X-ray activity, rotation, and angular momentum loss are coupled processes: it is known that rotation and activity are related, and since angular momentum transfer changes the stellar rotation, so it affects activity. To further complicate matters, that activity can be related to angular momentum loss. Understanding this complex system of interdependence and feedback is critical for understanding pre-main sequence stellar evolution.

In both the cases of accretion-driven X-ray emission and angular momentum loss, large scale magnetic loops could be ideal structures for facilitating angular momentum and material transfer in a protostellar environment. Our work here investigates the relationship of large scale loops to circumstellar material as well as additional roles which could be played by these structures in the cases where they are not disk-connected.

2. Large Flares on Young Stars: Star-Disk Interaction?

X-ray flares are observed frequently on young stars, and the morphology of their light curves imply that stellar flares are governed by the same physics as solar flares. The Chandra Orion Ultradeep Project observed hundreds of young, low-mass stars undergoing highly energetic X-ray flares in the Orion Nebula Cluster (ONC; Getman et al. 2005). Solar-calibrated models (i.e., the uniform cooling loop model, Reale & Micela 1998) were used to ascertain the size scale of the magnetic field confining the X-ray emitting plasma. The results were provocative: in the most energetic flares observed, magnetic loops on TTS of the ONC were inferred to be tens of stellar radii in size (Favata et al. 2005), potentially representing the type of large scale star-disk fields long envisioned in magnetospheric accretion scenarios.

In order to determine whether these large-scale loops are indeed intersecting with circumstellar disks, in Aarnio et al. (2010) we construct spectral energy distributions, fitting photometric data with star+disk radiative transfer models (Whitney et al. 2003; Robitaille et al. 2006) to quantitatively establish the location of the inner edge of a dust disk. Our photometric data span the wavelength range of 0.3-24μm, allowing us to determine the structure of protoplanetary disks out to >10 AU from the stellar surface in most cases. We then compare the inner disk radii determined from SED fitting to the observed magnetic loop sizes.

In ~20% of the stars studied, we do find close-in dust disks with inner radii within reach of the magnetic loop (see Figure 1). In ~60% of the stars studied, however, we find either no disk for the magnetic loop to intersect, or the dust disk is truncated far from the loop (Figure 2). Thus, we see that in the plurality of cases, these large magnetic loops are free-standing, anchored only to the stellar surface. If these extended structures are not participating in star-disk interaction, then how are they stabilized, and what happens when they de-stabilize? To address this, we look again to solar physics: solar flares occur due to magnetic field destabilization, as do other energetic solar phenomena such as solar energetic particle events (SEPs) and CMEs. In the same way that multiple energetic phenomena are often observed to occur simultaneously

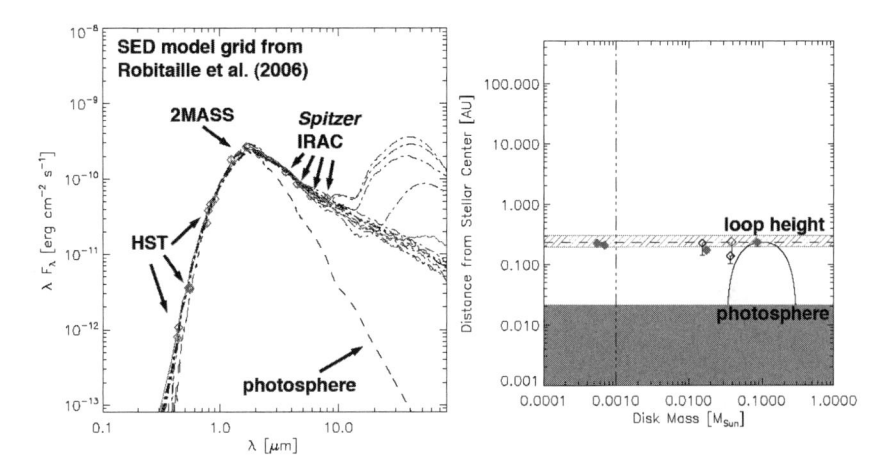

Figure 1. SED of COUP 454. Left panel: Diamonds are photometric data from HST, 2MASS, and the *Spitzer* Infrared Array Camera (IRAC). Black dash-dotted curves are best-fit model SEDs from the grid of Robitaille et al. (2006). The blue SED is a fiducial model with the inner disk radius set to the magnetic loop height (or dust destruction, if the loop is within the dust sublimation radius). The dashed curve is a model stellar atmosphere. Right panel: Comparing R_{loop} (hatched region, loop illustrated by cartoon parabola) and R_{trunc} for the best-fit SEDs models. Open diamonds show the location of the inner disk edge, with a vertical line extending downward to the dust destruction radius for that particular model. If the disk is truncated at dust destruction, the diamond is filled in red.

on the Sun, perhaps stellar flares could be used as tracers for other high-energy events which we cannot presently observe on TTS. Specifically, we seek to infer the properties of TTS coronal mass ejections using their observed X-ray flares.

3. Determining a Solar Flare Flux-CME Mass Relationship

To determine a stellar angular momentum loss rate for TTS, the following basic calculation could be done:

$$\frac{dJ}{dt} = F(E,t) \times \frac{\text{CME mass}}{\text{Flare energy}} \times \frac{\text{Angular momentum loss}}{\text{unit mass}} \times f, \qquad (1)$$

where $F(E, t)$ is a function describing a stellar flare energy and rate distribution, the second term relates flare flux and CME mass, and f is an efficiency factor (i.e., how often does a CME occur which is associated with a flare). We focus here on constraining the second term for the Sun, with the eventual goal of extending the analogy to the energy scales representative of TTS flares.

In Aarnio et al. (2011, in press), we use long-term observations of the Sun documenting flares, coronal mass ejections (CMEs), and their properties in an effort to assess

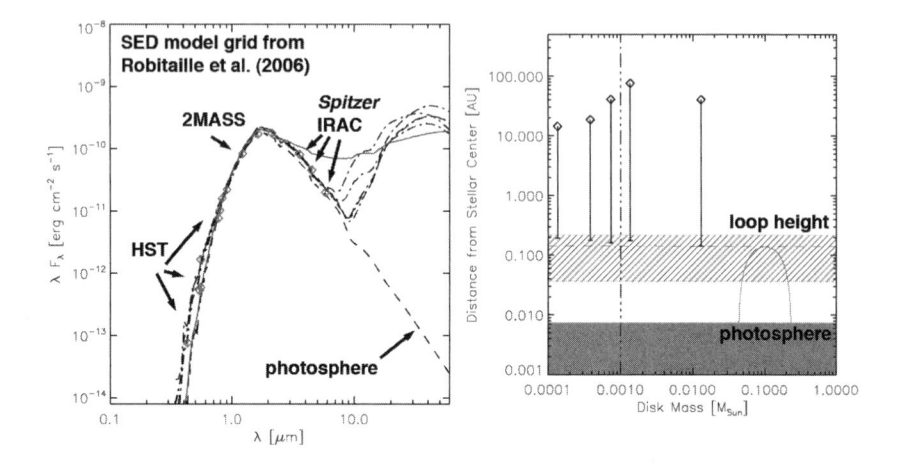

Figure 2. SED of COUP 262; all symbols as described in Figure 1. The IRAC photometry measured for COUP 262 does not indicate any close-in, dusty disk material. Thus, the SED fitter, even lacking photometric constraint beyond 5.8μm, results only in models with disks truncated at 10 AU or farther, well beyond reach of the flaring loop and beyond dust destruction for every model. We intepret this to be most consistent with a Class III T Tauri star.

the correlation of flare and CME properties. The LASCO CME Catalog[1] and GOES X-ray flare database[2] catalog observations from the Large Angle and Spectrometric Coronagraph and the Geostationary Operational Environmental Satellite; we correlate CME occurrence and X-ray flares from these databases over the period of 1996 to 2006. We compare ~7,000 CMEs with well-measured masses and ~12,000 flares with positions verified by optical counterparts. We require the events to be cospatial and cotemporal. Figure 3 shows our positional constraint: converting each flare's Stonyhurst (Cartesian) position to polar coordinates, we require the flare's polar angle and corresponding CME's central position angle to be coincident within ±45°.

For flares and CMEs which fit the spatial criterion, we then determined the temporal correlation window. Looking at a wide range of time offsets (CME − flare start times), we find a clearly correlated sample within the initial ±2 hour window. Figure 4 illustrates our choice of time window: we require any correlated CMEs to occur 10-80 minutes after their associated flare.

The final sample of associated flares and CMEs consists of 826 event pairs, the largest study of this nature to date. With such high number statistics, we discover that the flare fluxes and CME masses are indeed correlated (see black diamonds, Figure 5). The correlation holds over approximately 3 dex in flare flux, with CME mass increasing by an order of magnitude over this range. Interestingly, there appears to be a break in

---

[1] The CME catalog is generated and maintained at teh CDAW Data Center by NASA and The Catholic University of America in cooperation with the Naval Research Laboratory. SOHO is a project of international cooperation between ESA and NASA.

[2] http://www.ngdc.noaa.gov/stp/SOLAR/ftpsolarflares.html

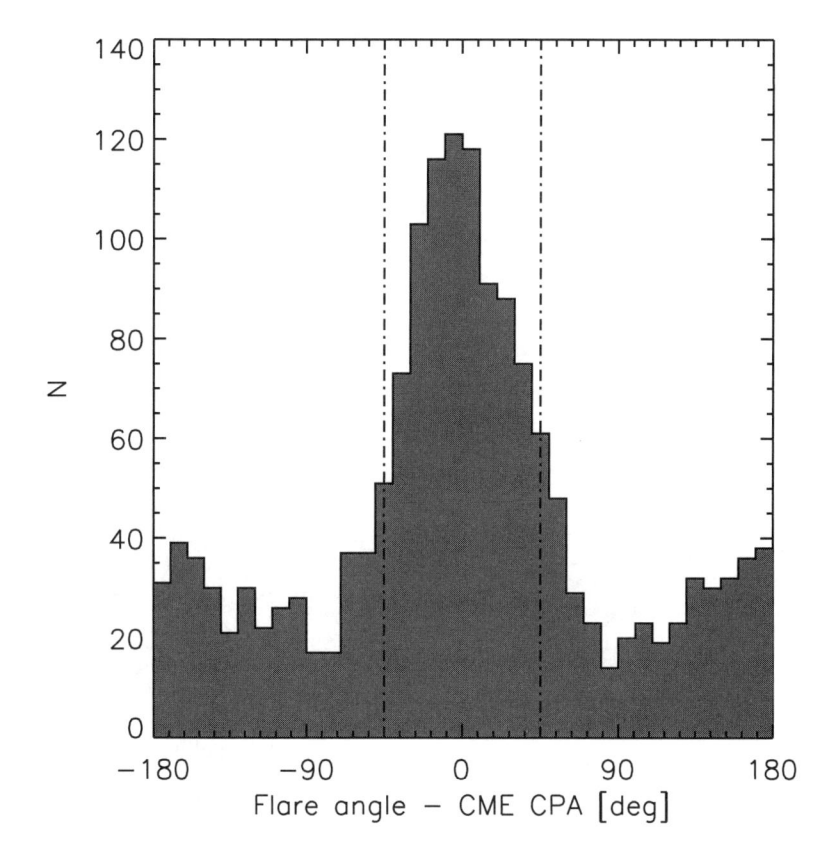

Figure 3. Using a loose time criterion of ±2 hours, a very distinct correlation in position angles of flares and CMEs is observed. Vertical lines indicate our final adopted angular separation range of 0±45°.

the function around log(flux)~ −4.5: at this point, the slope of the function becomes more shallow.

In our analysis, we excluded both flares which lacked measured masses and those which had poorly constrained masses; because of this, we had in fact excluded the widest, most massive CMEs observed, their large widths having created the measurement difficulty. We re-introduced these so-called halo CMEs to the analysis, and found that with increasing flare flux, the width of the correlated CME increases (Figure 6). Effectively, we had failed to recover the CMEs associated with the highest flux flares.

The functional form of our flare flux/CME mass relationship, including halo CMEs, is as follows:

$$\log(\text{CME mass}) = (18.67 \pm 0.27) + (0.70 \pm 0.05) \times \log(\text{flare flux}) \quad (2)$$

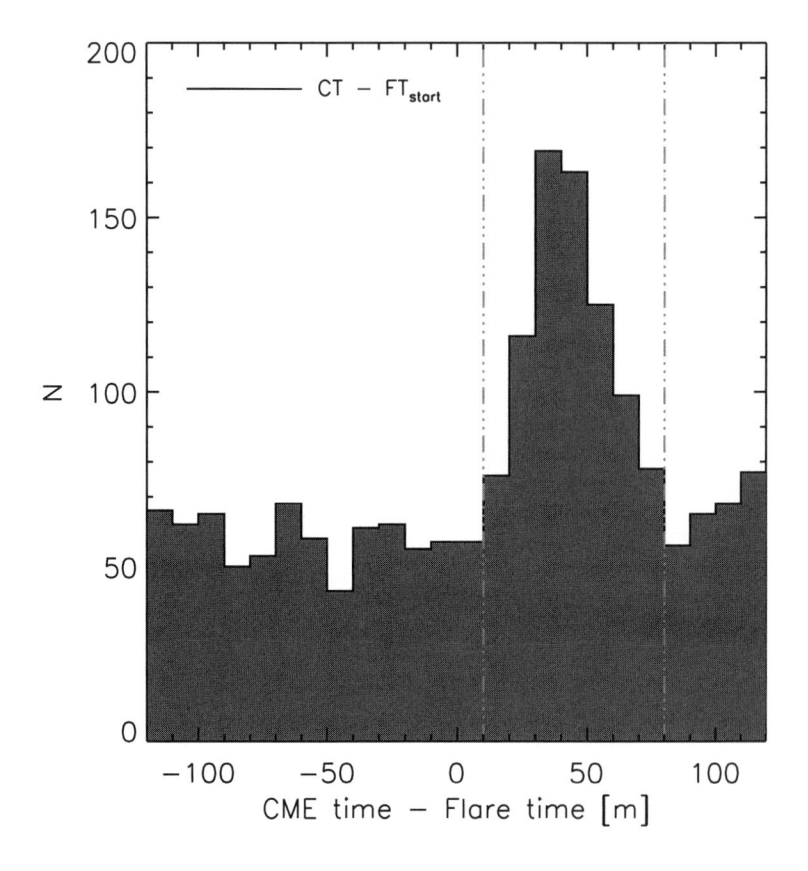

![Figure 4 histogram showing N vs CME time − Flare time in minutes, with peak near 20-60 minutes]

Figure 4. For CMEs and flares within 45° in position angle, here we show the number of paired flares and CMEs within a ±2 hour time window. The times used were the flare start time and the CME's first appearance in the LASCO C2 detector. The observed peak spanning the range of time offsets from 10-80 minutes (indicated by gray vertical lines) is a factor of ~3 above the apparent "background" event level.

4. Summary and Conclusions

In summary, we have assessed, for the 32-star sample of the most energetically flaring stars in the ONC, whether large-scale magnetic structures are sufficiently large and their circumstellar dust close enough for star-disk interaction to be presently ongoing. Solar models imply the flares have been caused by magnetic loops that are up to tens of stellar radii in extent. Finding that in most cases circumstellar dust material is truncated well beyond the reach of the inferred large-scale magnetic loops (Aarnio et al. 2010), we analyze other consequences of these large scale flaring magnetic structures. We invoke solar physics to relate solar flare flux to CME mass; this log-linear relationship can be then scaled up to the T Tauri parameter regime, making use of TTS X-ray flare observations to estimate mass losses via presently unobservable TTS CMEs (Aarnio

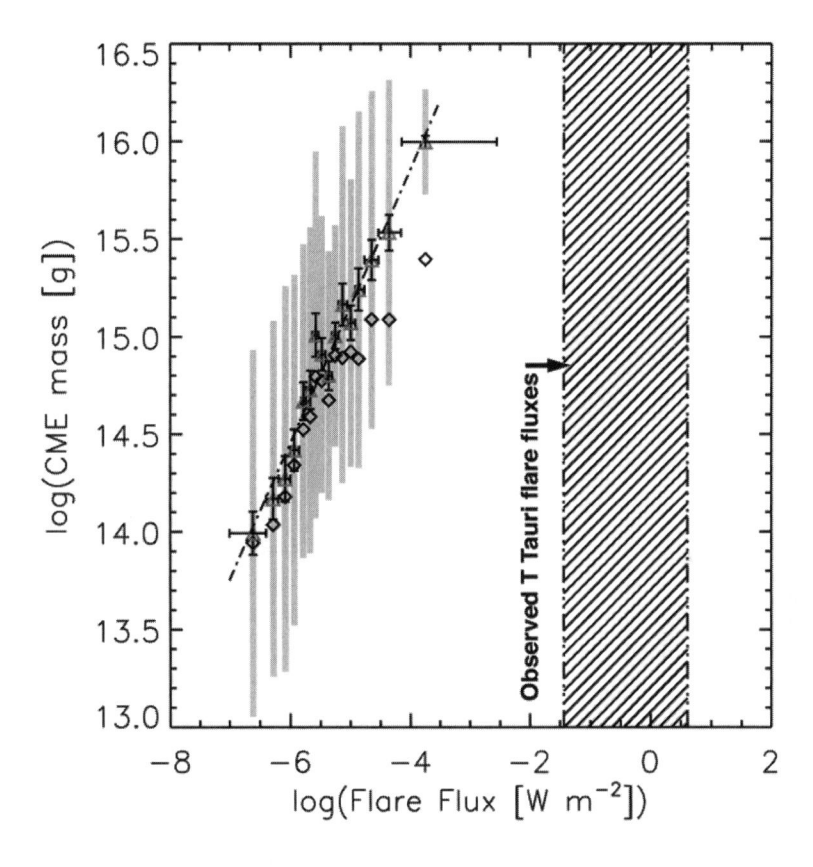

Figure 5. Relationship beween flare flux and CME mass. Black diamonds indicate correlated flares and CMEs for which the CMEs have well-measured masses. Red triangles indicate how the function would change were we to include halo CMEs, assigning them the maximum mass from the bin occupied by the associated flares. Vertical error bars show the standard deviation of the mean masses plotted with red triangles; shaded, gray boxes are of arbitrary width, their heights showing the standard deviation of the masses in each flare flux bin. Horizontal error bars indicate the with of each bin in flare flux. The black dot-dashed line represents the fit to the red triangles, indicating the steepest slope we could anticipate for the flare flux-CME mass relationship assuming that halo CMEs have masses which lie within the observed distributions. Finally, the hatched region represents the range of flare fluxes observed on TTS in the ONC.

et al. 2011, in press). Mass losses from effective lever arm heights of tens of stellar radii could shed substantial angular momentum (Aarnio et al. 2009).

Equation 1 outlines how to go about calculating an angular momentum loss rate via stellar CMEs. The first term, a stellar flare event rate, has already been measured for several young stellar clusters (e.g., Albacete Colombo et al. 2007). In this work, we have shown how to obtain the second term relating flare flux and mass lost via

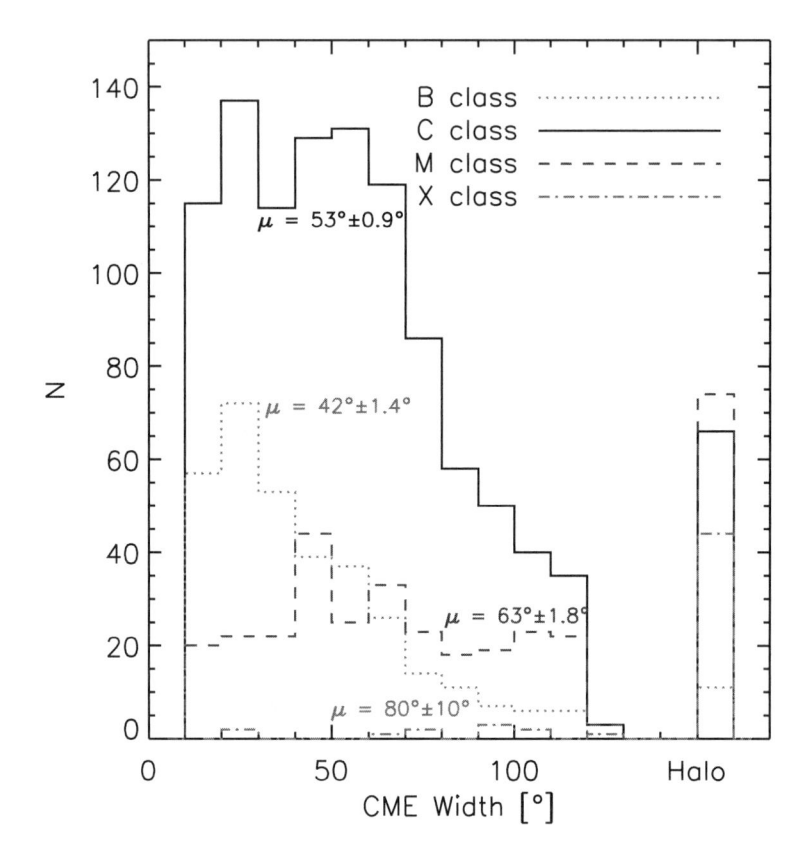

Figure 6. CME widths as a function of associated flare flux. The average width and the corresponding standard deviation of each distribution's mean are printed on the figure. Each distribution's mean is statistically significantly different from the others. The majority of X-class flares (80%) in our sample are associated with "halo" CMEs; this fraction decreases with decreasing flare flux.

associated CME for the solar case. As can be seen in Figure 5, there are several orders of magnitude in X-ray flare flux to be carefully traversed before arriving at a flare flux/CME mass distribution for TTS, but these steps represent a beginning point of a novel method for understanding episodic mass and angular momentum loss events in young stars.

Acknowledgments. This work was supported by NSF grant AST-0808072 (K. Stassun, P.I.). K. G. S. acknowledges a Cottrell Scholar award from the Research Corporation, and a Diversity Sabbatical Fellowship from the Ford Foundation. Work at Boston University was supported by CISM, which is funded by the STC Program of the National Science Foundation under Agreement Number ATM-0120950.

References

Aarnio, A. N., Stassun, K. G., Hughes, W. J., & McGregor, S. L. 2011, ArXiv e-prints. 1011.0424
Aarnio, A. N., Stassun, K. G., & Matt, S. P. 2009, in American Institute of Physics Conference Series, edited by E. Stempels, vol. 1094 of American Institute of Physics Conference Series, 337
— 2010, ApJ, 717, 93. 1005.2128
Albacete Colombo, J. F., Caramazza, M., Flaccomio, E., Micela, G., & Sciortino, S. 2007, A&A, 474, 495. 0708.2399
Favata, F., Flaccomio, E., Reale, F., Micela, G., Sciortino, S., Shang, H., Stassun, K. G., & Feigelson, E. D. 2005, ApJS, 160, 469. arXiv:astro-ph/0506134
Getman, K. V., Feigelson, E. D., Grosso, N., McCaughrean, M. J., Micela, G., Broos, P., Garmire, G., & Townsley, L. 2005, ApJS, 160, 353. arXiv:astro-ph/0504370
Kastner, J. H., Huenemoerder, D. P., Schulz, N. S., Canizares, C. R., & Weintraub, D. A. 2002, ApJ, 567, 434. arXiv:astro-ph/0111049
Reale, F., & Micela, G. 1998, A&A, 334, 1028
Robitaille, T. P., Whitney, B. A., Indebetouw, R., Wood, K., & Denzmore, P. 2006, ApJS, 167, 256. arXiv:astro-ph/0608234
Stassun, K. G., Ardila, D. R., Barsony, M., Basri, G., & Mathieu, R. D. 2004, AJ, 127, 3537. arXiv:astro-ph/0403159
Stassun, K. G., van den Berg, M., & Feigelson, E. 2007, ApJ, 660, 704. arXiv:astro-ph/0701735
Stassun, K. G., van den Berg, M., Feigelson, E., & Flaccomio, E. 2006, ApJ, 649, 914. arXiv:astro-ph/0606079
Whitney, B. A., Wood, K., Bjorkman, J. E., & Wolff, M. J. 2003, ApJ, 591, 1049. arXiv:astro-ph/0303479

Finding the Youngest Planets

Christopher J. Crockett,[1,2] Naved Mahmud,[3] Lisa Prato,[1] Christopher M. Johns–Krull,[3] Patrick Hartigan,[3] Daniel T. Jaffe,[4] and Charles A. Beichman[5,6]

[1]*Lowell Observatory, 1400 W Mars Hill Road, Flagstaff, AZ 86001, USA*

[2]*Department of Physics and Astronomy, University of California Los Angeles, 430 Portola Plaza, Box 951547, Los Angeles, CA 90095-1547, USA*

[3]*Department of Physics and Astronomy, Rice University, MS-108, 6100 Main Street, Houston, TX 77005, USA*

[4]*Department of Astronomy, University of Texas, R.L. Moore Hall, Austin, TX 78712, USA*

[5]*Jet Propulsion Laboratory, California Institute of Technology, 4800 Oak Grove Drive, Pasadena, CA 91109, USA*

[6]*NASA Exoplanet Science Institute (NExScI), California Institute of Technology, 770 S. Wilson Ave, Pasadena, CA 91125, USA*

Abstract. We are conducting a multiwavelength radial velocity (RV) survey of the Taurus-Auriga low-mass star forming region. The goal of this survey is to characterize the hot Jupiter population around a sample of 1−3 Myr old classical and weak-lined T Tauri stars. Given the young ages of these targets, a positive detection will help identify the planet formation timescale. The presence of large, cool star spots makes this a challenging environment in which to conduct an RV survey. To distinguish between spot-induced RV variability and true companions, we observe all of our targets in both visible light and K band: spot-induced RV variability exhibits a wavelength dependence that companion-induced variability does not. We present details on our methodology and analysis of several planet candidate targets.

1. Introduction

Since 1989, astronomers have discovered over 500 extrasolar planets, ~94% of which were identified using radial velocity (RV) techniques[1]. However, in spite of all these discoveries, a unified theory of planet formation is still elusive. While theorists propose a number of models, there exist few observational results with which to test current hypotheses.

Timescale is a critical test in evaluating planet formation models. In the core accretion model (e.g., Pollack et al. 1996), a planetary core is built up through the accretion

---
[1]Data from *The Extrasolar Planet Encyclopedia* (www.exoplanet.eu)

of ices, dust particles, and eventually planetesimals in the circumstellar disk. Recent results on core accretion predict planet formation in millions of years (e.g., Alibert et al. 2005; Hubickyj et al. 2005; Dodson-Robinson et al. 2008). Another avenue for giant planet formation is the gravitational instability model (e.g., Boss 1997) wherein instabilities in the circumstellar disk can lead to fragmentation and eventual collapse. These models predict very fast formation times for massive planets in long period orbits: 10^3 (Mayer et al. 2004) to 10^5 (Bodenheimer 2006) years. Clearly, any data which can narrow down the timescale will help constrain planet formation models. Ongoing campaigns to catalog the planets around main sequence stars can not directly provide this information. However, by observing late-type, pre-main sequence (PMS) T Tauri stars, one acquires a snapshot of the early stages of planet formation around solar analogue stars.

Low-mass young stars present challenging targets for traditional RV surveys. Late spectral types, large distances (>100 pc), and extinction from natal dust clouds make these targets faint at optical wavelengths. They also have strong magnetic fields (e.g., Johns-Krull 2007) that generate large, cool star spots. These spots impact RV surveys of young stars by introducing significant jitter which can mimic the velocity modulation imposed by a planet (Saar & Donahue 1997; Desort et al. 2007). Recent attempts at detecting substellar companions in young stellar populations (10–100 Myr) have generally been unsuccessful (e.g., Paulson et al. 2004; Paulson & Yelda 2006). These studies cite the small sample size and intrinsic RV variability of their targets as the main hinderance to such surveys. The youngest RV planet detected to date is a ∼6 M_{JUP} planet on a 850 day orbit around the 100 Myr old star HD 70573 (Setiawan et al. 2007).

Spectral line bisector analysis is typically used to distinguish spot-induced RV variations from companion-induced ones (e.g., Hatzes et al. 1997). Companion-induced reflex motion does not affect the shape of an absorption line's bisector whereas a spot distorts the line symmetry. The line bisector span, measured by the difference in bisector values at two different heights of the line profile, is a proxy for the average slope of the line bisector. A correlation between bisector span and RV variations suggests the RV fluctuation is spot-induced; otherwise, it may be caused by a companion. However, there may also be no correlation if the projected rotation velocity ($v \sin i$) of the star is comparable to or less than the velocity resolution of the spectrograph (Desort et al. 2007; Prato et al. 2008; Huerta et al. 2008). Therefore, bisector analysis is only a first step in identifying potential young planet hosts (e.g., Huélamo et al. 2008; Prato et al. 2008).

A more reliable method for distinguishing between spots and planets leverages the wavelength dependence of the spot-induced RV modulation amplitude. The reflex motion induced by a planet affects all wavelengths equally. However, the contrast between a photosphere and a cooler star spot decreases at longer wavelengths because of the flux-temperature scaling in the Rayleigh-Jeans limit of blackbody radiation (e.g., Vrba et al. 1986; Carpenter et al. 2001). As a result, the amplitude of any spot-induced RV variability will be smaller at longer wavelengths. By observing in the visible *and* NIR it is possible to distinguish between stellar activity and the presence of a companion by comparing the RV amplitudes at the two wavelengths. Reiners et al. (2010), however, present evidence that the while the effects of stellar activity are lessened at longer wavelengths, the magnitude of the decrease may be dependent on the temperature contrast between the photosphere and star spots; the RV jitter reduction in the NIR is more pronounced when the temperature contrast is small.

In this paper, we present our multiwavelength RV technique for detecting giant planets orbiting T Tauri stars. In §2 we discuss our observing strategy and data reduction algorithms. In §3 we present our methodology for bisector analysis and RV determination. We present results from a sample of observations in §4 and summarize our findings in §5.

2. Observing and Data Reduction

In 2004, we began the McDonald Observatory Young Planet Survey to monitor PMS stars in the nearby (~140 pc, Kenyon et al. 1994) Taurus-Auriga low-mass star forming region for evidence of substellar companions. Our sample of 143 classical and weak-lined T Tauri stars consists of V < 14 stars, most with $v \sin i$ < 20 km s^{-1} (Herbig & Bell 1988). Our observing strategy consists of obtaining every night observations of our targets for roughly a week at a time. The week-long observing window is chosen to closely match both the typical rotation periods of T Tauri stars and the companion orbital periods we are capable of detecting. Initial observations are conducted in visible light to identify potential planet hosts. Promising targets from these observations, based on measured RV variations and the results of bisector analysis, are selected for follow up K band observations to identify dependence of RV variability on wavelength.

2.1. Visible Light Spectroscopy

Visible light spectra were taken at the McDonald Observatory 2.7 meter Harlan J. Smith telescope with the Coudé echelle spectrograph (Tull et al. 1995) covering a wavelength range of 6233 Å – 7109 Å. A 1″.2 slit yielded a spectral resolution of $R \sim 60,000$. Integration times were typically 1800 s (depending on conditions) and average seeing was ~2″. We took Thorium-Argon (ThAr) lamp exposures before and after each target observation for wavelength calibration; typical RMS values for the dispersion solution were ~4 m s^{-1}. We observed RV standards on every night with an overall RMS scatter of ~140 m s^{-1}. We reduced our spectra with the REDUCE IDL echelle reduction package (Piskunov & Valenti 2002).

2.2. Near Infrared Spectroscopy

K band observations were taken at the 3 meter NASA Infrared Telescope Facility (IRTF) using the high-resolution Cassegrain-mounted echelle spectrograph, CSHELL (Tokunaga et al. 1990; Greene et al. 1993). We used the Circular Variable Filter (CVF) to isolate a 50 Å segment of spectrum centered at 2.298 μm onto the 256×256 InSb detector array. This region contains numerous deep photospheric absorption lines of the CO molecule as well as numerous telluric absorption features which we use as a wavelength reference. The 0″.5 slit yielded a typical FWHM of 2.6 pixels (~0.5 Å) corresponding to a spectral resolving power of R ~ 46,000. All of our target data were obtained using 10″ nodded pairs to enable subtraction of sky emission, dark current, and detector bias. Integration times for each nod were typically 600 seconds; for fainter targets we co-added multiple contiguous nod pairs. Total exposure times were therefore ~1 hour for our T Tauri targets and ~20 minutes for the RV standard. The signal-to-noise ratio (S/N) for all targets varied significantly depending on cloud cover, seeing, guiding errors, etc., with typical values ~140. We observed an RV standard (GJ 281) every night with an overall RMS scatter of ~70 m s^{-1} over two years. Our data reduction strategy closely follows that described in Johns-Krull et al. (1999).

3. Data Analysis

3.1. Visible Light Radial Velocities

We determined visible light RVs using a cross-correlation analysis of 8 echelle orders, each covering about 100 Å (Huerta et al. 2008) The orders were chosen for their high S/N, lack of stellar emission lines, and lack of telluric absorption lines. For the cross-correlation, we chose as our template spectrum the observation with the highest S/N. Using the target itself as a cross-correlation template prevents any spectral-type mismatching which would introduce additional uncertainty in the measurements. Each order was cross-correlated against the corresponding order in the template, thus our RVs are measured relative to one observation epoch. For each observation, the final RV and its internal uncertainty was taken to be the mean and standard deviations, respectively, of the RVs measured from the multiple echelle orders.

3.2. Visible Light Bisector Analysis

We performed bisector analysis on the visible light data to determine the origin of each target's RV variability. For each of the 8 echelle orders used to measure the RV for a given observation, we cross-correlated all absorption lines with the corresponding order in the template spectrum and measured the cross-correlation function (CCF). We used each CCF (one for each echelle order) to measure the bisector span (the inverse of the mean slope of the bisector) and calculated a mean bisector span for each observation.

Many spotted stars show a strong correlation between bisector span and RV (e.g., Huerta et al. 2008; Prato et al. 2008). However, Desort et al. (2007) showed that when $v \sin i$ approaches the resolution of the spectrograph, spot-induced RV and bisector span variations can mimic those expected from planetary companions. Because the resolution of our instrument, 5 km s^{-1}, is not much lower than the $v \sin i$ of our targets (~10 km s^{-1}), a lack of correlation between RV and the bisector span could be misleading. Huélamo et al. (2008) and Prato et al. (2008) show examples of young stars that do not display significant correlations between their bisector spans and RVs, yet their RV variations are shown to likely originate in spots. Thus, we cannot rely solely on bisector analysis to clarify the nature of the observed RV variability.

3.3. Near Infrared Radial Velocities

The telluric absorption features in the K-band provide an absolute wavelength and instrumental profile reference, similar in concept to the iodine gas cell technique used in high-precision optical RV exoplanet surveys (Butler et al. 1996). Using the atmosphere as a "gas cell" lets us superimpose a relatively stable wavelength reference onto our spectra which follows the same optical path as the light from the science target. This helps alleviate uncertainties introduced by variable slit illumination, changing optical path lengths, etc. We determined the RVs of our targets using a spectral modeling technique similar to the one presented in Blake et al. (2007) and Figueira et al. (2010). We modeled the stellar spectrum and the telluric features using two high-resolution template spectra. For the stellar spectrum, we employed NextGen stellar atmosphere models (Hauschildt et al. 1999) tailored to the T_{eff}, $\log g$, and metallicity of our targets. We used SYNTHMAG (Piskunov 1999) to generate spectra from the NextGen models along with atomic (Kupka et al. 2000) and CO (Goorvitch 1994) line lists. We modeled the telluric features using the NOAO telluric absorption spectrum (Livingston & Wallace 1991).

In order to match the model to our observations, we applied a number of transformations to the combination of the two templates: a velocity shift and a power law scaling factor, rotational and instrumental broadening, a second order continuum normalization, and a second order wavelength dispersion solution. The composite model is then binned down to the resolution of our CSHELL data. We used non-linear least squares fitting to determine the values of the model parameters which best reproduced our observations. Radial velocities of our targets were calculated by the difference in the velocity shifts between the best-fit stellar and telluric spectra.

4. Results

Figure 1 illustrates results for three of our targets which demonstrate three classes of optical and NIR RV variability: unresolved spot noise, resolved spot noise, and a potential planet host. For the first target, DN Tau (Figure 1, top panels), we obtained 44 optical and 8 K band spectra. A detailed analysis of this and two other targets is presented in Prato et al. (2008). We found an optical standard deviation of 438 m s^{-1} and a K band standard deviation of 144 m s^{-1}, a factor of ~3 difference between the two wavelengths. The K band variability is only ~30% greater than the 110 m s^{-1} uncertainty from their RV standards. The difference between the two data sets is similar to observations of the known spotted star V827 Tau (exhibiting a factor of 3.6 reduction in RV scatter). Despite the fact that our optical data show no correlation between RV and bisector span, we thus conclude that the RV variations of DN Tau result from star spots.

Hubble I 4 (Figure 1, middle panels) exhibits behavior similar to DN Tau. We obtained 27 optical and 19 K band spectra. The optical RVs show a 1051 m s^{-1} standard deviation compared to 426 m s^{-1} in the K band (Mahmud et al. in prep), a factor of 2.5 difference between the data sets. As with DN Tau, we interpret these results as indicative of star spots. Hubble I 4, however, is the first target for which we are able to resolve RV variability in the K band; the NIR RV scatter is 3.5 times greater than the 120 m s^{-1} uncertainty. DN Tau and Hubble I 4 therefore demonstrate a crucial lesson: *multiwavelength observations are an essential element for testing the planet hypothesis*, especially in young stars.

The third target (Figure 1, bottom panels) is our most intriguing target to date. We obtained 43 optical and 23 K band observations. This is the first star for which the RV scatter at two wavelengths is comparable: 584 m s^{-1} in the optical versus 461 m s^{-1} in the K band, a factor of only 1.3 difference. The lack of a strong dependance of RV amplitude with wavelength suggests that the RV variability is caused by a close-in giant planet. As of Nov 2009, the data were consistent with a ~6 M_{JUP} companion orbiting at ~0.06 AU. Recent IRTF observations have made this interpretation less clear. While the optical data continue to maintain phase coherence and a similar amplitude, the new K band RVs do not phase up as cleanly as the earlier data. This target is a classical T Tauri star with an actively accreting circumstellar disk and a K band veiling of ~0.8. Therefore, one possible explanation for the changing nature of the K band RVs

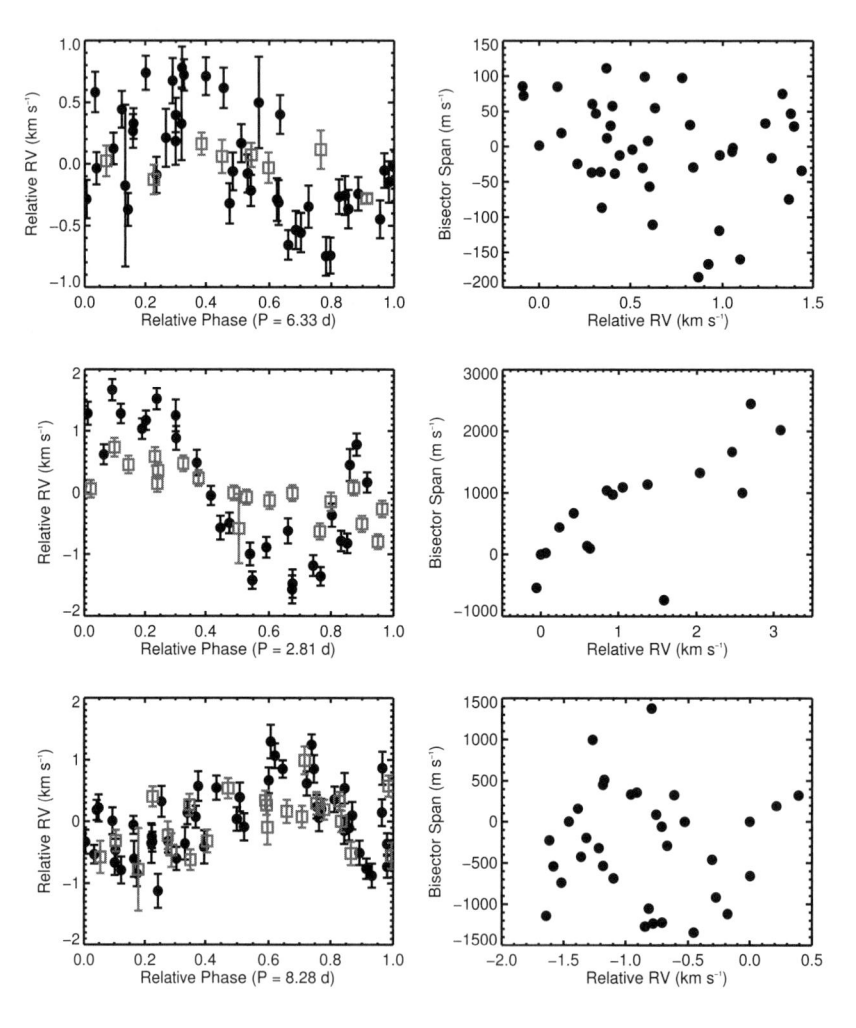

Figure 1. Results for three of our targets: DN Tau (top), Hubble I 4 (middle), and a potential exoplanet candidate (bottom). The left column shows phase-folded optical (filled circles) and NIR (open squares) RVs. The right column shows optical RV-bisector correlations.

is contamination of our spectra from the inner wall of the disk (Crockett et al. in prep). Whatever the underlying reason, these observations offer a reminder that T Tauri stars are dynamic, challenging environments in which to conduct such observations.

5. Summary

We are conducting a multiwavelength RV survey of 143 1–3 Myr old T Tauri stars in the Taurus-Auriga star forming region in order to catalog the young giant planet population around these stars. The goals of this survey are to place observational constraints on the planet formation timescale. Because these stars exhibit significant RV variability from large star spots, we conduct our observations at two widely separated wavelengths to distinguish stellar activity from orbiting companions. To date, all but one of our targets show no evidence of hot Jupiters. The one target which has the most potential as a planet host exhibits unusual behavior at longer wavelengths which may be caused by heating of an active circumstellar disk. While the discovery of any one planet is unlikely to discriminate between diverse planet formation models, the confirmation of a planet around any of our targets would represent the first solid data point of a hot Jupiter from the epoch of active planet formation and provide one of the first observational constraints on the formation timescale.

Acknowledgments. The authors acknowledge the *SIM* Young Planets Key Project for research support; funding was also provided by NASA Origins Grants 05-SSO05-86 and 07-SSO07-86. This work made use of the SIMBAD database, the NASA Astrophysics Data System, and the Two Micron All Sky Survey (2MASS), a joint project of the University of Massachusetts and IPAC/Caltech, funded by NASA and the NSF. We recognize the significant cultural role that Mauna Kea plays in the indigenous Hawaiian community and are grateful for the opportunity to observe there.

References

Alibert, Y., Mordasini, C., Benz, W., & Winisdoerffer, C. 2005, A&A, 434, 343
Blake, C. H., Charbonneau, D., White, R. J., Marley, M. S., & Saumon, D. 2007, ApJ, 666, 1198
Bodenheimer, P. 2006, Historical notes on planet formation (Cambridge University Press), chap. 1, 1
Boss, A. P. 1997, Science, 276, 1836
Butler, R. P., Marcy, G. W., Williams, E., McCarthy, C., Dosanjh, P., & Vogt, S. S. 1996, PASP, 108, 500
Carpenter, J. M., Hillenbrand, L. A., & Skrutskie, M. F. 2001, AJ, 121, 3160
Desort, M., Lagrange, A.-M., Galland, F., Udry, S., & Mayor, M. 2007, A&A, 473, 983
Dodson-Robinson, S. E., Bodenheimer, P., Laughlin, G., Willacy, K., Turner, N. J., & Beichman, C. A. 2008, ApJ, 688, L99
Figueira, P., Pepe, F., Melo, C. H. F., Santos, N. C., Lovis, C., Mayor, M., Queloz, D., Smette, A., et al. 2010, A&A, 511, 55
Goorvitch, D. 1994, ApJSS, 95, 535
Greene, T. P., Tokunaga, A. T., Toomey, D. W., & Carr, J. B. 1993, Proc. SPIE, 1946, 313
Hatzes, A. P., Cochran, W. D., & Johns-Krull, C. M. 1997, ApJ, 478, 374
Hauschildt, P. H., Allard, F., & Baron, E. 1999, ApJ, 512, 377
Herbig, G. H., & Bell, K. R. 1988, Third Catalog of Emission-Line Stars of the Orion Population (Lick Observatory Bulletin)
Hubickyj, O., Bodenheimer, P., & Lissauer, J. J. 2005, Icarus, 179, 415
Huélamo, N., Figueira, P., Bonfils, X., Santos, N. C., Pepe, F., Gillon, M., Azevedo, R., Barman, T., et al. 2008, A&A, 489, L9
Huerta, M., Johns-Krull, C. M., Prato, L., Hartigan, P., & Jaffe, D. T. 2008, ApJ, 678, 472
Johns-Krull, C. M. 2007, ApJ, 664, 975
Johns-Krull, C. M., Valenti, J. A., & Koresko, C. 1999, ApJ, 516, 900
Kenyon, S. J., Dobrzycka, D., & Hartmann, L. 1994, AJ, 108, 1872
Kupka, F. G., Ryabchikova, T. A., Piskunov, N. E., Stempels, H. C., & Weiss, W. W. 2000, Baltic Astronomy, 9, 590
Livingston, W., & Wallace, L. 1991, An atlas of the solar spectrum in the infrared from 1850 to 9000 cm-1 (1.1 to 5.4 micrometer) (National Solar Observatory)

Mayer, L., Quinn, T., Wadsley, J., & Stadel, J. 2004, ApJ, 609, 1045
Paulson, D. B., Cochran, W. D., & Hatzes, A. P. 2004, AJ, 127, 3579
Paulson, D. B., & Yelda, S. 2006, PASP, 118, 706
Piskunov, N. 1999, in Polarization, edited by K. N. Nagendra & J. O. Stenflo, vol. 243 of Astrophysics and Space Science Library, 515
Piskunov, N. E., & Valenti, J. A. 2002, A&A, 385, 1095
Pollack, J. B., Hubickyj, O., Bodenheimer, P., Lissauer, J. J., Podolak, M., & Greenzweig, Y. 1996, Icarus, 124, 62
Prato, L., Huerta, M., Johns-Krull, C. M., Mahmud, N., Jaffe, D. T., & Hartigan, P. 2008, ApJ, 687, L103
Reiners, A., Bean, J. L., Huber, K. F., Dreizler, S., Seifahrt, A., & Czesla, S. 2010, ApJ, 710, 432
Saar, S. H., & Donahue, R. A. 1997, ApJ, 485, 319
Setiawan, J., Weise, P., Henning, T., Launhardt, R., Müller, A., & Rodmann, J. 2007, ApJ, 660, L145
Tokunaga, A. T., Toomey, D. W., Carr, J., Hall, D. N. B., & Epps, H. W. 1990, Proc. SPIE, 1235, 131
Tull, R. G., MacQueen, P. J., Sneden, C., & Lambert, D. L. 1995, PASP, 107, 251
Vrba, F. J., Rydgren, A. E., Chugainov, P. F., Shakovskaia, N. I., & Zak, D. S. 1986, ApJ, 306, 199

The 16th Cambridge Workshop on Cool Stars, Stellar Systems and the Sun
ASP Conference Series, Vol. 448
Christopher M. Johns-Krull, Matthew K. Browning, and Andrew A. West, eds.
© 2011 Astronomical Society of the Pacific

Revealing the Chamaeleon: First Detection of a Low-mass Stellar Halo Around the Young Open Cluster η Chamaeleontis

Simon J. Murphy,[1,3] Warrick A. Lawson,[2] and Michael S. Bessell[1]

[1]Research School of Astronomy and Astrophyiscs, The Australian National University, Cotter Road, Weston Creek, ACT 2611, Australia

[2]School of PEMS, University of New South Wales, Australian Defence Force Academy, Canberra, ACT 2600, Australia

[3]murphysj@mso.anu.edu.au

Abstract. We have identified several lithium-rich low-mass ($0.08 < M < 0.3\ M_\odot$). stars within 5.5 deg of the young open cluster η Chamaeleontis, nearly four times the radius of previous search efforts. We propose 4 new probable cluster members and 3 possible members requiring further investigation. Candidates were selected on the basis of DENIS and 2MASS photometry, NOMAD astrometry and extensive follow-up spectroscopy. Several of these stars show substantial variation in their Hα emission line strengths on timescales of days to months, with at least one event attributable to accretion from a circumstellar disk. These findings are consistent with a dynamical origin for the current configuration of the cluster, without the need to invoke an abnormally top-heavy Initial Mass Function, as proposed by some authors.

1. Introduction

The open cluster η Chamaeleontis is one of the closest ($d \sim 94$ pc) and youngest ($t \sim 8$ Myr) stellar aggregates in the Solar neighbourhood. A census of its stellar population currently stands at 18 systems, covering spectral types B8–M5.5 (Mamajek et al. 1999, 2000; Lawson et al. 2002; Song et al. 2004; Luhman & Steeghs 2004; Lyo et al. 2004; Murphy et al. 2010a). At high and intermediate masses the cluster Initial Mass Function (IMF) follows that of other star-forming regions and young stellar groups, but there is a clear deficit of members at masses $<0.15\ M_\odot$. Comparing the observed mass function of the cluster to other young groups, Lyo et al. (2004) predict that an additional 20 stars and brown dwarfs in the mass range $0.025 < M < 0.15\ M_\odot$ remain to be discovered. Efforts to observe this hitherto unseen population have so far failed to find any additional members at either larger radii from the cluster core (Luhman 2004, to 1.5 deg, 4 times the radius of known membership) or to low masses in the cluster core (Lyo et al. 2006, to ~13 $M_{\rm Jup}$). Failure to find these low–mass members raises a fundamental question: has the cluster's evolution been driven by dynamical interactions which dispersed the stars into a diffuse halo at even larger radii, or does η Cha possess an abnormally top-heavy IMF deficient in low–mass objects? The latter result would seemingly be at odds with the growing body of evidence that suggests the IMF

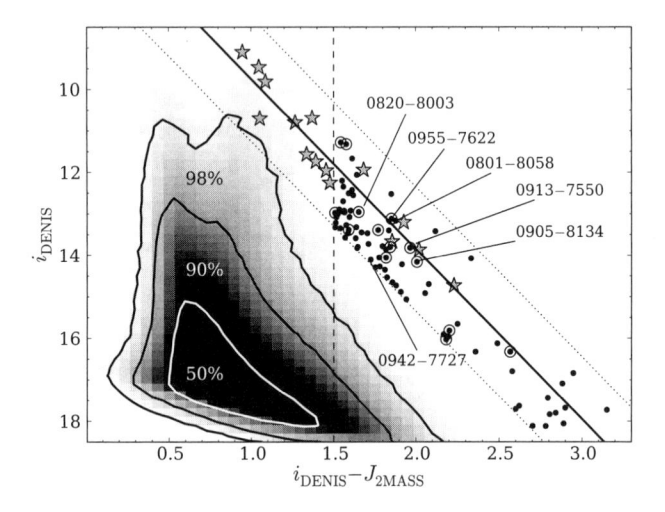

Figure 1. Selection criteria for the 81 photometric candidates from Murphy et al. (2010a). Contours show the cumulative total of stars enclosed. Known KM-type cluster members are shown as filled stars. We select candidates (filled circles) within 1.5 mag of the empirical cluster isochrone, having $i - J > 1.5$. Proper motion candidates are denoted by open circles. The 6 intermediate-gravity stars are labeled.

is universal and independent of initial star-forming conditions (for an excellent review see Bastian, Covey, & Meyer 2010).

Moraux, Lawson, & Clarke (2007) have attempted to model the observed properties of η Cha using N-body simulations of the cluster's dynamical evolution starting with standard initial conditions. They are able to replicate the current configuration of the cluster assuming a log-normal IMF and 30–70 initial members. New calculations incorporating binaries (Becker & Moraux, 2010, in prep.) show almost identical results can be obtained starting with ~20 binary systems. This suggests the deficit of low–mass objects seen in the present day cluster may not be due to a peculiar IMF but to dynamical evolution instead. These simulations predict there should exist a diffuse halo of cluster ejectees beyond the radius currently surveyed. To test the dynamical evolution hypothesis we have undertaken a detailed search for this putative halo of low-mass objects surrounding η Cha. Our survey methods and results are described in Murphy et al. (2010a), with an extensive follow-up spectroscopy campaign for two candidates thought to be harboring accretion disks presented in Murphy et al. (2010b). Definitive information can be found in these two papers – in the following contribution we give a precis of our work to date.

2. New Members of the Low-mass Halo Surrounding η Cha

From 2MASS and DENIS photometry of the 1.2×10^6 sources within 5.5 deg of η Cha we selected 81 photometric candidates having i_{DENIS} photometry within 1.5 mag of the empirical isochrone of confirmed η Cha members from the literature (Figure 1). Fourteen photometric candidates have NOMAD proper motions consistent with cluster membership. Naïvely one could expect to find proper motion candidates simply by se-

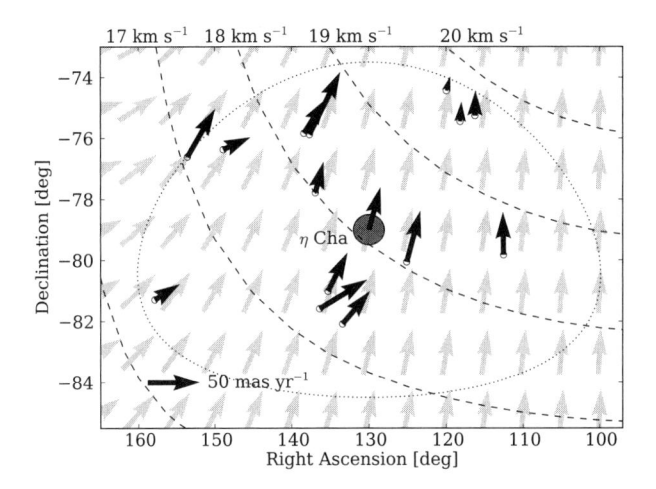

First Detection a Low-mass Stellar Halo Around η Chamaeleontis 63

Figure 2. The 14 proper motion candidates and their NOMAD proper motions (black arrows), compared to the cluster space motion projected onto the plane of the sky (gray arrows). Complementary radial velocity contours are shown by dashed lines. The 5.5 deg radius of our survey is given by the dotted border.

lecting objects with proper motions similar to the cluster mean. This would however bias any survey to candidates near the cluster on the sky. Figure 2 demonstrates the difficultly in using proper motion selection over such a large survey area. For a given *UVW* space motion the resultant proper motion vector (and radial velocity) depends on both sky position and distance. Given the large angular extent of the survey area, this can have a substantial effect on the expected proper motions and radial velocities of our candidates. We have corrected for this effect by comparing the observed proper motions to those expected at the position of each photometric candidate. The 14 resulting proper motion candidates are shown in Figure 2. Of these stars, four have intermediate gravities, as measured from the strength of the Na I 8200 Å absorption doublet. An intermediate gravity between dwarfs and giants is characteristic of Pre-Main Sequence (PMS) stars, which are still contracting towards their Main Sequence radii. Two other stars have intermediate gravities but are not proper motion members. All six stars have Li I 6708 Å equivalent widths consistent with a 5–10 Myr PMS population. In addition to the 2MASS/DENIS sources we also investigated a selection of stars from the literature. In particular several of the ROSAT-selected PMS stars from Covino et al. (1997) met the photometric, proper motion and Lithium requirements and were added to our final candidate list for dynamical modeling.

2.1. Dynamical Modeling

If the remaining candidates are in fact ejected members of η Cha we do not expect their space motions to be identical to that of the cluster proper. They will have each been imparted some ejection velocity which acts over time to disperse the star away from the cluster core. To determine the possible epoch and magnitude of this impulse we modeled the space motion of the candidates as a function of ejection time and current distance. Figure 3 shows the results of these simulations for several candidates. Similar plots for the other candidates and a more detailed explanation of the method used are presented in Murphy et al. (2010a). The colour-map in Figure 3 indicates the difference between the observed space motion and that expected from the current position of the star assuming ejection from the cluster. A similar map can be made for ejection speed. The kinematic distances estimated from the modeling can be checked against those expected from a star's position in the cluster CMD. Overall there is excellent agreement between the kinematic and photometric distances of the candidates. From such comparisons we have identified 4 new probable and 3 possible outlying members of η Cha, at radii of 1.5–5 deg (2.5–8 pc).

2.2. Dynamical Evolution or Abnormal IMF?

Are these numbers consistent with a dynamical origin for the paucity of low-mass objects currently observed in the cluster? To answer this we consider the radial distribution of ejectees presented in Moraux et al. (2007). Integrating the distribution out to 9 pc (5.5 deg) we could expect to find up to 6 stars across all masses in the survey area. Our blue photometry limit of $i - J > 1.5$ corresponds to an approximate spectral type of M3, whereas our spectroscopic campaign is complete to $i - J < 2$, approximately M5. Transforming these spectral types to temperatures and then to masses from an 8 Myr Baraffe & et al. (1998) isochrone gives an *approximate* surveyed mass range of $0.08 < M < 0.3\ M_\odot$. Moraux et al. find the mass distribution of ejectees is roughly constant with radius and consistent with the Chabrier (2003) input IMF. We therefore integrate this IMF over the above mass range to find the fraction of stars at those masses, this is ~40%. Hence we can expect to find 2–3 bona fide new η Cha members within the surveyed area and mass range. Our discovery of 3 probable 2MASS/DENIS members is consistent with this prediction. *We can therefore conclude that dynamical evolution is solely responsible for the current configuration of η Cha and it is not necessary to invoke an IMF deficient in low–mass objects.*

3. Episodic Disk Accretion at ~8 Myr

Nearby, isolated groups such as η Cha are ideal laboratories for investigating the dynamical evolution of young star clusters, in particular the influence that dynamics have had on the evolution of protoplanetary disks. Young clusters show a steady decline in the number of stars having disks and signatures of accretion with age (Mohanty et al. 2005; Jayawardhana et al. 2006). By an age of ~5 Myr, 90–95% of all young cluster members have stopped accreting material at a significant rate, yet ~20% of objects retain enough dust in their disks to produce a mid-IR excess (Fedele et al. 2010). By investigating the disk and accretion properties of any new dispersed members of η Cha, we can hope to gain a more unbiased view of the cluster as a whole, as well as addressing any influence dynamical interactions have had on disk evolution. For instance, we might expect any disks surrounding the outlying members to be truncated in mass and radius by the strong dynamical forces responsible for their dispersal, or show other systematic differences when compared to the evolution of the dynamically less-evolved inner members.

Two of our candidates showed strong and highly variable Hα emission over the six months of observations: the probable candidate 2MASS 0820–8003 and the possible candidate 2MASS 0801–8058. This star is a possible binary based on its elevated position in the cluster CMD and unusually large radial velocity variations. We obtained

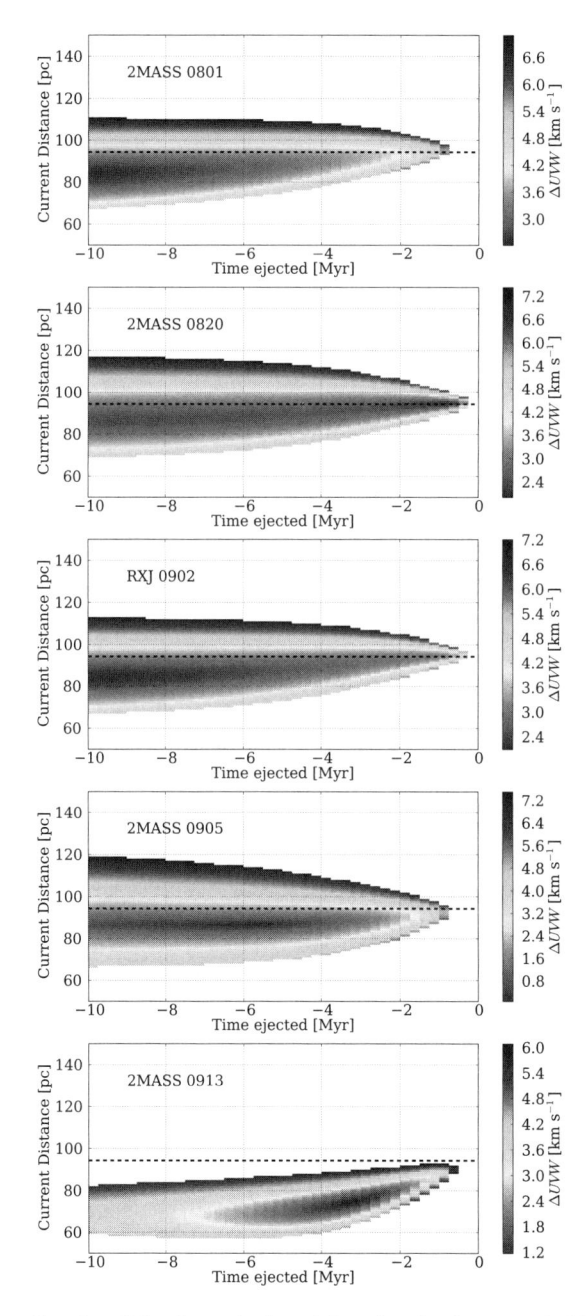

Figure 3. Results of the dynamical modeling for the 4 new probable members and the possible accretor 2MASS 0801. Based on these plots (and similar ones for ejection speed) we can estimate when and how fast the stars were ejected from the cluster and at what distances they currently lie. More information may be found in Murphy et al. (2010a).

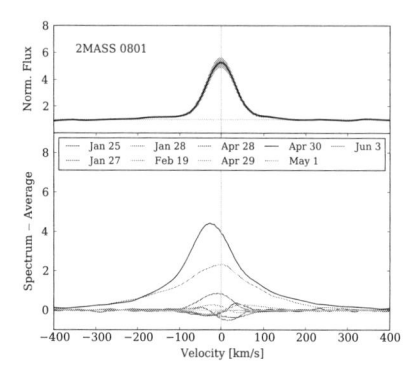

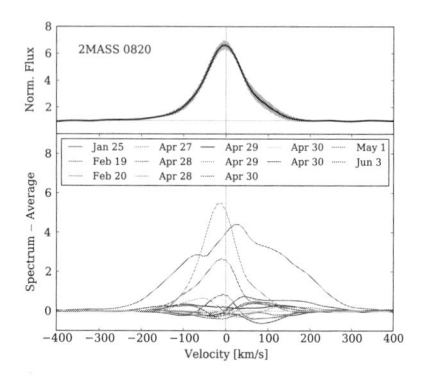

Figure 4. Hα velocity profiles for the 9 observations of 2MASS 0801 (left panel) and 13 observations of 2MASS 0820. In each plot the top panel shows the average quiescent spectrum spectrum and the standard deviation of quiescent spectra around the mean (shaded region). The bottom panel shows the variation around the mean for all epochs. Note the broad, multicomponent residual on Feb 19 for 2MASS 0820.

9 medium-resolution ($R \approx 7000$) observations of 2MASS 0801 over 2010 January–June and 13 epochs for 2MASS 0820. In addition to strong Hα emission both stars recurrently showed He I 5876 Å, 6678 Å and Na I D emission, often associated with accretion. Our full results can be found in Murphy et al. (2010b), including the variation of the Hα line in the Equivalent Width–10% Intensity Velocity Width plane (Lawson et al. 2004; Jayawardhana et al. 2006) and comparisons with other multi-epoch studies of η Cha members. We present here only the variation in the Hα line profile shapes (Figure 4). Immediately apparent are the broad residual profiles, tracing velocities up to ±200–300 km s^{-1}. The February 19 epoch of 2MASS 0820 shows a residual velocity profile reaching to ±300 km s^{-1}, with four distinct components visible and a large red asymmetry. Velocity shifts in the peaks of individual residual spectra are present at up to several tens of km s^{-1}.

Two possible mechanisms are usually invoked to explain such distinctive line profiles – accretion from a circumstellar disk or chromospheric activity. The work of Montes et al. (1998) has shown that in some older T Tauri stars the Hα line profile cannot be fitted by a single gaussian and two components are necessary: a narrow gaussian of FWHM <100 km s^{-1} and a much broader component with FWHM 100–500 km s^{-1}, sometimes offset in wavelength from the narrow component. They attribute these line profiles to chromospheric micro-flaring. Micro-flares are frequent, short duration events and have large-scale motions that could explain the broad wings observed in the lines and the residual spectra in Figure 4. Both stars also show He I 6678 Å in emission when Hα levels are strongest. Strong He I 6678 Å emission is an accretion diagnostic as it is only present in low-levels ($\ll 1$ Å) in older chromospherically active stars. While we do detect strong (1.5 Å) emission in the April 30 outburst spectrum of 2MASS 0801, at all other epochs where we detect the line it is weak (~ 0.5 Å). Given the weak He I line strengths generally observed in our stars and the simple gaussian-like profiles of the residual spectra we do not have strong evidence for ongoing accretion. Chromospheric activity is a much more likely explanation for the observed line profiles. Only the February 19 spectrum of 2MASS 0820 shows a broad, asymmetric residual characteristic of accretion. Multiple components are present at velocities up to ± 300 km s^{-1}, presumably tracing the ballistic infall of material from the inner edge of the disk onto the stellar surface.

Our results show that Hα variability in ~8 Myr PMS stars can be substantial on both short (hours–days) and long (months) timescales. This variation is probably driven primarily by chromospheric activity, which can generate broad Hα profiles mimicking accretion over short timescales. However we also have evidence for at least one accretion event in 2MASS 0820 which requires follow-up observations. Additional mid-IR observations will be necessary to detect the presence of any circumstellar disk around the star feeding the accretion. Assuming the duty-cycle of episodic accretion is low, single-epoch surveys of accreting objects, especially in the critical age range 5–10 Myr when inner disks are being cleared and giant planet formation takes place, are likely missing a large fraction of accreting objects. Gas depletion timescales derived from the fraction of accretors are therefore likely underestimated. A larger survey of the disk and accretion properties of outlying η Cha members, combined with more detailed investigation of the true accreting fraction of PMS clusters from multi-epoch surveys is needed to resolve the issue.

Acknowledgments. SJM acknowledges the generous support of the LOC and the receipt of an RSAA Alex Rodgers Traveling Scholarship and Joan Duffield Research Scholarship.

References

Baraffe, I., & et al. 1998, A&A, 337, 403
Bastian, N., Covey, K. R., & Meyer, M. R. 2010, ARA&A, 48, 339
Chabrier, G. 2003, PASP, 115, 763
Covino, E., Alcala, J. M., Allain, S., Bouvier, J., Terranegra, L., & Krautter, J. 1997, A&A, 328, 187
Fedele, D., van den Ancker, M. E., Henning, T., Jayawardhana, R., & Oliveira, J. M. 2010, A&A, 510, A72
Jayawardhana, R., Coffey, J., Scholz, A., Brandeker, A., & van Kerkwijk, M. H. 2006, ApJ, 648, 1206
Lawson, W. A., Crause, L. A., Mamajek, E. E., & Feigelson, E. D. 2002, MNRAS, 329, L29
Lawson, W. A., Lyo, A.-R., & Muzerolle, J. 2004, MNRAS, 351, L39
Luhman, K. L. 2004, ApJ, 616, 1033
Luhman, K. L., & Steeghs, D. 2004, ApJ, 609, 917
Lyo, A.-R., Lawson, W. A., Feigelson, E. D., & Crause, L. A. 2004, MNRAS, 347, 246
Lyo, A.-R., Song, I., Lawson, W. A., Bessell, M. S., & Zuckerman, B. 2006, MNRAS, 368, 1451
Mamajek, E. E., Lawson, W. A., & Feigelson, E. D. 1999, ApJ, 516, L77
— 2000, ApJ, 544, 356
Mohanty, S., Jayawardhana, R., & Basri, G. 2005, ApJ, 626, 498
Montes, D., Fernandez-Figueroa, M. J., de Castro, E., Cornide, M., Poncet, A., & Sanz-Forcada, J. 1998, in Cool Stars, Stellar Systems, and the Sun, edited by R. A. Donahue & J. A. Bookbinder, vol. 154 of ASP Conf. Series, 1516
Moraux, E., Lawson, W. A., & Clarke, C. 2007, A&A, 473, 163
Murphy, S. J., Lawson, W. A., & Bessell, M. S. 2010a, MNRAS, 406, L50
Murphy, S. J., Lawson, W. A., Bessell, M. S., & Bayliss, D. D. R. 2010b, MNRAS, in press
Song, I., Zuckerman, B., & Bessell, M. S. 2004, ApJ, 600, 1016

The 16th Cambridge Workshop on Cool Stars, Stellar Systems and the Sun
ASP Conference Series, Vol. 448
Christopher M. Johns-Krull, Matthew K. Browning, and Andrew A. West, eds.
© 2011 Astronomical Society of the Pacific

HST/COS Spectra of DF Tau and V4046 Sgr: First Detection of Molecular Hydrogen Absorption Against the Lyman α Emission Line[1]

Hao Yang,[2] Jeffrey L. Linsky,[2] and Kevin France[3]

[2]*JILA, University of Colorado and NIST, Boulder, CO, USA*

[3]*CASA, University of Colorado, Boulder, CO, USA*

Abstract. We present moderate-resolution (16,000 – 18,000) far-UV spectra of two classical T Tauri stars, DF Tau and V4046 Sgr, obtained with the Cosmic Origins Spectrograph (COS) on the Hubble Space Telescope (*HST*), and report for the first time detection of absorption in the Lyman-α profile produced by the H_2 pumping transitions. For most absorption features, the absorbed energy in the H_2 pumping transitions is significantly smaller than the amount of energy in the resulting fluorescent emission, indicative of additional absorption in the H I Lyman-α profile along our light of sight. We model the additional H I absorption and are able to correct the H_2 absorption/emission ratios close to unity. The required H I absorption for DF Tau is at a velocity close to the radial velocity of the star, consistent with H I absorption in the interstellar medium and the edge-on disk. For V4046 Sgr, a nearly face-on system, the required absorption is near +290 km s^{-1}, most likely resulting from H I gas in the accretion columns falling onto the star.

1. Introduction

In the far-ultraviolet (FUV) spectra of classical T Tauri stars (CTTSs), molecular hydrogen (H_2) emission lines are commonly detected (Herczeg et al. 2002, 2004, 2006; Ardila et al. 2002). These fluorescent H_2 lines are thought to be photoexcited from the ground electronic state to the B (or C) electronic state primarily by coincidence with hydrogen Lyman-α but also by other strong atomic emission lines (e.g., C II, C IV, and O IV) in the UV.

The H_2 fluorescence may arise from various locations in protostellar systems. In their analysis of *HST* STIS E140M spectra of 6 pre-main sequence (PMS) stars, Herczeg et al. (2006) found that blueshifted H_2 lines of RU Lupi, T Tau, and DG Tau are likely formed in outflows, while the H_2 lines of TW Hya, DF Tau, and V836 Tau show no radial velocity shifts from the photospheric lines and are likely formed in warm (~2500 K) surfaces of their circumstellar disks. Stars such as T Tau also show spatially extended H_2 fluorescent emission in associated nebulosity, which is likely pumped by local shocks and outflows rather than stellar Lyman-α emission (Walter et al. 2003).

---

[1]Based on observations made with the NASA/ESA *Hubble Space Telescope*, obtained from the data archive at the Space Telescope Science Institute. STScI is operated by the Association of Universities for Research in Astronomy, Inc. under NASA contract NAS 5-26555.

For the diskless counterparts of CTTSs, the naked T Tauri stars (NTTSs), H_2 features are not seen in their FUV spectra (Ingleby et al. 2009), indicating that H_2 fluorescent emission requires the presence of H_2 gas close to the central star. Studying the H_2 fluorescent emission therefore provides valuable information on the physical properties of protoplanetary disks.

While the H_2 fluorescent lines have been studied in a number of CTTSs, the pumping transitions had not been observed in absorption against the Lyman-α emission line. In this paper, we present new FUV spectroscopy of two CTTSs, DF Tau and V4046 Sgr, for which we detect such absorption for the first time as a result of the very low noise and high throughput of the Cosmic Origins Spectrograph (COS) on *HST*.

2. Observations and Data Reduction

We observed DF Tau (RA = 04:27:02.795, DEC = 57:12:35.38) and V4046 Sgr (RA = 18:14:10.466, DEC = -32:47:34.50) with the COS (J. C. Green et al. 2010, in preparation; S. N. Osterman et al. 2010, in preparation) on January 11 and April 27, 2010, respectively. COS is a high-throughput, moderate-resolution UV spectrograph installed on the *HST* in May 2009. During our *HST* GTO program 11533, we used both the G130M and G160M gratings of the COS FUV channel to cover the 1136 Å–1796 Å region. Since there is a small gap ($\sim$15 Å in wavelength coverage) between the two segments of the COS detector, we observed each star at 4 central wavelength settings for both gratings to provide continuous spectral coverage and minimize any fixed-pattern noise. The total exposure time for each star was about 10,000 sec during 4 *HST* orbits. The spectral resolution was approximately 17,000–18,000, with extended wings in the line-spread function. The extended wings are induced by polishing errors on the *HST* primary and secondary mirrors (see Ghavamian et al. 2010).

We reduced the DF Tau and V4046 Sgr spectra using the COS calibration pipeline, CALCOS (v2.12, March 19, 2010), and combined them with a custom IDL coaddition routine described by Danforth et al. (2010). In Figure 1, we show two portions of the coadded FUV spectra for both stars as examples.

3. Analysis and Results

DF Tau is a binary system consisting of two early M stars separated by 0.1" (Schaefer et al. 2006). Its disk is inclined by 80–85° (Johns-Krull & Valenti 2001; Ardila et al. 2002), i.e., the disk is seen nearly edge-on. The distance to DF Tau is generally adopted as 140 pc, the distance to the Taurus Molecular Cloud. Herczeg et al. (2006) analyzed in detail the H_2 fluorescent lines of DF Tau observed with the STIS E140M grating. Herczeg & Hillenbrand (2008) estimated a visual extinction (A_V) of 0.6 mag and accretion rates in the range of $2.3\text{--}4.6 \times 10^{-8} M_\odot$ yr^{-1} for DF Tau.

V4046 Sgr is a very close binary with a separation of 9 $R_\odot$ and an orbital period of 2.42 days (Stempels & Gahm 2004). The pair consists of a K7V and a K5V star (Quast et al. 2000). At a distance of $\sim$ 72 pc (Torres et al. 2008), it is an isolated system and the extinction is practically 0.0 mag (Stempels & Gahm 2004). The circumbinary disk is inclined at 35° (Quast et al. 2000; Kastner et al. 2008), close to face-on. V4046 Sgr may be a member of the β Pic Moving group and could be as old as 12 Myr (Ortega et al. 2002). We know of no previous detailed study of H_2 emission from V4046 Sgr.

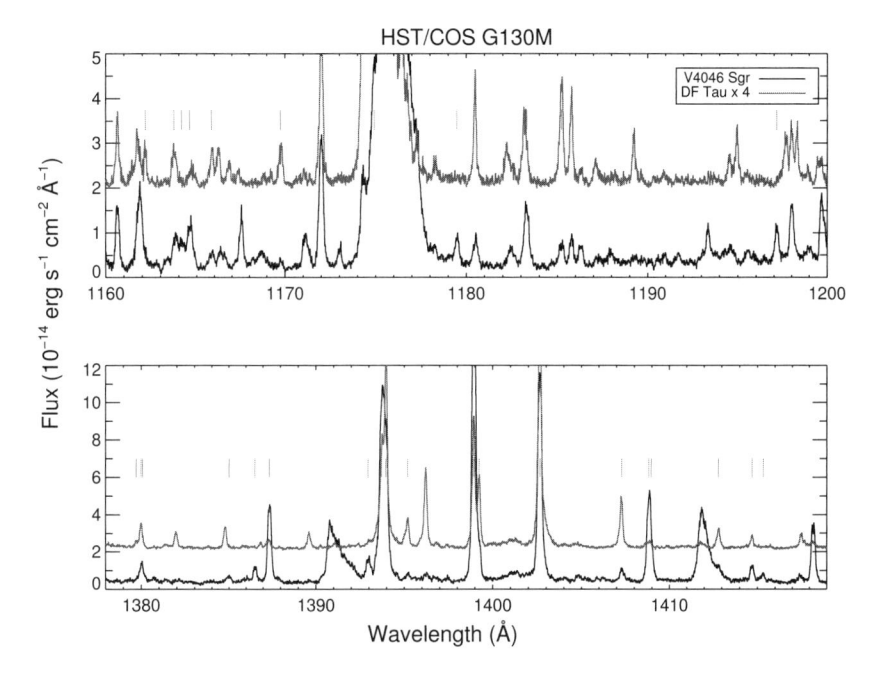

First Detection of H$_2$ Absorption Against Lyα Emission 71

Figure 1. Two portions of the COS spectra of DF Tau and V4046 Sgr. The red vertical ticks identify the Lyman-band H$_2$ emission lines.

In Figure 2 and 3, we show the Lyman-α profiles of DF Tau and V4046 Sgr, respectively. The apparent absorption features seen against the Lyman-α emission line have depths much greater than the noise at those wavelengths and coincide in wavelength with the Lyman-band H$_2$ pumping transitions (see Table 3 of Herczeg et al. 2006). We first mask out the absorption features and fit smooth curves to the Lyman-α profiles. To measure the amounts of absorbed energy, we integrate the area between the fitted curves and the observed spectra. The uncertainties in such measurements are mainly caused by the somewhat subjective determinations of the Lyman-α profiles without absorption. Some features result from the blended absorption of two or three pumping transitions, and the absorbed energy for these features represents the total absorption of the transitions.

We detect between 2 and 19 fluorescent H$_2$ lines in the progressions produced by each pumping transition observed in absorption against the Lyman-α lines. The H$_2$ fluorescent lines are identified based on the line list of Abgrall et al. (1993). To measure the line fluxes, we used a custom IDL fitting procedure (France et al. 2010) that convolves a Gaussian profile with the COS line spread function (LSF) to fit the observed H$_2$ line profiles. The convolution of a Gaussian profile with the COS LSF only changes the shape of the profile but not the total line flux. The uncertainties in the line fluxes are generally less than 5 % for unblended lines, indicative of the high signal-to-noise of the data. We next convert the line fluxes in each progression to the total energy emitted from the pumped upper level. Each H$_2$ line in a given progression yields an estimate of the total energy in the upper level from the line theoretical branching ratios. We average the estimated total energy emitted from each upper level using only the strong unblended

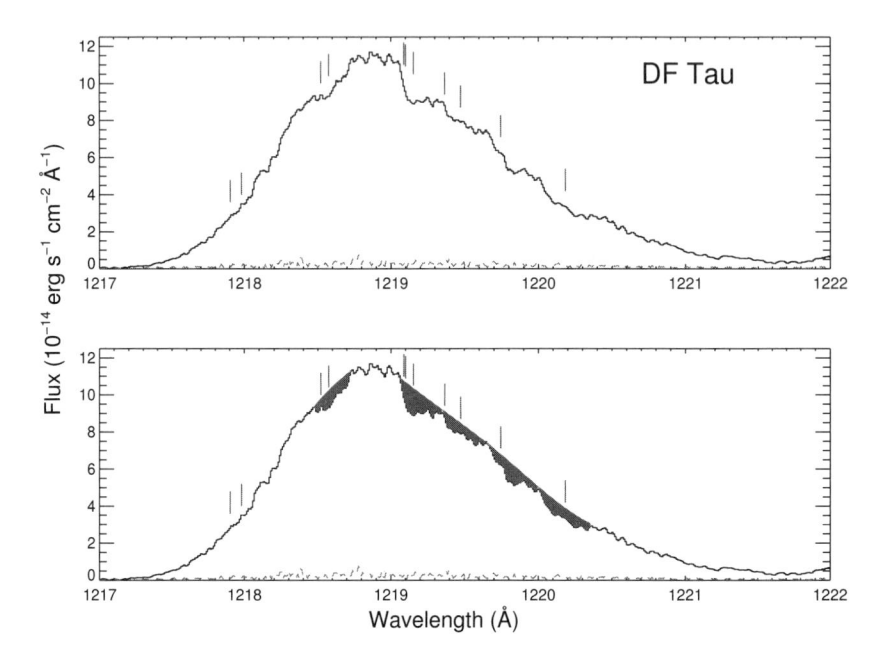

Figure 2. The Lyman-α emission profile of DF Tau. The red dashed lines indicate the uncertainty levels, and the vertical ticks mark the wavelengths of coincident Lyman-band H_2 transitions. In the bottom panel, the blue areas indicate our estimates of the absorption in the H_2 pumping transitions.

lines, and their standard deviations are $\leq 15\%$ of the mean values. The emission from each upper level is corrected for its dissociation probability as calculated by Abgrall et al. (2000). The dissociation probability is typically zero or only a few percent. The absorption and emission fluxes for DF Tau are also corrected for extinction using $A_V = 0.6$ mag and the Cardelli et al. (1989) extinction law.

For each absorption feature observed against the Lyman-α emission line, we have estimated the absorbed energy (E_{abs}) and the emitted energy (E_{em}) from the corresponding H_2 upper level. If there are no additional sources of absorption or emission, then the global average of E_{abs}/E_{em} should be unity. We begin the analysis by assuming that E_{abs}/E_{em} along our line of sight should be close to unity, but the measurements show otherwise. We do find that the three features in the Lyman-α blue wing for V4046 Sgr have E_{abs}/E_{em} within a factor of two of unity (1.38, 0.99, 0.56), but the E_{abs}/E_{em} ratios at longer wavelengths are smaller than one by factors of 4–60. We therefore propose that additional Lyman-α absorption in the line of sight between the location where H_2 is pumped and the observer has reduced the observed absorbed energy in the pumping line and thus the E_{abs}/E_{em} ratios. To model the additional hydrogen absorption, we calculate Lyman-α absorption profiles using the Voigt function for a range of hydrogen column densities (N_{HI}) and radial velocities (v_{rad}). Then for each combination of N_{HI} and v_{rad}, we calculate the optical depths, $\tau(\lambda)$, at the wavelengths of the pumping lines. We correct the observed E_{abs}/E_{em} ratios by multiplying by $e^{-\tau(\lambda)}$.

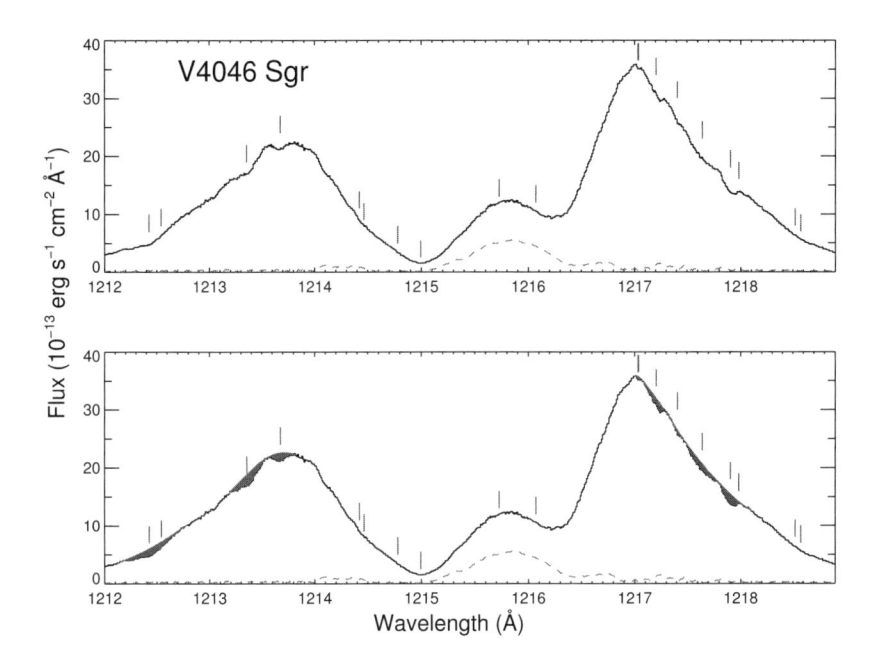

Figure 3. The Lyman-α emission profile of V4046 Sgr. The red dashed lines indicate the uncertainty levels, and the vertical ticks mark the wavelengths of coincident Lyman-band H$_2$ transitions. In the bottom panel, the blue areas indicate our estimates of the absorption in the H$_2$ pumping transitions.

In Figure 4 and 5, we show for DF Tau and V4046 Sgr the combinations of N_{HI} and v_{rad} that can correct the E_{abs}/E_{em} ratios to be close to unity. For DF Tau, if we assume that the extinction is all due to interstellar dust, we can convert $A_V = 0.6$ mag to a total hydrogen column density of $\log(N_H) = 21.03$ according to the relation in Predehl & Schmitt (1995). Note that this value assumes a standard gas-to-dust ratio for the interstellar medium. We mark in Figure 4 the corresponding N_{HI} values assuming 100% and 50% neutral hydrogen. Herczeg et al. (2006) estimated for DF Tau the absorption against the red wing of the Lyman-α emission line and measured a $\log(N_{HI})$ of 20.75. This value is close to the hydrogen column density converted from A_V with 50% neutral content and represents the neutral hydrogen in both the interstellar medium along the line of sight and possibly the edge-on disk of the system. Within reasonable ranges of N_{HI}, the additional absorption required to bring all of the E_{abs}/E_{em} ratios close to unity for DF Tau requires radial velocities close to zero. Given that the uncertainty in A_V could be as large as 0.4 mag, and the ionization fraction and the gas-to-dust ratio in the disk are unknown, we conclude that the absorption is likely caused by a combination of interstellar medium and neutral hydrogen in the edge-on disk.

For V4046 Sgr, which suffers negligible continuum extinction, the absorption required has a radial velocity between 100 and 290 km s^{-1}, as shown in Figure 4. If A_V is close to zero (Stempels & Gahm 2004), then the radial velocity is close to 290 km s^{-1}. This is consistent with a scenario in which H I in the accretion columns is absorbing

DF Tau

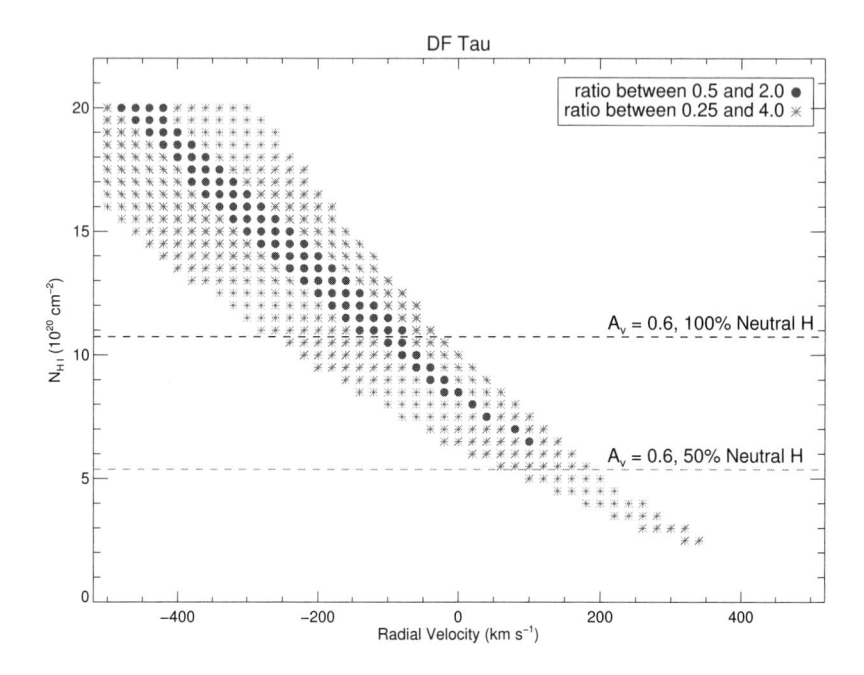

Figure 4. H$_2$ absorption/emission ratios of DF Tau, corrected for additional Lyman-α absorption for a range of H 1 column densities and radial velocities.

the red wings of the Lyman-α emission line in this system, which is oriented nearly face-on.

4. Discussion

Thanks to the excellent sensitivity and low background of COS, we were able to detect for the first time H$_2$ absorption against the Lyman-α emission line profiles in two CTTSs. Because of the large aperture of COS (2.5″), the center of the Lyman-α line is filled with geocoronal Lyman-α emission and not usable (see Figure 1). The STIS E140M spectrum of DF Tau reported by Herczeg et al. (2006) shows that the line center and blue wing of Lyman-α are completely absorbed. Interstellar absorption must be responsible for the disappearance of the line center, because we see many H$_2$ lines pumped by transitions coincident with the center of Lyman-α. On the other hand, we detect only a few H$_2$ lines pumped by the transitions blueward of line center, suggesting that the blue-wing emission of Lyman-α has been absorbed by the stellar wind before the blue-wing radiation reaches the molecular gas in the disk. Since the disk of DF Tau is viewed nearly edge-on, the stellar wind must be present near the stellar equator to absorb the blue wing of the Lyman-α emission line.

Understanding the geometry of the V4046 Sgr system requires more detailed consideration. The stellar wind must be weak for this nearly face-on system since the blue wing of Lyman-α is not totally absorbed. Our results show that there is additional absorption in the red wing of Lyman-α that could be explained by accretion with ve-

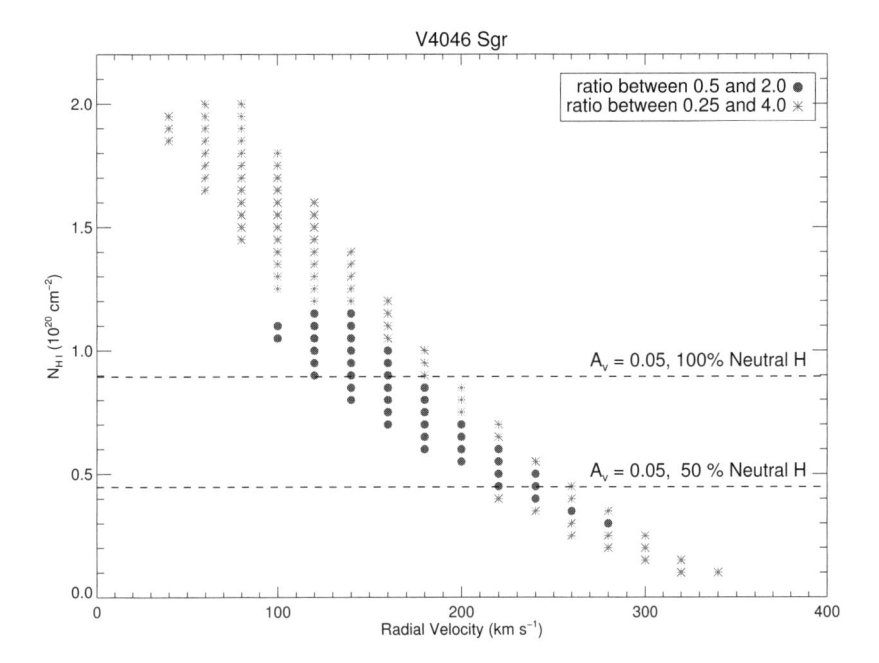

First Detection of H$_2$ Absorption Against Lyα Emission

Figure 5. H$_2$ absorption/emission ratios of V4046 Sgr, corrected for additional Lyman-α absorption for a range of H 1 column densities and radial velocities.

locities that are at least 100 km s^{-1} and are likely as large as 290 km s^{-1}. We envision a model in which a portion of the Lyman-α emission, likely formed near the accretion shocks, is reflected to the observer by neutral hydrogen in the inner disk. The H$_2$ pumping and fluorescence occurs where these Lyman-α photons are present in the inner disk, as described by the "thick disk" model in Herczeg et al. (2004). The reflected Lyman-α emission line, including the H$_2$ absorption at the pumping wavelengths, is then absorbed by infalling neutral hydrogen in large accretion funnels in our line of sight.

Accretion of gas from circumstellar disks onto CTTSs is generally thought to be controlled by stellar magnetic fields (Bouvier et al. 2007). Strong magnetic fields of a few kilogauss (e.g., Johns-Krull 2007) truncate the circumstellar gas disk at a few stellar radii and direct the accretion funnels onto the star near the magnetic poles. From models of the spectral energy distribution (SED) of V4046 Sgr, Jensen & Mathieu (1997) found that dust in the inner regions is cleared out to about 0.18 AU, which is 38.6 $R_\odot$. At this distance, the stellar magnetic fields are not strong enough to interact efficiently with the disk, and, more importantly, the temperature of molecular gas at the surface of the dusty disk may not be high enough for the hydrogen molecules to be electronically excited by the Lyman-α photons. For this transitional disk system, we think that both the fluorescent emission and accretion columns likely originate from the inner molecular gas disk (Muzerolle et al. 2003; Kastner et al. 2008) that is closer to the central star than the dust disk. The absence of any significant differences between the radial velocities of the star, pumping lines and fluorescent lines is consistent with the fluorescent H$_2$ gas lying in an inner gas disk seen nearly face-on.

Günther & Schmitt (2007) modeled the accretion shocks on V4046 Sgr, and their best-fit model of the X-ray observations yields an accretion rate of $3 \times 10^{-11} M_\odot$ yr^{-1} and a maximum infall velocity of 535 km s^{-1} where the accretion gas strikes the stellar surface. Since the H 1 absorption velocity is between +100 and +290 km s^{-1}, the absorbing H 1 gas is located between 2.3 R_* and 4.4 R_* above the accretion shock (cf. Eq. 1 in Calvet & Gullbring 1998) if we assume that the accretion columns have a constant cross-sectional area and the infalling gas sees only gravitational forces.

References

Abgrall, H., Roueff, E., & Drira, I. 2000, A&AS, 141, 297
Abgrall, H., Roueff, E., Launay, F., Roncin, J. Y., & Subtil, J. L. 1993, A&AS, 101, 273
Ardila, D. R., Basri, G., Walter, F. M., Valenti, J. A., & Johns-Krull, C. M. 2002, ApJ, 566, 1100
Bouvier, J., Alencar, S. H. P., Harries, T. J., Johns-Krull, C. M., & Romanova, M. M. 2007, Protostars and Planets V, 479. arXiv:astro-ph/0603498
Calvet, N., & Gullbring, E. 1998, ApJ, 509, 802
Cardelli, J. A., Clayton, G. C., & Mathis, J. S. 1989, ApJ, 345, 245
Danforth, C. W., Keeney, B. A., Stocke, J. T., Shull, J. M., & Yao, Y. 2010, ApJ, 720, 976
France, K., Linsky, J. L., Brown, A., Froning, C. S., & Béland, S. 2010, ApJ, 715, 596
Ghavamian, P., et al. 2010, in Bulletin of the American Astronomical Society, vol. 42 of Bulletin of the American Astronomical Society, 499
Günther, H. M., & Schmitt, J. H. M. M. 2007, Mem. Soc. Astron. Italiana, 78, 359
Herczeg, G. J., & Hillenbrand, L. A. 2008, ApJ, 681, 594. 0801.3525
Herczeg, G. J., Linsky, J. L., Valenti, J. A., Johns-Krull, C. M., & Wood, B. E. 2002, ApJ, 572, 310
Herczeg, G. J., Linsky, J. L., Walter, F. M., Gahm, G. F., & Johns-Krull, C. M. 2006, ApJS, 165, 256
Herczeg, G. J., Wood, B. E., Linsky, J. L., Valenti, J. A., & Johns-Krull, C. M. 2004, ApJ, 607, 369
Ingleby, L., et al. 2009, ApJ, 703, L137
Jensen, E. L. N., & Mathieu, R. D. 1997, AJ, 114, 301
Johns-Krull, C. M. 2007, ApJ, 664, 975
Johns-Krull, C. M., & Valenti, J. A. 2001, ApJ, 561, 1060
Kastner, J. H., Zuckerman, B., Hily-Blant, P., & Forveille, T. 2008, A&A, 492, 469
Muzerolle, J., Calvet, N., Hartmann, L., & D'Alessio, P. 2003, ApJ, 597, L149
Predehl, P., & Schmitt, J. H. M. M. 1995, A&A, 293, 889
Quast, G. R., Torres, C. A. O., de La Reza, R., da Silva, L., & Mayor, M. 2000, in IAU Symposium, edited by B. Reipurth, & H. Zinnecker, vol. 200 of IAU Symposium on The Formation of Binary Stars, held 10-15 April, 2000, in Potsdam, Germany, 28
Schaefer, G. H., Simon, M., Beck, T. L., Nelan, E., & Prato, L. 2006, AJ, 132, 2618
Stempels, H. C., & Gahm, G. F. 2004, A&A, 421, 1159
Walter, F. M., Herczeg, G., Brown, A., Ardila, D. R., Gahm, G. F., Johns-Krull, C. M., Lissauer, J. J., Simon, M., & Valenti, J. A. 2003, AJ, 126, 3076

Poster Sessions at Cool Stars 16

Part II
Fundamental Parameters

Lithium Depletion in Solar Type stars: Lithium and Planet Presence

S. G. Sousa,[1,2] N. C. Santos,[1,3] G. Israelian,[2] E. Delgado Mena,[2] J. Fernandes,[4] M. Mayor,[5] S. Udry,[5] C. Domínguez Cerdeña,[2] R. Rebolo,[2] and S. Randich[6]

[1] *Centro de Astrofisica, Universidade de Porto, Rua das Estrelas, 4150-762 Porto, Portugal*

[2] *Instituto de Astrofisíca de Canarias, Via Láctea s/n, E-38200 La Laguna*

[3] *Departamento de Física e Astronomia, Faculdade de Ciências, Universidade do Porto, Portugal*

[4] *Observatrio Astronómico e Departamento de Matemática, Universidade de Coimbra, Coimbra, Portugal*

[5] *Observatoire de Genève, Université de Genève, 51 ch des Maillettes, CH-1290 Versoix, Switzerland*

[6] *Instituto Nazionale di Astrofisica, Osservatorio di Arcetri, Largo Fermi 5, I59125 Firenze, Italy*

Abstract. The lithium (Li) abundance measured in the solar atmosphere is 140 times smaller than expected considering the proto-solar value. Furthermore, measurements of Li abundance made for many stars similar to the Sun reveal a large dispersion. These observations have defied the models of light element depletion for decades. We present a strong evidence for a correlation between Li depletion and the presence of planets. This result comes from the analysis of an unbiased sample of solar-analogue stars with and without planets detected, and for which precise spectroscopic stellar parameters were derived in an uniform way. Planet host stars are found to have typically only 1% of the primordial Li abundance while about 50% of the solar analogues without detected planets have on average ten times more Li. In addition, stellar evolutionary models were used to show that differences in stellar mass and age cannot be responsible for the observed correlation. These results suggest that the observed lithium difference is likely linked to some process related to the formation and evolution of planetary systems.

1. Introduction

The study of planet formation is being well fed by the new insights coming from the increasing number of discovered planets (e.g. Udry & Santos 2007). The first important insight coming from large uniform studies (Santos et al. 2001, 2004b; Fischer & Valenti 2005) was that the stars with giant planets have chemical abundances that are distinctly different from those found in "single" stars. This result provided significant constraints for modeling the formation and evolution of planets (e.g. Pollack et al. 1996; Mordasini et al. 2009; Boss 2002).

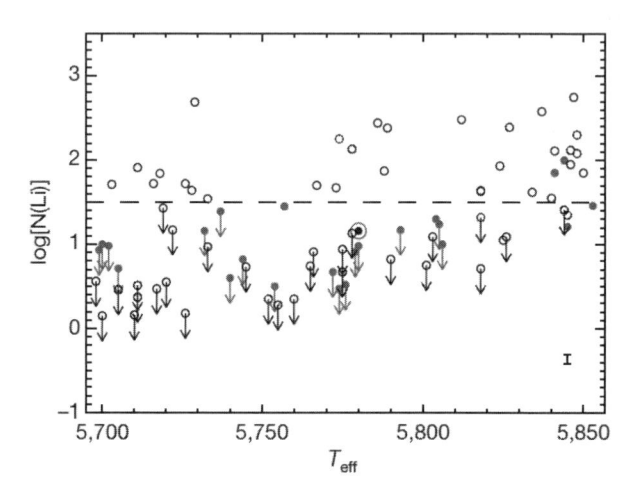

Figure 1. Lithium abundance plotted against effective temperature in solar-analogue stars with and without detected planets. The planethosting and single stars are shown by filled red and empty black circles, respectively. The red circle with the black point at its centre indicates the Sun. Figure from Israelian et al. (2009).

Most of the studies of chemical abundances in planet host stars were focus on iron as a metallicity proxy. However, there were several works analyzing a variety of refractory and volatile elements in planet host stars (e.g. Gilli et al. 2006; Ecuvillon et al. 2006; Takeda et al. 2007; Neves et al. 2009). Besides the general chemical enrichment found for stars with giant planets (interestingly not found for stars hosting very low mass planets, see: (Sousa et al. 2008)), none of these found compelling evidence for other chemical peculiarities.

In the last years it was proposed the possibility that the planet host stars present different abundances of the light elements Lithium and Beryllium (King et al. 1997; Garcia Lopez & Perez de Taoro 1998; Deliyannis et al. 2000; Cochran et al. 1997; Ryan 2000; Gonzalez & Laws 2000; Gonzalez 2008; Israelian et al. 2004; Santos et al. 2004a; Takeda & Kawanomoto 2005; Takeda et al. 2007; Chen & Zhao 2006; Luck & Heiter 2006). The different abundances could indicate pollution events caused by planets or planetary material that fell into the star during its lifetime (and in which quantity) (e.g. Israelian et al. 2001), or suggest that the rotational history of the star depends on the direct presence of planets or indirectly from the planet formation process (Bouvier 2008; Castro et al. 2009; Garaud & Bodenheimer 2010; Baraffe & Chabrier 2010; Eggenberger et al. 2010).

Israelian et al. (2009) presented a large uniform study of lithium abundances for a sample of stars from the HARPS planet search program (Mayor et al. 2003). In their work it was reported a significant difference regarding the depletion of lithium observed for planet host stars when comparing with the stars with no detected planet in the range 5700 K< T_{eff} < 5850 K. This was a confirmation for former suspicions (e.g. Israelian et al. 2004; Takeda et al. 2007). This result seems to indicate that stars with planets in the temperature range around the solar temperature (solar analogues) have significantly lower lithium abundances when compared with "single" stars (for

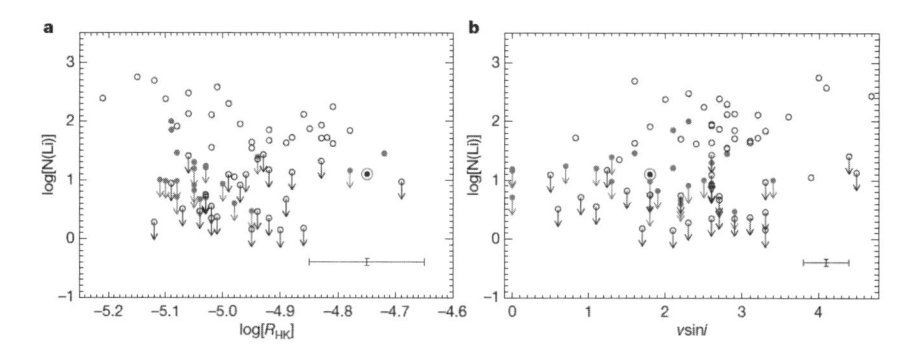

Figure 2. Lithium vs. chromospheric activity indices RHK and vsini of planet-hosting solar analogue and comparison stars. The solar symbol in the plot mark the activity index that corresponds to the solar maximum. Figure from Israelian et al. (2009).

which no planets were detected so far). The effects like stellar rotation, stellar activity, or chemical abundances were discarded as a cause for the observed difference since the study is based on an uniform HARPS sample composed mainly of old and inactive stars.

Nevertheless, Sousa et al. (2010) used stellar evolution models to explore the possibility that stellar age and mass could be responsible for the observed difference. In their work it could not be found any clear evidence for the lithium dependence on age for the selected stars.

More recently, after reanalyzing the published data from Israelian et al. (2009), Baumann et al. (2010) claimed that the analyzed stars are in fact following an age trend and that the presence of giant planets is not related to low lithium abundances.

The propose of this proceeding is to make a summary of the results presented in Israelian et al. (2009) and Sousa et al. (2010), and finally to discuss the conclusions of Baumann et al. (2010) about the systematic biases that led to the possible incorrect conclusion of an enhanced lithium depletion in planet-host stars.

2. Lithium and Planet Connection

The main result presented in Israelian et al. (2009) is shown here on Fig. 1. This plot shows the differences for Lithium abundances for both the planet host stars, and stars with no detected planets, studied in their work. The choice of the sample of stars for this kind of study is crucial. It is required to have a significant number of stars, both for the planet host and for the "single" stars. And, for all of these stars, it is also a requirement to have precise stellar parameters obtained in an homogeneous and systematics way. The big majority of the stars used by Israelian et al. (2009) were selected from the HARPS planet search program (Mayor et al. 2003). The spectroscopic parameters were derived in Sousa et al. (2008). To increase the number of stars with temperatures between 5,600 K and 5,900 K a few more stars were introduced. The full sample was composed of 162 stars with 46 being planet hosts. All of these stars were analyzed in the same way to obtain homogeneous Lithium abundance.

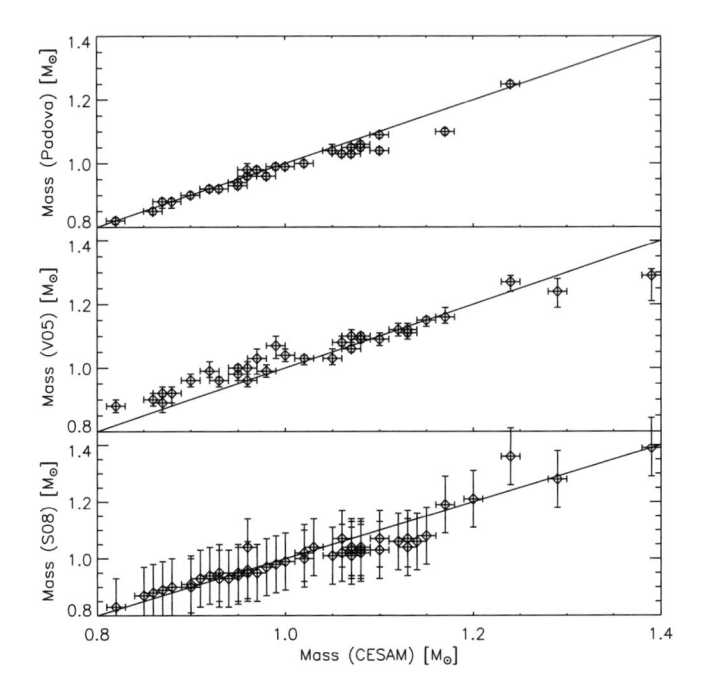

Figure 3. Comparison between masses determined in this work using CESAM with the ones presented in Sousa et al. (2008) (bottom), Fischer & Valenti (2005) (midle) and determined using the Padova models (top). The filled lines represent the identity line. Figure from Sousa et al. (2010).

The conclusion drawn from the study was that "planet-bearing stars have less than one per cent of the primordial Li abundance, while about 50 per cent of the solar analogues without detected planets have on average ten times more Li". One of the ideas pointed out to explain this difference was that "the presence of planets may increase the amount of mixing and deepen the convective zone to such an extent that the Li can be burned".

The dependence of the Lithium with age is known to exist for a long time. Therefore the plot shown in Fig. 2 was used to support the assumption that the stars in the sample have similar age and therefore eliminates any age effect in the result. This assumption is supported by the first selection made for the stars in the HARPS planet search program. With the purpose to search for smaller and with longer period planets, the selection of stars for the HARPS program was made in a way to only choose slow rotators stars with low activity levels. These constraints are required to permit high precision radial velocity measurements. This selection effect can be clearly seen in 2 where all the stars seem to have low activity, mostly lower than the one observed for the Sun and their projected rotational velocity is less than 5 Km/s. These characteristics are typical of old main sequence stars.

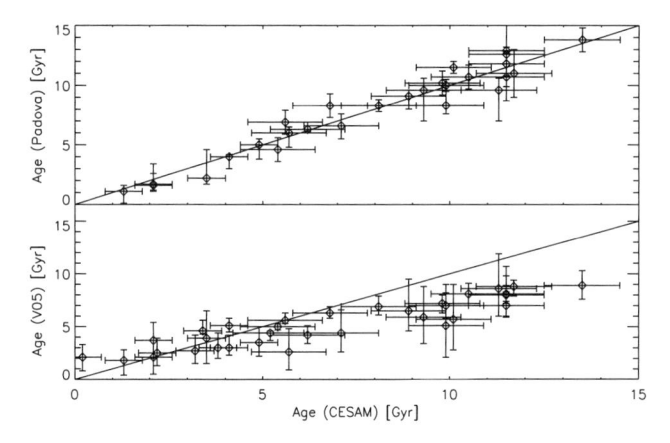

Figure 4. Same as Fig. 3, but for ages. The comparison with the values of Fischer & Valenti (2005) are here revised. An improvement is seen for younger stars with the respect with the original figure from Sousa et al. (2010).

3. Masses and Ages

In order to verify that the stars were in fact old and that there was no significant Lithium dependence with age for this sample, Sousa et al. (2010) used stellar evolution models to derive stellar age and mass. For this careful task only the stars from the HARPS planet search program were selected with effective temperature between 5700 K and 5850 K. The reason for this is to, first, make sure we eliminate any kind of systematics coming from instruments (e.g. spectral resolution) and, second, to be only focus on the interesting temperature range that is around to the solar temperature. The stellar age and mass were determined by means of comparison between stellar evolutionary models and observations. The stellar models were computed with the CESAM code version 3 (Morel 1997), running in the Coimbra Observatory. The procedure is described in more detail in the article. To check the values obtained we present in Fig. 3 the comparison between the obtained mass with others found in the literature. A very good agreement is observed for the three panels.

The same kind of comparison was performed for age and the result can be seen in Fig. 4. In this proceeding we are using a revised table for the values of age of the work Fischer & Valenti (2005) that was passed to us by Jeff Valenti during the Conference. The comparison is still similar with the original plots presented in Sousa et al. (2010) but we verify a slight improvement, specially for the younger stars, in the comparison with the values of Fischer & Valenti (2005).

From both Figure 1 and 2 we conclude that the masses and ages derived are consistent with the ones we can find using other procedures. With the masses and ages estimated in an homogeneous manner for the selected stars in the sample, we can now observe the results of the Lithium vs. temperature, age, and mass respectively in Fig. 5. From these plots we don't see any clear evidence for a correlation between Lithium with any of the presented parameters for the selected stars with a very restrict effective temperature. The higher depletion for the planet host stars is however clearly seen in the three panels of the figure confirming the results of Israelian et al. (2009).

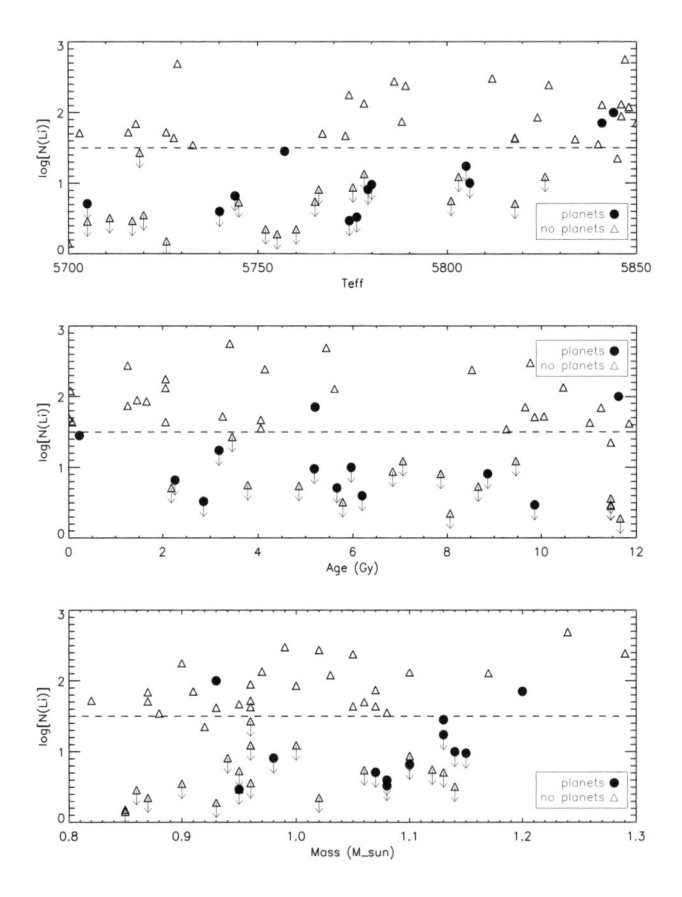

Figure 5. Lithium abundances as a function of effective temperature (top), age (middle), and mass (lower panel). Filled circles denote stars with planets, while open triangles denote single stars. Arrows indicate upper limits. Figure from Sousa et al. (2010).

4. Discussion

These new results give strength to the idea that there is a correlation between the presence of planets and the abundance of light elements in the stars. This discovery is important not only because it give us more insights about the stellar and planet formation, but it can also be useful to select stars for planet search, since it will be more likely to find planets hosts if the star suffers more lithium depletion.

However, this idea has been hardly debated in the last years by several groups, some confirming and some not confirming this idea. As time goes by, the recent works start to have more strength in their results because are making use of larger and more homogeneous samples with a very significant number of stars. Nevertheless a recent work by Baumann et al. (2010) does not confirm the idea presented in Israelian et al. (2009). In this new work they claim that there is still an age effect that is hidden in the

Lithium Depletion in Solar Type stars: Lithium and Planet Presence 87

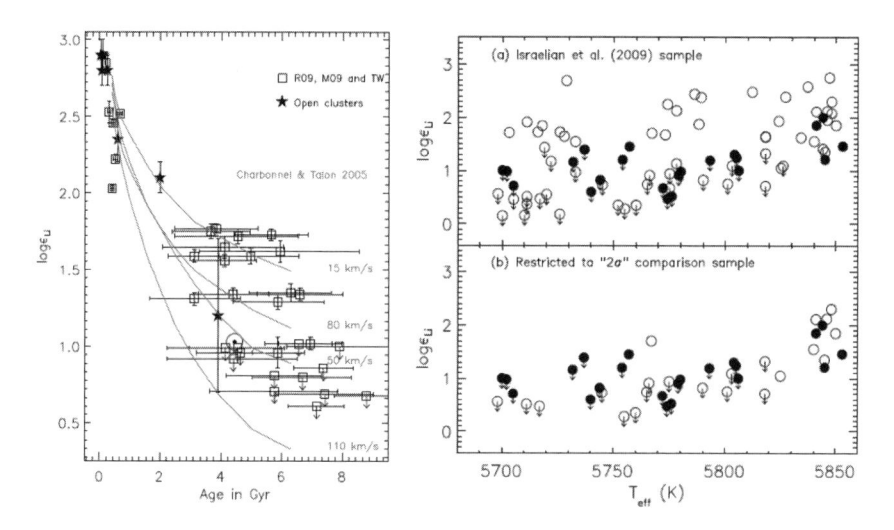

Figure 6. Left panel: Lithium vs. Age for solar twins. (Fig 3. from Baumann et al. (2010)); Right panel: Lithium abundance as a function of Teff in stars with and without detected planets from the Israelian et al. (2009) sample. (Fig 8. from Baumann et al. (2010))

data. Figure 6 shows two important plots taken from Baumann et al. (2010) regarding this subject.

In the left panel of Figure 6 we present the Lithium dependence on age for solar twins. For stars older than 2 Gyr it is clear that we observe an increase of the dispersion for the abundance of lithium in the stars. One possible explanation for this dispersion might be related with the rotational history of the stars. In this panel we can see the solid lines that show the predicted values from the models by Charbonnel & Talon (2005) for different initial rotational velocities. Comparing this result with the middle panel of Fig. 5 we can see a very similar dispersion for the stars older than 2 Gyr that can be also explained by the same reasoning. The only extra information here is that for some reason the planet host stars have on average less Lithium content than the rest of the stars. This may indicate that the planet presence may influence directly the lithium depletion or indirectly in the process of formation and evolution of these stellar systems.

In the same work it is presented a reanalysis of the data published in Israelian et al. (2009). In this case the choice of the comparison sample of stars was more selective. It is only plot stars without detected planets with stellar parameters (Teff, log g, [Fe/H]) within 2σ of the planet-hosting stars. In this case, when using stars with similar fundamental parameters, lithium seems to not be abnormally low in stars with detected giant planets (right panel of Fig. 6).

We think that this approach is very interesting indeed, however, the choice of the value 2σ when using the precision errors given for the parameters is, in our opinion, an excessive limitation, reducing significantly the number of comparison stars to just a few. Following the same idea we present a new plot in Figure 7 where we are assuming a more realistic accuracy error for the temperature and metallicity of 50 K and 0.05 dex, respectively, for all the stars. This will take into account any systematic effect coming

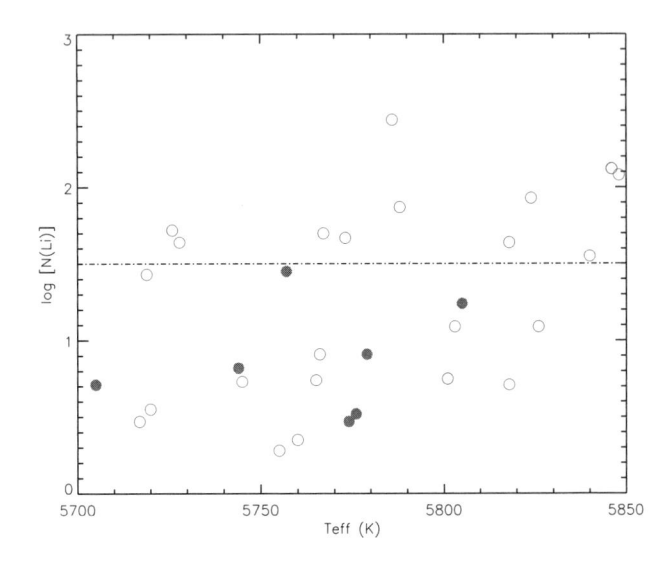

Figure 7. Same as middle panel of Figure 5 but with a more carefull selection of the comparison stars. See text for more details.

from the derived spectroscopic temperature. Moreover we have selected only the stars with surface gravity greater than 4.3, so we are sure that we are only considering dwarf and main sequence stars. Using this new selection of comparison stars we can see from the new plot(Fig. 7 that the planet host stars still have on average less Lithium than the rest of the stars and therefore confirming the result presented in Israelian et al. (2009).

The reader may now wonder that this is just a matter of playing with the numbers. Messing with the right numbers one can obtain any result that will satisfy him. If this is the case then the reader should think on raising the question with a different perspective. If we do not believe that there is in fact a correlation between the presence of giant planets and the high lithium depletion, then why didn't we found any planet for the stars with high lithium content? Looking to the right panel in Figure 6 consider the stars that just disappeared from the top right panel. All of these stars were followed in the same way to obtain radial velocity HARPS program. All of them have more than 20 precise radial velocities measures during the last years. This number is more than enough to detect any orbiting giant planets. Why is no detected planet in these higher Lithium content stars?

5. Summary

In this proceeding we summarized the results presented recently about a possible observational correlation between Lithium and the presence of planets. The main conclusions that came out of these work are as follow:

- Planet hosts stars seem to suffer for more severe lithium depletion.

- Lithium dependence on stellar age cannot completely explain this correlation.

- Initial rotation of stars may be influenced by the orbiting planets and explain the different rates of lithium depletion.
- Why don't we find planets in stars with higher lithium abundance?

Acknowledgments. We acknowledge the support by the European Research Council/European Community under the FP7 through Starting Grant agreement number 239953. S.G.S acknowledges the support from the Fundação para a Ciência e a Tecnologia (Portugal) in the form of a grant SFRH/BPD/47611/2008. NCS also acknowledges the support from Fundação para a Ciência e a Tecnologia (FCT) through program Ciência 2007 funded by FCT/MCTES (Portugal) and POPH/FSE (EC), and in the form of grant reference PTDC/CTE-AST/098528/2008.

References

Baraffe, I., & Chabrier, G. 2010, A&A, 521, A44+. 1008.4288
Baumann, P., Ramírez, I., Meléndez, J., Asplund, M., & Lind, K. 2010, A&A, 519, A87+. 1008.0575
Boss, A. P. 2002, ApJ, 567, L149
Bouvier, J. 2008, A&A, 489, L53. 0808.3917
Castro, M., Vauclair, S., Richard, O., & Santos, N. C. 2009, A&A, 494, 663. 0811.2906
Charbonnel, C., & Talon, S. 2005, Science, 309, 2189. arXiv:astro-ph/0511265
Chen, Y. Q., & Zhao, G. 2006, AJ, 131, 1816. arXiv:astro-ph/0607295
Cochran, W. D., Hatzes, A. P., Butler, R. P., & Marcy, G. W. 1997, ApJ, 483, 457. arXiv:astro-ph/9611230
Deliyannis, C. P., Cunha, K., King, J. R., & Boesgaard, A. M. 2000, AJ, 119, 2437
Ecuvillon, A., Israelian, G., Santos, N. C., Mayor, M., & Gilli, G. 2006, A&A, 449, 809. astro-ph/0512221
Eggenberger, P., Maeder, A., & Meynet, G. 2010, A&A, 519, L2+. 1010.5385
Fischer, D. A., & Valenti, J. 2005, ApJ, 622, 1102
Garaud, P., & Bodenheimer, P. 2010, ApJ, 719, 313. 1005.1618
Garcia Lopez, R. J., & Perez de Taoro, M. R. 1998, A&A, 334, 599. arXiv:astro-ph/9803016
Gilli, G., Israelian, G., Ecuvillon, A., Santos, N. C., & Mayor, M. 2006, A&A, 449, 723
Gonzalez, G. 2008, MNRAS, 386, 928. 0802.0434
Gonzalez, G., & Laws, C. 2000, AJ, 119, 390
Israelian, G., Delgado Mena, E., Santos, N. C., Sousa, S. G., Mayor, M., Udry, S., Domínguez Cerdeña, C., Rebolo, R., & Randich, S. 2009, Nat, 462, 189. 0911.4198
Israelian, G., Santos, N. C., Mayor, M., & Rebolo, R. 2001, Nature, 411, 163
— 2004, A&A, 414, 601. arXiv:astro-ph/0310378
King, J. R., Deliyannis, C. P., Hiltgen, D. D., Stephens, A., Cunha, K., & Boesgaard, A. M. 1997, AJ, 113, 1871
Luck, R. E., & Heiter, U. 2006, AJ, 131, 3069
Mayor, M., Pepe, F., Queloz, D., Bouchy, F., Rupprecht, G., Lo Curto, G., Avila, G., Benz, W., Bertaux, J.-L., Bonfils, X., dall, T., Dekker, H., Delabre, B., Eckert, W., Fleury, M., Gilliotte, A., Gojak, D., Guzman, J. C., Kohler, D., Lizon, J.-L., Longinotti, A., Lovis, C., Megevand, D., Pasquini, L., Reyes, J., Sivan, J.-P., Sosnowska, D., Soto, R., Udry, S., van Kesteren, A., Weber, L., & Weilenmann, U. 2003, The Messenger, 114, 20
Mordasini, C., Alibert, Y., & Benz, W. 2009, A&A, 501, 1139. 0904.2524
Morel, P. 1997, A&AS, 124, 597
Neves, V., Santos, N. C., Sousa, S. G., Correia, A. C. M., & Israelian, G. 2009, A&A, 497, 563. 0902.3374
Pollack, J. B., Hubickyj, O., Bodenheimer, P., Lissauer, J. J., Podolak, M., & Greenzweig, Y. 1996, Icarus, 124, 62

Ryan, S. G. 2000, MNRAS, 316, L35. arXiv:astro-ph/0007057
Santos, N. C., Israelian, G., García López, R. J., Mayor, M., Rebolo, R., Randich, S., Ecuvillon, A., & Domínguez Cerdeña, C. 2004a, A&A, 427, 1085
Santos, N. C., Israelian, G., & Mayor, M. 2001, A&A, 373, 1019. URL http://adsabs.harvard.edu/cgi\-bin/nph\-bib_query?bibcode=2000A\%26A... 363..228S\&db_key=AST
— 2004b, A&A, 415, 1153
Sousa, S. G., Fernandes, J., Israelian, G., & Santos, N. C. 2010, A&A, 512, L5+. 1003.0405
Sousa, S. G., Santos, N. C., Mayor, M., Udry, S., Casagrande, L., Israelian, G., Pepe, F., Queloz, D., & Monteiro, M. J. P. F. G. 2008, A&A, 487, 373. 0805.4826
Takeda, Y., & Kawanomoto, S. 2005, PASJ, 57, 45
Takeda, Y., Kawanomoto, S., Honda, S., Ando, H., & Sakurai, T. 2007, A&A, 468, 663
Udry, S., & Santos, N. C. 2007, ARA&A, 45, 397

Model Atmospheres From Very Low Mass Stars to Brown Dwarfs

F. Allard,[1] D. Homeier,[2] and B. Freytag[1,3]

[1]*Centre de Recherche Astrophysique de Lyon, UMR 5574: CNRS, Université de Lyon, École Normale Supérieure de Lyon, 46 Allée dItalie, F-69364 Lyon Cedex 07, France*

[2]*Institut für Astrophysik Göttingen, Georg-August-Universität, Friedrich-Hund-Platz 1, 37077 Göttingen, Germany*

[3]*Istituto Nazionale di Astrofisica, Osservatorio Astronomico di Capodimonte, Via Moiariello 16, I-80131 Naples, Italy*

Abstract. Since the discovery of brown dwarfs in 1994, and the discovery of dust cloud formation in the latest Very Low Mass Stars (VLMs) and Brown Dwarfs (BDs) in 1996, the most important challenge in modeling their atmospheres has become the understanding of cloud formation and advective mixing. For this purpose, we have developed radiation hydrodynamic 2D model atmosphere simulations to study the formation of forsterite dust in presence of advection, condensation, and sedimentation across the M-L-T VLMs to BDs sequence (T_{eff} = 2800 K to 900 K, Freytag et al. 2010). We discovered the formation of gravity waves as a driving mechanism for the formation of clouds in these atmospheres, and derived a rule for the velocity field versus atmospheric depth and T_{eff}, which is relatively insensitive to gravity. This rule has been used in the construction of the new model atmosphere grid, **BT-Settl**, to determine the microturbulence velocity, the diffusion coefficient, and the advective mixing of molecules as a function of depth. This new model grid of atmospheres and synthetic spectra has been computed for 100,000 K > T_{eff} > 400 K, 5.5 > logg > -0.5, and [M/H]= +0.5 to -1.5, and the reference solar abundances of Asplund et al. (2009). We found that the new solar abundances allow an improved (close to perfect) reproduction of the photometric and spectroscopic VLMs properties, and, for the first time, a smooth transition between stellar and substellar regimes — unlike the transition between the **NextGen** models from Hauschildt et al. 1999a,b, and the **AMES-Dusty** models from Allard et al. 2001). In the BDs regime, the **BT-Settl** models propose an improved explanation for the M-L-T spectral transition. In this paper, we therefore present the new **BT-Settl** model atmosphere grid, which explains the entire transition from the stellar to planetary mass regimes.

1. The Impact of the New Solar Abundances on VLMs Spectral Properties

The modeling of the atmospheres of very low mass stars (hereafter VLMs) has evolved with the development of computing capacities from an analytical treatment of the transfer equation using moments of the radiation field (Allard 1990), to a line-by-line opacity sampling in spherical symmetry (Allard et al. 1997; Hauschildt et al. 1999a,b), and finally 3D radiation transfer (Seelmann et al. 2010). In parallel to detailed radiative transfer in an assumed static environment, hydrodynamical simulations have been developed to reach a realistic representation of the granulation and the line profiles shifted and shaped by the hydrodynamical flow of the sun and sun-like stars (see for details the review in a special isssue of the Journal of Computational Physics by Freytag et al. 2011) by using a non-grey (multi-group binning of opacities) radiative transfer using a pure blackbody source function (scattering is neglected).

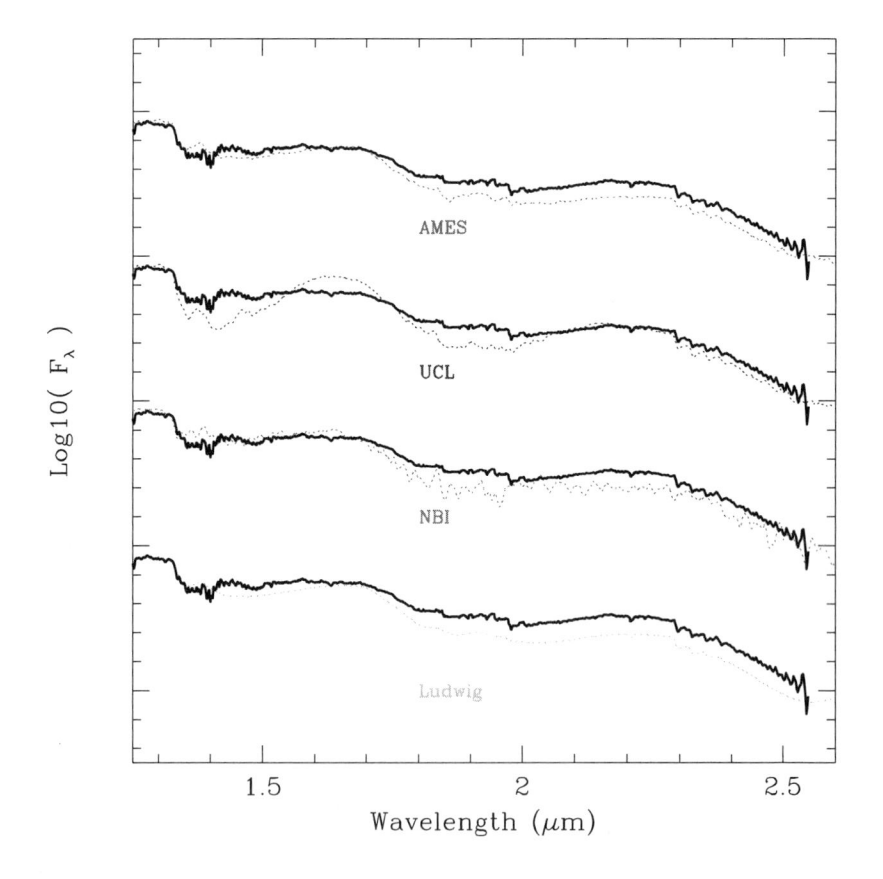

Figure 1. Model atmosphere synthetic spectrum versus VB10 IR SED, using different water opacity profiles over the years (Ludwig 1971; Jørgensen et al. 2001; Schryber et al. 1995; Partridge & Schwenke 1997).

The model atmospheres and synthetic spectra have also been made possible thanks to the development of realistic opacities calculated often ab initio for the needs of an accurate account of their cooling and heating effects in the internal atmospheric layers where temperatures close to 3000 K can prevail. For some time, the remaining discrepancies in the model synthetic spectra were believed to be due to incomplete water vapor line lists, both in temperature and in the rotation quantum number J of the molecular simulations. In Fig.1, model atmosphere synthetic spectra are shown using different published water vapor opacity profiles such as hot flames laboratory experiments (Ludwig 1971), empirical calculations (Jørgensen et al. 2001), and ab initio calculations

(Schryber et al. 1995; Partridge & Schwenke 1997) based on independently measured interaction potential surface. As can be seen from Fig.1 where the models are compared to the infrared spectrum of an M8 dwarf (VB10), the water vapor opacity profile which shape this part of the spectrum has strongly change over time with the improvement of computational capacities and a better knowledge of the interaction potential surface. But in general, all opacity profiles converge in predicting an over-opacity (or lack of flux in the model) in the K bandpass. The UCL opacity profile — likely because of its incompleteness — could allow a seemingly correct $J - K$ color while not allowing a detailed spectral comparison of the **NextGen** models (Allard et al. 1994). It became however clear that the more recent versions of the water vapor profile (Partridge & Schwenke 1997; Barber et al. 2006) all agree in establishing a systematic lack of flux in the K bandpass in the models. This is the reason why the Allard et al. (2001) **AMES-Cond/Dusty** model grids, based on the Partridge & Schwenke (1997) water vapor opacity profile were never proposed for the study of VLMs, and reserved for the study of the limiting properties of brown dwarfs.

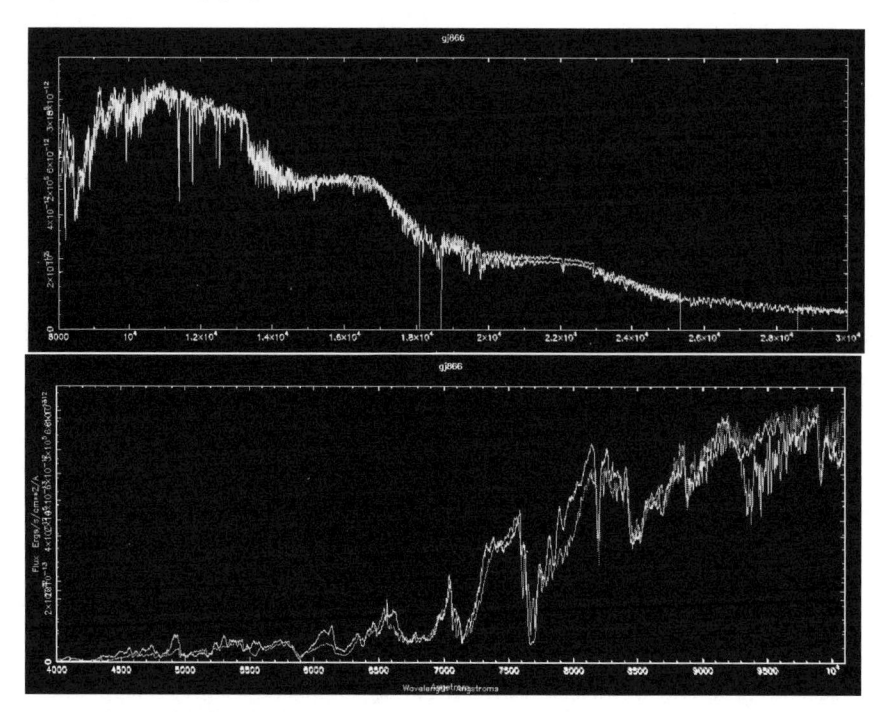

Figure 2. A **BT-Settl** synthetic spectrum with Teff=2900K, logg=5.0, and solar metallicity by Asplund et al. (2009) is shown in yellow compared to the SED of the red dwarf GJ 866 (cyan curve). Both are plotted in absolute flux, i.e no flux adjustment at a specific wavelength is needed. The infrared SED (top panel) is now perfectly reproduced by the model. The agreement is particularly good in the Wing Ford FeH bands near 0.99 μm while some discrepancy prevail in the missing CaOH bands around 6000Å. Telluric absorption is corrected from the observations in red.

In this regard, the new solar abundances based on radiation hydrodynamical simulations of the solar atmosphere (Asplund et al. 2009; Caffau et al. 2010) help as they predict an oxygen abundance of 0.3 dex (a factor of 2) lower than the previously used solar abundances of Grevesse et al. (1993). The results are shown in Fig.2, where the new **BT-Settl** model atmosphere synthetic spectra for a T_{eff} = 2900 K, logg=5.0 and solar abundances according Asplund et al. (2009) are compared to the spectrum of an M6 dwarfs, GJ866 (kindly provided by M. Bessel, Mt-Stromlo Obs.). For the first time, we find a perfectly fitting spectra distribution across the near-IR to infrared spectral region (the model is the yellow line). The agreement is also excellent in the optical to red part of the spectrum in particularly in the FeH Wing Ford bands near 0.99 μm, and in the VO bands thanks to line lists provided by B. Plez (GRAAL, Montpellier, France). Missing opacities are, however, affecting still the spectral distribution (CaOH bands), and the Allard et al. (2000) TiO line list becomes now too corse compared to progress made with water vapor.

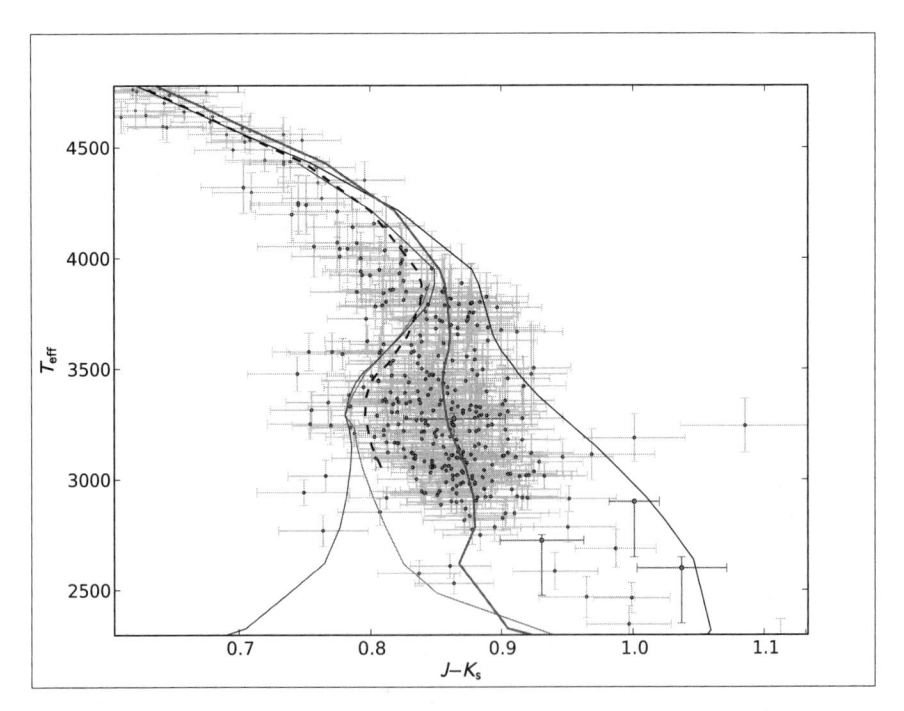

Figure 3. Estimated Teff for M dwarfs by Casagrande et al. (2008) and brown dwarfs by Golimowski et al. (2004) are reported as a function of J-K short. Overplotted are the **NextGen** model isochrones for 5 Gyrs (Baraffe et al. 1997, 1998) using varius generations of model atmospheres, starting with the **NextGen** (black line), pursuing with the limiting case **AMES-Cond/Dusty** grids by Allard et al. (2001) (blue and red line respectively), and finishing with the **BT-Settl** models using the Asplund et al. (2009) solar abundances (green line).

One can see from Fig.3 that the **NextGen** models systematically and increasingly overestimate T_{eff} through the lower main sequence, while the **AMES-Cond/Dusty** models were underestimating T_{eff} compared to the averaged empirical determinations of T_{eff} of individual stars (Casagrande et al. 2008). This situation is relieved when using the newer Asplund et al. (2009) abundances, and the **BT-Settl** models now agree fairly well with most of the empirical estimations of T_{eff}. Evolution models are currently being prepared using the **BT-Settl** model atmosphere grid.

2. Dust Formation in Late Type VLMs and Brown Dwarf Atmospheres

One of the most important challenge in modeling the atmospheres and spectral properties of VLMs and brown dwarfs is the formation of dust clouds and its associated greenhouse effects making the infrared colors of late M and early L dwarfs extremely red compared to colors of low mass stars. The cloud composition, according to equilibrium chemistry, is going from zirconium oxide (ZrO_2), to refractory ceramics (perovskite and corundum; $CaTiO_3$, Al_2O_3), to silicates (forsterite; Mg_2SiO_4), to salts (CsCl, RbCl, NaCl), and finally to ices (H_2O, NH_3, NH_4SH) as brown dwarfs cools down with time from M through L, T spectral types and beyond (Allard et al. 2001; Lodders & Fegley 2006). Many cloud models have been constructed to address this problem in brown dwarfs over the past decade (see Helling et al. 2008, for a review on the subject). However, none treated the mixing properties of the atmosphere, and the resulting diffusion mechanism realistically enough to reproduce the properties of the spectral transition from M through L and T spectral types without changing cloud parameters (for example Ackerman & Marley 2001). It is in this context that we have decided to address the issue of mixing and diffusion by 2D Radiation HydroDymanic (hereafter RHD) simulations of VLMs and brown dwarfs atmospheres, using the **Phoenix** opacities in a multi-group binning, and forsterite geometric cross-sections (Freytag et al. 2010). We found that gravity waves have a decisive role in clouds formation in brown dwarfs, while around $T_{\text{eff}} \leq 2200$ K the cloud layers become optically thick enough to initiate cloud convection, which participate in the global mixing. Overshoot is also important in the mixing of the largest dust particles (see paper by D. Homeier in this book). In Fig.4, 3D RHD simulations (a $350 \times 350 \times 170$ km^3 box at the surface of the star) are shown for dwarfs with $T_{\text{eff}} = 2600$ K, 2200 K, and 1500 K from top to bottom. For each simulation, two snapshots are shown side-by-side to illustrate the intensity variation due to cloud formation and granulation. The 2600 K case shows no or negligible dust formation, while dust formation progresses to reach optically thick density at around 2200K, before sedimenting out again towards the 1300K regime. The $T_{\text{eff}} = 1500$ K case illustrate the importance of gravity waves, where the minima of the waves reach condensation levels while the maxima remain in condensed phase. The box of simulations are too small compared to the radius of the star to show adequately the variability of these objects, however. See Freytag et al. (2010) and B. Freytag's poster paper in these proceedings for details.

These simulations permitted to account in **Phoenix** for the advective forces bringing fresh condensible material from the hotter lower layers to the cloud forming layers. Our cloud model is based on the condensation and sedimentation timescales from a study of Earth, Venus, Mars and Jupiter atmospheres by Rossow (1978). However we had to compute the supersaturation pressure from our pre-tabulated equilibrium chemistry in order to obtain the correct amount of dust formation (as opposed to use an approximate value cited in Rossow, 1978). We then solved the cloud model and equilibrium chemistry in turn layer by layer inside out to account for the sequence of grain species formation as a function of cooling of the gas. One can see from Fig.5 that the late-type M and early type L dwarfs behave as if dust is formed nearly in equilibrium with the gas phase with extremely red colors in some agreement with the **BT-Dusty**

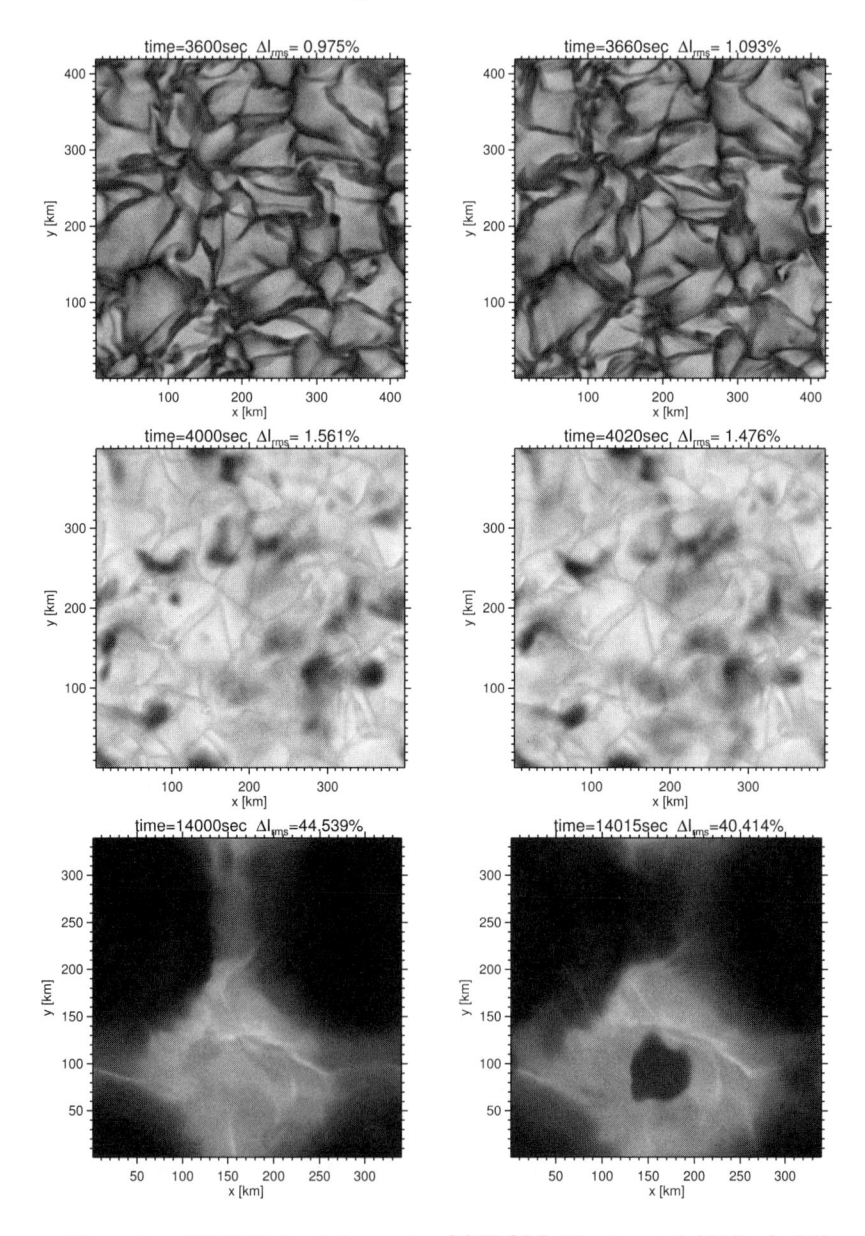

Figure 4. 3D HRD simulations using **CO5BOLD** (Freytag et al. 2010) of a $350 \times 350 \times 170$ km^3 of atmosphere at the surface of, from top to bottom, 2600K, 2200K and 1500K VLMs and brown dwarfs of logg=5.0, and solar metallicity. For each model two intensity snapshots are shown in time to illustrate the intensity variability.

models. At the low Teff regime dominated by T-type dwarfs, the **AMES-Cond** models also appear to provide a good limitation for the brown dwarf colors. However, the chemistry of T dwarfs can be very far from the equilibrium in the models. The **BT-Settl**

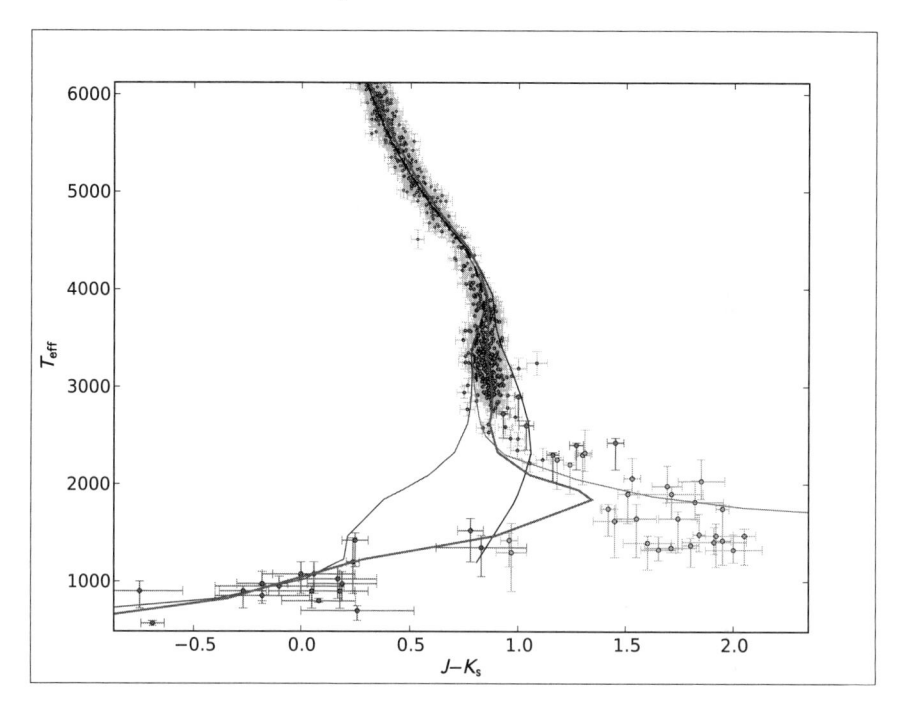

Figure 5. Same plot as Fig.3 but zooming out and extending into the brown dwarfs region of the diagram. This region below 2500K is dominated by dust formation (essentially forsterite and other silicates). The limiting case **AMES-Cond/Dusty** models atmosphere provide a description of the span in colors of the brown dwarfs in this diagram.

models, which account for a cloud model, dynamical mixing from RHD simulations, a supersaturation computed from pre-tabulated equilibrium chemistry calculations, and the Asplund et al. (2009) solar abundances, manage to reproduce the main sequence down to the L-type brown dwarf regime, before turning to the blue in the late-L and T dwarf regime. The models used an age of 5 Gyrs, and in the case of the **BT-Settl** models, a younger age of a few Gyrs would easily reproduce the reddest brown dwarfs.

3. Summary and Future Prospects

We propose, in this paper, a new model atmosphere grid, named **BT-Settl**, computed using the atmosphere code **Phoenix** which has been updated, compared to the Allard et al. (2001) **AMES-Cond/Dusty** models, for: i) the Barber et al. (2006) BT2 water opacity line list, ii) the solar abundances revised by Asplund et al. (2009), and iii) a cloud model accounting for supersaturation and RHD mixing. It is covering the whole range of VLMs and brown dwarfs and beyond: 1000,000 K < T_{eff} < 400 K; -0.5 < logg < 5.5; and +0.5 < [M/H] < -4.0, including various values of the alpha element enhancement. Only the confrontation of the models using spectral synthesis will allow to define the content of oxygen and alpha elements of VLMs and brown dwarfs. But it is clear that these objects are excellent constraint for the solar oxygen abundance. The models are available at the Phoenix simulator website "http://phoenix.ens-lyon.fr/simulator/" and are in preparation for publication. However, the interior and evolution models are expected for the second half of 2011.

In order to say something about the spectral variability of VLMs and brown dwarfs, 3D RHD simulations of "the star in the box" with rotation will be required. This is our current project supported by the French "Agence Nationale de la Recherche" for the period 2010-2015. Rotation is already modeled for a scaled down model of the Sun using **CO5BOLD** (Steffen & Freytag 2007) and can be applied to brown dwarf simulations.

Acknowledgments. We would like to thank specifically Mickael Bessel (Mount Stromlo Obs.) for his visit to CRAL and the fruitful discussions, as well as Robert Barber (UCL) for his generous support. We thank the french "Agence Nationale de la Recherche" (ANR) and "Programme National de Physique Stellaire" (PNPS) of CNRS (INSU) for their financial support. The computations of dusty M dwarf and brown dwarf models were performed at the *Pôle Scientifique de Modélisation Numérique* (PSMN) at the *École Normale Supérieure* (ENS) in Lyon.

References

Ackerman, A. S., & Marley, M. S. 2001, ApJ, 556, 872
Allard, F. 1990, Ph.D. thesis, PhD thesis. Ruprecht Karls Univ. Heidelberg, (1990)
Allard, F., Hauschildt, P. H., Alexander, D. R., & Starrfield, S. 1997, ARA&A, 35, 137
Allard, F., Hauschildt, P. H., Alexander, D. R., Tamanai, A., & Schweitzer, A. 2001, ApJ, 556, 357
Allard, F., Hauschildt, P. H., Miller, S., & Tennyson, J. 1994, ApJ, 426, L39
Allard, F., Hauschildt, P. H., & Schwenke, D. 2000, ApJ, 540, 1005
Asplund, M., Grevesse, N., Sauval, A. J., & Scott, P. 2009, ARA&A, 47, 481. 0909.0948
Baraffe, I., Chabrier, G., Allard, F., & Hauschildt, P. H. 1997, A&A, 327, 1054
— 1998, A&A, 337, 403
Barber, R. J., Tennyson, J., Harris, G. J., & Tolchenov, R. N. 2006, MNRAS, 368, 1087
Caffau, E., Ludwig, H., Steffen, M., Freytag, B., & Bonifacio, P. 2010, Solar Phys., 66. 1003.1190
Casagrande, L., Flynn, C., & Bessell, M. 2008, MNRAS, 389, 585. 0806.2471
Freytag, B., Allard, F., Ludwig, H., Homeier, D., & Steffen, M. 2010, A&A, 513, A19+. 1002.3437
Golimowski, D. A., Leggett, S. K., Marley, M. S., Fan, X., Geballe, T. R., Knapp, G. R., Vrba, F. J., Henden, A. A., Luginbuhl, C. B., Guetter, H. H., Munn, J. A., Canzian, B., Zheng, W., Tsvetanov, Z. I., Chiu, K., Glazebrook, K., Hoversten, E. A., Schneider, D. P., & Brinkmann, J. 2004, AJ, 127, 3516
Grevesse, N., Noels, A., & Sauval, A. J. 1993, A&A, 271, 587
Hauschildt, P. H., Allard, F., & Baron, E. 1999a, ApJ, 512, 377. arXiv:astro-ph/9807286
Hauschildt, P. H., Allard, F., Ferguson, J., Baron, E., & Alexander, D. R. 1999b, ApJ, 525, 871
Helling, C., Ackerman, A., Allard, F., Dehn, M., Hauschildt, P., Homeier, D., Lodders, K., Marley, M., Rietmeijer, F., Tsuji, T., & Woitke, P. 2008, MNRAS, 391, 1854. 0809.3657
Jørgensen, U. G., Jensen, P., Sørensen, G. O., & Aringer, B. 2001, A&A, 372, 249
Lodders, K., & Fegley, B. J. 2006, arXiv:astro-ph/0601381v1
Ludwig, C. B. 1971, Appl.Optics, 10, 1057
Partridge, H., & Schwenke, D. W. 1997, Journal for Computational Physics, 106, 4618
Rossow, W. B. 1978, ICARUS, 36, 1
Schryber, J. H., Miller, S., & Tennyson, J. 1995, JQSRT, 53, 373
Seelmann, A. M., Hauschildt, P. H., & Baron, E. 2010, A&A, 522, A102+. 1007.3419
Steffen, M., & Freytag, B. 2007, Astronomische Nachrichten, 328, 1054

Low-Mass Eclipsing Binaries: Observations vs. Theory

J. C. Morales,[1,2] I. Ribas,[1,2] and C. Jordi[2,3]

[1]*Institut de Ciències de l'Espai (CSIC-IEEC), Campus UAB-Facultat de Ciències, Torre C5 - parell 2a planta, 08193 Bellaterra, Spain*

[2]*Institut d'Estudis Espacials de Catalunya (IEEC), Edif. Nexus, C/ Gran Capità 2-4, 08034 Barcelona, Spain*

[3]*Departament d'Astronomia i Meteorologia, Institut de Ciències del Cosmos, Universitat de Barcelona, C/ Martí i Franquès 1, 08028 Barcelona, Spain*

Abstract. In recent years, analyses of eclipsing binary systems have unveiled differences between the observed fundamental properties of low-mass stars and those predicted by stellar structure models. Particularly, radii and effective temperatures computed from models are 5–10% lower and 3–5% higher than observed, respectively. Given the high accuracy of the empirical measurements (typically 1–2%), these differences are very significant. The discrepancies have been attributed to different factors, notably to the high levels of magnetic activity present on these stars. Here, we review the observational evidence on the fundamental properties of low-mass stars and discuss possible ways to reconcile observation and theory. In particular, we analyze the consequences of magnetic activity both on models and on the observational analysis of eclipsing binaries. With all evidence in hand, we propose a self-consistent scenario that explains the discrepancies by considering several factors, including an observational bias caused by polar spots, the radiative impact of spots themselves and the reduced efficiency of convective energy transport in the presence of magnetic fields.

1. Introduction

Knowing precisely the structure and evolution of low-mass stars ($M \lesssim 1$ M$_\odot$) is especially important since they share common characteristics with other objects such as brown dwarfs or gaseous exoplanets, and they also drive the properties and habitability of the exoplanets they may host. However, although low-mass stars constitute the most numerous stellar population in the Galaxy, their structure and evolution is not yet fully understood. This is a consequence of the relative lack of good calibrators because of their intrinsic faintness and the difficulty in determining their fundamental properties such as masses and radii. In the last 15 years, theoretical models of stellar interiors and atmospheres, together with an increasing number of accurate measurements of stellar properties, have brought great progress towards the understanding of low-mass stars. The crucial point to validate the models is a critical comparison of their predictions with observed fundamental properties such as masses, radii, effective temperatures, metal contents and ages.

Detached double-lined eclipsing binaries (hereafter DDLEBs) are the best objects to determine the fundamental properties of low-mass stars with the needed accuracy

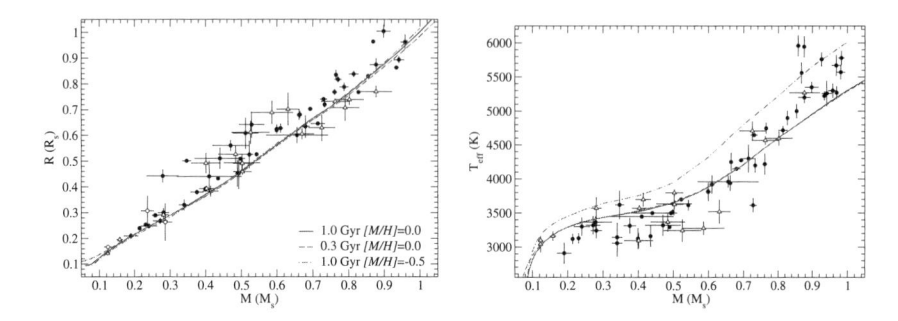

Figure 1. $M-R$ and $M-T_{\mathrm{eff}}$ diagrams observed for low-mass stars. Observations from double-lined eclipsing binaries (solid circles), single-lined eclipsing binaries (open circles), and stars with interferometric measurement of radius (triangles) are compared with different isochrones from Baraffe et al. (1998) as labeled. Error bars are plotted when available.

to test stellar models (Andersen 1991; Torres et al. 2010). Masses and radii of the components of these systems can be determined with 1–2% accuracy from the analysis of multi-band light curves and radial velocity curves. In the last decade, the results from DDLEBs have revealed that the observed radii of low-mass stars is ∼ 5 − 10% larger than predicted by models, while effective temperatures are ∼ 3−5% cooler. In contrast, luminosities are correctly reproduced by models (see Ribas 2006, for a review). As Figure 1 shows, these differences are very significant for the stars in DDLEBs, while interferometric measurements or stellar properties derived from single-lined eclipsing binaries do not have sufficient precision for a critical analysis. It is also apparent that the radii of stars below the fully convective limit ($M \lesssim 0.35$ M$_\odot$) is better reproduced by models, thus suggesting the possible existence of differences in the behavior of stars depending on the presence of a radiative core.

Several authors have argued that magnetic activity could be the cause of the discrepancies found between models and observations of DDLEBs although its mechanism is not yet fully understood. Here, we analyze the effect of activity on the stellar properties of low-mass stars both from the point of view of the observations and of the stellar models. In Section 2, we summarize properties regarding activity of these stars and the advances of theoretical stellar models in the treatment of magnetic activity. In Section 2, we analyze the effect of activity, in the form of stars spots, on the observations of low-mass DDLEBs. Section 4 presents an scenario to reconcile models with observations that is discussed in Section 4, and, finally, the conclusions are presented in Section 6.

2. Activity as the Cause of the Discrepancies

Analyses of activity-age relationships of stars (see Soderblom 2010, for a review) indicate that younger stars are more active than their older counterparts. This is a consequence of the spin-down of the star and the consequent loss of angular momentum during the main-sequence phase. Pizzolato et al. (2003), comparing the X-ray to bolometric luminosity of stars with different rotation periods, showed that stars with periods below 10 days present saturated levels of activity with $\log(L_{\mathrm{X}}/L_{\mathrm{bol}}) \sim -3$. In the case of DDLEBs, tidal forces tend to circularize the orbit of the system and to synchronize the rotation with the orbital motion (Mazeh 2008). The components of DDLEBs with orbital periods below ~ 10 days are synchronized and therefore they are fast rotators showing a saturated level of magnetic activity, regardless of their age. The observation of photospheric spots on the components of low-mass DDLEBs, through their imprints on the light curves in the form of photometric variations, clearly indicates that these are magnetically active systems. Given this evidence, several authors suggested that activity, i.e. the presence of star spots on the components of these systems, might be the cause of the discrepancies between models and observations (Torres & Ribas 2002; López-Morales & Ribas 2005; Torres et al. 2006; López-Morales & Shaw 2007; Ribas et al. 2008).

Other possible causes of these discrepancies that have been proposed include model parameters such as metallicity or opacity (Berger et al. 2006; Casagrande et al. 2008). However, extremely low metallicities, in conflict with observations, or unrealistically high opacities are needed to reconcile the observed mass-radius relationships with the model predictions (Morales et al. 2010). Besides, both metallicity and opacity effects should also be present on single stars, which are known to be well described by theoretical models (Demory et al. 2009). Further, similar radius differences as those between models and observations of DDLEBs where found when comparing single magnetically active stars and their inactive counterparts (Morales et al. 2008). This result is in accordance with previous color differences found between active and inactive stars (Stauffer & Hartmann 1986), thus strengthening the activity hypothesis.

With this evidence in hand, Chabrier et al. (2007) introduced the effects of activity in the Lyon stellar group models of low-mass stars (Baraffe et al. 1998). They considered two different scenarios:

a) the effect of magnetic activity and rotation alter the efficiency of convective energy transport by inhibiting convective motions. This can be modelled by setting the mixing length parameter (α) to lower values than those used for solar models.

b) the magnetic activity on low-mass stars is associated with the appearance of photospheric spots that block part of the outgoing flux, which the star compensates by increasing its size. This can be modeled assuming a new stellar luminosity given by

$$L' \propto (1 - \beta) R'^2 T'^4_{\text{eff}}, \qquad (1)$$

where R' and T'_{eff} are the new radius and luminosity and β is the fraction of stellar surface covered by dark spots.

The main conclusion of the work by Chabrier et al. (2007) is that the reduction of the mixing length parameter (α) does not have a significant impact for stars below the fully convective limit, thus leaving the remarkable case of CM Draconis (hereafter CM Dra, Morales et al. 2009a) unexplained. On the other hand, a dark spot coverage of between 30% and 50% of the stellar surface could roughly reproduce the mass-radius relationship of the overall sample of low-mass DDLEBs, including CM Dra, while conserving the stellar luminosity as observed.

3. Effect of Activity on DDLEBs

The results of Chabrier et al. (2007) reinforce the activity hypothesis suggested to explain the differences between models and observations, although it leaves some points to be thoroughly tested. In their modelization of spots, the authors considered spots completely dark, i.e. with $T_{\rm eff,s} = 0$. Actually, spots on low-mass stars, as determined from light curve analysis and Doppler tomography (see Strassmeier 2009, for a review), are generally a few hundreds of K cooler than the photosphere. Therefore, β does not represent the fraction of spotted surface but the fraction of the flux blocked by spots which is given by

$$\beta = \frac{S_s}{S}\left[1 - \left(\frac{T_{\rm eff,s}}{T_{\rm eff}}\right)^4\right], \qquad (2)$$

where S_s/S is the portion of surface covered by spots and $T_{\rm eff,s}/T_{\rm eff}$ the temperature contrast of the spots and the photosphere. Taking into account this relation, a β factor between 0.3 and 0.5, as that suggested by Chabrier et al. (2007), would translate into a surface coverage of about 65–100% with spots $\sim$ 500 K cooler than the photosphere. Such spot coverages seem to be too large compared with those found from the analysis of DDLEB light curves. However, it should be emphasized that light curve analysis is partly degenerate and does not provide absolute values of the spot coverage.

In Morales et al. (2010), we performed a blind test in order to check for the effect of spots on light curves, both to determine the modulation produced on the out-of-eclipse phases and the possible systematic effects on the derived parameters. To perform the test, we generated a series of spot patterns on stars using different distributions. We used spots with 10° in diameter and with a temperature contrast of 0.85. This constrains the possible values of β between 0 (no spots) and 0.5 (fully spotted star). The light curves were generated from the input parameters of GU Boo (Ribas 2003), used as template, with the Wilson-Devinney code (hereafter WD, Wilson & Devinney 1971). The left panel of Figure 2 shows the probability density functions over co-latitude of the distributions used to place spots on the stellar surface of DDLEB components. A uniform distribution was assumed on the longitudinal direction. Two distributions adapted from Granzer et al. (2000), referred to as Distribution 1 and Distribution 2, were used as well as a uniform distribution for comparison.

Regarding the amplitude of the out-of-eclipse modulations due to spots, these simulations indicate that the observed modulations in DDLEBs, up to a few tenths of magnitude, can be reproduced with any of the distributions, especially when taking into account the dependence of this modulation with properties of the spots such as the size and the temperature factor (see Morales et al. 2010, for further details).

To check for systematic effects on the light curve parameters, we added a random Gaussian noise to these curves and we used the same procedure as a normal light curve analysis to derive the fundamental parameters. We fitted a total of 33 light curves using the WD code, setting as free parameters the orbital inclination, the ratio of effective temperatures of the stars and the radii of the stars, as well as the longitude and the size of spots, testing for different values of latitude and temperature contrast. As usual, we tried a combination of a few bright and dark spots, typically less than four, because of limitations related to degeneracies and photometric noise. The right panel of Figure 2 displays an example of a DDLEB system spotted according to Distribution 2 with $\beta = 0.2$ and its corresponding simulated light curve with the best fit obtained.

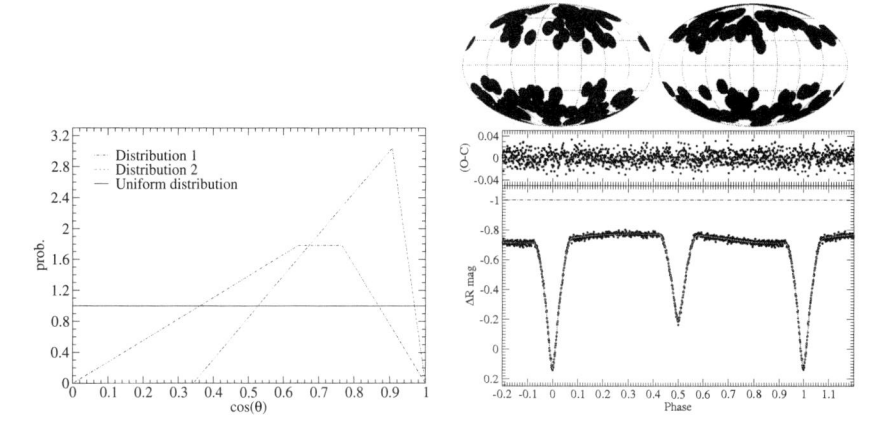

Figure 2. Left: Probability density functions over $\cos\theta$ used to simulate spotted stars. The equator corresponds to $\theta = 90°$. Right: Representation of a simulated spotted DDLEB system with the resulting light curve and its best fit. Star surfaces are represented in a Mollweide projection with the center of the each star being $0°$ in longitude and $90°$ in co-latitude. This case was simulated with Distribution 2 and $\beta = 0.2$.

The relevant parameters for the light curves were not found to have significant differences with respect the input parameters, except for the case of the radii of the components. Figure 3 plots the sum and the ratio of radii relative to to the input parameters of the simulation for different distributions and β. Each point is computed as the average of three simulations with the same configuration. For the case of Distribution 2 it is clear that the sum of radii is biased with respect to the input parameters, yielding radius that can be up to 6% larger than real. This indicates that when spots are located preferentially on the poles, the radii derived from light curves could be overestimated, just in the same direction than the differences found between DDLEB observations and stellar models. This effect is due to the loss of circular symmetry of the stars with polar spots. This causes the isophotes to be elongated, therefore eclipses are widened and the WD code, which assumes a Roche geometry, finds a best fit with a larger stellar radius.

The outcome of this battery of tests is that the presence of starspots on the stellar surface could bias the determination of the stellar radii when spots are concentrated toward the poles. Therefore, if polar spots are indeed present on low-mass stars, as found in some fast rotating stars with Doppler imaging (Strassmeier 2009), the stellar models should be compared with the observed values corrected for the systematic effect of polar spots.

4. Reconciling Models with Observations

The results in Chabrier et al. (2007) and the systematic effect that polar spots can have on DDLEB analysis make it worth revisiting the comparison between models and observations to derive a complete picture of the problem. We used the sample of best known DDLEBs to perform this test in Morales et al. (2010). The theoretical models indicate that activity effects may play a role on the convective transport, parameterized

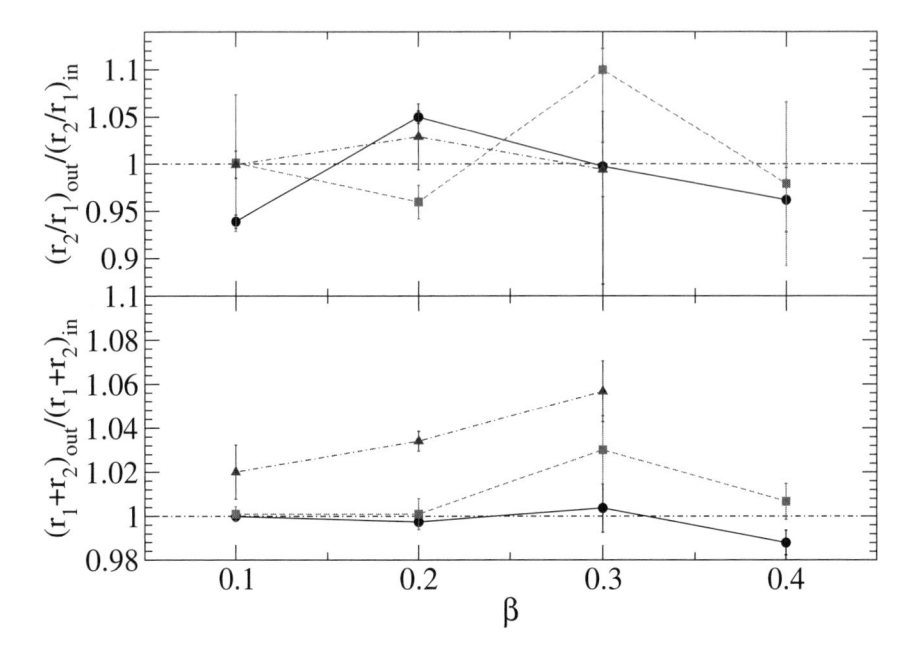

Figure 3. Differentials between the input parameters to the simulations and those recovered from the fits. Uniform distribution, Distribution 1 and Distribution 2 are labelled as solid, dashed and dot-dashed lines, respectively.

with α, or on the appearance of photospheric spots, parameterized with β. Both of these scenarios may explain the differences in radii between models and observations, but, as already mentioned, in the case of fully convective stars the first mechanism is not significant. This places the case of CM Dra as a unique system to disentangle the effect of activity on α and on β since its components are fully convective stars, i.e., their structure is not affected by a reduction of α. Thus, we used this system to find the β value that yields the best fit to the components of CM Dra when taking into account both the systematic effect of spots on the determination of the radii from the light curves and its effect on the stellar structure. We assumed the presence of polar spots on the components of the system, and for different β values we applied to the observed radii the corresponding correction according to the results shown on Figure 3. We subsequently compared with the model predictions for each β value. After iterating several times, we find that a model with $\beta = 0.17 \pm 0.03$ reproduces both components of the system when their radii are downward corrected by 3% (as a consequence of the bias induced by the polar spots). If spots are $\sim 15\%$ cooler than the photosphere, the resulting β translates into $(36 \pm 6)\%$ of the star surfaces covered by spots. This value is in agreement with findings from Doppler imaging for systems such as HK Aqr (Barnes & Collier Cameron 2001).

In order to explain the case of more massive, partially convective stars, some assumptions have to be made:

  a) that the spots on active DDLEB stars are preferentially concentrated toward the poles, maybe due to the fast rotation of the star,

b) that these stars have the same level of spot coverage as fully convective systems such as CM Dra.

The latter assumption is derived from the fact that all these systems show saturated levels of magnetic activity. The mechanism of saturation is not well understood yet, but one feasible explanation could be that saturation is reached when the entire stellar corona is full of active regions. This would imply that the β parameter is similar for all stars in the saturated regime.

Assuming these hypotheses, we found that the overall sample of best known low-mass DDLEBs is reproduced within 1σ of its error bars with $\beta = 0.20 \pm 0.04$, i.e., $(42 \pm 8)\%$ of spot coverage. Figure 4 shows the mass-radius and mass-effective temperature relationships for models with different values of α and β and the fundamental properties of the best known DDLEBs normalized to an age of 1 Gyr for consistent comparison (this implied very small corrections <2% to the radii to compensate for the effects of stellar evolution). It can be seen that the model that best reproduces the case of CM Dra ($\beta = 0.17$) does also explain well the radii of all the systems when they are downward corrected by the 3% systematic effect of polar spots, although the are still some significant differences for fast rotating systems such as GU Boo or YY Gem. If we set $\beta = 0$ and reduce the mixing length parameter, only the more massive systems can be roughly explained but not the case of CM Dra as already expected from the conclusions in Chabrier et al. (2007). It is also very interesting that the same model with $\beta = 0.17$ explains the mass-effective temperature relationships too, except for the cases of CU Cnc, whose temperatures may be affected by a circumstellar dust disk (Ribas 2003), and IM Vir, possibly due to its subsolar metallicity (Morales et al. 2009b). Again, a reduction of α can not reproduce the temperatures of CM Dra.

Figure 5 shows in depth the differences between the model with $\beta = 0.17$ and the best known DDLEBs. There is a conspicuous trend of an increasing radius difference with increasing rotational velocity (computed assuming synchronization with the orbital motion). The only exception is the components of NGC2204-S892 (two rightmost points on the plot), but this is the shortest period binary and there is no estimation of its age or the metallicity (Rozyczka et al. 2009), which could be affecting the model comparison. This tentative trend may indicate that, while the spot effect is a clear contributor to explain the differences between observation and theory, the loss of efficiency of convective energy transport may also be at play, explaining up to $\sim 4\%$ of the radius difference.

5. Discussion

The scenario presented here can explain the discrepancies between the observed fundamental properties on low-mass stars in DDLEBs and the theoretical predictions considering only the two assumptions described above, i.e., the existence of polar spots and its universal effect on the entire range of low-mass stars.

But, while the scenario seems to provide a reasonable framework, there are still some caveats that are worth pointing out. Regarding the spots, it is not clear that polar spots are ubiquitous on low-mass stars. Polar caps have been detected on rapidly rotating stars such as AG Dor, AB Dor or LO Peg (Strassmeier 2009). YY Gem is the only DDLEB in the sample used here to compare models with observations with published Doopler images, although no indication of polar spots was found (Hatzes

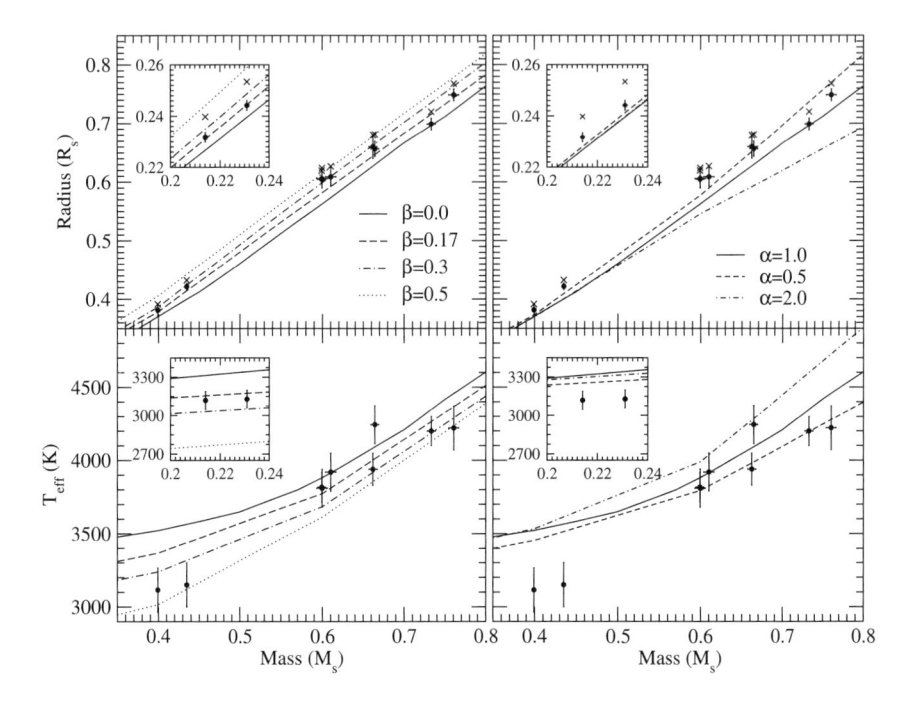

Figure 4. Comparison between models and observations for different β values and $\alpha = 1$ (left) and for different α values and $\beta = 0$ (right). Crosses indicate the measured radius of DDLEB components while circles represent the values corrected for the 3% systematic factor described in the text and further normalized to and age of 1 Gyr. The insets display the case of CM Dra.

1995). However, they are difficult to detect with Doppler tomography for near edge-on inclinations. Studies on the magnetic topology of low-mass stars could further hint the distribution of spots, but there are few cases studied that present some degeneracies, showing low-mass stars with both dipolar and toroidal fields (Morin et al. 2008, 2010). Besides, the connection between the fields and the location of spots is not clear yet. Note that the polar spot hypothesis on the systematic effect on radius determinations could be tested with light curve observations in the near-IR bands. The spot contrast is significantly reduced in the JHK bands with respect to the visible and thus the radius differential should be smaller. The value could be some 1–2% lower and so unveiling this requires of abundant high-precision photometric data that is not yet available for a sample of DDLEBs.

With respect to the *universality* in the effects of spots on the entire range of low-mass stars studied here (from 0.1 to 0.8 $M_\odot$), it is not known if this is really the case. This hypothesis allows us to minimize the number of free parameters at the expense of assuming that fully- and partially-convective stars behave the same and it may also be the case of brown dwarfs. Regading this later case, Chabrier et al. (2007) explained the puzzling case of a brown dwarf eclipsing binary showing a reversal on effective temperatures between the primary and secondary components (Stassun et al. 2006, 2007) assuming different values of α and β for each star. The necessity of different parameters for each star would be in contradiction with the *universality* in the effects of spots,

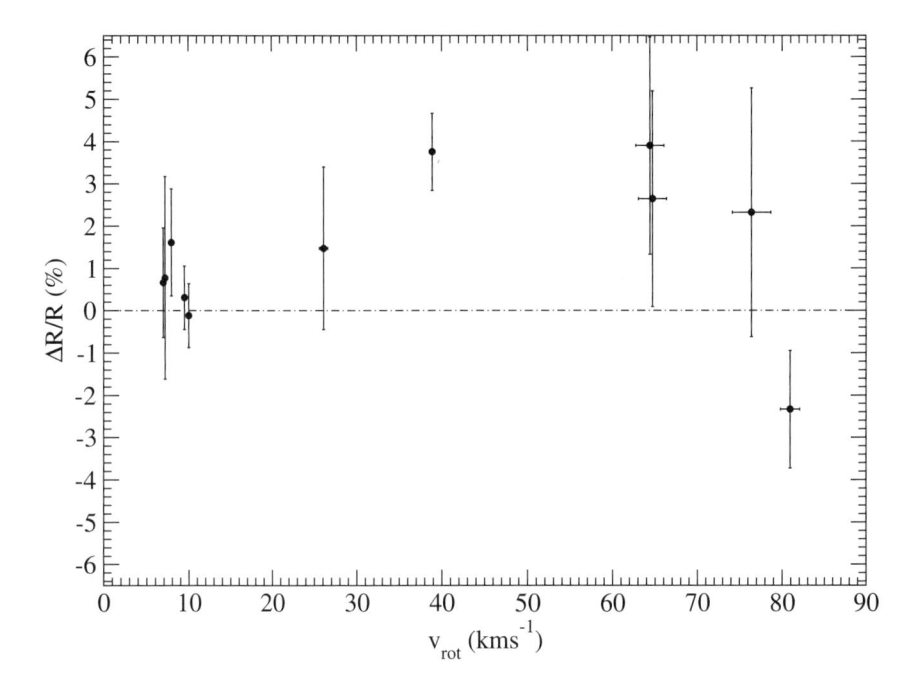

Figure 5. Residual discrepancy between observations and the model with $\beta = 0.17$ that best matches the case of CM Dra, which is not affected by variations of the mixing length parameter α.

however, coevality for this system, with an estimated age of about 1 Myr, is an important issue. Further, the luminosities of this system do not agree with standard models, thus indicating that they may be different from more massive stars whith luminosities in accordance with theoretical models.

Mullan & MacDonald (2001) also developed a set of theoretical stellar structure models considering the effect that magnetic fields have on the interior structure equations and studied this brown-dwarf system (MacDonald & Mullan 2009). They also concluded that high magnetic fields reduce the onset of convection and thus change the global structure of the star. Although they claim to be able to explain the case of the brown-dwarf binary (MacDonald & Mullan 2009; Mullan & MacDonald 2010), their magnetic models predict strong central magnetic fields, which are free parameters in the models, and they also predict different luminosities to those of standard models. This is in contradiction with the findings from low-mass DDLEBs, whose luminosities are known to be well reproduced by models, thus suggesting that the effect of spots is stronger on more massive systems.

6. Conclusions

In summary, the results of this analysis confirm that magnetic activity is the main responsible for the radius and effective temperature differences between observations and stellar models since metallicity or opacities have to be modified to unrealistic values to explain the same differences. Besides, isolated main sequence stars show the same pattern when comparing active and inactive stars. We have also found that the presence of spots has a significant impact both on observations and on theoretical models. In the case of observations, the presence of polar caps can introduce a systematic overestimation of the radii of DDLEB components through the light curve analysis. In the case of stellar models, according to Chabrier et al. (2007), introducing stellar spots on stellar models yields stars with larger radii.

The exact mechanism of the effect of activity on the stellar structure is not yet fully understood, but all the evidence points at the combined effects of spots and reduced convective efficiency. We have shown here that assuming that very active stars (such as those belonging to close binaries) have mostly polar spots and similar spot coverages, the best known DDLEB systems are well reproduced by models if:

a) 3% of the radius difference is due to the systematic effect of polar spots,

b) 2% of the radius difference is due to the radius increase predicted by models to compensate the loss of radiative efficiency due to spots (parameterized with β),

c) up to 4% of the radius difference is due to the lower convective efficiency in fast rotating stars (parameterized with α).

To fully vet the scenario, further observations of DDLEBs over a wide range of component masses (including down to the brown dwarf regime) are still needed. In particular, high-precision near-IR light curves will be of much help to disentangle some of the effects discussed above. Accurate fundamental properties of low-mass DDLEBs in long-period systems ($P_{\rm orb} > 10$ d) will also contribute to further understand the effect of activity on low-mass stars. The components of such long-period systems are not expected to be synchronized with the orbital motion, and therefore should retain their natural rotation velocities (that could be low for moderately old stars). Recent papers (Coughlin et al. 2010; Kraus et al. 2010) suggest that there is a clear trend of reducing the radius discrepancy between models and observations for longer period binaries. Missions such as *CoRoT* and *Kepler* can provide valuable samples of such systems for subsequent follow-up that will increase the statistical significance of the preliminary results. Finally, Doppler imaging of the DDLEBs analyzed or new methods of mapping stars using interferometry (Monnier et al. 2007) should also provide indications of the spot distribution on the stellar surfaces to better understand the structure and evolution of low-mass stars.

Acknowledgments. We acknowledge support from Spanish Ministerio de Ciencia e Innovación via grants AYA2006-15623-C02-01, AYA2006-15623-C02-02, AYA2009-06934, and AYA2009-14648-C02-01.

References

Andersen, J. 1991, A&A Rev., 3, 91
Baraffe, I., Chabrier, G., Allard, F., & Hauschildt, P. H. 1998, A&A, 337, 403
Barnes, J. R., & Collier Cameron, A. 2001, MNRAS, 326, 950
Berger, D. H., Gies, D. R., McAlister, H. A., Brummelaar, T. A. t., Henry, T. J., Sturmann, J., Sturmann, L., Turner, N. H., Ridgway, S. T., Aufdenberg, J. P., & Mérand, A. 2006, ApJ, 644, 475
Casagrande, L., Flynn, C., & Bessell, M. 2008, MNRAS, 389, 585

Chabrier, G., Gallardo, J., & Baraffe, I. 2007, A&A, 472, L17
Coughlin, J. L., Lopez-Morales, M., Harrison, T. E., Ule, N., & Hoffman, D. I. 2010, ArXiv e-prints. 1007.4295
Demory, B., Ségransan, D., Forveille, T., Queloz, D., Beuzit, J., Delfosse, X., di Folco, E., Kervella, P., Le Bouquin, J., Perrier, C., Benisty, M., Duvert, G., Hofmann, K., Lopez, B., & Petrov, R. 2009, A&A, 505, 205
Granzer, T., Schüssler, M., Caligari, P., & Strassmeier, K. G. 2000, A&A, 355, 1087
Hatzes, A. P. 1995, in IAU Symposium, vol. 176 of IAU Symposium, 90
Kraus, A. L., Tucker, R. A., Thompson, M. I., Craine, E. R., & Hillenbrand, L. A. 2010, ArXiv e-prints. 1011.2757
López-Morales, M., & Ribas, I. 2005, ApJ, 631, 1120
López-Morales, M., & Shaw, J. S. 2007, in The Seventh Pacific Rim Conference on Stellar Astrophysics, edited by Y. W. Kang, H.-W. Lee, K.-C. Leung, & K.-S. Cheng, vol. 362 of Astronomical Society of the Pacific Conference Series, 26
MacDonald, J., & Mullan, D. J. 2009, ApJ, 700, 387
Mazeh, T. 2008, in EAS Publications Series, edited by M.-J. Goupil & J.-P. Zahn, vol. 29 of EAS Publications Series, 1
Monnier, J. D., Zhao, M., Pedretti, E., Thureau, N., Ireland, M., Muirhead, P., Berger, J., Millan-Gabet, R., Van Belle, G., ten Brummelaar, T., McAlister, H., Ridgway, S., Turner, N., Sturmann, L., Sturmann, J., & Berger, D. 2007, Science, 317, 342. 0706.0867
Morales, J. C., Gallardo, J., Ribas, I., Jordi, C., Baraffe, I., & Chabrier, G. 2010, ApJ, 718, 502
Morales, J. C., Ribas, I., & Jordi, C. 2008, A&A, 478, 507
Morales, J. C., Ribas, I., Jordi, C., Torres, G., Gallardo, J., Guinan, E. F., Charbonneau, D., Wolf, M., Latham, D. W., Anglada-Escudé, G., Bradstreet, D. H., Everett, M. E., O'Donovan, F. T., Mandushev, G., & Mathieu, R. D. 2009a, ApJ, 691, 1400
Morales, J. C., Torres, G., Marschall, L. A., & Brehm, W. 2009b, ApJ, 707, 671
Morin, J., Donati, J., Petit, P., Delfosse, X., Forveille, T., Albert, L., Aurière, M., Cabanac, R., Dintrans, B., Fares, R., Gastine, T., Jardine, M. M., Lignières, F., Paletou, F., Ramirez Velez, J. C., & Théado, S. 2008, MNRAS, 390, 567. 0808.1423
Morin, J., Donati, J., Petit, P., Delfosse, X., Forveille, T., & Jardine, M. M. 2010, MNRAS, 407, 2269. 1005.5552
Mullan, D. J., & MacDonald, J. 2001, ApJ, 559, 353
— 2010, ApJ, 713, 1249
Pizzolato, N., Maggio, A., Micela, G., Sciortino, S., & Ventura, P. 2003, A&A, 397, 147
Ribas, I. 2003, A&A, 398, 239
— 2006, in Astrophysics of Variable Stars, edited by C. Sterken, & C. Aerts, vol. 349 of Astronomical Society of the Pacific Conference Series, 55
Ribas, I., Morales, J. C., Jordi, C., Baraffe, I., Chabrier, G., & Gallardo, J. 2008, Memorie della Societa Astronomica Italiana, 79, 562
Rozyczka, M., Kaluzny, J., Pietrukowicz, P., Pych, W., Mazur, B., Catelan, M., & Thompson, I. B. 2009, Acta Astronomica, 59, 385
Soderblom, D. R. 2010, ARA&A, 48, 581. 1003.6074
Stassun, K. G., Mathieu, R. D., & Valenti, J. A. 2006, Nat, 440, 311
— 2007, ApJ, 664, 1154. 0704.3106
Stauffer, J. R., & Hartmann, L. W. 1986, ApJS, 61, 531
Strassmeier, K. G. 2009, A&A Rev., 17, 251
Torres, G., Andersen, J., & Giménez, A. 2010, A&A Rev., 18, 67
Torres, G., Lacy, C. H., Marschall, L. A., Sheets, H. A., & Mader, J. A. 2006, ApJ, 640, 1018
Torres, G., & Ribas, I. 2002, ApJ, 567, 1140
Wilson, R. E., & Devinney, E. J. 1971, ApJ, 166, 605

Testing Theory with Dynamical Masses and Orbits of Ultracool Binaries

Trent J. Dupuy,[1] Michael C. Liu,[1] and Michael J. Ireland[2]

[1] *Institute for Astronomy, University of Hawai'i, 2680 Woodlawn Drive, Honolulu, HI 96822, USA*

[2] *Sydney Institute for Astronomy, School of Physics, University of Sydney NSW 2006, Australia*

Abstract. Mass is the fundamental parameter that governs the evolution of stars, brown dwarfs, and gas-giant planets. Thus, direct mass measurements are essential to test the evolutionary and atmospheric models that underpin studies of these objects. We present results from our program to test models using precise dynamical masses for visual binaries based on Keck laser guide star adaptive optics astrometric monitoring of a sample of over 30 ultracool (> M6) objects since 2005. In just the last 2 years, we have more than tripled the number of late-M, L, and T dwarf binaries with precise dynamical masses. For most field binaries, based on direct measurements of their luminosities and total masses, we find a "temperature problem" in that evolutionary model radii give effective temperatures that are inconsistent with those from model atmosphere fitting of observed spectra by 100–300 K. We also find a "luminosity problem" for the only binary with an independent age determination (from its solar-type primary via age–activity–rotation relations). Evolutionary models underpredict the luminosities of HD 130948BC by a factor of ≈ 2, implying that model-based substellar mass determinations (e.g., for directly imaged planets and cluster IMFs) may be systematically overestimating masses. Finally, we have employed the current sample of binary orbits to carry out a novel test of the earliest evolutionary stages, by using the distribution of orbital eccentricities to distinguish between competing models of brown dwarf formation.

1. Introduction

Binary systems have long been used to probe both the inner workings and the origins of stars. In multiple systems the laws of gravity can be used to infer the masses of stars. Mass is the fundamental property of any object as this largely determines its entire life history, and thus binaries with direct dynamical mass measurements are important benchmarks for testing theory. Binaries also record within their orbits a dynamical imprint of their formation and subsequent evolution. For a given mass, the total energy of the system is related solely to the semimajor axis of the orbit ($E = -GM/2a$), while the angular momentum depends on the orbital eccentricity ($L = \sqrt{GMa(1-e^2)}$). Decades before it was known that nucleosynthesis powers stars, well-determined binary orbits were commonplace (e.g., > 200 orbits were published in the compilation of Aitken 1918) and were, for example, used to argue for a dynamical constraint on the age of the Galaxy of $\lesssim 10$ Gyr (Ambartsumian 1937) as opposed to the ~10,000 Gyr age limit supported by Jeans (1935). Thus, the study of binary orbits has a rich heritage, predating and defining much of modern astrophysics.

It has only recently been possible to extend such studies below 0.1 $M_\odot$, to masses at the bottom of the main sequence and into the brown dwarf regime. The field population over this mass range encompasses spectral types $\gtrsim$ M7, collectively referred to as "ultracool" owing to their shared low-temperature atmospheric physics. Only ≈10 years ago were the first large samples of ultracool dwarfs identified via wide-field surveys such as 2MASS (e.g., Burgasser et al. 1999), DENIS (e.g., Delfosse et al. 1999), and SDSS (e.g., Hawley et al. 2002). The underlying binary population was subsequently uncovered by high-resolution imaging surveys using *HST* (e.g., Bouy et al. 2003; Burgasser et al. 2003) and adaptive optics (AO) from the ground (e.g., Close et al. 2002; Liu & Leggett 2005). In the first half of the last decade, orbital monitoring of a few select binaries yielded three dynamical mass measurements: LHS 1070BC (M8.5+M9.5; Leinert et al. 2001); Gl 569Bab (M8.5+M9; Lane et al. 2001); and 2MASS J0746+2000AB (L0+L1.5; Bouy et al. 2004). These first direct mass constraints on ultracool models probed relatively warm temperatures ($\gtrsim$ 2200 K) and high masses (only Gl 569Bb is likely to be a brown dwarf; Dupuy et al. 2010).

Over just the last three years the number of dynamical masses sufficiently precise for meaningful model tests ($\leq$ 30%) has tripled, with components now extending down to masses of ≈30 $M_{\rm Jup}$ and temperatures of ≈1000 K. This rapid expansion in dynamical masses has been driven by high-precision orbital monitoring programs that utilize the relatively new capability of laser guide star (LGS) AO to resolve the shortest period binaries from ground-based telescopes. Our program using Keck LGS AO has been ongoing since 2005 and has provided five of the six new high-precision dynamical masses. All nine ultracool binaries with high-precision dynamical masses are listed in Table 1. In the cases where multiple orbits have been published for a binary, the parameters from the highest quality orbit determination are given; previous orbit references are listed in parentheses.

Table 1. Ultracool visual binaries with precise dynamical masses ($\sigma_M/M \leq 30\%$)

Name	Component Sp. Types	Period [days]	Eccentricity	Total mass [$M_\odot$]	Ref.
Gl 569Bab	M8.5+M9	864.5 ± 1.1	0.316 ± 0.005	$0.140^{+0.009}_{-0.008}$	1 (2–5)
LP 349-25AB	M7.5+M8	2834 ± 15	0.051 ± 0.003	$0.120^{+0.008}_{-0.007}$	1 (5)
HD 130948BC	L4+L4	3760 ± 60	0.176 ± 0.006	0.109 ± 0.002	6 (5,7)
2M0746+20AB	L0+L1.5	4640 ± 30	0.487 ± 0.003	0.151 ± 0.003	5 (8)
LHS 2397aAB	M8+L7	5190 ± 40	0.350 ± 0.005	0.146 ± 0.014	9 (5)
2M1534−29AB	T5+T5.5	5500^{+800}_{-600}	$0.25^{+0.11}_{-0.13}$	0.056 ± 0.003	10 (5)
LHS 1901AB	M6.5+M6.5	5880 ± 180	0.830 ± 0.005	$0.194^{+0.025}_{-0.021}$	1
LHS 1070BC	M8.5+M9.5	6214.7 ± 0.4	0.034 ± 0.002	0.157 ± 0.009	11 (12)
2M2206−20AB	M8+M8	12800^{+2200}_{-1800}	0.25 ± 0.08	$0.15^{+0.05}_{-0.03}$	13 (5)

References. — (1) Dupuy et al. (2010); (2) Lane et al. (2001); (3) Zapatero Osorio et al. (2004); (4) Simon et al. (2006); (5) Konopacky et al. (2010); (6) Dupuy et al. (2011, revised); (7) Dupuy et al. (2009b); (8) Bouy et al. (2004); (9) Dupuy et al. (2009c); (10) Liu, Dupuy, & Ireland (2008); (11) Seifahrt et al. (2008); (12) Leinert et al. (2001); (13) Dupuy et al. (2009a).

2. Testing Evolutionary and Atmospheric Models

The growing sample of ultracool dwarf dynamical masses has enabled some of the strongest tests of theoretical models to date, particularly in the brown dwarf regime. Because brown dwarfs do not have a sustained source of internal energy generation, they simply grow colder and fainter over time. Thus, the sample of field brown dwarfs spans a wide range of masses at a given luminosity, and the only means of pinpointing a given brown dwarf's evolutionary status is to measure its mass or age. As shown by Liu, Dupuy, & Ireland (2008), fundamental properties such as $T_{\rm eff}$ and $\log(g)$ are $\approx 5\times$ better constrained (i.e., model tests are $\approx 5\times$ stronger) with dynamical mass measurements as compared to age determinations, which are ordinarily of lower precision.

One of the major components of our work has been to develop an analysis method that produces rigorous, quantitative tests of models based solely on directly measured properties (i.e., mass and luminosity). This is of fundamental importance, as other previous approaches were sometimes based on $T_{\rm eff}$ estimates or the model-predicted color–magnitude diagram, both of which are often prone to severe systematic biases (e.g., see discussions in Gizis & Reid 2006 and Section 4.5 of Dupuy et al. 2010). The discrepancies we have found between observations and models to date fall into two categories: (1) problems with the luminosities predicted by evolutionary models, (2) problems with temperature estimates derived from model atmospheres and/or from evolutionary model radii.

2.1. "Luminosity Problem"

Conventional wisdom is that the most robust predictions of substellar models are bulk properties such as the radius and bolometric luminosity ($L_{\rm bol}$) since independent models put forth by different groups produce nearly identical (to within a few percent) values at ages beyond a few Myr. Consequently, evolutionary models have become trusted to provide accurate mass estimates in the many situations when luminosity and age are well constrained, such as cluster mass functions (e.g., Lodieu et al. 2007) and directly imaged extrasolar planets (e.g., Marois et al. 2008). However, until recently no direct measurements of luminosity and mass were available for any brown dwarfs that also had well determined ages.

The first, and to date only, system that enables a direct test of substellar luminosity evolution is the L4+L4 binary HD 130948BC (Dupuy et al. 2009b). Its model-derived age (given its measured total mass and component luminosities) is significantly younger than the age of the primary G2V star in this triple system. This is a 2σ discrepancy when accounting for the errors in the binary's measured mass and luminosities and the gyrochronology relation used to age-date the primary star (Mamajek & Hillenbrand 2008). If we assume that the stellar age and its error are correct (i.e., HD 130948A is not simply an outlier as compared to the cluster rotation sequences), then the luminosities predicted by evolutionary models for HD 130948B and C are a factor of ≈ 2 lower than what is observed. In considering this potential luminosity problem, we will now address the following questions: (1) is there additional evidence for models underpredicting luminosities of substellar objects; and (2) what physical mechanism could be responsible for this effect?

(1) Is there additional evidence for the luminosity problem? If models underpredict luminosities, this would manifest as either model-derived masses (based on $L_{\rm bol}$ and age) that are $\approx 25\%$ too high or model-derived ages (based on mass and $L_{\rm bol}$) that are

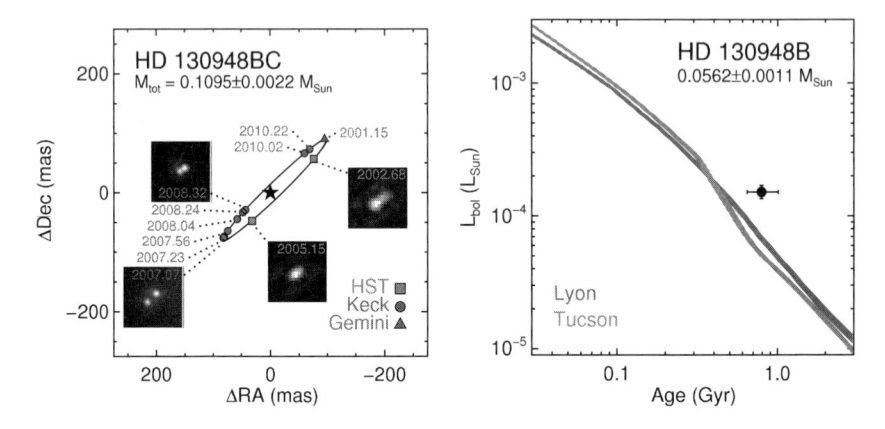

Figure 1. *Left:* Relative astrometry of the L4+L4 binary HD 130948BC from *HST*, Gemini, and Keck AO imaging along with the best-fit orbit. *Right:* Given the measured dynamical mass, evolutionary models predict the luminosity of the components of HD 130948BC as a function of age (colored lines; widths indicate the 2.0% mass error propagated through the models). The data point shows the measured luminosity of HD 130948B at the age determined from age–activity–rotation relations for the primary star in this triple system ($0.79^{+0.22}_{-0.15}$ Gyr). Both HD 130948B and HD 130948C are a factor of ≈ 2 more luminous than predicted by models.

too young. The former problem can be tested in the multiple planet system around HR 8799. Fabrycky & Murray-Clay (2010) have shown that there only two plausible scenarios for the three tightly packed, massive planets (model-derived masses of 7, 10, and 10 $M_{\rm Jup}$) to be dynamically stable: (1) the inner planets must be in 2:1 resonance, and all three planets must have masses lower than predicted by models by at least 20%–30%; or (2) both the inner pair and outer pair of planets must be in 2:1 resonances, in which case the masses of the planets could be as much as a factor of ~2 larger. The careful construction of the system required in the latter case may suggest that the first scenario is in fact more plausible, and the "luminosity problem" may also be seen here.

Further evidence for the luminosity problem can also come from objects with model-derived ages that appear to be too young. Dupuy et al. (2010) have observed this in the M7.5+M8 binary LP 349-25AB for which Lyon models give an age of 130 ± 20 Myr. This is essentially identical to the canonical Pleiades age of 125 Myr as determined from the lithium depletion boundary independent of model luminosity predictions (Stauffer et al. 1998). However, LP 349-25AB appears to be significantly older than the Pleiades given the strong upper limit on its Li I abundance (Reiners & Basri 2009), and it also does not show spectroscopic signatures of youth as is normally seen in young (~100 Myr) field dwarfs (e.g., McGovern et al. 2004; Allers et al. 2007).

(2) What physical mechanisms could explain the luminosity problem? After a few Myr of evolution, model cooling tracks that adopt very different initial conditions converge to essentially the same luminosity (e.g., Baraffe et al. 2002), so accounting for different formation histories such as in "cold start" models with radiative energy losses due to accretion (e.g., Marley et al. 2007) will not alter model cooling tracks sufficiently to explain our observed luminosity discrepancy. Rather, it is more likely that a process sustained over these objects' lifetimes, such as convection, is responsible for the problem. Mullan & MacDonald (2010) have recently constructed evolutionary

models tailored to the components of HD 130948BC in which the onset of convection is inhibited by very strong magnetic fields throughout the interior. This effect can substantially alter the radius evolution of the components of HD 130948BC and thus explain the observed discrepancy on the H-R diagram reported by Dupuy et al. (2009b) because their new models yield larger radii, which shifts model tracks to lower T_{eff}. However, at the same time these models also *lower* the model-predicted luminosities by ≈0.2 dex (≈50%), increasing the observed luminosity discrepancy.

If the inhibition of the onset of convection has the opposite effect on the luminosity evolution, then perhaps the efficiency of convection must be changed. Chabrier et al. (2007) consider the impact of strong interior magnetic fields and fast rotation on the convective energy transport in low-mass stars. They find that these processes result in lowered convective efficiency, which in turn results in bloated radii for ≈0.3–0.8 $M_\odot$ stars, while lower mass fully convective stars are much less affected. The effect of reduced convective energy transport for brown dwarfs is not investigated in detail, but Chabrier et al. (2007) point out that its main effect will be to slow their contraction rate. This may effectively "slow down the clock" of substellar evolution, i.e., for a given mass and age objects would be larger and more luminous than predicted by conventional evolutionary models that do not include the effects of convective inhibition due to rotation or strong interior magnetic fields.

2.2. "Temperature Problem"

Model atmospheres offer the possibility of determining the temperature and surface gravity of an object from a single spectrum. This approach is widely used in the study of very low-mass stars, brown dwarfs, and directly imaged planets. However, the reliability of such estimates (and the conclusions drawn from them) depend entirely the fidelity of the model atmospheres being used. This can only be assessed in special cases where independent constraints on the mass, age, or radius are available. We have utilized our measured dynamical masses of ultracool field dwarfs to provide some of the strongest tests of model atmospheres to date. This is made possible by the fact that direct mass measurements effectively pin down the evolutionary status of a field object to a degree of certainty commensurate with the precision of the measured mass and luminosity (both typically $\lesssim$ 10%). The values of T_{eff} and $\log(g)$ then inferred from evolutionary models are limited to a very narrow range, typically exceeding the precision of the model atmosphere grid steps of 50–100 K and 0.5 dex. This evolutionary model-derived T_{eff} is essentially a restatement of the model radius given the measured luminosity, following the definition of effective temperature, $T_{\text{eff}} \equiv (L_{\text{bol}}/4\pi R^2 \sigma)^{1/4}$. Therefore, these binaries provide precise temperature benchmarks that can be compared to results from model atmosphere fitting.

We have found that temperatures derived from model atmospheres are typically inconsistent with those from evolutionary models at the level of ≈100–300 K. The amplitude and the sign of this disagreement vary over the full range of our sample. For example, T dwarf model atmosphere fits give ≈100–200 K warmer temperatures than evolutionary model radii (Liu et al. 2008), but L dwarf model atmosphere fits give ≈200–300 K cooler temperatures (Dupuy et al. 2009b). The lack of a systematic offset for all objects indicates that the disagreement is not simply due to a uniform error in predicted radii at all masses and spectral types, but without directly measured radii, which would provide T_{eff} directly, it is impossible to distinguish whether evolutionary or atmospheric models are responsible for the observed discrepancies.

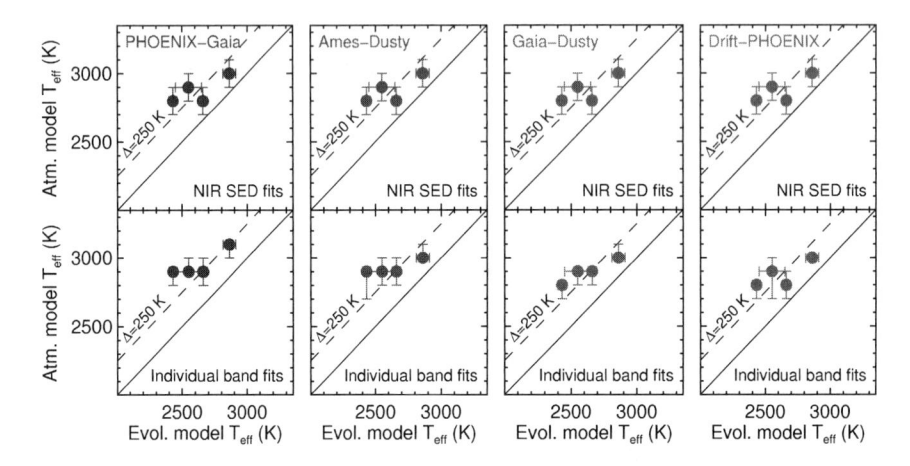

Figure 2. Late-M dwarf effective temperatures determined from model atmosphere fitting compared to Lyon Dusty evolutionary model-derived $T_{\rm eff}$, (i.e., from measured total mass and individual luminosities Dupuy et al. 2010). Four independent model atmosphere grids were tested (left to right) and two different methods of fitting atmospheres: fitting the full 0.95–2.42 μm near-infrared spectrum (top panels), and fitting the Y, J, H, and K bands individually (bottom panels). In all cases, an offset of 250 K between the two classes of models is observed (dashed line). This implies that either evolutionary model estimates are too cool (i.e., radii too large by 15%–20%) or that atmospheric model estimates are too warm by 250 K. The latter is more likely given that this sample spans a wide range of masses, ages, and activity levels, but a narrow range of $T_{\rm eff}$.

If we restrict our comparisons to a narrow range of temperatures, we can essentially remove one variable from the problem and focus on how the $T_{\rm eff}$ discrepancies vary with mass and age. To date, the largest subset of mass benchmarks at any given temperature are the late-M dwarfs (Table 1). We obtained integrated-light near-infrared spectroscopy of four nearly equal-flux late-M binaries with component mass determinations ranging from ≈0.06–0.10 $M_\odot$ (i.e., young brown dwarfs to old stars) and with a range of chromospheric activity levels (Dupuy et al. 2010). After fitting these spectra with four independent model atmosphere grids, we found that all best-fit temperatures were systematically ≈250 K warmer than the $T_{\rm eff}$ values derived from evolutionary models (Figure 2). Thus, the observed $T_{\rm eff}$ discrepancy is the same for objects with very similar spectra but with very different masses, ages, and activity levels, indicating that the model atmospheres are largely responsible for the observed $T_{\rm eff}$ discrepancy.

3. Testing Formation Models with Eccentricities

Some properties of very low-mass binaries, such as orbital semimajor axes, mass ratios, and multiplicity fractions, can be studied in a statistical fashion using a large sample without any knowledge of orbital motion (e.g., Allen 2007), and such studies have provided very important constraints on formation models at very low masses. However, only well-determined orbits can provide eccentricities (e), which are directly related to the angular momentum of the formation process. In fact, the only two theoretical

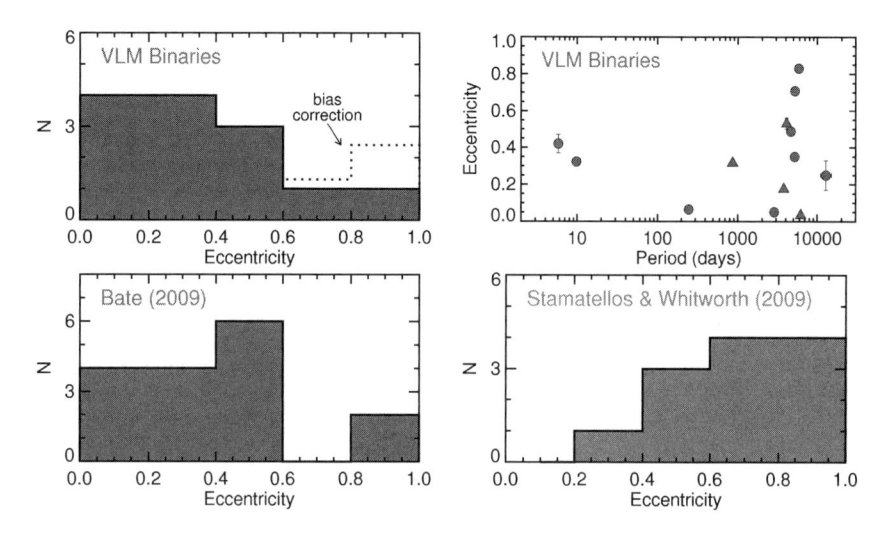

Figure 3. Eccentricity distribution of all very low-mass binaries with well-determined orbits, including both visual and spectroscopic binaries (top panels). The observed distribution is very similar to that predicted by the cluster formation model of Bate (2009, bottom left) but is inconsistent with the much higher eccentricities predicted by the gravitational instability model of Stamatellos & Whitworth (2009, bottom right), even after applying a bias correction to the observed distribution.

simulations that report eccentricities of very low-mass binaries predict very different types of orbits: (1) the cluster formation model of Bate (2009) yields a total of 16 binaries with typically modest eccentricities, and (2) the disk gravitational instability model of Stamatellos & Whitworth (2009) produces 12 binaries with typically high eccentricities (Figure 3).

The sample of very low-mass binaries with well-determined orbits is now of comparable size to the binaries produced in theoretical simulations, and thus it can be used to try to discriminate between the very different model predictions. Figure 3 shows the eccentricity distribution for all very low-mass binaries with eccentricity uncertainties less than 0.1. This excludes the visual binary 2MASS J1534−2952AB (Table 1) and includes three spectroscopic binaries: PPl 15 ($e = 0.42 \pm 0.05$; Basri & Martín 1999), 2MASS J0535−0546 ($e = 0.323 \pm 0.006$; Stassun et al. 2006), and 2MASS J0320−0446 ($e = 0.065 \pm 0.016$; Blake et al. 2008). This sample populates a broad range in eccentricity from nearly circular to highly eccentric ($e = 0.03$–0.83) and shows a strong preference for modest eccentricities ($\bar{e} = 0.35$, median of 0.32). The observed distribution of eccentricities is well-matched to that predicted by the cluster formation model of Bate (2009), although we note that this simulation is not in full agreement with observations as it produces much wider binaries than are observed in the field. In contrast, the gravitational instability model of Stamatellos & Whitworth (2009) produces too many high-e binaries, with an absence of *any* modest eccentricity ($e < 0.3$) orbits, which is highly inconsistent with the observed distribution.

References

Aitken, R. G. 1918, The Binary Stars
Allen, P. R. 2007, ApJ, 668, 492
Allers, K. N., Jaffe, D. T., Luhman, K. L., Liu, M. C., Wilson, J. C., Skrutskie, M. F., Nelson, M., Peterson, D. E., Smith, J. D., & Cushing, M. C. 2007, ApJ, 657, 511
Ambartsumian, V. A. 1937, Astron. Zhurn., 14, 207
Baraffe, I., Chabrier, G., Allard, F., & Hauschildt, P. H. 2002, A&A, 382, 563
Basri, G., & Martín, E. L. 1999, AJ, 118, 2460
Bate, M. R. 2009, MNRAS, 392, 590
Blake, C. H., Charbonneau, D., White, R. J., Torres, G., Marley, M. S., & Saumon, D. 2008, ApJ, 678, L125
Bouy, H., Brandner, W., Martín, E. L., Delfosse, X., Allard, F., & Basri, G. 2003, AJ, 126, 1526
Bouy, H., et al. 2004, A&A, 423, 341
Burgasser, A. J., Kirkpatrick, J. D., Reid, I. N., Brown, M. E., Miskey, C. L., & Gizis, J. E. 2003, ApJ, 586, 512
Burgasser, A. J., et al. 1999, ApJ, 522, L65
Chabrier, G., Gallardo, J., & Baraffe, I. 2007, A&A, 472, L17
Close, L. M., Siegler, N., Potter, D., Brandner, W., & Liebert, J. 2002, ApJ, 567, L53
Delfosse, X., et al. 1999, A&AS, 135, 41
Dupuy, T. J., Liu, M. C., & Bowler, B. P. 2009a, ApJ, 706, 328
Dupuy, T. J., Liu, M. C., Bowler, B. P., & Cushing, M. C. 2010, ApJ, 721, 1725
Dupuy, T. J., Liu, M. C., & Ireland, M. J. 2009b, ApJ, 692, 729
— 2009c, ApJ, 699, 168
Fabrycky, D. C., & Murray-Clay, R. A. 2010, ApJ, 710, 1408
Gizis, J. E., & Reid, I. N. 2006, AJ, 131, 638
Hawley, S. L., et al. 2002, AJ, 123, 3409
Jeans, J. H. 1935, Nat, 136, 432
Konopacky, Q. M., Ghez, A. M., Barman, T. S., Rice, E. L., Bailey, J. I., White, R. J., McLean, I. S., & Duchêne, G. 2010, ApJ, 711, 1087
Lane, B. F., Zapatero Osorio, M. R., Britton, M. C., Martín, E. L., & Kulkarni, S. R. 2001, ApJ, 560, 390
Leinert, C., Jahreiß, H., Woitas, J., Zucker, S., Mazeh, T., Eckart, A., & Köhler, R. 2001, A&A, 367, 183
Liu, M. C., Dupuy, T. J., & Ireland, M. J. 2008, ApJ, 689, 436
Liu, M. C., & Leggett, S. K. 2005, ApJ, 634, 616
Lodieu, N., Dobbie, P. D., Deacon, N. R., Hodgkin, S. T., Hambly, N. C., & Jameson, R. F. 2007, MNRAS, 380, 712
Mamajek, E. E., & Hillenbrand, L. A. 2008, ApJ, 687, 1264
Marley, M. S., Fortney, J. J., Hubickyj, O., Bodenheimer, P., & Lissauer, J. J. 2007, ApJ, 655, 541
Marois, C., Macintosh, B., Barman, T., Zuckerman, B., Song, I., Patience, J., Lafrenière, D., & Doyon, R. 2008, Science, 322, 1348
McGovern, M. R., Kirkpatrick, J. D., McLean, I. S., Burgasser, A. J., Prato, L., & Lowrance, P. J. 2004, ApJ, 600, 1020
Mullan, D. J., & MacDonald, J. 2010, ApJ, 713, 1249
Reiners, A., & Basri, G. 2009, ApJ, 705, 1416
Seifahrt, A., Röll, T., Neuhäuser, R., Reiners, A., Kerber, F., Käufl, H. U., Siebenmorgen, R., & Smette, A. 2008, A&A, 484, 429
Simon, M., Bender, C., & Prato, L. 2006, ApJ, 644, 1183
Stamatellos, D., & Whitworth, A. P. 2009, MNRAS, 392, 413
Stassun, K. G., Mathieu, R. D., & Valenti, J. A. 2006, Nat, 440, 311
Stauffer, J. R., Schultz, G., & Kirkpatrick, J. D. 1998, ApJ, 499, L199
Zapatero Osorio, M. R., Lane, B. F., Pavlenko, Y., Martín, E. L., Britton, M., & Kulkarni, S. R. 2004, ApJ, 615, 958

Cool Stars 16 Opening Reception (Top) and Physics/Astronomy Building at the University of Washington Campus (Bottom)

The 16th Cambridge Workshop on Cool Stars, Stellar Systems and the Sun
ASP Conference Series, Vol. 448
Christopher M. Johns-Krull, Matthew K. Browning, and Andrew A. West, eds.
© 2011 Astronomical Society of the Pacific

New Low-Mass Eclipsing Binaries from Kepler

Jeffrey L. Coughlin,[1,4] Mercedes López-Morales,[2] Thomas E. Harrison,[1] Nicholas Ule,[1] and Douglas I. Hoffman[3]

[1]*Department of Astronomy, New Mexico State University, P.O. Box 30001, MSC 4500, Las Cruces, New Mexico, USA 88003-8001*

[2]*Carnegie Institution of Washington, Department of Terrestrial Magnetism, 5241 Broad Branch Road NW, Washington, DC 20015, USA; Hubble Fellow*

[3]*California Institute of Technology, MC 249-17, 1200 East California Blvd, Pasadena, CA 91125*

[4]*NSF Graduate Research Fellow*

Abstract. We identify 231 objects in the newly released Cycle 0 dataset from the Kepler Mission as double-eclipse, detached eclipsing binary systems with $T_{\rm eff} < 5500$ K and orbital periods shorter than ~32 days. We model each light curve using the JK-TEBOP code with a genetic algorithm to obtain precise values for each system. We identify 95 new systems with both components below 1.0 $M_\odot$ and eclipses of at least 0.1 magnitudes, suitable for ground-based follow-up. Of these, 14 have periods less than 1.0 day, 52 have periods between 1.0 and 10.0 days, and 29 have periods greater than 10.0 days. This new sample of main-sequence, low-mass, double-eclipse, detached eclipsing binary candidates more than doubles the number of previously known systems, and extends the sample into the completely heretofore unexplored $P > 10.0$ day period regime. We find preliminary evidence from these systems that the radii of low-mass stars in binary systems decrease with period. This supports the theory that binary spin-up is the primary cause of inflated radii in low-mass binary systems, although a full analysis of each system with radial-velocity and multi-color light curves is needed to fully explore this hypothesis.

1. Introduction

A double-lined, detached, eclipsing binary (DDEB) is a system that contains two non-interacting, eclipsing stars, in which the spectra of both components can be clearly seen, allowing for the radial-velocity (RV) of each component to be obtained. In these systems, the mass and radius of each star can be determined with errors usually less than 1-2%, thus making DDEBs currently the most accurate method of obtaining masses and radii of stars. Models of main-sequence stars with masses similar to or greater than the Sun have been tested over the years using DDEBs (Andersen 1991; Torres et al. 2010). However, while observations of DDEBs have enhanced our understanding of stellar structure and evolution for stars with $M \geq 1.0$ $M_\odot$, low-mass, main-sequence (LMMS) stars, ($M < 1.0$ $M_\odot$ and $T_{\rm eff} < 5800$ K), have not been tested to the same extent.

Although a couple systems with late G or early K type components had been studied prior to 2000, (c.f. Popper 1980; Andersen 1991; Torres et al. 2006; Clausen et al.

2009, and references therein), only three LMMS DDEBs with late K or M type components were known (Lacy 1977; Leung & Schneider 1978; Delfosse et al. 1999). This number had only increased to nine by the beginning of 2007 (cf. López-Morales 2007, Table 1). Despite the fact that the majority of main-sequence stars are low-mass, these stars are both intrinsically fainter, and physically smaller, than their more massive counterparts. Therefore, they have a lower eclipse probability and are harder to discover and study. As outlined by López-Morales (2007), analysis of these systems showed that the observed radii for these stars are consistently ~10-20% larger than predicted by stellar models (Baraffe et al. 1998) for 0.3 $M_\odot \lesssim M \lesssim 0.8$ $M_\odot$.

Enhanced magnetic activity has been proposed as the principal cause of inflated radii (Chabrier et al. 2007; López-Morales 2007; Morales et al. 2008). Shorter-period binary systems, with the stellar rotation rate enhanced by the revolution of the system, are expected to show greater activity and thus larger radii than longer-period systems (Chabrier et al. 2007). Binary systems with component masses of 0.5 $M_\odot$ are expected to synchronize, and therefore be spun-up, in less than 0.1 Gyr for periods less than 4 days, and in less than 1 Gyr for periods less than 8 days (Zahn 1977, 1994). Thus, the discovery of LMMS DDEBs with $P \gtrsim 10$ days, where the binary components should have natural rotation rates, is crucial to probing if enhanced rotation due to binarity is the underlying cause of this phenomenon.

Though several more LMMS DDEB systems have been found since 2007, (Coughlin & Shaw 2007; Shaw & López-Morales 2007; Becker et al. 2008; Blake et al. 2008; Devor et al. 2008a,b; Shkolnik et al. 2008; Hoffman et al. 2008; Irwin et al. 2009; Dimitrov & Kjurkchieva 2010; Shkolnik et al. 2010), there are to-date only 7 well-studied systems with 1.0 < P < 3.0 days (López-Morales 2007, and references therein) (Becker et al. 2008; Shkolnik et al. 2008), and only one has a larger period, at P = 8.4 days (Devor et al. 2008b). This is mostly due to the fact that ground-based photometric surveys, such as NSVS, TrES, and OGLE, are either cadence, precision, magnitude, or number limited, and thus not sensitive to long periods. The Kepler mission, with 3 years of constant photometric monitoring of over 150,000 stars with $V \lesssim 17$, at 30-minute cadence and sub-millimagnitude precision, is the key to discovering a large number of long-period, LMMS DDEBs.

In this paper we present the results of our search through all the newly available Kepler Q0 and Q1 public data for LMMS DDEBs. Section 2 describes the data we use in this paper. Section 3 describes our binary identification technique, and section 4 describes how we model the light curves. Our selection of new LMMS DDEBs is presented in section 5, and in section 6 we compare the new LMMS DDEBs with theoretical models. We conclude with a summary of our results in section 7. Once accurate mass and radius values exist for a large range of both mass and period, our understanding of these objects should substantially improve, and we will be one step closer to extending to the lower-mass regime the advanced study of stellar structure and evolution that sun-like and high-mass stars have been a subject of for some time.

2. Observational Data

The data used in our analysis consists of the 201,631 light curves made public by the Kepler Mission[1] as of June 15, 2010 from Kepler Q0 and Q1 observations. All light curves can be accessed through the Multi-mission Archive at STScI (MAST)[2]. The data consist of 51,366 light curves from Kepler Q0, (observed from 2009-05-02 00:54:56 to 2009-05-11 17:51:31 UT), and 150,265 light curves from Kepler Q1, (observed from 2009-05-13 00:15:49 to 2009-06-15 11:32:57 UT), each at 29.43 minute cadence. Individual light curves for Q0 contain ~470 data points, and for Q1 contain ~1,600 data points. Targets range in Kepler magnitude from 17.0 to 5.0.

The Kepler team has performed pixel level calibrations, (including bias, dark current, flat-field, gain, and non-linearity corrections), identified and cleaned cosmic-ray events, estimated and removed background signal, and then extracted time-series photometry using an optimum photometric aperture. They have also removed systematic trends due to spacecraft pointing, temperature fluctuations, and other sources of systematic error, and corrected for excess flux in the optimal photometric aperture due to crowding (Van Cleve 2010). It is this final, "corrected" photometry that we have downloaded for use in our analysis.

3. Eclipsing Binary Identification

Our search consisted of two steps. The first was to identify variable stars, and to do so, we placed a light curve standard deviation limit above which the objects are classified as variables. We first subtracted an error-weighted, linear fit of flux versus time from all data, to remove any remaining linear systematic trends, and then plotted the standard deviation of each light curve versus its average flux and fit a power law. Next, we used the flux ratio (FR) measurement criterion, which we adapted from the magnitude ratio given in Kinemuchi et al. (2006), and is defined as

$$FR = \frac{\text{maximum flux} - \text{median flux}}{\text{maximum flux} - \text{minimum flux}} \quad (1)$$

as a measure of whether or not the variable spends most of its time above (low FR value) or below (high FR value) the median flux value. Perfectly sinusoidal variables have FR = 0.50, pulsating variables, such as RR Lyrae's, have FR > 0.5, and eclipsing binaries have FR < 0.5. As we are principally interested in finding well detached systems with relatively deep, narrow eclipses, which thus have low FR values, we make a further cut of the systems and only examine those variables with FR < 0.1.

The second step of the analysis was to determine the orbital period of each candidate. This was done using two independent techniques that are both well-suited for detached eclipsing binary systems. The first is Phase Dispersion Minimization (PDM) (Stellingwerf 1978), which attempts to find the period that best minimizes the variance in multiple phase bins of the folded light curve. This technique is not sensitive to the shape of the light curve, and thus is ideal for non-sinusoidal variables such as detached

---

[1] http://kepler.nasa.gov/

[2] http://archive.stsci.edu/kepler/

eclipsing binaries. The downside of this technique is that if strong periodic features exist in the light curve, which do not correspond to the period of eclipses, such as rapidly varying spots, stellar pulsations, or leftover systematics, they can weaken the signal of the eclipse period. We use the latest implementation given by Stellingwerf (2006), and determine the best three periods via this technique to ensure that the true period is found, and not just an integer multiple, or fraction, thereof.

The second technique we use is one we invented specifically for detached eclipsing binaries, and call Eclipse Phase Dispersion Minimization (EPDM). The idea behind EPDM is that we want to automatically identify and align the primary eclipses in an eclipsing binary, thus finding the period of the system. To accomplish this, EPDM finds the period that best minimizes the dispersion of the actual phase values of the faintest N points in a light curve, i.e. the very bottom of the eclipses. Since EPDM only selects the N faintest points in a light curve, it is not affected by systematics or periodic features that do not correspond to the period of eclipses, assuming the systematics do not extend below the depth of the eclipses. The technique works for all binary systems with equal or unequal eclipse depths, and transiting planets, both with either zero or non-zero eccentricity. Computationally, EPDM is significantly faster than traditional PDM techniques. We use EPDM to find the three best fit periods for each system as well, for the same reasons as we did with PDM.

We identified 577 EB candidates in the Q1 data, of which 486 were already listed by Prša et al. (2010) as detached eclipsing binaries, and 20 were identified as semi-detached eclipsing binaries. Of the 71 remaining candidates, 48 turned out to be false positives, but 23 turned out to be new or elsewhere listed eclipsing binary systems or transiting planet candidates.

4. Light Curve Modeling

We modeled each system using a robust global minimization scheme with a commonly used, physically detailed eclipsing binary modeling code. We took all 314 detached eclipsing binaries with T_{eff} < 5500K and that are publicly available, (5 systems are still proprietary), identified from both our search and the Prša et al. (2010) catalog, combined Q0 and Q1 data if available, and via manual inspection classified systems as double-eclipse (i.e. contained two visible eclipses), single-eclipse (i.e. only contained one eclipse), or as spurious results that were not recognizable as eclipsing systems. (Given the errors in the KIC temperature determination, and to ensure the primary is below 1.0 $M_\odot$, we used 5500K as our cutoff, instead of 5800K.

We then used the JKTEBOP eclipsing binary modeling program (Southworth et al. 2004a,b) to model every double-eclipse eclipsing binary system, of which there were 231, solving for the period, time of primary minimum, inclination, mass ratio, $e \cdot \cos(\omega)$, $e \cdot \sin(\omega)$, surface brightness ratio, sum of the fractional radii, ratio of the radii, and out of eclipse flux. In addition, we also solved for the amplitude and time of minimum of a sinusoidal term imposed on the luminosity of the primary component, with the period fixed to that of the binary, in order to account for spots. We used the quadratic limb darkening law, which works well for late-type stars (e.g. Manduca et al. 1977; Wade & Rucinski 1985; Claret & Gimenez 1990), with coefficients set to those found by Sing (2010) for the Kepler bandpass via interpolation given the systems' effective temperatures, surface gravities, and metallicities as listed in the Kepler Input Catalog

(KIC)[3]. We also fixed the gravity darkening exponent based on the effective temperature as prescribed by Claret (2000).

In order to model such a large number of systems over such a large solution space, and to ensure we have found the best global solution, we adapted the JKTEBOP code to use a modified version of the asexual genetic algorithm (AGA) described by Cantó et al. (2009), coupled with its standard Levenberg-Marquardt minimization algorithm. Genetic algorithms (GA) are an extremely efficient method of fitting computationally intensive, multi-parameter models over a large and potentially discontinuous parameter space, and thus ideal for this work. We found that our modified AGA does an excellent job of solving well-behaved light curves, simultaneously varying all 12 aforementioned parameters over the entire range of possible solutions. For some of the systems however, strong systematics and/or variable star spots introduced a significant amounts of noise, especially in systems with shallow eclipses, for which it was more difficult to arrive at a robust solution. For these systems we had to manually correct the systematics, often by either eliminating the Q0 or Q1 data, equalizing the base flux levels of Q0 and Q1 data, or subtracting out a quasi-sinusoidal variation in the base flux level due to remaining Kepler systematics. When possible we attempted to minimize the amount of manual interference. Hopefully this will become much less of a problem with subsequent data releases. We then re-ran the AGA using a larger initial population until a good solution was found. Every light curve in the end was visually inspected to be a good fit compared to the scatter of the data points, and the obtained parameters were confirmed to be reasonable when visually inspecting the light curves.

5. New Low-Mass Binary Candidates

In order to identify the main-sequence stars from our list of 231 candidates, and determine the best candidates for follow-up, we employ the following technique to estimate the temperature, mass, and radius of each star using the sum of the fractional radii, r_{sum}, and period, P, obtained from our JKTEBOP models, the luminosity ratio, L_r, (which is derived from the surface brightness ratio, J, and radii ratio, k, obtained from the models), and the effective temperature of the system, $T_{\rm eff}$, obtained from the KIC, with an estimated error of 200 K.

The value for $T_{\rm eff}$ given in the KIC was determined via interpolation of standard color magnitude relations as determined by ground-based, multi-wavelength photometry (Van Cleve 2010). Although in principle one might be able to deconvolve two separate spectral energy distributions from this photometry, in reality given the level of photometric error in the KIC and uncertainty at which binary phase the photometry was obtained, this is untenable. Instead, we assume the stars radiate as blackbodies, and that each star contributes to the determined $T_{\rm eff}$ in proportion to its luminosity. Thus, following our assumption, we obtain the following relation,

$$T_{\rm eff} = \frac{L_1 T_1 + L_2 T_2}{L_1 + L_2} \qquad (2)$$

where L_1, L_2, T_1 and T_2 are the luminosities and effective temperatures of star 1 and 2 respectively. Still assuming the stars radiate as blackbodies, the luminosity of

---

[3] http://archive.stsci.edu/kepler/kepler_fov/search.php

each star is proportional to its radius squared and temperature to the fourth power, with the temperature proportional to is surface brightness to the one-fourth power. Thus, we find that the luminosity ratio can be expressed as,

$$L_r = \frac{L_1}{L_2} = \frac{r_1^2 T_1^4}{r_2^2 T_2^4} = k^2 T_r^4 = k^2 \left[\left(\frac{SB_1}{SB_2}\right)^{1/4}\right]^4 = k^2 \left(J^{\frac{1}{4}}\right)^4 = k^2 J \quad (3)$$

where SB_1 and SB_2 are the surface brightnesses of star 1 and star 2 respectively, and r_1 and r_2 are the fractional radii of star 1 and 2 respectively, defined as R_1/a and R_2/a, where R_1 and R_2 are the physical radius of each star, and a is the semi-major axis of, or separation between, the components. Combining equations 2 and 3 yields the expression,

$$T_{\text{eff}} = \frac{L_r T_1 + T_2}{L_r + 1} \quad (4)$$

which has two known parameters, T_{eff} and L_r, and two unknown parameters, T_1 and T_2. To place a further constraint upon the values of T_1 and T_2, we make the assumption that both stars in the binary are on the main-sequence, and employ the mass, temperature, radius, and average of the V-band and R-band luminosity relations given in Baraffe et al. (1998) for $0.075 \leq M \leq 1.0$ $M_\odot$ and in Chabrier et al. (2000) for $M < 0.075$ $M_\odot$, both assuming an age of 5.0 Gyr and [M/H] = 0.0. (We average the V and R-band luminosities to obtain a very close approximation to the Kepler bandpass.) From these models, for a given value of T_1, there is only one value of T_2 which will reproduce the observed value of L_r. Thus, there only exists one set of unique values for T_1 and T_2 that reproduces both the observed T_{eff} and L_r values for the system.

For each T_1 and T_2 then, we obtain the absolute masses and radii, (M_1, M_2, R_1, and R_2), via interpolation from the Baraffe et al. (1998) and Chabrier et al. (2000) models. Then, utilizing Kepler's 3rd law, given the total mass of the system, we calculate the semi-major axis, a, via

$$a = (GM_{tot})^{\frac{1}{3}} \left(\frac{P}{2\pi}\right)^{\frac{2}{3}} \quad (5)$$

where M_{tot} is the total mass of the system, $M_1 + M_2$, and G is the gravitational constant. We then multiply each radius determined above by a constant so that the sum of the fractional radii derived from the JKTEBOP model, r_{sum}, is equal to $(R_1 + R_2)/a$, the sum of the fractional radii when using the physical values of M_1, M_2, R_1, R_2, and P. This technique is robust because while individual parameters such as i, J, and k can suffer from degeneracies, especially in systems with shallow eclipses, the values of r_{sum} and $L_r = k^2 J$, which we rely on, are firmly set by the width of the eclipses and the difference in their eclipse depths, respectively.

After estimating the individual mass, radius, and temperature for each component, we re-computed the gravity and limb-darkening coefficients for each individual star, and performed a Levenberg-Marquardt minimization starting from our previously best solutions, taking into account the 29.43 minute integration time as suggested by Kipping (2010). We then repeated the processes of deriving the physical values of the components, interpolating gravity and limb-darkening coefficients, and performing a Levenberg-Marquardt minimization several more times to ensure convergence. We find that 95 of the systems contain two main-sequence stars, which we define as having a radius less than 1.5 times the Baraffe et al. (1998) and Chabrier et al. (2000)

model relationships, and a light curve amplitude of at least 0.1 magnitudes, (suitable for ground-based follow-up and less likely to contain any third light). All of these 95 systems have both stars with masses less than 1.0 $M_\odot$. These results substantially increase the number of LMMS DDEB candidates in general, and provide 29 new LMMS DDEBs with both components below one solar mass, and at least 0.1 magnitude eclipse depths, in the heretofore unexplored period range of P > 10 days.

6. Comparison of the New Low-Mass Binary Candidates to Models

As described in the introduction, one of the current outstanding questions in the study of low-mass stars is whether the inflated radii observed in binaries is caused by their enhanced stellar rotation, and therefore enhanced magnetic activity. We explore this problem in this section using the 95 new LMMS DDEB candidates with estimated individual mass both below 1.0 $M_\odot$ and light curve amplitudes greater than 0.1 magnitudes. This sample, for the first time, provides a statistically significant number of systems with orbital periods larger than 10 days.

The left-side panels of Figure 1 show mass-radius diagrams using the mass and radius of each binary star component estimated in Section 5. The LMMS DDEB candidates have been separated into three diagrams, with orbital periods P < 1.0 day, 1.0 < P < 10 days, and P > 10 days. Each primary and secondary in a binary pair is traced by a connecting line. We also plot in each diagram the theoretical mass-radius relation predicted by the Baraffe et al. (1998) models for M ≥ 0.075 $M_\odot$, and the Chabrier et al. (2000) models for M < 0.075 $M_\odot$, both for [M/H] = 0.0, and an age of 5.0 Gyrs. We have also defined a main-sequence cutoff as 1.5 times the theoretical mass-radius relation, which is illustrated by the solid line in each diagram. In the models we have used an α = 1.0 for M ≤ 0.7 $M_\odot$ and interpolated the radius of the models for 0.7 $M_\odot$ < M ≤ 1.0 $M_\odot$ by fixing the radius of the 1.0 $M_\odot$ model to 1.0 $R_\odot$, therefore avoiding the dependence of the stellar radius with α between 0.7 $M_\odot$ and 1.0 $M_\odot$ (Baraffe et al. 1998). We also include in the mass-radius diagrams estimations of the error in our M and R values at several masses, computed by adding and subtracting 200 K, (the error in the T_{eff} determinations given by the KIC), from a given temperature and interpolating the mass and radius from the theoretical relations.

In the figure, many of the stellar radii of binaries with P < 1.0 days appear to fall above the model predictions, but as the orbital period increases, a larger fraction of the systems appear to have radii that are either consistent with or fall below the models. There certainly is a fair amount of scatter in these data introduced by the large error in the mass and radius estimations, but a histogram analysis of the radius distributions confirms these apparent trends. On the right-side panels of Figure 1 we show 5% bin-size histograms representing how many stars have a radius that deviates by a given percentage from the models. The average radius discrepancy is 13.0%, 7.5%, and 2.0% for the short (P < 1.0 days), medium (1.0 < P < 10.0 days), and long-period (P > 10.0 days) systems respectively. Although a full analysis of each system with multi-color light and radial-velocity curves is still needed, these preliminary estimates support the hypothesis that binary spin-up is the primary cause of inflated radii in short period LMMS DDEBs.

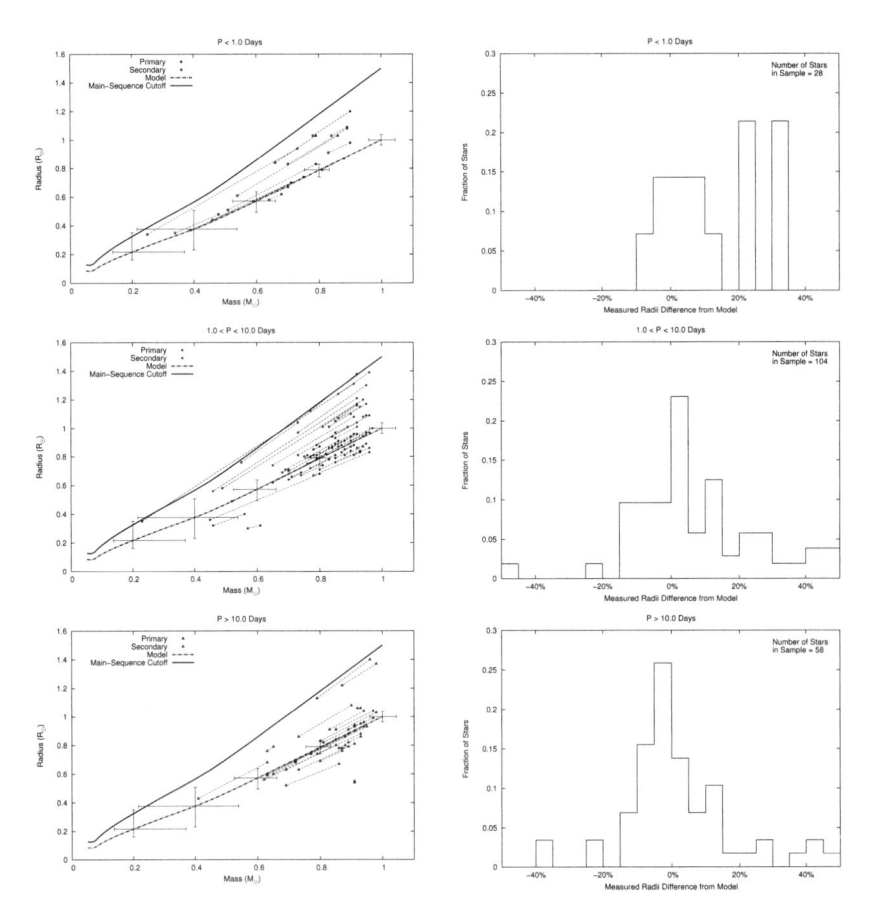

Figure 1. Left: Mass-radius diagrams for each binary with both components < 1.0 $M_\odot$ and photometric amplitudes greater than 0.1 mag, with systems connected by faint lines. The systems are sorted into short-period (P < 1.0 days, top panel), medium-period, (1.0 < P < 10.0 days, middle panel), and long-period groupings, (P > 10.0 days, bottom panel). The theoretical mass-radius relations of Baraffe et al. (1998) for 0.075 $M_\odot \leq M \leq 1.0\ M_\odot$, and of Chabrier et al. (2000) for M < 0.075 $M_\odot$, both for [M/H] = 0.0 and an age of 5.0 Gyr, are over-plotted. The solid line shows the main-sequence cutoff criterion. The error bars indicate the error in mass and radius obtained when interpolating from the mass-temperature-radius relations with an error of 200K. Right: Histograms of the fraction of stars in the sample versus their deviance from the models for each period grouping. As can be seen by both the mass-radius relation plots and the histograms, shorter period binaries in general appear to exhibit larger radii compared to the models than longer period systems.

7. Summary

We have identified and modeled 231 new double-eclipse, detached eclipsing binary systems with $T_{eff} < 5500$ K, found in the Cycle 0 data release of the Kepler mission. We estimate the masses and radii of the stars in these systems, and find that 95 of them contain two main-sequence stars with both components having $M < 1.0\ M_\odot$ and eclipse depths of at least 0.1 magnitude, and thus are suitable for ground-based follow-up. Of these 95 systems, 14 have periods less than 1.0 day, 52 have periods between 1.0 and 10.0 days, and 29 have periods greater than 10.0 days. This new sample of low-mass, double-eclipse, detached eclipsing binary candidates more than doubles the number of previously known systems, and extends the sample into the completely heretofore unexplored P > 10.0 day period range for LMMS DDEBs.

Comparison to the theoretical mass-radius relation models for stars below 1.0 $M_\odot$ by Baraffe et al. (1998) show preliminary evidence for better agreement with the models at longer periods, where the rotation rate of the stars is not expected to be spun-up by tidal locking, although, in the absence of radial-velocity measurements, the errors on the estimated mass and radius are still quite large. For systems with P < 1.0 days, the average radius discrepancy is 13.0%, whereas for 1.0 < P < 10.0 days and P > 10.0 days, the average radius discrepancy is 7.5% and 2.0%, respectively. Ground-based follow-up, in the form of radial velocity and multi-wavelength light curves, is needed to derive the mass and radius of each star in each system to ~1-2%, which we have already begun to acquire. With accurate masses and radii for multiple long-period systems, we should be able to definitively test the hypothesis that inflated radii in low-mass binaries are principally due to enhanced rotation rates.

Acknowledgments. We thank the entire Kepler team and all those who have contributed to the Kepler mission. We also thank the organizers of Cool Stars 16 for an exceptional conference, and Christopher Johns-Krull, Matthew Browning and Andrew West for editing these proceedings. J.L.C acknowledges support from a NSF Graduate Research Fellowship and the New Mexico Space Grant Consortium. J.L.C., M.L.M., & T.E.H. are grateful for funding from Kepler's Guest Observer Program. M.L.M. acknowledges support from NASA through Hubble Fellowship grant HF-01210.01-A/HF-51233.01 awarded by the STScI, which is operated by the AURA, Inc. for NASA, under contract NAS5-26555.

References

Andersen, J. 1991, A&A Rev., 3, 91
Baraffe, I., Chabrier, G., Allard, F., & Hauschildt, P. H. 1998, A&A, 337, 403
Becker, A. C., Agol, E., Silvestri, N. M., Bochanski, J. J., Laws, C., West, A. A., Basri, G., Belokurov, V., Bramich, D. M., Carpenter, J. M., Challis, P., Covey, K. R., Cutri, R. M., Evans, N. W., Fellhauer, M., Garg, A., Gilmore, G., Hewett, P., Plavchan, P., Schneider, D. P., Slesnick, C. L., Vidrih, S., Walkowicz, L. M., & Zucker, D. B. 2008, MNRAS, 386, 416
Blake, C. H., Torres, G., Bloom, J. S., & Gaudi, B. S. 2008, ApJ, 684, 635
Cantó, J., Curiel, S., & Martínez-Gómez, E. 2009, A&A, 501, 1259
Chabrier, G., Baraffe, I., Allard, F., & Hauschildt, P. 2000, ApJ, 542, 464
Chabrier, G., Gallardo, J., & Baraffe, I. 2007, A&A, 472, L17
Claret, A. 2000, A&A, 359, 289
Claret, A., & Gimenez, A. 1990, A&A, 230, 412

Clausen, J. V., Bruntt, H., Claret, A., Larsen, A., Andersen, J., Nordström, B., & Giménez, A. 2009, A&A, 502, 253
Coughlin, J. L., & Shaw, J. S. 2007, Journal of the Southeastern Association for Research in Astronomy, 1, 7
Delfosse, X., Forveille, T., Mayor, M., Burnet, M., & Perrier, C. 1999, A&A, 341, L63
Devor, J., Charbonneau, D., O'Donovan, F. T., Mandushev, G., & Torres, G. 2008a, AJ, 135, 850
Devor, J., Charbonneau, D., Torres, G., Blake, C. H., White, R. J., Rabus, M., O'Donovan, F. T., Mandushev, G., Bakos, G. Á., Fűrész, G., & Szentgyorgyi, A. 2008b, ApJ, 687, 1253
Dimitrov, D. P., & Kjurkchieva, D. P. 2010, MNRAS, 866
Hoffman, D. I., Harrison, T. E., Coughlin, J. L., McNamara, B. J., Holtzman, J. A., Taylor, G. E., & Vestrand, W. T. 2008, AJ, 136, 1067
Irwin, J., Charbonneau, D., Berta, Z. K., Quinn, S. N., Latham, D. W., Torres, G., Blake, C. H., Burke, C. J., Esquerdo, G. A., Fürész, G., Mink, D. J., Nutzman, P., Szentgyorgyi, A. H., Calkins, M. L., Falco, E. E., Bloom, J. S., & Starr, D. L. 2009, ApJ, 701, 1436
Kinemuchi, K., Smith, H. A., Woźniak, P. R., & McKay, T. A. 2006, AJ, 132, 1202
Kipping, D. M. 2010, MNRAS, 1184
Lacy, C. H. 1977, ApJ, 218, 444
Leung, K., & Schneider, D. P. 1978, AJ, 83, 618
López-Morales, M. 2007, ApJ, 660, 732
Manduca, A., Bell, R. A., & Gustafsson, B. 1977, A&A, 61, 809
Morales, J. C., Ribas, I., & Jordi, C. 2008, A&A, 478, 507
Popper, D. M. 1980, ARA&A, 18, 115
Prša, A., Batalha, N. M., Slawson, R. W., Doyle, L. R., Welsh, W. F., Orosz, J. A., Seager, S., Rucker, M., Mjaseth, K., Engle, S. G., Conroy, K., Jenkins, J. M., Caldwell, D. A., Koch, D. G., & Borucki, W. J. 2010, ArXiv e-prints, 1006.2815
Shaw, J. S., & López-Morales, M. 2007, in The Seventh Pacific Rim Conference on Stellar Astrophysics, edited by Y. W. Kang, H.-W. Lee, K.-C. Leung, & K.-S. Cheng, vol. 362 of Astronomical Society of the Pacific Conference Series, 15
Shkolnik, E., Liu, M. C., Reid, I. N., Hebb, L., Cameron, A. C., Torres, C. A., & Wilson, D. M. 2008, ApJ, 682, 1248
Shkolnik, E. L., Hebb, L., Liu, M. C., Reid, I. N., & Cameron, A. C. 2010, ApJ, 716, 1522
Sing, D. K. 2010, A&A, 510, A21
Southworth, J., Maxted, P. F. L., & Smalley, B. 2004a, MNRAS, 351, 1277
Southworth, J., Zucker, S., Maxted, P. F. L., & Smalley, B. 2004b, MNRAS, 355, 986
Stellingwerf, R. F. 1978, ApJ, 224, 953
— 2006, PDM2 Application, Technical Manual, and test data sets. URL http://www.stellingwerf.com/rfs-bin/index.cgi?action=PageView&id=29
Torres, G., Andersen, J., & Giménez, A. 2010, A&A Rev., 18, 67
Torres, G., Lacy, C. H., Marschall, L. A., Sheets, H. A., & Mader, J. A. 2006, ApJ, 640, 1018
Van Cleve, J. 2010, Kepler Data Release 5 Notes, NASA Ames Research Center, Moffett Field, CA
Wade, R. A., & Rucinski, S. M. 1985, A&AS, 60, 471
Zahn, J. 1977, A&A, 57, 383
— 1994, A&A, 288, 829

Dust and Chromospheres in M warfs

Michael S Bessell

Research School of Astronomy and Astrophysics, CMPS, The Australian National University, Australia

Abstract. Whilst impressive work has gone into modelling the spectra of M dwarfs over the past 10 years, there remain significant differences between the best model spectra and observations that indicate serious shortcomings in our understanding of M dwarf atmospheres. This contribution outlines the major observational differences and discusses the likelihood of dust opacity in the coolest M dwarfs and chromospheres in all M dwarfs as being explanations for the outstanding differences.

1. Introduction

M dwarfs are the most common stars in the galaxy falling near the maximum in the stellar luminosity function. They have masses between 0.5 and 0.08 M⊙ temperatures between 4000K and 2250K, and absolute V magnitudes between 9 and 20 mags. But although impressive work has gone into modelling the spectra of M dwarfs by incorporating millions of molecular lines (eg. MARCS[1] models (Gustafsson et al.(2008), PHOENIX models (Hauschildt, Allard & Baron 1999)) there remain significant differences between the best model spectra and observations that indicate serious shortcomings in our understanding of cool dwarf atmospheres.

2. Outstanding Problems

The first problem is that the spectra of late-M dwarfs are much brighter in K and redder in J-K than the models although the red continua and TiO bands are well fitted (Fig. 1). This was first noted by Kui (1991) who, although she had no explanation for it, was able to eliminate it by flattening the temperature structure in her models. It is still seen in the synthetic broad-band colors computed by Burrows, Sudarsky & Hubeny (2006, their Fig. 5) for modern dusty M and L dwarf atmospheres. This means that temperatures derived from the near-IR are uncertain and bolometric magnitudes for the late-M dwarfs suspect.

The second problem is the startling observation that the blue spectra of Proxima Cen (GJ551) and other mid-M dwarfs have metal lines hundreds of times weaker than the model spectra indicating the model temperature structure is seriously wrong. Fig. 2 shows an echelle blue spectrum of Proxima Cen compared to the MARCS model that well fits the red spectrum. Although there are few blue echelle spectra available

---

[1]http://marcs.astro.uu.se/

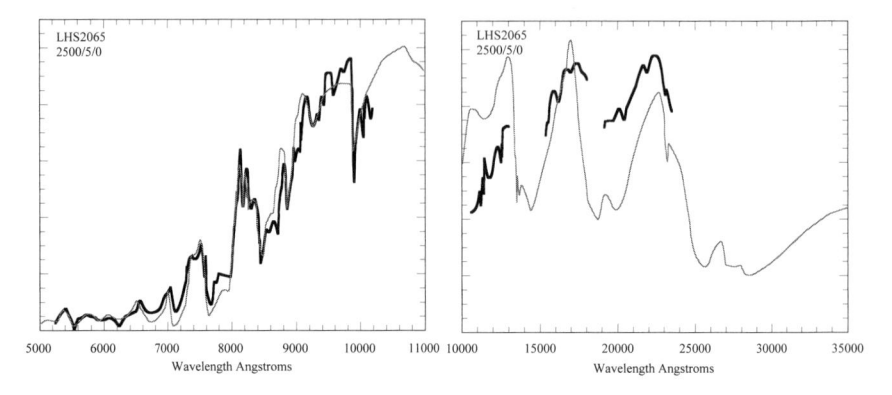

Figure 1. The observed (black) and model (red) spectrum of the late-M dwarf LHS 2065 (Kui 1991). Note the visible-red spectra agree well but the IR spectra are very different

for mid and late-M dwarfs, intermediate resolution spectra show the same large differences for all mid to late-M dwarfs observed. A broad blue depression centered about 4226Å is also seen in all mid to late-M dwarfs, including an extremely metal-deficient star. The echelle spectra rule out the depression resulting from veiling of molecular lines, although the bands of AlH were observed for the first time.

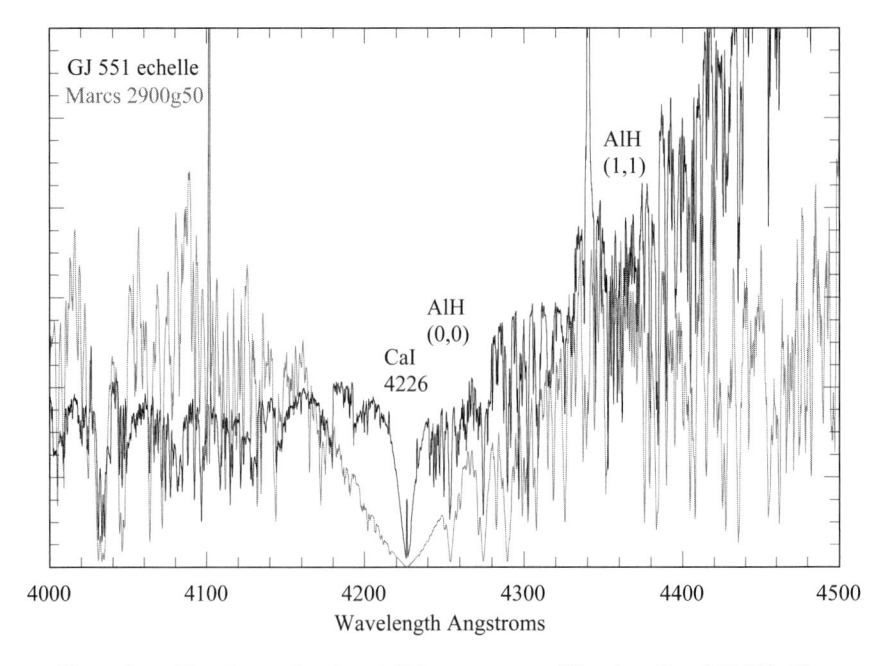

Figure 2. The observed and model blue spectrum of Proxima Cen (M5.5V). Note the huge difference in the strength of CaI 4226 and all the other lines of neutral metals and the difference in the shape of the underlying continua.

3. Dust in M Dwarf Atmospheres

As discussed by Tsuji, Ohnaka & Aoki(1996), if dust forms in any stellar atmospheres the most likely place is that of late-M dwarfs because of the low temperature and high density. Once dust is formed in the surface layers, the photosphere is greatly heated. In the following Fig. 3 and 4 from that paper are shown the temperature structures and resultant fluxes for two different kinds of model atmospheres: dust-free-models (dotted lines) where only molecules are considered as opacity sources although conditions for dust formation are satisfied in the surface layers and dusty models (solid lines) where dust opacities are added together with molecules.

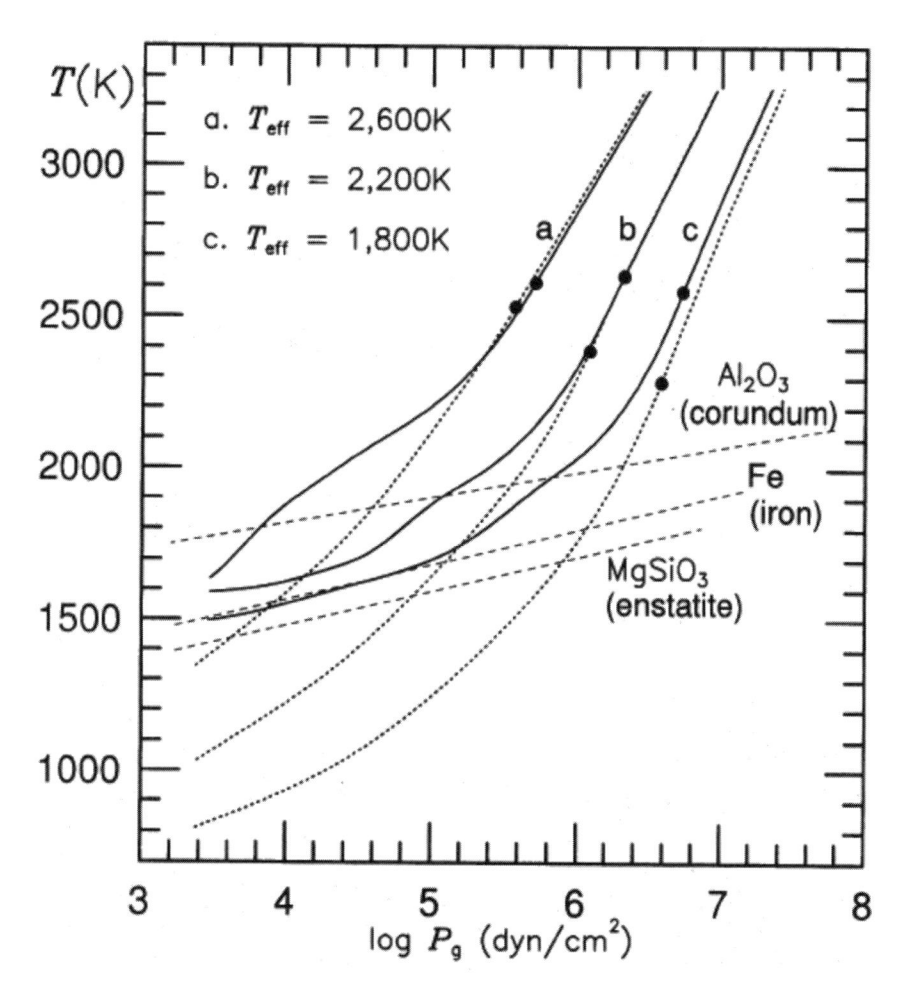

Figure 3. Model atmospheres for three different temperature dwarf models. Dust-free models (dotted lines) and dusty models (solid lines). The dashed curves are the dust condensation lines for corundum, Fe and enstatite. The filled circles indicate the onset of convection.

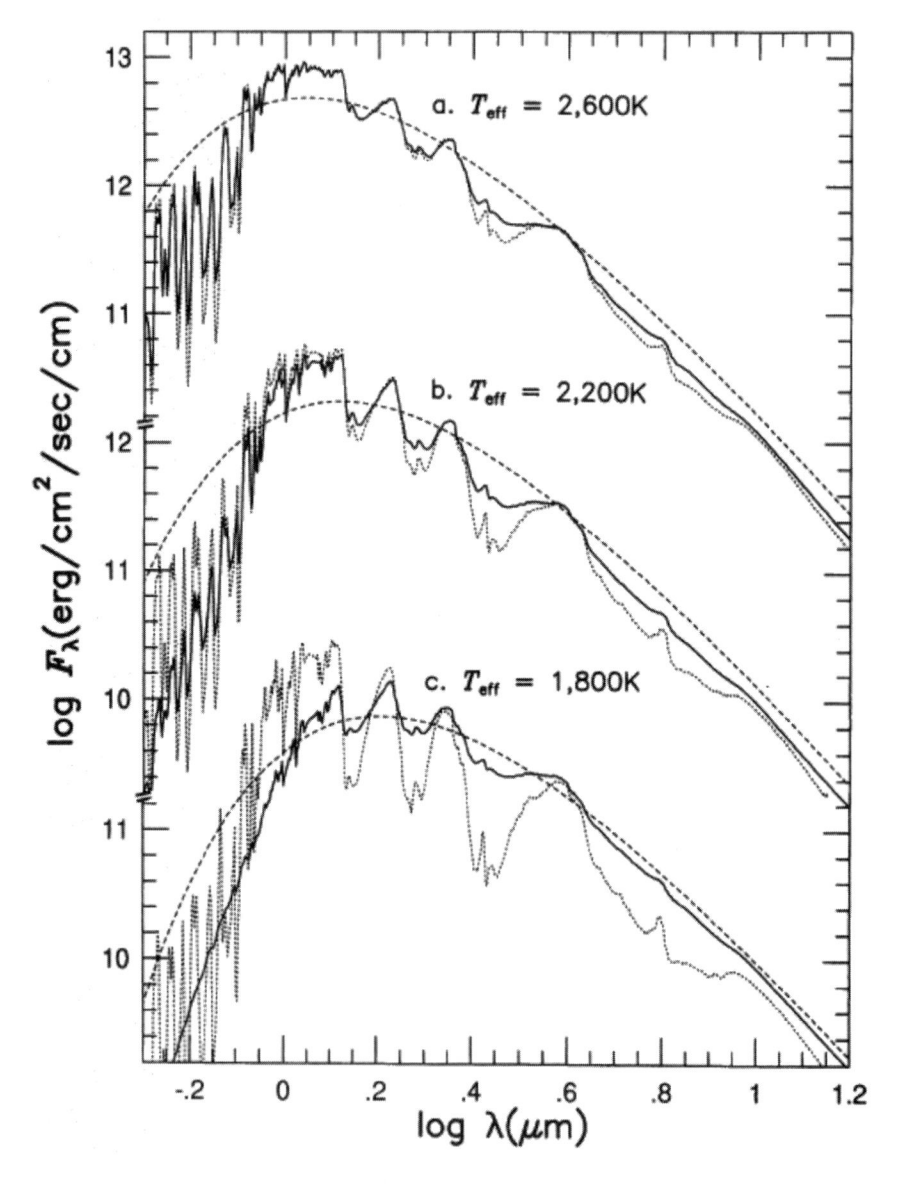

Figure 4. Predicted fluxes for the models shown in Fig 3. Note the weaker H_2O and TiO bands in the dusty models. The dashed curves indicate the black-body curves for the same temperatures.

Molecular absorption acts to cool the atmosphere while dust tends to heat it. Despite the negative feed-back effect of dust opacity, Al_2O_3 forms in all M dwarf models cooler than 2800K. Low temperature condensates such as Fe and enstatine ($MgSiO_3$) do not condense in the M dwarf regime ($T_e > 2000K$) (Tsuji et al. 1996). However, the more sophisticated DUSTY[23] models of Allard et al.(2001) show no such large effects from dust grains at the temperatures of the M dwarfs, as predicted by the preliminary models of Tsuji et al. (1996) and confirmed by observations (see also Leinert et al. 2000 Fig. 2 and 3) suggesting that a critical reexamination of dust opacity formulation at higher temperatures is necessary.

4. Convection in M Dwarfs

Hawley, Reid & Gizis(2000) discuss magnetic activity and physical attributes in M dwarfs. They show that the mid to late M dwarfs have the strongest convection and a well developed turbulent dynamo. This is the precisely the temperature regime where we are finding the biggest differences between predicted and observed spectra. Simulations suggest that nature of the magnetic fields changes from multiple small scale structures in the hotter or earlier M dwarfs to a large scale field in the cooler M dwarfs (Browning 2010). In addition, Reiners & Basri(2007) have found magnetic fields of up to 4kG in many M dwarfs. It is clear that the effects of complete convection and magnetic fields must be considered in modelling the photospheres of M dwarfs.

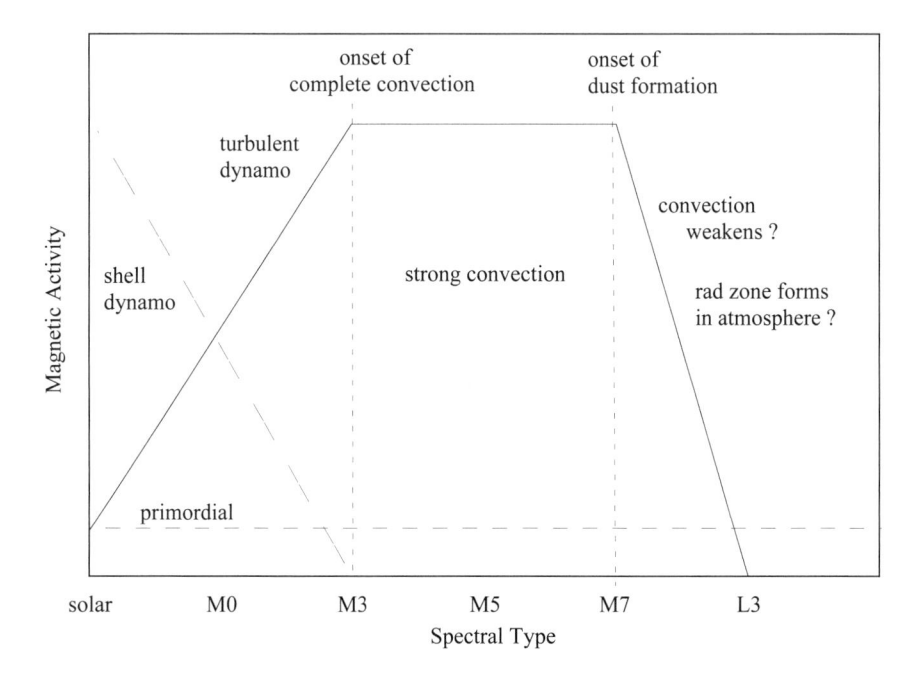

Figure 5. Stellar activity diagram from Hawley, Reid & Gizis.

---

[2]http://www.hs.uni-hamburg.de/EN/For/ThA/phoenix/index.html

[3]http://perso.ens-lyon.fr/france.allard/

5. Chromospheres in M Dwarfs

The temperature structure of normal atmospheres declines monotonically with height as does the gas pressure. The presence of a chromosphere is indicated by the temperature increasing in the outer layers due to presumed acoustic and magnetic heating. Most chromospheric modelling (e.g., Houdebine & Doyle 1995; Fuhrmeister, Schmitt & Hauschildt 2005) is done using empirical temperature-pressure relations dovetailed onto an underlying model photospheric temperature structure. The run of temperature and pressure through the upper layers is adjusted until a match is made with chromospheric observables such as CaII emission lines or Hα line profiles.

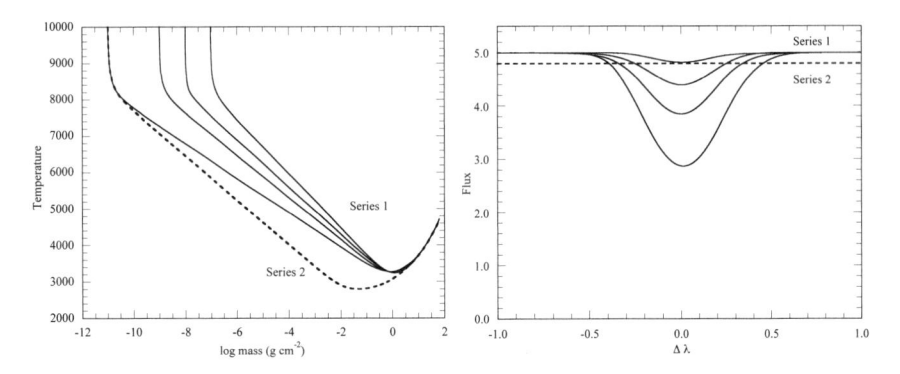

Figure 6. Several chromospheric temperature structures tested by Houdebine & Doyle and their effect on the CaIIK line profiles.

We intend to try and fit the blue metal absorption line strengths by arbitrarily increasing the run of temperature in the outer layers and positioning the temperature minimum at a range of depths as shown schematically in Fig. 6.

However, if the weak lines in the observed blue spectrum compared to the model spectrum result from the existence of a chromosphere, heated presumably by a large-scale magnetic field, we have in addition to explain why, as seen in Fig. 7, the underlying observed blue spectra of the most active (GJ1116B) and least active (GJ1002) M5.5 dwarfs are identical.

6. Summary

The stellar energy distributions of M dwarfs are not in good agreement with synthetic spectra computed using current 1D plane-parallel model atmospheres. This is especially obvious in the infrared, where the coolest M dwarfs are brighter in the K band and the H_2O bands are weaker, and in the blue, where the computed metal-line line strengths for mid-M dwarfs are much too strong. We believe that these discrepancies arise as a result of the inadequate treatment or neglect of dust in the cooler stars and because of the neglect of chromospheric heating from strong magnetic fields in the M dwarfs. Until these inadequacies are addressed, we should treat existing model atmospheres and effective temperatures derived using them, with reserve. We also need to understand why if magnetic heating is involved, this heating is independent of the activity levels of the stars.

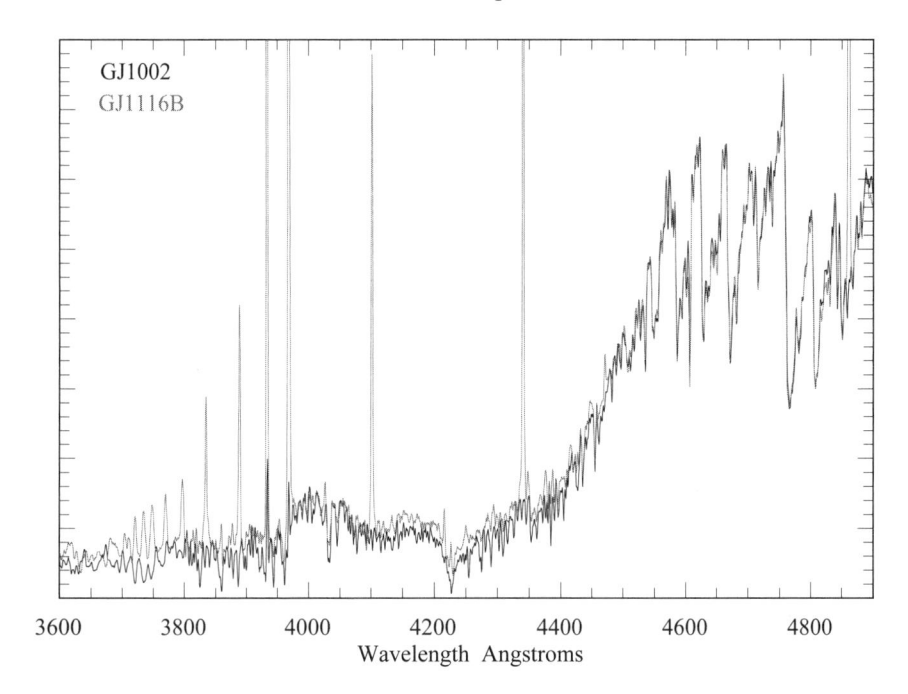

Figure 7. The blue WiFeS spectra of GJ1002, the least active M5.5V star and GJ1116B, the most active M5.5V star. Note also the broad blue depression.

Acknowledgments. I would like to thank France Allard and Bernt Freytag for discussions on dust and model atmospheres of M dwarfs; and Gerry Doyle, Eric Houdebine and Mihalis Mathioudakis for discussions on chromospheric heating and chromospheric modelling of the sun and M dwarfs. I would also like to especially thank Anna Frebel for obtaining the Magellan MIKE spectrum of Proxima Cen. Blue echelle spectra were also downloaded from the ESO UVES Archive. Medium dispersion optical spectra were obtained with the DBS and WiFeS spectrograph on the ANU 2.3m telescope at Siding Spring Observatory.

References

Allard, F., Hauschildt, P., Alexander, D., Tamanai, A., Schweitzer, A. 2001, ApJ, 556, 357
Browning, M. 2010 This conference proceedings
Burrows, A., Sudarsky, D., Hubeny, I., 2006, ApJ, 640, 1063
Fuhrmeister, B., Schmitt, J.H., Hauschildt, P. 2005, A&A, 439, 1137
Gustafsson, B., Edvardsson, B., Eriksson, K., Joegensen, U., Nordlund, A., Plez, B. 2008, A&A, 486, 951 (MARCS)
Hauschildt, P.H., Allard, F., Baron, E., 1999, ApJ, 512, 377 (NEXTGEN)
Hawley, S.L., Reid, I.N., Gizis, J.E. 2000, ASPC, 212, 252
Houdebine, E.R., Doyle, J.G. 1995, IrAJ, 22, 25
Kui, Ruan1991, Thesis ANU.
Leinert, C., Allard, F., Richichi, A., Hauschildt, P.H. 2000, A&A, 353, 691
Reiners, A., Basri, G., 2007, ApJ, 545, 1121
Tsuji, T., Ohnaka, K., Aoki, W., 1996, A&A 305, L1

Characterization of High-Energy Emissions of GKM Stars using Wide Binaries with White Dwarfs

S. Catalán,[1] A. Garcés,[2] and I. Ribas[2]

[1]*Centre for Astrophysics Research, University of Hertfordshire, Hatfield, AL10 9AB, UK*

[2]*Institute for Space Sciences (CSIC-IEEC), Campus UAB, 08193 Bellaterra, Spain*

Abstract. The definition of an age calibration for main-sequence late-type stars has multiple applications, e.g., in the fields of galactic evolution, stellar dynamos, theories of angular momentum loss and planetary atmospheres. In the latter, the characterization of the time-evolution of stellar high-energy emissions can help us understand the influence on planetary atmospheres and their potential habitability. A key element for this characterization is a reliable age determination. For this purpose we have studied a sample of late G, K, and M stars. To cover the age window up to 0.7 Gyr we have used stars belonging to open clusters, while for ages above this limit we use wide binaries containing white dwarfs. Since the evolution of white dwarfs can be understood as a cooling process, which is relatively well known at the moment, we can use them as age calibrators. Wide binary members are supposed to have been born simultaneously and with the same chemical composition. Since they are well separated (100-1000 AU aprox.) we can assume that no interaction has occurred between them in the past and they have evolved as single stars. So, from the white dwarf age we can infer the age of the system. We present our current results based in a sample of 30 binaries from the NLTT catalogue comprised by a DA white dwarf (showing only H absorption lines) and a G, K or M star.

1. Introduction

Exoplanet research has become a very dynamic and rapidly-advancing field in astronomy. In particular, studies of exoplanet atmospheres can tell us about the history of a planet and its potential habitability. Stellar astrophysics has become a critical element for these studies. The star is, by far, the main source of energy for an exoplanet's atmosphere and describing it in detail is of central importance. While the emissions of stars in the visible/IR are reasonably well understood, the same is not true for the high-energy part of the spectrum (from X-rays to the UV) that results from the magnetic activity of the star. It has been widely recognized that such high-energy emissions have a strong impact on planetary atmospheres, both in the Solar System planets (e.g., Bauer & Lammer 2004) and exoplanets (e.g., Lammer et al. 2003; Penz et al. 2008). However, the magnetic activity decays throughout the stellar lifetime in an as yet poorly understood manner, making it difficult to calculate the accumulated effects of the UV and X-ray emission on the planetary environments.

140 Catalán, Garcés, and Ribas

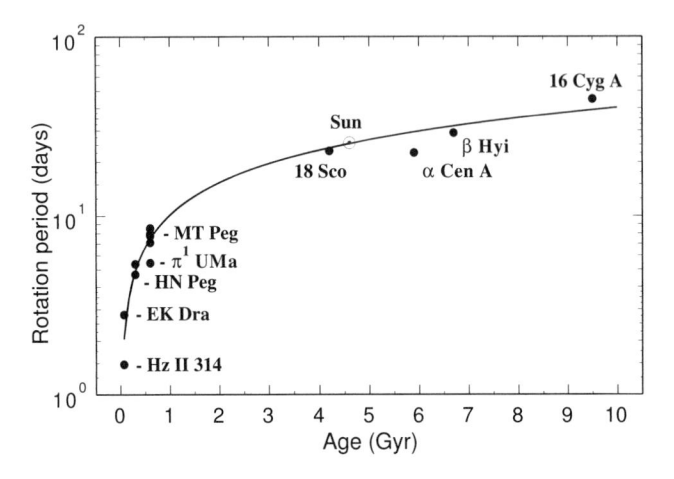

Figure 1. Rotation period as a function of age for a sample of solar-type stars. The solid line represents a power-law fit.

The decay of activity with stellar age is intimately linked to the rotational evolution of the stars. Compelling observational evidence (Guedel et al. 1997) shows that zero-age main sequence (ZAMS) solar-type stars rotate ~10 times faster than today's Sun. As a consequence, young solar-type stars, including the young Sun, have vigorous magnetic dynamos and correspondingly strong high-energy emissions. From the study of solar type stars with different ages, Skumanich (1972) showed that they lose angular momentum with time via magnetized winds thus leading to a secular increase of their rotation periods as a power law roughly proportional to $t^{1/2}$. This is illustrated in Fig. 1. In response to slower rotation, the dynamo strength diminishes with time causing high-energy emissions also to undergo significant decreases (Zahnle & Walker 1982; Ayres 1997).

1.1. Time Evolution of X-rays

The *Sun in Time* program was established to study the magnetic evolution of the Sun using a homogeneous sample of single, nearby G0-5 main sequence stars, which have known rotation periods and well-determined physical properties. Such stars could be used as proxies for the Sun at evolutionary stages covering from 0.1 up to about 7 Gyr. One of the primary goals of the *Sun in Time* program was to reconstruct the spectral irradiance evolution of the Sun. To this end, a large amount of multiwavelength (X-ray, EUV, FUV, UV, visible) data were collected covering a range between 0.1 and 330 nm. The results of this effort were published in Ribas et al. (2005) and show that the X-ray and EUV emissions of the young main-sequence Sun were about 100 to 1000 times stronger than those of the present Sun. Similarly, the FUV and UV emissions of the young Sun were 10 to 100 and 5 to 10 times stronger, respectively, than today. There is no doubt that such strong emissions should have had an important impact on the atmospheres of the Solar System planets (e.g., Kulikov et al. 2006).

While the long-term evolution of the high-energy emissions of solar-type stars is now well characterized, this is not the case for stars of lower mass, which are more active than their solar counterparts. The present program is a central part of the extension of the *Sun in Time* study to all stars of GKM spectral types owing to the differences in the high-energy emissions between, e.g., G-type and M-type stars, and to the interest of these stars as exoplanet hosts.

An analogous effort to that made for the main *Sun in Time* program (i.e., multi-wavelength observations) would require a *tour de force* that is prohibitive at this point. Instead, we prefer to take an optimized route by using some of the lessons learned. First, we have found that the overall stellar XUV emissions can be scaled from the X-ray flux (Ribas et al. 2005). This is understood if active stars share the same underlying physical mechanism responsible for the emission resulting in similar spectral energy distributions. Second, the X-ray flux is arguably the best proxy of stellar activity thanks to the large variations over time (1000× from the ZAMS to solar age) and the straightforward measurement. The approach we propose is to define X-ray vs. age relationships applicable to all GKM stars from which one could readily estimate the time-evolution of the overall XUV flux in a single step.

The bottleneck of this approach is the determination of stellar ages. It is well known that the ages of field GKM stars are extremely difficult to estimate and subject to large uncertainties. For young stars, cluster data provides relief and the ensemble age determination permits the study of the GKM population and activity/rotation relationships with limited age coverage. However, there are no nearby clusters with ages older than 0.65 Gyr (Hyades) and thus calibrations become unreliable. This is the case of recent calibrations proposed (Mamajek & Hillenbrand 2008) that, while providing good performance for young ages (15–20% uncertainty), do not yield accurate results beyond 0.5 Gyr because of the lack of calibrators and the increasing unreliability of rotation period estimates (differential rotation effects, weak signal from low-amplitude modulations). Further, the calibrations are not applicable to stars later than K2.

In a first attempt to establish the observational grounds we compiled X-ray data for a sample of clusters and calculated the mean X-ray luminosities for several spectral type intervals. We complemented the measurements with X-ray data of some particular objects, notably the α Cen system (which has a well-determined age of 6 Gyr from asteroseismology) and some field stars with large space velocities that were assigned a guessed age of 10 Gyr as members of the thick disk. The data are plotted in Fig. 2 and show a conspicuous lack of precise data between 0.6 and 6 Gyr for stars other than solar analogs. As a working hypothesis, we can assume that the evolution of L_x for a late-type star has a flat plateau from its arrival on the main sequence up to a certain age (end of saturation phase) and then decreases monotonically with age (e.g., Pizzolato et al. 2003). Such qualitative behavior can be used to obtain crude estimates of the X-ray flux of a star with a given age.

2. White Dwarfs to the Rescue

A reliable way to calculate stellar ages is via the use of wide binaries in which one of the components is a white dwarf (Silvestri et al. 2005). The members of a wide binary are assumed to have been born simultaneously and with the same chemical composition. Since they are well separated (100-1000 AU aprox.) we can assume that no interaction has occurred between them in the past and they have evolved as single stars (Oswalt et al. 1988). So, from the white dwarf age we can infer the age of the system. The evolution of a white dwarf can be described as a cooling process, which is rela-

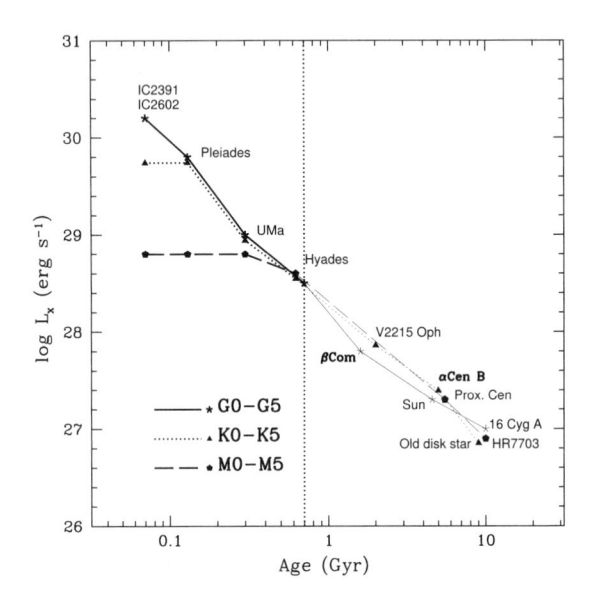

Figure 2. X-ray luminosity versus age for solar-like stars and for field and cluster K and M stars. Only a few K and M stars are available beyond 0.65 Gyr. Error bars in the cluster data correspond to the standard deviation of the mean.

tively well understood at the present moment (Salaris et al. 2000). The total age of a white dwarf can be expressed as the sum of its cooling time (t_{cool}) and the lifetime of its progenitor in the main-sequence (t_{prog}). Thus, accurate ages can be obtained from an initial-final mass relationship and stellar tracks to account for the pre-white dwarf lifetime. The initial-final mass relationship of white dwarfs still has moderate uncertainties but these are minimized for intermediate-age objects where the cooling age is the dominant factor.

2.1. The Sample Selection

Our first selection was done only considering the catalogues of Chanamé & Gould (2004) and Gould & Chaname (2004). We made a cautious selection of the binaries. Firstly, the white dwarf component should be classified as a DA (i.e., with the unique presence of Balmer lines), so that the fitting procedure is sufficiently accurate to derive realistic values for the effective temperature and surface gravity. Secondly, the other component of the pair should be a star of spectral type G, K or M. We revised all the wide binaries that satisfied the previous conditions, making sure the separation between the members was large enough (100-1000AU), in order to guarantee the no interaction between them. Information about the two proper motion (μ) components of both members of the wide binary is available in Chanamé & Gould (2004) and Gould & Chaname (2004) catalogues. We only considered those wide binaries with $\mu \geq 180$ mas yr^{-1} so that we can be sure they are physically bonded. The sample was completed by selecting some white dwarf + M wide binaries from Silvestri et al. (2005). Our final sample is formed by 30 pairs composed by a white dwarf and a GKM star companion.

3. Results

We have obtained low-medium resolution and high signal-to-noise spectroscopic data of the white dwarfs using ISIS and TWIN spectrographs, at WHT at Roque de los Muchachos Observatory (ORM) and the 3.5m telescope at CAHA Observatory, respectively. The white dwarf spectra were reduced using the standard procedures within single-slit tasks in IRAF[1]. First, the images were bias- and flatfield-corrected, and then, the spectra were extracted and wavelength calibrated using arc lamp observations. Finally, we normalized the spectra to the continuum so that we can compare them to synthetic models. Before this comparison the radial velocity of each star was obtained from a crosscorrelation with a reference model using the IRAF task fxcor. After correcting the spectra from their radial velocities, the atmospheric parameters (T_{eff} and $\log g$) were derived by performing fits to the observed Balmer lines with synthetic models (D. Koester, private communication) using the χ^2 minimization method Press et al. (1992). Balmer/Lyman lines in the models were calculated with the modified Stark broadening profiles of Tremblay & Bergeron (2009), kindly made available by the authors. In Fig. 3 we show the fits corresponding to 12 white dwarfs in our sample. From T_{eff} and $\log g$, the white dwarf mass and its cooling time can be determined via interpolation in the appropriate cooling sequences, for instance those of Salaris et al. (2000), which consider carbon-oxygen (C/O) core with a thick hydrogen envelope on top of a helium buffer. Although the initial-final mass relationship of white dwarfs still has large uncertainties, many improvements have been achieved during these last years, for instance with the coverage of the low-mass domain, Catalán et al. (2008), which is the most populated bin according to the initial mass function of Salpeter. So, using the initial-final mass relationship of Catalán et al. (2008) we can obtain the progenitor mass and then, considering the evolutionary tracks of Domínguez et al. (1999) derive the progenitors lifetime. This yields the total age of the white dwarf, which in turn, is the age of its companion. In Table 1 we show the results obtained for the 12 white dwarfs in Fig. 3. As can be noted, the total ages obtained (i.e., cooling ages plus progenitor lifetime) have precisions of 10–25%. This is sufficient in our context since magnetic activity is an intrinsically variable phenomenon and any relationship will have some inherent dispersion no matter how accurate the ages and coronal/chromospheric fluxes. So, with the use of wide binaries containing white dwarfs we will be able to obtain ages for stars that are impossible to calibrate using other methods. This will allow us to add valuable points to the not very populated region (above 0.7 Gyr) in the L_X vs age relation (Fig. 2).

---

[1]IRAF is distributed by the National Optical Astronomy observatories, which are operated by the Association of Universitites for Research in Astronomy, Inc., under cooperative agreement with the national Science Foundation (http://iraf.noao.edu).

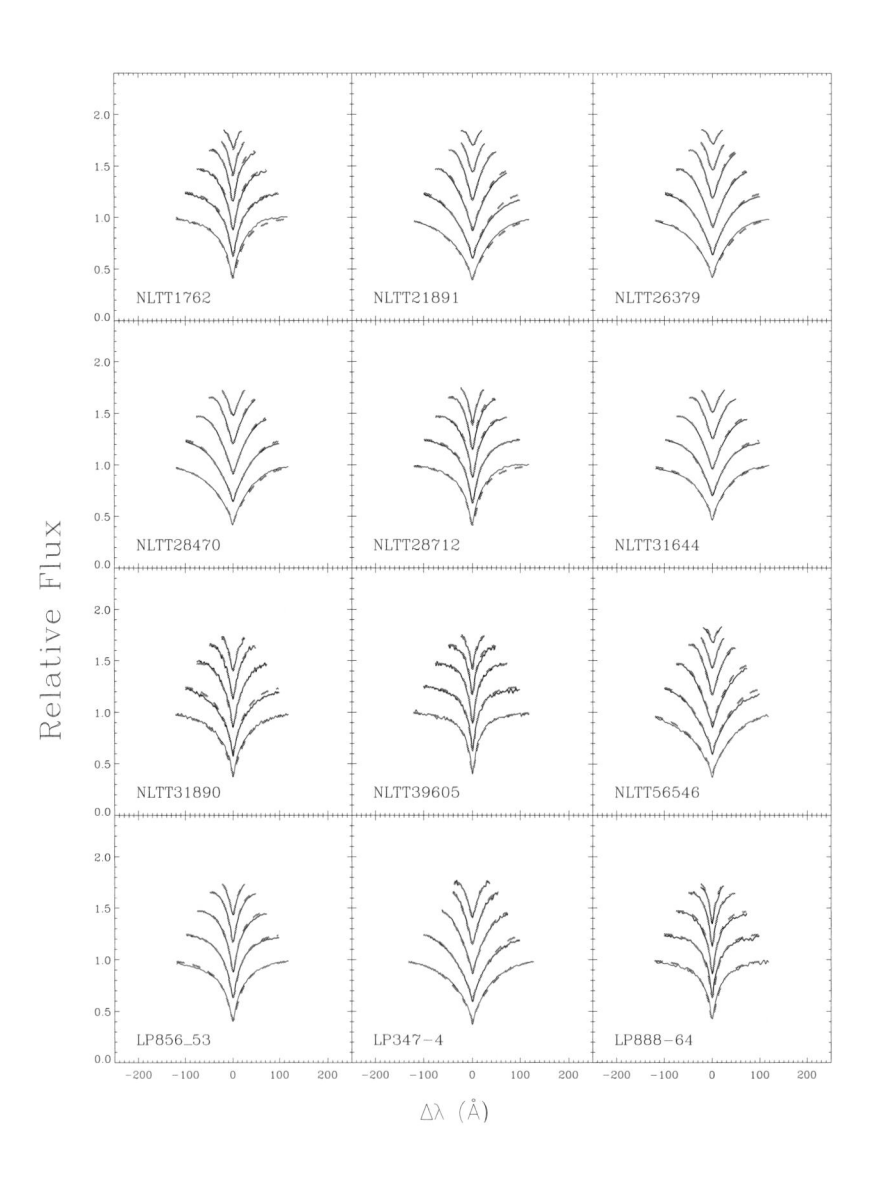

Figure 3. Fits of the observed Balmer lines (solid lines) to white dwarf models (dashed lines) for 12 white dwarfs representative of our sample.

Table 1. Parameters obtained for 12 white dwarfs (WDs) from our sample

WD Name	T_{eff} (K)	log g (dex)	M_{WD} ($M_\odot$)	t_{cool} (Gyr)	M_{prog} ($M_\odot$)	t_{prog} (Gyr)	Age (Gyr)
NLTT1762	10360±230	8.19±0.08	0.72±0.03	0.68±0.05	2.96±0.31	$0.49^{+0.07}_{-0.06}$	$1.2^{+0.1}_{-0.1}$
NLTT21891	13520±190	7.92±0.02	0.57±0.01	0.23±0.01	1.45±0.19	$3.36^{+0.70}_{-0.78}$	$3.6^{+0.7}_{-0.8}$
NLTT26379	15150±220	7.91±0.03	0.56±0.01	0.16±0.01	1.41±0.21	$3.76^{+0.88}_{-0.77}$	$3.9^{+0.9}_{-0.8}$
NLTT28470	15290±450	7.99±0.05	0.61±0.02	0.18±0.01	1.86±0.26	$1.49^{+0.32}_{-0.22}$	$1.7^{+0.3}_{-0.2}$
NLTT28712	9910±70	8.04±0.03	0.63±0.01	0.62±0.02	2.05±0.24	$1.15^{+0.17}_{-0.12}$	$1.8^{+0.2}_{-0.1}$
NLTT31644	19170±250	7.89±0.08	0.57±0.01	0.069±0.01	1.44±0.35	$3.46^{+1.20}_{-1.09}$	$3.5^{+1.2}_{-1.1}$
NLTT31890	10790±120	8.17±0.07	0.71±0.03	0.60±0.03	2.88±0.29	$0.53^{+0.07}_{-0.06}$	$1.1^{+0.1}_{-0.1}$
NLTT39605	9280±80	8.12±0.07	0.67±0.03	0.82±0.04	2.52±0.40	$0.76^{+0.11}_{-0.13}$	$1.6^{+0.1}_{-0.1}$
NLTT56546	13660±100	7.97±0.02	0.59±0.01	0.24±0.01	1.72±0.26	$1.93^{+0.32}_{-0.28}$	$2.2^{+0.3}_{-0.3}$
LP856_53	10200±100	8.26±0.06	0.76±0.02	0.80±0.04	3.26±0.28	$0.37^{+0.04}_{-0.04}$	$1.2^{+0.1}_{-0.1}$
LP347_4	13300±80	7.88±0.05	0.55±0.02	0.23±0.01	1.22±0.29	$6.12^{+2.23}_{-1.93}$	$6.3^{+2.2}_{-1.9}$
LP888_64	9525±100	7.92±0.05	0.56±0.02	0.58±0.02	1.34±0.27	$4.47^{+1.20}_{-1.22}$	$5.1^{+1.2}_{-1.2}$

4. Future Work

After determining the age of all the GKM stars in our sample the next step of this project is to obtain information about their activity. We have performed observations for the GKM companions in order to collect all their possible activity indicators to calibrate the activity-age relationship. We have already obtained high resolution spectroscopic observations of the companions with SARG and Stella echelle spectrographs at TNG in ORM and Observatorio del Teide, respectively. At the moment we are working in the data analysis of these companions.

X-ray observations are the most powerful means to obtain the activity information of late-type stars since their time variations are of several orders of magnitude (young stars emit thousands of times more X-ray flux than their older counterparts). The large age-related variations revert in a higher reliability of the relationships with respect to chromospheric activity diagnostics and minimize errors caused by intrinsic variations (e.g., no more than a factor of a few during the activity cycle). On the other hand, visible spectroscopy to measure chromospheric activity is easier to collect for a large sample.

We have also granted time for some of our targets with the Chandra satellite, and we have submitted new proposals for the XMM satellite. We will make efficient use of both chromospheric and coronal activity diagnostics by combining the Chandra data with optical spectroscopic data that we will collect shortly (3.5-m TNG in La Palma) for the same targets. Our strategy is to use X-ray observations of the Chandra sample to calibrate visible spectroscopic indicators (chromospheric flux measurements from Ca II H & K, H Balmer, Ca II IRT, etc). We will then define flux-flux relationships (e.g., Ayres 1999) applicable to the rest of the stars.

The results from our work should provide critical input information to any effort of understanding the long-term evolution of the atmospheres of planets around low-mass stars, including their potential habitability. It will also allow a better understanding of the evolution of high energy emissions of low-mass stars. And since age determination in the age domain > 0.7 Gyr remains difficult it could be used as an age calibrator when stellar high energy emissions can be measured.

Acknowledgments. We thank D. Koester for providing us with his white dwarf models. S.C. is supported by a Marie Curie Intra-European Fellowship within the 7th

European Community Framework Programme. A.G. and I.R. acknowledge the support of the Spanish MCINN grant AYA2009-06934.

References

Ayres, T. R. 1997, J. Geophys. Res., 102, 1641
— 1999, ApJ, 525, 240
Bauer, S. J., & Lammer, H. 2004, Planetary aeronomy : atmosphere environments in planetary systems (Lammer, H. and Bauer, S. J.)
Catalán, S., Isern, J., García-Berro, E., & Ribas, I. 2008, MNRAS, 387, 1693
Chanamé, J., & Gould, A. 2004, ApJ, 601, 289
Domínguez, I., Chieffi, A., Limongi, M., & Straniero, O. 1999, ApJ, 524, 226
Gould, A., & Chaname, J. 2004, ApJS, 150, 455
Guedel, M., Guinan, E. F., & Skinner, S. L. 1997, ApJ, 483, 947
Kulikov, Y. N., Lammer, H., Lichtenegger, H. I. M., Terada, N., Ribas, I., Kolb, C., Langmayr, D., Lundin, R., Guinan, E. F., Barabash, S., & Biernat, H. K. 2006, planss, 54, 1425
Lammer, H., Selsis, F., Ribas, I., Guinan, E. F., Bauer, S. J., & Weiss, W. W. 2003, ApJ, 598, L121
Mamajek, E. E., & Hillenbrand, L. A. 2008, ApJ, 687, 1264
Oswalt, T. D., Hintzen, P., & Luyten, W. 1988, ApJS, 66, 391
Penz, T., Micela, G., & Lammer, H. 2008, A&A, 477, 309
Pizzolato, N., Maggio, A., Micela, G., Sciortino, S., & Ventura, P. 2003, A&A, 397, 147
Press, W. H., Teukolsky, S. A., Vetterling, W. T., & Flannery, B. P. 1992, Numerical recipes in FORTRAN. The art of scientific computing (Cambridge: University Press, —c1992, 2nd ed.)
Ribas, I., Guinan, E. F., Güdel, M., & Audard, M. 2005, ApJ, 622, 680
Salaris, M., García-Berro, E., Hernanz, M., Isern, J., & Saumon, D. 2000, ApJ, 544, 1036
Silvestri, N. M., Hawley, S. L., & Oswalt, T. D. 2005, AJ, 129, 2428
Skumanich, A. 1972, ApJ, 171, 565
Tremblay, P., & Bergeron, P. 2009, ApJ, 696, 1755
Zahnle, K. J., & Walker, J. C. G. 1982, Reviews of Geophysics and Space Physics, 20, 280

Rotational Velocities of Very Low Mass Binaries

Q.M. Konopacky,[1] A.M. Ghez,[2] B.A. Macintosh,[1] R.J. White,[3] T.S. Barman,[4] E.L. Rice,[5] and G. Hallinan[6]

[1]*Lawrence Livermore National Laboratory, 7000 East Avenue, Livermore, CA 94550, USA*

[2]*Department of Physics and Astronomy, Uinversity of California, Los Angeles, Los Angeles, CA 90095, USA*

[3]*Department of Physics and Astronomy, Georgia State University, Atlanta, GA 30303, USA*

[4]*Lowell Observatory, 1400 West Mars Hill Road, Flagstaff, AZ 86001, USA*

[5]*Department of Astrophysics, American Museum of Natural History, New York, NY 10024, USA*

[6]*Department of Astronomy, University of California, Berkeley, Berkeley, CA 94720, USA*

Abstract. We present rotational velocities for individual components of eleven very low mass (VLM) binaries with spectral types between M7.5 and L4. These results are based on observations taken with the near-infrared spectrograph, NIRSPEC, and the Keck II laser guide star adaptive optics (LGS AO) system. The binaries were targeted as part of a dynamical mass program, and their orbital inclinations are used to translate $v\sin i$ into a rotational velocity for each component. We find that the observed sources tend to be rapid rotators ($v\sin i > 10$ km s^{-1}), consistent with previous measurements for ultracool objects. Five systems have component $v\sin i$'s that are statistically different, with three binaries having velocity differences greater than 25 km s^{-1}. To bring these discrepant rotational velocities into agreement would require their rotational axes to be inclined between 10 to 40° with respect to each other, and that at least one component has a significant inclination with respect to the orbital plane. Alternatively, each component could be rotating at a different rate, even though they have similar spectral types. Both differing rotational velocities and inclinations have interesting implications for binary star formation. Two of the binaries with large differences in rotational velocity are also known radio sources, LP 349-25AB and 2MASS 0746+20AB. LP 349-25B is rotating at ~95 km s^{-1}, within a factor of ~3 of the break up speed, and is one of the most rapidly rotating VLM objects known.

1. Introduction

Rotational velocity is an important diagnostic parameter for stellar objects, offering a window into the angular momentum evolution of a given source. Measurements of rotational velocity have been shown to correlate strongly with stellar activity, possibly driving the magnetic dynamo responsible for generating this activity (Browning 2008).

In addition, rotational velocities have been shown to correlate with the age of a system, offering a tool for estimating stellar ages (Delfosse et al. 1998).

The rotational behavior of very low mass stars and brown dwarfs has been studied by a number of authors in recent years (Mohanty & Basri 2003; Bailer-Jones 2004; Zapatero Osorio et al. 2006; Reiners & Basri 2008, 2010). It has been shown that the brown dwarfs tend to be rapid rotators, and that the minimum rotation rate is a function of spectral type (i.e., Zapatero Osorio et al. 2006; Reiners & Basri 2008). It has also been determined that the rotational velocities of brown dwarfs correlate with age and that their rotational evolution is probably dominated primarily by magnetic braking (Reiners & Basri 2008). However, it appears that the activity-rotation relationship that is very strong amongst M dwarfs tends to break down at these low masses (Mohanty & Basri 2003).

The majority of previous studies have been performed with seeing-limited observations, and most sources targeted are thought to be single. Known binaries have been included in various samples, and the rotational velocities derived for these objects have been from the combined light of both components. The rotational velocities of individual binary components can potentially provide a unique look at the rotational evolution of VLM objects. If the rotational velocities differ substantially from those of single objects, it could imply that VLM binaries evolve differently than single systems. Additionally, if any differences are seen between the velocities of the binary components, it could have implications for the way in which these binaries formed and their early accretion history. Finally, if the orbits of these binaries are known, it allows for the translation of projected rotational velocity into the true rotational velocity via the assumption that the rotation axis is perpendicular to the orbital plane.

We present here rotational velocity measurements for the components of a sample of tight, visual VLM binaries. The measurements of these spatially-resolved velocities are enabled by the W.M. Keck Observatory laser guide star adaptive optics (LGS AO) system, which provides high spatial resolution observations of optically faint targets (Wizinowich et al. 2006). This study is the first to examine systematically the rotational velocities of individual VLM objects that reside in binary systems.

2. Sample and Observations

2.1. Sample

Our sample is comprised of eleven VLM binaries that were targeted as part of an ongoing program to measure their dynamical masses. These objects have been observed both astrometrically and spectroscopically since 2006, and initial estimates of their orbital properties have been obtained (Konopacky et al. 2010). Their spectral types range from M7.5 to L4, and their separations range from $0\rlap{.}''07$ to $0\rlap{.}''35$. Because we are able to spatially resolve the components before obtaining high resolution spectroscopy (see section 2.2), our total sample consists of 22 VLM objects.

2.2. Observations

The eleven binaries were observed using the NIR spectrograph NIRSPEC (McLean et al. 2000) on Keck II 10 m in conjunction with the facility LGS AO system (NIRSPAO). These observations, taken between 2006 December and 2010 June, are described in detail in Konopacky et al. (2010). Briefly, we used the instrument in its high spectral resolution mode, selecting a slit 0.″041 in width and 2.″26 in length in AO mode. We observed in the K band in order to obtain data in the CO bandhead region (2.291 - 2.325 μm, dispersion order 33). Due to the dense population of lines in this region, our analysis for this work was done only in order 33, although the cross-dispersed data ranged from 2.044 - 2.382 μm.

A rotation was applied so that both components of each binary fell simultaneously on the high-resolution slit, which is at an angle of 105.9^o with respect to north. Typical observations consisted of four spectra of both components, each with 1200 second integration times, taken in an ABBA dither pattern along the length of the slit. Each target observation was accompanied by the observation of a nearby A0V star to measure the telluric absorption in the target spectra.

3. Analysis

Data reduction of our NIRSPAO data was performed as described in Konopacky et al. (2010). The basic reduction was done with REDSPEC, a software package designed for NIRSPEC[1]. As these systems are fairly tight binaries, cross-contamination can be an issue when extracting the spectra. We therefore use a Gaussian extraction method, fitting each spectral trace with a Gaussian of variable FWHM. In this way, one binary component's trace can be subtracted from a given frame before the other component's is extracted, minimizing contamination (Konopacky et al. 2010). The average signal-to-noise ratio of our individual spectra is ~50.

It has been demonstrated the CO bandhead line widths in order 33 are primarily a function of temperature and the projected rotational velocity ($v$sin$i$) for VLM objects, with an additional moderate dependence on surface gravity (Blake et al. 2007). With some knowledge of the temperature of a given object and an allowance for varying surface gravity, $v$sin$i$ measurements can be estimated from our extracted spectra. An example of the impact of $v$sin$i$ on spectral morphology of the CO bandhead is shown in Figure 1. These three sources, which have roughly the same spectral type, look quite different spectrally due to their different rotational velocities.

Our extracted spectra are not corrected for telluric absorption because these features not only provide a stable reference for absolute wavelength calibration, but also provide a means of estimating the instrumental PSF. We therefore model each spectrum as a combination of a KPNO/FTS telluric spectrum (Livingston & Wallace 1991) and a synthetically generated spectrum derived from the PHOENIX atmosphere models (Hauschildt et al. 1999). The model spectrum is parameterized to account for the wavelength solution, continuum normalization, instrumental profile (assumed to be Gaussian), $v$sin$i$, and radial velocity. To accurately estimate $v$sin$i$, the instrumental PSF is measured on our A0V calibrator stars, which by design are a clean measure of the actual telluric spectrum. We hold the instrumental profile fixed while fitting our actual VLM spectra. The best-fit model is determined by minimizing the variance-weighted reduced χ^2 of the difference between the model and the extracted spectrum, once this difference has been Fourier filtered to remove the fringing present in NIRSPEC K-band spectra (J. Bailey, in prep). This model therefore provides our $v$sin$i$ estimates.

---

[1] http://www2.keck.hawaii.edu/inst/nirspec/redspec/index.html

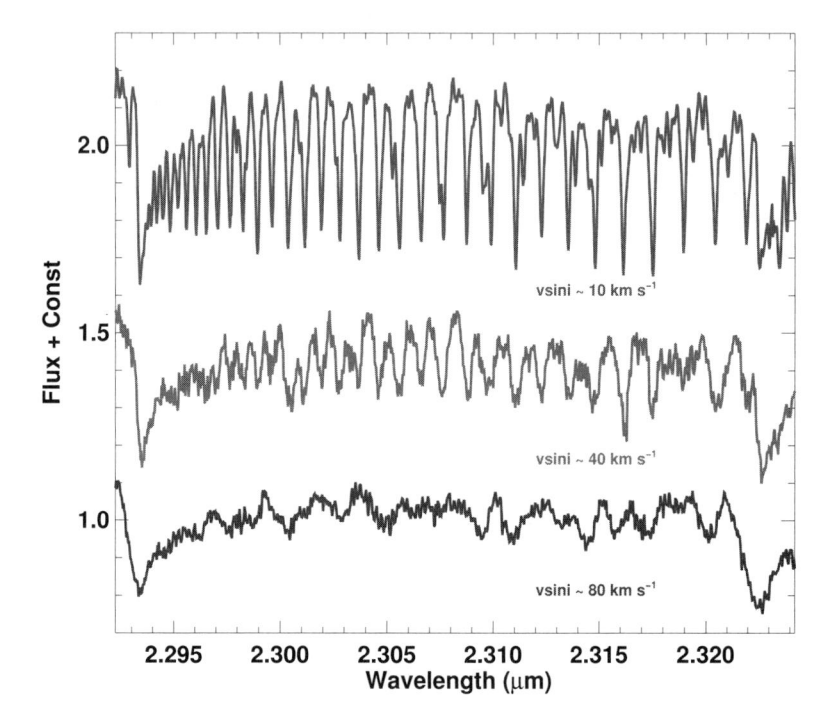

Figure 1. Three extracted and telluric corrected spectra from our sample. These spectra are from NIRSPAO dispersion order 33, and show strong CO features. All sources shown in this figure have a spectra type of ~M8, but have widely different morphologies due to varying $v\sin i$. The measured $v\sin i$'s for each source are given for reference.

The PHOENIX templates are generated at a fixed temperature and surface gravity. Our primary templates for each source have a temperature as measured in Konopacky et al. (2010) and a log(g) of 5.0. Statistical uncertainties are assigned by fitting each individual spectrum separately and taking the RMS of the values derived for each case. We also need to account for systematic uncertainties due to both the temperature and surface gravity dependence of our spectra. We fit each spectrum with templates spanning ±100 K in temperature and ranging from 3.0 - 4.5 dex in log(g). We then use the spread in these values around our best-fit value as our systematic uncertainty, and add these in quadrature with our statistical uncertainties.

We also estimate the lowest measureable value of $v\sin i$ in our spectra. To do this, we took our PHOENIX templates and broadened them first to the correct instrumental PSF and then to different values of $v\sin i$. We fit these spectra using the methodology described above. We find that the limiting value for which we could accurately measure $v\sin i$ is 3 km s^{-1}.

4. Results and Discussion

We average the measured $v\sin i$'s for each source across all epochs of data. The final uncertainty for each source is the error on the mean. The measured $v\sin i$'s for all sources are above the 3 km s^{-1} limit. Our typical uncertainties are of order 10%. We plot all of our values on the left panel of Figure 2 as a function of spectral type. Also included on this plot are values for other VLM sources from previous surveys (Mohanty & Basri 2003; Reiners & Basri 2008, 2010). It is apparent that our values are consistent with previous results. Our sources tend to be relatively rapid rotators, with 17 sources having $v\sin i > 10$ km s^{-1}. This implies that the rotational evolution of objects in binary systems is not drastically different than that of single VLM objects.

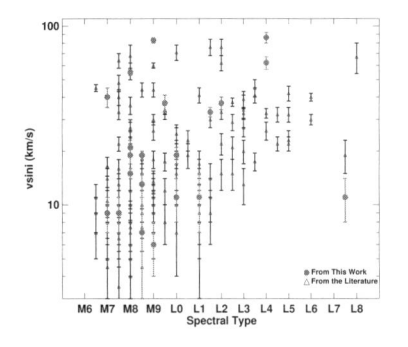

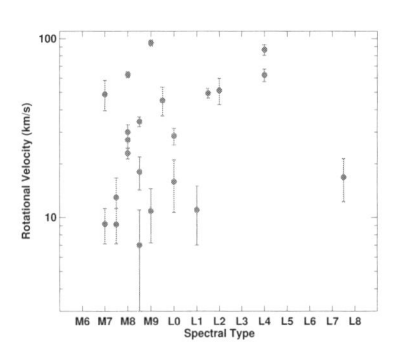

Figure 2. **Left**: Projected rotational velocity ($v\sin i$) versus spectral type for each of the binary components in our sample (red circles). Also plotted (open triangles) are previously measured values (seeing limited observations, so binaries are not resolved) of $v\sin i$ for VLM objects in the literature (Mohanty & Basri 2003; Reiners & Basri 2008, 2010). The values we measure for our sources are consistent with other VLM objects, with our sources tending towards rapid rotation. **Right**: True rotational velocities for our sources, assuming that the equatorial and orbital planes are aligned. Several of our sources are among the most rapidly rotating VLM objects yet observed.

Because these binary systems have been monitored in order to obtain their orbital solutions, the inclination of the orbital planes are measured for all sources (Konopacky et al. 2010). If we make the assumption that the inclination of the orbital plane is the same as the equatorial inclination of each source (meaning that the rotation axis is perpendicular to the orbital plane), we can turn our measured *projected* rotational velocities into *true* rotational velocities. The result of this deprojection is shown in the right panel of Figure 2.

We also examined whether the binary components have similar rotational velocities. Figure 3 shows the velocity of the secondary component of each binary plotted against that of the primary. It is immediately apparent that five sources in our sample have statistically different rotational velocities under the assumption that rotation axis is perpendicular to the orbital plane. In the case of three sources, the differences in velocity are >25 km s^{-1}. In most cases, it is the secondary component that is rotating more rapidly, although in one case the primary is rotating more rapidly.

There are two possible scenarios that could explain the results shown in Figure 3. The first is that our assumption that the equatorial planes of both components align with

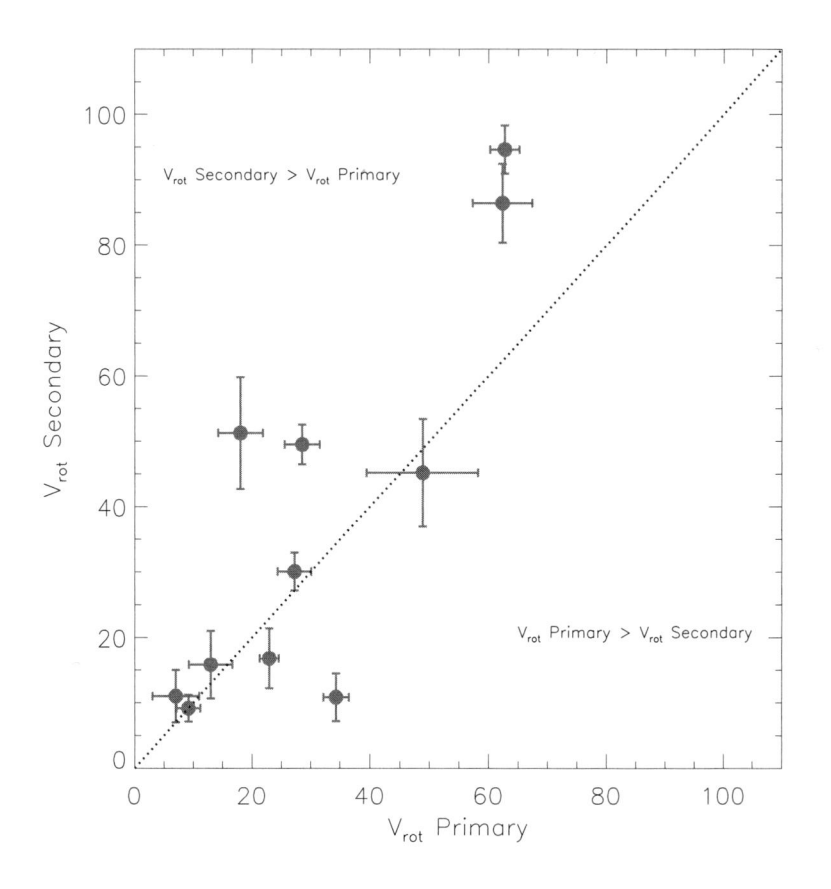

Figure 3. The rotational velocities of each secondary component plotted against its primary. Sources with consistent velocities should fall on the dotted line. Five of our eleven systems show components with different velocities, with three of these having differences >25 km s^{-1}

the orbital plane is incorrect. If one or both components is tilted with respect to the orbital plane, they could have the same true rotational velocity. All of the binaries in our sample have semimajor axes between 1 AU and 80 AU, beyond the separation regime where tidal circularization is expected. However, given the results of Hale (1994), one might expect binaries to have aligned equatorial and orbital planes, as substantial misalignment tends not to be observed for sources with separations $\lesssim$100 AU. Additional work on T Tauri binary systems have shown both slight and substantial planar misalignment via observations of disk orientation (Jensen et al. 2004; Monin et al. 2006), with the relative inclinations often of order 20° or less. To bring the rotational velocities our sources into agreement would require a differential inclination of 10 - 40°, consistent with these previous measurements. However, the T Tauri binaries have wider separations (>70 AU) than most of our systems, and also generally higher masses. In spite of this, the implication of such a result could be a consistent mode of formation between the higher mass and VLM systems. Turbulent core fragmentation theories of binary formation predict random component orientations (Bate 2000), whereas theories in which binaries form via gravitational instability in a disk predict rotational alignment (Bonnell & Bate 1994). In addition, interactions of components in multibody systems could produce misaligned axes. At least two of the objects in our sample with differing velocities are part of known triple systems with wide, higher mass components.

The other scenario is that the two components truly have different rotational velocities. One might basically expect binary components to have similar velocities if they formed out of the same cloud of material, with the cloud imprinting its angular momentum on the binary. However, differing component velocities have been observed in the case of the very young eclipsing binary brown dwarf, 2MASS0535-05AB (Gómez Maqueo Chew et al. 2009), making this a plausible scenario for the binaries observed in this study. A difference in rotational evolution could potentially come after initial formation, due to different amounts of accretion onto each component. In addition, it is possible that differing disk dissipation timescales for each component could have this effect. In terms of longer term evolution, the components of the binaries with discrepant $v\sin i$'s have spectral types that differ by at most 1.5 spectral subclasses. Thus it is unlikely, given the coevality of these binary components, that the known trend of increasing spindown time with spectral type (or roughly, mass) is the cause of the differing velocities. New simulations of VLM binary formation could shed light on all of these possibilities.

Two of our binary sources (2MASS0746+20AB and LP 349-25AB) are known radio sources (Antonova et al. 2007; Phan-Bao et al. 2007). The components of both of these binaries exhibit rapid rotation (>20 km s^{-1}). This suggests that the rapid rotation helps drive the activity of these sources. Both of these systems also have components with different $v\sin i$'s. This has interesting implications for determining which of the binary components is exhibiting the radio emission, as the radio measurements are typically unresolved. For instance, Berger et al. (2009) used the average $v\sin i$ of the two binary components from unresolved measurements to derive a radius for the radio emitting component, which they assumed to be the primary. However, the secondary is rotating more rapidly and thus is more likely to be the active source. Using the spatially resolved measurement of $v\sin i$ for the secondary rather than the combined measurement gives a different value for the radius, one that is actually consistent with expected values (G. Hallinan et al., in prep). This highlights the importance of obtaining fundamental parameters of binary components individually. Furthermore, it is likely that the secondary component of LP 349-25AB is the radio source, since it is rotating at the extremely rapid rate of ~95 km s^{-1}. This is within a factor of 3 of the breakup speed of this object, and is the most rapid rotator in our sample. Future VLBI observations of both sources will determine conclusively which component is generating the radio emission in these sources. In addition, LP349-25AB would be an excellent candidate for polarimetric observations (i.e., Tata et al. 2009).

With our new measurements, we will be able to compare $v\sin i$ to the masses of these objects. This may shed light on the rotational evolution as a function of mass, rather than spectral type. In addition, these observations can provide an further constraints on the ages of these systems, which in many cases is completely unknown or very uncertain. Age estimates can aid in the comparison of fundamental parameters to evolutionary models. Finally, we aim to obtain $v\sin i$ measurements for a greater sample of VLM binaries and push down to objects with later spectral types, allowing for the characterization of binaries across the entire substellar regime.

Acknowledgments. This work was performed under the auspices of the U.S. Department of Energy by Lawrence Livermore National Laboratory under Contract DE-AC52-07NA27344. This work is support by NASA Origins Grant NNX1 OAH39G. The W.M. Keck Observatory is operated as a scientific partnership among the California Institute of Technology, the University of California and the National Aeronautics and Space Administration. The Observatory was made possible by the generous financial support of the W.M. Keck foundation.

References

Antonova, A., Doyle, J. G., Hallinan, G., Golden, A., & Koen, C. 2007, A&A, 472, 257
Bailer-Jones, C. A. L. 2004, A&A, 419, 703
Bate, M. R. 2000, MNRAS, 314, 33
Berger, E., Rutledge, R. E., Phan-Bao, N., Basri, G., Giampapa, M. S., Gizis, J. E., Liebert, J., Martín, E., & Fleming, T. A. 2009, ApJ, 695, 310
Blake, C. H., Charbonneau, D., White, R. J., Marley, M. S., & Saumon, D. 2007, ApJ, 666, 1198
Bonnell, I. A., & Bate, M. R. 1994, MNRAS, 269, L45
Browning, M. K. 2008, ApJ, 676, 1262. 0712.1603
Delfosse, X., Forveille, T., Perrier, C., & Mayor, M. 1998, A&A, 331, 581
Gómez Maqueo Chew, Y., Stassun, K. G., Prša, A., & Mathieu, R. D. 2009, ApJ, 699, 1196
Hale, A. 1994, AJ, 107, 306
Hauschildt, P. H., Allard, F., & Baron, E. 1999, ApJ, 512, 377
Jensen, E. L. N., Mathieu, R. D., Donar, A. X., & Dullighan, A. 2004, ApJ, 600, 789
Konopacky, Q. M., Ghez, A. M., Barman, T. S., Rice, E. L., Bailey, J. I., White, R. J., McLean, I. S., & Duchêne, G. 2010, ApJ, 711, 1087
Livingston, W., & Wallace, L. 1991, An atlas of the solar spectrum in the infrared from 1850 to 9000 cm-1 (1.1 to 5.4 micrometer)
McLean, I. S., Graham, J. R., Becklin, E. E., Figer, D. F., Larkin, J. E., Levenson, N. A., & Teplitz, H. I. 2000, in Society of Photo-Optical Instrumentation Engineers (SPIE) Conference Series, edited by M. Iye & A. F. Moorwood, vol. 4008 of Society of Photo-Optical Instrumentation Engineers (SPIE) Conference Series, 1048
Mohanty, S., & Basri, G. 2003, ApJ, 583, 451
Monin, J., Ménard, F., & Peretto, N. 2006, A&A, 446, 201
Phan-Bao, N., Osten, R. A., Lim, J., Martín, E. L., & Ho, P. T. P. 2007, ApJ, 658, 553
Reiners, A., & Basri, G. 2008, ApJ, 684, 1390
— 2010, ApJ, 710, 924
Tata, R., Martín, E. L., Sengupta, S., Phan-Bao, N., Zapatero Osorio, M. R., & Bouy, H. 2009, A&A, 508, 1423
Wizinowich, P. L., Le Mignant, D., Bouchez, A. H., Campbell, R. D., Chin, J. C. Y., Contos, A. R., van Dam, M. A., Hartman, S. K., Johansson, E. M., Lafon, R. E., Lewis, H., Stomski, P. J., Summers, D. M., Brown, C. G., Danforth, P. M., Max, C. E., & Pennington, D. M. 2006, PASP, 118, 297
Zapatero Osorio, M. R., Martín, E. L., Bouy, H., Tata, R., Deshpande, R., & Wainscoat, R. J. 2006, ApJ, 647, 1405

Convective Core Overshoot and Mass Loss in Classical Cepheids: A Solution to the Mass Discrepancy?

Hilding R. Neilson,[1] Matteo Cantiello,[1] and Norbert Langer[1]

[1]*Argelander Institut für Astronomie, Universität Bonn*

Abstract. We explore the role of mass loss and convective core overshoot in the evolution of Classical Cepheids. Stellar evolution models are computed with a recipe for pulsation-driven mass loss and it is found that mass loss alone is unable to account for the long-standing Cepheid mass discrepancy. However, the combination of mass loss and moderate convective core overshooting does provide a solution, bringing the amount of convective core overshooting in Cepheids closer to that found for other stars.

1. Introduction

Classical Cepheids are powerful laboratories for understanding stellar astrophysics, yet an important unanswered question is what are the masses of Cepheids. Cepheid masses are determined using multiple different methods, including stellar evolution models matching measured effective temperatures and luminosities, stellar pulsation models matching pulsation periods and amplitudes, and from observations of binaries where one component is a Cepheid. However, these three methods do not agree such that masses based on stellar pulsation models tend to be smaller than masses based on stellar evolution models. Furthermore, dynamic masses are consistent with stellar pulsation masses (e.g. Evans et al. 2008).

This mass difference, called the Cepheid Mass Discrepancy, is a long-standing problem (Cox 1980) that is important not just for understanding Cepheids themselves but how stars evolve in general. Because dynamic masses agree with the pulsation masses but not the stellar evolution masses then there must be physics missing from the stellar evolution models leading to Cepheid masses being overestimated (Keller & Wood 2006; Keller 2008). By finding the underlying source of the mass discrepancy, we can constrain the evolution of massive stars in general, leading to a better understanding of their later phases, such as asymptotic giant branch stars, white dwarfs and supernovae.

Historically, the Cepheid Mass Discrepancy was found to be about 40% in general and about a factor of two for beat Cepheids (Cox 1980). However, Moskalik et al. (1992) reduced the mass discrepancy substantially when the OPAL opacities were included in stellar evolution models. This result was a tremendous step forward in resolving the mass discrepancy. However, Caputo et al. (2005) and Keller (2008) found that the mass discrepancy for Galactic Cepheids is still present, with typical values of 10 - 20%. Furthermore, Keller & Wood (2006) modeled stellar pulsation of Galactic, and Large and Small Magellenic Cloud Cepheids and found that the mass discrepancy is a function of metallicity, increasing with decreasing metallicity. The purpose of this work is to explore some of the physics used in stellar evolution calculations and test the impact of different input physics on the Cepheid mass discrepancy.

2. Possible Causes of the Mass Discrepancy

There have been a number of possible solutions suggested to resolve the mass discrepancy (see Bono et al. 2006, for more details). The four most probable causes are: missing opacity, rotational mixing, convective core overshooting (CCO), and mass loss.

Bono et al. (2006) argued that for opacity changes to account for the discrepancy, they would need to differ by a factor of two. The difference between the Opacity Project (Badnell et al. 2005) and previous opacities is at most 10%, suggesting that an opacity revision of a factor of two appears unlikely. Therefore new opacities would only have a small impact on the mass discrepancy, and thus missing opacity is not a plausible solution.

Rotational mixing is a second possible solution. A star that rotates rapidly during its main sequence evolution will mix hydrogen into the core. Thus its post main sequence helium core is more massive than if the same star evolved with negligible rotation. The more massive helium core would cause the star to be more luminous when the model crosses the Cepheid Instability Strip, and thus would predict a smaller mass for a Cepheid model with the same luminosity as a Cepheid model that does not include rotational mixing. Furthermore, rotational mixing will change the surface abundances of Cepheids as well as the structure of the post main sequence blue loop evolution where stars undergo their second and third crossings of the Instability Strip (Maeder & Meynet 2001). Rotational mixing is a possible solution to the mass discrepancy that should be explored further, however, it is beyond the scope of this work.

Keller (2008, and references therein) suggested convective core overshoot (CCO) as yet another possible solution. Convective core overshooting during the main-sequence evolution mixes hydrogen into the nuclear-burning region. Just like rotational mixing, this will lead to a more massive helium core and a more luminous Cepheid. The amount of CCO in stellar evolution models is often determined from the parameterization $\Lambda_c = \alpha_c H_P$, where H_P is the pressure scale height, α_c is a free parameter, and Λ_c is the distance above the convective region that overshooting penetrates. Keller & Wood (2006) and Keller (2008) found that to solve the mass discrepancy α_c must equal 0.5 - 1. However, this range of α_c is larger than that found for other stars.

Mass Loss occurs throughout the entire evolution of a star and hence suggests another possible solution to the mass discrepancy. However, mass loss during the main sequence and red giant stage of evolution of intermediate mass stars appears to negligibly contribute to the mass discrepancy (Lanz & Catala 1992; Willson 2000) with the exception of the Be stars. Neilson & Lester (2008, 2009) developed a prescription for pulsation-enhanced mass loss during the Cepheid stage of evolution and found that the predicted Cepheid mass-loss rates are large enough to contribute to the mass discrepancy. However, it is uncertain if mass loss completely solves the mass discrepancy. Further evidence of Cepheid mass loss includes observations of asymmetry of $H\alpha$ line profiles of Galactic Cepheids (Nardetto et al. 2008) as well as from modelling the infrared excess of Large Magellanic Cloud Cepheids (Neilson et al. 2010). However, Keller (2008) noted that mass loss needs to be as efficient for low-mass Cepheids as for large-mass Cepheids, which would be seem unlikely.

3. Methodology

We test if mass loss during the Cepheid stage of stellar evolution is a plausible solution to the Cepheid mass discrepancy, either on its own or when coupled with various values for convective core overshoot, α_c. We compute grids of stellar evolution models using a one-dimensional hydrodynamic stellar evolution code (Heger et al. 2000) for stars with initial mass $M = 4, 6, 8$, and $9\ M_\odot$. These models are evolved with four different sets of input parameters:

1. no pulsation-driven mass loss and no CCO,
2. pulsation-driven mass loss, using the prescription from Neilson & Lester (2008) and no CCO,
3. pulsation-driven mass loss and CCO, with $\alpha_c = 0.1$, and
4. pulsation-driven mass loss and CCO, with $\alpha_c = 0.335$, the value found by Brott et al. (in prep) (see Vink et al. 2010, for details).

We compute the contribution of mass loss to the mass discrepancy ($\Delta M/M$) by subtracting the initial mass of the stellar model from the mass at the end of Cepheid evolution, and divide by the initial mass. We calculate the contribution of CCO towards the mass discrepancy to be that $\alpha_c = 0.1$ leads to $\Delta M/M = 2.5\%$, $\alpha_c = 0.2$ is $\Delta M/M = 5\%$ and $\alpha_c = 0.335$ is $\Delta M/M = 8.375\%$ because of the change of luminosity due to increasing the amount of CCO. The pulsation mass is computed from the luminosity, period, and other pulsation quantities; therefore we need to include the contribution of CCO in our computed mass difference. We adopt the mass of the stellar model at the end of Cepheid evolution as equivalent to the pulsation mass that was found by Keller (2008) and the initial stellar mass with zero CCO is equivalent to the evolution mass based on the definition used by Caputo et al. (2005) and Keller (2008). Therefore, this measurement is a prediction of the maximum contribution of the chosen input physics on the Cepheid mass discrepancy. Note that we assume pulsation-driven mass loss occurs only on the Cepheid Instability Strip.

4. Results

The stellar evolution tracks for each case and mass are shown in Figure 1 and the computed contribution to the mass discrepancy is shown in Table 1. For the $\alpha_c = 0.335$ case, the $6, 8$, and $9\ M_\odot$ stars lose enough mass during the Cepheid stage of evolution to be consistent with the mass discrepancy determined by Keller (2008). Every stellar model undergoes blue loop evolution, where the width of the blue loop is determined by effective temperature range spanned, increases with initial mass implying that the stars undergo the Cepheid stage of evolution for each case. We do not model the radial pulsations but assume that the star is undergoing Cepheid evolution if the effective temperature and luminosity of the stellar model is inside the Cepheid Instability Strip.

Once a stellar model enters the Instability Strip, we use the Neilson & Lester (2008) prescription for mass loss where radial pulsation in the star generates shocks that enhance the wind. In this model, the mass-loss rate is determined by the stellar mass, luminosity, effective temperature, and pulsation period, where the period is computed from a Period-Mass-Radius relation (Gieren 1989). This relation has a predicted

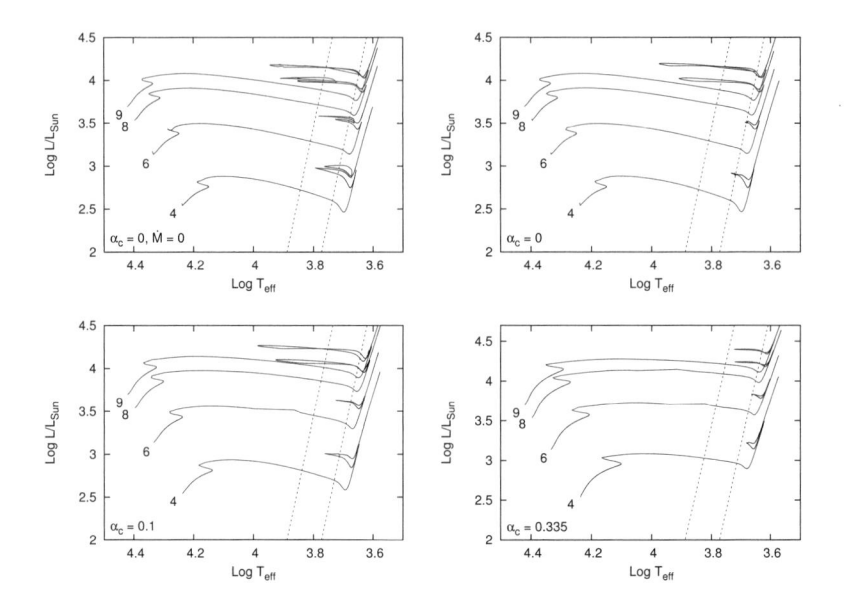

Figure 1. Stellar evolutionary tracks for models with initial mass $M = 4, 6, 8, 9\ M_\odot$ and for the four scenarios discussed in the text. (Upper Left) no pulsation-driven mass loss nor CCO, (Upper Right) with pulsation-driven mass loss and zero CCO, (Lower Left) with pulsation-driven mass loss and CCO, $\alpha_c = 0.1$, and (Lower Right) with pulsation-driven mass loss and CCO, $\alpha_c = 0.335$. The dashed lines are the borders of the Cepheid Instability Strip.

period error of approximately 25% which may lead to a mass-loss rate uncertainty of approximately a few hundred percent. Neilson & Lester (2008) found that mass-loss rates of Galactic Cepheids ranges from $10^{-11} - 10^{-7}\ M_\odot yr^{-1}$ and that mass-loss rates are potentially increasing with lower metallicity (Neilson & Lester 2009). We apply this prescription to stellar evolution models and explore how mass loss affects blue loop evolution.

Consider first the 8 $M_\odot$ stellar evolution models that are shown in Figure 2. The tracks for models with and without Cepheid mass loss and zero CCO have blue loops spanning identical ranges of effective temperature and luminosity suggesting that mass loss alone has little affect on the evolution of the star. When CCO is considered then the location of the blue loop shifts to larger luminosities. For example, the blue loop of the stellar evolution model with $\alpha_c = 0.1$ has a luminosity that is about 0.1 dex larger than the models without CCO while the model with $\alpha_c = 0.335$ is about 0.25 dex more luminous than the models without CCO.

However, it is not clear how mass loss changes with different amounts of CCO and thus how mass loss affects structure of the blue loop. For instance, the mass-loss rate depends on the ratio of luminosity and mass, L/M, which increases with α_c. On the other hand, the predicted mass-loss rate increases with decreasing pulsation period which in turn increases with stellar radius. The radius increases with luminosity and thus with increasing α_c. Thus a change of CCO leads to changes in fundamental

Table 1. The contribution of mass loss and CCO to the Cepheid Mass Discrepancy.

Initial Mass ($M_\odot$)	Case 1	Case 2	Case 3	Case 4
4	0%	6.25%	6.00%	8.72%
6	0%	3.00%	8.30%	18.54%
8	0%	1.75%	4.50%	16.24%
9	0%	1.67%	4.40%	15.00%

parameters that work against each other in determining the mass-loss rate. In Figure 2, the only instance where mass loss appears to affect the evolution of the star is for the case where $\alpha_c = 0.335$.

We explore this further in Figure 3, where we compare the predicted mass-loss rates for the 8 $M_\odot$ stellar evolution models for three cases where pulsation-driven mass loss is assumed for the second and third crossings of the Instability Strip; the first crossing is too short to contribute a significant mass change in the 8 ⊙ models. During the second crossing, the timescale is about twice as long for $\alpha_c = 0.335$ as opposed to $\alpha_c = 0$ and 0.1, yet there are little difference in mass-loss rates. However, during the third crossing, the $\alpha_c = 0.335$ model has a crossing time that is almost an order-of-magnitude longer than the cases with less CCO and has an average mass-loss rate that is a factor of 2-3 larger than for the other cases.

The predicted average mass-loss rate tends to increase with α_c when the star is evolving along the Instability Strip. For increasing CCO, then the ratio of the luminosity and mass, L/M, increases for Cepheids. Hence the amount of radiation pressure acting on a wind is larger and plays a bigger role in driving the wind than the change of pulsation period. This is consistent with the results of Neilson & Lester (2009). The change in timescale that the 8 $M_\odot$ stars spend on the Cepheid Instability Strip is also not directly due to the amount of CCO but on the mass-loss rate and its dependence on CCO. We hypothesize that the width of the blue loop decreases if the mass-loss rate is large, similar to the result found by Brunish & Willson (1987), thus "trapping" the star in the Cepheid Instability Strip. During the second crossing, a Cepheid's effective temperature is increasing and its radius is decreasing with time. However, mass loss acts against the contraction of the stellar envelope, and "puffs up" the envelope. The more inflated envelope prevents the effective temperature from increasing at the same rate as the cases with smaller CCO and mass loss.

We find that the 9 $M_\odot$ models appear to behave similarly as the 8 $M_\odot$ models. The width of the blue loop of the 9 $M_\odot$ models is smaller for the $\alpha_c = 0.335$ case and is also trapped in the Instability Strip. The stellar evolution tracks appear unchanged for smaller values of α_c. The 4 and 6 $M_\odot$ stellar models behave differently. The models with $\alpha_c = 0$ and zero pulsation-driven mass loss appear to undergo multiple blue loops and when pulsation mass loss is included the multiple loops disappear and the one blue loop has a smaller width for both models. The 4 and 6 $M_\odot$ blue loops are more sensitive to mass loss relative to the 8 and 9 $M_\odot$.

The 4 and 6 $M_\odot$ model blue loops are also sensitive to CCO. The blue loops for the $\alpha_c = 0.1$ models span a larger range of effective temperatures than the models with $\alpha_c = 0$. However, the $\alpha_c = 0.335$ models have a smaller blue loop, so much so that the 4 $M_\odot$ model blue loop does not enter the Cepheid Instability Strip. This behavior is associated with the minimum mass at which stellar evolution models can form a

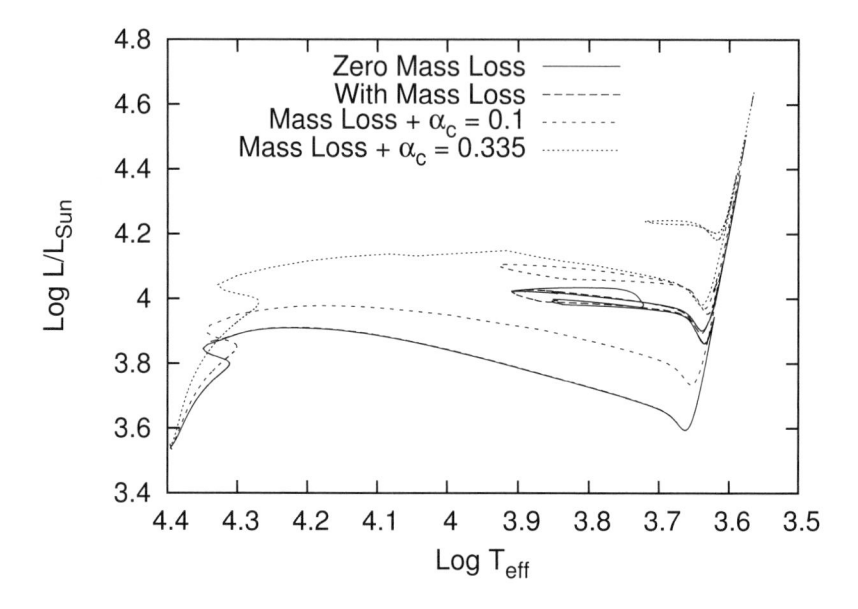

Figure 2. Comparison of the 8 $M_\odot$ stellar evolution tracks for the four scenarios explored in this work. The location and structure of the blue loops change due to the different input parameters.

blue loop. The minimum mass increases with increasing α_c, such that at $\alpha_c = 0.335$ the 4 $M_\odot$ model is approximately the minimum mass. The 6 $M_\odot$ model blue loop appears truncated at $\alpha_c = 0.335$ due to the combination of mass loss and convective core overshooting just like the 8 and 9 $M_\odot$ models.

5. Summary

The results of this work suggest that the Cepheid mass discrepancy can be resolved by the combination of pulsation-enhanced mass loss (Neilson & Lester 2008) on the Cepheid Instability Strip and convective core overshoot in the main sequence progenitors. On its own, CCO is not a plausible solution because of measurements of the amount of CCO in other stars. While this would require $\alpha_c > 0.5$ (Keller 2008), from eclipsing binaries, the value of $\alpha_c = 0.2 - 0.4$ (Clausen et al. 2010; Sandberg Lacy et al. 2010), from β Cephei stars $\alpha_c = 0.28 \pm 0.1$ (Lovekin & Goupil 2010) while from early B-type stars $\alpha_c = 0.335$ (Hunter et al. 2008; Vink et al. 2010). Furthermore, large values of α_c may act to suppress the blue loop evolution in the most massive Cepheids. Kippenhahn & Weigert (1990) noted that whether stars go through blue loop evolution depends on the mass and radius of the stellar core, the mass and change of hydrogen abundance of the hydrogen-burning shell. A more massive stellar helium core, such as that produced when CCO is included, acts to quench blue loop evolution.

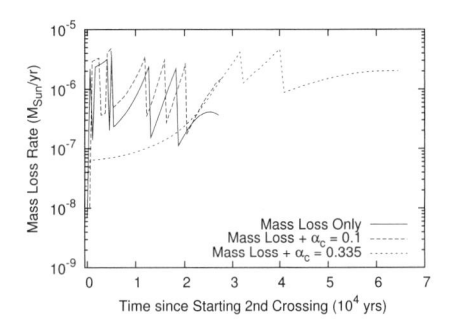

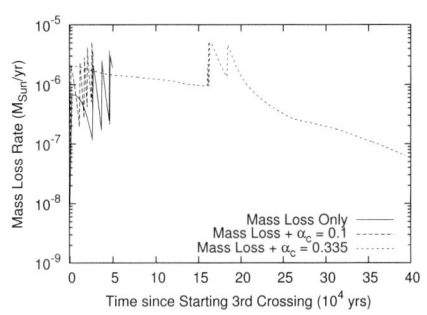

Figure 3. The pulsation-driven mass-loss rates predicted for the 8 $M_\odot$ models for the second and third crossings of the Cepheid Instability Strip and for the three scenarios that include pulsation-driven mass loss. The pulsation-driven mass-loss rate for the 8 $M_\odot$ during its first crossing is not shown because the change of mass is too small to contribute significantly.

Mass loss, on its own, also is not a plausible solution because the Cepheid crossing timescales are too short, and/or the mass-loss rates are too small. However, the combination of CCO, with $\alpha_c = 0.2 - 0.4$, and pulsation-driven mass loss together provide a plausible solution. This work is preliminary and we intend to explore a larger mass range up to $M = 15 M_\odot$ as well as Small and Large Magellanic Cloud metallicities to test the metallicity dependence of the mass discrepancy found by Keller & Wood (2006).

Acknowledgments. HRN is grateful for financial support from the Alexander von Humboldt Foundation.

References

Badnell, N. R., Bautista, M. A., Butler, K., Delahaye, F., Mendoza, C., Palmeri, P., Zeippen, C. J., & Seaton, M. J. 2005, MNRAS, 360, 458
Bono, G., Caputo, F., & Castellani, V. 2006, MmSAI, 77, 207
Brunish, W. M., & Willson, L. A. 1987, in Stellar Pulsation, edited by A. N. Cox, W. M. Sparks, & S. G. Starrfield, vol. 274 of Lecture Notes in Physics, Berlin Springer Verlag, 27
Caputo, F., Bono, G., Fiorentino, G., Marconi, M., & Musella, I. 2005, ApJ, 629, 1021. arXiv: astro-ph/0505149
Clausen, J. V., Frandsen, S., Bruntt, H., Olsen, E. H., Helt, B. E., Gregersen, K., Juncher, D., & Krogstrup, P. 2010, A&A, 516, A42+
Cox, A. N. 1980, ARA&A, 18, 15
Evans, N. R., Schaefer, G. H., Bond, H. E., Bono, G., Karovska, M., Nelan, E., Sasselov, D., & Mason, B. D. 2008, AJ, 136, 1137
Gieren, W. P. 1989, A&A, 225, 381
Heger, A., Langer, N., & Woosley, S. E. 2000, ApJ, 528, 368
Hunter, I., Brott, I., Lennon, D. J., Langer, N., Dufton, P. L., Trundle, C., Smartt, S. J., de Koter, A., Evans, C. J., & Ryans, R. S. I. 2008, ApJ, 676, L29. 0711.2267
Keller, S. C. 2008, ApJ, 677, 483
Keller, S. C., & Wood, P. R. 2006, ApJ, 642, 834
Kippenhahn, R., & Weigert, A. 1990, Stellar Structure and Evolution
Lanz, T., & Catala, C. 1992, A&A, 257, 663

Lovekin, C. C., & Goupil, M. 2010, A&A, 515, A58+
Maeder, A., & Meynet, G. 2001, A&A, 373, 555
Moskalik, P., Buchler, J. R., & Marom, A. 1992, ApJ, 385, 685
Nardetto, N., Groh, J. H., Kraus, S., Millour, F., & Gillet, D. 2008, A&A, 489, 1263
Neilson, H. R., & Lester, J. B. 2008, ApJ, 684, 569
— 2009, ApJ, 690, 1829
Neilson, H. R., Ngeow, C., Kanbur, S. M., & Lester, J. B. 2010, ApJ, 716, 1136
Sandberg Lacy, C. H., Torres, G., Claret, A., Charbonneau, D., O'Donovan, F. T., & Mandushev, G. 2010, AJ, 139, 2347
Vink, J. S., Brott, I., Gräfener, G., Langer, N., de Koter, A., & Lennon, D. J. 2010, A&A, 512, L7+
Willson, L. A. 2000, ARA&A, 38, 573

Cool Stars 16 banquet at Tillicum Village

Part III

Time Domain

The 16th Cambridge Workshop on Cool Stars, Stellar Systems and the Sun
ASP Conference Series, Vol. 448
Christopher M. Johns-Krull, Matthew K. Browning, and Andrew A. West, eds.
© 2011 Astronomical Society of the Pacific

Asteroseismology of Cool Dwarfs and Giants with Kepler

Ronald L. Gilliland

Space Telescope Science Institute, 3700 San Martin Drive, Baltimore, MD 21218

Abstract. The primary science goal of the *Kepler Mission* is to detect planets in the habitable zones of their host stars where liquid water could exist on the planet surface, and to determine the intrinsic frequency of these and other exoplanets. The technique employed by *Kepler* is to search for shallow transits, that in the case of a true Earth analog would be 85 parts per million deep, lasting an average of 10 hours, and happening once per year. This combination of small, short, and rare events drives mission development and operations to support ultra-high precision photometry on one field of 150,000 largely solar-like stars with nearly continuous exposures covering at least 3.5 years. Since the transit depth returns only the size of a candidate planet relative to the size of its host star, fulfillment of an additional *Kepler* science goal of determining physical properties of the discovered planets requires us to also determine stellar properties, with stellar radius being the most important. For the latter asteroseismology is a particularly important tool that allows stellar radii to be determined to ~1% in favorable cases.

The *Kepler* asteroseismology program is organized into a large international collaboration – the *Kepler Asteroseismic Science Consortium* consisting of a dozen working groups and some 300 members. The observations to date in the area of cool giant stars consist of 30 minute integrations for some 1,200 stars spanning a full year. For cool dwarfs the observations to date using 1 minute integrations aimed at asteroseismology consist of three types: (1) A few thousand stars observed for one month each in a "survey" mode to identify the best prospects for lengthier observations. (2) About 200 stars hosting planet candidates observed for periods now reaching up to a year for which asteroseismology is desired whenever possible. (3) A set of about 100 stars for which much more extended observations are possible that have been selected purely on the basis of asteroseismic interest.

The red giants have typical magnitudes of about 12th, with a couple of hundred selected at brighter magnitudes. In nearly all cases the astrophysical signatures of oscillations and granulation dominate noise sources.

Cool dwarfs studied for asteroseismology range from 16 Cyg A and B near 6th magnitude, to planet host stars of 13th magnitude for which extensive observations may allow detection of at least the asteroseismic large separation to constrain the stellar mean density and hence radius.

I review the target selections, the instrumental capabilities relevant for cool-star asteroseismology, and the basic approaches and promises of asteroseismology. Emphasis is given to reviewing the extensive new returns from the *Kepler Mission* that are providing new insights into cool dwarfs and giants from asteroseismic analyses.

1. *Kepler Mission* Goals, Team, Role of KASC

Kepler's multi-year mission is to go where no mission has gone before and discover strange new worlds like our own orbiting distant stars. The necessity of understanding the host stars of planets provides the rationale for including asteroseismology as a prime mission goal with *Kepler*. Simple analysis of the transit light curves provided by *Kepler* yields R_p/R_*, the relative radii of planet and star. With a robust means of setting R_* which asteroseismology provides to highest precision we then know the planet radius for cases allowing detection and study of stellar oscillations. More generally asteroseismology returns the most robust information available for obtaining secure knowledge of host star parameters allowing the *Kepler* planet discoveries to be placed in proper context.

Of greater relevance for the Cool Star community, *Kepler* has an extensive set of observations conducted for the express purpose of asteroseismology on a broad range of stellar types, with cool dwarfs and giants forming the core of these investigations. Asteroseismology with *Kepler* is conducted through the Kepler Asteroseismic Science Consortium (KASC) led by this author, Jørgen Christensen-Dalsgaard, Hans Kjeldsen, and Steven Kawaler. KASC is divided into a total of 13 working groups, now comprising over 300 individual members. Of most relevance to Cool Stars are the Solar-like p-mode oscillations group led by Bill Chaplin, Oscillation in Clusters, led by Sarbani Basu, and Red Giants led by Joris De Ridder. KASC is a very open collaboration, new members are welcome to join.

2. Spacecraft, Observations, Asteroseismology Target Selection

The hardware, operations, and observation planning for *Kepler* are much simpler than for many space missions, since the science goals are best served by simple, repetitive observations continuing for years. Koch, D. et al. (2010) provides a detailed overview of the mission design, performance, and outline of early science results across the planet and asteroseismology programs. Asteroseismology is introduced in the review paper by Gilliland, R.L. et al. (2010a).

2.1. Spacecraft

The *Kepler* spacecraft consists of a 1.4 m primary mirror and 0.95 m Schmidt corrector feeding a focal plane array of 42 CCDs each of 1024x2200 27 μ pixels. A unique aspect of the *Kepler* CCDs are full well depths of some 1.2×10^6 e- supporting the collection of large numbers of electrons. The point spread function delivered to the CCDs is quite sharp, with a FWHM before detector sampling that is only 1/3 of a pixel. This results in undersampled PSFs, with up to 64% of the total flux captured by the central pixel for stars centered on a pixel on the CCDs with sharpest PSFs. A motivation for using such sharp PSFs on a mission with a goal of returning ultra-precise photometry is to minimize the area subtended by stellar apertures to minimize the numbers of false positives for the transit detection program arising from blended background eclipsing binaries. Since the *Kepler* spacecraft is a compact design, in an Earth-trailing orbit, and maintains pointing at one field (except for monthly breaks to telemeter data back, and quarterly rolls to keep the solar panels aligned with the Sun) the guiding is exquisitely steady with line of sight jitter motion at the milli-pixel level supporting excellent time-series relative photometry even with the sharp PSFs.

2.2. Observations

Kepler has only two observing modes: long cadence consisting of 270 on-board summations of underlying 6 second exposures, and short cadence consisting of 9 such summations. The LC interval of 29.4 minutes is used for the bulk of planet program observations of some 156,000 targets (Jenkins, J.M. et al. 2010) as well as asteroseismic targets like red giants where this sampling suffices. The SC interval of 58.8 seconds is used for solar-like oscillations in dwarf stars (Gilliland, R.L. et al. 2010b). Although saturation of the CCD occurs for stars with $Kp \sim 11.5$, the CCD gains used support full sampling through saturation with the A/D and photometry on strongly saturated stars is very successful even as bright as 16 Cyg A & B near 6th magnitude. Asteroseismology on solar-like stars typically needs noise levels in power spectra near 1 ppm, this can be reached in one month near $Kp = 11.3$.

2.3. Asteroseismology Target Selection

Asteroseismic targets for *Kepler* are drawn from several categories. A set of 1,000 red giants just fainter than saturation ($Kp = 11.5 - 12.0$) and evenly distributed over the focal plane serve as controls for the mission astrometry goals and are shared with KASC. KASC is allowed to select an additional about 1,000 targets (red giants, RR Lyraes, etc.) for which LC data suffice. Since only 512 targets can be observed at SC at any time, and this mode is required for solar-like oscillations SC target selection is the most critical and competitive. Early in the mission KASC was allowed to use most of the 512 target slots and chose to rotate through a fresh set once per month which has ultimately led to a survey of the brightest ~3,000 targets in the field of view for which detection of oscillations was deemed feasible in one month.

After the first year of operations the target selection for solar-like stars has been adjusted in two primary directions, both intended to provide extensive, unique multi-year coverage of many targets for which successful asteroseismology is expected. From the large number of planet candidates the 100 for which the host star is most likely to support asteroseismic detections have been moved to Short Cadence observations. From the KASC survey program a comparably large number of about 100 (usually much brighter than the planet hosts) stars deemed particularly exciting on the basis of the survey results have been selected for long term monitoring. A small number of short cadence slots are also available to the annual call for General Observer program proposals.

3. Solar-like Oscillations

Since most small planets can be detected via transits only for dwarf host stars, the synergy between the asteroseismology program and the planet program occurs primarily in this realm. Asteroseismology has been used to refine knowledge of planets through improving knowledge of the host star, has demonstrated the promise of returning precise ages, and promises detailed population synthesis studies enabled by having precise stellar parameters known for order 1,000 stars.

The oscillation modes which can be studied with photometric time series consist of low-degree, high overtone p-modes which are well approximated by a regular series of peaks for which the oscillation frequencies are given by the approximate asymptotic relation:

$$\nu_{nl} \approx \Delta\nu_0(n + l/2 + \epsilon) - D_0 l(l+1) \quad (1)$$

where $\Delta\nu_0 = (2\int_0^R dr/c)^{-1}$ corresponds to the inverse of the sound travel time across the stellar diameter, and closely relates to the stellar mean density via:

$$\Delta\nu_0 \approx 135(M_*/R_*^3)^{1/2} \,\mu\text{Hz} \quad (2)$$

where M_* and R_* are the stellar mass and radius in solar units, and the large separation for the Sun is approximately 135 μHz. In this notation l is the number of zero crossings in the spherical harmonic expansions of the modes along a stellar circumference, and n the number of crossings in the radial direction defining the mode order. This results in a very characteristic signal of solar-like oscillations – near-equally spaced peaks in a power spectrum of oscillation strength plotted against frequency.

3.1. HAT-P-7

HAT-P-7 was a previously detected planet host (Pál, A. et al. 2008) in the *Kepler* field of view and served as an early example (Christensen-Dalsgaard, J. et al. 2010) of the promise of asteroseismology in refining exoplanet properties through improved knowledge of the host star. At a *Kepler* magnitude of 10.46, a T_{eff} near 6500 K, and a radius about twice solar this star provides an excellent opportunity for asteroseismology. In the first 45 days of data a robust detection of some 33 individual mode frequencies extending over about 800 - 1500 μHz (periods of 10 - 20 minutes) was possible. The power spectrum for HAT-P-7 with *Kepler* is shown in Fig. 1. Comparing these frequencies with the asymptotic formula allowed individual mode identification. And comparison with detailed stellar evolution model results and eigenfrequency analyses supported determination of $M = 1.52 \pm 0.04 \, M_\odot$, $R = 1.99 \pm 0.02 \, R_\odot$, and an age of 2.1 ± 0.3 Gyr.

HAT-P-7 was a well observed target from the ground before *Kepler*. Detailed spectroscopic studies had been conducted, as well as utilization of the transit light curve to constrain stellar host parameters (light curve shape during transit can be used to determine the mean stellar density – Seager & Mullén-Ornelas (2003)). From the ground based studies, coupled with radial velocity measurements to determine the planet mass the mean planet density was known to ~50%. With the order of magnitude improvement in basic stellar parameters resulting from *Kepler* asteroseismology the planetary mean density improves to ~5%. This is the level of planet knowledge needed to enable detailed considerations of planet composition and formation.

3.2. KIC 11026764

Another early target for detailed asteroseismic study was the slightly evolved G dwarf KIC 11026764 at $Kp = 9.3$. As with HAT-P-7 the first 45 days of *Kepler* observations allowed a robust detection (Chaplin, W.J. et al. 2010) of p-modes in the 700 - 1100 μHz range. Figure 2 contrasts this star (bottom panel) with two other less evolved G dwarf power spectra. Of particular interest in this case is presence of one $l = 1$ mode frequency that does not fit well the simple asymptotic relation given in Eq. (1). The deviation of the observed p-mode frequency from the simple relation results from an "avoided crossing" of eigenfunctions for the p-modes in the outer layers of the star, with eigenfunctions of g-modes unstable in the stellar core. As evolution proceeds the coincidence of p- and g- mode intrinsic frequencies can "bump" the observed p-mode

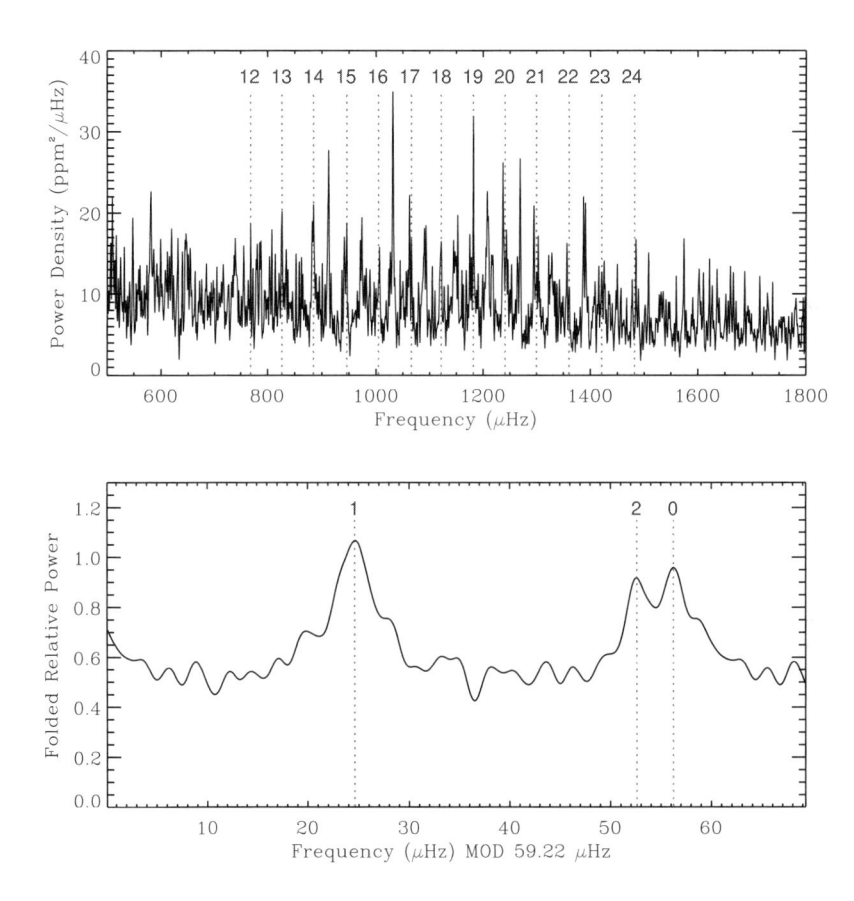

Figure 1. Upper panel shows the power spectrum for HAT-P-7 as observed by *Kepler* with mode frequencies evenly separated as expected by Eq. (1), and n values for radial modes indicated. The lower panel shows the power spectrum after folding by the large separation, averaging and normalizing to a maximum of unity. The characteristic single peak corresponding to radial degree, $l = 1$ modes and a doubled peak from $l = 2,0$ are clearly evident. (Courtesy of J. Christensen-Dalsgaard from Gilliland, R.L. et al. (2010a).)

frequency by an easily detectable amount. Furthermore the precise p-mode degree suffering this perturbation is an exquisitely sensitive function of the stellar evolutionary state. Metcalfe, T.S. et al. (2010) have shown that the age of KIC 11026764 can be determined to a precision near 1%. (Two solution branches with slightly different masses support equally good matches to the current observations; these result in differences of 15% for the age. Continued observation resulting in detection of higher frequency modes may well be able to break this degeneracy bringing the "large" systematic error down to the precision level achieved.)

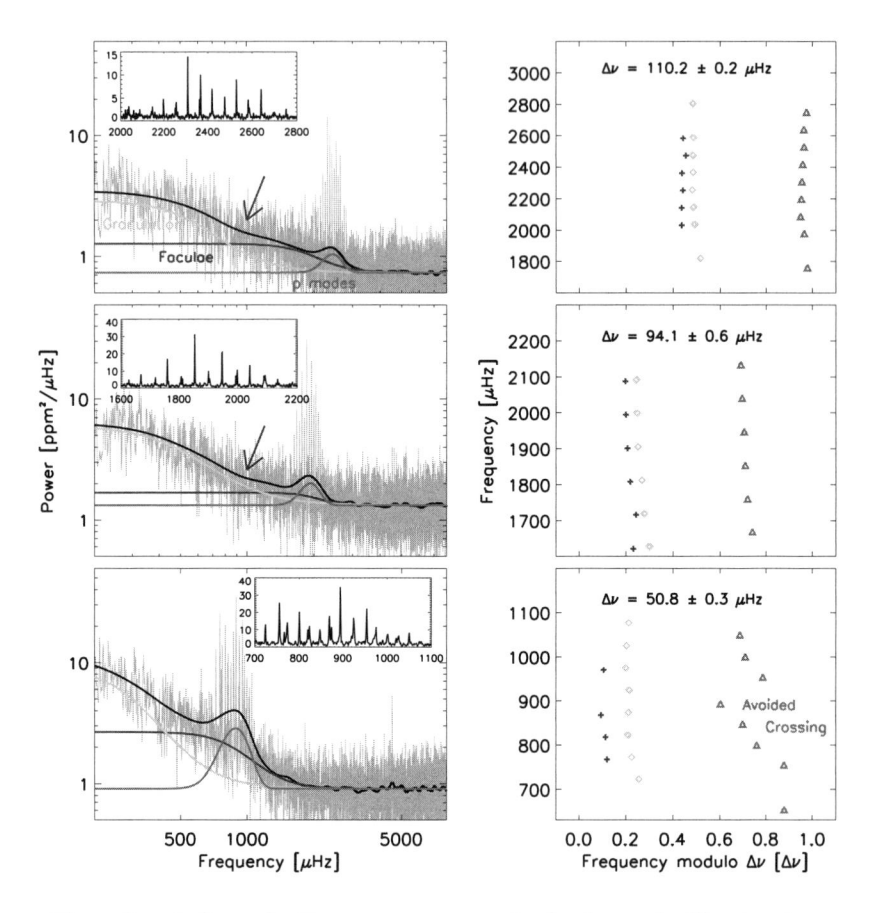

Figure 2. *Left panels:* Frequency-power spectra for three solar-like stars (*gray* over 200-8000 μHz. *Black lines* are from heavy smoothing of the spectra. Estimates of the underlying power spectral density contributions of p modes, bright faculae, and granulation as labeled in the top left panel are also shown. At higher frequencies the flat underlying contribution from photon shot noise is visible. The arrows mark a kink in the background power that is caused by the flattening toward lower frequencies of the facular component. The most prominent modes are shown in the insets. *Right panels:* So-called échelle plots of individual mode frequencies. Individual oscillation frequencies have been plotted against the frequencies modulo the average large frequency spacings (with the abscissa scaled to units of the large spacing of each star). The frequencies align in three vertical ridges that correspond to radial modes ($l = 0$, diamonds), dipole modes ($l = 1$, triangles) and quadrupole modes ($l = 2$, crosses). KIC 11026764 appears at the bottom – see text for discussion of the "Avoided Crossing". (Courtesy of W.J. Chaplin from Gilliland, R.L. et al. (2010a).)

3.3. Survey Program

Probably the most stunning achievement to date from *Kepler* in the realm of asteroseismology is simply in the number of successful detections, and the promise this holds for stellar astrophysics of cool stars. Despite being a primary goal for decades, the number of stars with asteroseismic detections before *Kepler* numbered only about 20. From the ground it has been necessary to dedicate some 7 - 10 straight nights of observations with a large (4-m class) telescope and a state of the art spectrograph for the resulting radial velocities (precisions of about 1 m/s are needed, with a sampling rate of once per minute for solar analogs even when combining data from many dedicated nights) to provide detections. Often ultimate success has only followed after organizing campaigns involving telescopes distributed in longitude to minimize aliases in power spectra resulting from the diurnal sampling of a single site. More recently CoRoT (Michel, E. et al. 2008) has provided excellent results, but only for a handful of actual near solar stars. Within the first year the number of detections, often of superior quality with *Kepler* has reached 500. The asteroseismology of solar-like stars has moved from a field starved for results, to one in which there is a momentary embarrassment of riches in which the community is now faced with the challenge of utilizing these exquisite data.

Primary applications are expected to include: (1) Detailed comparisons of eigenfrequencies with those from stellar evolution models will allow determination of stellar parameters such as He abundance and convection zone depth. (2) Similar comparisons will allow challenging assumptions made in generating stellar structure models, e.g. limits on mixing and convective overshoot. (3) Accurate knowledge of basic parameters for mass, radius and luminosity for order 1,000 stars will support detailed comparison with predictions of galactic population synthesis models.

4. Red Giants

Only a few years ago it was still a matter of discussion whether red giants would support coherent p-mode oscillations with a sufficiently long lifetime for good asteroseismology. Many ground-based attempts met with limited success despite large dedicated observing campaigns. CoRoT firmly answered this matter in the affirmative with successful detections of oscillations on several hundred large red giants (see, e.g. De Ridder, J. et al. (2009)). Mode lifetimes can be 10's to even 100 days with periods for the oscillations of hours to tens of hours allowing precise determination of individual frequencies. *Kepler* data are dramatically extending progress in obtaining and interpreting oscillations in red giants.

4.1. Clusters

With its large field of view there are serendipitously 4 open clusters (NGC 6811, NGC 6866, NGC 6819 and NGC 6791) observable by *Kepler*. Clusters provide particularly attractive targets for asteroseismology since one may assume that all of the stars have the same age and initial compositions. And the age and composition have usually been well fixed through matching CMDs to stellar evolution models, and detailed spectroscopic observations respectively. This leaves the stellar mass as the primary parameter leading to differences in individual cluster members, and through the detection of many individual modes of oscillation for each of many cluster members it should be possible to reach a point at which many more observables exist, than model parameters. In such a scenario one may hope to challenge fine details of stellar structure and evolution associated for instance with the treatment of convection and mixing processes.

Stello, D. et al. (2010) present results for NGC 6819 for which detections have been obtained for some 30 red giants along the red giant branch spanning two orders of magnitude in luminosity. As expected from Eq. (2) the oscillation frequencies are

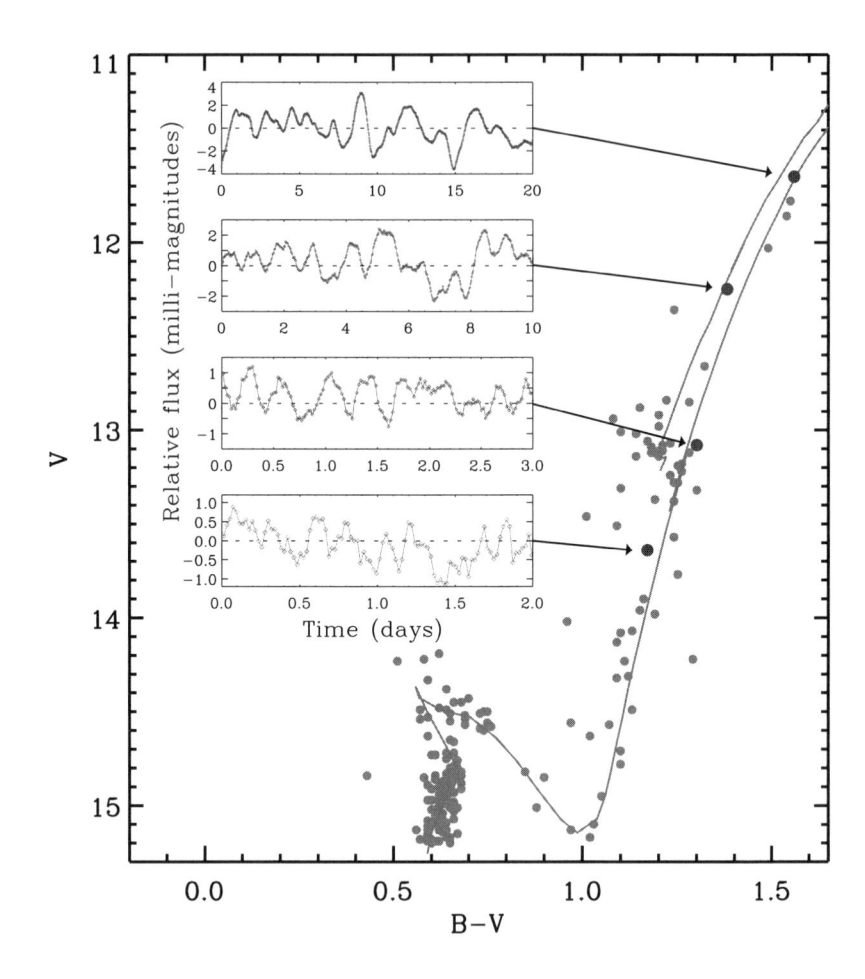

Figure 3. Color-magnitude diagram for NGC 6819. The grey points show stars that have high radial-velocity probabilities from Hole et al. (2009). The curve (red in on-line version) is a solar metallicity, 2.5 Gyr isochrone of Marigo et al. (2008). The larger, dark-grey (blue) points are four of the *Kepler* targets; we show the time series of their brightness fluctuation in the insets. As can be seen from the inset, the timescale of the oscillations decreases with increasing apparent magnitude of the stars. (Courtesy of D. Stello in Gilliland, R.L. et al. (2010a).)

smaller (timescales longer) for the oscillations for more evolved (larger) giants. Fig. 3 shows the NGC 6819 CMD along with sample *Kepler* time series for Red Giants at different evolutionary stages. From analysis of the first 33 days of *Kepler* data the oscillations proved particularly useful to test cluster membership, since the oscillation properties follow a tight sequence in frequency and amplitude along the red giant branch. Interlopers that happen to have the same V, $B - V$ in the CMD, are unlikely to also happen to have oscillations matching the cluster properties.

Future applications utilizing the many cluster giants will be particularly important for refining dependencies on age, mass and composition.

4.2. Universal Nature

When the individual mode frequencies for red giants are shown with reduced frequencies (i.e. scaled by the large separation) along the x-axis as shown in Fig. 4 (Bedding, T.R. et al. 2010) the sequences form nearly vertical ridges. This universal nature of the red giant frequencies considerably eases the task of mode identification needed for detailed interpretations.

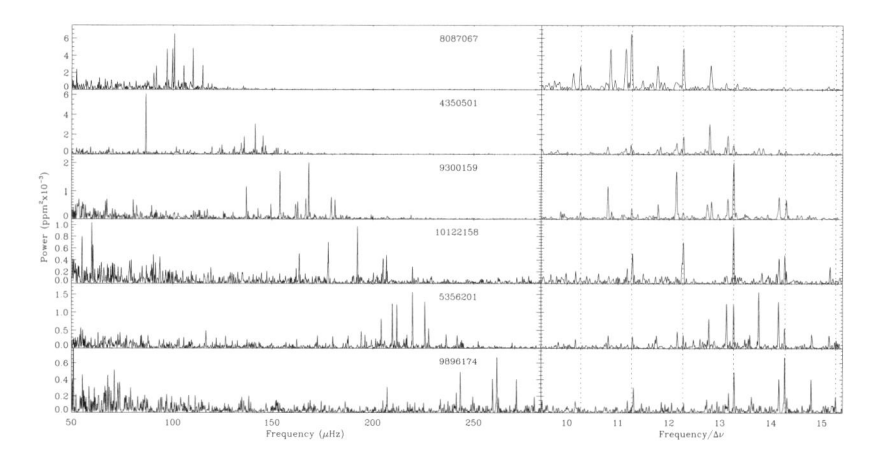

Figure 4. *Left:* distribution of power spectra for six representative low-luminosity red giants. *Right:* the same, but plotted after scaling by the large separation $\Delta \nu$. $l = 0$ modes are flagged by the equally spaced dotted lines. (Courtesy of Bedding, T.R. et al. (2010).)

4.3. Mode Lifetimes

Huber, D. et al. (2010) have found that the line widths of modes in power spectra from the first 138 days of *Kepler* data show evidence for lifetimes of $l = 1$ modes that vary from about 50 days for relatively large red giants, to only about 8 days for giants near the base of the red giant branch with radii of perhaps five times the Sun. The $l = 0$, radial modes, show the longest lifetimes of typically 50 - 100 days (good measurement will thus require use of longer observation sets which are only now becoming available), while $l = 2$ have lifetimes of 20-50 days. These long lifetimes are sufficient for the red giants to show the distinct "picket fence" distribution of coherent oscillations as predicted by the asymptotic equation (Eq. 1). Dupret, M.-A. et al. (2009) had provided predictions for mode characteristics based on detailed theory which are in surprisingly good agreement with the observations, thus providing a sound basis for utilization of the newly abundant data to challenge details of our understanding.

4.4. Population Synthesis

With stellar parameters of M, R, T_{eff} and L well determined for hundreds of red giants it is possible to make comparisons with expectations from models of galactic stellar distributions. A first application for Red Giants was by Mosser, B. et al. (2010) using CoRoT data for which accurate parameters were determined for about 930 stars. *Kepler* is returning similarly precise knowledge for a factor of twenty larger ensemble of giants promising strong constraints on star formation history along the *Kepler* sightline.

5. Conclusions

Asteroseismology is proving to be a useful tool in support of the planetary program in providing more accurate stellar parameters with radii to ~1% often possible, as well as uniquely accurate determination of stellar ages. The field of asteroseismology has moved from being starved for secure observational detections to having a momentary embarrassment of riches. After a couple of decades in which asteroseismology stubbornly remained largely a future promise, the present and future are bright. Returns are expected that will significantly boost our understanding of cool stars in most of the theme areas of Cool Stars. The data returned by *Kepler* in support of asteroseismology are providing advances far beyond anything previously available.

Acknowledgments. I would like to thank the original members of the *Kepler* asteroseismology program, Tim Brown, Jørgen Christensen-Dalsgaard and Hans Kjeldsen whose extensive contributions have made possible the impressive array of *Kepler* asteroseismology results. I also acknowledge the entire *Kepler* science team for their efforts that have made the mission possible. Funding for this Discovery mission is provided by NASA's Science Mission Directorate.

References

Bedding, T.R. et al. 2010, ApJL, 713, 176
Chaplin, W.J. et al. 2010, ApJL, 713, 169
Christensen-Dalsgaard, J. et al. 2010, ApJL, 713, 164
De Ridder, J. et al. 2009, Nature, 459, 398
Dupret, M.-A. et al. 2009, A&A, 506, 57
Gilliland, R.L. et al. 2010a, PASP, 122, 131
— 2010b, ApJL, 713, 160
Hole, K., Geller, A., Mathieu, R., Platais, I., Meibom, S., & Latham, D. 2009, AJ, 138, 159
Huber, D. et al. 2010, ApJ, 723, 1607
Jenkins, J.M. et al. 2010, ApJL, 713, 87
Koch, D. et al. 2010, ApJL, 713, 79
Marigo, P., Girardi, L., Bressan, A., Groenewegen, M., Silva, L., & Granato, G. 2008, A&A, 482, 883
Metcalfe, T.S. et al. 2010, ApJ, 723, 1583
Michel, E. et al. 2008, Science, 322, 558
Mosser, B. et al. 2010, A&A, 517, 22
Pál, A. et al. 2008, ApJ, 680, 1450
Seager, S., & Mullén-Ornelas, G. 2003, ApJ, 585, 1038
Stello, D. et al. 2010, ApJL, 713, 182

Starspots and Stellar Rotation: Stellar Activity with *Kepler*

Lucianne M. Walkowicz and Gibor S. Basri

Astronomy Department, UC Berkeley, 601 Campbell Hall, Berkeley CA 94720 USA

Abstract. While the telescopic study of starspots dates back to Galileos observations of our own Sun, recent space-borne photometric missions (such as MOST, CoRoT, and *Kepler*) are opening a new window into understanding these ubiquitous manifestations of stellar activity. Because of the intimate link between stellar rotation and the generation of the magnetic eld, starspots cause a modulation in the lightcurve at the rate of stellar rotation. To complicate matters, stars rotate differentially, so the stellar rotation rate is not really best characterized by a single value but rather by a range of rotation rates. Through high- precision, long-term photometric monitoring of stars of different spectral types and activity strengths, it is possible to determine stellar rotation rates and differential rotation measures. In addition, modeling these lightcurves can tell us about the properties of stellar spots, such as location, areal coverage, and lifetime. New observations provide precision photometry for a large cohort of stars, ranging from Sun-like to rather different stellar properties, at a spread of ages, making these lightcurves a powerful tool for understanding magnetic activity for stars of all activity levels. Here, I will discuss how *Kepler* can provide new insight into the continuum of stellar activity and our own Suns place amongst the stars.

1. Introduction

A new era of stellar astrophysics exploration has begun, enabled by a slew of space-based precision photometry missions: MOST (Walker et al. 2003), CoRoT (Baglin 2003), and most recently, *Kepler* (Koch et al. 2010). Although its primary purpose is the discovery of transiting exoplanets, the high time cadence, precise photometry of *Kepler* makes it an excellent source of data for detailed study of stellar atmospheres. Here, I focus in particular on the early results on stellar activity from the first thirty days of *Kepler* data, known as *Kepler* Quarter 1.

Stellar magnetic activity leads to a plethora of observable effects, from star spots that modulate the stellar brightness on the order of weeks, to flares, which release highly energetic radiation over a few hours. In solar-type stars (late-F, G, K and early-M dwarfs), the stellar magnetic field operates via the alpha-omega dynamo, driven by rotational shearing at the boundary between the radiative zone and convective zone. The magnetic field generation in stars is therefore intimately linked with stellar rotation (Pizzolato et al. 2003). For this reason, most photometric variability due to starspots and active regions occurs near the rotational timescale of the star, and provides a natural way to trace its rotation period. Different magnetic features dominate on different timescales: starspots are visible throughout their passage across the stellar disk, but are most visible at disk center, while bright faculae are most visible while crossing the stellar limb, but are a minimal effect at disk center. By measuring rotation periods for a large sample of stars, we can better constrain the relationship of rotation, activity, and age, as well as inform models of the magnetic field production.

2. Quantifying Variability

The first step in our analysis of the *Kepler* lightcurves was to develop a quantitative way to characterize the variability of the target stars and relate their variability to their physical properties where possible. Stellar physical parameters, such as gravity and effective temperature, are available from the Kepler Input Catalogue (Batalha et al. 2010) for most of the stars in the *Kepler* exoplanet search sample. The variability of the target stars is measured from the lightcurves themselves: the *Kepler* long cadence data comprise photometry for $\sim 150,000$ stars measured at a 30 minute cadence. With such a large dataset, however, it is inefficient to examine each lightcurve by eye to determine its variability characteristics. We therefore developed a set of metrics that can be calculated from the lightcurves in an automated way (Basri et al. 2010b,a). In the results I show here, I primarily use the variability range, or the amplitude between the 5th and 95th percentile of intensity in the lightcurve (which helps to avoid large positive or negative excursions, therefore better characterizing the basic level of photometric variability). In the following discussion, I also refer to the number of "zero crossings", the number of time a differential lightcurve crosses zero. The number of zero crossings is essentially a proxy for the characteristic timescale of variability, where slowly varying lightcurves only cross their midpoint perhaps a few times in 30 days, but quickly varying things cross many times.

Along with the lightcurve variability characteristics and stellar properties, we also compute Lomb-Scargle periodograms for the targets in our sample (Horne & Baliunas 1986). For each lightcurve we store the location and power in the top 20 strongest peaks, as well as various metadata regarding the periodograms– for example, the number of significant peaks, the location and power of the strongest peaks, etc. The periodograms for many stars are quite complex, consisting not of a single strong signal but of multiple signals. In addition, one must bear in mind that the strongest peak may not be the true period of the star but rather an alias– for example, for stars with symmetric spots on either hemisphere, the periodogram may identify the half period rather than the true period of the star. The half period may also be found when one searches a lightcurve that covers less than a full period in duration. Having longer lightcurves improves this problem in one sense and makes it more complicated in another: longer lightcurves are subject to spot evolution, dissipation and reemergence, as well as the effects of differential rotation, and may therefore have power on a multitude of timescales.

3. The Sun Among Stars

Our initial work with the *Kepler* Quarter 1 lightcurves dealt with the variability of the *Kepler* target stars in comparison with the total solar irradiance from the 2001 active Sun, and was originally reported in (Basri et al. 2010b). Active Sun lightcurves were constructed using the SOHO Virgo Instrument g+r lightcurves, which combined form a reasonable approximation to the *Kepler* white light bandpass. We then subdivided the SOHO lightcurve into ~ 30 day long chunks sampled at the *Kepler* 30 minute cadence, such that they were reasonable analogs to the *Kepler* Quarter 1 observations of other stars. We then calculated our variability characteristics on these chunks, and compared them to the *Kepler* data.

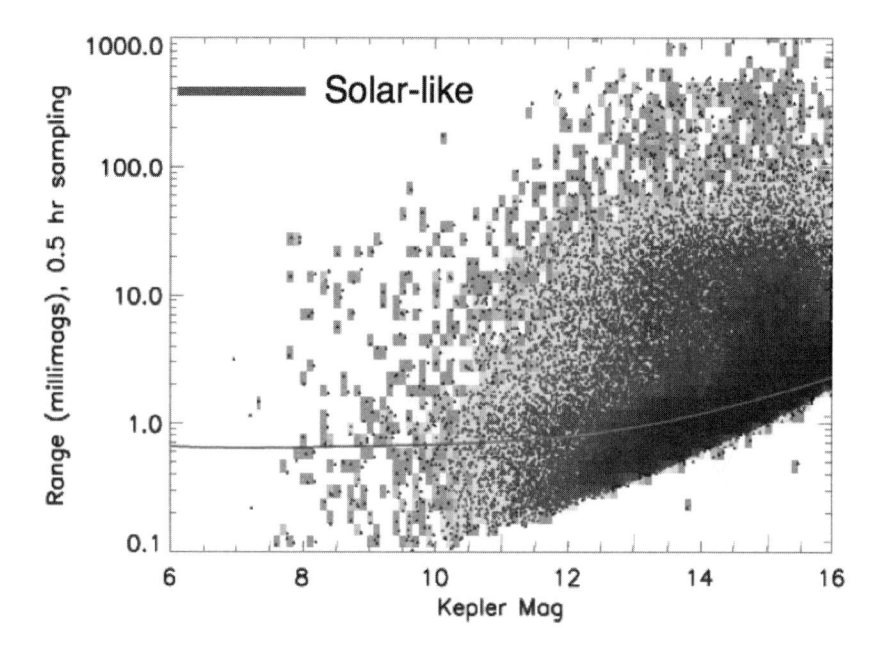

Figure 1. The distribution of variability range as a function of *Kepler* magnitude for the *Kepler* dwarf stars. The red line shows the variability range of the 2001 active Sun as determined from the SOHO lightcurves. Figure adapted from Basri et al. (2010b).

Through an extensive by-eye examination, we compared the SOHO active Sun lightcurves against lightcurves of *Kepler* dwarfs, and determined the locus of stars showing similar variability to the active Sun. We then used this locus as a dividing point of comparison to determine which stars were as or less variable than the active Sun, and which were more variable than the active Sun. In Figure 1, we show the distribution of variability range in the *Kepler* targets, with the locus of the active Sun indicated in red. Of the total sample of *Kepler* target stars, we found that ~46% of the *Kepler* dwarfs were more variable than the active Sun, in keeping with previous results from smaller surveys of solar-type stars (Batalha et al. 2002; Lockwood et al. 2007). Of these, only ~18% were twice as variable as the active Sun. This result is excellent news for Kepler's primary application as a planet hunting instrument, as the target selection and survey design were based on these approximate ratios of variable and quiet stars.

The comparison with our Sun also yielded some interesting trends in variability with effective temperature. Most dwarfs of similar effective temperature as our Sun lie close to the solar value of variability range or below it, but of the stars cooler than ~5200K (the K and M dwarfs in our sample), most stars were more variable than the Sun. While one might expect this trend to be due to cool stars being preferentially fainter (and therefore having higher variability caused by the rising noise floor

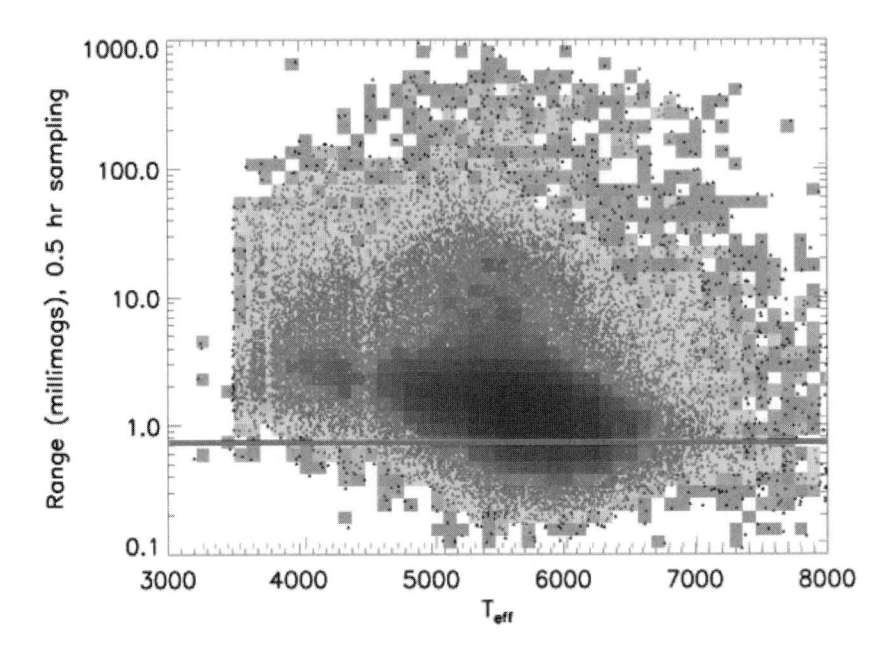

Figure 2. The distribution of variability range with effective temperature for *Kepler* dwarfs. The red line shows the variability range of the 2001 active Sun as determined from the SOHO lightcurves. Figure adapted from Basri et al. (2010b).

for fainter stars), the trend of cooler stars being more variable is true for both bright stars (those with a *Kepler* magnitude brighter than 13.5) and faint ones (see Basri et al. 2010b). In Table 1, we provide the fraction of stars more variable than the active Sun for the entire sample, the bright stars and faint stars, broken into temperature bins. Below each fraction, we show the number of stars in each bin for reference.

Table 1. Fraction of stars more variable than the 2001 active Sun

	$T_{eff} > 6000K$	$6000K > T_{eff} > 5500K$	$5500K > T_{eff} > 4500K$	$T_{eff} < 4500K$
All stars	0.33	0.37	0.57	0.87
	21023	42832	33288	6522
Bright stars	0.41	0.31	0.46	0.84
	8747	6164	2613	369
Faint stars	0.33	0.38	0.57	0.88
	12276	36668	30675	6153

4. Rotation Analysis

As the *Kepler* Quarter 1 observations span only a month, they are suitable for finding periodicities in the data on timescales shorter than ~16 days. In the case of short periods these may actually be the rotation periods of the stars, but some may be fractional periods for those stars where a full rotation was not observed in the course of a month. A thorough analysis of periodicities in the *Kepler* Quarter 1 data appears in Basri et al. (2010a)– I outline the major results here.

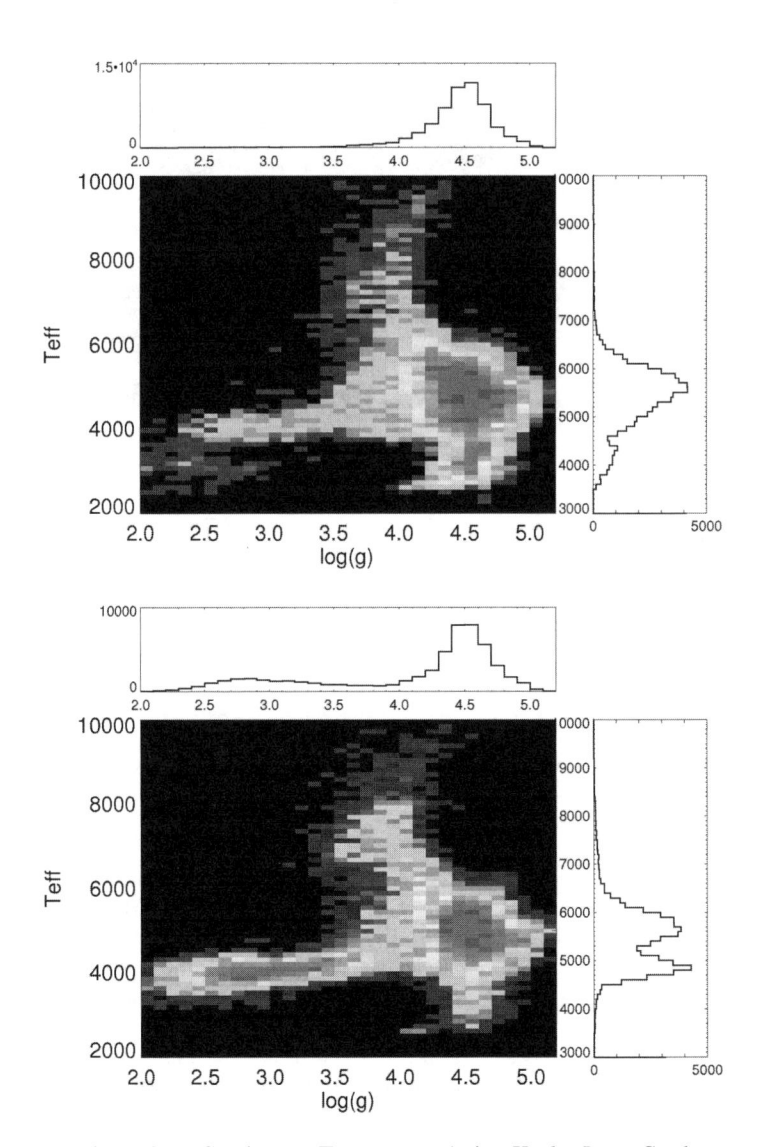

Figure 3. Gravity vs. Temperature (using Kepler Input Catalogue parameters). The upper sample has stars with clear periodicity, while the lower sample is non-periodic. The periodic sample shows a stronger component of cool dwarfs. The non-periodic sample has post-main sequence stars (with low gravity); the hotter stars there tend to be pulsators with too high a frequency to trigger our filter for periods. Figure reprinted from Basri et al. (2010a).

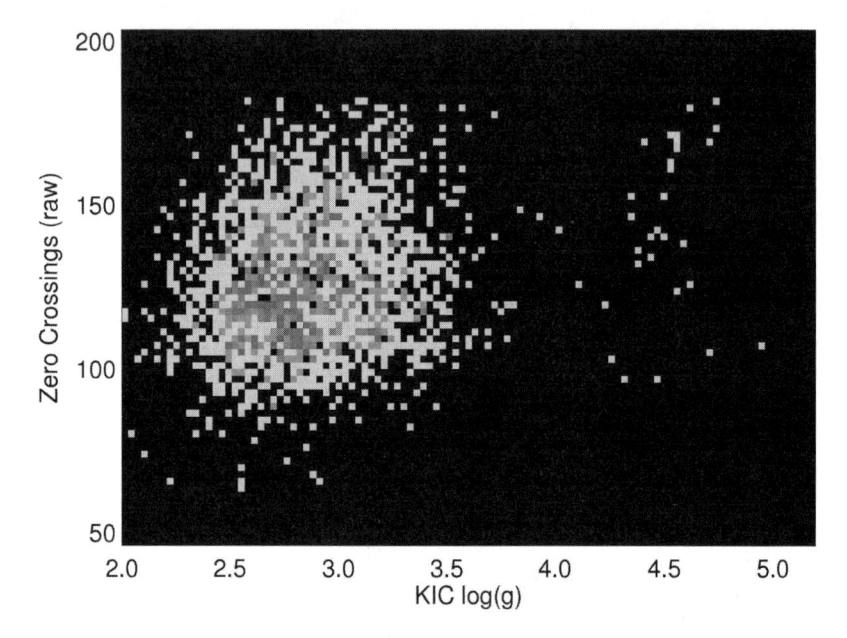

Figure 4. The results of photometric selection (see text) of giants. The ordinate is the number of times the differential lightcurve crossed zero. The KIC gravities were not used in choosing the stars, but afterwards to see whether giants were chosen. The high gravity cases look very much like giants photometrically (and it is possible that some were misclassified by the KIC). Figure reprinted from Basri et al. (2010a).

We used the results of our Lomb-Scargle periodograms to identify the sample of periodic stars in the *Kepler* data. We examined the resulting periodograms extensively by eye to determine what power threshold corresponded to obvious periodicity in the lightcurves, versus marginal periodicity or aperiodic variability. We refer to stars as part of the periodic sample if they had power greater than 60 in the strongest identified periodogram peak, marginal if they had power between 35 and 60 in their strongest peak, and nonperiodic for stars with no peaks stronger than 35. Of the ~150,000 stars measured, roughly 60,000 stars were identified as having some periodicity in the Q1 observations. Given that these lightcurves are only a month in duration, this initial sample of periodic stars is a testament to how powerful *Kepler* will be in identifying stellar rotation periods.

We can use the results of the periodogram analysis in combination with stellar physical parameters from the Kepler Input Catalogue and the previously discussed variability characteristics to identify interesting trends in stellar behavior with physical properties. In Figure 3, we show the distribution in effective temperature and gravity of the periodic and nonperiodic samples. While both periodic and nonperiodic dwarfs can be found in large number, the vast amount of low gravity stars (giants and subgiants) are aperiodically variable. The lightcurves of these stars are dominated by small amplitude oscillations that result in periodogram power being distributed over a wide range of periods, with no one period dominating the spectrum (e.g. Gilliland 2008; Bedding

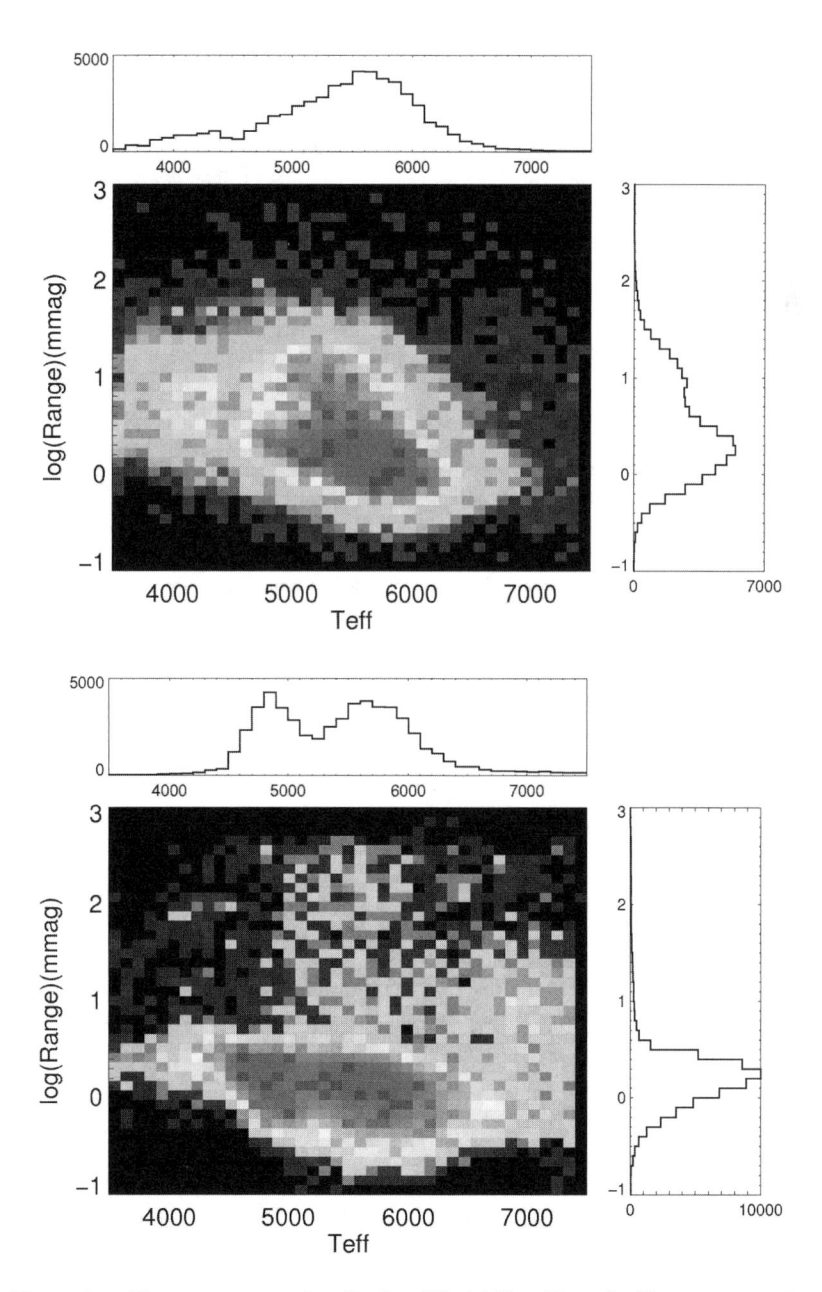

Figure 5. Temperature vs. Amplitude of Variability (Range). The upper panel shows periodic stars and the lower one non-periodic stars. The periodic stars have a cloud of higher amplitude objects at all temperatures, although there are many dwarfs with lower amplitudes for which as many are non-periodic. One can also see a tendency for cooler periodic stars to have greater amplitudes. Figure reprinted from Basri et al. (2010a).

et al. 2010). We further explored how dwarfs and giants may be separated by their photometric characteristics alone. Using a combination of the number of zero crossings, the variability range, and lack of a strong period to identify stars having "giant-like" photometric characteristics, we selected a subset of stars from the data. We then used the Kepler Input Catalogue gravities to check whether these stars were in fact classified as having low gravities. As can be seen in Figure 4, most of the stars identified by these means were indeed low gravity. The resulting few that have high gravity in the catalogue may be the rare misclassification, or dwarf stars that happen to possess similar variability morphology.

In Figure 5, we examine the distribute of variability range as a function of effective temperature for both the periodic (top) and non-periodic (bottom) samples. The middle panel of each of these figures shows the density distribution between the range and effective temperature, while the side panels show histograms of the range and effective temperature, respectively. Both the periodic and non-periodic populations include stars with relatively low amplitude variability– amongst the solar-type stars, low amplitude variability is associated with both periodic and non-periodic variability. However, one can also see that the periodic sample (upper plot) generally has higher variability, and in particular that the periodic cool dwarfs tend to have higher variability than their warmer counterparts. Longer time coverage will help to discern whether this effect is due primarily to our sampling shorter period variability in this first month of data, or whether this is a general property of the cool stars in the *Kepler* sample.

Finally, Figure 6 shows the relationship between the range and the distribution of periods found in our Quarter 1 sample. We show periods between 4 and 16 days due to (on the short period end) the extremely short period sample is dominated by pulsators, rather than spotted stars, and (on the long period end) the fact that the data duration does not make it suitable for detecting longer periods. Indeed, some of the periodicities found in our sample may in fact be fractions of the true periods for some stars (i.e. P/2, P/3, etc.), so this distribution should be viewed as preliminary until longer lightcurves become available.

5. Conclusions

Although its primary mission goal is the detection of exoplanets, *Kepler* provides a rich and unique dataset for studies of stellar activity. In this initial examination of activity and rotation in the first 30 days of *Kepler* data, we have already identified a number of intriguing trends with stellar properties and a host of periodic variables that may yield rotation periods in the near future. We find that overall, cool stars tend to have larger amplitude variability than solar-type stars, and that in particular the periodic cool stars tend to have larger amplitude variability. We also find that both periodic and non-periodic variability are common at all amplitudes, which may indicate that we are seeing a population of some stars whose surfaces are mottled with spots, such that the amplitude of their variability is relatively small but periodic on <30 day timescales. While it is too early at this point to make definitive statements about the true distribution of rotation periods in the *Kepler* sample, this early analysis is indicative of the power of *Kepler* photometry to open a new window on stellar magnetic activity.

Acknowledgments. LMW thanks the Scientific Organizing Committee of Cool Stars 16 for inviting her to present this work, and the Local Organizing Committee for facilitating the workshop's great success.

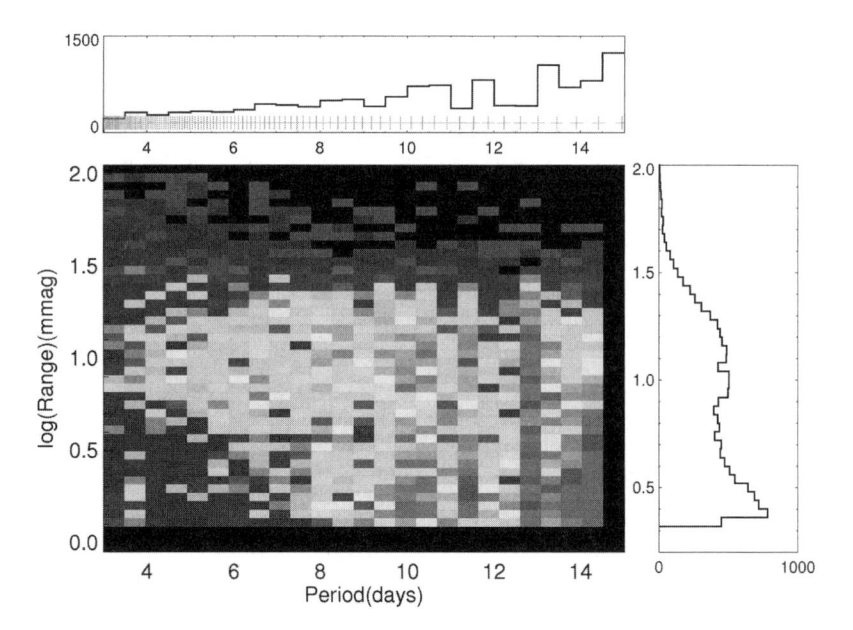

Figure 6. The distribution of periods found for a selected sample of periodic main sequence stars. The actual sampling of the periods search is given by the row of plusses in the upper histogram. The period histogram rises toward our limit of validity of 16 days. The distribution of variability range peaks just above the active Sun value, with a substantial tail at high variability (which extends to shorter periods). Figure reprinted from Basri et al. (2010a).

References

Baglin, A. 2003, Advances in Space Research, 31, 345
Basri, G., Walkowicz, L. M., Batalha, N., Gilliland, R. L., Jenkins, J., Borucki, W. J., Koch, D., Caldwell, D., Dupree, A. K., Latham, D. W., Marcy, G. W., Meibom, S., & Brown, T. 2010a, ArXiv e-prints. 1008.1092
Basri, G., Walkowicz, L. M., Batalha, N., Gilliland, R. L., Jenkins, J., Borucki, W. J., Koch, D., Caldwell, D., Dupree, A. K., Latham, D. W., Meibom, S., Howell, S., & Brown, T. 2010b, ApJ, 713, L155. 1001.0414
Batalha, N. M., Borucki, W. J., Koch, D. G., Bryson, S. T., Haas, M. R., Brown, T. M., Caldwell, D. A., Hall, J. R., Gilliland, R. L., Latham, D. W., Meibom, S., & Monet, D. G. 2010, ApJ, 713, L109. 1001.0349
Batalha, N. M., Jenkins, J., Basri, G. S., Borucki, W. J., & Koch, D. G. 2002, in Stellar Structure and Habitable Planet Finding, edited by B. Battrick, F. Favata, I. W. Roxburgh, & D. Galadi, vol. 485 of ESA Special Publication, 35
Bedding, T. R., Huber, D., Stello, D., Elsworth, Y. P., Hekker, S., Kallinger, T., Mathur, S., Mosser, B., Preston, H. L., Ballot, J., Barban, C., Broomhall, A. M., Buzasi, D. L., Chaplin, W. J., García, R. A., Gruberbauer, M., Hale, S. J., De Ridder, J., Frandsen, S., Borucki, W. J., Brown, T., Christensen-Dalsgaard, J., Gilliland, R. L., Jenkins, J. M., Kjeldsen, H., Koch, D., Belkacem, K., Bildsten, L., Bruntt, H., Campante, T. L., Deheuvels, S., Derekas, A., Dupret, M., Goupil, M., Hatzes, A., Houdek, G., Ireland, M. J., Jiang, C., Karoff, C., Kiss, L. L., Lebreton, Y., Miglio, A., Montalbán, J., Noels, A., Roxburgh, I. W., Sangaralingam, V., Stevens, I. R., Suran, M. D., Tarrant, N. J., & Weiss, A. 2010, ApJ, 713, L176. 1001.0229
Gilliland, R. L. 2008, AJ, 136, 566. 0806.1497
Horne, J. H., & Baliunas, S. L. 1986, ApJ, 302, 757
Koch, D. G., Borucki, W. J., Basri, G., Batalha, N. M., Brown, T. M., Caldwell, D., Christensen-Dalsgaard, J., Cochran, W. D., DeVore, E., Dunham, E. W., Gautier, T. N., Geary, J. C., Gilliland, R. L., Gould, A., Jenkins, J., Kondo, Y., Latham, D. W., Lissauer, J. J., Marcy, G., Monet, D., Sasselov, D., Boss, A., Brownlee, D., Caldwell, J., Dupree, A. K., Howell, S. B., Kjeldsen, H., Meibom, S., Morrison, D., Owen, T., Reitsema, H., Tarter, J., Bryson, S. T., Dotson, J. L., Gazis, P., Haas, M. R., Kolodziejczak, J., Rowe, J. F., Van Cleve, J. E., Allen, C., Chandrasekaran, H., Clarke, B. D., Li, J., Quintana, E. V., Tenenbaum, P., Twicken, J. D., & Wu, H. 2010, ApJ, 713, L79. 1001.0268
Lockwood, G. W., Skiff, B. A., Henry, G. W., Henry, S., Radick, R. R., Baliunas, S. L., Donahue, R. A., & Soon, W. 2007, ApJS, 171, 260. arXiv:astro-ph/0703408
Pizzolato, N., Maggio, A., Micela, G., Sciortino, S., & Ventura, P. 2003, A&A, 397, 147
Walker, G., Matthews, J., Kuschnig, R., Johnson, R., Rucinski, S., Pazder, J., Burley, G., Walker, A., Skaret, K., Zee, R., Grocott, S., Carroll, K., Sinclair, P., Sturgeon, D., & Harron, J. 2003, PASP, 115, 1023

Weather at the L/T Transition: A Large J-Band Survey for Variability of Cool Brown Dwarfs

Jacqueline Radigan,[1] Ray Jayawardhana,[1] David Lafrenière,[2] and Étienne Artigau[2]

[1] Department of Astronomy and Astrophysics, University of Toronto, 50 St. George Street, Toronto, ON, M5S3H4, Canada M5S3H4

[2] Département de physique and Observatoire du mont Mégantic, Université de Montréal, C.P. 6128, Succ. Centre-Ville, Montréal, QC, H3C 3J7, Canada

Abstract. An outstanding issue in the understanding of ultracool dwarf atmospheres is the abrupt shift from red ($J - K$ ~2) to blue ($J - K$ ~0) near-infrared colors, accompanied by a J-band brightening, at the boundary between L and T spectral types, marking the transition from cloudy to cloud-free atmospheres. While current 1-dimensional cloud models generally fail to capture the main features of the transition, most notably the narrow temperature range over which it occurs, it has been hypothesized that a disruption of the cloud layer as it sits progressively lower in the photosphere could contribute to a more abrupt decrease in dust opacity in this regime. If present, such patchy cloud coverage should give rise to photometric variability on rotational timescales. Motivated in part by the recent discovery of ~50 mmag periodic variability of the T2 dwarf SIMP0136 by Artigau et al. 2009, we have undertaken the most comprehensive near-infrared variability survey of cool brown dwarfs to date, designed to test for heterogeneous cloud coverage in a sample of brown dwarfs spanning mid-L to late-T spectral types. Our J-band search has targeted 56 objects with high cadence, high precision, photometric sequences, and is complemented by follow-up observations in H and K_s bands to characterize the nature of the variations. Here we describe our large survey, and showcase the first, preliminary results. Our data set hints at a higher incidence of variability within the the L/T transition regime.

1. Introduction

The transition between predominantly cloudy late-L and predominantly clear mid-T spectral types occurs at temperatures of ~1200 K, and is characterized by a significant decrease in condensate opacity over *a relatively small range of effective temperatures*. For a population of ultracool dwarfs on a color magnitude diagram, the transition is evident as an abrupt blue-ward shift in $J - K_s$ color by ~2 magnitudes, accompanied by a brightening in the J-band. This dramatic blue-ward evolution and J-band brightening are not well-reproduced by 1D cloud models which predict a much more gradual turn around of $J - K_s$ color with decreasing effective temperature (e.g. Marley et al. 2002;

Tsuji & Nakajima 2003; Allard et al. 2003; Burrows et al. 2006)[1]. One solution, put forward by Burgasser et al. (2002b) is that condensate clouds are disrupted at the L/T transition, opening up holes in the cloud layer. This scenario would account for the abrupt decrease in dust opacity and temporary increase in J-band flux, as photons from deeper, warmer atmospheric layers are suddenly able to emerge. If this scenario is true, one might expect that heterogeneous clouds at the L/T transition will lead to rotationally modulated variability, as surface features rotating with the BD pass in and out of the observer's line of sight[2]. In particular, a high contrast between dust clouds and gaps is expected in the J-band (~1.2 μm), where the underlying gaseous photosphere forms at depths several hundred degrees hotter than the dust layer (Marley et al. 2002).

Here we describe our large J-band survey for weather-induced variability of cool brown dwarfs (BDs).

2. Motivation

Although the combination of condensate clouds and rapid rotation has long-motivated searches for variability of ultracool dwarfs (e.g. Tinney & Tolley 1999), comparatively little attention has been paid to the L/T transition regime. Surveys in the I-band (e.g. Bailer-Jones & Mundt 2001; Gelino et al. 2002; Koen 2003, 2004, 2005a,b) have focussed on late-M and early-L targets, due to a significant drop-off in I-band flux for later spectral types. Near-infrared (NIR) surveys (e.g. Enoch et al. 2003; Koen et al. 2004, 2005; Goldman et al. 2008; Bailer-Jones 2008; Clarke et al. 2008), while sampling a wide range of spectral types, have sampled few late-L and T dwarfs with intermediate NIR colors ($0.4 \lesssim J - K \lesssim 1.6$) indicative of atmospheres that are in the process of cloud clearing (see figure 1). Most recently, in a single targeted study Artigau et al. (2009) discovered convincing J-band variability of the T2.5 dwarf SIMP J013656.57+093347.3 (SIMP0136 hereafter; Artigau et al. 2006) with a peak-to-peak amplitude of ~50 mmag and a period of ~2.3 hr. Notably, with $J - K_s = 0.89$, SIMP0136 falls directly within the L/T transition regime (see figure 1). Since previous NIR variability studies have only sparsely sampled the L/T transition, it is impossible to infer how common objects like SIMP0136 may be. Therefore, the question of ultracool dwarf variability at the L/T transition merits further investigation.

3. Observations and Sample

Here we describe our large J-band survey for variability of cool BDs using the Wide Field Infrared Camera (WIRC) on the 2.5-m Du Pont telescope at Las Campanas (48 targets with $J < 17$ and $\delta < 15$ degrees), supplemented with additional observations of northern targets observed using the Wide-field InfraRed Camera (WIRCam) on the 3.6-m Canada-France-Hawaii Telescope (CFHT; 5 targets with $J < 16.5$), and the Camera PAnoramique Proche InfraRouge (CPAPIR) on the 1.6-m Observatoire du Mont Mégantic (OMM) telescope in Québec, Canada (3 targets with $J < 15.5$).

---

[1] We note that a much improved 1D model of the L/T transition is presented in this proceeding by Allard et al. (2010).

[2] However, we note that heterogeneities that are spatially small and/or distributed axisymmetrically may not produce a measurable signal

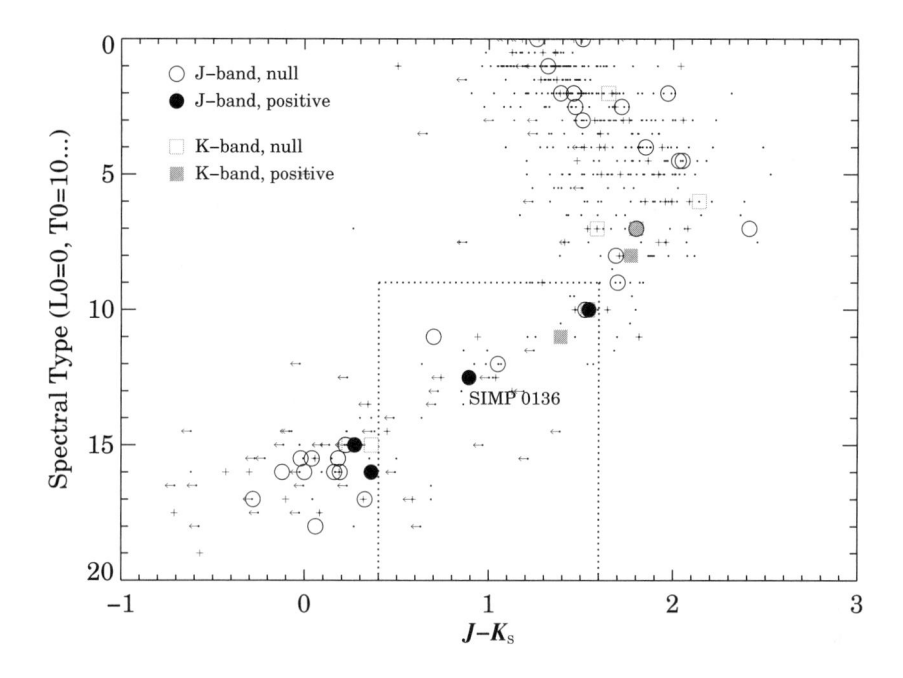

J-Band Variability Survey of Cool Brown Dwarfs 189

Figure 1. Summary of previous NIR photometric monitoring surveys of field L and T dwarfs. Spectral type versus $J - K_s$ color is plotted as black dots for all L and T dwarfs in the DwarfArchives (http://DwarfArchives.org, Maintained by Chris Gelino, Davy Kirkpatrick and Adam Burgasser) that have *J*-band detections in the Two Micron All Sky Survey (2MASS; Skrutskie et al. 2006). Known binaries are marked with crosses, and objects that have no K_s-band detection in 2MASS are marked with left-pointing arrows. A rough outline of the L/T transition regime is shown with dotted lines. Overplotted on the overall population are results from the *J*-band variability surveys of Koen et al. (2004, 2005); Clarke et al. (2008). Detection limits for different targets vary widely, but are generally $\lesssim$20 mmag (peak-to-peak). Filled circles represent claimed detections of periodic variability (all from Clarke et al. 2008), while open symbols show null or inconclusive results (possible detections that were not ascribed high significance by their respective authors were interpreted here as null results; please refer to the references provided for a more complete picture). The *J*-band detection of 50 mmag variability for SIMP0136 (Artigau et al. 2009) is labelled (also a filled circle). The K_s-band survey of Enoch et al. (2003) is overplotted in light grey. Positive peridic detections and null/inconclusive detections are shown as open and closed squares respectively. We note that this latter study should be viewed cautiously, due to large photometric uncertainties, and hence lower sensitivities compared to other work.

Our sample consists mostly of late-L and T-dwarfs, with a focus on the elusive L/T transition regime (see figure 3). In total we have monitored 56 late-L and T-dwarf targets for variability in the *J* band.

Our survey was designed to achieve high precision (<1%), high-cadence *J*-band differential photometry of cool BDs over rotational timescales. Each target was monitored continuously over a ~3-5 hr time span during a single epoch. Since ultracool dwarfs are known to be rapid rotators with periods ranging from ~1.5-12 hr (e.g. Reiners & Basri 2008), the chosen time baseline ensures that most targets were monitored over a large fraction of a rotation period. Continuous, high cadence monitoring was chosen instead of a sparser time sampling over weeks or months because (i)it leads to more stable differential photometry, and (ii)the evolution of surface features (e.g. clouds) on a BD's surface may prevent observations taken over multiple epochs from being phased together in a coherent way. In addition, during the Du Pont observing runs new data were reduced nightly, and target light curves checked for variability. This allowed us to immediately respond to positive detections, and obtain follow-up observations on subsequent nights and in additional (H and K_s) bandpasses.

3.1. Du Pont Observations

The WIRC camera consists of 4 HAWAII-I arrays, each with a 3.2′ field of view and a pixel scale of 0.2″. New J-band filters closely matching the Mauna Kea Observatory (MKO) system (Tokunaga & Vacca 2005) were purchased and installed for our survey in order to minimize the effects of differential telluric extinction (which can mimic variability), by cutting off telluric water absorption bands redward of 1.35 μm (e.g Artigau 2006).

Observations were obtained either in staring mode (i.e. keeping the target fixed on the same detector pixel, without dithering) or by using a random dither pattern confined to an area within a 15″ radius on the detector. Dome flats and dark frames and were taken on the evening preceeding or morning following each night of observations.

3.1.1. Data Reduction and Photometry

All raw images were corrected for non-linearity, dark-subtracted, and flat-fielded. For dithered sequences, a running sky frame was subtracted from all science images. Faint stars were identified in an initial first-pass reduction, and subsequently masked for the second pass so as not to bias the sky frames. Hot or dead pixels, as well as pixels having more than 35,000 counts (>3% non-linear) were flagged. Except for sky subtraction, reduction of the staring sequences is almost identical to the procedure described above, however in these cases sky frames obtained before and after the staring sequence were used for rough sky subtraction.

For each monitoring sequence, aperture photometry was performed on the target and a set of reference stars, using a circular aperture of radius 1-1.5 times the median full width at half maximum of all stars in each image. Residual sky levels in the vicinity of each star were measured inside an annulus centered on each source. Differential photometry was performed with respect to a set of bright reference stars (typically several in the 3.2′×3.2′ WIRC field of view). One or more comparison stars of similar brightness to the target were chosen as controls.

An example of the photometric precision achieved for the T5.5 target 2MASS J15462718-3325111 (Burgasser et al. 2002a), with J=15.6, is shown in figure 2. In general, a relative photometric precision of $\lesssim$10 mmag every few minutes is typical for targets with $J < 16$ in clear conditions and seeing $\lesssim$0.8″.

3.2. CFHT and OMM Observations

Observations of 5 targets were carried out during the 2009B semester at the CFHT, using WIRCam in queue mode. A staring strategy was employed with a small defocus

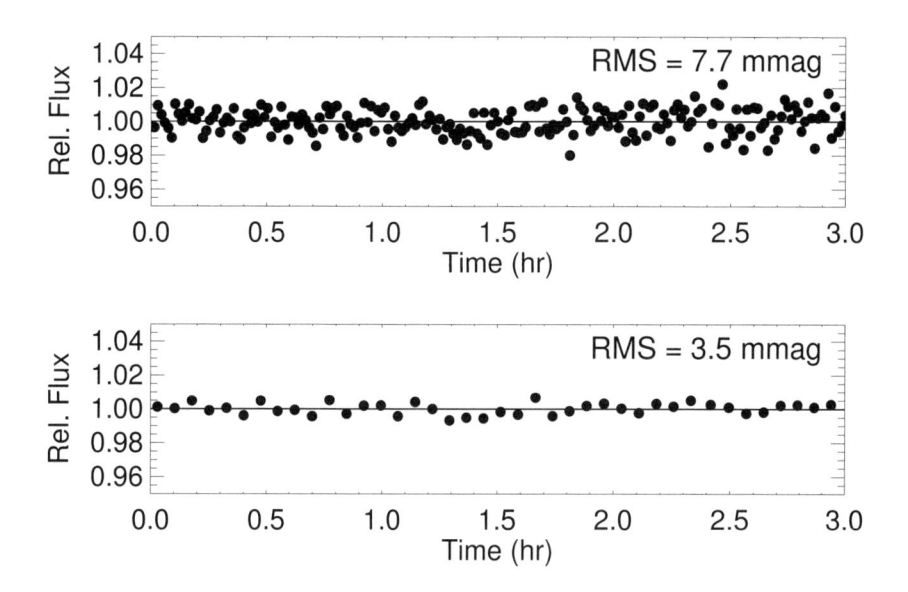

Figure 2. A demonstration of photometric sensitivity that can be achieved using WIRC on the Du Pont 2.5-m telescope at Las Campanas for a J=15.6 magnitude target in photometric conditions and with 0.7″ seeing. Both panels show the same *J*-band lightcurve of the *non-variable* T5.5 brown dwarf 2MASS J15462718-3325111 obtained using a staring strategy. This target has a large number of bright reference stars in its field of view that can be used to preform differential photometry. The top panel shows the precision reached with unbinned 40 s exposures, while the bottom pannel shows the precision reached after binning the observations by a factor of 5. Approximately 4 mmag precision is reached every 4.3 min.

for bright targets. The data were processed using the standard WIRCam pipeline, and aperture photometry was performed in a similar manner as described above for the WIRC data. With a 20′×20′ field of view, and pixel scale of 0.3″, WIRCam typically provides several tens of good reference stars for differential photometry.

To date, 3 targets have been observed using CPAPIR (30′×30′ field of view and a pixel scale of 0.89″) on the OMM telescope. Images were obtained using standard dither sequences and processed by the CPAPIR pipeline. Aperture photometry was performed as described above.

4. Preliminary Results and Highlights

Here we describe some *preliminary* results and highlights of our variability search.

Of 56 targets observed, 6 show clear evidence of variability upon visual examination of their lightcurves, with peak-to-peak amplitudes ranging from ∼1.5-25%. In all cases, the variability does not appear to be random but follows a smooth trend on timescales consistent with rotation. For some targets we have indentified a stable periodicity, while other targets will require further monitoring. Of the 6 detections, 1 has yet to be followed up at additional epochs and therefore remain promising, but tentative.

We note that there may be additional, low-amplitude detections in our dataset that are not immediately obvious by eye, but may become apparent in a more rigorous analysis.

Four of 6 detections fall directly within the L/T transition regime, roughly indicated by the dotted box in figure 1. The $J - K_s$ and spectral type distributions of (i)our entire sample and (ii)of positive detections are shown in figure 3. A higher incidence of variability is hinted at within the L/T transition, especially compared to later spectral types ($\gtrsim$T4) with bluer $J - K_s$ colors. For earlier spectral types ($\lesssim$L9) with red $J - K_s$ colors, there also appears to be a decline in variability, however a trend cannot be claimed with great significane due to the low number of statistics in this regime. Nonetheless, the four variable targets falling directly within the L/T transition regime are also the most variable objects in our sample (all with peak-to-peak amplitudes $\gtrsim$ 30 mmag, and photometric errors $\lesssim$ 5 mmag). Previous J-band studies (see figure 1) have detected exclusively null results at the ~30 mmag level. Thus a careful analysis of both the frequencies and amplitudes of variability as a function of spectral type and color is required. Preliminary results are intriguing, to say the least.

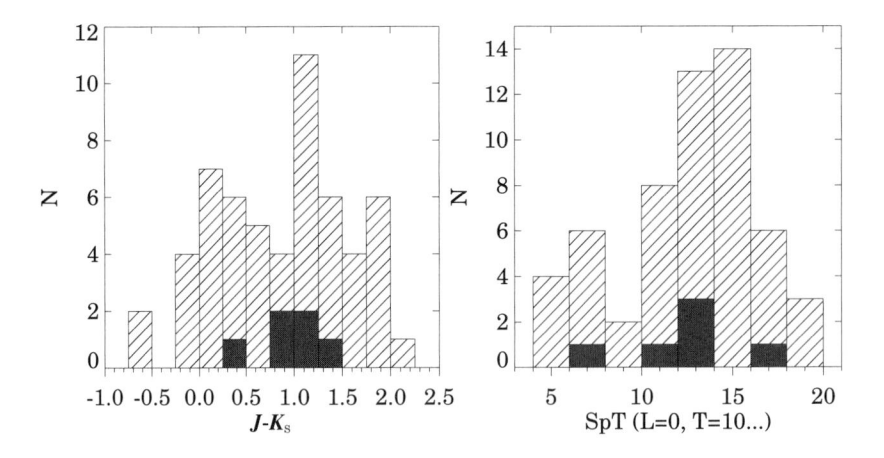

Figure 3. Distribution of our sample by $J - K_s$ color and spectral type. Hatched regions represent all 56 targets monitored for variability. Solid grey regions represent the population found to be variable in our preliminary analysis.

We confirm the variability of two objects previously classified as such in the literature: SIMP0136 and 2MASS J22282889-431026 (2M2228 hereafter; Burgasser et al. 2003). While Artigau et al. (2009) observed a peak-to-peak amplitude of ~50 mmag for SIMP0136, in our observations the amplitude had dropped to ~30 mmag. Clarke et al. (2008) found 2M2228 to be variable with a 1.43±0.16 hr period, and peak-to-peak amplitude of 15 mmag. We monitored 2M2228 for nearly 6 consective hours (over 4 periods) and derive a period of 1.43±0.02 hr and peak-to-peak amplitude of 14 mmag by fitting a sinusoid to the data (figure 4).

Perhaps the most notable and surprising highlight of our survey is the detection of ~25% variability of an early T-dwarf (figure 5). The variability is periodic with a ~7.7 hr period, consistent with ultracool dwarf rotation rates. The lightcurve amplitude and shape appear to change with time, and observations over long timescales cannot be phased together, implying the evolution of surface features. It remains to be seen if holes in the dust clouds of this BD are capable of producing such large amplitude vari-

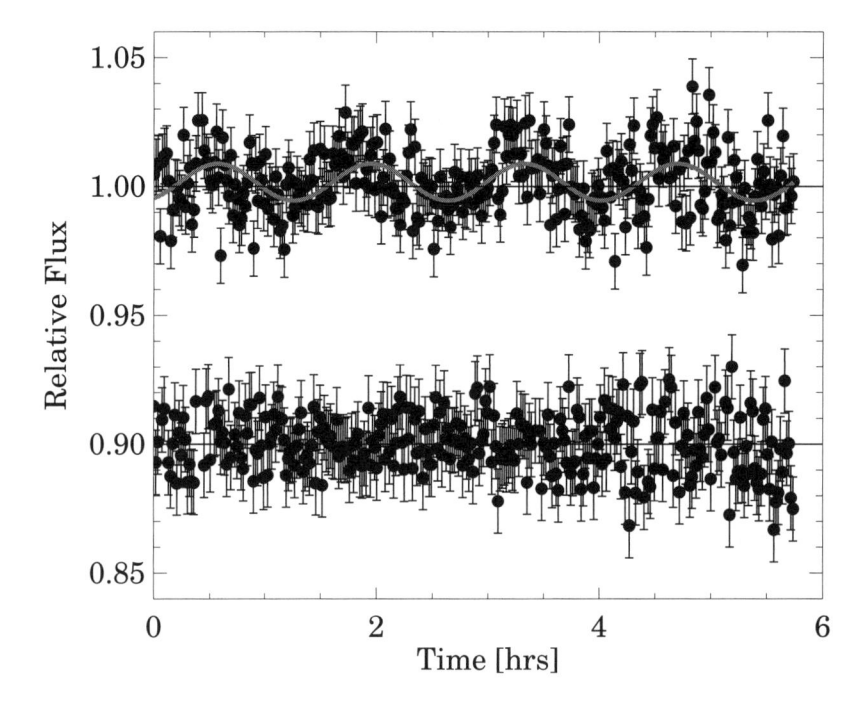

Figure 4. A *J*-band lightcurve of the T6 brown dwarf 2M2228 (top) and a comparison star of similar brightness (bottom) obtained using WIRC on the Du Pont 2.5-m telescope at Las Campanas. A best-fitting sinusoid is overplotted (red solid curve in the online version) with a period of 1.43±0.02 hr and a peak-to-peak amplitude of 14 mmag.

ability, or if some other explanation is required. If the 7.7 hr period is indeed a rotation period, then the leading explanation would be surface features; either weather-induced or magnetic in origin. The latter explanation seems unlikely due to the expectation that magnetic Reynolds numbers will be $\ll 1$ throughout the photospheres of T-dwarfs (e.g. Gelino et al. 2002; Mohanty et al. 2002).

Detections of large-amplitude variability on the scale of that shown in figure 5 will permit detailed photometric and spectroscopic follow-up. Such follow-up can potentially constrain the compositions, temperature contrasts, and fractional coverage of clouds and clearings in ultracool dwarf atmospheres. Furthermore, our observations suggest that heterogeneous clouds may account for key observed characteristics of the transition between L and T spectral types, and may indicate the need to account for patchy cloud coverage when modelling the L/T transition (e.g. Marley et al. 2010). These conclusions are based on a work in progress and further analysis is required to better understand the significance of trends in our data. In addition, there remain additional questions about what distinguishes variable from non-variable L/T transition

194   Radigan et al.

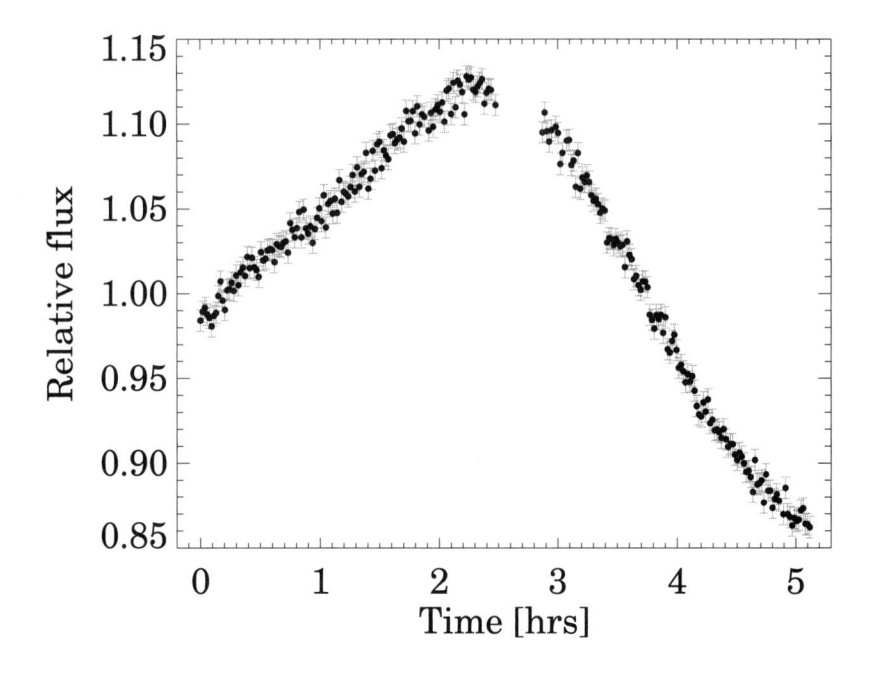

Figure 5. A J-band lightcurve from a single monitoring sequence of the most variable target in our sample (Radigan et al., in preparation) obtained using WIRC on the Du Pont 2.5-m telescope at Las Campanas. The lightcurve presented shows about two-thirds of a full period. If measured as a brightening from minimum light, the peak-to-peak variability is 30%, while a dimming from maximum light would correspond to a 23% change.

objects (binarity, axisymmetric cloud distributions, etc.) and whether a single mechanism is responsible for the variability in all cases[3].

Acknowledgments. The authors wish to thank Ian Thompson of Carnegie Observatories and the staff of the Las Campanas Observatory for their help in scheduling and carrying out many of the observations presented here. JR and RJ also wish to acknowledge grant support from the Natural Sciences and Engineering Research Council of Canada.

References

Allard, F., Guillot, T., Ludwig, H., Hauschildt, P. H., Schweitzer, A., Alexander, D. R., & Ferguson, J. W. 2003, in Brown Dwarfs, edited by E. Martín, vol. 211 of IAU Symposium, 325

---

[3]Observations of broadband optical variability for radio-pulsating late-M and early-L dwarfs is presented in the contribution of L. Harding et al. to these proceedings, and is possibly linked to auroral emmissions, discussed in the contribution of G. Hallinan et al.

Allard, F., Homeier, D., & Freytag, B. 2010, ArXiv e-prints. 1011.5405
Artigau, E. 2006, Ph.D. thesis, Universite de Montreal (Canada
Artigau, É., Bouchard, S., Doyon, R., & Lafrenière, D. 2009, ApJ, 701, 1534. 0906.3514
Artigau, É., Doyon, R., Lafrenière, D., Nadeau, D., Robert, J., & Albert, L. 2006, ApJ, 651, L57. arXiv:astro-ph/0609419
Bailer-Jones, C. A. L. 2008, MNRAS, 384, 1145
Bailer-Jones, C. A. L., & Mundt, R. 2001, A&A, 367, 218. arXiv:astro-ph/0012224
Burgasser, A. J., Kirkpatrick, J. D., Brown, M. E., Reid, I. N., Burrows, A., Liebert, J., Matthews, K., Gizis, J. E., Dahn, C. C., Monet, D. G., Cutri, R. M., & Skrutskie, M. F. 2002a, ApJ, 564, 421. arXiv:astro-ph/0108452
Burgasser, A. J., Marley, M. S., Ackerman, A. S., Saumon, D., Lodders, K., Dahn, C. C., Harris, H. C., & Kirkpatrick, J. D. 2002b, ApJ, 571, L151. arXiv:astro-ph/0205051
Burgasser, A. J., McElwain, M. W., & Kirkpatrick, J. D. 2003, AJ, 126, 2487. arXiv:astro-ph/0307374
Burrows, A., Sudarsky, D., & Hubeny, I. 2006, ApJ, 640, 1063. arXiv:astro-ph/0509066
Clarke, F. J., Hodgkin, S. T., Oppenheimer, B. R., Robertson, J., & Haubois, X. 2008, MNRAS, 386, 2009
Enoch, M. L., Brown, M. E., & Burgasser, A. J. 2003, AJ, 126, 1006. arXiv:astro-ph/0305048
Gelino, C. R., Marley, M. S., Holtzman, J. A., Ackerman, A. S., & Lodders, K. 2002, ApJ, 577, 433. arXiv:astro-ph/0205305
Goldman, B., Cushing, M. C., Marley, M. S., Artigau, É., Baliyan, K. S., Béjar, V. J. S., Caballero, J. A., Chanover, N., Connelley, M., Doyon, R., Forveille, T., Ganesh, S., Gelino, C. R., Hammel, H. B., Holtzman, J., Joshi, S., Joshi, U. C., Leggett, S. K., Liu, M. C., Martín, E. L., Mohan, V., Nadeau, D., Sagar, R., & Stephens, D. 2008, A&A, 487, 277. 0801.2371
Koen, C. 2003, MNRAS, 346, 473
— 2004, MNRAS, 354, 378
— 2005a, MNRAS, 360, 1132
— 2005b, MNRAS, 357, 1151
Koen, C., Matsunaga, N., & Menzies, J. 2004, MNRAS, 354, 466
Koen, C., Tanabé, T., Tamura, M., & Kusakabe, N. 2005, MNRAS, 362, 727
Marley, M. S., Saumon, D., & Goldblatt, C. 2010, ApJ, 723, L117. 1009.6217
Marley, M. S., Seager, S., Saumon, D., Lodders, K., Ackerman, A. S., Freedman, R. S., & Fan, X. 2002, ApJ, 568, 335. arXiv:astro-ph/0105438
Mohanty, S., Basri, G., Shu, F., Allard, F., & Chabrier, G. 2002, ApJ, 571, 469. arXiv:astro-ph/0201518
Reiners, A., & Basri, G. 2008, ApJ, 684, 1390
Skrutskie, M. F., et al. 2006, AJ, 131, 1163
Tinney, C. G., & Tolley, A. J. 1999, MNRAS, 304, 119. arXiv:astro-ph/9809165
Tokunaga, A. T., & Vacca, W. D. 2005, PASP, 117, 421. arXiv:astro-ph/0502120
Tsuji, T., & Nakajima, T. 2003, ApJ, 585, L151. arXiv:astro-ph/0302169

The Galactic M Dwarf Flare Rate

Eric J. Hilton,[1] Suzanne L. Hawley,[1] Adam F. Kowalski,[1] and Jon Holtzman[2]

[1]*University of Washington, Box 351580, UW, Seattle WA, 98195, USA*

[2]*New Mexico State University, P. O. Box 30001, MSC 4500, Las Cruces NM, 88003, USA*

Abstract. M dwarfs are known to flare on timescales from minutes to hours, with flux increases of several magnitudes in the blue/near-UV. These frequent, powerful events, which are caused by magnetic reconnection, will have a strong observational signature in large, time-domain surveys. The radiation and particle fluxes from flares may also exert a significant influence on the atmospheres of orbiting planets, and affect their habitability. We present a statistical model of flaring M dwarfs in the Galaxy that allows us to predict the observed flare rate along a given line of sight for a particular survey depth and cadence. The parameters that enter the model are the Galactic structure, the distribution of magnetically active and inactive M dwarfs, and the flare frequency distribution (FFD) of both populations. The FFD is a function of spectral type, activity, and Galactic height. Although inactive M dwarfs make up the majority of stars in a magnitude-limited survey, the FFD of inactive stars is very poorly constrained. We have organized a flare monitoring campaign comprising hundreds of hours of new observations from both the ground and space to better constrain flare rates. Incorporating the new observations into our model provides more accurate predictions of stellar variability caused by flares on M dwarfs. We pay particular attention to the likelihood of flares appearing as optical transients (i.e., host star not seen in quiescent data).

1. Introduction

Stellar flares are powered by magnetic reconnection events. Although most last for only a few minutes, they have been observed to last for up to 8 hours (Kowalski et al. 2010). Their strongest observational signal is a flux increase in the blue/near-UV, which can be several magnitudes for the biggest flares. Because M dwarfs account for such a large fraction of the stars in the Galaxy, photometric variability on these stars will have an significant signature in time-domain surveys such as Pan-STARRS (Kaiser 2004), PTF (Law et al. 2009), and LSST (LSST Science Collaborations: Paul A. Abell 2009).

M dwarfs are an excellent place to look for extrasolar planets because planets have relatively large RV effects and transit depths on low-mass stars. Many planets, including several super-Earths, have been discovered orbiting M dwarfs (e.g., Udry 2007; Charbonneau 2009; Correia et al. 2010). However, magnetic activity on M dwarfs can hamper planet searches, since starspots and flares introduce both photometric noise and radial velocity jitter (Wright 2005; Basri et al. 2010; López-Santiago et al. 2010). At the same time, the intense flux of high energy photons and particles caused by flares may have a significant impact on the atmospheres of Earth-like planets in the close-in habitable zones of M dwarfs (Buccino et al. 2007; Segura et al. 2010).

Our group has used the serendipitous observations of flares in surveys to make estimates of the flare rate. Kowalski et al. (2009) searched for flares in the ~80 photometric epochs from the SDSS Stripe 82 survey (Ivezić et al. 2007), while Hilton et al. (2010) identified flares in the sample of M dwarfs with SDSS spectroscopy (West et al. 2008). These studies confirmed that while the flare rate increases with later spectral subtype, the flare energy on earlier type stars was higher. We also found that the flare stars were preferentially closer to the Galactic midplane, which we interpret as an age effect, with younger stars flaring more frequently.

Although the occurrence of an individual flare cannot be predicted, dedicated photometric monitoring campaigns of individual stars can yield measurements of the frequency of flares of various energies, also called the flare frequency distribution (FFD). Moffett (1974) and Lacy et al. (1976) obtained FFDs on several of the most well-known and most prolific flare stars in the Solar neighborhood, nearly all of which are active mid-type M dwarfs. We have extended these measurements of the FFD to inactive stars, as well as both early- and late-type M dwarfs.

2. Observations

Measuring the FFD of a star requires time-resolved photometric monitoring over the course of many flares. We have used four telescopes (described in Table 1) to obtain almost 500 hours of flare monitoring of our targets.

Table 1. The telescopes used to collect flare-monitoring data

Location	Size	Operator	Filters
Apache Point Observatory	3.5m	Astrophysical Research Consortium	B
Apache Point Observatory	1.0m	New Mexico State University	U
Manastash Ridge Observatory	0.8m	University of Washington	u,g
Apache Point Observatory	0.5m	Astrophysical Research Consortium	u

We chose bright targets in order to have higher time resolution while maintaining high signal-to-noise. The cadence of our observations was typically about ten seconds, except for the NSMU 1.0m telescope, which has a longer readout time and a cadence of approximately 60 seconds. The data reduction and aperture photometry was done with standard IRAF tasks. Because we are only concerned about variation in the lightcurve, we used differential photometry. We use the equivalent width of Hα to classify each of our targets as either magnetically active or inactive. Following the criteria of West et al. (2008), stars with Hα equivalent width of more than 1Å in emission are considered active.

3. Flare Frequency Distributions

In order to derive the flare frequency distributions for each bin of spectral type and activity level, we first need to identify flares, calculate the flare energies, transform each of these flare energies to a common filter, and then combine the observations from different nights and different stars.

3.1. Flare Identification and Energies

For each night of observations, we first define a period of quiescence. This is identified by eye and is at least an hour in duration. We use this quiescent period to define the quiescent flux (taken as the median flux value) and the precision of our data, taken as the standard deviation of the data points during the quiescent period. We define a flare as having occurred if there are at least three consecutive measurements more than 2.5 σ above the quiescent value, with at least one of those measurements being $\geq 5\sigma$ above the quiescent value. The flare start and stop times are identified as when a running average of several measurements is equal to the quiescent value. Since this procedure requires the flux to return to the quiescent value before the flare is determined to have ended, groups of flares or subpeaks within a flare are typically identified as one flare.

We follow the method of Gershberg (1972); Moffett (1974); Lacy et al. (1976) to find the flare energies in the observed filter. We first calculate the flare flux as

$$I_f(t) = \frac{I_{Flare}(t) + I_0}{I_0} - 1 \tag{1}$$

where I_0 is the quiescent flux. The equivalent duration, P, is simply the time integral between the time of flare start and stop (described above).

$$P = \int I_f(t)\, dt \tag{2}$$

The total flare energy in filter x is

$$E_x = q_x P \tag{3}$$

where q_x is the quiescent flare luminosity in filter x. The quiescent flare luminosity is easily calculated from the known absolute magnitude of the star. The absolute magnitudes come from the Palomar/MSU Nearby Star Spectroscopic Survey (Reid et al. 1996; Hawley et al. 1997; Gizis et al. 2002; Reid et al. 2002), or the Gleise catalogue (Gliese 1969). We transform the energies of all flares into SDSS u using our own simultaneous observations and a relationship from Lacy et al. (1976). The total observing time, number of stars, and number of flares in each bin is given in Table 2.

Table 2. Flare Monitoring Targets

Bin	Total (hrs)	# of Flares	# Stars/Bin
Inactive M0-2	201.05	5	9
Inactive M3-5	107.96	4	3
Active M3-5	123.15	43	7
Active M6-9	60.27	29	4
Total	492.43 hours	81 flares	23 stars

3.2. Constructing the FFD

The FFD is a cumulative distribution of flare energies such that the FFD evaluated at energy E is the number of flares per unit time with energy greater than E. To construct the FFD, we need not only the energy of each flare, but also the duration of observations during which a flare of that energy is able to be detected. Because conditions change from night to night, and because we must bin together stars of different magnitudes with different comparison stars, the minimum detectable flare energy is different for each night on each star.

For each bin, the FFD is constructed from the list of flare energies in that bin and the duration (τ) and minimum observable flare energy (E^{min}) for each observing period for stars in that bin. The differential flare frequency is the inverse of the sum of all observing periods whose minimum observable energy is less than the flare energy.

$$FF(E_i) = 1/\sum_{E_i^{min}}^{E_{max}^{min}} \tau_i \qquad (4)$$

where E_i^{min} is the minimum observable flare energy on a night with duration τ_i, and E_{max}^{min} is the night with the largest threshold for detecting a flare. The cumulative FFD is simply the cumulative sum of the flare frequencies.

The cumulative FFDs for each of our bins are shown in Figure 1. Previous studies (e.g., Lacy et al. 1976; Audard et al. 2000) have described the FFD as a linear relation between the log of the flare energy and the log of the frequency

$$\log \nu = \alpha + \beta \log E_U \qquad (5)$$

Overplotted in Figure 1 is our best fit for each FFD.

The FFD for the active M3-M5 stars is consistent with the measured FFDs of several of the most well-known flare stars (all active and between M3 and M5) from Lacy et al. (1976). The FFDs for the other three bins represent the first measurement for these bins. We note that inactive stars flare much less frequently than active stars, and with less energy. For the inactive stars, flares on earlier type stars are more energetic than on later type stars. This trend is also apparent for the active stars, if the FFD of the only measured active early M star, Au Mic (Lacy et al. 1976, not shown), is representative of all active early M stars. Although it may be that flares on earlier type stars are intrinsically more energetic, this is at least partially a selection effect, since a flare on an earlier type star will cause a smaller change in apparent magnitude than a flare of equal energy on a later type star, the so called "contrast effect."

4. Galactic Flare Model

We would like to predict the number and size of flares that will be seen in time domain surveys. To do this, we create a model of flaring in the Galaxy. The first step in modeling the Galactic flare rate is modeling the light curves of individual stars. We adopt the classic flare shape (see e.g., Moffett 1974; Hawley & Pettersen 1991; Kowalski et al. 2010) of an impulsive phase with a linear flux rise and partial linear decrease, followed by a gradual phase consisting of an exponential flux decay. We insert flares into a light curve by drawing from the measured FFD for a particular type of star. Using the model light curves, we calculate the fraction of time a star is seen at increased brightness. The black line in Figure 2 (top panel) shows the FFD for the active M3-M5 bin, while the black line in the bottom panel shows the fraction of time a hypothetical dM3.5e star is seen at increased brightness from flares. The uncertainty in the measurement of the

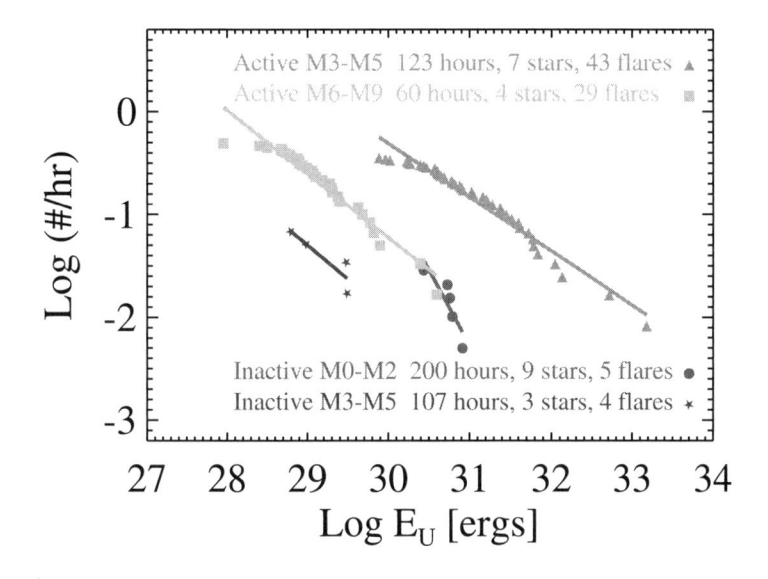

Figure 1. Using nearly 500 hours of flare monitoring observations, we calculate the FFDs of four bins of differing spectral type and activity status are shown here, along with the best fit model. We find that inactive stars flare less frequently than active stars.

FFD (shaded gray) is dominated by our choice of the flare definition, especially the start and stop times, and the assumption that all stars within a particular bin have the same intrinsic FFD. Different input FFDs result in significant changes in the amount of time spent at increased brightness. In principle, we can use time domain surveys, which provide measurements of the fraction of time spent at increased brightness, to determine which FFD is the best fit to the survey data.

We have created a model of flaring on M dwarfs in the Galaxy to predict the number and magnitude of flares that will be observed in time domain surveys, with a particular focus on LSST. We generate a model Galaxy with thin disk, thick disk, and halo components (Juric et al. 2010). The scale heights, lengths, and densities of these components have been determined from SDSS star counts (Jurić et al. 2008; Bochanski et al. 2010). The active fractions of each spectral subtype as a function of Galactic height are extrapolated from West et al. (2008).

For a given pointing and field of view size, the Galactic model determines the number, color, distance, and activity level of M dwarfs along that line of sight. The activity status and spectral type determine the FFD used to generate a light curve for each star in the model.

To verify the utility of this model, we compared our model results to the measured flare rates (Kowalski et al. 2009) from the SDSS Stripe 82 photometric catalogue (Ivezić et al. 2007). Each of the ~50,000 M dwarfs in Stripe 82 was observed ~80 times, with 271 epochs meeting stringent flare criteria (Kowalski et al. 2009). The magnitude

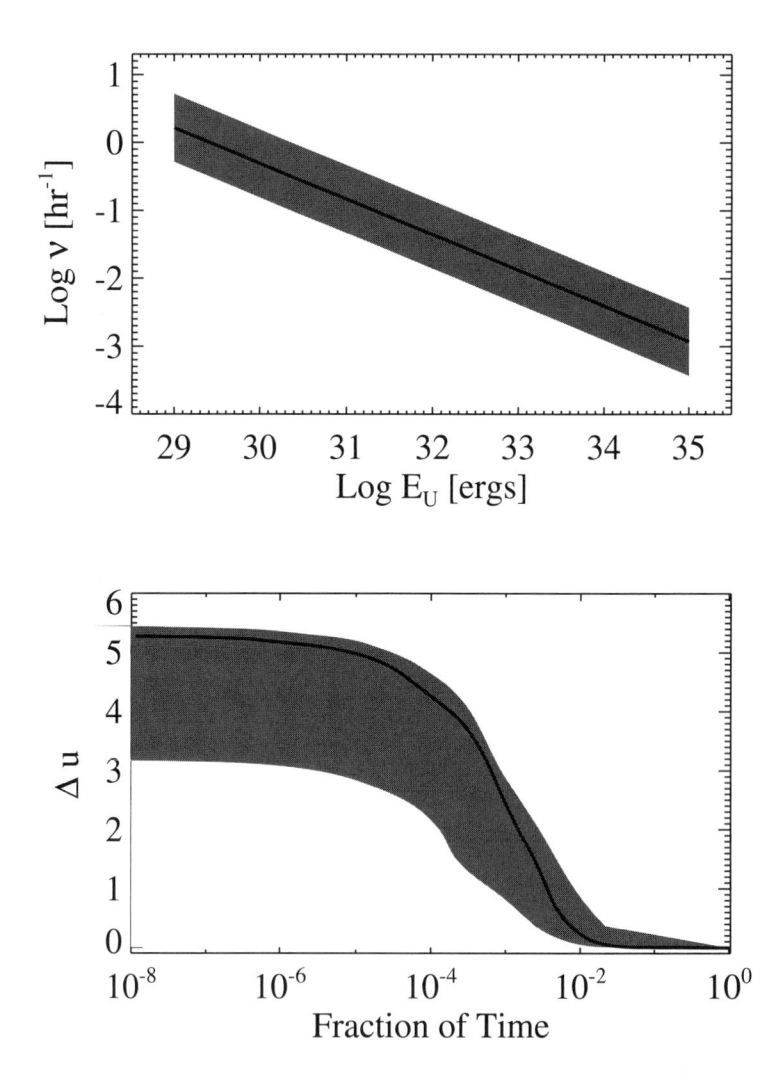

Figure 2. Top: The measured FFD of an active M3.5, with the uncertainty shown as shaded gray. Bottom: The fraction of time an active M3.5 spends at increased brightness. Stars with different FFDs spend vastly different amounts of time at highly enhanced magnitudes. Future studies may be able to use the large number of flare observations from time domain surveys to determine the FFD more precisely.

distribution of these flares is shown in Figure 3 (red line). Adopting the same number of epochs, field of view, flare criteria, and survey depth of the Stripe 82 data in our flare model, we predict 283 flares, which are overplotted in Figure 3 (blue line). Our model not only matches the total number of flares, but independently reproduces the distribution of flare magnitudes remarkably well.

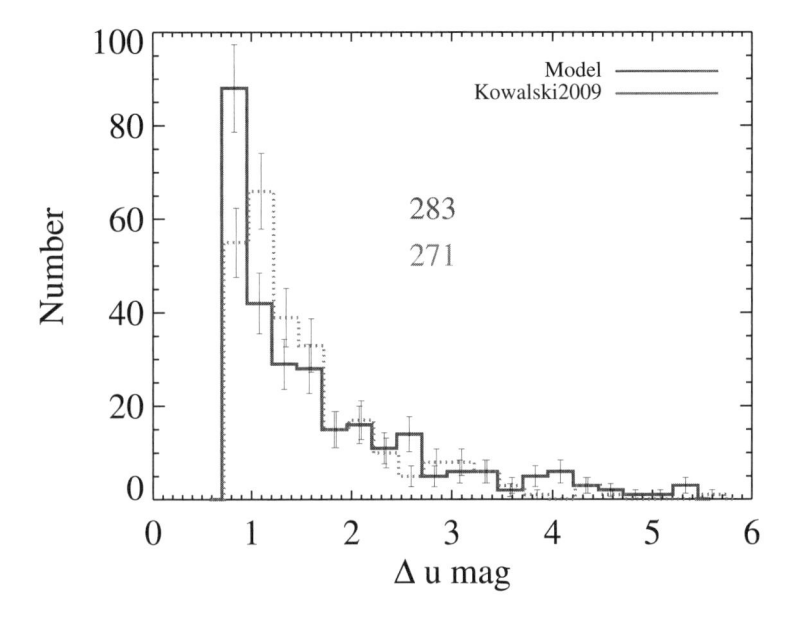

Figure 3. The magnitude distribution of flares measured in SDSS Stripe 82 from Kowalski et al. (2009) (red line) and the model prediction using the same survey parameters (blue line).

4.1. Predicting Flare Rates in Time Domain Surveys

The good agreement between our flare model and the measured Stripe 82 flare rates gives us confidence that we can predict the frequency of flares in future time domain surveys. However, there are three areas where limited data may affect the model predictions. These limitations are especially important to the prediction of transients — flares that are bright enough to make a star briefly visible that is otherwise too faint to be seen in the survey.

The first uncertainty is the active fraction of stars as a function of spectral type and distance from the Galactic midplane. The West et al. (2008) results are based on SDSS spectroscopic data, which are limited to bright objects. The active fraction at late spectral types and at far distances, which will certainly be visible to LSST, are currently unconstrained, and will likely remain so in the near-term. The second reason is that a study of flares in the SDSS spectroscopic survey finds that flare stars are preferentially closer to the Galactic plane than the active stars (Hilton et al. 2010), indicating that the FFD of active stars may be a function of distance from the Galactic plane. The third limitation to our model inputs is in the rate of flares at the high energy, low frequency end of the distribution. Because large flares happen infrequently, we have limited information on the frequency of the highest energy flares. We don't know if there is a break in the distribution, or simply a cut-off.

We have used the expected survey parameters for LSST to predict the number of flares seen in a single 15-second u-band exposure. Figure 4 shows the number of flares seen as a function of Galactic latitude. The decrease with increasing latitude is primarily caused by the smaller number of stars in sightlines looking out of the Galactic plane. As expected, there are more small flares than large flares. The error bars are calculated using the uncertainties in the FFDs (see Figure 2). The number of flares seen in a single night of LSST observing will be larger than the entirety of the Kowalski et al. (2009) SDSS Stripe 82 results. In the future, we can determine the FFD of stars to a much greater precision, and for much smaller spectral type and Galactic height bins, by varying our input FFDs in the model to fit the large number of LSST flare observations.

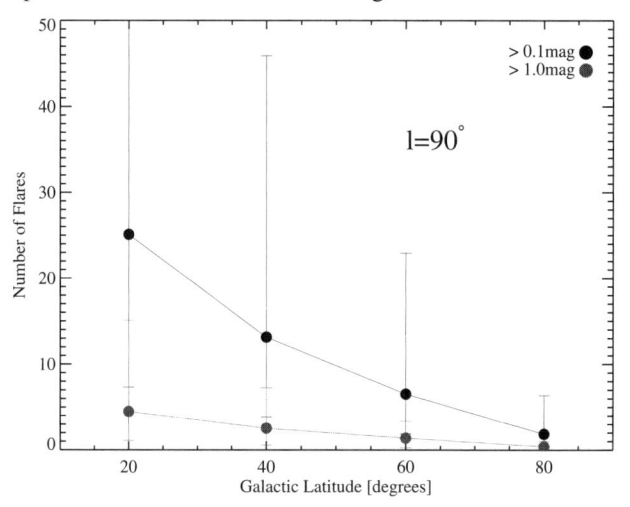

Figure 4. The predicted number of flares that will be seen in a single LSST u-band exposure. The decrease with Galactic latitude is caused by a decrease in the number of M dwarfs in the sightline.

Astronomers interested in exotic transients can neglect most stellar flares, since the source of the flare is known from the quiescent photometry. However, it is certainly possible for an M dwarf to have such a large flare that it will appear as an optical transient. In this case, the star is too faint to be visible in quiescence in any filter, but is briefly bright enough that it is seen during the flare, and will thus appear as a blue point source with no observed host. To meet this criteria, a flare must have a large change in magnitude:

$$\Delta u > (u_q - z_q) + (z_q - z_{lim}) + (z_{lim} - u_{lim}) = u_q - u_{lim} \qquad (6)$$
$$\text{where } z_q < z_{lim}$$

where the subscript q represents quiescent measurements and the subscript lim represents the survey limiting magnitude in that filter. The first term is the $u - z$ color of the star, which is several magnitudes for late M dwarfs. The second term means that stars close to the survey limit more likely to appear as transients, and the final term is the difference between the survey limits in the two filters. Note that although we use u and z filters in this example, we can make similar predictions for any other filter.

Using the LSST survey criteria, we predict that the probability of M dwarf flares appearing as optical transients in a single u-band exposure is ∼10% at low Galactic latitude, decreasing to just a few percent at high latitudes. Only the largest flares will meet the transient criteria, and as mentioned above, the frequency of very large flares is not well measured. The error bars are therefore quite large.

5. Conclusions

We have collected nearly 500 hours of flare monitoring observations to make the first measurements of the FFD of inactive early- and mid-type M dwarfs and for active late-type M dwarfs. The FFD is best represented as a linear relation between the log of the quiescent flare energy and the log of the frequency. We find that flares on earlier type stars are more energetic than on later type stars, and that inactive stars flare less frequently than active stars.

We incorporate the measured FFDs into our Galactic M dwarf flare model, which can be used to predict the number and magnitude of flares that will be seen in any survey. To make this prediction, we also adopt a flare shape, an active fraction (which is a function of spectral type and distance from the Galactic plane), a luminosity function, and a Galactic model. We draw from the appropriate FFD to generate a light curve for each star along a given sightline. Using the survey parameters from SDSS Stripe 82, our model was able to reproduce both the number and the magnitude distribution of flares found by Kowalski et al. (2009).

Applying our model to the survey parameters of LSST predicts tens of flares > 0.1 magnitudes and several flares > 1.0 magnitudes in each u-band exposure. We can also predict the probability of observing a large flare occurring on a star that is too faint to be seen during quiescence. These large flares, which appear as optical transients in the survey, have a few percent probability of being observed in each u-band exposure, although the uncertainties are large.

Finally, when large numbers of flares are measured in time domain surveys such as Pan-STARRS, PTF, and LSST, our model can be used to precisely determine the FFD of stars in very small bins of spectral type or color and distance from the Galactic plane. This will inform our knowledge of how the magnetic field of M dwarfs changes with spectral type (in particular how it changes across the fully convective boundary), with age, and with activity level.

References

Audard, M., Güdel, M., Drake, J. J., & Kashyap, V. L. 2000, ApJ, 541, 396
Basri, G., Walkowicz, L. M., Batalha, N., Gilliland, R. L., Jenkins, J., Borucki, W. J., Koch, D., Caldwell, D., Dupree, A. K., Latham, D. W., Meibom, S., Howell, S., & Brown, T. 2010, ApJ, 713, L155. 1001.0414
Bochanski, J. J., Hawley, S. L., Covey, K. R., West, A. A., Reid, I. N., Golimowski, D. A., & Ivezić, Ž. 2010, AJ, 139, 2679. 1004.4002
Buccino, A. P., Lemarchand, G. A., & Mauas, P. J. D. 2007, Icarus, 192, 582. arXiv: astro-ph/0701330
Charbonneau, e. a., D. 2009, Nat, 462, 891
Correia, A. C. M., Couetdic, J., Laskar, J., Bonfils, X., Mayor, M., Bertaux, J., Bouchy, F., Delfosse, X., Forveille, T., Lovis, C., Pepe, F., Perrier, C., Queloz, D., & Udry, S. 2010, A&A, 511, A21+. 1001.4774
Gershberg, R. E. 1972, Ap&SS, 19, 75
Gizis, J. E., Reid, I. N., & Hawley, S. L. 2002, AJ, 123, 3356. arXiv:astro-ph/0203499

Gliese, W. 1969, Veröeffentlichungen des Astronomischen Rechen-Instituts Heidelberg, 22, 1
Hawley, S. L., Gizis, J. E., & Reid, N. I. 1997, AJ, 113, 1458
Hawley, S. L., & Pettersen, B. R. 1991, ApJ, 378, 725
Hilton, E. J., West, A. A., Hawley, S. L., & Kowalski, A. F. 2010, AJ, 140, 1402
Ivezić, Ž., Smith, J. A., Miknaitis, G., Lin, H., Tucker, D., Lupton, R. H., Gunn, J. E., Knapp, G. R., Strauss, M. A., Sesar, B., Doi, M., Tanaka, M., Fukugita, M., Holtzman, J., Kent, S., Yanny, B., Schlegel, D., Finkbeiner, D., Padmanabhan, N., Rockosi, C. M., Jurić, M., Bond, N., Lee, B., Stoughton, C., Jester, S., Harris, H., Harding, P., Morrison, H., Brinkmann, J., Schneider, D. P., & York, D. 2007, AJ, 134, 973. arXiv:astro-ph/0703157
Juric, M., Cosic, K., Vinkovic, D., & Ivezic, Z. 2010, BAAS, 42, 222
Jurić, M., Ivezić, Ž., Brooks, A., Lupton, R. H., Schlegel, D., Finkbeiner, D., Padmanabhan, N., Bond, N., Sesar, B., Rockosi, C. M., Knapp, G. R., Gunn, J. E., Sumi, T., Schneider, D. P., Barentine, J. C., Brewington, H. J., Brinkmann, J., Fukugita, M., Harvanek, M., Kleinman, S. J., Krzesinski, J., Long, D., Neilsen, E. H., Jr., Nitta, A., Snedden, S. A., & York, D. G. 2008, ApJ, 673, 864. arXiv:astro-ph/0510520
Kaiser, N. 2004, in Ground-based Telescopes. Edited by Oschmann, Jacobus M., Jr. Proceedings of the SPIE, Volume 5489, pp. 11-22 (2004)., edited by J. M. Oschmann, Jr., vol. 5489 of Presented at the Society of Photo-Optical Instrumentation Engineers (SPIE) Conference, 11
Kowalski, A. F., Hawley, S. L., Hilton, E. J., Becker, A. C., West, A. A., Bochanski, J. J., & Sesar, B. 2009, AJ, 138, 633. 0906.2030
Kowalski, A. F., Hawley, S. L., Holtzman, J. A., Wisniewski, J. P., & Hilton, E. J. 2010, ApJ, 714, L98. 1003.3057
Lacy, C. H., Moffett, T. J., & Evans, D. S. 1976, ApJS, 30, 85
Law, N. M., Kulkarni, S. R., Dekany, R. G., Ofek, E. O., Quimby, R. M., Nugent, P. E., Surace, J., Grillmair, C. C., Bloom, J. S., Kasliwal, M. M., Bildsten, L., Brown, T., Cenko, S. B., Ciardi, D., Croner, E., Djorgovski, S. G., van Eyken, J., Filippenko, A. V., Fox, D. B., Gal-Yam, A., Hale, D., Hamam, N., Helou, G., Henning, J., Howell, D. A., Jacobsen, J., Laher, R., Mattingly, S., McKenna, D., Pickles, A., Poznanski, D., Rahmer, G., Rau, A., Rosing, W., Shara, M., Smith, R., Starr, D., Sullivan, M., Velur, V., Walters, R., & Zolkower, J. 2009, PASP, 121, 1395. 0906.5350
López-Santiago, J., Montes, D., Gálvez-Ortiz, M. C., Crespo-Chacón, I., Martínez-Arnáiz, R. M., Fernández-Figueroa, M. J., de Castro, E., & Cornide, M. 2010, A&A, 514, A97+. 1002.1663
LSST Science Collaborations: Paul A. Abell 2009, ArXiv e-prints. 0912.0201
Moffett, T. J. 1974, ApJS, 29, 1
Reid, I. N., Gizis, J. E., & Hawley, S. L. 2002, AJ, 124, 2721
Reid, I. N., Hawley, S. L., & Gizis, J. E. 1996, AJ, 111, 2469
Segura, A., Walkowicz, L. M., Meadows, V., Kasting, J., & Hawley, S. 2010, Astrobiology, 10, 751. 1006.0022
Udry, e. a., S. 2007, A&A, 469, L43
West, A. A., Hawley, S. L., Bochanski, J. J., Covey, K. R., Reid, I. N., Dhital, S., Hilton, E. J., & Masuda, M. 2008, AJ, 135, 785. 0712.1590
Wright, J. T. 2005, PASP, 117, 657. arXiv:astro-ph/0505214

Observations of Late-Type Stars with the Infrared Spatial Interferometer

Edward Wishnow, Charles Townes, Vikram Ravi, Sean Lockwood,
Hemma Mistry, Walt Fitelson, William Mallard, and Dan Werthimer

Space Sciences Laboratory and Department of Physics, University of California, Berkeley, CA 94720, USA

Abstract. The Infrared Spatial Interferometer (ISI) has been conducting mid-IR measurements of red supergiant and AGB stars for about 20 years. This paper reviews the ISI system and results with an emphasis on measurements of changes in stellar sizes and shapes, and those of the surrounding dust shells, over time scales of weeks to decades. The long-term measurement record provides important new observables for stellar theory. A new spectrometer-correlator system is discussed where this system will measure visibilities on-and-off individual molecular spectral lines.

1. Introduction

The ISI is a three-telescope interferometer that operates in the mid-infrared spectral region (9–11 μm) and it is located at the Mt. Wilson Observatory. The telescopes have a Pfund optical design with 1.65 m diameter primary mirrors and they are mounted in movable semi-trailers. The trailers can be placed at various locations at the ISI site giving a range of baseline separations of 4 to 85 meters. Figure 1 shows a photograph of the ISI in a short baseline linear array configuration. The system is currently in a triangular configuration with ~35 m baselines.

The ISI uses heterodyne detection where radiation from CO_2 lasers is mixed with starlight at each telescope, and the resulting intermediate frequency (IF) "beats" are correlated and detected. In this regard the system is like a radio telescope array, but it operates near 27 THz. Early CO_2 laser heterodyne spectroscopy studies measured the CO_2 abundance and winds of Venus and Mars (Betz et al. 1976; Johnson et al. 1976). Early heterodyne interferometry was performed using the auxillary siderostats at the McMath-Pierce telescope at Kitt Peak (Johnson et al. 1974) and circumstellar dust shells around a number of bright stars were observed (Sutton et al. 1977). A dedicated two-telescope ISI system obtained first fringes in 1989 and the third telescope was added in 2004. A more complete description of the ISI system is given in Hale et al. (2000) and at the ISI web site http://isi.ssl.berkeley.edu/.

Mid-infrared wavelengths are scattered less by dust than visible or near-IR wavelengths and hence stars can be observed even when they are embedded within dust shells. The dust shells can also be measured since, even though their emissivity is low, they are at a temperature where the Planck function peaks in the mid-infrared. The ISI has a narrow detection bandwidth with a spectral resolution of ~5000 that is tuned by choosing CO_2 laser lines to observe, or avoid, spectral lines due to molecular gases

Figure 1. The ISI telescope array at Mt. Wilson. The baselines shown are 4, 8 and 12 m, along an E-W line. Starlight is reflected from a steering flat mirror to a vertical parabola on the right. Light is then focused through the hole in the flat mirror to the detection system in the cabin behind the flat mirror. The master laser is in the trailer with the three periscopes and the control room trailer is at the left edge of the photo.

surrounding these stars. Many ISI studies of stellar sizes have thereby been made of the continnum, free of confusion due to gases. A new spectrometer-correlator system is being developed to measure visibilities at frequencies of spectral lines and the continuum simultaneously.

The ISI observational program is directed towards precision observations of red supergiant stars and AGB stars. The measurements have provided new information on stellar properties. Among these are the sizes and shapes of stars, the sizes and shapes of dust shells surrounding stars, and importantly, the changes of these properties with time.

2. Stellar Interferometry

Optical-IR interferometry has advanced sufficiently that it is making contributions to stellar astrophysics in many areas; this work is covered in a recent review by Ten Brummelaar et al. (2009). The ISI has contributed to the development of interferometry and some discussion of interferometers is appropriate for understanding ISI results. In a typical optical-IR interferometer, beams are combined together and detected to form interference fringes. In the ISI, the intermediate frequency signals derived at each telescope are brought together and correlated electronically. The quantity measured is the "Michelson visibility", or the normalized, square of the complex visibility, which according to the van Cittert-Zernike theorem, is the two-dimensional Fourier transform of the source intensity distribution. The ISI measures the visibility squared, $V^2 = (FP)^2/P_1 P_2$ where FP is the power in the interference fringe and P_1 and P_2 are the individual powers measured at each telescope. The visibility is sampled at various points in the spatial frequency (U, V) plane, given by $\mathbf{B}/\lambda$ where $\mathbf{B}$ is the projection of the interferometer baseline vector perpendicular to the direction to the star, and λ is the wavelength. As the Earth rotates, or if the telescope positions are moved, a set of measurements is assembled that samples the visibility function.

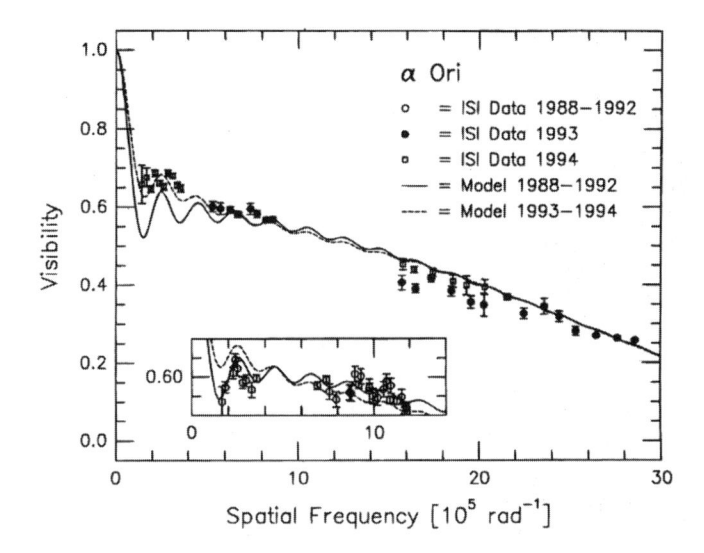

Figure 2. ISI visibility measurements of Betelgeuse at 11.15 μm. The low spatial frequency ripples correspond to a ~ 1"radius dust shell, and the high spatial frequencies match a uniform disk of 23 mas radius. The variation of low frequency points over time, shown in the inset, indicates the shell is hollow during 1988–1992, and then dust is present near the star in the 1993-1994 period. From Bester et al. (1996).

The classic Young's double-slit experiment is an analog of stellar interferometry. For a point source, the fringe pattern observed at a screen behind the slits is determined by the separation (baseline) of the slits. If the slit separation is increased, the fringe frequency increases, but the contrast does not vary. If, however, the source has a finite size, then as the slit separation increases, there will be a separation for which the fringe contrast is zero. The variation in fringe contrast with slit separation is very similar to the experiments conducted at the 100" telescope by Michelson & Pease (1921).

Visibility measurements and their interpretation are best illustrated by example using ISI measurements of Betelgeuse shown in Figure 2. The measurements cover spatial frequencies from about 2 to 30 SFU, or spatial frequency units (1 SFU= 1×10^5 cycles/radian) where these correspond to telescope baseline separations of ~2 to 32 m. The high spatial points are associated with the star itself and the lower spatial frequency ripples correspond to a shell of dust surrounding the star. The solid curve is a model fitted to data from 1988–1992 which consists of a uniform stellar disk of 23 mas radius plus a thin dusk shell of 1"radius. The dashed curve is a model fitted to later data where the star is again a uniform disk, but the low frequency ripples must be fitted with shell of material closer to the star (0.01") in addition to the large shell of material at 1". This figure shows an early measurement of changes of material surrounding supergiant stars on year time scales.

The ISI cannot measure the phase of an interference fringe, only the magnitude, due to atmospheric turbulence. If the phase is not known, the positions of features are not determined and the measured visibilities only have information related to a symmetric source intensity distribution. With three telescopes a quantity called the closure phase, Φ, can be measured where $\Phi = \phi_{12} + \phi_{23} + \phi_{31}$ Here, the ϕ_{ij}'s are the phases of the individual fringes of the three telescope baselines summed around a closed triangle. This quantity is not distorted by atmospheric turbulence and it provides information related to stellar asymmetries. For example, the closure phase is zero for sources which are centro-symmetric, like circles and ellipses, but non-zero for egg-shapes or a pair of disks where one is brighter than the other.

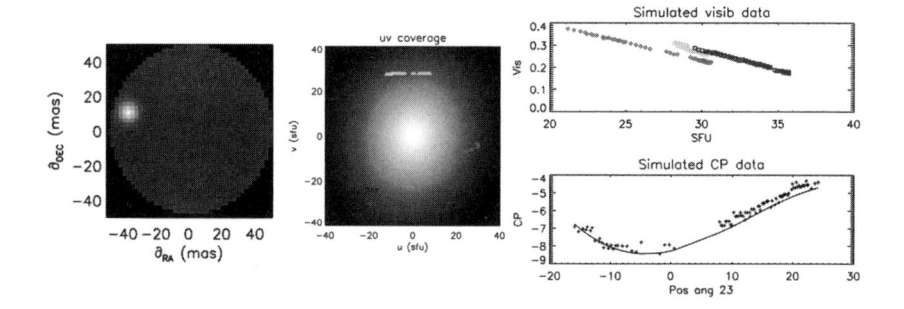

Figure 3. Simulation of an asymmetric source illustrating a non-zero closure phase. Left panel, source with hot spot; middle panel, 2D FFT of source image with UV samples from ISI observations. Red, green and blue colors indicate the telescope pairs, 1-2, 2-3 and 3-1. The upper right panel shows the visibility magnitudes at the sampled points. The lower right panel shows the closure phases obtained from the sampling complex 2D FFT. The visibilities could be due to an ellipse, but the closure phase indicates the size and orientation of the source asymmetry.

Figure 3 shows an example of a simulated two dimensional source distribution with a bright spot placed off-center on a stellar disk. The magnitude of the two-dimensional Fourier transform of this distribution, the complex visibility, is shown in the middle panel and the colored points show UV samples of the visibility. The upper right panel plots the sampled visibility magnitudes as a function of spatial frequency. The lower right panel shows the closure phases obtained from summing the phases of the UV samples of the complex visibility; the closure phases are plotted as a function of the position angle of the telescope 2-3 baseline (green points). The variation of the closure phase can then be used to determine that there is an asymmetry in the source, where this may not be apparent from the measured visibilities. Modeling of visibility and closure phase measurements of Betelgeuse over the period 2006-2009 are presented by Ravi et al. (2010) in these proceedings.

3. Heterodyne Interferometry System

A schematic diagram of the ISI system is shown in Figure 4 below. A more complete description of this diagram is given in Wishnow et al. (2010a).

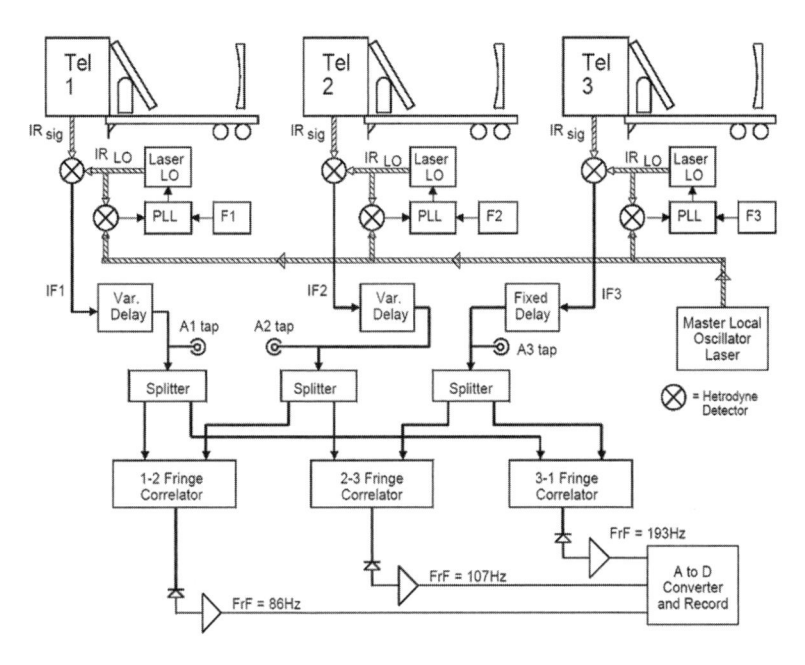

Figure 4. ISI system diagram. Radiation from a star is combined with that from CO_2 laser local oscillators (LO). The resulting IF signals are delayed, correlated and low-pass filtered. The interference fringes occur at frequencies chosen to be 86, 107 and 193 Hz, by locking the beat between the LOs and the master oscillator to computer controlled synthesized frequencies F1, F2 and F3. The delay lines compensate for differential geometric, and signal, pathlengths through the telescopes and cables to the correlators with the required accuracy of 5 mm (this is set by the bandwidth).

4. Measurements Over Time: AGB Stars

The ISI has measured the dust distribution surrounding a number of supergiant, carbon and Mira stars (Danchi et al. 1994). The ISI has also made precise measurements of changes in stellar sizes over time and Figure 5 shows the size variation of Mira over three stellar luminosity cycles. The maximum mid-IR size lags the maximum visible brightness by ~0.135 of the luminosity period.

Recent measurements of Mira and R Leo, using three telescopes and closure phase, have been fitted with uniform disk+hot spot models. Some variations in stellar sizes have also been examined over a time period of a few months (Tatebe et al. 2008). The ISI size variations over time form a set of observables that can be compared to stellar models, such as those of Freytag & Höfner (2008) and Chiavassa et al. (2010b).

The ISI has also observed variations in circumstellar material as it flows outwards from the star (Tatebe et al. 2006; Chandler et al. 2007). Figure 6 shows the measured visibilities, closure phases, and one-dimensional integrated intensities of the dust shell that surrounds Mira obtained over four observing periods. This work was obtained using the short-baseline configuration shown in Figure 1. The notable feature of this series of measurements is the formation of a fairly large amount of dust, with an asym-

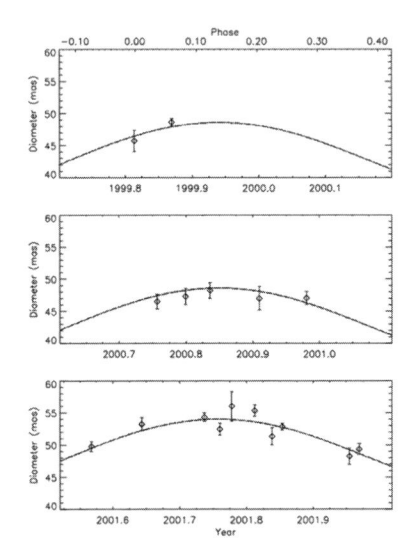

Figure 5. ISI diameter measurements of Mira measured over three stellar periods. The average diameter is 48 mas and the fitted curve has a peak-to-peak amplitude of 12 mas. Note the maximum size varies across stellar cycles. From Weiner et al. (2003a).

metric distribution, in fall of 2003 and the expansion and dissipation of this dust shell in subsequent years. Mass-loss and dust formation processes are apparently episodic. In addition to observing material in circumstellar regions fairly far away from the star, measurements of the flow of material (presumably dust) just above the stellar photosphere have been measured for the star W Hya (Wishnow et al. 2010b).

5. Measurements Over Time: Red Supergiants

ISI studies of Betelgeuse show very interesting variations of the stellar size and shape over relatively short time intervals. Tatebe et al. (2007a) observed an asymmetry in 2006 which was fitted with a uniform disk+hot spot model, and Ravi et al. (2010) in an accompanying paper discuss the change of model images over the time period 2006–2009. It would be very valuable to model these variations using an approach similar to that used by Chiavassa et al. (2010a) to interpret the H band interferometric measurements of Betelgeuse (Haubois et al. 2009)

ISI observations of the diameter of Betelgeuse over the period 1994 to 2009 are shown below in Figure 7. A recent paper showed that the star diminished in size over the first 16 years of observation (Townes et al. 2009). More recently, the star's size has increased. The sizes shown in this plot differ slightly from those in the Ravi stellar model images because the data for 2007 and 2008, has been fitted with simple uniform disks, and the analysis did not take into account the small fraction of flux associated with asymmetries. There is no obvious correlation between the stellar size and the visible photometry, but it is tempting to associate the recent size increase with a recent brightness increase. Given these recent variations, it will be quite interesting to continue

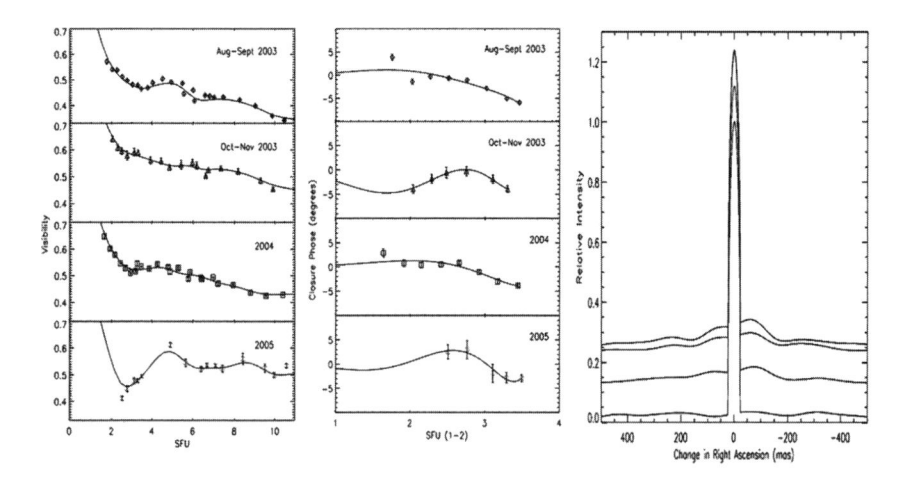

Figure 6. ISI measurements of dust surrounding Mira over time. Left to right: visibility vs. spatial frequency; closure phase vs. spatial frequency; 1D integrated intensity profile (Fourier transform of the visibility), The curves from top to bottom are: Aug-Sept 2003, Oct-Nov 2003, 2004, 2005 and they are displaced upwards from the baseline of 2005 by an amount proportional to the time difference. From Chandler et al. (2007).

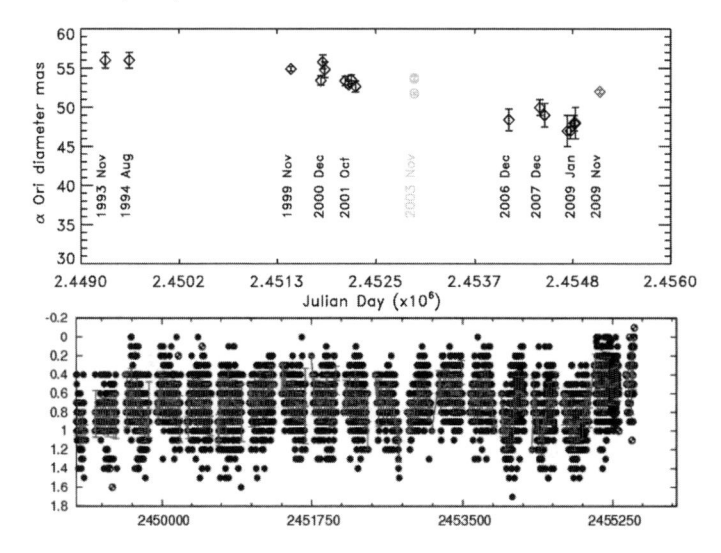

Figure 7. Upper panel, mid-IR diameter measurements of Betelgeuse vs. time. ISI measurements are in black (Townes et al. 2009) with a recent one in red; VLTI mid-IR measurements are in green (Perrin et al. 2007). Lower panel, visible photometry vs. time. AAVSO data is shown by black dots and 30 day averages are in red.

monitoring this star. It will also be important to compare the mid-IR measurements to those made in other wavelength bands over similar time periods.

Another long-term ISI supergiant study examined the behavior of the dust surrounding α Her. The star was surrounded by a moderately dense shell of dust in 1992, but by 1993 it had begun to dissipate and since 1996 there is relatively little dust around this source (Tatebe et al. 2007b).

6. Spectroscopic Interferometry

Evolved giant stars do not have well defined surfaces and their appearances are further affected by limb darkening, or by emission from overlying material. As a result, the diameter of a star varies as a function of wavelength due to the various contributing factors to the stellar opacity. Near-IR Keck aperture masking interferometry of Mira shows diameter variations of greater than 2 across the band, where large diameters are observed in spectral regions of water vapor emission (Woodruff et al. 2009). Variations in the apparent diameter of Betelgeuse of 30% were observed across the mid-IR band using the VLT interferomter (Perrin et al. 2007). The contributions of various molecular gases, dust types or free electron scattering, to the mid-IR opacity of extended stellar atmospheres is not completely understood.

Mid-IR interferometric measurements of Betelgeuse were well modeled by a detached molecular shell, often called a "MOLsphere," by Perrin et al. (2007). On the other hand, spectroscopic measurements of OH and H_2O spectral lines have been interpreted using a higher temperature region in close proximity to the stellar surface (Jennings & Sada 1998; Ryde et al. 2006).

In order to better understand the distribution of molecules surrounding red supergiant and AGB stars, and the nature of the mid-IR opacity, we are currently building a spectrometer-correlator system to observe visibilities on-and-off individual molecular spectral lines. The spectrometer-correlator is being built in collaboration with the Center for Astronomy Signal Processing and Electronics Research (CASPER) at UC Berkeley. It is based on very high speed Field Programmable Gate Array (FPGA) processors and 6 Giga-samples/sec analog-to-digital converter; the system is described in more detail in Wishnow et al. (2010a).

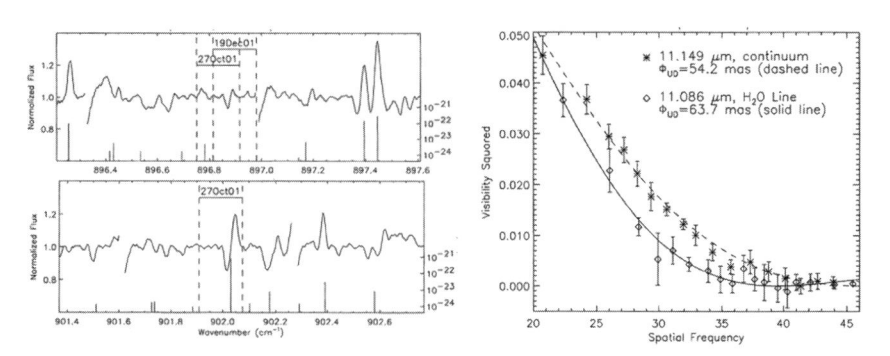

Figure 8. Left panels: two spectral regions of Mira. Upper plot continuum, lower plot H_2O line region (11.149 μm=896.94 cm^{-1}, 11.086 μm=902.04 cm^{-1}), dashed lines show ISI detection bands. Right panel: visibility squared of Mira on-and-off the H_2O line. The apparent size on-line is larger than the continuum size. From Weiner et al. (2003b).

Previous ISI spectroscopy studies used an analog filterbank system to show that ammonia (NH_3) and silane (SiH_4) form at large distances from the star, 20 and 80 stellar radii, respectively, in the case of IRC+10216 (Monnier et al. 2000). In the case of VY Cma, ammonia was observed at a distance of 40 stellar radii. Figure 8 shows ISI visibility measurements of Mira made by Weiner et al. (2003b) The left panels panels show spectra obtained using the TEXES high resolution spectrometer (Lacy et al. 2002) and the "stick spectra" are predicted intensities of H_2O lines at 1000 K. The ISI detection bands are shown with dashed lines where the apparent shifts of the bands in the upper plot are due to the barycentric motion of the Earth (in fact the spectra shift, but the frequency axes are in the star's reference frame). The right panel shows visibility measurements in spectral regions with, and without, a water line. The visibility measurements were obtained using the entire RF detection band so the H_2O line intensity is diluted and the diameter associated with the line may also be reduced. Notice that the water line near 902 cm^{-1} has an "inverse P-Cygni" lineshape. This arises when absorbing material along the line of sight is moving towards the star and emitting gas is flowing outwards. This spectral shape may indicate a velocity differential on either side of an outward propogating shock.

Initial spectroscopic-interferometry studies will measure visibilities on H_2O and OH lines which are present in atmospheres of red supergiant and AGB stars. H_2O, OH lines, and continuum regions of interest are selected by predicting the molecular spectrum based on a stellar atmospheric temperature and the recessional velocity. The region of interest must then fall within the ± 3 GHz spectrometer bandwidth centered on a CO_2 laser line. Figure 9 shows a transparent spectral region of Earth's atmosphere and a set of laser lines for two isotopologues of CO_2 which are chosen to avoid absorption by telluric $^{16}O^{12}C^{16}O$.

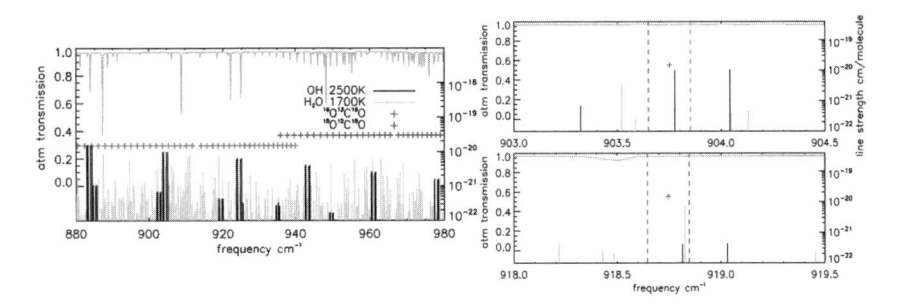

Figure 9. Left panel, atmospheric transmission vs. frequency (upper blue curve). Predicted intensities of OH and H_2O lines are shown as "stick spectra"; laser lines are indicated by red and blue + symbols. Right panels, two expanded regions of predicted stick spectra of α Ori for 8 Nov. ISI detection bands are indicated by dashed red lines which are centered on CO_2 laser lines. For this star, barycentric motion shifts the stellar lines with respect to the detection band by about ± 0.17 cm^{-1} over a year.

ISI interferometric-spectroscopy measurements will determine the abundance and temperatures of molecules in the regions just beyond the continuum photosphere. Studies of spectral line profiles, and hence the gas kinematics of stellar atmospheres, may also be possible with the spectrometer-correlator. The system is being tested and observations will begin in the summer of 2011.

7. Conclusion

The ISI has been performing precision measurements of evolved giant stars and the material that surrounds them. These measurements of stellar features and behavior can be used to test and motivate stellar theory. A new spectroscopic-interferometry capability will give information on the distribution of molecules and their temperatures in extended stellar atmospheres.

The ISI capabilities are unique compared to other mid-IR interferometers (Keck, VLT and LBT) as it performs measurements with a cadence that is not practical at large facilities; this allows studies of stars on both short- and long-time scales. The new spectrometer-correlator system has a spectral resolution of ~600,000 (although both sidebands fold on top of each other) and it is the only mid-IR interferometer capable of visibility measurements on individual spectral lines. ISI high spectral and high spatial resolution capabilities complement far-IR spectroscopy by the Herschel Space Observatory and sub-mm spectroscopy that will be performed by ALMA.

Acknowledgments. This work made use of the SIMBAD and AAVSO databases. We appreciate support from the Office of Naval Research, the Gordon and Betty Moore Foundation and the National Science Foundation.

References

Bester, M., Danchi, W. C., Hale, D., Townes, C. H., Degiacomi, C. G., Mekarnia, D., & Geballe, T. R. 1996, ApJ, 463, 336
Betz, A. L., Johnson, M. A., McLaren, R. A., & Sutton, E. C. 1976, ApJ, 208, L141
Chandler, A. A., Tatebe, K., Wishnow, E. H., Hale, D. D. S., & Townes, C. H. 2007, ApJ, 670, 1347
Chiavassa, A., Haubois, X., Young, J. S., Plez, B., Josselin, E., Perrin, G., & Freytag, B. 2010a, A&A, 515, A12+. 1003.1407
Chiavassa, A., Lacour, S., Millour, F., Driebe, T., Wittkowski, M., Plez, B., Thiébaut, E., Josselin, E., Freytag, B., Scholz, M., & Haubois, X. 2010b, A&A, 511, A51+. 0911.4422
Danchi, W. C., Bester, M., Degiacomi, C. G., Greenhill, L. J., & Townes, C. H. 1994, AJ, 107, 1469
Freytag, B., & Höfner, S. 2008, A&A, 483, 571
Hale, D. D. S., Bester, M., Danchi, W. C., Fitelson, W., Hoss, S., Lipman, E. A., Monnier, J. D., Tuthill, P. G., & Townes, C. H. 2000, ApJ, 537, 998
Haubois, X., Perrin, G., Lacour, S., Verhoelst, T., Meimon, S., Mugnier, L., Thiébaut, E., Berger, J. P., Ridgway, S. T., Monnier, J. D., Millan-Gabet, R., & Traub, W. 2009, A&A, 508, 923. 0910.4167
Jennings, D. E., & Sada, P. V. 1998, Science, 279, 844
Johnson, M. A., Betz, A. L., McLaren, R. A., Townes, C. H., & Sutton, E. C. 1976, ApJ, 208, L145
Johnson, M. A., Betz, A. L., & Townes, C. H. 1974, Physical Review Letters, 33, 1617
Lacy, J. H., Richter, M. J., Greathouse, T. K., Jaffe, D. T., & Zhu, Q. 2002, PASP, 114, 153. arXiv:astro-ph/0110521
Michelson, A. A., & Pease, F. G. 1921, ApJ, 53, 249
Monnier, J. D., Danchi, W. C., Hale, D. S., Tuthill, P. G., & Townes, C. H. 2000, ApJ, 543, 868. arXiv:astro-ph/0007418
Perrin, G., Verhoelst, T., Ridgway, S. T., Cami, J., Nguyen, Q. N., Chesneau, O., Lopez, B., Leinert, C., & Richichi, A. 2007, A&A, 474, 599. 0709.0356
Ravi, V., Wishnow, E. H., Lockwood, S., & Townes, C. H. 2010, ArXiv e-prints. 1012.0377
Ryde, N., Harper, G. M., Richter, M. J., Greathouse, T. K., & Lacy, J. H. 2006, ApJ, 637, 1040. arXiv:astro-ph/0510177

Sutton, E. C., Storey, J. W. V., Betz, A. L., Townes, C. H., & Spears, D. L. 1977, ApJ, 217, L97
Tatebe, K., Chandler, A. A., Hale, D. D. S., & Townes, C. H. 2006, ApJ, 652, 666
Tatebe, K., Chandler, A. A., Wishnow, E. H., Hale, D. D. S., & Townes, C. H. 2007a, ApJ, 670, L21
Tatebe, K., Hale, D. D. S., Wishnow, E. H., & Townes, C. H. 2007b, ApJ, 658, L103
Tatebe, K., Wishnow, E. H., Ryan, C. S., Hale, D. D. S., Griffith, R. L., & Townes, C. H. 2008, ApJ, 689, 1289
Ten Brummelaar, T., Creech-Eakman, M., & Monnier, J. 2009, Physics Today, 62, 060000
Townes, C. H., Wishnow, E. H., Hale, D. D. S., & Walp, B. 2009, ApJ, 697, L127
Weiner, J., Hale, D. D. S., & Townes, C. H. 2003a, ApJ, 588, 1064
— 2003b, in Society of Photo-Optical Instrumentation Engineers (SPIE) Conference Series, edited by W. A. Traub, vol. 4838 of Presented at the Society of Photo-Optical Instrumentation Engineers (SPIE) Conference, 172
Wishnow, E. H., Mallard, W., Ravi, V., Lockwood, S., Fitelson, W., Wertheimer, D., & Townes, C. H. 2010a, in Society of Photo-Optical Instrumentation Engineers (SPIE) Conference Series, vol. 7734 of Presented at the Society of Photo-Optical Instrumentation Engineers (SPIE) Conference
Wishnow, E. H., Townes, C. H., Walp, B., & Lockwood, S. 2010b, ApJ, 712, L135
Woodruff, H. C., Ireland, M. J., Tuthill, P. G., Monnier, J. D., Bedding, T. R., Danchi, W. C., Scholz, M., Townes, C. H., & Wood, P. R. 2009, ApJ, 691, 1328. 0811.1642

A Search for Periodic Optical Variability in Radio Detected Ultracool Dwarfs: A Consequence of a Magnetically-Driven Auroral Process?

Leon K. Harding,[1] Gregg Hallinan,[2,3] Richard P. Boyle,[4] Ray F. Butler,[1] Brendan Sheehan,[1] and Aaron Golden[1]

[1]*Centre for Astronomy, National University of Ireland Galway, University Road, Galway, Ireland*

[2]*National Radio Astronomy Observatory, 1003 Lopezville Road, NM 87801-0387, USA*

[3]*Astronomy Department, University of California at Berkeley, CA 94720-3411, USA*

[4]*Vatican Observatory Research Group, Steward Observatory, University of Arizona, Tucson, AZ 85721, USA*

Abstract. A number of ultracool dwarfs have been unexpectedly detected as radio sources in the last decade, four of which have been found to be producing periodic pulses. More recently, two of these pulsing dwarfs have also been found to be periodically variable in broadband optical photometry. The detected periods match the periods of the radio pulses which have previously been associated with the rotation period of the dwarf. For one of these objects, it has also been established that the optical and radio periodic variability are possibly linked, being a consequence of magnetically-driven auroral processes. In order to investigate the ubiquity of the periodic optical variability in radio detected sources, the GUFI instrument (Galway Ultra Fast Imager) was commissioned on the 1.8m Vatican Advanced Technology Telescope, on Mt. Graham, Arizona, and has been obtaining data for the past eighteen months. More than two hundred hours of multi-epoch photometric monitoring observations of radio detected ultracool dwarfs have been completed. We present initial results confirming optical periodic variability for four of this sample, three of which have been newly confirmed using GUFI.

1. Introduction

Ultracool dwarfs are fully convective stars which occupy the spectral range late-M, L and T (<2500 K). A number of investigations undertaken in the last decade have found optical and infrared variability in this class of object. In most cases, this variability has been attributed to the expected presence of atmospheric dust (Bailer-Jones & Mundt 2001; Martin et al. 2001; Gelino et al. 2002; Enoch et al. 2003; Maiti 2007; Littlefair et al. 2008; Goldman et al. 2008; Clarke et al. 2008; Artigau et al. 2009); however, for higher temperature objects (specifically late-M dwarfs), the presence of magnetic spots and other magnetic related activity, as seen for earlier M-dwarfs, may also be present (Rockenfeller et al. 2005; Lane et al. 2007). Modulation at the expected rotation period has been found in some cases (Clarke et al. 2002; Koen 2006; Lane et al. 2007), whereas in other cases aperiodic variability, or periodic modulations on time-scales not associated with rotation have been inferred (Gelino et al. 2002; Lane et al. 2007; Maiti 2007).

The association of magnetic activity with the observed optical variability in these investigations was prompted as a result of the detection of both quiescent and time-variable radio emission in the late-M and L dwarf range (Berger et al. 2001; Berger 2002; Berger et al. 2005; Burgasser & Putman 2005; Osten et al. 2006; Berger 2006; Phan-Bao et al. 2007; Hallinan et al. 2006, 2007; Antonova et al. 2007; Berger et al. 2009). To date, periodic pulses have been detected from four ultracool dwarfs, and this emission has been associated with the period of rotation of the dwarf (Hallinan et al. 2006, 2007, 2008; Berger et al. 2009). Observations by Hallinan et al. (2008) of the M9 dwarf TVLM 513-46546 (henceforth TVLM 513), reveal electron cyclotron maser emission as the mechanism responsible for the observed 100% circularly polarised periodic pulses, where magnetic field strengths of at least 3 kG are present. Indeed, such a model requires kG magnetic field strengths and infers that such fields are large-scale and stable in configuration. More recently, for TVLM 513 in particular, we have evidence that the optical and radio emissions are produced by the same population of electrons in the magnetosphere, and may be auroral in nature (Hallinan et al., this volume).

We therefore undertook a campaign to investigate the ubiquity of optical periodic variability for known radio detected ultracool dwarfs using the GUFI photometer, on the 1.8m Vatican Advanced Technology Telescope (VATT). These data will provide an insight into the cause of this optical emission, its connection to the radio processes, and most importantly determine whether optical periodic signals are present *only* in radio pulsing dwarfs. We present initial results of periodic variability, from multiple epoch observations, for the following dwarfs: the M8V tight binary dwarf LP 349-25; the M8.5 dwarf LSR J1835+3259 (henceforth LSR J1835); the M9 dwarf TVLM 513; the L dwarf binary 2MASS J0746+2000 (henceforth 2M J0746). We have also observed the M9.5 dwarf BRI 0021-0214 and the L3.5 dwarf 2MASS J0036+1821 and detected periodicity, however these are single epochs and a second set of observations are required. Each confirmed period has multiple epochs where the periodic variability is categorically present in the data.

2. Optical Observations

2.1. GUFI mk.II - the Galway Ultra Fast Imager Photometer

The GUFI instrument was originally commissioned by astronomers in NUI Galway as an optical photometer capable of high-time resolution imaging (Sheehan & Butler, in prep.). We modified the GUFI mk.II system to be compatible with the 1.8m VATT on Mt. Graham, Arizona, where it is currently stationed as a visitor instrument. The system uses the Andor iXon DV887 EM-CCD camera from e2v technologies, which has a CCD97 thinned back-illuminated sensor hosting >90% quantum efficiency with a native 512×512 frame transfer sensor (~2ms transfer rate). It offers variable readout rates up to 10 MHz and can operate full-frame at 32 frames per second (fps) and up to 526 fps in a windowed configuration. The native field of view (FOV) of GUFI at the VATT Cassegrain focus is ~ $1.7' \times 1.7'$ with a corresponding plate scale of $0.2''$/pixel. Focal reducer options for wider fields are limited by the short VATT back focal distance of 50.8mm, but GUFI provides near-infrared (NIR) and visible-optimised focal reducers, offering a FOV of ~ 3′ × 3′ and a larger plate scale of 0.35″/pixel.

2.2. Observations and Data Reduction

Observations of selected targets were carried out between May 2009 - December 2010. Due to the red nature of these objects, the NIR and visible-optimised focal reducers were used in order to maximise flux sensitivity during exposures. Consequently, we used the VATT I-Arizona (~7200-9100 Å) and R-Harris (~5600-8800 Å) broadband filters. The campaign encompassed observations of all radio detected dwarfs to search for periodic variability, that were visible from the VATT observatory site (32°42′4.78″ 109°53′32.5″W). Typical acquisition parameters employed 5 second exposure times and 1 MHz readout rates using the conventional amplifer (EM gain at unity).

Data reduction was carried out using the in-house GUFI L3 Pipeline (Sheehan & Butler, in prep.). Standard data reduction techniques were employed where the data were bias subtracted using zero-integration frames and flat fielded using twilight flat fields. Twilight flat fields for any given observation consisted of >100 median-combined dithered frames taken from a blank part of the sky. Frames were summed in image space to increase the signal-to-noise ratio (SNR) and differential photometry was carried out on all science data in order to achieve milli-magnitude photometric precision. The ~ 3′ × 3′ FOV of GUFI provided between 3-30 reference stars for a given field. Photometry for all reference stars was also obtained as a measure of their stability in order to ensure variability was intrinsic to the target star. These stars were chosen on the basis of their stability, position, isolation, the properties of their seeing profiles, and comparable magnitudes and colour to that of the target. Photometric apertures (in pixels) which provided the highest SNR for the target star were selected for aperture photometry; however aperture and sky annulus diameters varied from night to night depending on the average seeing conditions, which typically ranged from 0.7 to 1.6 arcseconds. Differential photometry was obtained by dividing the target flux by the mean flux of the reference stars. Although changing seeing (varying point spread function) can ultimately introduce photometric errors, for a given observation the photometric parameters remained constant for all stars - this ensured that the same fraction of total flux was being observed in the aperture of each source.

3. Discussion of Individual Targets and Results

Each consecutive target is discussed in the following sections and its GUFI photometry is shown in Fig. 1 and Fig. 2. In order to detect periodic variability, and assess its significance, the Lomb-Scargle (LS) Periodogram (Lomb 1976; Scargle 1982) was calculated and the power spectra were analysed for significant peaks, which correspond to periodic variability (Fig. 1).

3.1. LP 349-25

LP 349-25 is a tight M8V binary that is located at a distance of ~7.8 pc. It has an inferred mass of m~ $0.12^{+0.008}_{-0.007} M_\odot$, and evolutionary models estimate the binary's age to be 140±30 Myr (Dupuy et al. 2010), which infers it should have an abundance of lithium. However, Reiners & Basri (2009) detect no lithium in the binary dwarf's atmosphere, which suggests that the age derived from modelling is perhaps an underes-

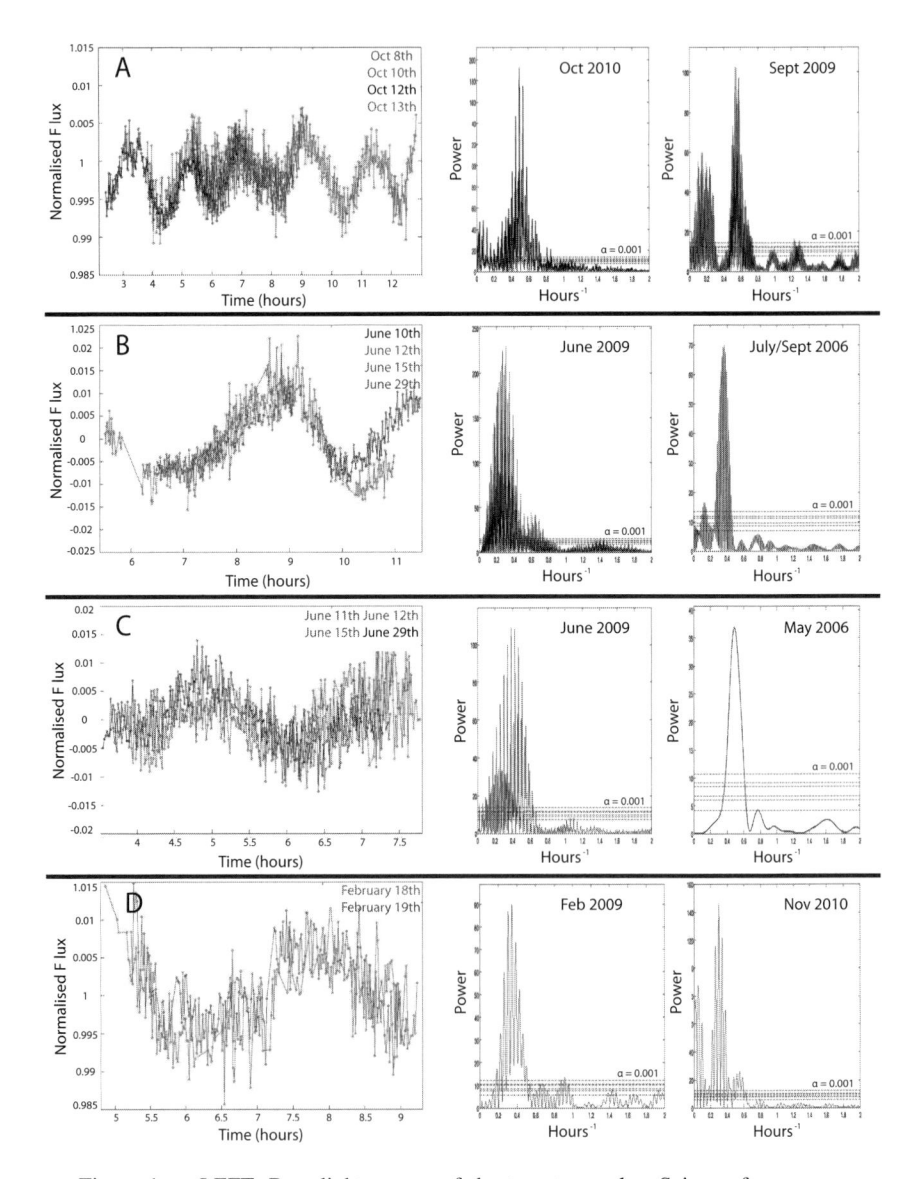

Figure 1. LEFT: Raw light curves of the target sample. Science frames were summed to 1 minute to increase SNR (raw 5 sec). Each consecutive night of a given observation has been shifted in phase by the detected period of rotation. RIGHT: Corresponding LS Periodograms showing multiple epochs, and the significance of the detection. *A:* LP 349-25, Oct 10, 12, 15 & 29 2010, $p \approx 1.86$ hrs; *B:* LSR J1835, Jun 8, 10, 12 & 13 2009, $p \approx 2.84$ hrs; *C:* TVLM 513, Jun 11, 12, 13 & 15 2009, $p \approx 1.96$ hrs; *D:* 2M J0746, Feb 18 & 19 2010, $p \approx 3.3$ hrs.

timate of the system's true age. It was reported as the most radio luminous ultracool dwarf yet detected by Phan-Bao et al. (2007), however no pulsing was found during the observation. They argued that either gyrosynchrotron or electron cyclotron maser emission could be the mechanism responsible. We observed LP 349-25 in I-band and R-band broadband photometry to investigate the optical variable nature of the source. We detect the binary as a periodically varying source in both bands over the course of three separate epochs of observations for a total of 60 hours, spanning fourteen months. We show periodic sinusoidal variability with a period of ~1.86 hours for one component of the tight binary, present in each band and varying with a peak to peak amplitude of ~1.5% in I and ~3% in R respectively.

3.2. LSR J1835+3259

LSR J1835 was reported by Hallinan et al. (2008) to produce persistent pulses of radio emission with a period of 2.84±0.01 hours, which they attributed to the dwarf's rotation period. They selected this dwarf because of its close proximity (~6.2 pc), similar spectral type (M8.5) to the previously detected pulsating dwarf TVLM 513 and the detection of radio emission by Berger (2006). Since optical periodic variability was also present for TVLM 513 (Lane et al. 2007), we decided to investigate the presence of such periodic variability for LSR J1835. We conducted observations over a period of three years encompassing three separate epochs. Initial epochs were taken as test data for the GUFI mk.I system, and we observed the dwarf in I-band to confirm its periodic nature for a total of 18 hours in June 2009. We determined an optical period of ~2.84 hours, consistent with the VLA radio observations of Hallinan et al. (2008). These data reveal stable periodic sinusoidal variability with peak to peak amplitude variations of ~3%.

3.3. TVLM 513-46546

TVLM 513 is an M9 ultracool dwarf which is at a distance of ~10.6 pc. Based on its mass ($m<0.8M_\odot$) and age (<10 Gyr), it is located at the substellar boundary and is one of the most rapidly rotating dwarfs discovered thus far with a rotation rate of ~60 km s^{-1} (Basri 2001). Periodic pulses of ~1.96 hours were detected by Hallinan et al. (2006, 2007), which they attributed to coherent electron cyclotron maser emission and argued that large-scale, stable magnetic fields of strength ~3 kG were present. Lane et al. (2007) detected a periodic signal of ~1.96 hours in photometric I-band data, which matched the period of the radio pulses of Hallinan et al. (2008). We observed TVLM 513 with GUFI on VATT for 28 hrs in June 2009. These observations were performed to 1) confirm the stability of the optical period over time-scales of years, and 2) determine the period of rotation to a greater level of accuracy via an increased temporal baseline. We confirm consistent periodic variability of ~1.96 hours with peak to peak amplitude variations of ~2.5%.

3.4. 2MASS J0746+2000

2M J0746 is an L dwarf binary (L0+L1.5) possessing an integrated rotation velocity for both components of ~31 km s^{-1} (Reiners & Basri 2008) at a distance of ~12.2 pc. The first detection of radio emission was reported by Antonova et al. (2008), where they reveal 100% circularly polarised emission at 4.9 GHz. Berger et al. (2009) presented periodic radio emission at 4.86 GHz of 2.07±0.002 hours, as well as periodic Hα emission and an inferred magnetic field strength of ~1.7 kG. They proposed that the radio and spectroscopic periodic emission was emanating from the primary binary member. We obtained multiple epoch I-band observations of 2M J0746 to investigate if the same periodic emission was present at these wavelengths as was observed at radio and spectroscopic bands. Observations were taken over three separate epochs which were nine months apart, for a total of 25 hours. Most intriguingly, we show optical periodic modulation of ~3.3 hours, where the sinusoidal variability exhibits peak to peak amplitude variations of ~2%. Therefore this optical variability originates from the other component to that producing the radio emission. The optical periodicity is categorically present in the data, and therefore we conclude that the period of rotation of ~3.3 hours is that of the slower component of the binary dwarf.

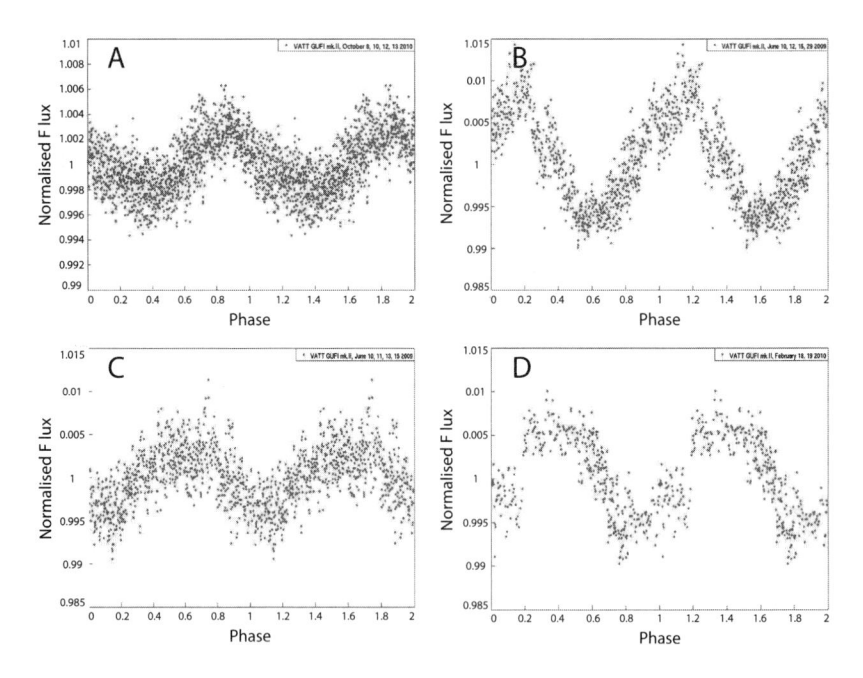

Figure 2. Initial results of binned, phase-folded light curves of each consecutive target taken with GUFI mk.II on VATT is presented here. *A:* LP 349-25, Oct 2010; *B:* LSR J1835, Jun 2009; *C:* TVLM 513, Jun 2009; *D:* 2M J0746, Feb 2010.

Table 1. Results from the radio detected ultracool dwarf sample, listed in order of increasing spectral type. † = Lane et al. (2007) and confirmed here.

Target	SpT	Observations [hours$_{total}$]	Rotation period [hours]	Significance [%]
LP 349-25	M8V	60	~1.86	>99.99
LSR J1835	M8.5	18	~2.84	>99.99
TVLM 513	M9	28	~1.96†	>99.99
2M J0746	L0+L1.5	25	~3.3	>99.99

4. Summary and Conclusions

We have reported on optical observations of four ultracool dwarfs - the M8V tight binary LP 349-25, the M8.5 dwarf LSR J1835, the M9 dwarf TVLM 513, and the L dwarf binary 2M J0746, and we detect periodic variability for three of these dwarfs for the first time. Lane et al. (2007) presented a period of rotation for TVLM 513 of ~1.96 hours and we confirm this value in our data. We found the M8.5 dwarf LSR J1835 to exhibit I-band periodic modulation of ~2.84 hours (~3% peak to peak variations), a periodicity that has been established over three separate epochs, from 2006 - 2009. Similarly to TVLM 513, the optical periodic signals for LSR J1835 are consistent with the radio pulses detected by Hallinan et al. (2008), which they argue is due to the rotation of the dwarf - we also attribute the optical periodicity to the dwarf's rotation. They also argue that the electron cyclotron maser instability is the process responsible for the detected periodic radio pulses. For TVLM 513 in particular, these emissions have recently been attributed to auroral emissions in the magnetosphere of the dwarf (Hallinan et al., this volume).

In the case of the tight binary dwarfs, we present I-band periodic sinusoidal variability of ~1.86 hours (~1.5% variations) and ~3.3 hours (~2% variations) for LP 349-25 and 2M J0746 respectively. We also conducted R-band observations of LP 349-25 and found the same periodic behaviour, however these signals were varying at the ~4% level. Although LP 349-25 was detected as an extremely bright radio source by Phan-Bao et al. (2007), this detection was not transient in nature. Therefore we present these data as the first periodically modulated detection from one component of the system. We find a most intriguing result in the case of 2M J0746 however. This dwarf was detected by Berger et al. (2009) to exhibit periodic radio pulses with a period of 2.07±0.002 hours. However, we report the non-radio detected dwarf to be periodically varying in optical bands with a period of ~3.3 hours. The detected optical periodic variability of these targets may be due to auroral emissions as supported by Hallinan et al. (this volume) for TVLM 513. Other possibilities such as magnetic spots still remain. Further simultaneous radio and spectrophotometric observations will distinguish between these possibilities.

Acknowledgments. LKH gratefully acknowledges the support of the Science Foundation Ireland (Grant Number 07/RFP/PHYF553). We thank the VATT team for their help and guidance, especially Dave Harvey, Ned Franz and Ken Duffek. LKH would also like to personally thank Dr. Mark Lang for his constant assistance, in addition to the staff of the NRAO who provided significant support when LKH was conducting some of his work during his appointment as an NRAO graduate intern.

References

Antonova, A., Doyle, J. G., Hallinan, G., Bourke, S., & Golden, A. 2008, ApJ, 487, 6
Antonova, A., Doyle, J. G., Hallinan, G., Golden, A., & Koen, C. 2007, A&A, 472, 4
Artigau, E., Bouchard, S., Doyon, R., & Lafreniere, D. 2009, ApJ, 701, 5
Bailer-Jones, C. A. L., & Mundt, R. 2001, ApJ, 367, 17
Basri, G. 2001, in 11th Cambridge Workshop on Cool Stars, Stellar Systems and the Sun, Astronomical Society of the Pacific Conference Series, 261
Berger, E. 2002, ApJ, 572, 11
— 2006, ApJ, 648, 8

Berger, E., Ball, S., Becker, K. M., Clarke, M., Frail, D. A., Fukuda, T. A., Hoffman, I. M., Mellon, R., Momjian, E., Murphy, N. W., Teng, S. H., Woodruff, T., Zauderer, B. A., & Zavala, R. T. 2001, Nature, 410, 3
Berger, E., Rutledge, R. E., Phan-Bao, N., Basri, G., Giampapa, M. S., Gizis, J. E., Liebert, J., Martin, E., & Fleming, T. A. 2009, ApJ, 695, 10
Berger, E., Rutledge, R. E., Reid, I. N., Bildsten, L., Gizis, J. E., Liebert, J., Martin, E., Basri, G., Jayawardhana, R., Brandeker, A., Fleming, T., Johns-Krull, C. M., Giampapa, M. S., Hawley, S. L., & Schmitt, J. H. M. M. 2005, ApJ, 627, 14
Burgasser, A. J., & Putman, M. E. 2005, ApJ, 626, 12
Clarke, F. J., Hodgkin, S., Oppenheimer, B. R., Robertson, J., & Haubois, X. 2008, MNRAS, 386, 5
Clarke, F. J., Tinney, C. G., & Covey, K. R. 2002, MNRAS, 332, 5
Dupuy, T. J., Liu, M. C., Bowler, B. P., Cushing, M. C., Helling, C., Witte, S., & Hauschildt, P. 2010, ApJ, 721, 23
Enoch, M. L., Brown, M. E., & Burgasser, A. J. 2003, AJ, 126, 10
Gelino, C. R., Marley, M. S., Holtzman, J. A., Ackerman, A. S., & Lodders, K. 2002, ApJ, 557, 13
Goldman, B., Cushing, M. C., Marley, M. S., Artigau, E., Baliyan, K. S., Bejar, V. J. S., Caballero, J. A., Chanover, N., Connelley, M., Doyon, R., Forveille, T., Ganesh, S., Gelino, C. R., Hammel, H. B., Holtzman, J., Joshi, S., Joshi, U. C., Leggett, S. K., Liu, M. C., Martin, E. L., Mohan, V., Nadeau, D., Sagar, R., & Stephens, D. 2008, A&A, 487, 15
Hallinan, G., Antonova, A., Doyle, J. G., Bourke, S., Brisken, W. F., & Golden, A. 2006, ApJ, 653, 10
Hallinan, G., Antonova, A., Doyle, J. G., Bourke, S., Lane, C., & Golden, A. 2008, ApJ, 684, 10
Hallinan, G., Bourke, S., Lane, C., Antonova, A., Zavala, R. T., Brisken, W. F., Boyle, R. P., Vrba, F. J., Doyle, J. G., & Golden, A. 2007, ApJ, 663, 4
Koen, C. 2006, MNRAS, 367, 4
Lane, C., Hallinan, G., Zavala, R. T., Butler, R. F., Boyle, R. P., Bourke, S., Antonova, A., Doyle, J. G., Vrba, F. J., & Golden, A. 2007, ApJL, 668, 4
Littlefair, S. P., Dhillon, V. S., Marsh, T. R., Shahbaz, T., Martin, E. L., & Copperwheat, C. 2008, MNRAS, 391, 4
Lomb, N. R. 1976, Ap&SS, 39, 16
Maiti, M. 2007, AJ, 133, 11
Martin, E. L., Zapatero Osorio, M. R., & Lehto, H. J. 2001, ApJ, 557, 8
Osten, R. A., Hawley, S. L., Bastian, T. S., & Reid, I. N. 2006, ApJ, 637, 4
Phan-Bao, N., Osten, R. A., Lim, J., Martin, E. L., & Ho, P. T. P. 2007, ApJ, 658, 4
Reiners, A., & Basri, G. 2008, ApJ, 684, 14
— 2009, ApJ, 705, 9
Rockenfeller, B., Bailer-Jones, C. A. L., & Mundt, R. 2005, A&A, 448, 13
Scargle, J. D. 1982, ApJ, 263, 19

Cool Stars 16 participants at a Seattle Mariners baseball game

Part IV

Magnetic Fields and Magnetic Activity

Solar Energetic Events, the Solar-Stellar Connection, and Statistics of Extreme Space Weather

C. J. Schrijver

Lockheed Martin Advanced Technology Center, 3251 Hanover Street, Palo Alto, CA 94304, U.S.A.

Abstract. Observations of the Sun and of Sun-like stars provide access to different aspects of stellar magnetic activity that, when combined, help us piece together a more comprehensive picture than can be achieved from only the solar or the stellar perspective. Where the Sun provides us with decent spatial resolution of, e.g., magnetic bipoles and the overlying dynamic, hot atmosphere, the ensemble of stars enables us to see rare events on at least some occasions. Where the Sun shows us how flux emergence, dispersal, and disappearance occur in the complex mix of polarities on the surface, only stellar observations can show us the activity of the ancient or future Sun. In this review, I focus on a comparison of statistical properties, from bipolar-region emergence to flare energies, and from heliospheric events to solar energetic particle impacts on Earth. In doing so, I point out some intriguing correspondences as well as areas where our knowledge falls short of reaching unambiguous conclusions on, for example, the most extreme space-weather events that we can expect from the present-day Sun. The difficulties of interpreting stellar coronal light curves in terms of energetic events are illustrated with some examples provided by the SDO, STEREO, and GOES spacecraft.

1. Introduction

Magnetic activity of Sun and Sun-like or "cool" stars results in a rich variety of observable phenomena that range from the asterospheres that surround these stars down to the stellar surfaces and – with rapid advances expected in asteroseismology – below. Many of these phenomena are directly observable on Sun and stars alike, including the outer-atmospheric phenomena of persistent chromospheres and coronae, as well as their perturbations in the form of short-lived light-curve perturbations that are the signature of energetic events. It is on the latter that I focus here.

The proximity of the Sun enables us to see details in the evolving magnetic field and associated atmospheric phenomena that are simply impossible to infer from stellar observations. Small wonder, then, that many of the names for stellar phenomena are taken from the solar dictionary: active regions, spots, flares, eruptions, and even the processes such as differential rotation and meridional advection that are part of the equivalent of the magnetic activity cycle, all the way to the loss of angular momentum associated with the gusty outflow of hot, magnetized plasma. Ensembles of observations taken over periods of years to centuries are revealing the statistical properties of some of these phenomena, even as state-of-the art observatories in space and on the ground are revealing physical processes and the interconnectedness of the global outer atmosphere.

But recent observations of the Sun only provide a very limited view of what its magnetic activity has on offer, for at least two reasons. First, our Sun has a magnetic activity cycle that is relatively long compared to researchers' careers as well as to the era of advanced technology that has aided us in our observations. Consequently, we can expect even the 'present-day Sun' to have surprises in store for us that we have not yet observed simply because we have not been looking long enough. Some of these surprises may lie hidden in records such as polar ice sheets, while others lie embedded in rocks from outer space. But the lessons that can be learned from these invaluable and, as yet, under-explored archives are limited by access to these resources, by the limited temporal resolution of such records, and by the long chains of processes that sit between a solar phenomenon like the sunspot cycle or its largest flares and the 'recording physics' for the archive from which we are attempting to learn about them. In addition to learning about the Sun from such 'geological' records of its activity, one can also perform an ensemble study of states of infrequent extreme solar activity by looking at a sample of stars like it. This can provide us with a large enough sample of Sun-like stars that we can begin to assess how frequently the Sun may subject us to rare but high-impact events such as dangerous superflares and disruptive geomagnetic storms: although rare, the damage that may be inflicted to our global society and its safety and economy by extreme events is of such a magnitude that in-depth study of their properties and likelihood is prudent.[1]

A second reason why stellar studies are crucial to understanding of solar activity is that only stellar observations allow us to explore what the Sun's activity has been in the very distant past or what it will be in the very distant future (measured on time scales up to billions of years) by selecting stars of a wide range of ages.

In this review, I discuss a sampling of the results coming out of the study of what has been termed the solar-stellar connection. I focus, in particular, on lessons that we are learning about what could be called 'space climate', i.e., the characteristic state of activity of a star like the Sun including the fluctuations about the mean in the form of energetic events like flares and coronal mass ejections (CMEs). Consequently, one of the topics selected for this review is a comparison of frequency distributions of bipolar regions, flares, CMEs, and solar energetic particle (SEP) events, the possible relationships between them, and the lessons learned by combining solar and stellar observations in our quest to establish the 'laws' of astro-magnetohydrodynamics. Another topic is that of light curves, which touches on the need to have pan-chromatic knowledge to guide our interpretation of stellar observations as well as on the long-standing concept of 'sympathetic events' in stellar magnetic activity.

2. The Flux Spectrum of Bipolar Regions

Magnetic flux emerges from the solar interior as flux bundles that shape themselves into bipolar regions. Those regions large enough to contain spots during at least some of their mature, coherent phase, are called active regions. Their emergence frequency for sunspot cycle 22 was found to be characterized by a power-law distribution with an index $-\alpha_\Phi = 2.0 \pm 0.1$ (Harvey & Zwaan 1993; Schrijver & Harvey 1994).

---

[1] See the NRC report on "Severe Space Weather Events – Understanding Societal and Economic Impacts" at http://www.nap.edu/catalog.php?record_id=12507.

The flux bundles that are too small to contain spots, typically with fluxes below about 10^{20} Mx, but larger than about 10^{18} Mx, are called ephemeral regions. The flux spectrum of the ephemeral regions appears to smooth transition into that of the active region spectrum. When the frequency distributions for active and ephemeral regions are combined into an approximate power-law distribution, one finds

$$N(\Phi)d\Phi \propto \Phi^{-\alpha_\Phi}d\Phi \qquad (1)$$

for $\alpha_\Phi = 2.69$ (Thornton & Parnell 2011). The difference between the above values of α_Φ indicates that a single power-law is too simple as an approximation, with the flux spectrum somewhat less steep with increasing size. For the purpose of the discussion below, I will take the range as a measure of uncertainty, i.e. $\alpha_\Phi = 2.3 \pm 0.3$.

The active-region spectrum for cycle 22 appeared to be roughly fixed in shape, going up and down by a multiplicative factor over the cycle with a power-law index changing by at most a few tenths of its value (Harvey & Zwaan 1993, Schrijver & Harvey 1994). During the recent cycle minimum (late 2008 and early 2009), active regions were essentially absent for a long time, with the 3-month average sunspot number hovering around an unusually low value of 1.5, down from around 150 during characteristic sunspot maxima. The ephemeral-region population, in contrast, remained essentially unchanged (Schrijver et al. 2011). Hence, the bipolar-region spectrum is likely not a single, time-independent power law. Yet, at least during relatively active phases, on which I shall focus below, this simple approximation appears warranted.

Eq. (1) characterizes the frequency of emerging bipoles. In a study of bipolar regions existing on the solar surface during 1996 to 2008, Zhang et al. (2010) find a power law with an index of $\alpha_{\text{exist}} \sim 1.63$; its lower (absolute) value than α_Φ in Eq. (1) is caused by the longer life time of larger active regions.

The flux distribution reported on by Zhang et al. (2010) exhibits a marked drop below the power law for fluxes exceeding $\Phi \sim 6 \times 10^{23}$ Mx, and they find no regions above $\Phi_{\text{max}} \sim 2 \times 10^{24}$ Mx. Historically, the largest sunspot group recorded occurred in April of 1946, with a value of 6 milliHemispheres (Taylor 1989); for an estimated field strength of 3 kG, that amounts to a flux in the spot group alone of $\Phi_{\text{spots}} \sim 6 \times 10^{23}$ Mx. The total flux in this spot group was likely larger, but perhaps within a factor of 2 – 3 of that in the spots, and thus of the same order of magnitude as the upper limit to the distribution found by Zhang et al. (2010).

During their mature, coherent phase solar active regions are characterized by a remarkably similar flux density, $\langle B \rangle$ of about 100 Mx/cm^2 to 150 Mx/cm^2 (Schrijver & Harvey 1994) regardless of region size. This allows us to perform a transformation of the frequency distribution of fluxes to one of total energies in the atmospheric field (to be discussed below), using a simple approximation that the energy E_B contained in the bipolar-region field for an active region with size ℓ scales as

$$E_B \propto \langle B \rangle^2 \ell^3 \propto (\langle B \rangle \ell^2)^{3/2} = \Phi^{3/2}. \qquad (2)$$

Thus, when rewriting Eq. (1) in terms of energies one finds:

$$N(\Phi)d\Phi = n(E_B)\frac{d\Phi}{dE_B}dE_B \propto E_B^{-\frac{2}{3}\alpha_\Phi - \frac{1}{3}}dE_B \equiv E_B^{-\mu}dE_B = E_B^{-1.9\pm 0.2}dE_B. \qquad (3)$$

For the above value of Φ_{max}, the corresponding energy is $E_{B,\text{max}} \sim 10^{37}$ ergs.

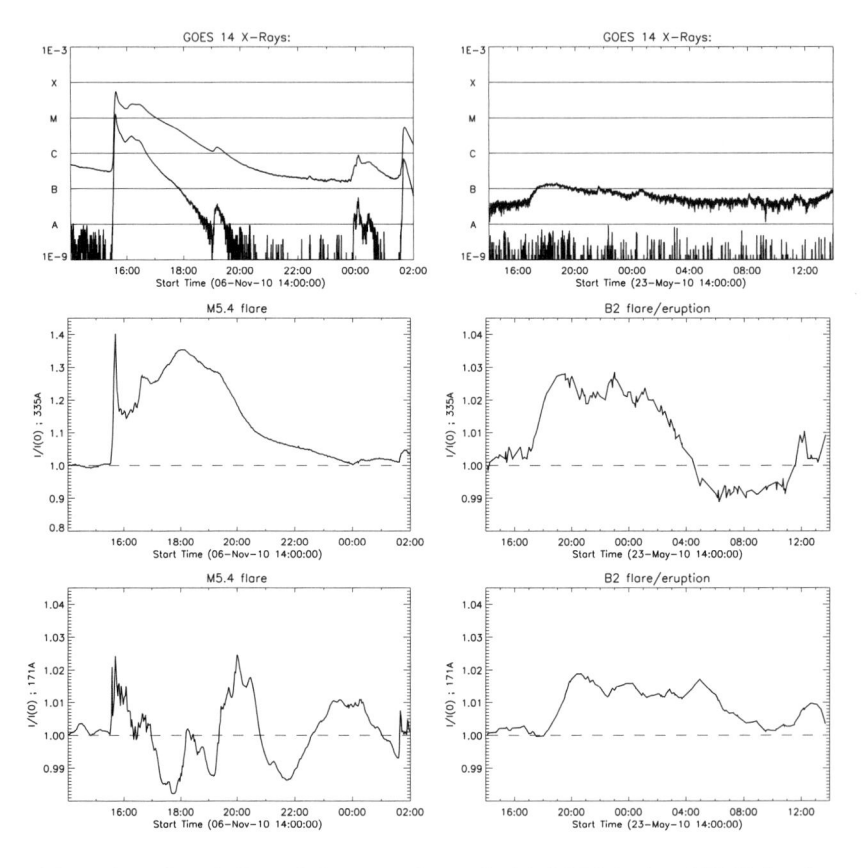

Figure 1. Sample X-ray lightcurves observed with GOES (top) and EUV signals observed with SDO/AIA in the Fe XVI 335 Å channel (center) and the Fe IX/X 171 Å channel (bottom), for an M5.4 active-region flare (left) and for the eruption of a large filament from a quiet-Sun region (right). The intensities were scaled to a pre-event reference value (at the dashed line). Note that the vertical scale of the center-left panel is 10× that of the other AIA panels.

3. The Spectral Appearance of Flares and Eruptions

Solar flares span an astonishing range of energies: from the largest flares, emitting brightly in hard X-rays and even γ rays, down to the weakest EUV flares barely detectable against the persistent background glow of the quiet-Sun corona. The fact that flares shift in color from EUV for $\sim 10^{24}$-erg nanoflares to hard X-rays for $\sim 10^{32}$-erg X-class events makes a statistical comparison difficult: small and large flares are generally observed with different instruments, exacerbating the difficulties in estimating total energies in the absence of bolometric observables (solar flares are too faint to be picked up by total irradiance monitors, with very few exceptions). This is compounded by problems in estimating the strong background in the case of the weakest events. For the largest flares, there are statistical limitations owing to their low frequency.

The launch of the Solar Dynamics Observatory (SDO) in the spring of 2010 has opened a window onto the global Sun that enables a direct comparison of solar and stellar observations in terms of light curves and their interpretation. The Atmospheric Imaging Assembly (AIA; Lemen et al. 2010) on SDO is a full-disk EUV imaging telescope array that provides high-resolution (~ 1.4 arcsec), high-cadence (~ 12 s) coronal images in 7 narrow-band spectral windows (in addition to a few UV channels). SDO's Extreme-ultraviolet Variability Experiment (EVE) measures the X-ray and EUV spectral irradiance (e.g., Woods et al. 2010). The comparison of the data from these two instruments is just starting, but already some unanticipated findings are emerging.

One of these surprises has to do with the relationship between flares and CMEs. With the Sun in a moderately active state through much of 2010, SDO has thus far observed primarily C-class flares and only a few low M-class flares. Woods et al. (2010) note that the AIA observations show that about 80% of the C-class flares are associated with CMEs. Earlier work had suggested numbers as low as ~ 20% for C-class flares, increasing to ~ 40% and ~ 100% for M- and X-class flares, respectively (see summary and references in Schrijver 2009). The reason for the relatively large fraction of eruptive events remains subject to investigation; it may reflect a selection bias in earlier studies, or may suggest that flaring during low-activity states more readily breaks through the relatively weak overlying field than during higher activity states.

On another front, even the Sun still has surprises in store as to how flares show themselves in different pass bands. Associated with the large fraction of eruptive C-class flares is a characteristic signature seen in the EVE spectral irradiance measurements: Woods et al. (2010) point out that these eruptive flares have late-phase emissions in, e.g., Fe XV and Fe XVI that AIA data show to be associated with relatively high post-eruption loop systems, apparently reconnecting and cooling after the eruptive flare. That late-phase emission was not observed for C-class solar flares until now.

Another example of what we are learning about solar energetic events is shown in Fig. 1. The top-left panel shows the GOES $0.5 - 4.0$ Å (lower curve) and $1.0 - 8.0$ Å (upper curve) signals for the 2010/11/05 M5.4 flare. The AIA lightcurves below it are for the Fe XVI 335 Å and Fe IX/X 171 Å channels. This eruptive flare shows a spike in the Fe XVI 335 Å channel followed by a long-duration signal indicative of the cooling of post-eruption loops (as discussed above for the less energetic C-class flares). The much cooler Fe IX/X 171 Å signal shows a mixture of brightenings and darkenings that reflect heating, cooling, and expansion-related coronal dimmings. For an analogous discussion on stellar flaring, see, e.g., Osten & Brown (1999).

The right-hand panels in Fig. 1 show the signals in these same pass bands associated with the eruption of a large quiet-Sun filament from an otherwise quiet-Sun disk. The X-ray brightening barely registers as a flare, reaching no more than the B2 level. The Fe XVI 335 Å signal shows a long-lived brightening followed by a dimming before the signal recovers to pre-eruption levels some 18 h after the onset, while the Fe IX/X 171 Å signal persists as a brightening associated with quiet-Sun post-eruption arcades for that entire period, starting about an hour after the onset of the B2 flare. The total energy (fluence) associated with this event (estimated from EVE measurements by R. Hock, priv. comm.) is $\sim 2 \times 10^{30}$ ergs, i.e. comparable to a low M-class flare, not dissimilar from that shown on the left. Despite the comparable energies involved in these two events, the softness of the spectrum and the duration of this very extended event, spanning a good fraction of a solar radius, lead to an entirely different observational appearance if observed through lightcurves as can be obtained for stars.

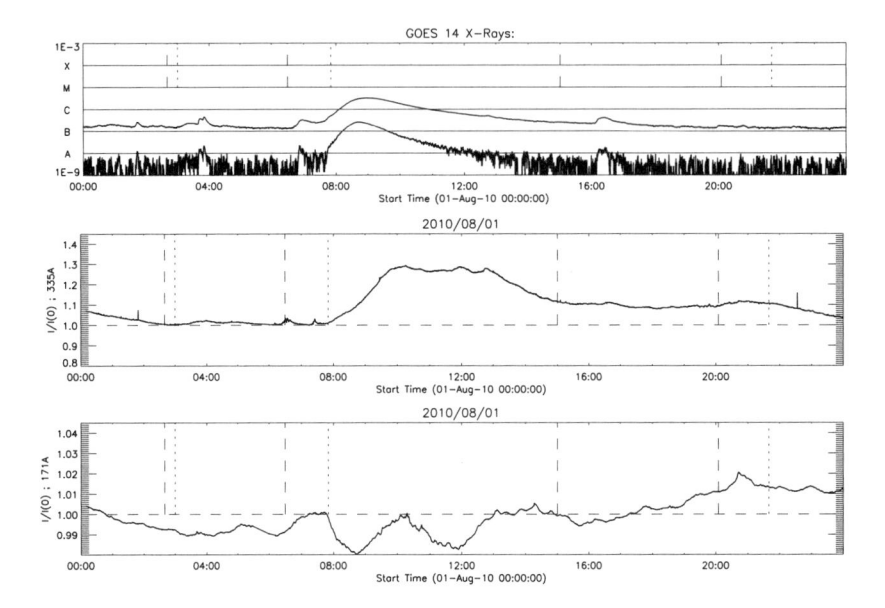

Figure 2. Sample lightcurves observed with GOES (top) and with SDO/AIA in the Fe XVI 335 Å channel (center) and the Fe IX/X 171 Å channel (bottom), for a series of related coronal mass ejections and flares (as discussed by Schrijver & Title 2010). The intensities were scaled to the time of minimum brightness in the 335 Å channel (at the dashed line). Vertical dashed lines show approximate times of the start of filament eruptions seen in AIA's 304 Å channel, and dotted lines show when the first signs of CMEs were seen in STEREO's COR1 images.

4. Local and Global Influences on Eruptions

High-resolution solar instruments typically have a limited field of view, while telemetry constraints on, e.g., the full-Sun X-ray imager on YOHKOH often led observers to down-select regions around a flare site to bring down higher-cadence imaging. From that perspective, it is not surprising that many studies looking for conditions leading to the explosive/eruptive release of magnetic energy focus on the conditions within or nearby an active region. This has led to the recognition of certain patterns in the magnetic field generally associated with flaring regions, in particular neutral lines, often with high shear or high gradients (as summarized, e.g., by Schrijver 2009).

The capabilities of SDO's AIA, particularly when combined with EUV observations from the two STEREO spacecraft that both approach quadrature relative to the Sun-Earth line, are revealing that long-range interactions play an important part in flares and eruptions in addition to the 'internal conditioning'. Schrijver & Title (2010) discuss one particular set of flares and eruptions in detail, but many more similar connections have been inferred from other observations. They use full-sphere magnetic field maps in combination with the STEREO and SDO observations to show that a series of events on 2010/08/01 are directly connected by magnetic field, in particular by field lines that are part of a web of topological fault zones, i.e., separatrices, separators, and quasi-separatrix layers. These long-range connections span across half a hemisphere, part of which was invisible from Earth, but covered by one of the STEREO spacecraft.

These observations reveal that in many cases, what may look like a single event is instead a complex of simultaneous events, or events that occur in close succession, across a large area of a stellar surface. The solar events of 2010/08/01 are dominated by short active-region flare loops in the X-ray part of the spectrum and by long post-eruption loops over quiet-Sun filaments; the combination of these wavelengths to estimate the properties of the field involved, as one might do if these were stellar observations, would lead to fundamentally erroneous conclusions. This is illustrated by the lightcurves in Fig. 2 (showing the same set of signals as Fig. 1): the flares are identifiable in the GOES X-ray signal (top panel), but already of a substantially different character in the Fe XVI 335 Å passband shown below it. Around 1 MK, in the Fe IX/X 171 Å channel, the flares are absent, while the dimmings are not obviously connected to the filament eruptions. Where multi-wavelength coverage would help interpret stellar observations, separation of the multitude of events on that day using only light curves and then estimating total energies from available limited pass bands is obviously an exercise fraught with large ambiguities and uncertainties. The repercussions on such linked (or "sympathetic") events on the hotly-debated problem of the causal links between flares and CMEs continue to be studied; we may need to follow Harrison (1996) who acknowledges the complex linkages between these event classes and suggests that we refer to their composites as 'coronal magnetic storms,' a term equally suitable for the stellar arena (where sympathetic flaring is discussed, e.g., by Osten & Brown 1999).

5. Flare Energy Spectrum

Despite the difficulties that arise in establishing total energies for the variety of flares even for solar events, the comparison from nanoflares to X-class flares has been made, with the intriguing result that flares span a range of at least 8 orders of magnitude with a frequency distribution that can be approximated by a power-law spectrum.

Statistics on stellar data are even harder to assemble and interpret, but over the years lessons have been learned and numbers extracted. One such study, by Audard et al. (2000), assembles EUVE observations on 10 cool stars into downward-cumulative frequency distributions (to suppress the problem of low-number statistics) of estimated energies. Here, too, power-law spectra arise, in fact with slopes very much like those reported for solar flares. In another study, Wolk et al. (2005) look at young, active stars in the Orion Nebula Cluster, just a few million years old; their flare energy distribution yield a power-law index of $\alpha_f \approx 1.7$.

Osten & Brown (1999) – for over 140 d of observations on 16 tidally interacting RS CVn binaries – and Audard et al. (2000) – for some 74 d of observing of 10 single and binary stars – both conclude that the overall flaring rate increases essentially linearly with the background stellar X-ray luminosity, with flares reaching energies of over 10^{34} ergs, roughly comparable to a solar X500 flare. If we use the inferred proportionality of flare frequency with X-ray brightness to normalize the observed stellar frequency distribution for flare energies to the Sun characteristic of cycle maximum, the combined solar and stellar flare statistics (Fig. 3, which shows the downward-cumulative frequency distribution) reveals a frequency distribution for flare energies of

$$N(E_f)dE_f \propto E_f^{-\alpha_f} dE_f, \tag{4}$$

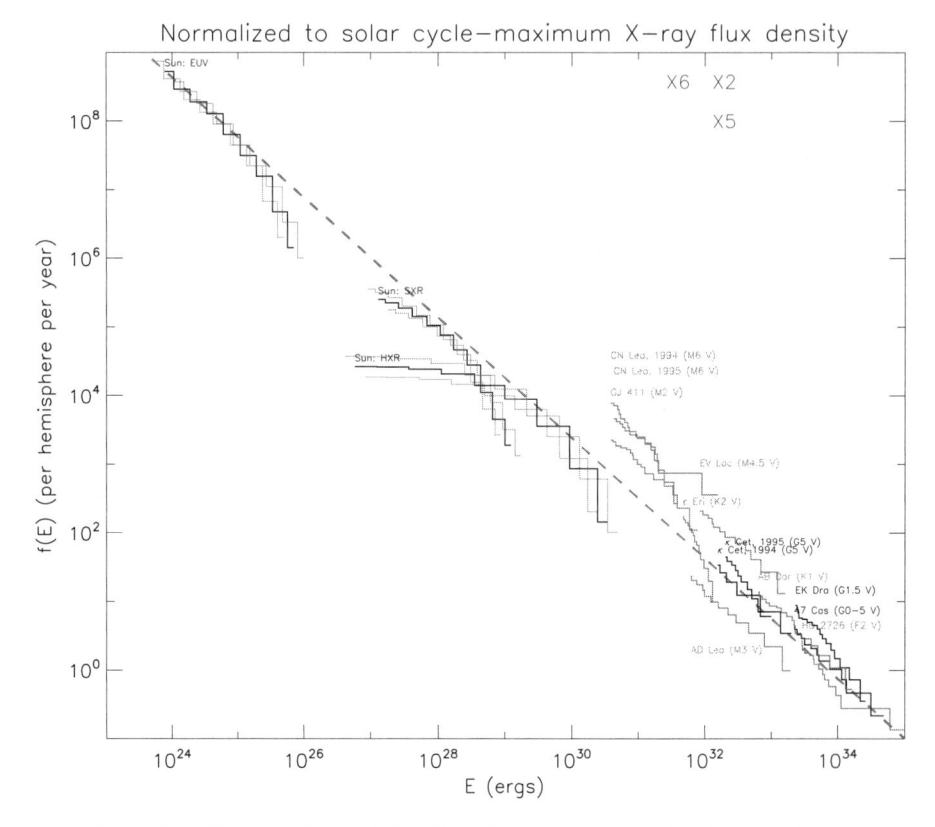

Figure 3. Downward-cumulative flare frequency distribution for energies exceeding E, normalized to approximate solar-maximum X-ray flux density levels (1.3×10^5 erg cm^{-2} s^{-1}; Judge et al. (2003) assuming a linear dependence of the X-ray surface flux density. Flare spectra for Sun-like G-type stars are shown in black, for warmer and cooler stars in grey. The grey dashed power-law fit has an index of $\alpha_f + 1 = -0.87$. EUV data from Aschwanden (2000), soft X-ray data from Shimizu (1997), hard X-ray data > 8 keV from Lin et al. (2001), and stellar data from Audard et al. (2000). The grey histograms for solar data bracket a conservative energy uncertainty of a factor of 2. Three estimates of flare energies for GOES X flares are shown near the top, from Aschwanden & Alexander (2001) and Benz (2008).

with $\alpha_f = -1.87 \pm 0.10$ (rough uncertainty estimate). The solar data align remarkably well with those of the G-type main-sequence stars in the sample by Audard et al. (2000).

One property of note here is that the slopes α_f in Eq. (4) and μ in Eq. (3) agree within their uncertainties. What can we infer from this? The scaling of the flare frequency spectra observed for stars is essentially linear with the X-ray brightness of the stars (Audard et al. 2000), which, in turn, is essentially linearly dependent on the unsigned magnetic flux threading the stellar surface (Schrijver 2000), and modeling with a surface flux-transport model suggests that that scales linearly with the rate of emergence of bipolar active regions (Schrijver 2000).

With this chain of linear scalings, it is very tempting to conclude that the flare frequency is proportional to the emergence frequency of bipolar regions. One possible inference from this is that flares draw their power from the same source as the active region magnetic configurations. In that context, it is interesting to note that the flare energy distribution in Fig. 3 does not suggest a cutoff in energy up to values approaching 10^{35} ergs, which is comfortably below the maximum available energy $E_{B,\text{max}}$ estimated above for historically observed solar active regions.

Interestingly, recent Kepler observations of white light flares reported at the 16th workshop on "Cool Stars, Stellar Systems, and the Sun" suggest flares can occur up to at least 10^{37} ergs, although these most energetic flares are (at least at present) reported for stars significantly cooler than the Sun, see, e.g., Walkowicz et al. (2010). What does it take to power such large flares? If we assume that a fraction of $f = 0.01 - 0.1$ of the magnetic energy density in a volume with a characteristic mean field strength of B_0 can be converted into what eventually is radiated from the flare site, the typical dimension d_0 and magnetic flux $\Phi_0 = B_0 d_0^2$ in such a flaring region are given by

$$d_0 = \left(\frac{4\pi E_f/f}{B_0^2}\right)^{1/3} \; ; \; \Phi_o = B_0 d_0^2 = \frac{(4\pi E_f/f)^{2/3}}{B_o^{1/3}}. \tag{5}$$

At $B_0 = 300$ G, for flares with $E_f = 10^{35}$ ergs, $d_0 \approx (0.7 - 1.6)R_\odot$ and $\Phi_0 \approx (0.8 - 4) \, 10^{24}$ Mx. Although very sizeable, these numbers are still compatible with the largest fluxes, Φ_{max}, discussed in Sect. 2; note that the value of B_0 here was chosen 3 times higher than characteristic of solar regions to illustrate how challenging it is to fit the flaring region on the available solar surface (see also, e.g., Osten & Brown 1999 for evidence of extended flaring regions in some RS CVn stars). For $E_f = 10^{37}$ ergs, on the other hand, $d_0 \approx (3 - 7)R_\odot$ and $\Phi_0 \approx (20 - 80) \, 10^{24}$ Mx, which simply would not fit on the Sun, and involves fluxes well above Φ_{max}; conditions for such stellar flares must differ from those on the present-day Sun, and perhaps – as I discuss next – these very extreme events no longer occur on the aged star next door.

6. Spectrum of Solar Energetic Particle Events

The frequency distribution in Fig. 3 suggests that if we estimate an X1 flare to have an energy of about 10^{32} ergs, we should see about 30 flares per year of that magnitude or larger during active phases in the solar cycle. That number compares quite well with the average frequency of 25 per year for X-class flares over the past three solar cycles, when counting only during the active half of the cycles (see the compilation by Hudson 2007). We would expect an X10 or larger about 4 times per year, which is high by a factor of about three given the 21 observed X-class flares since 1976 with the Sun in an active state for about 15 years within that interval. Extending that spectrum even further, we would expect an X100 flare or larger once every other year for the Sun near cycle maximum. As we have not experienced such large flares in at least half a century, we face the possibility that despite the intriguing alignment of solar and stellar flare data after normalization to the mean coronal brightness, there may in fact be an upper limit to stellar flare energies that may shift to lower values with increasing age even as the spectrum below that value is left unchanged in slope: the solar flare energy distribution appears to drop below the power-law fit in Fig. 3 above X10, and the largest flare observed to date has been estimated to be an X45±5 (Thomson et al. 2004).

Are there constraints that can be set on the occurrence of extreme flares that can be found in 'archives' on Earth other than those compiled by mankind? Solar eruptive events are frequently associated with solar energetic particle (SEP) events that modulate the galactic cosmic ray (GCR) background which itself varies on time scales of years and longer (Schrijver & Siscoe 2010a, 2010b). SEP events can accelerate particles to energies up to several GeV which suffices to initiate a nuclear cascade in the upper layers of the Earth's atmosphere and in lunar or meteoric rocks. Some constraints on large SEP events can be recovered by studying the mix of radioactive nuclides or in radioactive decay products as a function of penetration depth in lunar and meteoric rock samples, or by studying the so-called odd nitrogen concentrations that are modulated by reactions high in the Earth's atmosphere that are sensitive to ionization states. The combination of space-age measurements with such geological archival information led Usoskin (2008) to conclude that the downward-cumulative probabilities of large solar proton events scaled with the fluence F_p as a power law with a slope of $\delta + 1 \sim 0.4$ for relatively small events, then turning towards a steeper slope with a lower limit of $\delta + 1 \leq 0.9$, with the 'break' occurring around fluences of 10^{10} cm^{-2} for protons above 10 MeV.

One might deduce from the inferred break in the SEP fluence distribution that the behavior of solar flares changes at high energies. That conclusion is ambiguous based on this argument only, however, as argued by combining several observed frequency distributions. Let us start from the particle fluence distribution

$$N_p dF_p \propto F_p^{-\delta} dF_p, \qquad (6)$$

with the above value for δ as estimated from the cumulative distribution function of the fluences. In order to relate the fluences at Earth to those originating from the solar flares and associated coronal mass ejections (both of which appear to contribute to the SEPs in a mixture that continues to be debated), let us assume that the particles are emitted from their source region in the corona or inner heliosphere into a solid angle that scales with the total energy E_f of the eruptive event as

$$\Omega \propto E_f^\gamma. \qquad (7)$$

The value of γ can be estimated by comparing the flare energy distribution in Eq. (4) and the distribution of opening angles, a (in degrees), for eruptions from very large CMEs seen by SOHO's LASCO to small fibril eruptions observed by TRACE and perhaps for even smaller events seen in STEREO data (summarized by Schrijver 2010)

$$N_a da = b a^{-\beta} da, \qquad (8)$$

with $\beta = 2.0 \pm 0.3$ (estimated from Fig. 2 in Schrijver (2010); $b \approx 1.1$ for $\beta = 2$). For given a (in radians), the corresponding fractional solid angle is given by

$$\frac{\Omega}{4\pi} = \frac{1}{2}(1 - \cos a) \approx \frac{1}{4}a^2, \qquad (9)$$

where the first term in the Taylor expansion shown on the right approximates the solid angle only for sufficiently small values of a, otherwise saturating when the hemisphere above the event is filled at $\Omega = 2\pi$. Using the right-hand expression in Eq. (9) together with Eq. (7), we can rewrite Eq. (8) as

$$N_a da \propto \Omega^{-\frac{1}{2}(1+\beta)} d\Omega \propto E_f^{-\frac{\gamma}{2}(1+\beta)+\gamma-1} dE_f. \qquad (10)$$

With Eq. (4), $\alpha_f = \frac{\gamma}{2}(1+\beta) - \gamma + 1$ one finds $\gamma = 2(\alpha_f - 1)/(\beta - 1)$.

Let us further assume that the particle fluence at Earth, F_p is a fixed fraction f of E_f, diluted by expanding over a solid angle Ω, i.e., that with Eq. (7),

$$F_p \propto f \frac{E_f}{\Omega} \propto E_f^{1-\gamma}. \tag{11}$$

With this, Eq. (6) can be transformed to read

$$N_p dF_p \propto E_f^{-\delta(1-\gamma)-\gamma} dE_f \propto E_f^{-\delta+2(\alpha-1)(\delta-1)/(\beta-1)} dE_f. \tag{12}$$

The assumption that the SEPs are emitted within a solid angle Ω implies that only those for which the direction of the Earth (mapped through the curved path of the Parker spiral of the heliospheric magnetic field) is included within that solid angle can be detected (here, I ignore the fact that relatively small flares often do not connect to the heliosphere, see Sect. 3). This means that only a fraction

$$p = \frac{\Omega}{4\pi} \propto E_f^\gamma \tag{13}$$

of the total number of events is detected. Hence, to recover the flare energy distribution from the observed SEP fluences, the distribution in Eq. (12) has to be divided by p:

$$N_f dE_f \propto E_f^{-\delta+2(\alpha-1)(\delta-2)/(\beta-1)} dE_f \equiv E_f^{-\epsilon} dE_f \stackrel{?}{=} E_f^{-\alpha_f} dE_f. \tag{14}$$

With the values of the exponents above, one finds $\epsilon = 2.4 \pm 0.5$, consistent with the value of α_f in the flare energy distribution of Eq. (4) within $\sim 1.1\sigma$, provided we limit the comparison to events for which the SEPs are spread over a solid angle a small compared to 2π steradians. This slope holds up to an SEP frequency at Earth of about 1/yr; if we assume for those events $\Omega \approx 2\pi$, $p \approx 1/2$, so ≈ 2/yr for the full Sun; Eq. (8) has that downward cumulative frequency for $a \approx 200°$, which is slightly high, but roughly consistent with our hypothetical argumentation. For relatively large opening angles a, the expression in Eq. (9) saturates to a constant value, as does the detection probability p. In that case, one would expect $\epsilon \approx \delta \approx \alpha$. Such a value is just allowed by the empirical upper limits to the energy distributions from Moon rocks. On the other hand, such a slope may be too shallow relative to what is suggested around the 'break' in the spectrum from NO_3 data from ice cores as summarized by Usoskin (2008).

From this, we can tentatively conclude that at least the frequency distributions for solar flares, for CME opening angles, and for SEP fluences are consistent for proton events up to the largest observed in the instrumental era. The break in the fluence spectrum above those values might reflect the saturation of the spreading of the SEPs over essentially a full hemisphere over the solar source region. But it is possible that a true saturation occurs somewhere along the chain of events from Sun to Earth: flare energies may have an upper cutoff for the present-day Sun (with flare probabilities dropping significantly below the power-law fit in Fig. 3 for flares above X10) that is not readily inferred by looking at samples of young Suns or the generation of energetic particles may be limited (either at the flare site or within heliospheric shocks).

7. In Conclusion

The sample observations discussed above demonstrate that the combination of geophysical, heliophysical, and astrophysical data can teach us much about the Sun's magnetic climate, up to the most energetic of events. The interpretations outlined are, of course, to be tested and alternatives, that doubtlessly exist, are to be explored. Despite the speculative nature of the scenarios sketched above, it appears that we are close to having the material available to learn where the solar flare-energy spectrum drops below a solar-stellar power law: the combination of the study of archives in ice and of stellar flare statistics should be able to provide us with an answer. On a less positive note, the solar observations discussed above demonstrate that measuring the energies involved in explosive and eruptive events is difficult, that separating events based on lightcurves alone is an ambiguous exercise, and that broad wavelength coverage is essential to both of these objectives: to learn about the most severe space weather, we have to accept that long-duration, large-sample, pan-chromatic (and thus often multi-observatory) stellar observations are needed because they can provide crucial information that can otherwise only be gathered by observing the Sun for a very long time and undergoing the detrimental effects of extreme magnetic storms on the Sun and around the Earth.

Acknowledgments. I thank M. Aschwanden, J. Beer, R. Osten, C. Parnell, and A. Title for discussions, suggestions, and pointers to the literature.

References

Aschwanden, M. J. 2000, Solar Phys., 190, 233
Aschwanden, M. J., & Alexander, D. 2001, Solar Phys., 204, 91
Audard, M., Güdel, M., Drake, J. J., & Kashyap, V. L. 2000, ApJ, 541, 396
Benz, A. O. 2008, Living Reviews in Solar Physics, 5, 1
Harrison, R. A. 1996, Solar Phys., 166, 441
Harvey, K. L., & Zwaan, C. 1993, Solar Phys., 148, 85
Hudson, H. S. 2007, ApJL, 663, 45.
Judge, P. G., Solomon, S. C., & Ayres, T. R., 2003 ApJ, 593, 534.
Lemen, J. R., Title, A. M., Akin, D. J., et al. 2010, submitted to Solar Phys.
Lin, R. P., Feffer, P. T., & Schwartz, R. A. 2001, ApJL, 557, L125
Osten, R. A., & Brown, A. 1999, ApJ, 515, 746
Schrijver, C., & Siscoe, G. 2010a, Heliophysics. Evolving solar activity and the climates of space and Earth (Cambridge, U.K.: Cambridge University Press)
— 2010b, Heliophysics. Space storms and radiation: Causes and Effects (Cambridge, U.K.: Cambridge University Press)
Schrijver, C. J. 2001, ApJ, 547, 475
Schrijver, C. J. 2009, Advances in Space Research, 43, 739.
— 2010, ApJ, 710, 1480.
Schrijver, C. J., & Harvey, K. L. 1994, Solar Phys., 150, 1
Schrijver, C. J., Livingston, W. C., Woods, T. N., & Mewaldt, R. A. 2010, in preparation.
Schrijver, C. J., & Title, A. M. 2010, JGR, in press
Shimizu, T. 1997, in Magnetic Reconnection in the Solar Atmosphere, edited by R. D. Bentley & J. T. Mariska, vol. 111 of Astronomical Society of the Pacific Conference Series, 59
Taylor, P. O. 1989, J. of the American Assoc. of Variable Star Observers, 18, 65
Thomson, N. R., Rodger, C. J., & Dowden, R. L. 2004, Geophys. Res. Lett., 31, 6803
Thornton, L. M., & Parnell, C. E. 2011, Solar Phys., in press
Usoskin, I. G. 2008, Living Reviews in Solar Physics, 5, 3.
Walkowicz, L. M., Basri, G., Batalha, N., et al. 2010, ArXiv e-prints 1008.0853

Wolk, S. J., Harnden, F. R., Jr., Flaccomio, E., Micela, G., Favata, F., Shang, H., & Feigelson, E. D. 2005, ApJSS, 160, 423.
Woods, T. N., Eparvier, F., Jones, A. R., Hock, R., Chamberlin, P. C., Klimchuk, J. A., Didkovsky, L., Judge, D., Mariska, J., Tobiska, W. K., Schrijver, C. J., & Webb, D. F. 2010
Zhang, J., Wang, Y., & Liu, Y. 2010, ApJ, 723, 1006

New Insights into Stellar Magnetism from the Spectropolarimetry in All Four Stokes Parameters

O. Kochukhov,[1] F. Snik,[2] N. Piskunov,[1] S. V. Jeffers,[2] C. U. Keller,[2] V. Makaganiuk,[1] J. A. Valenti,[3] C. M. Johns-Krull,[4] M. Rodenhuis,[2] and H. C. Stempels[1]

[1]*Department of Physics and Astronomy, Uppsala University, Box 516, 751 20 Uppsala, Sweden*

[2]*Sterrekundig Instituut, Universiteit Utrecht, P.O. Box 80000, NL-3508 TA Utrecht, The Netherlands*

[3]*Space Telescope Science Institute, 3700 San Martin Dr, Baltimore MD 21211, USA*

[4]*Department of Physics and Astronomy, Rice University, 6100 Main Street, Houston, TX 77005, USA*

Abstract. Development of high-resolution spectropolarimetry has stimulated a major progress in our understanding of the magnetism and activity of late-type stars. During the last decade magnetic fields were discovered and mapped for various types of active stars using spectropolarimetric methods. However, these observations and modeling attempts are inherently incomplete since they are based on the interpretation of the stellar circular polarization alone. Taking advantage of the recently commissioned HARPS polarimeter, we obtained the first systematic observations of cool active stars in all four Stokes parameters. Here we report detection of the magnetically induced linear polarization in the RS CVn binary HR 1099 and phase-resolved full Stokes vector observations of ε Eri. For the latter star we measured the field strength with the precision of ~0.1 G over a complete rotation cycle and reconstructed the global field topology with the help of magnetic Doppler imaging. Our observations of the inactive solar-like star α Cen A indicate the absence of the global field stronger than 0.2 G.

1. Introduction

Magnetic fields play a central role in the formation of stars and their evolution. The processes of the generation, transformation and decay of magnetic fields are responsible for most of the short-term stellar variability and give rise to a wide range of non-equilibrium time-dependent phenomena on the stellar surfaces (non-thermal emission, flares, star spots, activity cycles). But despite the prominent role of magnetism in stellar physics, very little is known about the actual strength and topology of magnetic fields in stars other than the Sun. Due to the difficulty of direct detection of stellar magnetic fields, the studies of "magnetic activity" commonly rely on proxy indicators (X-ray emission, chromospheric lines), which are interpreted with a blunt extrapolation of the solar magnetism-activity paradigm.

Recent development of the high-resolution spectropolarimeters capable of recording polarization in spectral lines over a wide wavelength regions (Donati et al. 1999, Donati et al. 2006) added a significant new momentum to the studies of stellar magnetism. It has become possible to detect very weak fields (down to 1 G in best cases) in essentially all types of stars and even to obtain detailed maps of the magnetic field distributions over stellar surfaces with the help of Zeeman Doppler imaging (Donati et al. 2003, Kochukhov & Wade 2010). This led to a qualitatively new picture of the stellar magnetism and showed that the solar magnetic activity paradigm is of limited use when other stars are concerned (Donati & Landstreet 2009).

Spectacular results of recent investigations notwithstanding, spectropolarimetry of stellar magnetic fields remains to be fundamentally limited, both regarding the type of commonly available observational data and the methods used to interpret these data. Most importantly, with an exception of several studies of strongly magnetic Ap stars based on full Stokes vector observations (Wade et al. 2000), the current stellar magnetism investigations are limited to circular polarization. Such incomplete Stokes parameter data sets do not allow an unambiguous mapping of the stellar magnetic field topologies (Kochukhov & Piskunov 2002) and tend to miss the small-scale magnetic features (Kochukhov & Wade 2010) . The failure of the circular polarimetry to retrieve complete information about the stellar magnetic fields is illustrated, for example, by a major discrepancy between the results of the Zeeman broadening and circular polarization analyses of low-mass stars (Reiners & Basri 2009).

Full four Stokes parameter observations of moderately active cool stars represent the critical step towards more robust and complete analysis of stellar magnetic topologies. Here we present the first results of the simultaneous linear *and* circular polarization observations of active stars with a new, very high resolution spectropolarimeter at ESO.

2. The HARPS Polarimeter

The HARPS spectrometer (Mayor et al. 2003) attached to the ESO 3.6-m telescope is one of the most powerful astronomical tools for precision studies of stellar radial velocities variations due to planetary companions and stellar pulsations. It is also the highest resolution ESO spectrograph, providing coverage of the 380–690 nm wavelength region in a single exposure with the resolving power of $\lambda/\Delta\lambda = 115{,}000$. Recently the capabilities of HARPS have been extended to the polarimetric measurements in all four Stokes parameters.

The new polarimeter (HARPSpol, see Snik et al. 2008, 2010) occupies the place of the decommissioned iodine cell unit in the Cassegrain focus of the 3.6-m telescope, feeding the two existing HARPS optical fibers. The polarimeter consists of two separate polarimetric units (Fig. 1): one for circular and another for linear polarization measurements. Each unit is equipped with the modulator – a superachromatic quarter-wave or half-wave plate (QWP, HWP), based on a multi-layer zero-order birefringent polymer. This design yields a much lower fringing amplitude compared to the off-the-shelf quartz-based superachromatic wave plates. The light is split into two orthogonally polarized beams by a polarizing beam-splitter (Foster prism). The corresponding spectra, i.e. $I \pm X$ with $X = QUV$, are recorded simultaneously on the 4×4K CCD mosaic. A slider inserts the polarimeter from a completely retracted position, and allows for selection of either the circular or the linear polarimeter.

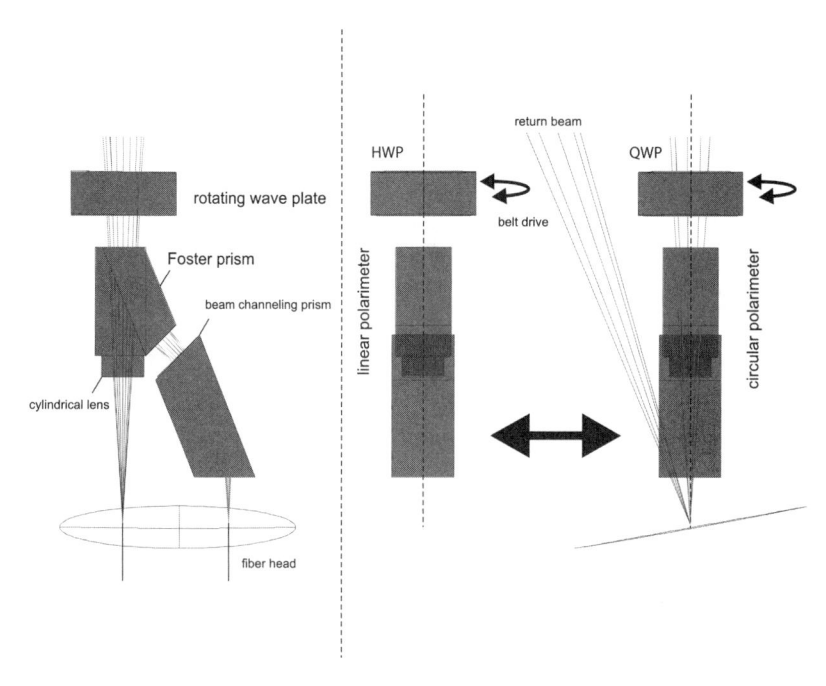

Figure 1. Design of the HARPS polarimeter. The converging beam from the telescope is split by a custom polarizing beam-splitter and fed into the two HARPS fibers. The modulation is performed by superachromatic QWP and HWP, respectively, for the two separate polarimetric units that can be positioned into the beam.

The two orthogonally polarized beams injected into the fibers can be exchanged by rotating the half-wave plate with the increment steps of 22.5° and the quarter-wave plate with 90° steps. A combination of exposures obtained with different modulator orientations (at least two for each of the Stokes parameters) allows to compensate, to first order, flat-fielding errors and other artifacts, reaching the polarimetric sensitivity of 10^{-4}–10^{-5} (Semel et al. 1993).

HARPSpol is fully integrated with the ESO control electronics and software. The standard echelle spectra-reduction and calibration procedures are performed by the improved HARPS pipeline. The polarimetric demodulation of the resulting one dimensional spectra is based on the "double ratio" method described by Bagnulo et al. (2009).

3. First Scientific Results

The HARPS polarimeter was commissioned at the 3.6-m ESO telescope in May 2009. Additional alignments were performed in December 2009. Since January 2010 our consortium carries out a comprehensive program of science verification of the new instrument by observing magnetic stars over the entire H-R diagram. Here we present preliminary results from our first two observing runs, focusing on the observations of bright late-type stars in all four Stokes parameters. Results of the HARPSpol circular polarization monitoring of T Tauri stars are reported separately (Johns-Krull, this meet-

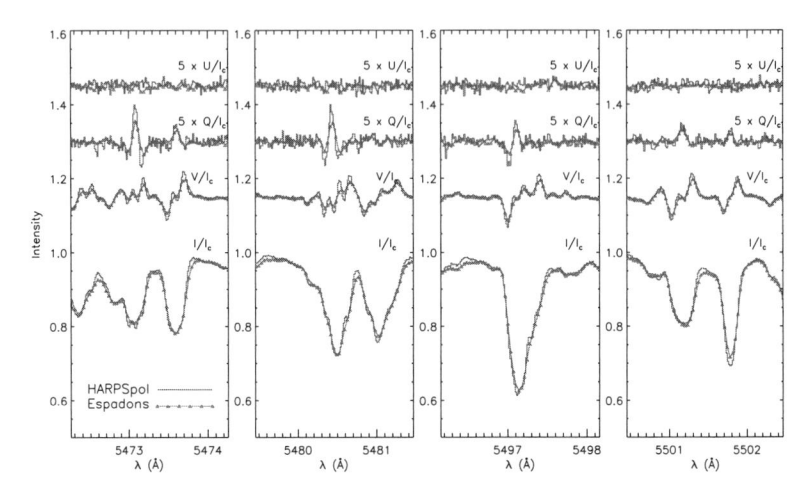

Figure 2. Comparison of the HARPSpol four Stokes parameter spectra of the Ap star γ Equ with the (specially calibrated) ESPaDOnS data for the same star. The polarization spectra are shifted vertically with respect to Stokes I. The Stokes Q and U spectra are magnified by a factor of 5.

ing). In addition, a sensitive HARPSpol search of weak magnetic fields in chemically peculiar B-type stars is discussed by Makaganiuk et al. (2010).

3.1. Spectropolarimetry of Magnetic Ap Stars

Owing to their strong fossil magnetic fields, slow rotation, and enhanced line blanketing due to chemical overabundances, Ap/Bp stars are excellent targets for stellar high-resolution spectropolarimetry in all four Stokes parameters (Wade et al. 2000). They also represent the only class of magnetic stars other than the Sun that currently allows detection and modeling of polarization signals in individual spectral lines (e.g., Kochukhov & Wade 2010).

A number of bright magnetic Ap stars is included in our HARPSpol science verification observations. In particular, we have obtained partial spectropolarimetric coverage of the rotation cycles for the hot magnetic chemically peculiar stars HD 75049 and HD 137509, which both have extremely strong magnetic fields, as well as for the bright cool Ap stars α Cir, HR 1217, and γ Equ. The latter star is especially useful for verification of the high-resolution linear polarization observations due to its relatively strong magnetic field and rotation period exceeding 90 yr (Bychkov et al. 2006). This star is systematically observed with the ESPaDOnS spectropolarimeter at CFHT with the aim to characterize significant circular to linear polarization cross-talk produced in this instrument by the telescope optics and ADC. A special sequence of exposures at different orientations of the rotating CFHT Cassegrain adaptor fully compensates this cross-talk. In Fig. 2 we compare these specially calibrated ESPaDOnS Stokes profiles of γ Equ with regular HARPSpol observations of this star obtained during the first commissioning of the instrument (May 2009).

It is clear that standard HARPSpol spectra match well the specially calibrated ESPaDOnS data, indicating the absence of significant cross-talk in our instrument. In general HARPSpol allows to resolve more details in circular polarized line profiles

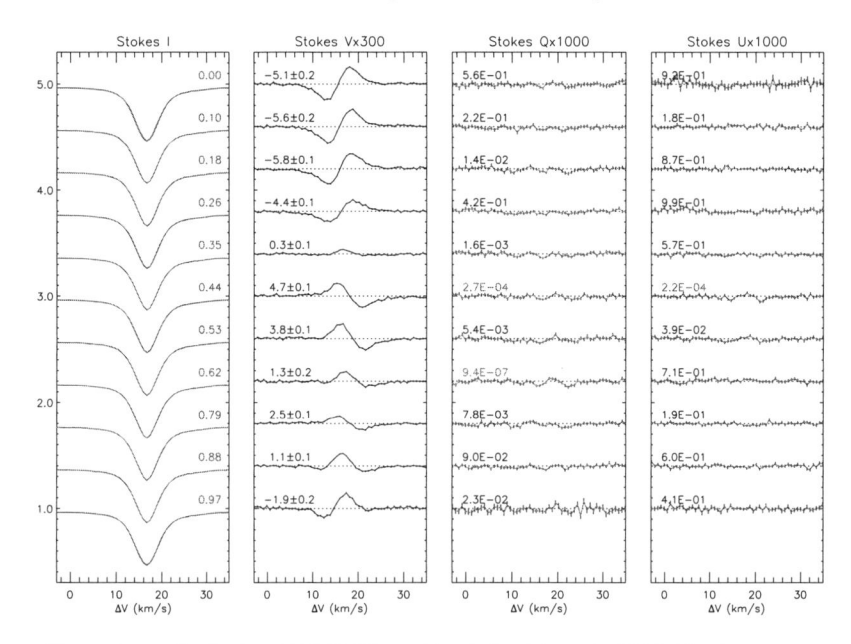

Figure 3. The Stokes $IQUV$ LSD profiles derived from the HARPSpol spectropolarimetric monitoring of ε Eri. The LSD profiles for different observing nights are shifted vertically. The numbers in the Stokes I panel give rotational phase. The numbers in the Stokes V panel indicate the longitudinal magnetic field in gauss. The False Alarm Probability estimate of the signal detection is given for Stokes Q and U. Note that the Stokes V and QU profiles are magnified by factors 300 and 10^3, respectively, relative to Stokes I.

and yields significantly higher amplitude of the linear polarization signals, owing to its larger spectral resolution. Four Stokes parameter observations of several slowly rotating Ap stars show that an increase of the resolving power from $\lambda/\Delta\lambda = 65{,}000$ (ESPaDOnS, NARVAL) to 115,000 (HARPSpol) results in roughly factor of two higher amplitude of the Stokes Q and U profiles.

At the same time, a higher pixel-to-pixel scatter evident in Fig. 2 for the HARPSpol spectra compared to the ESPaDOnS observations with similar effective exposure time reflects a combination of a somewhat lower overall efficiency of HARPS, its higher dispersion, and light losses in the second, lower quality HARPS fiber, which is normally used to record simultaneous ThAr calibration but not for science observations. An imperfect alignment of the linear polarimeter at the time of the HARPSpol observations of γ Equ also contributed to the noise in Q and U profiles. An upgrade of the HARPS fibers is ongoing. Meanwhile, the current efficiency reduction of the HARPSpol observations relative to the standard mode of HARPS is estimated to be $\approx 30\%$.

3.2. Ultra-precise Magnetic Observations of ε Eri

The young K2 dwarf ε Eri (HR 1084) is one of the nearest solar-type stars and the closest exoplanet host (Benedict et al. 2006). This star is surrounded by the dust disk and shows a high level of magnetic activity. Previous studies of the Zeeman broadening

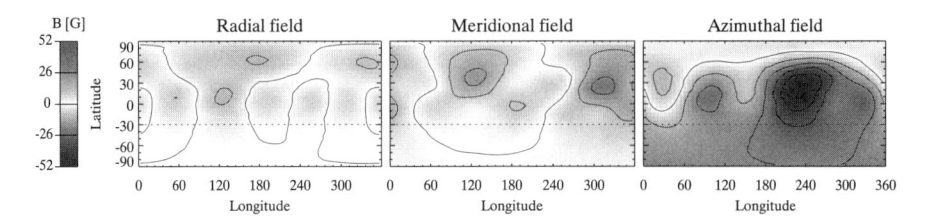

Figure 4. Magnetic field topology of ε Eri reconstructed from the HARPSpol spectropolarimetric observations using magnetic Doppler imaging technique. The panels show rectangular projections of the radial, meridional, and azimuthal field components. The contours of equal field strength are plotted with a step of 10 G.

in optical and IR magnetically sensitive lines suggested the presence of fields with spatially averaged strength of 100–200 G (Valenti et al. 1995, Rüedi et al. 1997).

We observed ε Eri in all four Stokes parameters during 11 nights, fully covering ≈ 11.3 d rotation period of the star (Frölich 2007). Similar to previous studies of moderately active solar-type stars (Petit et al. 2008), we could not find polarimetric signal in individual spectral lines and resorted to the Least Squares Deconvolution (LSD) method (Donati et al. 1997) for the spectropolarimetric diagnostic. Using the code by Kochukhov et al. (2010), we derived the mean line profiles from a sample of ≈ 7500 lines. Fig. 3 illustrates the LSD intensity and polarization spectra of ε Eri. There is no evidence of the variability in Stokes I, but variable magnetic field is clearly detected in the circular polarization. The mean line of sight field component varies between −6 and 5 G. The HARPSpol spectra allow us to measure the longitudinal magnetic field with the precision of 0.1–0.2 G. Such polarimetric accuracy was previously achieved for a few brightest giants (Auriére et al. 2009), but never for a dwarf solar-like object.

Despite the high signal-to-noise ratio of our polarimetric data, the LSD Q and U profiles of ε Eri generally do not reveal magnetic signatures. However, on one particular night the Q signature is detected with a False Alarm Probability (FAP) of $< 10^{-5}$, which is usually interpreted as a definite detection (Donati et al. 1997). On another night, a marginal signal ($10^{-5} <$ FAP $< 10^{-3}$) is present in both Q and U profiles.

Taking advantage of the very precise polarimetric measurements of ε Eri, we reconstructed the global magnetic field topology in this star using magnetic Doppler imaging. We modeled the LSD V profiles with the code described by Kochukhov & Piskunov (2009), assuming that the LSD profile behaves as a real spectral line with the average parameters. Preliminary results of the magnetic inversion are presented in Fig. 4. We find the field strength of up to ≈ 50 G on the surface of ε Eri and infer that the global field topology is dominated by the toroidal magnetic component. In this respect ε Eri appears to be similar to the solar twins rotating faster than 12 d (Petit et al. 2008).

3.3. Detection of Linear Polarization in HR 1099

The K1 subgiant primary of the RS CVn binary HR 1099 (V711 Tau) is one of the most magnetically active late-type stars. It was frequently studied with Doppler imaging and circular spectropolarimetry (Petit et al. 2004). Previous studies found evolving star spots and large-scale azimuthal magnetic field structures, stable over several years. We observed HR 1099 in all four Stokes parameters at two different nights in January 2010.

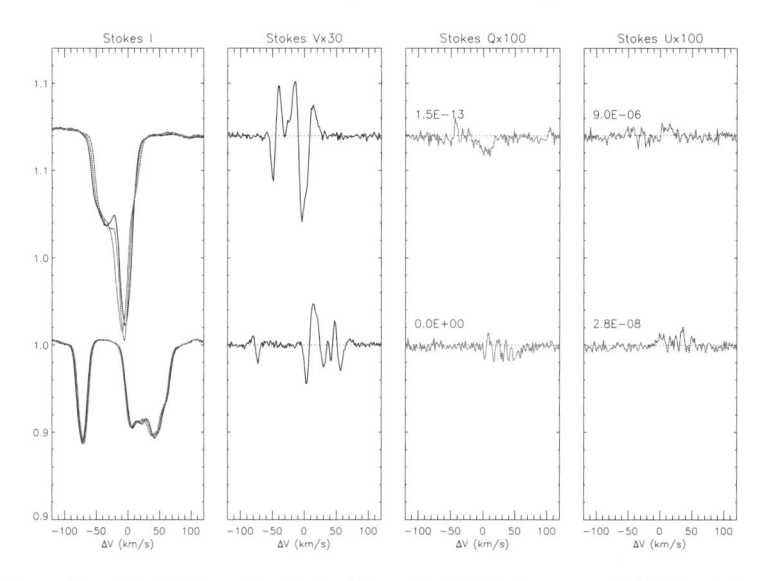

Figure 5. The LSD profiles derived from the full Stokes vector HARPSpol observations of the RS CVn-type binary star HR 1099. The LSD Stokes profiles for two different observing nights are shifted vertically. The Stokes V and QU profiles are magnified by factors 30 and 100, respectively, relative to Stokes I. The numbers in the Stokes Q and U panels give the False Alarm Probability estimate of the signal detection. Definite detection of the linear polarization inside line profiles of a cool star is reported here for the first time.

Fig. 5 shows the LSD Stokes $IQUV$ profiles derived by combining polarimetric signal in 8300 lines. The distortions in the mean intensity profile of the primary indicate significant spot coverage at the time of our observations. The Stokes V profile of the primary shows a more complex morphology than in previous spectropolarimetric studies of this star (Donati et al. 2003,Petit et al. 2004), probably indicating a more complex field configuration. We also clearly detect magnetic field signature in the Stokes V profile of the main sequence secondary component.

The Q and U signatures of the primary are detected in both HARPSpol observations with a very high confidence (FAP < 10^{-5}). *This is the first ever definite detection of the linear polarization inside line profiles of a cool active star.* We find that the LSD Stokes QU profiles are significantly more complex than the corresponding Stokes V spectra. The amplitude of the linear polarization LSD profiles is ~10^{-4}, which is approximately an order of magnitude lower in comparison to the mean Stokes V spectrum.

3.4. Null Result for α Cen A

The solar analogue α Cen A was observed during the first commissioning of HARPSpol, during partly cloudy conditions. Nevertheless, sufficient photons were collected to reach a polarimetric sensitivity of ~10^{-5} with the help of LSD based on 9000 metal lines. As demonstrated by Fig. 6, no clear signals are detected in either V, Q or U profiles. The corresponding null spectra are also featureless, confirming the lack of polarization artifacts down to the level of ~10^{-5}.

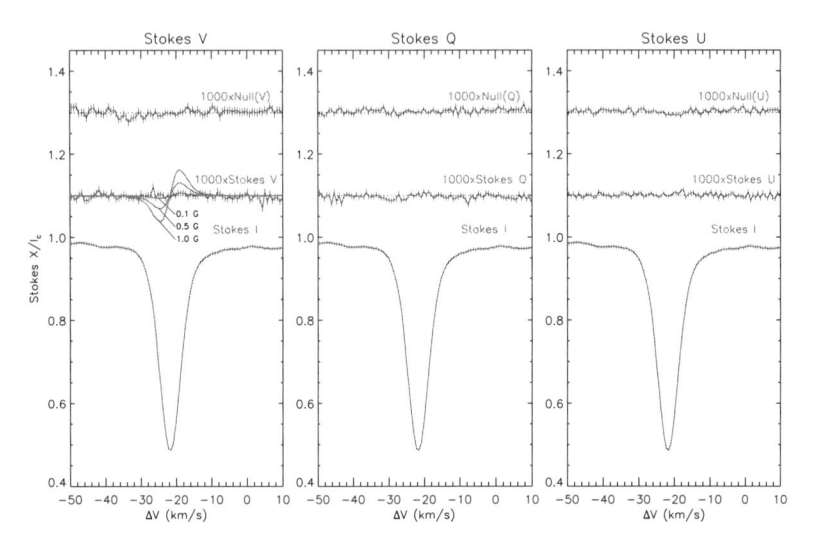

Figure 6. The LSD profiles derived from the full Stokes vector HARPSpol observations of α Cen A. The LSD Stokes I is compared with the LSD VQU and corresponding null profiles. Polarization profiles are shifted vertically and magnified by a factor of 10^3. The curves plotted on top of the Stokes V spectrum illustrate the circular polarization signal expected for 0.1, 0.5, and 1.0 G longitudinal magnetic field.

The non-detection of the Stokes V signature corresponds to the 3σ upper limit of 0.2 G for the disk-averaged line of sight magnetic field. In comparison, the longitudinal field of Sun as a star reaches 1 G during activity maximum and is below 0.1 G during the minimum (Demidov et al. 2002). We therefore conclude that HARPSpol can detect global magnetic fields of active solar twins. In its present low activity state, α Cen A appears to be reminiscent of the Sun at cycle minimum, in agreement with recent X-ray observations (Ayres 2009).

4. Conclusions

The HARPS polarimeter is now available to the astronomical community as a standard ESO instrument at the 3.6-m telescope at La Silla. It offers similar capabilities as the ESPaDOnS instrument at CFHT (Donati et al. 2006), although with a significantly higher spectral resolution, more reliable linear polarization mode and unique radial velocity stability of HARPS. HARPSpol is currently the only high-resolution four Stokes parameter spectropolarimeter in the southern hemisphere.

Acknowledgments. OK is a Royal Swedish Academy of Sciences Research Fellow supported by grants from the Knut and Alice Wallenberg Foundation and the Swedish Research Council. We thank Gregg Wade for providing us the ESPaDOnS data of γ Equ.

References

Aurière, M., Wade, G. A., Konstantinova-Antova, R., Charbonnel, C., Catala, C., Weiss, W. W., Roudier, T., Petit, P., Donati, J., Alecian, E., Cabanac, R., van Eck, S., Folsom, C. P., & Power, J. 2009, A&A, 504, 231
Ayres, T. R. 2009, ApJ, 696, 1931
Bagnulo, S., Landolfi, M., Landstreet, J. D., Landi Degl'Innocenti, E., Fossati, L., & Sterzik, M. 2009, PASP, 121, 993
Benedict, G. F., McArthur, B. E., Gatewood, G., Nelan, E., Cochran, W. D., Hatzes, A., Endl, M., Wittenmyer, R., Baliunas, S. L., Walker, G. A. H., Yang, S., Kürster, M., Els, S., & Paulson, D. B. 2006, AJ, 132, 2206
Bychkov, V. D., Bychkova, L. V., & Madej, J. 2006, MNRAS, 365, 585
Demidov, M. L., Zhigalov, V. V., Peshcherov, V. S., & Grigoryev, V. M. 2002, Solar Phys., 209, 217
Donati, J., Catala, C., Landstreet, J. D., & Petit, P. 2006, in Astronomical Society of the Pacific Conference Series, edited by R. Casini & B. W. Lites, vol. 358 of Astronomical Society of the Pacific Conference Series, 362
Donati, J., & Landstreet, J. D. 2009, ARA&A, 47, 333
Donati, J.-F., Cameron, A. C., Semel, M., Hussain, G. A. J., Petit, P., Carter, B. D., Marsden, S. C., Mengel, M., López Ariste, A., Jeffers, S. V., & Rees, D. E. 2003, MNRAS, 345, 1145
Donati, J.-F., Catala, C., Wade, G. A., Gallou, G., Delaigue, G., & Rabou, P. 1999, A&AS, 134, 149
Donati, J.-F., Semel, M., Carter, B. D., Rees, D. E., & Collier Cameron, A. 1997, MNRAS, 291, 658
Fröhlich, H. 2007, Astronomische Nachrichten, 328, 1037
Kochukhov, O., Makaganiuk, V., & Piskunov, N. 2010, A&A, 524, A5
Kochukhov, O., & Piskunov, N. 2002, A&A, 388, 868
— 2009, in Astronomical Society of the Pacific Conference Series, edited by S. V. Berdyugina, K. N. Nagendra, & R. Ramelli, vol. 405 of Astronomical Society of the Pacific Conference Series, 539
Kochukhov, O., & Wade, G. A. 2010, A&A, 513, A13
Makaganiuk, V., Kochukhov, O., Piskunov, N., Jeffers, S. V., Johns-Krull, C. M., Keller, C. U., Rodenhuis, M., Snik, F., Stempels, H. C., & Valenti, J. A. 2010, A&A, in press. arXiv: 1010.3931
Mayor, M., Pepe, F., Queloz, D., Bouchy, F., Rupprecht, G., Lo Curto, G., Avila, G., Benz, W., Bertaux, J.-L., Bonfils, X., Dall, T., Dekker, H., Delabre, B., Eckert, W., Fleury, M., Gilliotte, A., Gojak, D., Guzman, J. C., Kohler, D., Lizon, J.-L., Longinotti, A., Lovis, C., Megevand, D., Pasquini, L., Reyes, J., Sivan, J.-P., Sosnowska, D., Soto, R., Udry, S., van Kesteren, A., Weber, L., & Weilenmann, U. 2003, The Messenger, 114, 20
Petit, P., Dintrans, B., Solanki, S. K., Donati, J.-F., Aurière, M., Lignières, F., Morin, J., Paletou, F., Ramirez Velez, J., Catala, C., & Fares, R. 2008, MNRAS, 388, 80
Petit, P., Donati, J., Wade, G. A., Landstreet, J. D., Bagnulo, S., Lüftinger, T., Sigut, T. A. A., Shorlin, S. L. S., Strasser, S., Aurière, M., & Oliveira, J. M. 2004, MNRAS, 348, 1175
Reiners, A., & Basri, G. 2009, A&A, 496, 787
Rüedi, I., Solanki, S. K., Mathys, G., & Saar, S. H. 1997, A&A, 318, 429
Semel, M., Donati, J.-F., & Rees, D. E. 1993, A&A, 278, 231
Snik, F., Jeffers, S., Keller, C., Piskunov, N., Kochukhov, O., Valenti, J., & Johns-Krull, C. 2008, SPIE, 7014
Snik, F., Kochukhov, O., Piskunov, N., Rodenhuis, M., Jeffers, S., Keller, C., Dolgopolov, A., Stempels, E., Makaganiuk, V., Valenti, J., & Johns-Krull, C. 2010, ArXiv e-prints. 1010.0397
Valenti, J. A., Marcy, G. W., & Basri, G. 1995, ApJ, 439, 939
Wade, G. A., Donati, J.-F., Landstreet, J. D., & Shorlin, S. L. S. 2000, MNRAS, 313, 823

Magnetic Fields on Cool Stars

Ansgar Reiners

Universität Göttingen, Institut für Astrophysik, Friedrich-Hund-Platz 1, D-37077 Göttingen, Germany

Abstract. Magnetic fields are an important ingredient to cool star physics, and there is great interest in measuring fields and their geometry in order to understand stellar dynamos and their influence on star formation and stellar evolution. During the last few years, a large number of magnetic field measurements became available. Two main approaches are being followed to measure the Zeeman effect in cool stars; 1) the measurement of polarized light, for example to produce magnetic maps, and 2) the measurement of integrated Zeeman broadening to measure the average magnetic field strength on the stellar surface. This article briefly reviews the two methods and compares results between them that are now available for about a dozen M-type stars. It seems that we see a great variety of magnetic geometries and field strengths with typical average fields of a few kG in active M-type stars. The interpretation of geometries, however, has not yet led to a clear picture of magnetic dynamos and field configuration, and work is needed on more observational data but also on the fundamental understanding of our measurements.

1. Introduction

The Sun and cool stars are known to harbor magnetic fields leading to all phenomena summarized under the term *stellar activity*. It may be debatable whether magnetic fields are actually the most interesting aspect of cool star and solar physics (cp. Moore & Rabin 1985), but it is certainly an exciting field that brings together a large variety of physical mechanisms and subtle analysis techniques. This makes it sometimes difficult to interpret observational results and compare them to theoretical expectations – even if both are available, or even if one compares observations achieved from different techniques.

A particularly interesting class of stars are cool stars of spectral type M. Covering the mass spectrum between ~ 0.6 and $0.1\,M_\odot$, M dwarfs are the most frequent type of stars. Within this mass range, objects can have very different physical properties. The very important transition from partly convective (sun-like) to fully convective stars happens in the M dwarf regime, probably around spectral type M3/M4. Furthermore, atmospheres of M dwarfs can be very different and both molecules and dust gain importance as the temperature drops toward late spectral types.

In this article, I will concentrate on measurements of magnetic fields on M dwarfs because most of the currently available measurements of cool star magnetic fields are from M dwarfs, which is because sun-like (field) stars tend to be less active (less rapidly rotating because of shorter braking timescales) implying lower average magnetic fields that are more difficult to detect.

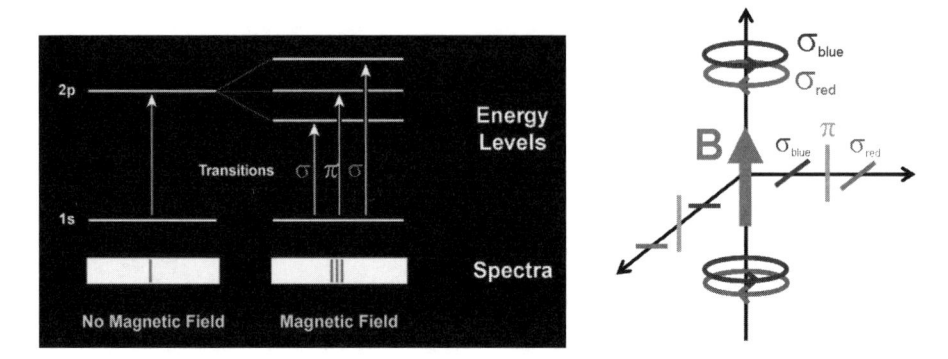

Figure 1. Schematic view of Zeeman splitting. The upper level in the example (left panel) is split into three levels producing three spectral lines that are separated. The polarization of the π and σ components are shown in the right panel.

2. Technical Aspects

To measure the magnetic fields of stars, determination of spectral line splitting due to the Zeeman effect is the most commonly used method. In this article, I will not give a comprehensive overview of the Zeeman technique but try to emphasize a few technical aspects that are particularly important for the interpretation of the currently available measurements. More comprehensive discussions and reviews on magnetic fields and their measurement in cool stars are, e.g., Landstreet (1992); Valenti et al. (1995); Mestel & Landstreet (2005); Donati & Landstreet (2009); Miesch & Toomre (2009), or the conference reviews by Saar (1996); Johns-Krull (2008).

In short, dipole transitions obey the selection rule $\Delta M = -1, 0$ or $+1$. The three groups of lines according to different ΔM are degenerate if no magnetic field is present, but separated in the presence of a magnetic field. Transitions with $\Delta M = 0$ are called π-components, $\Delta M = \pm 1$ are called σ-components. If a magnetic field is present, the energy (wavelength) of the σ-components are shifted according to the sensitivity of the transition (summarized in the *Landé*-factor g), the strength of the magnetic field B, and the wavelength of the transition itself (λ_0). The average velocity displacement of the spectral line components can be written as

$$\Delta v = 1.4 \lambda_0 g B, \qquad (1)$$

with B in kG, λ_0 in μm, and Δv in km s^{-1}. Fig. 1 shows a simplified scheme of the splitting (left panel) and of the polarization (right) of the π and σ components; the π component is always linearly polarized while the σ components can be linearly or circularly polarized depening on the viewing angle.

In order to measure a magnetic field, the measurement of different polarization states can be of great advantage. This is immediately clear from the right panel of Fig. 1 since the different Zeeman components are polarized in a characteristic fashion. A commonly used system are the Stokes components I, Q, U, and V, which are defined in the following sense:

$$I = \updownarrow + \leftrightarrow$$
$$Q = \updownarrow - \leftrightarrow$$
$$U = \nwarrow - \nearrow$$
$$V = \circlearrowleft - \circlearrowright$$

Stokes I is just the integrated light (unpolarized light), Stokes Q and U measure the two directions of linear polarization, and Stokes V measures circular polarization. It has been shown that the magnetic field distribution of a star can be reconstructed from observations of all four Stokes parameters (Kochukhov & Piskunov 2002; Kochukhov et al. 2010), and that using only a subset of Stokes vectors leads to ambiguities that should be interpreted with caution. Unfortunately, measurements of linear polarization are extremely challenging in cool stars so that typically only Stokes I and (sometimes) Stokes V are available.

2.1. Field, Flux, and Filling Factor

The situation shown in Fig. 1 is simplified. In most cases, line splitting is a bit more complicated, but the main difficulty in measuring the splitting is the small value of Δv compared to other broadening agents like intrinsic temperature and pressure broadening, and rotational broadening. In a kG-magnetic field, typical splitting at optical wavelengths is below 1 km s^{-1}, which is well below intrinsic line-widths of several km s^{-1} and also below the spectral resolving power of typical high-resolution spectrographs. Thus, individual components of a spectral line can normally not be resolved even if the star only had one well-defined magnetic field component. Real stars, however, can be expected to harbor a magnetic field distribution that is much more complex than this. Thus, even if spectral lines were intrinsically very narrow and spectral resolving power infinitely high, we would expect the Zeeman-broadened lines to look smeared out since in our observations we integrate over all magnetic field components on the entire visible hemisphere.

An important consequence of the fact that individual Zeeman-components are usually not resolved is the degeneracy between magnetic field B and filling factor f. A strong magnetic field covering a small portion of the star looks similar to a weaker field covering a larger portion of the star. An often used way around this ambiguity is to specify the value Bf, i.e., the product of the magnetic field and the filling factor (if more than one magnetic component is considered, Bf is the weighted sum over all components). Products of B with some power of f, for example $Bf^{0.8}$ are also considered because they seem to be better defined by observations (see, e.g., Valenti et al. 1995, for a deeper discussion). One important point to observe is that Bf is often called the "flux" – because it is the product of a magnetic field and an area – but it has the unit of a magnetic field. In fact, the term flux is very misleading since 1) Bf is identical to the average magnetic field on the stellar surface, i.e., $Bf \equiv <B>$, and 2) the total magnetic flux of two stars with the same values of Bf can be extremely different according to their radii because the actual flux is proportional to the radius squared; $\mathcal{F} \propto Bfr^2$. As a consequence, the value Bf will be much lower in a young, contracting star compared to an older (smaller) one if flux is conserved.

A related source of confusion is the difference between the signed magnetic field (or flux), and the unsigned values or the square of the fields (used to calculate magnetic energy). With Stokes I, both polarities produce the same signal and only the unsigned flux is measured. This implies that Stokes I carries only partial information about field geometry, but it also means that Stokes I always probes the entire magnetic flux of the star. On the other hand, Stokes V can provide information on the sign of the magnetic fields, but this comes with another serious caveat: Since we cannot resolve the stellar surface, and Stokes V measures the signed field, regions of opposite magnetic fields on the stellar surface cancel out and can become invisible to the Stokes V signal.

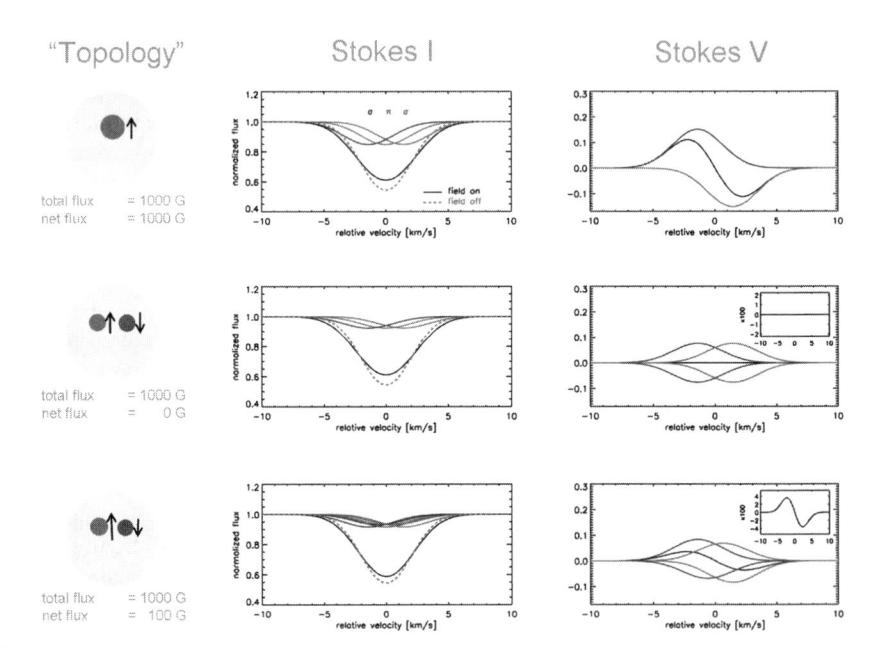

Figure 2. Three examples of simplified field topologies and their signals in Stokes I and V.

Examples of signatures of magnetic fields in Stokes I and V are sketched in Fig. 2. The left panel shows the "topology", which is actually not the topology of a stellar magnetic field, but nothing else than two areas of radial magnetic fields put on a flat surface (the spherical shape of a star has not been taken into account in this example). In the top row, a simple magnetic field region with only one direction is shown; signed and unsigned "net" flux both are 1 kG in this example. The line in Stokes I is effectively broadened, and Stokes V shows a clear signal on the order of 10%. Note that the direction of the Stokes V signal indicates the direction of the magnetic field. In the second row, two magnetic regions with only half the size as in the first example are observed. Both regions have the same absolute field strength and area but with opposite polarity. In this example, the Stokes I signal is identical to the first example, but the signal in Stokes V entirely vanishes because the net (signed) flux of this configuration is exactly zero; any field strength in this cancelling configuration is invisible to Stokes V. The last row shows a case in which one of the two areas is slightly larger than the other, the total flux is still 1000 G, but the net flux is only 100 G. The amplitude in Stokes V is ca. 4%.

2.2. Viewing Angle

Another thing that becomes immediately clear is that the geometric interpretation of Zeeman splitting on an unresolved stellar disk can be arbitrarily complicated, no matter if polarized or unpolarized light is used. In addition to the ambiguity between magnetic field strength and filling factor (which includes our ignorance about the number and distribution of magnetic components), the signature of a magnetic field region in stellar spectra depends on the angle between the magnetic field lines and the line of sight (Fig. 1). In reality, again, a distribution of angles will be present because field lines are probably bent on the stellar surface, and because the stellar surface is spherical. As a result, even geometrically relatively simple field distributions will lead to complicated splitting patterns. If the star is rotating (as most stars probably are) that pattern again depends a lot on the time a star is observed – which in turn can be utilized by observing the change of observed spectra with rotation phase.

3. Measurements of Cool Star Magnetic Fields

Reliable detections of magnetic fields in cool stars of spectral type F–K are relatively rare, see, e.g., Saar (1996, 2001). The challenges of detecting Zeeman splitting in these stars using atomic lines at optical wavelengths have recently been revisited by Anderson et al. (2010). As realized earlier (e.g., Basri et al. 1990), magnetic field signatures can often be mimicked by the signatures of cool spots so that a definite measurement of magnetic fields, and in particular their distribution and geometry, is a delicate task.

M dwarfs, however, are a bit more cooperative. Although their spectra exhibit a much denser forest of absorption lines rendering it difficult to investigate isolated spectral lines, their magnetic fields can be higher than fields in hotter stars at given rotational velocity $v \sin i$. An important step in our understanding of M dwarf magnetic fields was the investigation of seven M dwarfs by Johns-Krull & Valenti (1996). Line shapes of an atomic Fe I line at 846.7 nm, hidden in a forest of TiO molecular lines, were analyzed and average magnetic field strengths of several kG were reported in a few stars that are probably fully convective (the results of their analysis were updated by those reported in Johns-Krull & Valenti (2000) using a multi-component approach). Complementary work looking for magnetic field signatures in Stokes V was done in T Tauri stars very successfully. Using polarized light, signatures of kG-strength magnetic fields could be discovered (see Johns-Krull 2007).

During the last years, two main routes have been followed searching for direct detections of magnetic fields in M dwarfs. One method is to employ near-infrared molecular absorption lines of FeH to search for Zeeman broadening in Stokes I, another is to extract the polarization signal from several hundred lines at optical wavelengths in order to search for field signatures in Stokes V. Both methods proved to be successful down to the latest M dwarfs, and it is possible to compare the results of both methods in a small number of stars. In the following, an overview on both methods is given and the results are compared and interpreted.

3.1. Results from Stokes I

The spectra of M dwarfs are dominated by the presence of molecular absorption bands. At optical wavelengths, bands from TiO and VO appear. Analysis of Zeeman broadening in these bands, however, is difficult because the lines are not individually resolved.

According to Eq. 1, Zeeman broadening (in terms of velocity shift) is proportional to wavelength, which implies that it is easier to detect at longer wavelength, in particular because other Doppler broadening agents like rotation do not depend on wavelength. At near-infrared wavelengths, a molecular band of FeH can be found in the spectra of M dwarfs, and Wallace et al. (1999) showed that this band is well suited for the measurement of magnetic fields. An observational problem of FeH is that its most suitable band is located at around $\lambda = 1\mu$m, which is too red for most CCDs and too blue for most astronomically used infrared spectrographs. As a consequence, only very few high-resolution spectrographs can provide spectra of this wavelength, and the efficiencies are typically ridiculously low. On the other hand, M dwarfs emit much of their flux at near-infrared wavelengths so that in comparison to optical wavelengths, the signal quality around 1μm is not much lower than around 700 nm if the spectra are obtained with an optical/near-IR echelle spectrograph like HIRES (Keck observatory) or UVES (ESO VLT). The 2004-upgrade of the HIRES spectrograph allowed a thorough test of the FeH spectral range. We developed a method to semi-empirically determine the magnetic fields of M dwarfs comparing FeH spectra of our targets to spectra taken of two template stars; one with no magnetic field and one with a known, strong magnetic field (Reiners & Basri 2006). As reference, spectra of stars with magnetic fields previously measured in a detailed analysis of atomic lines are used (Johns-Krull & Valenti 2000). Thus, all magnetic field measurements are relative to the reference star (Gl 873, $< B >= 3.9$ kG), and magnetic fields higher than this value cannot be quantified.

Obviously, systematic uncertainties of the measurements are quite large, typically several hundred Gauss. Unfortunately, Zeeman splitting of the FeH molecule is complicated and cannot entirely be described so far (see Berdyugina & Solanki 2002). Meanwhile, progress has been made using an empirical approach to model Zeeman splitting in FeH lines, and this approachs suggests that the fields determined semi-empirically may be overestimated by some $\sim 20\%$[1] (Shulyak et al. 2010).

3.1.1. Average Fields in M-type Stars

Magnetic field measurements in M dwarfs are reported in Saar (1994); Johns-Krull (2007); Reiners & Basri (2007, 2009, 2010); Reiners et al. (2009b,a). This is a non-exhaustive list of publications, and a number of magnetic field measurements of stars re-analysed here can be found in earlier literature (with more or less consistent results); see Saar (1996, 2001). Measurements of M dwarf magnetic fields for spectral types M1–M9 are shown in Fig. 3.

It is well known that early-M dwarfs (M0–M3) in the field are much less active and rotating slower than later, fully convective M stars (e.g., Reiners & Basri 2008). Early-M dwarfs appear to suffer much more effective rotational braking so that their activity lifetime is shorter than in later M dwarfs. Whether this is an effect of different magnetic field topologies in partially and fully convective stars is not clear – this question is one of the basic motivations for the comparison of field measurement results summarized here. It is important to realize that at spectral type M3/M4, several parameters of the stars change dramatically so that the reason for a change in braking timescales may actually be much more fundamental than magnetic field topology.

---

[1]This could be due to an overestimate of the reference magnetic field measurement derived from the atomic line analysis.

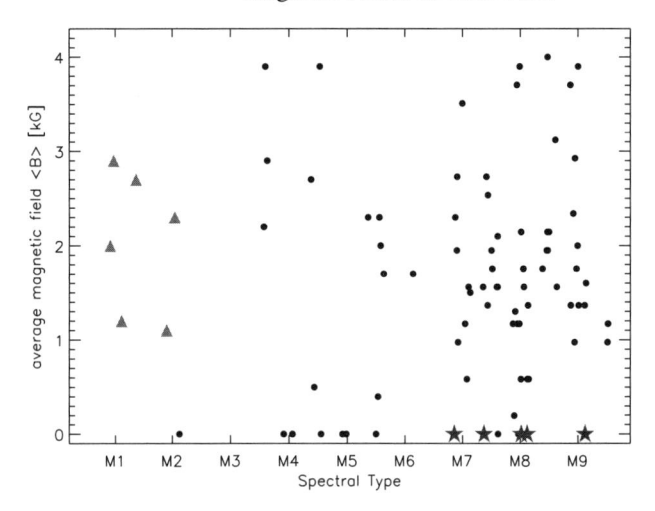

Figure 3. Measurements of M dwarf magnetic fields from Stokes I. Red triangles: young, early M-stars; blue stars: young accreting brown dwarfs; black circles: field M dwarfs. See text for references. Spectral types are offset by a small amount to enhance visibility of different objects.

The field strengths of young, early-M and field mid- and late-M dwarfs are on the order of a few kG. This is the main results from Zeeman analysis and consistently found using different indicators (at least in mid-M dwarfs). Compared to the Sun, the average magnetic field hence is larger by two to three orders of magnitude, an observational result that must have severe implications for our understanding of low-mass stellar activity. It is not clear whether our picture of a star with spots more or less distributed over the stellar surface is actually valid in M dwarfs. If, for example, 50 % of the surface of a star with a mean magnetic field of 4 kG is covered with a "quiet" photosphere and low magnetic field, the other half of the star must have a field strength as large as ∼8 kG. The two components of the stellar surface on such a star probably have very different temperatures and properties, and the definition of effective temperature must be considerably different from the temperature of the "quiet" photosphere.

In early-M dwarfs (M3 and earlier), magnetic fields were found in young stars that are still rapidly rotating. Since old, early-M dwarfs in the field are generally slowly rotating and inactive there has been no search for magnetic fields in any large sample of them. Typical field values can be expected to be on the level of a few hundred Gauss and less, which is difficult to detect with Stokes I Zeeman measurements.

3.1.2. Young Stars and Young Brown Dwarfs

While young, early-M stars exhibit magnetic fields of several kG, which is consistent with the magnetic fields of older M stars rotating at comparable pace and young stars of earlier spectral type, it is surprising that in young brown dwarfs of spectral types M7–M9 only upper limits of a few hundred Gauss could be determined for their magnetic fields (Reiners et al. 2009b) (blue stars in Figs. 3 and 4). Interestingly, all five young brown dwarfs seem to be accretors implying that they still have a disk, and a magnetic field of a few hundred Gauss appears to be enough for magnetospheric accretion in such

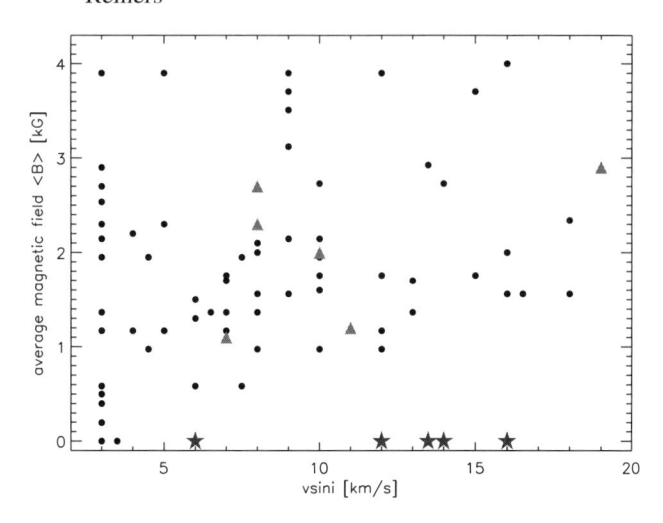

Figure 4. Measurements of M dwarf magnetic fields from Stokes I as a function of projected rotation velocity, $v \sin i$. Symbols are the same as in Fig. 3.

a low-mass object. So far, no direct magnetic field measurement could be performed in non-accreting, old, field brown dwarfs, but it is expected that they also harbor substantial magnetic fields (Reiners & Christensen 2010). Observations of radio-emission indicate that fields of a few kG strength are in fact present on some L-type field brown dwarfs (Hallinan et al. 2008; Berger et al. 2009).

It is an open question whether the non-detection of magnetic fields in brown dwarfs is due to the presence of accretion disks around the objects observed so far. If this is the case, there ought to be some mechanism for the disk to regulate the magnetic field of the central object, which is not easily understood. Alternatively, large difference in radius may be of importance in this context because the surface area of young brown dwarfs is about an order of magnitude larger than the surface of old brown dwarfs. If magnetic flux is approximately conserved during its evolution, the average magnetic field would be an order of magnitude lower in young, large brown dwarfs than in old, small, field brown dwarfs.

3.1.3. Rotation and Magnetic Fields

Stokes I measurements of magnetic fields are shown as a function of projected surface velocity $v \sin i$ in Fig. 4. The typical signature of a Γ-shaped rotation-activity relation is visible, which means that the lower (left) end of the relation is not resolved because the non-saturated part of the rotation-activity relation falls below the detection limit in rotation velocities. On the other hand, all rapidly rotating ($v \sin i \gtrsim 3 \, \mathrm{km \, s^{-1}}$) field stars show detectable magnetic fields. However, again, the young brown dwarfs violate this relation since they are rapidly rotating but do not show detectable fields.

The rotation-activity relation describes the fact that slowly rotating stars are less active than rapid rotators. In terms of Rossby number, $Ro = P/\tau_{\mathrm{conv}}$, with P the rotation period and τ_{conv} the convective overturn time, stars with $Ro \lesssim 0.1$ are saturated in activity. Stars with larger values of Ro show activity proportional to Ro. We know from the Sun that activity is caused by magnetic fields, and a rotation-magnetic field

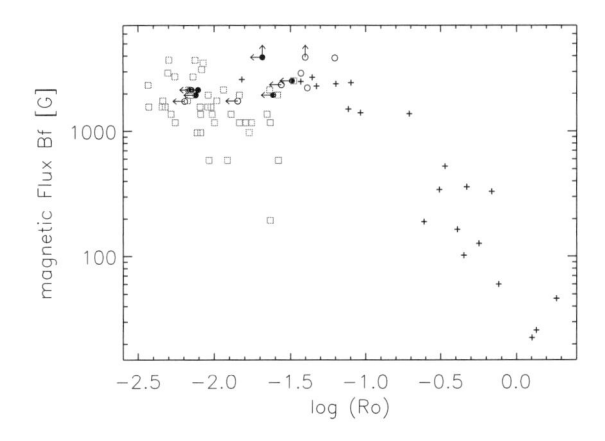

Figure 5. Magnetic fields as a function of Rossby number. Crosses and circles are stars M6 and earlier (see Reiners et al. 2009a, and references therein). Red squares are objects of spectral type M7–M9.

relation was shown by Saar (1996). At that time, however, magnetic fields could not be measured in stars with very low Rossby numbers (saturated regime) because spectral line widths are too broad due to the rotational broadening occuring at these velocities. M dwarfs, on the other hand, have very small radii (and long overturn times) so that for low Rossby numbers the corresponding surface velocities are relatively low. This allows measuring magnetic fields of stars well within the saturated regime. For M dwarfs of spectral type M6 and earlier, it was found that average magnetic fields indeed saturate (Reiners et al. 2009a). This implies that the saturation of the rotation-activity relation is due to a saturation of the average magnetic field and that B itself is limited (in contrast to a limit in the filling factor f only).

Recently, we have measured magnetic fields in a sample of M7–M9 stars (Reiners & Basri 2010). These measurements are shown as red squares in Fig. 5. Stars as cool as M7 seem to deviate from the rotation-activity relation; they still show higher magnetic fields at lower Ro, but saturation seems to occur at much lower values of Ro.

3.2. Results from Stokes V

Maps of magnetic fields in M stars derived from time-series of Stokes V measurements are presented in Donati et al. (2008); Morin et al. (2008, 2010). The Stokes V profiles are derived simultaneously from several hundred lines through least-squares-deconvolution (LSD). LSD makes use of the weak-field approximation so that very strong magnetic fields (several kG) may not be detectable with this method. As mentioned above, the use of only Stokes V means that the magnetic geometry will not be fully characterized. In particular, magnetic regions of opposite polarity can cancel each other so that they remain undetected. A way to overcome this issue is to take observations at different times so that magnetic regions may remain visible when they appear at the limb of the star (Zeeman Doppler Imaging, ZDI). The resolution of such a Doppler image critically depends on the rotation velocity of the star and the exposure time required to obtain the necessary data quality.

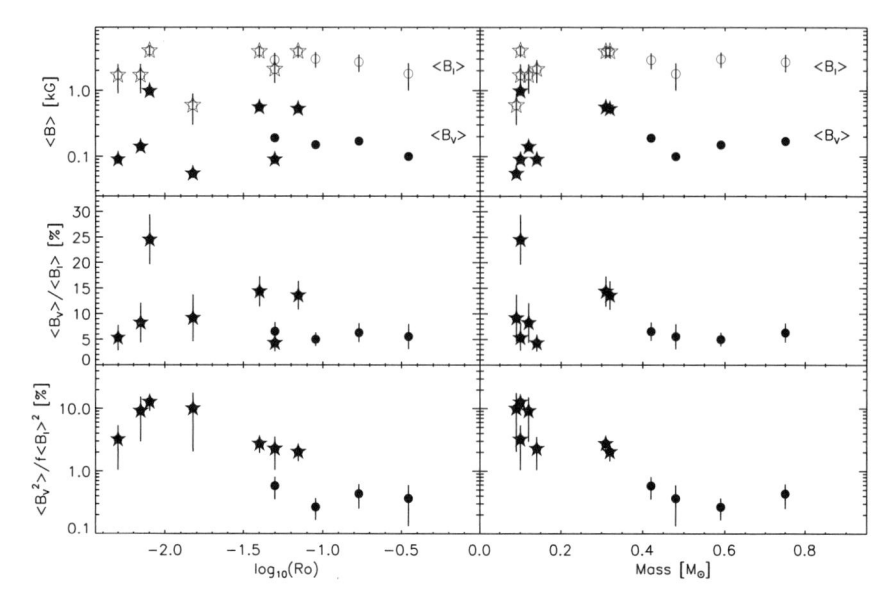

Figure 6. Measurements of M dwarf magnetic fields from Stokes I and Stokes V. *Top panel:* Average magnetic field – Open symbols: measurements from Stokes I; Filled symbols: measurements from Stokes V. *Center panel:* Ratio between Stokes V and Stokes I measurements. *Bottom panel:* Ratio between magnetic energies detected in Stokes V and Stokes I. Circles show objects more massive than $0.4 M_\odot$, stars show objects less massive than that.

A tremendous amount of work was put into the analysis of magnetic geometries in stars through ZDI, and the possibility of reconstructing magnetic fields on stellar surfaces is truly amazing. However, the interpretation of the field maps is very difficult, and conclusions have to be drawn with great care.

3.2.1. Magnetic Field Strengths

Typical average magnetic field strengths found in Stokes V measurements of M dwarfs are about a few hundred Gauss. Note that this is the average value for the detected unsigned magnetic field, $|B|$, the same as in Stokes I measurements; the average value of the signed magnetic field is zero by construction. An average field strength of a few hundred Gauss is much lower than average field strengths of magnetically active stars observed in Stokes I that are a typically few kG. The literature today contains eleven M dwarfs for which magnetic field measurements were carried out independently both in Stokes V and Stokes I. I compare the results from the mentioned papers in Fig. 6, which is an update of Fig. 2 in Reiners & Basri (2009).

Fig. 6 shows the average magnetic fields from Stokes I and V, their ratios, and the ratios of magnetic energies as a function of Rossby number and stellar mass. In the top panel, the measurements are shown directly, the center panel shows the ratio between the average magnetic fields $<B_V>/<B_I>$. For the majority of stars, the ratio is on the order of ten percent or less, which means that <10 % of the full magnetic field is detected in the Stokes V map. In other words, more than 90 % of the field is invisible to this method. As discussed above, this is probably a consequence of cancellation between field components of different polarity. One notable exeption is the M6 star WX Uma, which has an average field of approximately 1 kG in Stokes V (Gl 51 shows an even higher field but has not yet been investigated with the Stokes I method).

A second observable that comes out of the Stokes V maps is the average squared magnetic field, $<B^2>$, which is proportional to the magnetic energy of the star. Under some basic assumptions, this value can be approximated from the Stokes I measurement, too (see Reiners & Basri 2009). The ratio between approximate magnetic energies detected in Stokes V and I is shown in the bottom panel of Fig. 6, it is between 0.3 and 10 % for the stars considered.

3.2.2. Topologies of the Detected Fields

In contrast to the conclusions suggested in Donati et al. (2008) and Reiners & Basri (2009), evidence for a change is magnetic topologies at the boundary between partial and complete convection is not very obvious when the new results are included. Four of the late-M dwarfs have ratios $<B_V>/<B_I>$ below 10 % while earlier results suggested that more flux is detectably in Stokes V in fully convective stars. On the other hand, the ratio of detectable magnetic energies stays rather high in this regime ($\gtrsim$ 2%), which may reflect an influence of the convective nature of the star. The main problem, however, seems to be why some stars have very different ratios $<B^2>/<B>^2$. This may well be an effect of different magnetic topologies but large differences occur even within the group of fully convective stars (see Morin et al. 2010).

4. Summary

Our knowledge on magnetic fields in cool stars, particularly in M stars across the full convection boundary, has seen enormous progress during the last few years. Intensive observations of many M dwarfs led to the construction of Stokes V Doppler maps, and the exploitation of the FeH molecular spectra allow a determination of the entire field from Stokes I. We can now start to compare results from independent methods and search for the influence of stellar parameters including convective nature. The interpretation of results from different methods opens a parameter space that certainly contains deep information about the fields and their topology, but it is not yet clear what our measurements are actually telling us. Field strengths and topologies have ramifications to a broad range of astrophysics, and at spectral type late-M, we are approaching the brown dwarf regime. More field measurements, determination of molecular constants, and fundamental investigation of detectabilities are required to push the field forward, and to understand the many facets of magnetic fields in cool stars, brown dwarfs, star formation, and their links to exoplanets.

Acknowledgments. It is a pleasure to thank my main collaborators in the work on Stokes I magnetic fields, Gibor Basri, Denis Shulyak, and Andreas Seifahrt, and I thank Julien Morin for insightful discussions. I want to thank the organizers of CS16 for a very fruitful and extremely well-organized meeting. My work is funded through a DFG Emmy-Noether fellowship under RE 1664/4-1.

References

Anderson, R. I., Reiners, A., & Solanki, S. K. 2010, ArXiv e-prints. 1008.2213
Basri, G., Valenti, J. A., & Marcy, G. W. 1990, ApJ, 360, 650
Berdyugina, S. V., & Solanki, S. K. 2002, A&A, 385, 701
Berger, E., Rutledge, R. E., Phan-Bao, N., Basri, G., Giampapa, M. S., Gizis, J. E., Liebert, J., Martín, E., & Fleming, T. A. 2009, ApJ, 695, 310. 0809.0001
Donati, J., & Landstreet, J. D. 2009, ARA&A, 47, 333. 0904.1938
Donati, J., Morin, J., Petit, P., Delfosse, X., Forveille, T., Aurière, M., Cabanac, R., Dintrans, B., Fares, R., Gastine, T., Jardine, M. M., Lignières, F., Paletou, F., Velez, J. C. R., & Théado, S. 2008, MNRAS, 390, 545. 0809.0269
Hallinan, G., Antonova, A., Doyle, J. G., Bourke, S., Lane, C., & Golden, A. 2008, ApJ, 684, 644. 0805.4010
Johns-Krull, C. M. 2007, ApJ, 664, 975. 0704.2923
— 2008, in 14th Cambridge Workshop on Cool Stars, Stellar Systems, and the Sun, edited by G. van Belle, vol. 384 of Astronomical Society of the Pacific Conference Series, 145
Johns-Krull, C. M., & Valenti, J. A. 1996, ApJ, 459, L95+
— 2000, in Stellar Clusters and Associations: Convection, Rotation, and Dynamos, edited by R. Pallavicini, G. Micela, & S. Sciortino, vol. 198 of Astronomical Society of the Pacific Conference Series, 371
Kochukhov, O., Makaganiuk, V., & Piskunov, N. 2010, ArXiv e-prints. 1008.5115
Kochukhov, O., & Piskunov, N. 2002, A&A, 388, 868
Landstreet, J. D. 1992, A&A Rev., 4, 35
Mestel, L., & Landstreet, J. D. 2005, in Cosmic Magnetic Fields, edited by R. Wielebinski & R. Beck, vol. 664 of Lecture Notes in Physics, Berlin Springer Verlag, 183
Miesch, M. S., & Toomre, J. 2009, Annual Review of Fluid Mechanics, 41, 317
Moore, R., & Rabin, D. 1985, ARA&A, 23, 239
Morin, J., Donati, J., Petit, P., Delfosse, X., Forveille, T., Albert, L., Aurière, M., Cabanac, R., Dintrans, B., Fares, R., Gastine, T., Jardine, M. M., Lignières, F., Paletou, F., Ramirez Velez, J. C., & Théado, S. 2008, MNRAS, 390, 567. 0808.1423
Morin, J., Donati, J., Petit, P., Delfosse, X., Forveille, T., & Jardine, M. M. 2010, MNRAS, 407, 2269. 1005.5552
Reiners, A., & Basri, G. 2006, ApJ, 644, 497. arXiv:astro-ph/0602221
— 2007, ApJ, 656, 1121. arXiv:astro-ph/0610365
— 2008, ApJ, 684, 1390. 0805.1059
— 2009, A&A, 496, 787. 0901.1659
— 2010, ApJ, 710, 924. 0912.4259
Reiners, A., Basri, G., & Browning, M. 2009a, ApJ, 692, 538. 0810.5139
Reiners, A., Basri, G., & Christensen, U. R. 2009b, ApJ, 697, 373. 0903.0857
Reiners, A., & Christensen, U. R. 2010, A&A, 522, A13+. 1007.1514
Saar, S. H. 1994, in Infrared Solar Physics, edited by D. M. Rabin, J. T. Jefferies, & C. Lindsey, vol. 154 of IAU Symposium, 493
— 1996, in Stellar Surface Structure, edited by K. G. Strassmeier & J. L. Linsky, vol. 176 of IAU Symposium, 237
— 2001, in 11th Cambridge Workshop on Cool Stars, Stellar Systems and the Sun, edited by R. J. Garcia Lopez, R. Rebolo, & M. R. Zapaterio Osorio, vol. 223 of Astronomical Society of the Pacific Conference Series, 292
Shulyak, D., Reiners, A., Wende, S., Kochukhov, O., Piskunov, N., & Seifahrt, A. 2010, ArXiv e-prints. 1008.2512
Valenti, J. A., Marcy, G. W., & Basri, G. 1995, ApJ, 439, 939
Wallace, L., Livingston, W. C., Bernath, P. F., & Ram, R. S. 1999, An atlas of the sunspot umbral spectrum in the red and infrared from 8900 to 15,050 cm(-1) (6642 to 11,230 Å), revised

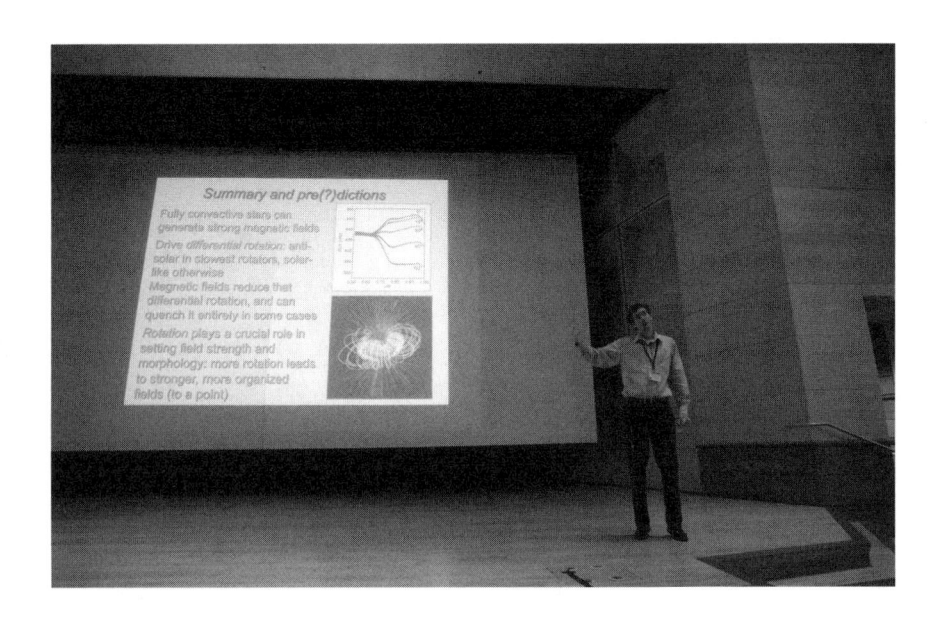

Cool Stars 16 Plenary Talks

The Age-Rotation-Activity Relation: From Myrs to Gyrs

Kevin R. Covey,[1,2] Marcel A. Agüeros,[3] Jenna J. Lemonias,[3]
Nicholas M. Law,[4] Adam L. Kraus,[5] and the Palomar Transient
Factory Collaboration

[1]*Hubble Fellow; Cornell University, Department of Astronomy, 226 Space Sciences Building, Ithaca, NY 14853, USA*

[2]*Visiting Researcher, Department of Astronomy, Boston University, 725 Commonwealth Avenue, Boston, MA 02215, USA*

[3]*Department of Astronomy, Columbia University, 550 West 120th Street, New York, NY 10027, USA*

[4]*Dunlap Institute for Astronomy and Astrophysics, University of Toronto, 50 St. George Street, Toronto M5S 3H4, Ontario, Canada*

[5]*Hubble Fellow; University of Hawaii-IfA, 2680 Woodlawn Drive, Honolulu, HI 96822, USA*

Abstract. Over the past 40 years, observational surveys have established the existence of a tight relationship between a star's age, rotation period, and magnetic activity. The age-rotation-activity relation is essential for understanding the interplay between, and evolution of, a star's angular momentum content and magnetic dynamo. It also provides a valuable age estimator for isolated field stars. While the age-rotation-activity relation has been studied extensively in clusters younger than 500 Myr, its subsequent evolution is less constrained. Empirically measured rotation periods are scarce at intermediate ages (i.e., Hyades or older), complicating attempts to test reports of a break in the age-activity relation near 1 Gyr (e.g., Pace & Pasquini 2004; Giardino et al. 2008).

Using the Palomar Transient Factory (PTF), we have begun a survey of stellar rotation to map out the late-stage evolution of the age-rotation-activity relation: the Columbia/Cornell/Caltech PTF (CCCP) survey of open clusters. The first CCCP target is the nearby ~600 Myr Hyades-analog Praesepe; we have constructed PTF light curves containing >150 measurements spanning more than three months for ~650 cluster members. We measure rotation periods for 40 K & M cluster members, filling the gap between the periods previously reported for solar-type Hyads (Radick et al. 1987; Prosser et al. 1995) and for a handful of low-mass Praesepe members (Scholz & Eislöffel 2007).

Our measurements indicate that Praesepe's period-color relation undergoes a transition at a characteristic spectral type of ~M1 — from a well-defined singular relation at higher mass, to a more scattered distribution of both fast and slow-rotators at lower masses. The location of this transition is broadly consistent with expectations based on observations of younger clusters and the assumption that stellar-spin down is the dominant mechanism influencing angular momentum evolution at ~600 Myr. Combining these data with archival X-ray observations and Hα measurements provides a portrait of the ~600 Myr age-rotation-activity relation (see contribution by Lemonias et al. in these proceedings). In addition to presenting the results of our photometric monitoring of Praesepe, we summarize the status and future of the CCCP survey.

1. Introduction

In his seminal 1972 paper, Andrew Skumanich showed that stellar rotation decreases over time such that $v_{rot} \propto t^{-0.5}$ — as does CaII emission, a measure of chromospheric activity and proxy for magnetic field strength. This relationship between age, rotation, and activity has been a cornerstone of stellar evolution work over the past 40 years, and has generated almost as many questions as applications. For example, angular momentum loss due to stellar winds is generally thought to be responsible for the Skumanich (1972) law, but the exact dependence of v_{rot} on age is not entirely clear, and relies on the assumed stellar magnetic field geometry and degree of core-envelope coupling (Kawaler 1988; Krishnamurthi et al. 1997). Furthermore, later-type, fully convective stars appear to have longer active lifetimes than their early-type brethren (e.g., West et al. 2008), indicating that the lowest mass stars are capable of generating significant magnetic fields even in the absence of a standard solar-type dynamo (Browning 2008). The lack of a comprehensive theoretical understanding of the age-rotation-activity relation has not prevented the development and use of gyrochronology, however, which attempts to determine the precise ages of field stars based on a presumed age-rotation relation (e.g., Barnes 2007; Mamajek & Hillenbrand 2008; Collier Cameron et al. 2009; Barnes 2010), nor of empirical age-activity relations, which do not always find activity decaying with time quite as simply as predicted by the Skumanich law (e.g., Feigelson et al. 2004; Pace & Pasquini 2004; Giampapa et al. 2006).

Fully mapping out the dependence of stellar rotation and activity on age requires the study of stars ranging in both mass and age. Statistical constraints on the age-rotation-activity relation can be derived via analysis of Galactic field stars (e.g., Feigelson et al. 2004; Covey et al. 2008; Irwin et al. 2010), but the homogeneous, coeval populations in open clusters provide an ideal environment for studying time-dependent stellar properties. There are relatively few nearby open clusters, however, and fewer still have the high quality optical data needed to characterize their rotations — in part because of the sheer difficulty involved in systematically monitoring a large number of stars over several months or more. As a result, our current view of the age-rotation-activity relation depends on observations of handfuls of stars in the field and in a small number of well-studied clusters (with the Hyades being a particularly key cluster, e.g., Radick et al. 1987; Jones et al. 1996; Stauffer et al. 1997; Terndrup et al. 2000).

The advent of time-domain surveys, with their emphasis on wide-field, automated, high-cadence observing, makes it possible to monitor stellar rotation in clusters on an entirely new scale (e.g., Irwin et al. 2007; Meibom et al. 2009; Hartman et al. 2010). Many ongoing time-domain surveys are primarily designed to identify transiting exoplanets, however, and thus aim to cover the widest area possible to a relatively modest depth. Deep targeted surveys of open clusters are therefore still required to measure rotation periods, particularly for the lowest mass stars in older, more distant open clusters. The Palomar Transient Factory provides deep, multi-epoch photometry over a wide field-of-view, and our Columbia/Cornell/Caltech PTF survey is leveraging this capability to map $v_{rot}(t)$ in open clusters of different ages. Our first CCCP target, Praesepe (the Beehive Cluster, M44, $08^h 40^m 24^s + 19°41'$), is a nearby (~170 pc), intermediate-age

(~600 Myr), and rich (its membership has recently been expanded to ~1200 stars; Kraus & Hillenbrand 2007) open cluster that shares many characteristics with the Hyades.

Here we report the stellar rotation periods for Praesepe members derived from our first season of PTF observations. Our campaign produced ~200 distinct observations of four overlapping fields covering a 3.75 × 3.30 deg area designed to include a large number of Praesepe members identified by Kraus & Hillenbrand (2007). In Section 2 we describe our PTF observations. We discuss our period-finding algorithm in Section 3 and our results in Section 4. Finally, in Section 5 we outline the current status and future of the CCCP survey.

2. Observations

PTF is a transient detection system comprised of a wide-field survey camera mounted on the automated Samuel Oschin 48 inch telescope at Palomar Observatory, CA (known as the P48), an automated real time data reduction pipeline, a dedicated photometric follow-up telescope (the Palomar 60 inch), and an archive of all detected sources (for details, see Law et al. 2009; Rau et al. 2009). The survey camera has 101 megapixels, 1″ sampling and a 8.1 deg^2 field-of-view covered by an array of 12 CCD chips, one of which is now inoperative. Currently, observations are performed in one of two filters: SDSS-g or Mould-R. Under the median 1.1″ seeing conditions the camera produces 2.0″ FWHM images and reaches 5σ limiting AB magnitudes of $m_g \approx 21.3$ and $m_R \approx 20.6$ mag in its standard 60 s exposures.

Praesepe was monitored by PTF for 3.5 months in early 2010, beginning on 2010 Feb 2 and ending on 2010 May 19. Our observing cadence was sensitive to P_{rot} from a few to a few hundred hours, covering the range occupied by the few cluster members with known periods (Scholz & Eislöffel 2007). PTF monitored Praesepe by imaging four overlapping 3.5 x 2.31 deg fields, which together cover ~18 deg^2 in the cluster's center. The extent of the CCCP footprint is shown in Fig. 1; >80,000 individual objects were detected in our observations.

In developing their cluster catalog, Kraus & Hillenbrand (2007) combined archival data from multiple surveys to calculate proper motions and photometry for several million sources within 7 deg of Praesepe's center. Their census covers a larger area of sky and is deeper than any previous proper motion study, and extends to near the stellar/substellar boundary. The resulting catalog includes 1169 candidate members with membership probability >50% (hereafter referred to as the P50 stars); 442 are identified as high-probability candidates for the first time. Kraus & Hillenbrand (2007) estimate that their survey is >90% complete across a wide range of spectral types, from F0 to M5.

Of the 1169 P50 members in the Kraus & Hillenbrand (2007) catalog, 923, or close to 80%, lie within the CCCP footprint. Of these, 661 are fainter than the PTF saturation limit (~14 mag): PTF detected 534 (or 81%) of these candidate members, with the rest falling within chip gaps or on the dead chip. Kraus & Hillenbrand (2007) provide spectral types for Praesepe stars based on spectral energy distribution fitting; the PTF-detected members are late K through early M stars, which is as expected given the distance to Praesepe and the PTF exposure time. Aperture photometry was measured for each candidate member at each epoch using the SExtractor software (Bertin & Arnouts 1996). Positional matching was then used to merge detections across epochs, producing a single light curve for sources. The photometric zeropoints were adjusted on a per chip basis to minimize the median photometric variability of the detected sources.

272   Covey et al.

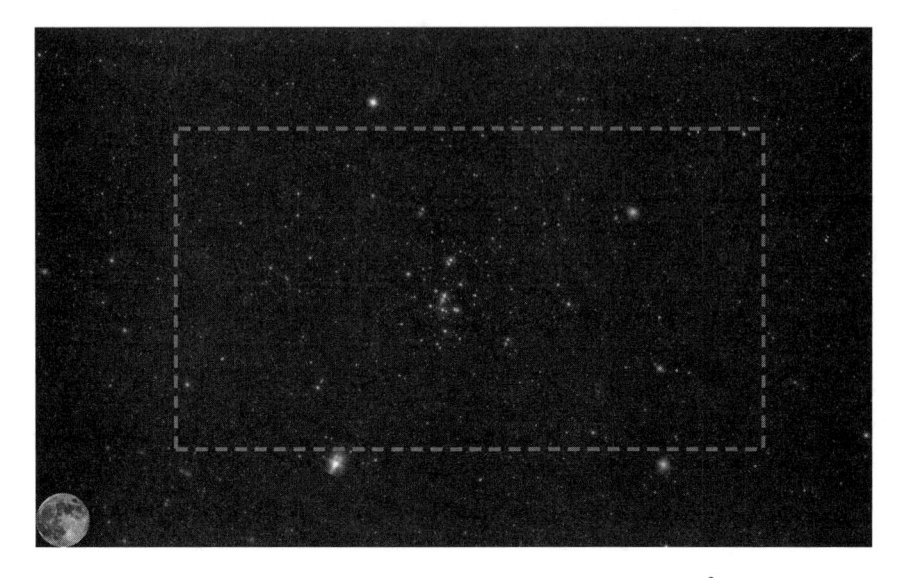

Figure 1. SDSS image of the Praesepe open cluster. The ~18 deg^2 monitored by the CCCP survey is shown by the blue dashed rectangle; the Moon is shown to scale in the lower left corner of the image.

For more information on this reduction process, see Law et al. 2010 elsewhere in these proceedings

3. Period Measurements

We use a modified version of the Lomb-Scargle algorithm to search our light curves for periodic signals due to rotation. The properties of each star's light curve defined the range of frequencies that were searched for periodic signals: we followed Eq. 11 of Frescura et al. (2008) in calculating the frequency grid from the number of individual measurements obtained for each star, as well as the total duration of the light curve. We oversampled this grid by a factor of five to ensure maximum sensitivity to any periodic behavior sampled by our monitoring. Potential beat frequencies between the primary periodogram peak and a possible one-day alias, typical for ground-based, nightly observing campaigns, were flagged following Eq. 1 of Messina et al. (2010).

To test the significance of the periods identified by our modified Lomb-Scargle algorithm, we performed a Monte Carlo analysis of our light curves (for a similar analysis, see Sturrock & Scargle 2010). Having measured potential rotation periods for all cluster members detected by PTF, we then conducted an identical analysis on each light curve after randomly scrambling the magnitudes measured at each epoch. Repeating this test 100 times on each scrambled light curve, we were able to identify the maximum measured periodogram peak as the power threshold corresponding to a <1% false alarm probability (FAP) in the absence of ordered variations. This analysis established that, across our entire sample, a periodogram peak with power >25 corresponded to a FAP <1%; indeed, for only three of the 534 stars analyzed here did the 1% FAP correspond to a periodogram power threshold >20.

Adopting a conservative power threshold of 30 to select potentially periodic cluster members, we then visually inspected the output of our search for each candidate. Periodograms were checked to confirm the presence of a single narrow peak, well separated from the underlying background power; further scrutiny established that the periodic behavior was visible and stable across the full light curve, well sampled in phase, and of an amplitude at least comparable to the observational noise. Representative periodograms and phased light curves for the resulting high-confidence detections are shown in Fig. 2.

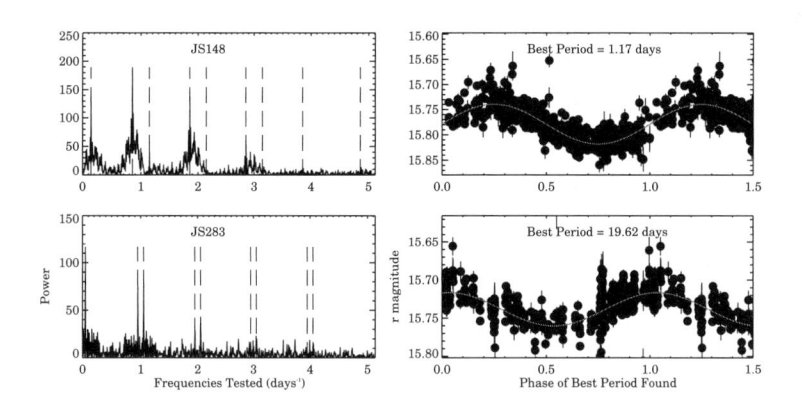

Figure 2. Example periodograms (left panels) and phased light curves (right panels) for two P50 Praesepe stars with high-confidence rotation period measurements. A fast and slow rotator are shown in the top and bottom panels, respectively.

4. Results

The analysis described above produces high-confidence measurements of rotation periods ranging from 0.45 to 35.85 days for a total of 40 stars. The CCCP data clearly indicate the presence of both fast and slow rotators in Praesepe; examples of a fast and slow rotator are shown in Fig. 3, and the location of the full sample in color/period space is shown in Fig.4, along with additional stars from comparably aged clusters (i.e., Hyades & Coma Ber). These observations, along with complementary measurements of low-mass Hyades members (see contribution by Delorme et al. elsewhere in this proceedings), establish that the ~600 Myr period-color relation is single-valued for colors blueward of $r-K_s \sim 3$ and $J-K_s \sim 0.825$, corresponding to a spectral type of ~M1 and a mass of $M \sim 0.6\ M_\odot$ (Bochanski et al. 2007; Kraus & Hillenbrand 2007). Redward of this color, however, the distribution of stars in color-period space is scattered, with populations of both fast and slow rotators.

The morphology of the color-period relation we measure in Praesepe is broadly consistent with observations of somewhat younger clusters, assuming that stellar spin-down is the dominant mechanism governing the stellar angular momentum at these ages. The color-period relation observed in the slightly younger M37, for example, departs from a single-valued relation at $M \sim 0.8\ M_\odot$ (Hartman et al. 2009), suggesting that stars with $0.8\ M_\odot < M < 0.6\ M_\odot$ possess spin-down timescales of ~500-650 Myr.

To go beyond this crude comparison, we are currently analyzing the color-period relation we measure in Praesepe in light of the gyrochronology relations presented in the literature (e.g., Mamajek & Hillenbrand 2008; Barnes 2010).

This spin-down timescale also agrees well with the ~600 Myr magnetic activity lifetime predicted for ~M1 stars based on statistical analyses of low-mass stars in the Galactic field (West et al. 2008). We are currently conducting a comprehensive census of magnetic activity in Praesepe to test directly the equivalence of the activity lifetime and spin-down timescale. For more on our activity analysis, see the presentation by Lemonias et al. elsewhere in these proceedings.

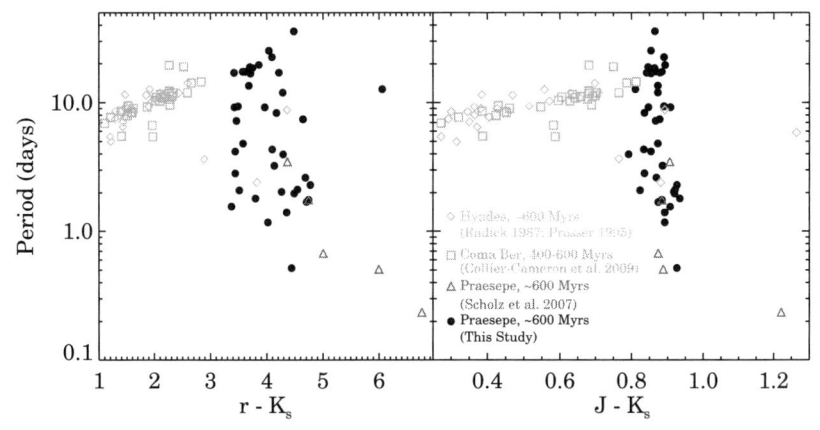

Figure 3. Color-period relations for stars in moderately old (t~600 Myr) open clusters. Stellar periods are plotted as a function of SDSS r - 2MASS K_s color (left panel) and 2MASS J-K_s color (right panel). All stars possess native 2MASS J-K_s colors. r-K_s colors were constructed for Praesepe stars from native SDSS r and 2MASS K_s magnitudes. For Hyades stars, r-K_s colors were determined using synthetic r magnitudes calculated from observed B and V magnitudes (Joner et al. 2006) and the r(B,V) transformation defined by Jester et al. (2005). For Coma Ber stars, synthetic r-K_s colors were estimated from observed J-K_s colors using the median SDSS-2MASS stellar locus defined by Covey et al. (2007).

5. Current Status and Future Plans

We recently began our second season of observations of Praesepe. Our observing strategy has changed somewhat, and the CCCP footprint now features less overlap and covers a total area of ~40 deg^2. In parallel to this continued photometric monitoring, we are conducting a spectroscopic campaign to confirm the membership and measure the chromospheric activity of the Kraus & Hillenbrand (2007) P50 stars. Our observations with the 2.4-m Hiltner Telescope at the MDM Observatory and using the Hydra multi-object spectrograph on the WIYN 3.5-m telescope, both on Kitt Peak, AZ, will enable us to confirm our targets as members based on their radial velocities, and to obtain Hα line measurements to diagnose their levels of chromospheric activity. Moreover, we have proposed to obtain new X-ray imaging of Praesepe to expand the sample of cluster members with measured coronal activity, and to establish once and for all if

Praesepe's X-ray luminosity function can be reconciled with that of the Hyades (see additional discussion in Lemonias et al., this volume).

Our next CCCP target, NGC 752, is a ~1.1 Gyr cluster at a distance of 450 pc, and is the best studied, closest old cluster to the Sun. Stellar rotation at this age is generally not well constrained, and no rotation periods that we know of exist for NGC 752 members. The CCCP footprint for this cluster is comprised of two overlapping PTF fields such that the cluster core (which is roughly 0.5 deg across) is contained almost entirely on one chip in each field. Our observations began on 2010 Aug 22, and to date we have >300 individual observations of each field, so that the cluster core has been visited >600 times. As the NGC 752 low-mass population is poorly defined, we are also beginning a spectroscopic campaign on this cluster this winter to determine the membership and activity level of candidate low-mass members in the cluster field.

Acknowledgments. This research has made use of NASA's Astrophysics Data System Bibliographic Services, the SIMBAD database, operated at CDS, Strasbourg, France, the NASA/IPAC Extragalactic Database, operated by the Jet Propulsion Laboratory, California Institute of Technology, under contract with the National Aeronautics and Space Administration, and the VizieR database of astronomical catalogs (Ochsenbein et al. 2000). IRAF (Image Reduction and Analysis Facility) is distributed by the National Optical Astronomy Observatories, which are operated by the Association of Universities for Research in Astronomy, Inc., under cooperative agreement with the National Science Foundation.

The Two Micron All Sky Survey was a joint project of the University of Massachusetts and the Infrared Processing and Analysis Center (California Institute of Technology). The University of Massachusetts was responsible for the overall management of the project, the observing facilities and the data acquisition. The Infrared Processing and Analysis Center was responsible for data processing, data distribution and data archiving.

References

Barnes, S. A. 2007, ApJ, 669, 1167. 0704.3068
— 2010, ApJ, 722, 222
Bertin, E., & Arnouts, S. 1996, A&AS, 117, 393
Bochanski, J. J., West, A. A., Hawley, S. L., & Covey, K. R. 2007, AJ, 133, 531. arXiv: astro-ph/0610639
Browning, M. K. 2008, ApJ, 676, 1262. 0712.1603
Collier Cameron, A., Davidson, V. A., Hebb, L., Skinner, G., Anderson, D. R., Christian, D. J., Clarkson, W. I., Enoch, B., Irwin, J., Joshi, Y., Haswell, C. A., Hellier, C., Horne, K. D., Kane, S. R., Lister, T. A., Maxted, P. F. L., Norton, A. J., Parley, N., Pollacco, D., Ryans, R., Scholz, A., Skillen, I., Smalley, B., Street, R. A., West, R. G., Wilson, D. M., & Wheatley, P. J. 2009, MNRAS, 400, 451. 0908.0189
Covey, K. R., Agüeros, M. A., Green, P. J., Haggard, D., Barkhouse, W. A., Drake, J., Evans, N., Kashyap, V., Kim, D., Mossman, A., Pease, D. O., & Silverman, J. D. 2008, ApJS, 178, 339. 0805.2615
Covey, K. R., Ivezić, Ž., Schlegel, D., Finkbeiner, D., Padmanabhan, N., Lupton, R. H., Agüeros, M. A., Bochanski, J. J., Hawley, S. L., West, A. A., Seth, A., Kimball, A., Gogarten, S. M., Claire, M., Haggard, D., Kaib, N., Schneider, D. P., & Sesar, B. 2007, AJ, 134, 2398. 0707.4473
Feigelson, E. D., et al. 2004, ApJ, 611, 1107
Frescura, F. A. M., Engelbrecht, C. A., & Frank, B. S. 2008, MNRAS, 388, 1693

Giampapa, M. S., Hall, J. C., Radick, R. R., & Baliunas, S. L. 2006, ApJ, 651, 444. arXiv: astro-ph/0607313
Giardino, G., Pillitteri, I., Favata, F., & Micela, G. 2008, A&A, 490, 113. 0808.3451
Hartman, J. D., Bakos, G. Á., Kovács, G., & Noyes, R. W. 2010, MNRAS, 408, 475
Hartman, J. D., Gaudi, B. S., Pinsonneault, M. H., Stanek, K. Z., Holman, M. J., McLeod, B. A., Meibom, S., Barranco, J. A., & Kalirai, J. S. 2009, ApJ, 691, 342. 0803.1488
Irwin, J., Berta, Z. K., Burke, C. J., Charbonneau, D., Nutzman, P., West, A. A., & Falco, E. E. 2010, ArXiv e-prints. 1011.4909
Irwin, J., Hodgkin, S., Aigrain, S., Hebb, L., Bouvier, J., Clarke, C., Moraux, E., & Bramich, D. M. 2007, MNRAS, 377, 741. arXiv:astro-ph/0702518
Jester, S., Schneider, D. P., Richards, G. T., Green, R. F., Schmidt, M., Hall, P. B., Strauss, M. A., Vanden Berk, D. E., Stoughton, C., Gunn, J. E., Brinkmann, J., Kent, S. M., Smith, J. A., Tucker, D. L., & Yanny, B. 2005, AJ, 130, 873. arXiv:astro-ph/0506022
Joner, M. D., Taylor, B. J., Laney, C. D., & van Wyk, F. 2006, AJ, 132, 111
Jones, B. F., Fischer, D. A., & Stauffer, J. R. 1996, AJ, 112, 1562
Kawaler, S. D. 1988, ApJ, 333, 236
Kraus, A. L., & Hillenbrand, L. A. 2007, AJ, 134, 2340
Krishnamurthi, A., Pinsonneault, M. H., Barnes, S., & Sofia, S. 1997, ApJ, 480, 303
Law, N. M., Kulkarni, S. R., Dekany, R. G., Ofek, E. O., Quimby, R. M., Nugent, P. E., Surace, J., Grillmair, C. C., Bloom, J. S., Kasliwal, M. M., Bildsten, L., Brown, T., Cenko, S. B., Ciardi, D., Croner, E., Djorgovski, S. G., van Eyken, J., Filippenko, A. V., Fox, D. B., Gal-Yam, A., Hale, D., Hamam, N., Helou, G., Henning, J., Howell, D. A., Jacobsen, J., Laher, R., Mattingly, S., McKenna, D., Pickles, A., Poznanski, D., Rahmer, G., Rau, A., Rosing, W., Shara, M., Smith, R., Starr, D., Sullivan, M., Velur, V., Walters, R., & Zolkower, J. 2009, PASP, 121, 1395. 0906.5350
Mamajek, E. E., & Hillenbrand, L. A. 2008, ApJ, 687, 1264. 0807.1686
Meibom, S., Mathieu, R. D., & Stassun, K. G. 2009, ApJ, 695, 679. 0805.1040
Messina, S., Parihar, P., Koo, J., Kim, S., Rey, S., & Lee, C. 2010, A&A, 513, A29+. 0912.4131
Ochsenbein, F., Bauer, P., & Marcout, J. 2000, A&AS, 143, 23
Pace, G., & Pasquini, L. 2004, A&A, 426, 1021. arXiv:astro-ph/0406651
Prosser, C. F., Shetrone, M. D., Dasgupta, A., Backman, D. E., Laaksonen, B. D., Baker, S. W., Marschall, L. A., Whitney, B. A., Kuijken, K., & Stauffer, J. R. 1995, PASP, 107, 211
Radick, R. R., Thompson, D. T., Lockwood, G. W., Duncan, D. K., & Baggett, W. E. 1987, ApJ, 321, 459
Rau, A., Kulkarni, S. R., Law, N. M., Bloom, J. S., Ciardi, D., Djorgovski, G. S., Fox, D. B., Gal-Yam, A., Grillmair, C. C., Kasliwal, M. M., Nugent, P. E., Ofek, E. O., Quimby, R. M., Reach, W. T., Shara, M., Bildsten, L., Cenko, S. B., Drake, A. J., Filippenko, A. V., Helfand, D. J., Helou, G., Howell, D. A., Poznanski, D., & Sullivan, M. 2009, PASP, 121, 1334. 0906.5355
Scholz, A., & Eislöffel, J. 2007, MNRAS, 381, 1638. 0708.2274
Skumanich, A. 1972, ApJ, 171, 565
Stauffer, J. R., Balachandran, S. C., Krishnamurthi, A., Pinsonneault, M., Terndrup, D. M., & Stern, R. A. 1997, ApJ, 475, 604
Sturrock, P. A., & Scargle, J. D. 2010, ApJ, 718, 527. 1006.0546
Terndrup, D. M., Stauffer, J. R., Pinsonneault, M. H., Sills, A., Yuan, Y., Jones, B. F., Fischer, D., & Krishnamurthi, A. 2000, AJ, 119, 1303. arXiv:astro-ph/9911507
West, A. A., Hawley, S. L., Bochanski, J. J., Covey, K. R., Reid, I. N., Dhital, S., Hilton, E. J., & Masuda, M. 2008, AJ, 135, 785. 0712.1590

Global-scale Magnetism (and Cycles) in Dynamo Simulations of Stellar Convection Zones

Benjamin P. Brown,[1,2] Matthew K. Browning,[3] Allan Sacha Brun,[4] Mark S. Miesch,[5] and Juri Toomre[6]

[1]*Dept. Astronomy, University of Wisconsin, Madison, WI 53706-1582*

[2]*Center for Magnetic Self Organization in Laboratory and Astrophysical Plasmas, University of Wisconsin, Madison, WI 537066-1582*

[3]*Canadian Institute for Theoretical Astrophysics, University of Toronto, Toronto, ON M5S3H8 Canada*

[4]*DSM/IRFU/SAp, CEA-Saclay and UMR AIM, CEA-CNRS-Université Paris 7, 91191 Gif-sur-Yvette, France*

[5]*High Altitude Observatory, NCAR, Boulder, CO 80307-3000*

[6]*JILA and Dept. Astrophysical & Planetary Sciences, University of Colorado, Boulder, CO 80309-0440*

Abstract. Young solar-type stars rotate rapidly and are very magnetically active. The magnetic fields at their surfaces likely originate in their convective envelopes where convection and rotation can drive strong dynamo action. Here we explore simulations of global-scale stellar convection in rapidly rotating suns using the 3-D MHD anelastic spherical harmonic (ASH) code. The magnetic fields built in these dynamos are organized on global-scales into wreath-like structures that span the convection zone. We explore one case rotates five times faster than the Sun in detail. This dynamo simulation, called case D5, has repeated quasi-cyclic reversals of global-scale polarity. We compare this case D5 to the broader family of simulations we have been able to explore and discuss how future simulations and observations can advance our understanding of stellar dynamos and magnetism.

1. Introduction

Magnetism is a ubiquitous feature of stars like our Sun. The magnetism we see at the surface probably has its origin in stellar dynamo action arising in the convective envelopes beneath the photosphere. There, turbulent plasma motions couple with rotation to build organized fields on global-scales. These processes occur in the Sun as well and are probably the source of the 11-year activity cycle. Despite intense study, solar and stellar dynamos are poorly understood, and at present we are unable to reliably predict even large-scale features of the solar cycle.

Observations of young, rapidly rotating stars indicate that they have strong magnetic fields at their surfaces. There are clearly observed correlations between rotation and activity which appear to hold generally for stars on the lower main sequence (e.g.,

Pizzolato et al. 2003). Many of these stars show cycles of activity as well, though here the dependence on rotation rate, stellar mass and other fundamental parameters is less clear (e.g., Saar & Brandenburg 1999; Oláh et al. 2009). At present even from a theoretical perspective we do not understand how the stellar dynamo process depends in detail on rotation.

Motivated by this rich observational landscape, we have explored the effects of more rapid rotation on 3-D convection and dynamo action in simulations of stellar convection zones. These simulations have been conducted using the anelastic spherical harmonic (ASH) code to study global-scale magnetohydrodynamic convection and dynamo action in stellar convection zones (e.g., Clune et al. 1999; Miesch et al. 2000; Brun et al. 2004). In the past, global-scale convective dynamo simulations have focused primarily on the Sun, but now explorations are beginning for a variety of stars, ranging from A-type (e.g., Brun et al. 2005; Featherstone et al. 2009) to the M-type dwarfs (Browning 2008).

Here we will discuss simulations of G-type stars that rotate more rapidly than the Sun. We began these explorations by exploring convection in hydrodynamic simulations at a variety of rotation rates (Brown et al. 2008). These simulations capture the convection zone only, spanning from $0.72\,R_\odot$ to $0.97\,R_\odot$, and take solar values for luminosity and stratification but the rotation rate is more rapid. The total density contrast across such shells is about 25. In those simulations we found that the differential rotation generally becomes stronger as the rotation rate increases, while the meridional circulations appear to become weaker and multi-celled in both radius and latitude.

These rapidly rotating stars have vigorous dynamos, and the magnetic fields created in the dynamos are often organized on global-scales into banded wreath-like structures (Brown et al. 2010a). Surprisingly, this organization occurs in the middle of the convection zone itself, rather than in a tachocline of penetration and shear between the convection zone and stable radiative zone beneath. Many of the wreath-building undergo quasi-cyclic reversals of magnetic polarity. Here we explore one of these cyclic dynamos (§2), before putting it in context with other such dynamos (§3).

2. Wreaths and Cycles in a Stellar Convection Zone

Our main focus here is on a convective dynamo in a star rotating five times faster than our Sun currently does, which we call case D5 (Brown et al. 2010b). Vigorous convection in this simulation drives a strong differential rotation, which in turns fuels a strong dynamo. The magnetic fields created in this dynamo are organized on global-scales into banded wreath-like structures, as shown in Figure 1a. Two wreaths are visible near the equator, spanning the depth of the convection zone and latitudes from roughly ±30°. The longitudinal field B_ϕ dominates the magnetic structures, and the two wreaths have opposite polarities (positive in northern hemisphere, negative in the southern). Magnetic fields meander in and out of each wreath, connecting them to one another across the equator where small knots of alternating polarity are visible throughout. The wreaths are also connected to high latitudes, where magnetic structures of opposite polarity are visible; these polar structures are relic wreaths from the previous global-scale reversal.

We follow one such reversal in Figure 1. During the reversal (Fig. 1b), new wreaths of opposite polarity form near the equator and begin to grow in strength. After a reversal (Fig. 1c) the new magnetic wreaths dominate the equatorial region, while the old wreaths propagate towards the poles. The origin of this poleward propagation appears to be a combination of a nonlinear dynamo wave, arising from systematic spatial offsets between the generation terms for mean poloidal and toroidal magnetic field, and possibly a poleward-slip instability arising from magnetic stresses within the wreaths. Life near the equator can be quite complex, and at times during the middle of the cycle states with substantial non-axisymmetry are realized (Fig. 1d). In the polar regions, convection begins to unravel the wreaths from the previous cycle, reconnecting them with the pre-existing flux there.

3. Wreath-building Dynamos

Case D5 is part of a much larger family of simulations that we have conducted exploring convection and dynamo action in younger suns. The properties of this broad family are summarized in Figure 2a. Indicated here are 26 simulations at rotation rates ranging from $0.5\,\Omega_\odot$ to $15\,\Omega_\odot$. At individual rotation rates (e.g., $3\,\Omega_\odot$), further simulations explore the effects of lower magnetic diffusivity η and hence higher magnetic Reynolds numbers. Some of these follow a path where the magnetic Prandtl number Pm is fixed at 0.5 (triangles) while others sample up to Pm=4 (diamonds). The most turbulent simulations have fluctuating magnetic Reynolds numbers of about 500 at mid-convection zone. Wreath-building dynamos are achieved in most simulations (17), though a smaller number do not successfully regenerate their mean poloidal fields (9, indicated with crosses). Very approximate regimes of dynamo behavior are indicated, based on the time variations shown by the different classes of dynamos.

Detailed studies of cases D3 and D5 indicate that the magnetic wreaths are built by both the global-scale differential rotation and by the turbulent emf arising from correlations in the convection (Brown et al. 2010a,b). Generally, the mean longitudinal magnetic field $\langle B_\phi \rangle$ in the wreaths is generated by the Ω-effect: the stretching of mean poloidal field by the shear of differential rotation into mean toroidal field. Production of $\langle B_\phi \rangle$ by the differential rotation is typically balanced by turbulent shear and advection, and by ohmic diffusion on the largest scales.

The mean poloidal field in these simulations is generated by the turbulent emf $E_\mathrm{FI} = \langle \vec{u'} \times \vec{B'} \rangle$, where the fluctuating velocity is $\vec{u'} = \vec{u} - \langle \vec{u} \rangle$ and the fluctuating magnetic fields are $\vec{B'} = \vec{B} - \langle \vec{B} \rangle$. In cases D3 and D5, E_FI is generally strongest at the poleward edge of the wreaths, centered at approximately $\pm 20°$ latitude, whereas the Ω-effect and $\langle B_\phi \rangle$ peak at roughly $\pm 15°$ latitude. This spatial offset between E_FI and $\langle B_\phi \rangle$ means that the turbulent emf is not generally well represented by a simple α-effect description, e.g.,

$$E_\mathrm{FI} = \langle \vec{u'} \times \vec{B'} \rangle|_\phi \neq \alpha \langle B_\phi \rangle \tag{1}$$

when α is a scalar quantity. This is true even when α is estimated from the kinetic and magnetic helicities present in the simulation. More sophisticated mean-field models may do much better at matching the observed emf E_FI, and other terms in the mean-field expansion may play a significant role; in particular, the gradient of $\langle B_\phi \rangle$ is large on the poleward edges of the wreaths where E_FI is significant. During reversals in case D5, both E_FI and the production of $\langle B_\phi \rangle$ associated with the Ω-effect surf on the poleward edge of the wreaths as those structures move poleward. This systematic phase shift appears to contribute to that propagation.

280   Brown et al.

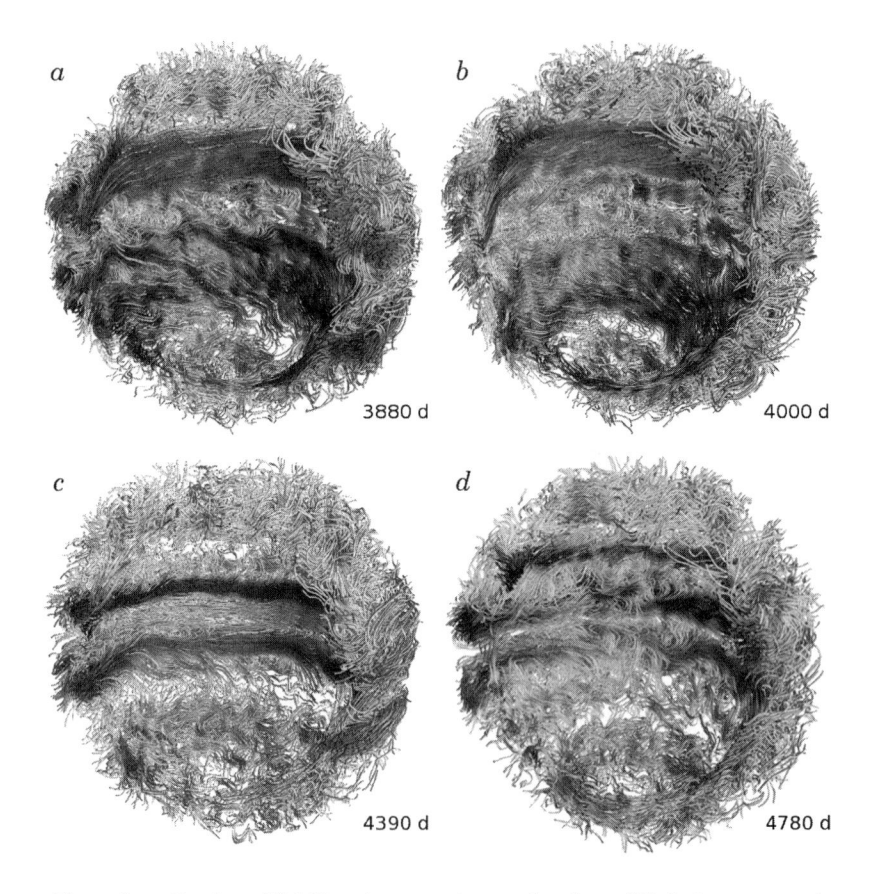

Figure 1. Tracing of fieldlines in magnetic wreaths of case D5 during a magnetic reversal; volume shown spans slightly more than a full hemisphere. (a) Shortly before a reversal, with a positive polarity wreath above the equator (red tones) and negative polarity below (blue tones). Relic wreaths from the previous cycle remain visible in the polar caps. (b) During a reversal, new wreaths with opposite polarity form at the equator. (c) When the reversal completes, the polarity of the wreaths have flipped, with negative polarity wreath above the equator and positive below. The old wreaths propagate towards the poles where they slowly dissipate. (d) Mid-cycle, complex non-axisymmetric states are sometimes attained but do not always trigger reversals. Times of snapshots are labeled, and color tables range from ±25kG, with peak fields reaching ±40kG.

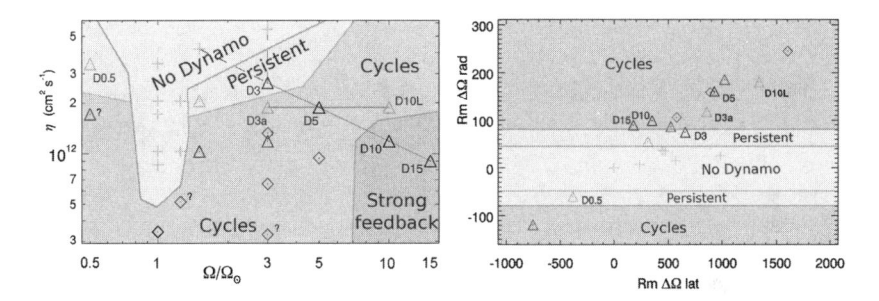

Figure 2. Parameter space explored by wreath-building dynamos. (*a*) Primary control parameters magnetic diffusivity η and rotation rate Ω are shown for dynamo simulations at rotation rates ranging sampling 0.5–15 $\Omega_\odot$, with very approximate dynamo regimes shown and with some cases labeled. In some regions, magnetic Reynolds numbers are too low to sustain dynamo action, while in other regions persistent magnetic wreaths form which do not show evidence for cycles. At higher magnetic Reynolds numbers (occurring here at low η or high Ω), wreaths typically undergo quasi-cyclic reversals. At the highest rotation rates the Lorentz force can substantially modify the differential rotation, but dynamo action is still achieved. Cases marked with question marks show significant time-variation but have not been evolved for long enough to establish cyclic behavior. (*b*) Plot of magnetic Reynolds numbers associated with global-scale differential rotation. The dynamos are largely driven by differential rotation, and the radial shear (vertical axis) appears to discriminate between the different dynamo regimes.

As the differential rotation plays a crucial role in these dynamos, we define magnetic Reynolds numbers associated with the latitudinal shear at mid-convection zone and the radial shear across the convection zone:

$$\text{Rm } \Delta\Omega \text{ lat} = \frac{\Delta\Omega_{\text{lat}} RD}{\eta} \qquad (2)$$

$$\text{Rm } \Delta\Omega \text{ rad} = \frac{\Delta\Omega_r D^2}{\eta} \qquad (3)$$

were $R = 0.85 R_\odot$ is the radial location of the mid-convection zone, $D = 0.3 R_\odot$ is the depth of the convection zone, $\Delta\Omega_{\text{lat,r}}$ are the angular velocity contrasts in latitude and radius respectively (e.g., Brown et al. 2010a) and η is the magnetic diffusivity at mid-convection zone. These magnetic Reynolds numbers are shown in Figure 2*b* for many of the dynamos, neglecting some cases with very high magnetic Reynolds numbers.

The latitudinal shear is generally large in all of these dynamos (horizontal axis), and all of the simulations, including those that fail to sustain dynamo action, succeed in initially producing global-scale toroidal magnetic structures. The radial shear near the equator is relatively weaker (vertical axis), and this quantity more clearly separates those dynamos that succeed from those that fail. The radial differential rotation also discriminates the cyclic dynamos from those that build persistent fields. Somewhat surprisingly, the magnetic Reynolds number associated with the fluctuating convection does not provide as good of a discriminant between dynamos that succeed or fail. The simulations that fail to sustain dynamo action are those that do not regenerate their

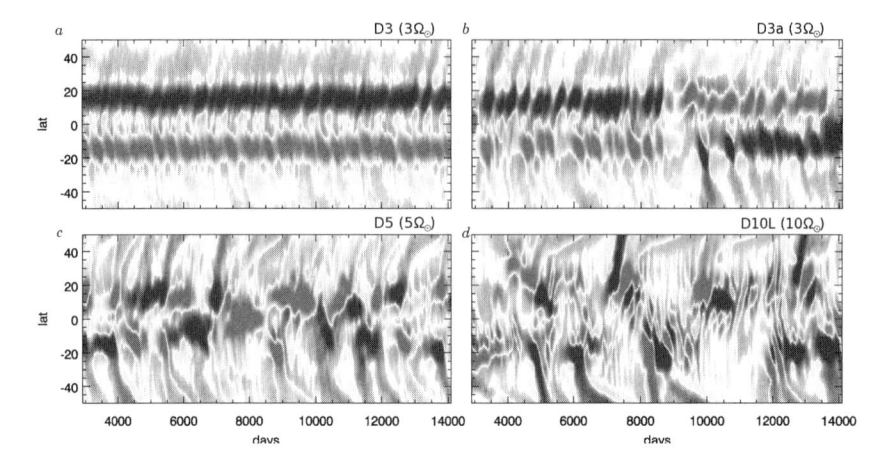

Figure 3. Mean toroidal magnetic field in four wreath-building dynamos at mid-convection zone, shown as time latitude maps. (*a*) Persistent case D3 with fields that do not change in sense. (*b*) More turbulent companion case D3a shows long cycles. (*c*) Cyclic case D5. (*d*) Case D10L.

poloidal fields quickly enough. The clear dependence on $\Delta\Omega_r$ and the weak dependence on the properties of the fluctuating convection suggest again that these wreath-building dynamos may rely on effects other than a classical α-effect to build their turbulent emf E_{FI} which generates the global-scale poloidal fields.

Two of the dynamos shown in Figure 2*b* have negative magnetic Reynolds numbers. These are the two slowly-spinning simulations which rotate half as quickly as our Sun currently does (e.g., case D0.5). In these simulations, the differential rotation is anti-solar in nature and opposite in sense to that of the Sun, with rapidly spinning pole and a more slowly spinning equator. Despite this fundamental difference, these simulations drive strong dynamos and build magnetic wreath-like structures in their convection zones. Anti-solar differential rotation appears to arise when convection is only slightly constrained by rotation (e.g., when the Rossby number is large), while solar-like differential rotation arises in rapidly rotating stars (when the Rossby number is small). The angular velocity shear associated with the differential rotation increases as the Rossby number becomes either large or small. The Sun itself appears to be very near Rossby number unity, and this partially explains the difficulty in attaining wreath-building dynamos in previous solar dynamo simulations: the angular velocity contrast in the Sun is smaller than that realized in the rapidly rotating dynamos. As a consequence, the solar dynamo simulations require low values of η to build wreaths, which in turn calls for high resolutions and that exacts a large computational cost.

Many of these dynamos show global-scale reversals, but the dependence of this phenomena on rotation rate or Reynolds number are somewhat unclear. Near the onset of wreath-building dynamo action we generally find little time variation in the axisymmetric magnetic fields associated with the wreaths. This is illustrated for case D3 in Figure 3*a*, where the mean longitudinal field $\langle B_\phi \rangle$ is shown at mid-convection zone over an interval of nearly 10,000 days. Though small variations are visible on a roughly 500 day timescale, the two wreaths retain their polarities for the entire time simulated (more than 20,000 days), which is significantly longer than the convective overturn time (roughly 10–30 days), the rotation period (9.3 days), or the ohmic diffusion time (about 1300 days at mid-convection zone). We refer to the dynamos in this regime as persistent wreath-builders.

Generally, we find that wreath-building dynamos begin to show large time dependence as the magnetic diffusivity η decreases and as the rotation rate Ω increases. Case D3a, rotating three times the solar rate but with lower diffusivities than case D3, is an example of the first behavior and undergoes reversals even though D3 did not (Fig. 3b). To explore the dependence on rotation rate, we compare cases D3a, D5 and D10L, which have the same magnetic, momentum and thermal diffusivities but rotate at three, five and ten times the solar rate respectively. In case D5 the global-scale reversals are more frequent, occurring with a roughly 1500 day timescale (Fig. 3c), though during some intervals the dynamo can fall into other states. When we explore case D10L rotating ten times the solar rate, we find that the cycles are somewhat harder to define, with the northern and southern hemispheres showing distinctly different behavior (Fig. 3d). It is unclear at present how the cycle period depends on the rotation rate of the star; case D3a and D5 would imply that faster rotation leads to shorter cycles in general agreement with observations, but all of these simulations are highly variable and actual cycle periods are difficult to quantitatively define.

The magnetic wreaths act back strongly on the differential rotation that feeds their generation, and the global-scale shear is much weaker in dynamo simulations than in corresponding hydrodynamic simulations. Individual convective structures are largely unaffected by the magnetic wreaths except when the fields reach very large amplitudes; in case D5 this occurs when B_ϕ exceeds values of roughly 35 kG at mid-convection zone. At the highest rotation rates the Lorentz force of the axisymmetric magnetic fields becomes strong enough to substantially modify the differential rotation, largely wiping out the latitudinal and radial shear (e.g., cases D10 and D15 in Fig. 2a). In these cases, wreath-like structures can still form though they typically have more complex structure and are less axisymmetric.

4. Overview

Advances in massively parallel supercomputers are now permitting simulations that can capture global-scale convection and dynamo action in stars like our Sun. Dynamo simulations of solar-type stars are revealing that organized magnetic fields can be built in the convection zone itself, without necessarily relying on a tachocline in between the convection zone and radiative zone for this organization. This is a marked departure from many solar dynamo theories, where the tachocline plays a vital role. Many of these simulations show quasi-cyclic reversals of magnetic polarity. These cycles are not yet like the solar cycle: namely, they are typically too short and generally the magnetic fields migrate towards the poles, rather than towards the equator as observed at the solar surface (though see Ghizaru et al. 2010). Simulations remain well separated in parameter space from real stellar convection, which remains humblingly out of reach for the foreseeable future, but these global-scale simulations are entering a regime where resolved turbulence plays a larger role than explicit diffusion. Thus they are beginning to capture in a self-consistent fashion the processes which likely contribute most directly to stellar dynamo action.

In the future we will be exploring convection and dynamo action in K- and F-type stars, to understand how stellar mass and convection zone depth affect the global-scale dynamo. This work will complement ongoing work exploring dynamo action in fully convective M-dwarfs (Browning 2008). As simulations move away from the Sun, we need further constraints from stellar observations. In particular, measurements of stellar differential rotation are vitally important, given the role of that global-scale flow in the wreath building dynamos. An observational understanding of how dynamo properties including magnetic activity and cyclic period scale with differential rotation would be of great utility.

Acknowledgments. We thank Nicholas Nelson for his excellent and continuing work on dynamo case D3a (Figure 3*b*) and its relatives. This research is supported by NASA through Heliophysics Theory Program grants NNG05G124G and NNX08AI57G, with additional support for Brown through the NSF Astronomy and Astrophysics postdoctoral fellowship AST 09-02004. CMSO is supported by NSF grant PHY 08-21899. Miesch was supported by NASA SR&T grant NNH09AK14I. NCAR is sponsored by the National Science Foundation. Browning is supported by the Jeffrey L. Bishop fellowship at CITA. Brun was partly supported by the Programme National Soleil-Terre of CNRS/INSU (France), and by the STARS2 grant from the European Research Council. The simulations were carried out with NSF PACI support of PSC, SDSC, TACC and NCSA. Field line tracings shown in Figure 1 were produced using VAPOR (Clyne et al. 2007).

References

Brown, B. P., Browning, M. K., Brun, A. S., Miesch, M. S., & Toomre, J. 2008, ApJ, 689, 1354. 0808.1716
— 2010a, ApJ, 711, 424
Brown, B. P., Miesch, M. S., Browning, M. K., Brun, A. S., & Toomre, J. 2010b, ApJ. submitted
Browning, M. K. 2008, ApJ, 676, 1262. arXiv:0712.1603
Brun, A. S., Browning, M. K., & Toomre, J. 2005, ApJ, 629, 461. arXiv:astro-ph/0610072
Brun, A. S., Miesch, M. S., & Toomre, J. 2004, ApJ, 614, 1073
Clune, T. L., Elliott, J. R., Glatzmaier, G. A., Miesch, M. S., & Toomre, J. 1999, Parallel Computing, 25, 361
Clyne, J., Mininni, P., Norton, A., & Rast, M. 2007, New Journal of Physics, 9, 301
Featherstone, N. A., Browning, M. K., Brun, A. S., & Toomre, J. 2009, ApJ, 328, 1126
Ghizaru, M., Charbonneau, P., & Smolarkiewicz, P. K. 2010, ApJ, 715, L133
Miesch, M. S., Elliott, J. R., Toomre, J., Clune, T. L., Glatzmaier, G. A., & Gilman, P. A. 2000, ApJ, 532, 593
Oláh, K., Kolláth, Z., Granzer, T., Strassmeier, K. G., Lanza, A. F., Järvinen, S., Korhonen, H., Baliunas, S. L., Soon, W., Messina, S., & Cutispoto, G. 2009, A&A, 501, 703. 0904.1747
Pizzolato, N., Maggio, A., Micela, G., Sciortino, S., & Ventura, P. 2003, A&A, 397, 147
Saar, S. H., & Brandenburg, A. 1999, ApJ, 524, 295

Magnetic Field Measurements on the Classical T Tauri Star BP Tauri

Wei Chen and Christopher M. Johns-Krull

Physics and Astronomy Department, Rice University, 6100 Main Street, MS-108, Houston, TX 77005; wc2@rice.edu, cmj@rice.edu

Abstract. We apply our newly developed least-squares deconvolution (LSD) code to several solar-like stars as tests of the code and to 6 nights of observations of BP Tau to measure the mean longitudinal magnetic field on the stellar surface. Null results or weak fields are detected on the solar-like stars as is expected, while a maximum of 180 ± 30 G is detected from the photospheric lines of BP Tau. A 400 G dipole tilted at 113° with respect to the rotation axis fits our photospheric measurements well. Measurements of several emision lines (He I 5876 Å, Ca II 8498 Å and 8542 Å) show the presence of strong magnetic fields in the line formation regions, which are believed to be the base of the accretion footpoints.

1. Introduction

Classical T Tauri Stars (CTTSs) are young low-mass stars at a typical age of several million years. They are still accreting material from their circumstellar disks, whose existence is usually inferred by an infrared (IR) continuum excess. Magnetic fields are believed to play a fundamental role in this stage of mass accretion onto the star. The most popular model describing CTTSs is the so-called "magnetospheric accretion" model (e.g., Shu et al. 1994,), which posits that strong stellar magnetic fields truncate the inner disk at a radius slightly smaller than the co-rotation radius, and the material from the disk attaches to the field lines and is directed gravitationally and centrifugally onto the central star. Magnetic fields also provide braking torques to prevent the stellar rotation from being increased to the breakup velocity or close to this velocity by the angular momentum transport from the accreting material. This explains why observed rotation velocities are usually an order of magnitude smaller than the breakup velocity (Hartmann & Stauffer 1989). Therefore, measuring the magnetic fields on CTTSs and probing the field geometry are of great importance in understanding the general picture of how a CTTS interacts with its surrounding disk.

The Zeeman effect is widely used to detect and measure magnetic fields. Generally, for late type stars measuring Zeeman broadening from several magnetically sensitive spectral lines in unpolarized spectra is the best choice to get the field strength (Valenti et al. 1995; Johns-Krull & Valenti 1996). However, in CTTS because of the rapid rotation, Zeeman broadening is usually coupled with Doppler broadening, which makes this kind of measurements more difficult. Johns-Krull et al. (1999a) measured a field of 2.6 ± 0.3 kG on BP Tau using Zeeman broadening. Spectropolarimetry is usually used to measure the mean longitudinal magnetic field, B_z, by measuring line shifts in right and left circularly polarized spectra (Valenti & Johns-Krull 2004; Johns-

Krull et al. 1999b, hereafter JK99). Due to the shift, Zeeman signatures are present in Stokes V spectra; however they are usually buried in the noise since they are often quite weak. Multiple-line techniques (e.g. Least-squares Deconvolution - LSD) have been developed to extract these weak signatures by averaging over many lines (Donati et al. 1997; Donati 2008, hereafter D08).

2. Observations and Data Reduction

All data were obtained from the 2.7m Harlan J. Smith Telescope at McDonald Observatory using the Zeeman analyzer (ZA) and cross-dispersed coude echelle spectrometer. This system is described by Tull et al. (1995) and Vogt et al. (1980). The incoming light beam, first corrected by a Babinet-Soleil phase compensator to reduce the spurious linear polarization produced by the coude mirrors, goes through the ZA and is split into two beams, a left circularly polarized component and a right circularly polarized one. The two parallel beams enter the spectrometer slit. The two copies of the spectrum are recorded on a 2048 × 2048 CCD with a two pixel resolution of $R = \lambda/\Delta\lambda \approx 60,000$. For each night, one pair of exposures were taken for each object. Between each pair of exposures, a 1/2 wave plate was used to reverse the sense of circular polarization of the two beams. From UT date Nov 26 1998 to Dec 1 1998 we observed BP Tau and Babcock's Star; from Apr 20 1999 to Apr 26 1999, ξ Boo A and 61 Cyg A were observed. We used the echelle reduction package of Valenti (1994) to reduce all the spectra and the spectra of a thorium-argon lamp to determine the wavelength solution.

3. Analysis

3.1. Least-Squares Deconvolution and Photospheric Lines

Least-Squares Deconvolution (LSD) is widely used to measure weak magnetic fields on different type stars along the Hertzprung-Russell diagram, such as Ap and Bp stars (Wade et al. 2000), late type stars (Barnes 1999), and T Tauri Stars (Donati 2007; Hussain 2009), including BP Tau (Donati 2008). Mean longitudinal fields on these stars are usually of the order of several hundred Gauss and can be as weak as several Gauss. The Zeeman signatures are usually embedded in the noise, even in spectra with S/N of 100, which is a little greater than the signal-to-noise of our BP Tau spectra. It is very difficult to measure these fields directly from individual lines, which is why LSD is used. It has the advantage of greatly enhancing the signal-to-noise of the obtained intrinsic Stokes I and V profiles by averaging over hundreds to thousands of spectral lines, permitting a sensitive measurement of the weak mean longitudinal magnetic fields (Donati et al. 1997).

3.1.1. Line Selection

The first step using the LSD method is to create a line list containing the wavelength, the mean Lande g factors and the central depths of the lines to be used. We use the Vienna Atomic Line Database (VALD) and pick lines in the following manner: line depths are greater than d_{min} and smaller than d_{max} (d_{min} and d_{max} vary from star to star). Lines should not be in regions where strong lines (such as Balmer and Ca II H & K and Na I D lines) or telluric lines appear. Spectral lines of rotating stars such as BP Tau are rotationally broadened. VALD doesn't have an option to include rotation to calculate the actual central depth. Therefore, we run the SYNTHMAG (Piskunov 1999) on each line and rotationally broaden them and find the actual central depth for each line. Since the weakest lines with $d < d_{min}$ do not contribute much to the spectrum, including them only consumes more computer time and does not add to the final measurement. The reason for setting an upper limit for line central depth, that is, using only weaker lines (different from lines used in other LSD codes), is that the assumptions described in Section 3.1.2 apply reasonably well only to the weaker lines. The advantage of doing so is obvious: we get a nice gaussian shaped Stokes I profile and no Lorentzian wings. The tradeoff is also apparent: if there are stronger lines blended with the weaker lines we used, the resultant Stokes profiles will have larger uncertainties starting on the wings and going outwards. We assume that the advantage overcomes the shortcoming.

3.1.2. Least-Squares Deconvolution

The LSD method makes several quite strong assumptions: (1) all spectral lines have more or less the same shape; (2) the local intensity of the lines is proportional to the local line central depth, d; (3) intensities of blended lines add up linearly; (4) limb darkening is independent of wavelength; (5) the lines satisfy the weak field approximation (see below). Assumptions (1), (2) and (3) are obviously not applicable to optically thick lines but are reasonably good approximations for relatively optically thin lines; assumption (4) guarantees that line ratios between any two lines hold after integration of the intensity over the whole star. We follow the same procedures as described in Donati et al. (1997); the solution for this method gives the intrinsic Stokes profile as:

$$Z = (M^t \cdot S^2 \cdot M)^{-1}(M^t \cdot S^2 \cdot Y), \quad (1)$$

where M is the weight matrix, weighted by $\lambda g d$ (λ is the line wave length, g is the mean lande g factor, and d is the line central depth) when calculating Stokes V, and the weights are just d for Stokes I. The vector Y is the observed Stokes V or I (actually 1-I) spectrum. The matrix S is diagonal and its elements are the reciprocals of the uncertainties for the elements of Y. The diagnal elements of $(M^t \cdot S^2 \cdot M)^{-1}$ are the variances of the resultant vector Z. Hence the average Stokes I and V spectra will be

$$I = 1 - \bar{d} * Z_I, \quad V = \bar{\lambda} \bar{g} \bar{d} * Z_V, \quad (2)$$

where bar on the top means taking average values over the line list used in the LSD method. The vector Z_I is the intrinsic Stokes I profile and the vector Z_V is the intrinsic Stokes V profile calculated in equation (1). Figure 1 shows a pair of sample LSD Stokes I and V profiles.

3.1.3. Weak Field Approximation Method for Measuring B_z

The Weak field approximation (WFA) method uses an approximate relation between Stokes V and the derivative of Stokes I to calculate the field strength. The relation is described in Brown & Landstreet (1981) as follows:

$$\frac{V}{I} \approx \frac{e}{4\pi mc} B_z \bar{g} \bar{\lambda}^2 \frac{dI/d\lambda}{I}, \quad (3)$$

where all values are in cgs.

A fitting between $\frac{V}{I}$ and $\frac{dI/d\lambda}{I}$ gives the products of the terms in front of $\frac{dI/d\lambda}{I}$ on the right hand side, thus the longitudinal field strength B_z.

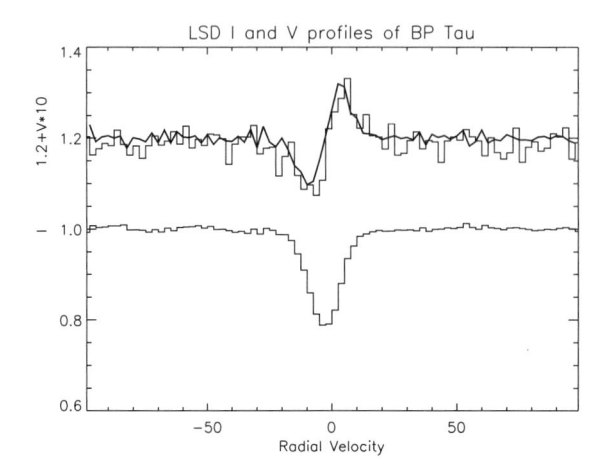

Figure 1. LSD Stokes I (bottom) and V (top) profiles (histograms). The Stokes V profiles is multiplied by 10 and shifted by 1.2. The dark solid line is a -433 Gauss (B_z) fit using WFA.

The purpose of taking pairs of exposures for our objects is to cancel out the possible existence of spacial shifts on the detector due to a tilt between the slit and the detector CCD. Such a shift may result in an artificial contribution to the mean longitudinal fields that have magnitudes comparable to the fields that are being measured. Therefore, we average the resulting pair of B_z measurements.

3.1.4. Uncertainties

Uncertainties propagating through from the uncertainties in the observed spectra are much smaller than the standard deviation of the noise in the final LSD Stokes spectra. In the continuum the Stokes V spectra are expected to be zero while the Stokes I spectra should be unity. This indicates that either the uncertainties in the observed spectra are underestimated and/or there are other sources of uncertainties. First, an incomplete line list is used. Though modern atmosphere models are already very sophisticated and robust, they are not perfect. There are lines that are in the actual spectra but not in the ones from models and vice versa; and the lines we adopted have both lower and upper limits, which exclude the weakest and strongest lines that have a chance of blending with the lines adopted. Second, line lists from VALD and SYNTHMAG do not reproduce perfectly the line depths in the observed spectra. Third, the assumptions we made are not strictly valid: line profiles can differ from one line to another; limb darkening can vary with different wavelength, etc. Fourth, instrument resolution is not infinite. Although we are using a high resolution instrument with a two-pixel resolution of $R = \lambda/\delta\lambda \approx 60,000$, about 5.0 km/s in velocity space. Therefore, we use the standard deviation of the noise in the final LSD Stokes spectra as the uncertainties when determining the field strength. This enables us to include uncertainties from the sources mentioned above.

Additionally, to conservatively estimate our error bars on the measurements, we examine potential systematic errors using tests on several solar-like stars, whose mean longitudinal magnetic fields are expected to be zero or very weak: a few to a few tens of

Gauss. We follow the procedures described above and apply our LSD code to spectra from the Sun as a star (by observing the asteroid Vesta), Arcturus, 61 Cyg A and ξ Boo A. Table 1 shows the results for these tests.

Table 1. LSD Results for Sun, Arcturus, 61 Cyg A and ξ Boo A

Star	UT Date	Time	B_z	Star	UT Date	Time	B_z
Sun	11/24/97	03:28	-2.6±17.1	ξ Boo A	04/21/99	07:04	13.9±17.2
Arcturus	12/01/98	12:42	-13.4±17.5		04/21/99	10:24	14.6±18.8
61 Cyg A	04/20/99	11:23	3.2±17.3		04/22/99	11:08	4.9±17.5
	04/20/99	11:39	-17.5±17.3		04/22/99	11:25	-0.2±18.0
	04/25/99	11:22	-13.0±18.7		04/23/99	10:01	6.5±17.3
	04/25/99	11:39	-27.1±17.8		04/23/99	10:25	-1.5±17.3
	04/26/99	11:39	-15.5±17.7		04/25/99	03:36	-8.7±17.7
ξ Boo A	04/20/99	07:41	3.1±17.2		04/25/99	04:07	11.8±18.5
	04/20/99	07:58	4.9±17.5		04/26/99	09:36	-22.7±17.3
	04/21/99	07:04	-6.7±17.2		04/26/99	11:11	-30.6±17.5

All the field strengths obtained here are less than 30 Gauss, which is consistent with what we are expecting. The errors bars above are modified by adding a systematic error. The original error bars are around 4 Gauss. Back-to-back observations were taken for 61 Cyg A and ξ Boo A at some nights, such as the night of UT Date Apr 20 1999. The same field measurements are expected for these back-to-back observations, but this is not the case here, which indicates that there exists sources of uncertainties other than errors propagated down from the original spectra. Since ξ Boo A is among the best studied solar-type stars in this area (Borra et al. 1984; Hubrig et al. 1994; Plachinda & Tarasova 2000), and it is also the test star that we have the most observations on, we consider it as a standard star and try to estimate the systematic errors by fitting our observations to previous studies. A second order polynomial is fitted to the data of Plachinda & Tarasova (2000), and we then calculate the standard deviation of our observations from this polynomial as our systematic error. Doing so gives 17.0 G.

3.1.5. Results

For BP Tau a total of 365 lines with central depths greater than 0.2 and smaller than 0.55 are picked. Using lines deeper than 0.55 yields Lorentzian wings in the resultant Stokes I profiles. The results are included in Table 2 and Figure 2.

3.2. Emission Lines

We studied three emission lines covered by our observations: He I (5876 Å) and Ca II (8498 Å and 8542 Å). The observed spectra of these three lines (no LSD) and the center-of-gravity method (Mathys & Lanz 1992) were used to calculate the mean longitudinal fields in the regions where these lines are formed. Results are shown in Table 2 and Figure 2.

Table 2. Mean Longitudinal Magnetic Fields on BP Tau

UT Date	Time	Photospherical lines	He I 5876 Å	Ca II (average) 8498 Å and 8542 Å
26-Nov-1998	05:19	-181.9 ± 30.9	2720.3 ± 264.4	2348.8 ± 243.5
27-Nov-1998	03:05	-32.7 ± 31.1	1338.6 ± 237.8	1291.4 ± 193.1
28-Nov-1998	03:21	-3.7 ± 26.3	590.7 ± 231.5	846.9 ± 165.9
29-Nov-1998	03:15	-10.8 ± 37.7	640.8 ± 349.8	550.0 ± 258.1
30-Nov-1998	04:05	17.4 ± 35.5	390.2 ± 398.6	-213.0 ± 316.6
01-Dec-1998	03:27	24.0 ± 22.5	751.1 ± 211.8	417.5 ± 151.2

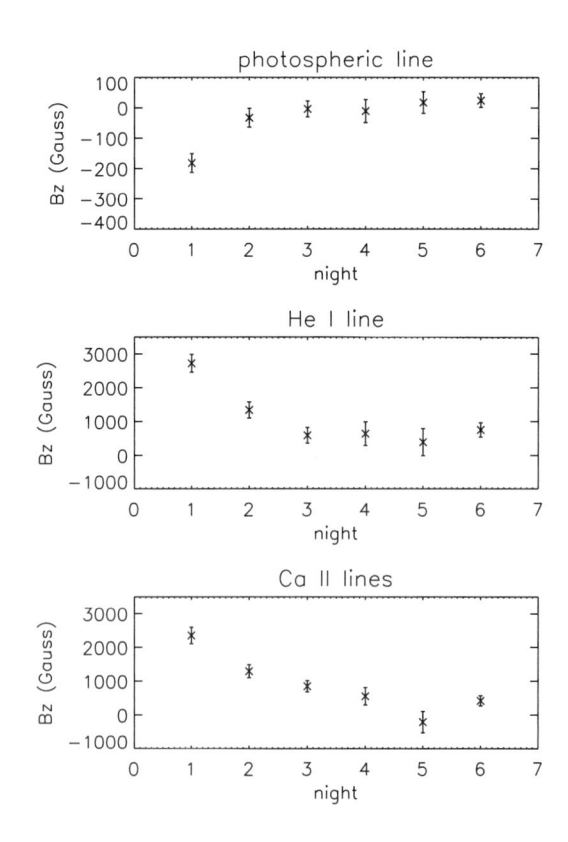

Figure 2. Mean longitudinal magnetic field measurements on BP Tau

4. Discussion

Spectropolarimetry studies on the mean longitudinal magnetic fields of BP Tau have been carried out in the past. A common conclusion from all these studies is that fields measured from emissions lines are very strong, at the order of several kG, while measurements from photospheric lines are significantly weaker. An interpretation is that small scale strong magnetic fields are present on the surface of BP Tau; however, integrating over the whole stellar surface cancels out these small scale strong fields, leaving only a significantly reduced average field. The regions where these emission lines forms are believed to be the base of the accretion footpoints and cover only a small fraction of the stellar surface, therefore these emission lines show very strong polarization, indicative of the small scale field.

JK99 studied the He I 5876 Å emission line and measured a field stronger than 2 kG. Further, Valenti & Johns-Krull (2004) used the same line and measured the mean longitudinal fields on six consecutive nights and found that the measurements fit well to a model where a single magnetic spot rotates with the star at latitude 48°, with a stellar inclination of 38° with respect to the line of sight, and a field strength of about 2.1 kG.

From photospheric lines, JK99 used four unblended magnetically sensitive (longer wavelength and greater Lande g factor) Fe I lines and got a mean longitudinal field of −40 ± 50 G so fields larger than 200 G are ruled out at the 3 σ level; Valenti & Johns-Krull (2004) obtained similar results and found the surface fields are not globally dipolar. On the other hand, D08 measured photospheric fields ranging from 60 G to 600 G with uncertainties ranging from 40 G to 140 G, using the same LSD technique described in this paper and a total of 9400 lines with central depths greater than 0.4 when no non-thermal broadening mechanism is present, and they claimed that the field of BP Tau consists primarily of a 1.2 kG dipole and 1.6 kG octupole, both slightly tilted from the rotation axis which they took to be inclined at an angle of 45° with respect to the line of sight. There is a discrepancy between the interpretation of D08 and JK99 though the observations need not be inconsistent with each other given the quoted uncertainties. D08 argued that this might come from either variations of BP Tau throughout the years (data for JK99 were taken in 1997 and the two sets of data for D08 were taken in 2006, with one set in February and the other in November and December) or the use of cross-correlation methods by JK99, which tend to underestimate both the mean longitudinal fields and the error bars, instead of the center of gravity method or the equivalents as used in D08. Here our observations were taken on six consecutive nights in 1999, covering almost a full rotation, and the same method LSD is used as D08. We find mean longitudinal fields that are no stronger than 200 G and go down to almost zero. These results tend to match with JK99 better than they do with D08. Generally, higher order components from a spherical harmonic expansion contribute much less to the longitudinal component than lower orders. For example, assuming their polar fields are all 1, a dipole field reaches its maximum mean longitudinal field strength of 0.324 when viewed pole on, while the maximum a quadrupole can get is 0.058 and an octupole 0.022. Therefore, if not at some extreme angle the information from these mean longitudinal field measurements will be dominantly from the dipole component. A dipole fitting to our measurements yields a dipole field at a tilted angle of 113° with respect to the rotation axis with its polar field of about 400 G (conservatively with an uncertainty of 200 G).

5. Summary

With our newly developed LSD code, we are able to extract the weak Zeeman signatures from the right and left polarized spectra. Tests on several solar-like stars demonstrate the functionality of this LSD code. Applying this code to spectra of BP Tau using 365 photospheric lines yields measurements of mean longitudinal fields with a maximum of 180 ± 30 G. After a dipole fitting to these results it turns out that the best fit is a dipole tilted at an angle of 113° with respect to the rotation axis, whose inclination is 45° with respect to the line of sight, and its polar field strength is about 400 G. This is not consistent with the result from D08. The discrepancies may rise from the intrinsic variations on the star itself. To show this we need more observations on BP Tau. Studies of the emission lines gave us measurements at the order of kG, which is consistent with previous studies mentioned above. In the future our targets will and need to be extended to other T Tauri Stars to investigate the processes involved under the control of different magnetic field configurations.

After the first introduction of the least-square deconvolution method more than a decade ago, many modified or improved methods have been developed. Sennhauser & Berdyugina (2010) developed Zeeman Component Decomposition (ZCD), and claimed that ZCD overcomes limitations of the weak-field and weak-line approximations and has the advantage of applying simultaneously to all Stokes parameters. Kochukhov et al. (2010) described their iLSD method, which uses multiple average profiles instead of one single profile as assumed in a regular LSD code, and showed the increase in the quality of extracted intrinsic profiles. It's worthwhile to try out these new techniques on our data.

Acknowledgments. We acknowledge the support from the NASA Origins of Solar Systems grant NNX10AI53G to Rice University. This work also made use of the VALD line database and the NASA Astrophysics Data System.

References

Barnes, J. R. 1999, Ph.D. thesis, University of St. Andrews
Borra, E. F., Edwards, G., & Mayor, M. 1984, ApJ, 284, 211
Brown, D. N., & Landstreet, J. D. 1981, ApJ, 246, 899
Donati, J., Semel, M., Carter, B. D., Rees, D. E., & Collier Cameron, A. 1997, MNRAS, 291, 658
Donati, J.-F. e. a. 2007, MNRAS, 380, 1297
— 2008, MNRAS, 386, 1234
Hartmann, L., & Stauffer, J. R. 1989, AJ, 97, 873
Hubrig, S., Plachinda, S. I., Hunsch, M., & Schroder, K. P. 1994, A & A, 291, 890
Hussain, G. A. J. e. a. 2009, MNRAS, 398, 189
Johns-Krull, C. M., & Valenti, J. A. 1996, ApJ, 459, 95
Johns-Krull, C. M., Valenti, J. A., & Koresko, C. 1999a, ApJ, 516, 900
Johns-Krull, C. M., Valenti, J. A., P., H. A., & Kanaan, A. 1999b, ApJ, 510, 41
Kochukhov, O., Makaganiuk, V., & Piskunov, N. 2010, A & A, 524, 5
Mathys, G., & Lanz, T. 1992, A & A, 256, 169
Piskunov, N. 1999, Ap & SS, 243, 515
Plachinda, S. I., & Tarasova, T. N. 2000, ApJ, 533, 1016
Sennhauser, C., & Berdyugina, S. V. 2010, A & A, 522, 57
Shu, F. H., Najita, J., Ostriker, E., Wilkin, F., Ruden, S., & Lizano, S. 1994, ApJ, 429, 781
Tull, R. G., MacQueen, P. J., Sneden, C., & Lambert, D. L. 1995, PASP, 107, 251
Valenti, J. A. 1994, Ph.D. thesis, University of California Berkeley
Valenti, J. A., & Johns-Krull, C. M. 2004, Ap & SS, 292, 619
Valenti, J. A., Marcy, G. W., & Basri, G. 1995, ApJ, 439, 939
Vogt, S. S., Tull, R. G., & Kelton, P. W. 1980, ApJ, 236, 308
Wade, G. A., Donati, J.-F., Landstreet, J. D., & Shorlin, S. L. 2000, MNRAS, 313, 823

The 16th Cambridge Workshop on Cool Stars, Stellar Systems and the Sun
ASP Conference Series, Vol. 448
Christopher M. Johns-Krull, Matthew K. Browning, and Andrew A. West, eds.
© 2011 Astronomical Society of the Pacific

The Mouse that Roared: A SuperFlare from the dMe Flare Star EV Lac Detected by Swift and Konus-Wind

Rachel A. Osten,[1] Olivier Godet,[2] Stephen Drake,[3] Jack Tueller,[3] Jay Cummings,[3] Hans Krimm,[3] John Pye,[4] Valentin Pal'shin,[5] Sergei Golenetskii,[5] Fabio Reale,[6] Samantha R. Oates,[7] Mat J. Page,[7] and Andrea Melandri[8]

[1] *Space Telescope Science Institute 3700 San Martin Drive, Baltimore, MD 21218 USA*

[2] *Université de Toulouse, UPS, CESR, 9 avenue du Colonel Roche, 31028 Toulouse Cedex 9, France*

[3] *NASA Goddard Space Flight Center, Greenbelt MD USA*

[4] *Department of Physics and Astronomy, University of Leicester, University Road, Leicester LE1 7RH UK*

[5] *Ioffe Physico-Technical Institute, Laboratory for Experimental Astrophysics, 26 Polytekhnicheskaya, St Petersburg 194021, Russian Federation*

[6] *Dip. Scienze Fis. & Astron., Sez. Astron.,Universitá di Palermo,P.za Parlamento 1, 90134 Palermo, Italy*

[7] *Mullard Space Science Laboratory, University College London, Holmbury St. Mary, Dorking, Surrey, RH5 6NT, UK*

[8] *Astrophysics Research Institute, Liverpool John Moores University, Twelve Quays House, Birkenhead CH41 1LD UK*

Abstract. We report on a large stellar flare from the nearby dMe flare star EV Lac observed by the Swift and Konus-Wind satellites and the Liverpool Telescope. It is the first large stellar flare from a dMe flare star to result in a Swift trigger based on its hard X-ray intensity. Its peak f_X from 0.3–100 keV of 5.3×10^{-8} erg cm^{-2} s^{-1} is nearly 7000 times larger than the star's quiescent coronal flux, and the change in magnitude in the white filter is ≥ 4.7. This flare also caused a transient increase in EV Lac's bolometric luminosity (L_{bol}) during the early stages of the flare, with a peak estimated L_X/L_{bol} ~3.1. We apply flare loop hydrodynamic modeling to the plasma parameter temporal changes to derive a loop semi-length of $l/R_\star = 0.37\pm0.07$. The soft X-ray spectrum of the flare reveals evidence of iron Kα emission at 6.4 keV. We model the Kα emission as fluorescence from the hot flare source irradiating the photospheric iron, and derive loop heights of $h/R_\star$=0.1, consistent within factors of a few with the heights inferred from hydrodynamic modeling. The Kα emission feature shows variability on time scales of ~200 s which is difficult to interpret using the pure fluorescence hypothesis. We examine Kα emission produced by collisional ionization from accelerated particles, and find parameter values for the spectrum of accelerated particles which can accommodate the increased amount of Kα flux and the lack of observed nonthermal emission in the 20-50 keV spectral region.

1. Introduction

EV Lac has a record of frequent (Huenemoerder et al. 2009) and large outbursts, from optical flaring (Kodaira et al. 1976; Roizman & Shevchenko 1982) to radio (Osten et al. 2005) to X-ray (Schmitt 1994; Favata et al. 2000) wavelengths. On April 25, 2008, EV Lac underwent an outburst which was serendipitously detected by two Gamma-Ray Burst satellites and had supporting ground-based observations. This flare was detected by its transient *hard X-ray* flux, belonging to a small handful of stellar flares which have been triggered using this method. These results have been published in Osten et al. (2010); we provide highlights of the results in this conference proceedings.

2. Light Curves

With the exception of the Konus data, there is no information on the flux variation of EV Lac prior to the trigger by the Burst Alert Telescope (BAT). The trigger time was 2008 April 25 T05:13:57. Figure 1 displays the temporal behavior of the flare as it was caught by the different satellites/instruments involved. The left panel details evolution over the first ~hour, while the right panel details evolution on longer timescales, up to 40,000 s after the trigger. The Konus data indicate that the peak of the flare was achieved within a minute of the source entering the BAT field of view and causing the trigger. The following instruments/telescopes recorded information on the flare: Konus/Wind, Swift/BAT, Swift/X-ray Telescope (XRT), Swift/UV Optical Telescope (UVOT), and Liverpool Telescope (LT).

The blue horizontal lines in Figure 1 indicate the times over which X-ray spectra were extracted. Because of the high count rates, annular regions in the wings of the PSF had to be used to avoid pile-up. Sparse photometric coverage was obtained, indicated in the bottom panel (v); the flare seen in the white filter was brighter than the count rate limit for the UV Optical Telescope, causing a safing event; thus only a lower limit is obtained. Nevertheless, this allows a lower limit on the optical area involved, under the assumption that the white light flare can be described by a black-body with temperature near 10^4K.

3. Time Variation of Plasma Parameters and Spectral Modelling

Time-resolved spectroscopy was performed on data from the XRT and BAT, with different time intervals. The blue horizontal lines in Figure 1 indicate the times when spectra were extracted. APEC model fits were performed to the X-ray spectra. For BAT spectra in the 14-100 keV range, a single temperature only was needed, while for the XRT spectra in the 0.3-10 keV range, two temperatures were used. The variation of temperature and volume emission measure (VEM) with time is smooth, and there are no apparent abundance (Z) variations. In the early stages of the flare, the X-ray luminosity exceeds the bolometric luminosity, by a factor of ~3 at the largest. The right figure shows a fit to both the XRT and BAT spectra over the time interval T0+171 s:T0+961 s. Two temperatures are adequate to fit both soft and hard X-ray emission; no additional thermal or nonthermal component is needed.

By using Reale et al. (1997)'s hydrodynamic modelling method applied to Swift data, we can estimate the coronal loop length during the decay phase as:

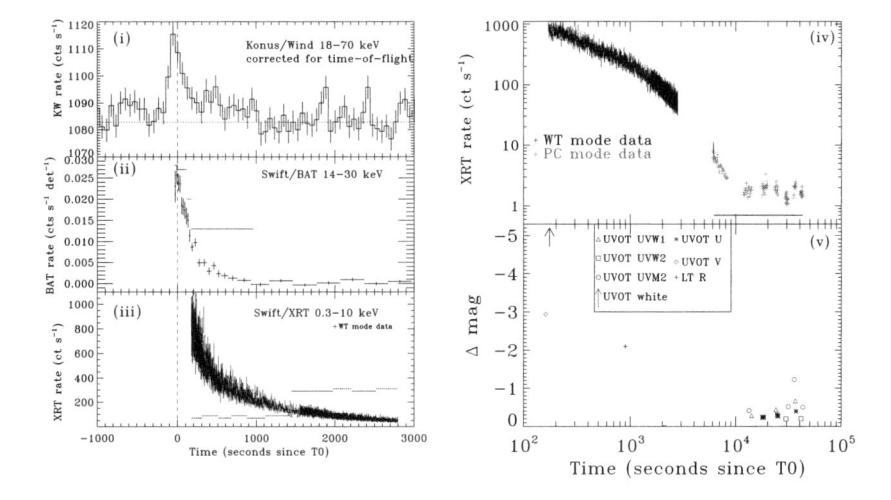

Figure 1. Left panel shows light curves in the initial stages of the flare decay of EV Lac, right panel shows light curves at later times.

$$L = \frac{\tau_{lc} \sqrt{(T)}}{\alpha F(\zeta)} \quad (1)$$

where τ_{lc} is the light curve exponential decay time, T is the flare maximum temperature, related to the maximum best-fit temperature through calibration of this method to the instrument under consideration, $\alpha = 3.7 \times 10^{-4}$ cm^{-1} s^{-1} K$^{1/2}$, and $F(\zeta)$ is a function which quantifies the amount of heating occurring during the decay phase of the flare. The variable ζ is the slope in the log(density-temperature) plane, or equivalently, 0.5*log(VEM)-log(T) plane. The result is $l/R_\star = 0.37 \pm 0.07$, or for a vertical semi-circular loop, $h/R_\star = 0.24 \pm 0.04$.

4. Kα Emission

One of the most surprising findings of this investigation was the clear detection of the Iron Kα emission line at 6.4 keV in the initial stages of the flare decay. Figure 2 shows a close-up of a spectrum in the 5–8 keV region detailing this. When fitting the data with only APEC plasma components, there is a clear excess around 6.4 keV which we attribute to emission from the Iron Kα line. Besides the Sun, there are only two other cool stars which have shown evidence for this feature during large stellar flares: the active binary system II Peg (Osten et al. 2007), and the single active evolved star HR 9024 (Testa et al. 2008). The line, at 6.4 keV, is formed from neutral or nearly neutral iron after a K shell electron is removed. The conventional mechanism as to what causes this is photoionization by X-rays above the photoionization edge of 7.11 keV. In the stellar context, it is cold iron which produces the emission, but the source of the radiation is the high energy photons from coronal plasma. Thus, in one X-ray spectrum, information about both coronal and photospheric material is obtained. The utility of the emission line as a diagnostic in the stellar context stems from the dependence of the line strength on the height of the coronal (X-ray-emitting) material above the neutral (photospheric) material. Drake et al. (2008) have explored the emission strength of Iron Kα as a function of height for different scenarios, and we use their results to interpret our line strengths. The luminosity above 7.11 keV is required, as is the fluorescent efficiency, which depends on both the plasma temperature and the height of the X-ray source above the fluorescent source. The interpretation is done self-consistently for each time interval, using the results of the spectral modelling for that time interval. The source heights are in general agreement with the one derived using the hydrodynamic modelling from investigation of the plasma parameters, and indicates a fairly compact loop, with $h/R_\star \lesssim 0.3$.

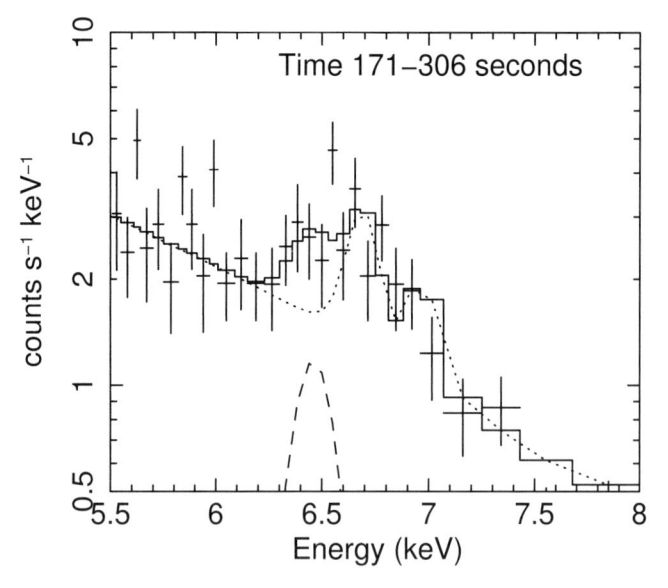

Figure 2. X-ray spectrum of EV Lac during the time period listed. The times refer to seconds since the BAT trigger time. The pluses indicate data and errors, the histogram is the fit to the data including all model components. The dotted line shows the thermal plasma components, while the dashed line shows the Gaussian component at 6.4 keV.

5. Kα Variability

Because there were a large number of time intervals in the initial decay of the flare, we explored the intensity of the Kα emission line in each of the time intervals. Table 1 lists the emission line parameters derived for the first ten time intervals. We only considered time intervals in which the probability was more than 97%. This results in five time intervals in the beginning stages of the flare decay having Iron Kα detections, but with variability detected on quite short timescales: in the time interval T0+306:T0+508, the line is not detected, and then in the next time interval, T0+508:T0+671, the line is detected with a strength much larger than what would be predicted for a fluorescent

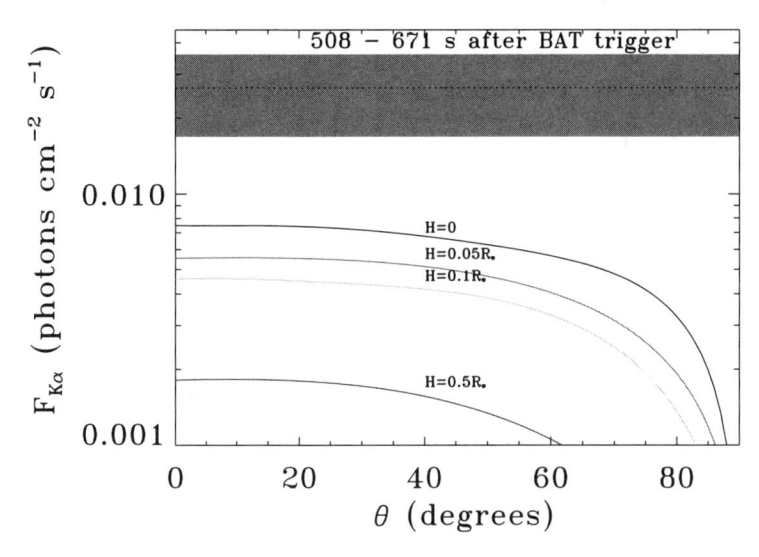

Figure 3. Modelling of the time interval 508–671 s after the BAT trigger, when the Iron Kα line showed an increase in line strength compared with time intervals near it. The dotted line shows the measured value of the line strength, and the magenta curve shows the 3σ error range. The curves below indicate the expected value of the Kα flux for different heights of the X-ray emitting source above the photosphere. Even for the most compact source, the observed flux cannot be reproduced with a purely fluorescent process.

Table 1. Iron Kα Emission Line Parameters

Time Interval (s)	E_0^1 (keV)	EW^2 (eV)	F^3 (10^{-3} photons cm^{-2} s^{-1})	P^4
T0+171:T0+306	6.44 ± 0.08	$146^{+67}_{-72.5}$	$27.3^{+14.7}_{-12.9}$	99.94
T0+306:T0+508	$6.42^{+0.31}_{-0.41}$	$52.2^{+64.8}_{-52.2}$	$6.96^{+8.23}_{-6.96}$	55.42
T0+508:T0+671	6.41 ± 0.04	320 ± 102	$26.4^{+9.6}_{-9.5}$	99.97
T0+671:T0+874	$6.42^{+0.07}_{-0.10}$	103^{+95}_{-81}	$6.51^{+4.67}_{-4.37}$	99.19
T0+874:T0+1118	$6.46^{+0.05}_{-0.06}$	143^{+106}_{-114}	$6.86^{+4.18}_{-4.04}$	99.92
T0+1118:T0+1451	$6.41^{+0.15}_{-0.41}$	$84.3^{+155.7}_{-84.3}$	$2.65^{+2.35}_{-2.33}$	97.69
T0+1451:T0+1971	$6.45^{+0.10}_{-0.09}$	91^{+139}_{-91}	$2.12^{+2.02}_{-1.63}$	...
T0+1971:T0+2215	$6.38^{+0.11}_{-0.12}$	$94.8^{+288.2}_{-94.8}$	$1.20^{+0.94}_{-1.06}$	...
T0+2215:T0+2517	$6.11^{+0.20}_{-0.11}$	$331^{+52}_{-281.3}$	1.65 ± 0.85	...
T0+2517:T0+2798	$6.22^{+0.13}_{-0.22}$	295^{+302}_{-295}	$0.95^{+0.77}_{-0.71}$	...

[1] center of Gaussian fitted to data
[2] equivalent width of emission line
[3] flux in emission line
[4] probability from Monte Carlo simulations for significance of detection

process using the plasma parameters. Figure 3 shows the observed line strength as a function of astrocentric angle θ (defined so that $\theta = 0$ is on the center of the stellar disk and $\theta = 90$ is on the limb). Also plotted are the expected values of the line strength for different heights of the X-ray emitting source above the photosphere. Even for the most compact X-ray source, the Iron Kα emission is too strong to be explained by a purely fluorescence mechanism.

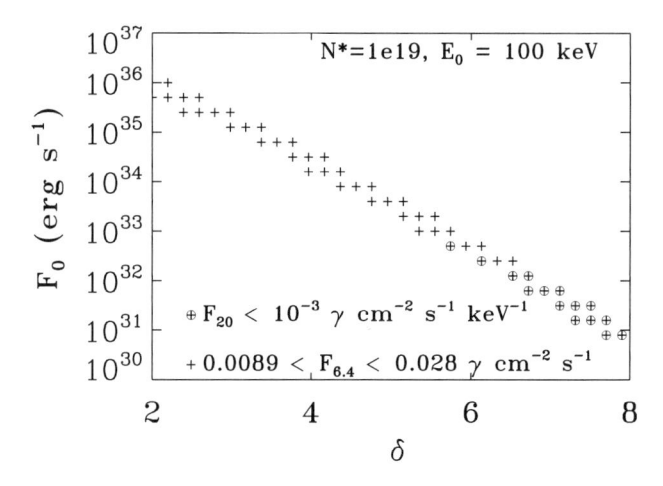

Figure 4. Parameters describing accelerated electrons within the context of explaining the enhanced level of Kα line flux. Pluses indicate values of total power and index δ which can produce the excess Kα emission, while pluses with circles around them satisfy the additional constraint that the nonthermal bremmstrahlung radiation at 20 keV is less than 10^{-3} photons cm^{-2} s^{-1} keV^{-1}.

Motivated by a few solar studies that showed flares which had excess Iron Kα line strengths above that which could be predicted from fluorescence mechanisms, we explored the addition of another Kα-producing mechanism. Figure 4 shows our attempt to interpret the excess Kα emission observed from T0+508 s: T0+671 s as collisional ionization from accelerated electrons. This is based upon the observation of excess Kα emission in a few solar flares above that accounted for by fluorescence (Emslie et al. 1986; Zarro et al. 1992). We use two constraints: the spectrum and amount of accelerated electrons must be enough to produce the excess Kα emission, but produce a nonthermal bremsstrahlung flux below the hard X-ray photon flux level observed (10^{-3} photons cm^{-2} s^{-1} keV^{-1} at 20 keV). Parameter values with encircled pluses fit this constraint. With this interpretation, and the lower limit on photospheric flare area of 2×10^{19} cm^2, we can estimate the nonthermal beam flux of 10^{11}-10^{14} erg cm^{-2} s^{-1}, which overlaps the range seen and modeled in solar and stellar flares, respectively (Allred et al. 2005, 2006).

6. Conclusions

The flare studied in this paper is unique in two main aspects: the level of X-ray radiation marks it as one of the most extreme flares yet observed in terms of the enhancement of ~7000 compared to the usual emission levels, and it is one of only a handful of stellar flares from normal stars without disks to exhibit the Fe Kα line, with a maximum equivalent width of >200 eV.

References

Allred, J. C., Hawley, S. L., Abbett, W. P., & Carlsson, M. 2005, ApJ, 630, 573. arXiv: astro-ph/0507335
— 2006, ApJ, 644, 484. arXiv:astro-ph/0603195
Drake, J. J., Ercolano, B., & Swartz, D. A. 2008, ApJ, 678, 385. 0710.0621
Emslie, A. G., Phillips, K. J. H., & Dennis, B. R. 1986, Solar Phys., 103, 89
Favata, F., Reale, F., Micela, G., Sciortino, S., Maggio, A., & Matsumoto, H. 2000, A&A, 353, 987. arXiv:astro-ph/9909491
Kodaira, K., Ichimura, K., & Nishimura, S. 1976, PASJ, 28, 665
Osten, R. A., Drake, S., Tueller, J., Cummings, J., Perri, M., Moretti, A., & Covino, S. 2007, ApJ, 654, 1052. arXiv:astro-ph/0609205
Osten, R. A., Godet, O., Drake, S., Tueller, J., Cummings, J., Krimm, H., Pye, J., Pal'shin, V., Golenetskii, S., Reale, F., Oates, S. R., Page, M. J., & Melandri, A. 2010, ApJ, 721, 785. 1007.5300
Osten, R. A., Hawley, S. L., Allred, J. C., Johns-Krull, C. M., & Roark, C. 2005, ApJ, 621, 398. arXiv:astro-ph/0411236
Reale, F., Betta, R., Peres, G., Serio, S., & McTiernan, J. 1997, A&A, 325, 782
Roizman, G. S., & Shevchenko, V. S. 1982, Soviet Astronomy Letters, 8, 85
Schmitt, J. H. M. M. 1994, ApJS, 90, 735
Testa, P., Drake, J. J., Ercolano, B., Reale, F., Huenemoerder, D. P., Affer, L., Micela, G., & Garcia-Alvarez, D. 2008, ApJ, 675, L97. 0801.3857
Zarro, D. M., Dennis, B. R., & Slater, G. L. 1992, ApJ, 391, 865

The 16th Cambridge Workshop on Cool Stars, Stellar Systems and the Sun
ASP Conference Series, Vol. 448
Christopher M. Johns-Krull, Matthew K. Browning, and Andrew A. West, eds.
© 2011 Astronomical Society of the Pacific

Observations of X-ray Flares in G-K Dwarfs by XMM-Newton

J. C. Pandey[1] and K. P. Singh[2]

[1]*Aryabhatta Research Institute of Observational Sciences, Nainital -263 129, India*

[2]*Tata Institute of Fundamental Research, Mumbai -400 005, India*

Abstract. The temporal and spectral analysis of seventeen flares observed in six G-K dwarfs by XMM-Newton is presented here. The exponential decay of the X-ray light curves, and time evolution of the plasma temperature and emission measure are similar to those observed in compact solar flares. The loop length of flare is derived using various loop models including state-of-the-art hydrodynamic flare model and found to be much less than the stellar radius. For most of the flares, we have also found that the decay path of the flare is driven by sustained heating.

1. Introduction

Stellar flares are features in which a star's intensity increases rapidly to a peak value followed by more gradual decay. Flares are frequently observed in late-type active stars from the X-ray to the radio regime. In a standard model, flares are known to be a manifestation of the reconnection of magnetic loops, accompanied by particle beams, chromospheric evaporation, rapid bulk flows or mass ejection, and heating of plasma confined in loops. Flares from active stars present many analogies with the solar flares. However, stellar flares could be radiating several orders of magnitude more energy than a solar flare. Even though stellar flares are spatially unresolved, a great deal of information on the coronal heating and on the plasma structure morphology can be inferred from a detailed modeling of stellar flares. Analysis of light curves during flares can provide us with insights into the characteristics of the coronal structures and, therefore, of the magnetic field (Osten et al. 2010, Pandey & Srivastava 2009, Reale et al. 2004).

Active solar-type (G-K) dwarfs are known to possess magnetic fields on their surfaces, and the field strengths are as large as several kilogauss, i.e. stronger than the strongest field observed on the Sun. Cyclic behavior has been identified in several late-type active stars (Olah & Strassmeier 2002), suggesting that dynamos similar to the solar dynamos are also operative in solar-type stars. These solar-type stars are less active than the dMe stars. However, they show more frequent flaring activity than the late-type evolved stars. In this paper, we present temporal and spectral analysis of archival data obtained from XMM-Newton Observations of six G-K dwarfs (V268 Cep, XI Boo, IMVir, V471 Tau, EP Eri and CCEri).

2. Observations and Data Reductions

Star in the sample were observed with XMM-Newton satellite using the EPIC PN, MOS detectors and Reflection Grating Spectrometer (RGS) between 2001 to 2004 with exposure times 30 to 60 ks . The details on the XMM-Newton satellite and its detectors are given in Janesen et al. (2001) and Strüder et al. (2001). The data were reduced using the Science Analysis System (SAS) version 8.0 with updated calibration files. We have checked data for background proton flaring and pileup effects. These effects were removed using the standard processes given in handbook of SAS, where necessary. X-ray light curves and spectra of all target stars were generated from on-source counts obtained from circular regions with a radius ~40-50 arcsec around each source. However, for the sources which are affected by pile up, an annular region was taken to generate light curve and spectra. We have used, PN data for further analysis. However, for ξ Boo, data with MOS detector was used as no observation from PN detector was made.

3. X-ray Light Curves

The X-ray light curves observed with PN detector of the stars in the energy band 0.3-10 keV are shown in top panels of Figures 1 (a) - (e). Seventeen flare like features, marked as Fi, where i=1,2,...,17, are clearly seen. The quiescent states in the light curves are marked by Q. Flares F6, F7,F9, F14 and F16 were found to be long lasting (>2.3 h) flares. In fact, the flare F7 of the ξ Boo is longest flare observed with total duration of 3.1 h. The peak flux in these long duration flares was found to be more than two times than the quiescent state. Other detected flares were shorter than 1.3 h. The shortest duration flare was detected one in V368 Cep (F3) and other in CCEri (F12). In these short duration flares, peak X-ray flux was 1.3-1.8 times more than that of the quiescent state. The e-folding decay and rise times, and count rates at peak of the flare and quiescent states are given in table 1. The e-folding decay times were found to be in range to ks. Among the 17 flares presented here, only four flares (F6 and F7 of ξ Boo, F9 of IMVir and F14 of CC Eri) appear to be long decay flare ($\tau_d > 1$ h). In fact, the flare F7 of XI Boo was one of the longest duration flares observed with ($\tau_d \sim 10$ ks). This flare classification is purely based on the decay time. It has been shown that even much longer flares can occur in a single loop (Favata et al. 2005), while relatively short flares with heating dominated decay are probably arcade flares. For the flares F6, F9 and F14, the decay path is driven by the sustained heating (see §6 and Table 2), therefore, these flares cannot be classified as long decay flares. However, the sustained heating during the decay of the flare F7 is negligible. Thus, these flares can be classified as arcade. The flare decay time for the remaining 12 flares is similar to that of the solar compact flares. The morphological differences in these different types of flares indicate the different processes of energy released. In compact flares energy is probably released only during an impulsive phase, whereas in the two-ribbon flares a prolonged energy release is apparently required to explain their long decay time (Pallavicini et al. 1988). The rise time of these flares has been found to be less than 1 ks, which is similar to the rise time of impulsive flares observed in the M dwarfs (Pallavicini et al. 1990). However, for some flares (F6, F7 and F9) the rise time was found to be more than 1 ks.

Flares F2-F3-F4, F6-F7 and F10-F11 show the similar structure i.e. before ending the first flare the next flare starts. Similar loop systems have been observed to the flare

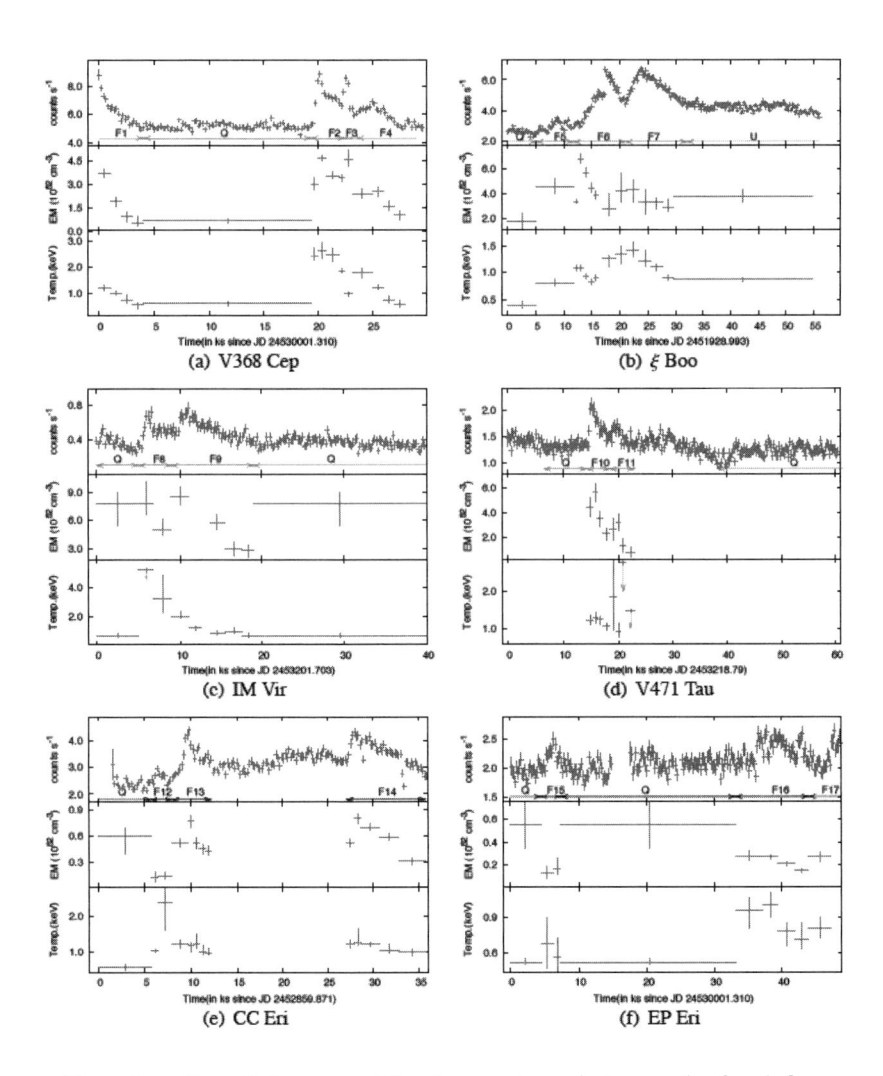

Figure 1. X-ray light curves of the stars are shown in top panels of each figure. The time evolution of temperature and corresponding emission measure are in bottom and middle panels of each figure.

on the Sun (e.g. so-called Bastille Day flare; Aschwanden & Alexander 2001), and in a stellar analogue dMe star Proxima Centauri (Reale et al. 2004). In these two events, a double ignition in nearby loops was observed or suggested, and the delay between the ignitions appears to scale with the loop sizes.

Table 1. Flare parameters from light curve. Here A and q are count rate at flare peak and quiescent state in counts s^{-1}, τ_d and τ_r are flare decay and rises times in s, and L_X and L_{Xq} are X-ray luminosities at flare peak and quiescent state in erg s^{-1}

Star	Flare	A	q	τ_d	τ_r	L_X	L_{Xq}
V368 Cep	F1	9.0 ± 0.2	5.1	1226 ± 86	...	9.4	6.8
	F2	9.1 ± 0.3		2387 ± 341	347 ± 53	11.8	
	F3	8.7 ± 0.2		496 ± 36	515 ± 79	9.5	
	F4	7.0 ± 0.2		1232 ± 133	681 ± 91	8.2	
XI Boo	F5	3.42 ± 0.12	2.66	2829 ± 950	879 ± 213	0.54	0.5
	F6	6.9 ± 0.1		4393 ± 333	3169 ± 259	1.02	
	F7	6.78 ± 0.06		9885 ± 362	3351 ± 244	1.11	
IM Vir	F8	1.02 ± 0.05	0.64	2025 ± 603	311 ± 121	14.6	8.3
	F9	1.27 ± 0.04		3450 ± 289	2041 ± 401	15.9	
V471 Tau	F10	2.12 ± 0.05	1.28	2135 ± 286	580 ± 109	12.3	7.5
	F11	1.70 ± 0.02		1847 ± 519	827 ± 307	10.8	
CC Eri	F12	4.6 ± 0.2	2.35	781 ± 98	335 ± 52	1.1	0.9
	F13	6.4 ± 0.3		2844 ± 312	722 ± 61	1.5	
	F14	7.4 ± 0.2		5992 ± 231	915 ± 51	1.5	
EP Eri	F15	2.48 ± 0.10	1.90	1094 ± 293	967 ± 201	0.7	0.6
	F16	2.54 ± 0.04		4052 ± 514	3200 ± 780	0.8	
	F17	2.51 ± 0.03			1014 ± 597	0.7	

4. Quiescent State X-ray Emission

X-ray spectra for each star during the quiescent states were analysed using XSPEC, version 12.3 (Arnaud 1996). Spectral analysis of EPIC data was performed in the energy band between 0.3 and 10.0 keV. The EPIC-PN spectra of the stars were fitted with a single (1T) and two (2T) temperature collisional plasma model known as APEC (Smith et al. 2001), with variable elemental abundances. Abundances for all the elements in the APEC were varied together. The interstellar hydrogen column density (NH) was left free to vary. Acceptable APEC 2T fits were achieved only when the abundances were allowed to depart from the solar values. For each star the value of N_H was found to be negligible small at the value of $\sim 10^{20}$ cm^{-2}. Abundance was found to be ~ 0.2 times of the solar values. The temperatures of cool component were found in between 0.2 to 0.49 keV, whereas temperatures of hot component were found in the range of 0.83 to 0.97 kev. The X-ray luminosities during the quiescent state for the stars are given table 1.

5. Spectral Evolution of X-ray Flares

In order to trace the spectral changes during the flares, we have binned flare spectra into different time segments covering their rising and decay phase. Most of the flare were binned into 4 - 5 time segments. For example the spectra of flare F2 of the star V368 Cep were binned into four time segments (one segment in rise phase and three segments in decay phase), and the corresponding different time segments spectra are shown in Fig. 1. Quiescent state spectra is also shown in this figure for a comparison. Because of a limited count statistics only one or two segment were made for few flares (e.g. F3, F5, F8, F12 and F15). To study the flare emission only, we have performed 1T spectral fits of the data, with the quiescent emission taken into account by including its best-fitting 2T model as a frozen background contribution. This is equivalent to consider the flare emission subtracted of the quiescent level, allows us to derive one effective temperature and one EM of the flaring plasma. The abundances were kept fix to that of the quiescent emission. Bottom and middle panels of Figs. 1 (a) - 1 (e) show the temporal evolution of the temperature and corresponding emission measure (EM) of the flares. Both the temperature and EM show the well-defined trends i.e. the changes in the temperature and the EM are correlated with the variations observed in the light curves during the flares . During the flares F1, F9 and F11 the EM was increased by a factor of 5 or more. For rest of the flares EM was increased by a factor less than 5. The peak flare temperature is found in the range of 10-80 MK for all the flares. These values are intermediate between those of flares observed on active dMe stars and young stellar objects (Reale & Micela 1998; Favata et al. 2000; Briggs & Pye 2003; Favata, Micela & Reale 2001; Favata et al. 2005). For most of flares both EM and temperature peaked simultaneously. However, for few flares (F9, F11, F13), it appears that the temperature was evolved before the EM. This is probably due to an impulsive flare event, in which loop does not reach equilibrium conditions, the density begins to decay later than the temperature.

6. Loop Modeling

In a star, flares cannot be resolved spatially. However, by an analogy with solar flares and using flare loop models, it is possible to infer the physical size and morphology of the loop structures involved in a stellar flare. Based on quasi static radiative and conductive cooling during the early phase of the decay Haisch (1983) suggested an approach to model a loop. Given an estimate of two measured quantities, the emission measure (EM) and the decay timescale of the flare (τ_d), this approach leads to the following expression for the loop length (L_{ha}):

$$L_{ha}(cm) = 5 \times 10^{-6}(EM)^{1/4}\tau_d^{3/4} \qquad (1)$$

After Haisch (1983) various stellar flare models came up. These are (i) two-ribbon flare method (Kopp & Poletto 1984), (ii) quasi-static cooling method (van den Oord & Mewe 1989), (iii) pure radiation cooling method (Pallavicini et al. 1990), (iv) rise and decay time method (Hawley et al. 1995), and (v) hydrodynamic method (Reale et al. 1997, 2007). The two-ribbon flare model assumes that the flare decay is entirely driven by heating released by magnetic reconnection of higher and higher loops and neglects completely the effect of plasma cooling. The other four methods are based on the cooling of plasma confined in a single flaring loop. The hydrodynamic model includes

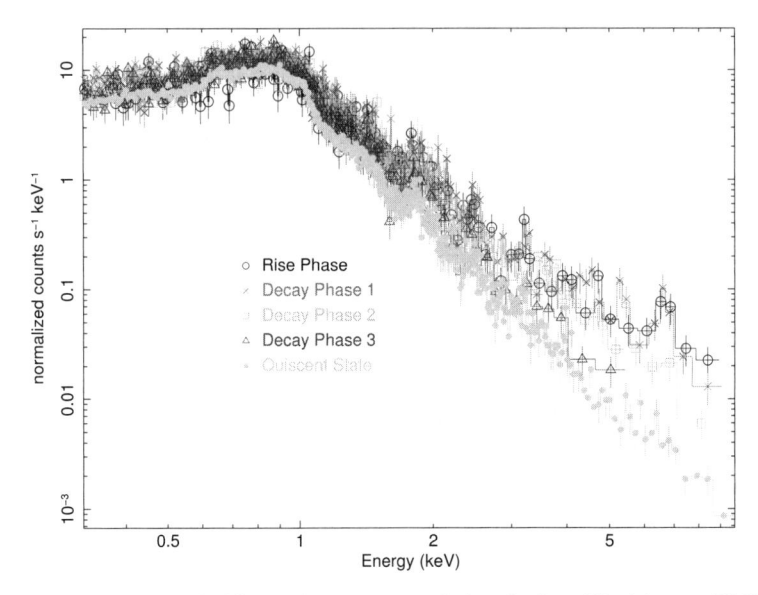

Figure 2. Spectra of different time segments during the flare F2 of the star V368 Cep

both plasma cooling and the effect of heating during flare decay. Reale et al. (1997) presented a method to infer the geometrical size and other relevant physical parameters of the flaring loops, based on the decay time and on evolution of temperature and the EM during the flare decay. The thermodynamic loop decay time can be expressed (Serio et al. 1991) as

$$\tau_{th} = \frac{3.7 \times 10^{-4} L}{\sqrt{T_{max}}} \qquad (2)$$

where L is the loop half-length in cm, and T_{max} is the flare maximum temperature (K) calibrated for EPIC instruments as,

$$T_{max} = 0.13 T_{obs}^{1.16} \qquad (3)$$

T_{obs} is the maximum best-fit temperature derived from single temperature fitting of the data. The average temperatures of the loops, usually lower than the real loop maximum temperatures, are found from the spectral analysis of the data. According to equation (3), the loop maximum temperatures for the flares are given in table 2.

The ratio of the observed exponential light curve decay time τ_d to the thermodynamic decay time τ_{th} can be written as a function which depends on the slope ζ of the decay in the density-temperature plane. The value of ζ is maximum ($\sim$ 2) if heating is negligible and minimum ($\sim$0.5) if heating dominates the decay (Sylwester et al. 1993). Considering the volume is constant throughout the loop, the $EM^{1/2}$ can be taken as a proxy of density. Fig. 3 shows the diagram density ($EM^{1/2}$) - temperature diagram for the flares F1, F2 and F4 in the stars V368 Cep. The solid lines represent the best linear fit to the corresponding data, providing the slope ζ. The values of ζ in density-temperature diagram for the observed flares are given table 2. The best-fitting slopes (ζ

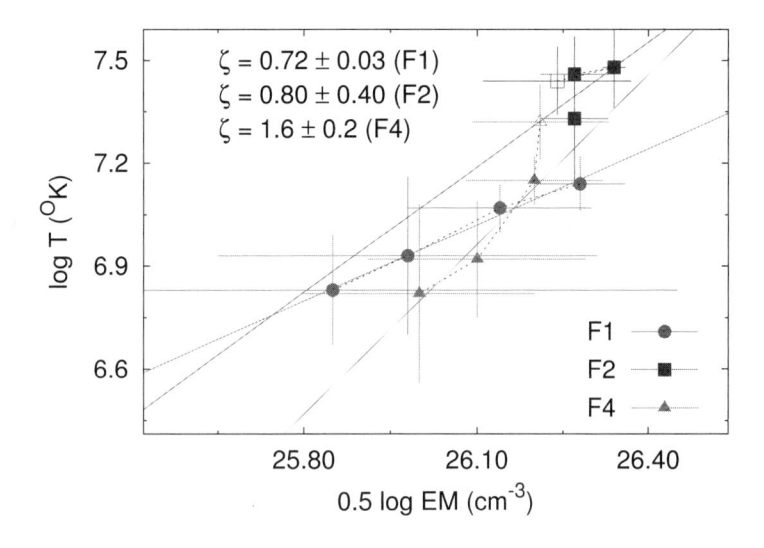

Figure 3. The density-temperature diagram for the three flares (F1, F2 and F4) of the star V368 Cep, where EM$^{1/2}$ has been used as a proxy of density. Solid symbols represents the decay phase while open symbol represents the rise phase of the flare. Dotted lines represent the locus of the density-temperature diagram. The continuous lines represent the best fit straight line to the decay phase of density-temperature diagram.

) for the decaying phase of flares indicate that the flares F1, F2, F6, F9, F10, F13, F14 and F16 are driven by the time-scale of the heating process, whereas sustained heating is negligible during the decay of the flares F4 and F7. For seven flares (F3,F5, F11, F12, F15 and F17), we could not determine the value of ζ either due to limited statistics or due to un-observed decay phase.

For the XMM-Newton EPIC spectral response the ratio $\tau_d/\tau_{th} = F(\zeta)$ is given as (Reale 2007)

$$\frac{\tau_d}{\tau_{th}} = F(\zeta) = \frac{0.51}{\zeta - 0.35} + 1.36 \qquad \text{(for } 0.35 < \zeta \leq 1.6\text{)} \qquad (4)$$

Combining above equations the expression for semi-loop length is (Reale et al. 1997)

$$L = L_{hyd} = \frac{\tau_d \sqrt{T_{max}}}{3.7 \times 10^{-4} F(\zeta)} \qquad (5)$$

Alternatively, Reale (2007) derive the semi-loop length from the rise phase and peak phase of the flare as

$$L_{hyr} = 3 \times 10^{5/2} t_M \frac{T_{max}^{5/2}}{T_M^2} \qquad (6)$$

T_M and t_M are temperature and time at which density peaks. In most of the cases the variation in EM was correlated with light curve during the flare (see Fig. 1, therefore, the e-folding rise time was used as a value of t_M in the estimation of loop length using equation 6.

Shibata & Yokoyama (1999, 2002) developed a method, which is based on a balance between heating due to magnetic reconnection and chromospheric evaporation. They assumed that, to maintain stable flare loops, the gas pressure of the evaporated plasma must be smaller than the magnetic pressure. The derived equations for a magnetic loop length L_p is given in equation 7. We did not apply this method to determine the loop-length because electron density outside the flare (n_e) is not known prior.

$$L_p = 10^9 \left(\frac{EM}{10^{48} \text{ cm}^{-3}} \right)^{3/5} \left(\frac{n_e}{10^9 \text{ cm}^{-3}} \right)^{-2/5} \left(\frac{T}{10^7 \text{ K}} \right)^{-8/5} \text{ cm} \qquad (7)$$

Table 2. Loop length derived from various loop models and other parameters.

Star Name	Flare	T_{max} (MK)	ζ	L_{ha} (10^9 cm)	L_{hyd} (10^9 cm)	L_{hyr} (10^9 cm)	$R_\star$ (10^9 cm)
V368 Cep	F1	25 ± 3	0.72 ± 0.03	11.9 ± 0.7	6.0 ± 0.6	...	56
	F2	62 ± 8	0.80 ± 0.40	23.6 ± 2.5	19.8 ± 8.6	9.0 ± 2.0	
	F3	20 ± 2	...	07.7 ± 0.4	....	...	
	F4	40 ± 2	1.6 ± 0.2	12.3 ± 1.0	11.7 ± 1.3	13.6 ± 3.2	
ξ Boo	F5	16 ± 3	...	10.9 ± 2.8	....	...	56
	F6	19 ± 1	0.73 ± 0.30	22.6 ± 1.3	18.6 ± 7.5	16.8 ± 1.8	
	F7	25 ± 3	2.1 ± 0.4	38.1 ± 1.3	79.4 ± 6.1	36.0 ± 8.0	
IM Vir	F8	79 ± 35	...	24.0 ± 5.4	...	...	56/46
	F9	47 ± 4	0.60 ± 0.20	35.6 ± 2.3	18.3 ± 8.9	41.0 ± 13.0	
V471 Tau	F10	34 ± 3	1.16 ± 0.07	22.1 ± 2.2	21.7 ± 2.5	7 ± 2	54
	F11	42 ± 36	...	16.6 ± 3.5	...	6.2	
CC Eri	F12	55 ± 28	...	04.4 ± 0.5	...	...	41/17
	F13	26 ± 3	0.51 ± 0.21	16.7 ± 1.3	< 14	6.6 ± 1.4	
	F14	27 ± 2	0.54 ± 0.15	29.7 ± 0.8	20.0 ± 10.0	11.4 ± 3.1	
EP Eri	F15	13 ± 4	...	05.9 ± 1.2	...	2.1	51
	F16	20 ± 2	1.10 ± 0.30	17.6 ± 1.6	23.5 ± 4.5	20.8 ± 5.8	
	F17	16 ± 2	...	...	...	3.8	

The loop length obtained from above three methods are given in Table2. For most of the flares the semi-loop lengths (L_{hyr}) estimated from rise approach for the flares are found to be consistent with that of estimated from the decay phase analysis. The estimated loop length for the flares F2, F7 and F10 are found to be less than that of determined from decay phase. However, for the flare F9 of IM Vir the estimated loop length is two times more than that of estimated from decay phase. The inconsistency in the estimation of the loop length from the two approaches is probably due to the involvement of other different coronal loops during the decay of the flare or the heat pulse triggering the flare is not a top-hat function. In general, the loop length derived from Haisch approach L_{ha}, is found to be more than that derived from the hydrodynamic method, but most of the cases the differences is well within 1σ level. In case of the flare F2, F4, F6, F10, F13 and F16, the L_{ha} is consistent with L_{hyd}. However, for the flare F4, F7 and F9 L_{ha} is consistent with L_{hyr}. We have also noticed that for the flares F4, F7, F10 and F16, the sustained heating during the decay phase was negligible. This implies that if the sustained heating during the decay phase was negligible, both hydrodynamic and Haisch methods derive the consistent loop lengths. Fig. 4 shows the plot between loop lengths derived from hydrodynamic method and Haisch approach. In this plot

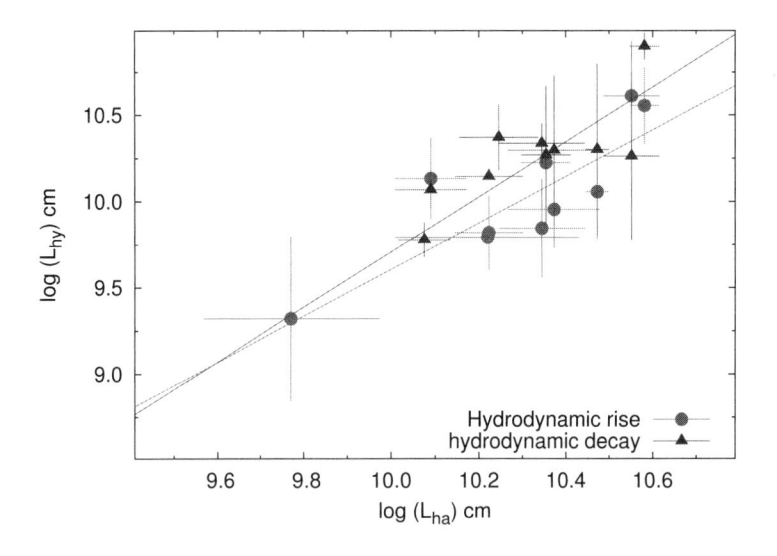

Figure 4. Plot of loop lengths : hydrodynamic method (L_{hyd}, L_{hyr}) versus Haisch approach (L_{ha}).

continuous lines show the best fit straight lines between L_{hyd} and L_{ha}, and L_{hyr} and L_{ha}. The best fit straight line show the relation $L_{ha} \propto L_{hyd}^{1.5\pm0.4}$ and $L_{ha} \propto L_{hyr}^{1.3\pm0.4}$. However, more sample is needed to make any conclusive relation between two approaches.

The loop lengths derived from any of the method was found to be less than the pressure scale height and also found to be smaller than the stellar radii.

7. Loop Parameters

Physical properties of the flaring plasma can be inferred from the analysis of flare data. The analysis of X-ray spectra provides values of the temperature and EM. From the EM and the plasma density (n_e), the volume (V) of the flaring loop is estimated as

$$EM = n_e^2 V \tag{8}$$

According to RTV scaling laws (Rosner, Tucker & Vaiana 1978), the maximum pressure, temperature, loop length and heating rate per unit volume are related as

$$T_{max} = 1.4 \times 10^3 (pL)^{1/3} \tag{9}$$

$$E_H = 10^{-6} T_{max}^{3.5} L^{-2} \text{ erg s}^{-1} \text{ cm}^{-3} \tag{10}$$

The minimum magnetic field necessary to confine the flaring plasma can be estimated as

$$p = \frac{B^2}{8\pi} \tag{11}$$

Using the above relations, the estimated loop parameters are given in table 3, where column 1 contains flare name, column 2 consists adopted loop length, columns 3, 4, 5, 6 and 7 consist maximum pressure in the loop, loop density, volume of flaring plasma, heating rate per unit volume at the flare peak and minimum magnetic field necessary to confine the plasma, respectively. Column 8 consists the loop aspect ratio (β), and is determined by RTV scaling law $V = 2\pi\beta^2 L^3$. Column 8 consists number loop in a arcade. To determine the flare parameters, we have adopted the loop length from hydrodynamic decay method. However, if loop lengths from hydrodynamic methods are consistent with Haisch method then loop length with less error was adopted.

Table 3. Loop parameters

Flare	L	p	n_e	V	E_H	B	β	NL
(1)	(2)	(3)	(4)	(5)	(6)	(7)	(8)	(9)
F1	6.0 ± 0.6^a	1007 ± 312	1.4 ± 0.3	1.8	2.32	159	1.15	..
F2	23.6 ± 2.5^c	3679 ± 1476	2.1 ± 0.6	1.0	3.37	304	0.11	~ 1
F3	7.7 ± 0.4^c	378 ± 92	0.7 ± 0.1	9.9	0.60	97	1.86	..
F4	11.7 ± 1.3^a	2038 ± 354	1.8 ± 0.3	0.8	3.03	226	0.28	~ 7
F5	10.9 ± 2.8^c	136 ± 94	0.3 ± 0.2	1.1	0.14	58	0.37	~ 13
F6	16.8 ± 1.8^b	146 ± 24	0.3 ± 0.1	8.8	0.10	60	0.54	~ 29
F7	38.1 ± 1.3^c	154 ± 49	0.2 ± 0.1	8.7	0.06	62	0.16	~ 2
F8	24.0 ± 5.4^c	7600	3.4	0.7	7.74	436	0.09	~ 1
F9	18.3 ± 8.9^a	2015 ± 1091	1.6 ± 0.8	4.6	2.06	224	0.35	~ 11
F10	21.7 ± 2.5^a	631 ± 200	0.7 ± 0.2	12.2	0.46	125	0.44	~ 19
F11	6.2^b	4292	3.7	0.2	12.27	328	0.37	~ 13
F12	4.4 ± 0.5^c	14081	9.1	0.0	65.32	594	0.01	~ 1
F13	16.7 ± 1.3^c	370 ± 128	0.5 ± 0.1	2.9	0.31	96	0.31	~ 9
F14	20.0 ± 10.0^a	358 ± 197	0.5 ± 0.3	3.5	0.26	94	0.26	~ 6
F15	2.1^b	381	1.1	0.2	1.79	97	1.85	..
F16	23.5 ± 4.5^a	127 ± 48	0.2 ± 0.1	5.4	0.07	56	0.26	~ 6
F17	3.8^b	392	0.9	0.4	1.13	99	1.08	..

For all the flares, the estimated maximum electron density under assumption of a totally ionized hydrogen plasma ($p = 2n_e k T_{max}$) is found in the order of 10^{10-11} cm^{-3}. This is compatible with the values expected for the plasma in coronal condition (Wu et al. 1986). To satisfy the energy balance relation for the flaring as a whole, the maximum X-ray luminosity must be lower than the total energy rate ($H = E_H.V$) at the flare peak. The rest of the input energy is used for thermal conduction, kinetic energy and radiation at lower frequencies. For the most of flares 11-26 % of H is observed as a peak luminosity. However, for the flares F2, F4, F5 and F11 the maximum value of X-ray luminosity is about 33-44% of H. These values are in agreement with those reported for the solar flares, where the soft X-ray radiation only accounts for up to 20 per cent of total energy (Wu et al. 1986). In comparison, the fraction of X-ray radiation to the total energy has been found to be 15 and 35 % for the M dwarfs Proxima Centauri and EV Lac, respectively (Favata et al. 2000; Reale et al. 2004). Applying loop scaling law ($V = 2\pi\beta^2 L^3$), and if the detected flares are produced by a single loop, their aspect ratio (β) were estimated in the range of 0.1 to 0.5 for the most of flares. Similar cross-section was also observed for the solar coronal loops for which typical values of are in the range of 0.1-0.3. If we assume $\beta = 0.1$, these flares occurred in the arcades, which are composed 2 to 29 loops.

8. Summary

A total 17 flares were observed in six G-K dwarfs. Decay time of flares were found from 0.5 to 10.0 ks. Quiescent state luminosity of these dwarfs varies from 0.5 to 8.3 $\times 10^{29}$ erg s^{-1}. Quiescent state corona of these dwarfs contain two temperature plasma with variable abundances of ~0.2 of the solar abundances. The average values of quiescent state temperatures of cool and hot components were 0.35 and 0.81 keV, respectively. Flare peak luminosities were found to be 1.3 to 2.5 times more than that of the quiescent state luminosity. The flaring corona was defined by 3 temperature plasma out of which cool two temperatures represent quiescent corona. The maximum temperature during the observed flares were found to be in the range 10 to 80 MK. The derived loop lengths were found to be of the order of 10^{10} cm, which is less than the stellar radii. In conclusion, flares observed in G-K dwarfs are similar to solar arced flare, which are as strong as M dwarfs and are much smaller than giants and pre-main sequence stars.

Acknowledgments. This research has made use of data obtained from HEASARC, provided by the NASA Godard Space Flight Center.

References

Arnaud K. A., 1996, ASPC, 101, 17
Aschwanden M. J., Alexander, D., 2001, Sol. Phys, 204, 91
Briggs K. R., Pye J. P., 2003, MNRAS, 345, 714
Favata F., Flaccomio E., Reale F., Micela G., Sciortino S., Shang H., Stassun K. G., Feigelson E. D., 2005, ApJS, 160, 469
Favata F., Micela G., Reale F., 2001, A&A, 375, 485
Favata F., Micela G., Reale F., 2000, A&A, 354, 1021
Haisch B. M., 1983, in IAU Colloq. 71, Activity in Red-Dwarf Stars, Vol. 102, ed. P. B. Byrne & M. Rodono (Dordrecht: Reidel), 255
Hawley S. L. et al., 1995, ApJ, 453, 464
Jansen F. et al., 2001, A&A, L365, 1
Kopp R. A., Poletto G., 1984, SoPh, 93, 351
Olah K., Strassmeier K. G., 2002, AN, 323, 361O
Osten, R. A, et al., 2010, ApJ, 721, 785
Pallavicini R., Tagliaferri G., Stella L., 1990, A&A, 228, 403
Pallavicini R., Monsignori-Fossi B. C., Landini M., Schmitt, J. H. M. M., 1988, A&A, 191, 109
Pandey J. C., Srivastava, A. K., 2009ApJ, 697L, 153
Reale F., 2007, A&A, 471, 271
Reale F., Güdel M., Peres G., Audard M., 2004, A&A, 416, 733
Reale F., Micela G., 1998, A&A, 334, 1028
Reale F., Betta R., Peres G., Serio S., McTiernan J., 1997, A&A, 325, 782
Rosner R., Tucker W. H., Vaiana G. S., 1978, ApJ, 220, 643
Smith R. K., Brickhouse N. S., Liedahl D. A., Raymond J. C., 2001, ApJ, 556, L91
Serio S., Reale F., Jakimiec J., Sylwester B., Sylwester J., 1991, A&A, 241, 197
Shibata, K., Yokoyama, T., 2002, ApJ, 577, 422
Shibata, K., Yokoyama, T., 1999, ApJ, 526L, 49
Strüder L. et al., 2001, A&A, 365, L18
Sylwester B., Sylwester J., Serio S., Reale F., Bentley R. D., Fludra A., 1993, A&A, 267, 586

van den Oord G. H. J., Mewe R., 1989, A&A, 213, 245
Wu S. T. et al., 1986, in Energetic Phenomenon on the Sun, eds. M. Kundu& B. Woodgate, No. 2439 in NASA Conference Publication, NASA, p. 5

The 16th Cambridge Workshop on Cool Stars, Stellar Systems and the Sun
ASP Conference Series, Vol. 448
Christopher M. Johns-Krull, Matthew K. Browning, and Andrew A. West, eds.
© 2011 Astronomical Society of the Pacific

The Activity and Rotation Limit in the Hyades

U. Seemann,[1,2] A. Reiners,[2] A. Seifahrt,[2,3] and M. Kürster[4]

[1]*European Southern Observatory, Karl-Schwarzschild-Straße 2, 85748 Garching, Germany; useemann@eso.org*

[2]*Institut für Astrophysik, Georg-August-Universität Göttingen, Friedrich-Hund-Platz 1, 37077 Göttingen, Germany*

[3]*Department of Physics, University of California, One Shields Avenue, Davis, CA 95616, USA*

[4]*Max-Planck-Institut für Astronomie, Königstuhl 17, 69117 Heidelberg, Germany*

Abstract. We conduct a study of K to M type stars to investigate the activity and the rotation limit in the Hyades. We use a sample of 40 stars in this intermediate-age cluster ($\approx$625 Myr) to probe stellar rotation in the threshold region where stellar activity becomes prevalent. Here we present projected equatorial velocities ($v_{rot} \sin i$) and chromospheric activity measurements (H_α) that indicate the existence of fast rotators in the Hyades at spectral types where also the fraction of stars with H_α emission shows a rapid increase ("H_α limit"). The locus of enhanced rotation (and activity) thus seems to be shifted to earlier types in contrast to what is seen as the rotation limit in field stars. The relation between activity and rotation appears to be similar to the one observed in field stars.

1. Introduction

Solar-type stars are mostly fast rotators and magnetically active when they are young. Their magnetic fields drive stellar winds, which rotationally slow-down the star by means of angular momentum transfer. The stellar spin-down over time is empirically quantified by the so-called "Skumanich-law" as $\omega \propto t^{-\frac{1}{2}}$ (Skumanich 1972). This relation, however, becomes invalid at very low masses. Among the field stars, it is observed that at the transition to fully convective stars at early M-type ($\approx 0.3\,M_\odot$), the rotational braking efficiency changes, and fast rotation ($v_{rot} > 3$ km/s) becomes predominant (Delfosse et al. 1998; Mohanty & Basri 2003; Reiners & Basri 2008). The threshold between slow and rapid rotation is thought to be age-dependent (Hawley et al. 1999), so that young cluster stars are expected to show a rotation limit shifted towards higher masses or earlier spectral types, compared to (old) field stars.

Stellar rotation and magnetic activity are tightly linked by the underlying dynamo processes. In (young) clusters, it is observed that the fraction of active stars (eg. with chromospheric H_α emission) sharply increases at different masses depending on the cluster age ("H_α limit", Hawley et al. 1999). At younger age, enhanced magnetic activity is seen at higher masses (ie. earlier spectral types) than it is for older clusters.

However, it is elusive whether this change in the locus of the H_α limit is also due to an increase of the rapid rotation rate (Radick et al. 1987; Stauffer et al. 1987). Previous studies have focussed on the evolution of the H_α limit in young and intermediate age clusters (Stauffer et al. 1997; Hawley et al. 1999; Reid et al. 1995) or field stars (West et al. 2004), but rotational velocities have only been measured extensively for field stars all across the main sequence (Delfosse et al. 1998; Mohanty & Basri 2003; West et al. 2008; Reiners & Basri 2008). However, for open clusters such as the Hyades, $v_{rot} \sin i$ measurements have concentrated on earlier spectral types F to K (eg. Radick et al. 1987), and on the very low-mass regime (M-type and below; eg. Reid & Mahoney 2000), so that in the mid-K to early M-type range (hence in the range of the H_α-limit) rotational velocities are scarce for the Hyades.

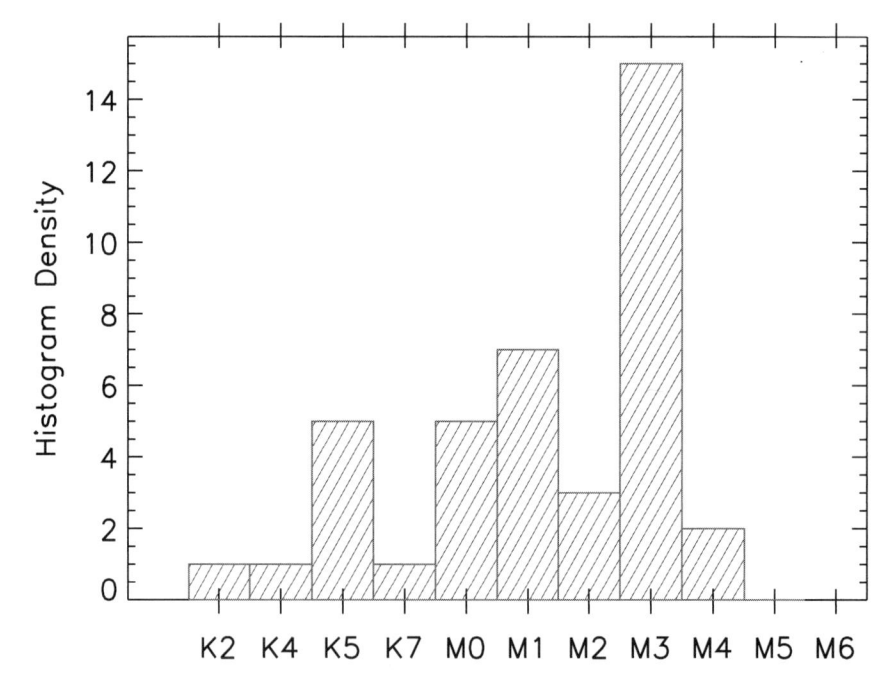

Figure 1. Distribution of spectral types in the Hyades sample. The total sample size is 40. The K-star bins are two spectral sub-types wide.

The present work adresses this scarcity between spectral classes K and M, and probes the coexistence of enhanced activity at the H_α limit and rapid rotation for young stars in the case of the intermediate aged Hyades. We thus aim to determine if rapid rotation occurs at a different threshold in the Hyades with respect to field stars.

2. Sample Selection and Observations

Our sample of low-mass stars comprises 40 members of the Hyades open cluster. Selection is based on color, and proper motions where available from the literature. All stars are drawn from spectral types early K to mid M, and thus bracket both sides of the H_α limit observed for the Hyades. The age of the Hyades open cluster has been es-

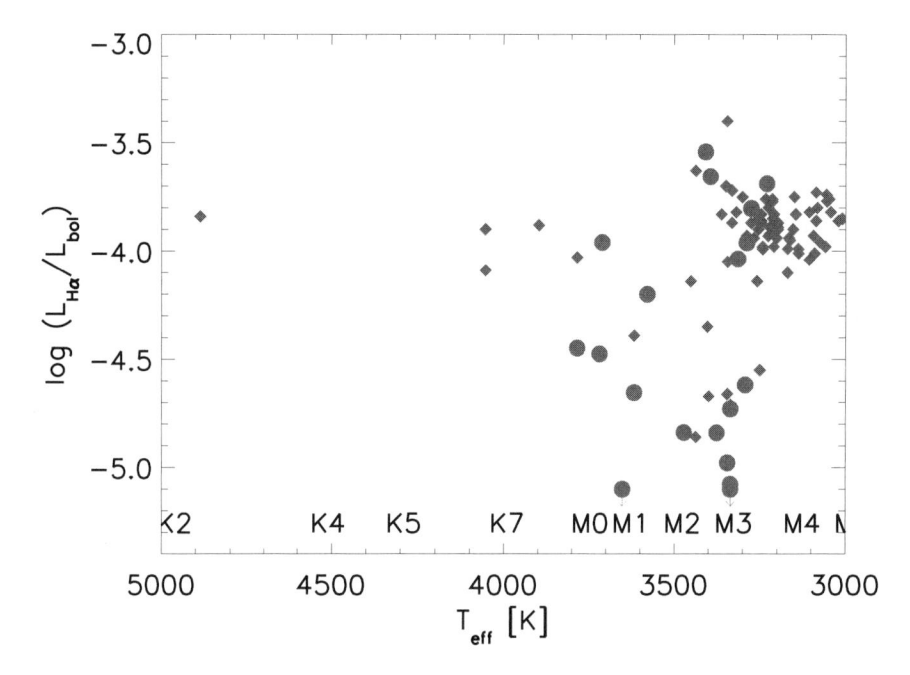

Figure 2. Normalized H$_\alpha$ activity vs. spectral type for our sample Hyads (red points). Data from Reid & Mahoney (2000) is overplotted as diamonds. The combined data indicates that chromospheric magnetic activity (H$_\alpha$ emission) for the Hyades arises at higher masses (M0–M2) than for (old) field stars ($\approx$M3). No predominant activity is seen in K-stars due to magnetic braking (slow rotators, cf. Fig. 3).

timated as 625 ± 50 Myr (Perryman et al. 1998), so that also the later M-type members have already settled on the zero age main sequence and are expected to spin down.

We obtained high-resolution (R = 48 000) optical spectra between 360 nm and 920 nm for all our sample stars using the Fiber-fed Extended Range Optical Spectrograph (FEROS), mounted on the ESO/MPG 2.2 m telescope on La Silla. The signal-to-noise ratio exceeds 60 around 800 nm for the K-type objects. For the M-dwarfs, the SED falls off rapidly towards the blue, so that for these intrinsically faint objects (V = 12..15) we still achieve a signal-to-noise ratio higher than 30 at 800 nm.

3. Data Analysis

Data reduction of the echelle spectra follows standard procedures employing the FEROS pipeline package within MIDAS. From the extracted spectra, H$_\alpha$ (λ656.3 nm) equivalent widths are measured as a proxy of chromospheric activity. We express the H$_\alpha$ emission strength relative to the bolometric luminosity, ie. L_{H_α}/L_{bol} to account for the steeply decreasing luminosity within the spectral range K–M, which would otherwise overestimate H$_\alpha$ emission for earlier spectral types. L_{bol} is computed from synthetic Phoenix spectra (Hauschildt et al. 1999) of the same spectral type.

Rotational velocities $v_{rot} \sin i$ are determined by a cross-correlation technique, similar to the methods used by Browning et al. (2010). For bins of adjacent spectral types, template spectra are constructed from slowly rotating stars of very similar spectral type as that of the sample stars. The template stars were also observed with FEROS to minimize the effects of instrumental profile, and have known $v_{rot} \sin i < 2.5$ km/s, which is our detection limit. We consider their rotation as negligible. The templates are then artificially broadened employing a line-broadening kernel, and cross-correlated against the object spectra to construct a line broadening curve, from which $v_{rot} \sin i$ is derived. Cross-correlation is performed in about 20 selected wavelength-ranges, typically 0.2 – 0.5 nm wide, between 500 – 900 nm that contain moderately deep, isolated photospheric lines. For each object, $v_{rot} \sin i$ is derived from a set of template stars that bracket the object in spectral type.

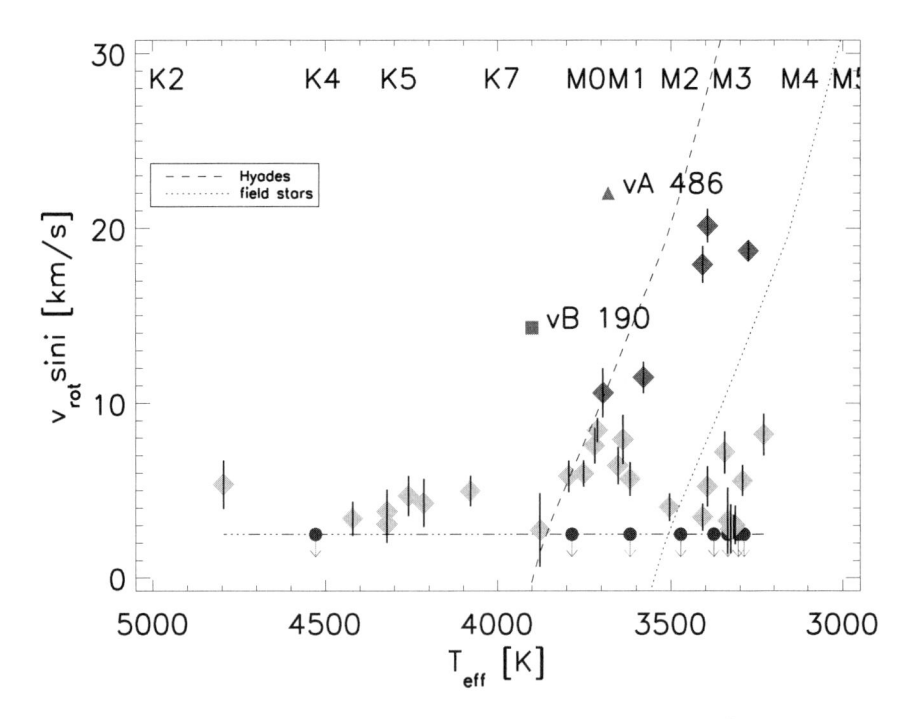

Figure 3. Measured rotational velocities $v_{rot} \sin i$ as a function of temperature (with corresponding spectral types) for the Hyades sample. Non-detections (below our detection limit; blue points), rotators (green diamonds) and rapid rotators (red diamonds) indicate an increase in $v_{rot} \sin i$ towards higher rotation rates in the Hyades around ≈M0 (dashed line for illustration), in contrast to field stars where this threshold kicks in at ≈M3 (dotted line). We thus see activity and rotation in the 625 Myr young Hyades at earlier spectral types than in old (≈5 Gyr) field stars. VA486 (red triangle) from Seifahrt et al. (2009), vB190 (red square) from Radick et al. (1987).

Spectral classes of the sample objects are determined from the photometric $H - K$ color (2MASS, Skrutskie et al. 2006), which gives a more reliable temperature indicator for the K and M-type stars than optical colors. The H and K magnitudes were also measured simultanously, and hence are free of an activity bias that sequentially taken magnitudes might suffer from. The distribution of spectral types covered by the sample is shown in Fig. 1.

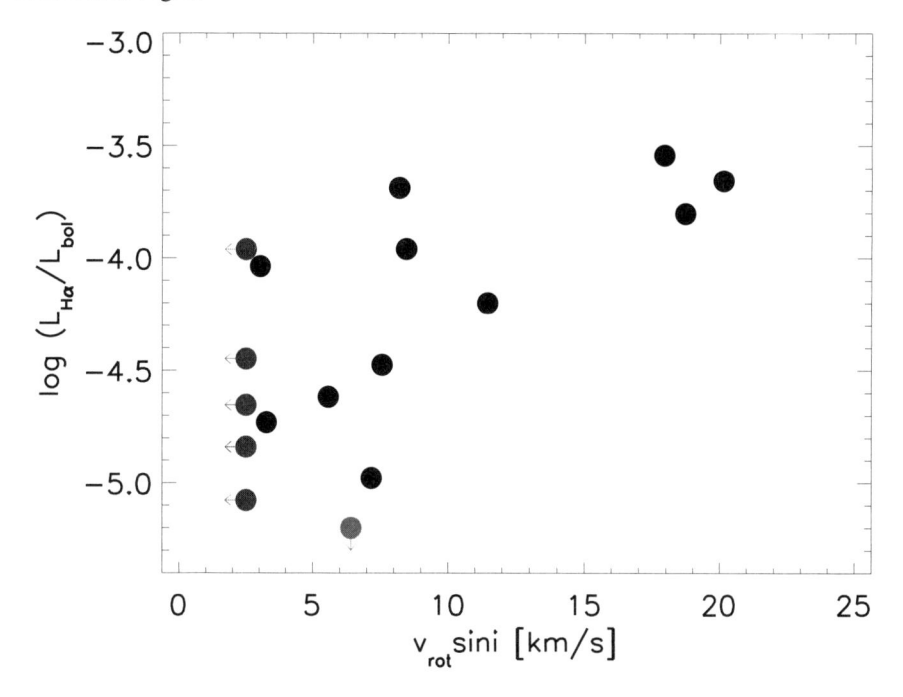

Figure 4. Normalized H_α activity vs. measured $v_{rot} \sin i$ for our sample Hyads (black points; gray and blue points mark upper limits in activity and rotational velocity, respectively). The most active objects are also the most rapid rotators, which is consistent with the rotation-activity relation seen in field stars.

4. Results

4.1. Activity

From the 40 cluster stars, we find significant H_α emission in 18 objects. Their normalized H_α luminosity distribution is plotted in Fig. 2. All of the active stars we find within spectral range M0–M4. The strongest emission is detected at spectral type M3, in agreement with data from Reid & Mahoney (2000), who studied H_α activity in a large sample of mostly M-type Hyades stars. The onset of activity in our data confirms an H_α limit in the Hyades shifted to earlier spectral type when compared to field stars.

4.2. Rotational Velocities

We detect rotation above the detection limit of $v_{rot} \sin i > 2.5$ km/s in 30 stars, spread over all spectral types in the sample. We find rapid rotation with $v_{rot} \sin i > 10$ km/s among the M dwarfs, with an increase in $v_{rot} \sin i$ towards higher rotation rates around M0, while we do not see any fast rotators ($v_{rot} \sin i > 5$ km/s) among the K stars. The latter is likely due to the stronger rotational braking in the K-type regime. We find that

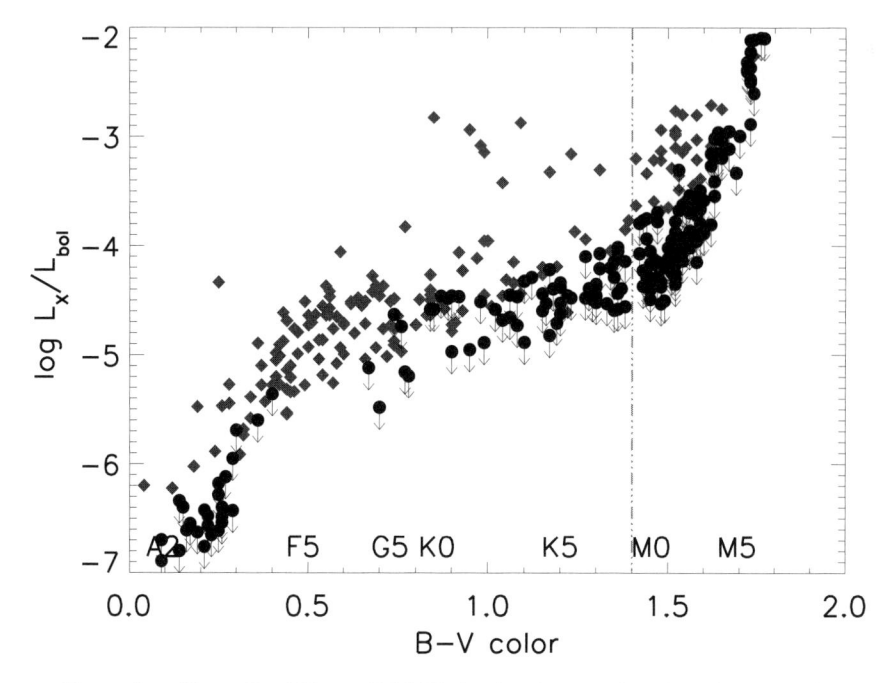

Figure 5. Normalized X-ray (ROSAT) luminosity as a function of (B-V) color (blue diamonds, upper limits as black points; with correspondence to spectral types) shows a strong increase in coronal activity at spectral type ≈M0 (indicated by dashed line) for the Hyades, analogous to chromospheric activity. Data from the catalog in Stern et al. (1995).

the locus of the rotation limit at ≈M0 in the 625 Myr aged Hyades data is shifted to higher masses, compared to the rotation limit at spectral class ≈M3 found in old field stars (see Joshi et al, these proceedings).

Our findings in H_α are in good agreement with previous chromospheric H_α (Hawley & Reid 1994) and coronal activity measurements from X-rays (Fig. 5; Stern et al. 1995; Güdel 2004). In both activity proxies, a threshold in normalized activity luminosity is observed at spectral type ≈M0 in the Hyades. The behaviour in $v_{rot} \sin i$ seems to coincide, suggesting that the rotation-activity relation known for field stars is still in place in a 625 Myr cluster (Fig. 4). This is also supported by recent photometric studies of the Hyades (see Delorme et al, these proceedings) that show a clear drop-off in rotational period at $J - K_s \approx 0.85$ (spectral type M0).

5. Conclusions

Our sample of Hyads shows increased rotation alongside with H_α activity at spectral types ≈M0 and later. We see evidence that young, early M- type stars rotate faster than field stars do. This is possibly influenced by the different contraction timescales present in these young stars over this range of spectral types, giving rise to different braking efficiencies and histories.

Activity and Rotation in the Hyades 319

Acknowledgments. We thank Ulli Käufl and Nandan Joshi for helpful discussions and advice. US and AR acknowledge research funding from the DFG under grant RE 1664/4-1. This research has made use of the Simbad and Vizier databases, operated at CDS, Strasbourg, France, and NASA's Astrophysis Data System Bibliographic Services.

References

Browning, M. K., Basri, G., Marcy, G. W., West, A. A., & Zhang, J. 2010, AJ, 139, 504
Delfosse, X., Forveille, T., Perrier, C., & Mayor, M. 1998, A&A, 331, 581
Güdel, M. 2004, A&A Rev., 12, 71. arXiv:astro-ph/0406661
Hauschildt, P. H., Allard, F., & Baron, E. 1999, ApJ, 512, 377. arXiv:astro-ph/9807286
Hawley, S. L., & Reid, I. N. 1994, in Bulletin of the American Astronomical Society, vol. 26 of Bulletin of the American Astronomical Society, 930
Hawley, S. L., Reid, I. N., Gizis, J. E., & Byrne, P. B. 1999, in Solar and Stellar Activity: Similarities and Differences, edited by C. J. Butler & J. G. Doyle, vol. 158 of Astronomical Society of the Pacific Conference Series, 63
Mohanty, S., & Basri, G. 2003, ApJ, 583, 451. arXiv:astro-ph/0201455
Perryman, M. A. C., Brown, A. G. A., Lebreton, Y., Gomez, A., Turon, C., Cayrel de Strobel, G., Mermilliod, J. C., Robichon, N., Kovalevsky, J., & Crifo, F. 1998, A&A, 331, 81. arXiv:astro-ph/9707253
Radick, R. R., Thompson, D. T., Lockwood, G. W., Duncan, D. K., & Baggett, W. E. 1987, ApJ, 321, 459
Reid, I. N., & Mahoney, S. 2000, MNRAS, 316, 827. arXiv:astro-ph/0007235
Reid, N., Hawley, S. L., & Mateo, M. 1995, MNRAS, 272, 828
Reiners, A., & Basri, G. 2008, ApJ, 684, 1390. 0805.1059
Seifahrt, A., Reiners, A., Scholz, A., & Basri, G. 2009, in American Institute of Physics Conference Series, edited by E. Stempels, vol. 1094 of American Institute of Physics Conference Series, 373. 0811.2485
Skrutskie, M. F., Cutri, R. M., Stiening, R., Weinberg, M. D., Schneider, S., Carpenter, J. M., Beichman, C., Capps, R., Chester, T., Elias, J., Huchra, J., Liebert, J., Lonsdale, C., Monet, D. G., Price, S., Seitzer, P., Jarrett, T., Kirkpatrick, J. D., Gizis, J. E., Howard, E., Evans, T., Fowler, J., Fullmer, L., Hurt, R., Light, R., Kopan, E. L., Marsh, K. A., McCallon, H. L., Tam, R., Van Dyk, S., & Wheelock, S. 2006, AJ, 131, 1163
Skumanich, A. 1972, ApJ, 171, 565
Stauffer, J. R., Hartmann, L. W., & Latham, D. W. 1987, ApJ, 320, L51
Stauffer, J. R., Hartmann, L. W., Prosser, C. F., Randich, S., Balachandran, S., Patten, B. M., Simon, T., & Giampapa, M. 1997, ApJ, 479, 776
Stern, R. A., Schmitt, J. H. M. M., & Kahabka, P. T. 1995, ApJ, 448, 683
West, A. A., Hawley, S. L., Bochanski, J. J., Covey, K. R., Reid, I. N., Dhital, S., Hilton, E. J., & Masuda, M. 2008, AJ, 135, 785. 0712.1590
West, A. A., Hawley, S. L., Walkowicz, L. M., Covey, K. R., Silvestri, N. M., Raymond, S. N., Harris, H. C., Munn, J. A., McGehee, P. M., Ivezić, Ž., & Brinkmann, J. 2004, AJ, 128, 426. arXiv:astro-ph/0403486

Part V

Galactic Context

New Surveys for Brown Dwarfs and Their Impact on the IMF

J. Davy Kirkpatrick

Infrared Processing and Analysis Center, California Institute of Technology, Pasadena, CA 91125, USA

Abstract. Although first predicted to exist in the early 1960s, brown dwarfs were not discovered in bulk until decades later by deep, large-area, infrared-capable surveys such as 2MASS, SDSS, and DENIS. Hundreds of examples are now known, enabling the study of brown dwarfs as a population in their own right. Despite these successes, only the warmest brown dwarfs have so far been identified. The coolest brown dwarfs currently known are field late-T dwarfs with $T_{eff} \approx 500 - 600K$ and implied masses of around 5-35 M_{Jup} for assumed ages of 1-10 Gyr. Foremost is the question of what cooler objects will look like spectroscopically and whether a new spectral class beyond T, dubbed "Y", will be needed. These cooler field brown dwarfs must exist, as studies of young star formation regions have revealed objects even lower in mass, which, at the age of the field population, will have cooled to temperatures well below 500K. Finding and characterizing such cold objects will set important boundary conditions on the shape of the initial mass function at the lowest masses and determine what the low-mass cut-off for star formation is. Here I highlight discoveries from the latest generation of brown dwarf surveys – UKIDSS, CFBDS, and the recently launched WISE – and discuss their impact on our understanding of the field mass function at very low masses.

1. Why Study Brown Dwarfs?

Brown dwarfs are objects formed during the star formation process but too low in mass to ignite sustained nuclear burning. Although fascinating to explore in their own right, brown dwarfs enable us to address several "Big Picture" issues of relevance to researchers in other fields:

First, field brown dwarfs provide valuable insights into the efficiency of star formation. For convenience, the shape of the mass function (MF) is usually described as a power law, $\Psi(M) \propto M^{-\alpha}$, where M is mass and α is the measure of efficiency, larger values of α indicating more low mass objects being formed relative to their higher mass counterparts. Figure 5 of Burgasser (2004), which simulates the way in which various forms of the initial mass function (IMF) would manifest themselves observationally in a Solar Neighborhood sample, demonstrates that the efficiency of star formation at low masses is most easily measurable at the temperatures below 1,000K. The makeup of the simulated observational sample is also critically dependent on the value of the low-mass cut-off for star formation. Empirical evidence from young clusters suggests masses as low as $\sim 10 M_{Jup}$ can be formed; theory has suggested several mechanisms to produce objects this low in mass, among them turbulent fragmentation, magnetic field confinement of low-mass cores, dynamical ejection of stellar embryos, and photo-evaporation of cores by hot stars. Figure 12 of Burgasser (2004), which shows another set of simulations where the low-mass cut-off is varied between 1 and 15 M_{Jup}, demonstrates that this cut-off formation mass is measurable only at extremely low temperatures, $T_{eff} < 400K$, in a sample of old, field brown dwarfs.

Second, brown dwarfs can be thought of as time capsules across the age of the Galaxy. These objects represent fossilized records of star formation at all Galactic epochs. Their mass, unlike that of high-mass main sequence stars, is never ejected back into the insterstellar medium. Because brown dwarfs of lowest mass never even have a period of heavy hydrogen burning, the material from which they formed is never further enriched. Hence, the objects preserve information on Galactic metallicity enrichment at the time of their births.

Third, brown dwarfs show what low-temperature atmospheres look like. Because they are often found in isolation, brown dwarfs are much easier to study than planets circling host stars, and hence they serve as valuable laboratories for low-temperature atmospheric physics, such as that found in gas giants (e.g., Marois et al. 2008). A gap of ~400K still exists between the temperature of the coldest known brown dwarf and that of the planet Jupiter. One open issue is whether or not the near-infrared spectral morphology of brown dwarfs in the $100K < T_{eff} < 500K$ range changes markedly enough from the methane-dominant T dwarfs now known, perhaps because of the appearance of ammonia shortward of 2.5 μm, to warrant another spectral class. This class has tentatively been denoted as "Y" (Kirkpatrick 2003).

Fourth, brown dwarfs may be some of our closest neighbors in space. Outside of the objects in our own Solar System, these brown dwarfs are also the nearest, brightest examples of cool atmospheres that we have to study. Circumstellar disks have been observed around many brown dwarfs (e.g., Apai et al. 2005), raising the intriguing possibility that the nearest brown dwarfs may host some of the closest planetary systems to our own.

2. Ongoing Surveys for Cold Brown Dwarfs

The first generation of large-area surveys responsible for detecting vast numbers of brown dwarfs in the Solar Neighborhood consisted of the Two-Micron All-Sky Survey (2MASS; Skrutskie et al. 2006), the Sloan Digital Sky Survey (SDSS; York et al. 2000), and the Deep Near-Infrared Survey of the Southern Sky (DENIS; Epchtein et al. 1997). Follow-up of candidates selected from these data sets revealed many hundreds of L dwarfs, with 2MASS and SDSS also enabling the discovery of ~100 T dwarfs. (See DwarfArchives.org for a list of known L and T dwarfs.) Despite this success, the survey depth even of 2MASS meant that the volume of space surveyed was too small to detect more than two T8 dwarfs, which were the coldest objects discovered until 2007. In order to address many of the "Big Picture" topics discussed above, colder objects were required, and thus a more sensitive set of second generation surveys was needed.

2.1. UKIDSS

The United Kingdom Infrared Deep Sky Survey (UKIDSS) uses the Wide Field Camera on the United Kingdom Infrared Telescope (UKIRT) on Mauna Kea, Hawai'i, to survey large tracts of sky at Y, J, H, and K bands. An overview of the goals of the brown dwarf portion of the survey can be found in Warren et al. (2007). There are several sub-surveys that comprise UKIDSS, the one best suited to brown dwarf hunting being the Large Area Survey. (Lucas et al. in this volume present the discovery of a nearby brown dwarf in another of the UKIDSS sub-surveys, the Galactic Plane Survey.)

Burningham et al. (2010) describe the survey selection and results thus far. Roughly 700 square degrees of the Large Area Survey has been searched to date. The selection of brown dwarf candidates imposes photometric criteria of $J < 19.3$ mag (essentially the completeness limit of the survey), $J - H < 0.1$ mag, $J - K < 0.1$ mag, and $Y - J > 0.5$ mag. To cull contaminants, overlap with the SDSS is essential, and a further constraint of $z' - J > 3.0$ mag is also imposed. Figure 1 of Burningham et al. (2010) shows the rationale behind the near-infrared color selections, which are driven by the loci in color space where atmospheric models predict cold brown dwarfs to fall.

The UKIDSS brown dwarf survey has been successful in finding brown dwarfs later in type and cooler than those discovered previously. Among these successes are the objects ULAS J003402.77−005206.7 (type T9, temperature estimate of $630K < T_{eff} < 690K$; Warren et al. 2007), Wolf 940B (T8.5, $545K < T_{eff} < 595K$; Burningham et al. 2009), and ULAS J133553.45+113005.2 (T9, $550K < T_{eff} < 600K$; Burningham et al. 2008). The implications of these discoveries will be discussed below.

2.2. CFBDS

The Canada-France Brown Dwarf Survey (CFBDS) uses MegaCam on the Canada-France-Hawai'i Telescope (CFHT) on Mauna Kea, Hawai'i, to survey large tracts of sky at i' and z' bands. An overview of the survey can be found in Delorme et al. (2008b).

Reylé et al. (2010) describe the current state of the survey. Most of the 780 square degree area has been searched fully using brown dwarf selection criteria of $z' < 22.5$ mag (AB system) and $i' - z' > 1.7$ mag. Candidates thus identified are then followed up with J-band imaging to cull contaminant objects. A plot of the $z' - J$ vs. $i' - z'$ color plane can be found as Figure 1 of Reylé et al. (2010). An additional survey, the CFBDS-Infrared (CFBDSIR), aims to acquire complete J-band imaging of approximately 335 square degrees so that individual J-band follow-up is no longer needed. Delorme et al. (2010) report results from the first 66 square degrees of the CFBDSIR.

As with the UKIDSS brown dwarf searches, the CFBDS has been successful in finding objects cooler that the T dwarfs previously identified. These cold objects are CFBDSIR J145829+101343 (T8+, $550K < T_{eff} < 600K$; Delorme et al. 2010) and CFBDS J005910.90−011401.3 (T9, $T_{eff} \approx 620K$; Delorme et al. 2008a). The implications of these discoveries will be discussed below.

2.3. WISE

The Wide-field Infrared Survey Explorer (WISE) is an Earth-orbiting NASA mission to survey the entire sky at wavelengths of 3.4, 4.6, 12, and 22 μm, hereafter referred to as bands W1, W2, W3, and W4, respectively. In unconfused regions, the 5σ point source senstivities are 80, 110, 1000, and 6000 μJy at W1, W2, W3, and W4 respectively. An overview of the mission can be found in Wright et al. (2010). The two shortest wavelength bands of WISE were specifically designed for optimum brown dwarf detection, as Figure 2 of Mainzer et al. (2010) shows. The W1 band falls squarely on the deep CH_4 fundamental absorption feature. The W2 band covers a region in the spectrum where, according to atmospheric models, cold brown dwarfs emit excess flux because of a lack of significant absorbers in that wavelength region. As a result, cold brown dwarfs stand out as having large W1-W2 colors, making them easy to identify relative to other astrophysical sources.

WISE completed it first full pass of the sky on 17 Jul 2010 and is now completing a second pass. This second set of observations is critical for identifying brown dwarfs because (a) we can perform a first-pass to second-pass proper motion survey to identify extremely high moving brown dwarfs, independent of preconceived notions regarding their colors, and (b) we can verify that the faintest candidates identified in the first pass are still viable candidates in the second. Because cryogen in the outer, secondary tank depleted on 10 Aug 2010, and cryogen in the inner, primary tank depleted on 30 Sep 2010, the second full pass of the sky is partly missing bands W4 and W3, which saturated at secondary and primary tank exhaustion, respectively. However, W1 and W2 were little affected and those are the bands most crucial for brown dwarf selection. WISE is scheduled to continue its data collection until 31 Jan 2011.

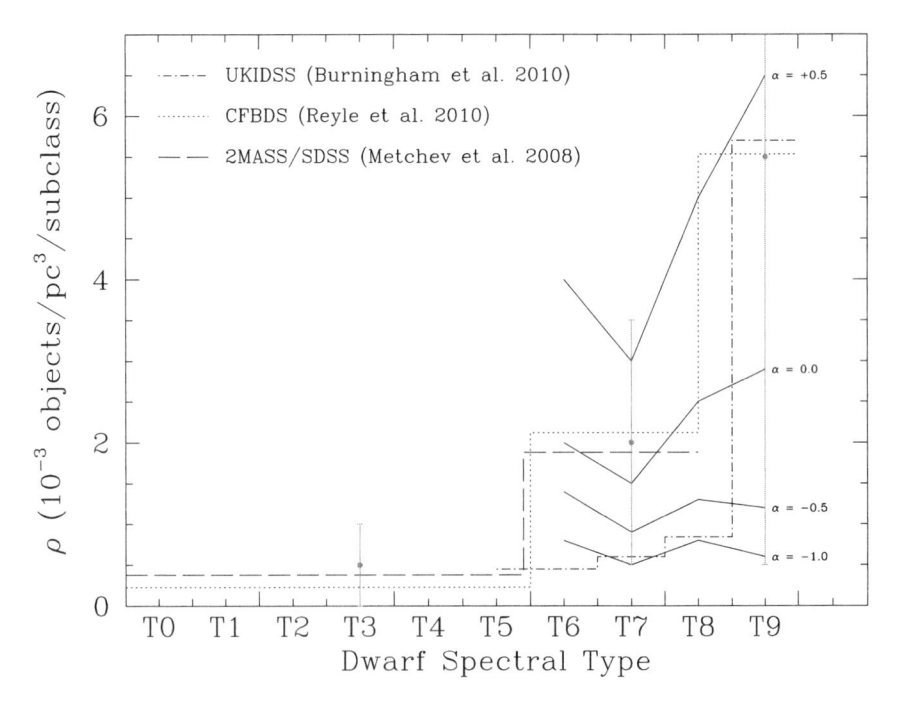

Figure 1. Space density of T dwarfs as a function of T spectral type for three recent measurements: UKIDSS (dash-dotted line; Burningham et al. 2010), CFBDS (dotted line; Reylé et al. 2010), and 2MASS/SDSS (dashed line; Metchev et al. 2008). Shown by the solid lines are averages of the simluations from Burningham et al. (2010) showing how various forms of the mass function, with α values ranging from -1.0 to +0.5, would manifest themselves on this same observational diagram. Typical error bars for the measured space density at three representative spectral type bins are shown in grey. At the latest types, the space density measurements have very large uncertainties because of a paucity of objects and small volumes probed.

3. Our Current Understanding of the Mass Function at Low Masses

Preliminary results from UKIDSS (Burningham et al. 2010) and CFBDS (Reylé et al. 2010) are summarized in Figure 1, along with a comparison of the 2MASS/SDSS results of Metchev et al. (2008). By simulating the star formation history of the Solar Neighborhood sample using various assumed forms of the mass function and translating the results into an observational diagram of space density vs. spectral type, Burningham et al. (2010) conclude that values of $\alpha \approx 0$ or lower are preferred. Nonetheless, at the latest types, T7 - T9, the error bars on the observed space density are still enormous, so values of α as large as 0.5 are not completely ruled out.

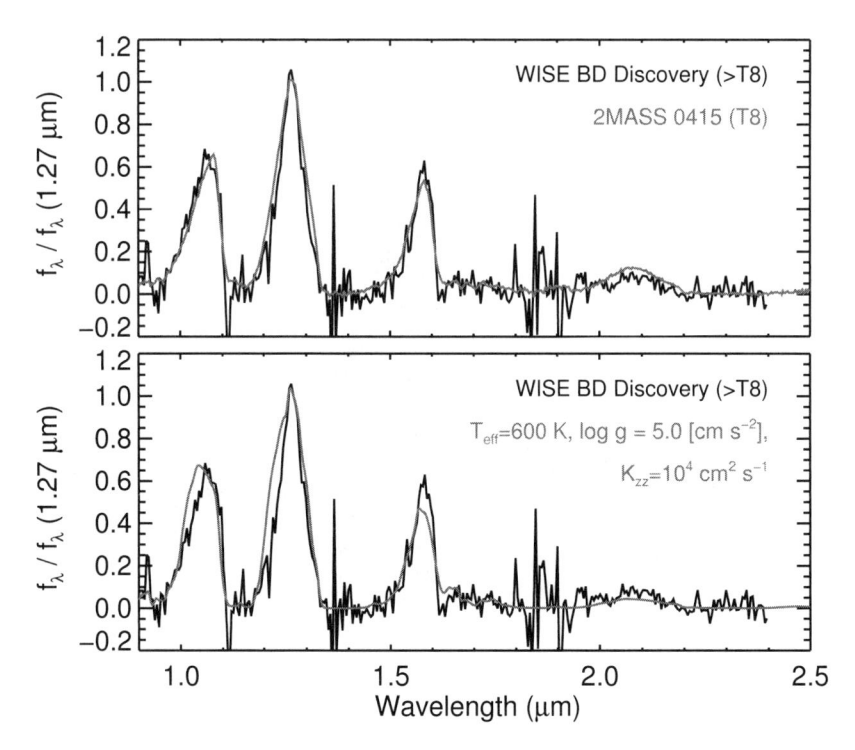

Figure 2. Follow-up spectrum (black) of a late-type WISE discovery compared with the T8 spectral standard (upper panel, grey; Burgasser et al. 2006). The narrow J-band peak of the WISE object suggests a spectral type later than T8. The observed spectrum is also compared to a 600K model (lower panel, grey) from Marley et al. (2002) and Saumon & Marley (2008). The spectrum of the WISE source was taken with the Folded-port Infrared Echellete (FIRE) on Magellan; for more on this instrument, see the contribution in this volume from Burgasser et al.

4. Preliminary Results from WISE

As Figure 1 demonstrates, the space density of late-T dwarfs is still poorly constrained. Prior to WISE, only seven T dwarfs with types >T8 were known, a significantly larger sample (and one with even cooler objects in it) being needed to determine the MF with more confidence. Uncovering these larger samples is one of the main reasons why WISE was launched. As explained in Wright et al. (2010), predictions from the Spitzer survey of Eisenhardt et al. (2010) can be used to predict the numbers of brown dwarfs that WISE will find. Those results showed that ~400 T (and Y) dwarfs are predicted with $T_{eff} < 750K$ assuming a W2 flux limit of 160 μJy; WISE's achieved 5σ sensitivity on orbit is 110 μJy at W2, so this constraint is very conservative. Within a narrower range, $T_{eff} < 500K$, a few dozen WISE objects are expected. In the ultra-cold regime, $T_{eff} < 300K$, only a few objects are expected, although this prediction is highly dependent on the assumed low-mass cut-off for star formation, which Wright et al. (2010) took to be 1 M_{Jup}.

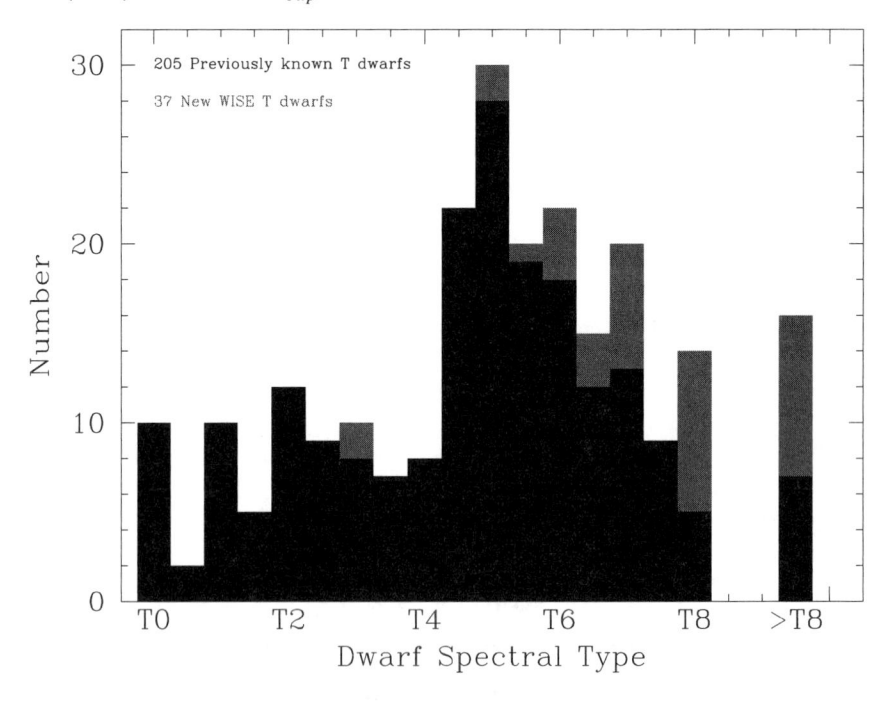

Figure 3. Distribution of the number of T dwarfs known prior to WISE (black) and new additions from WISE (grey) as a function of spectral type. Note that WISE has already more than doubled the number of T dwarfs known with types later than T8.

Follow-up of WISE candidates has already begun. The first ultra-cool T dwarf discovered by WISE was WISEPC J045853.90+643451.9 from Mainzer et al. (2010). This object is tentatively typed as T9, is estimated to lie at a distance of ~9 pc, and has been confirmed as a binary based on high-resolution, adaptive-optics imaging (see Gelino et al. paper, this volume). Other discoveries, some later in type than the Mainzer et al. (2010) object, are being written up by Burgasser et al. (in prep.) and Kirkpatrick et al. (in prep.). An example of one of these late-type WISE discoveries is shown in Figure 2.

WISE follow-up is still in its early stages. The summary listed below, which represents a snapshot of results available at the time of the Cool Stars 16 meeting in early

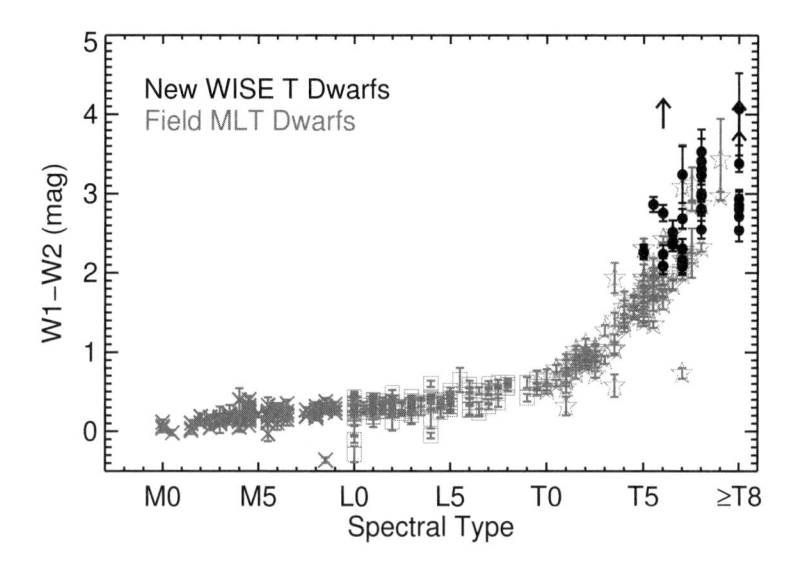

Figure 4. W1-W2 color as a function of spectral type for known M dwarfs (grey x's), L dwarfs (open grey squares), and T dwarfs (open grey stars). Also shown are the W1-W2 colors and spectral types of new WISE discoveries (solid black dots). Objects with color limits are denoted with arrows.

Sep 2010, are very encouraging. Figure 3 illustrates the spectral type distribution of WISE discoveries for which spectroscopic verification has been acquired. WISE has already more than doubled the number of brown dwarfs known with spectral types later than T8. Figures that follow demonstrate where these discoveries fall in color space.

Figure 4 shows the trend of W1-W2 color vs. spectral type for known M, L, and T dwarfs, along with the colors of some new, late-T dwarfs uncovered by WISE. Figure 5 shows these same objects on the W1-W2 vs. W2-W3 color plane. Although there are few astrophysical contaminants at colors W1-W2 > 2.0 mag, imposing an additional color cut of W2-W3 < 2.5 mag for earlier T dwarfs with colors of 1.0 mag < W1-W2 < 2.0 mag is beneficial, as shown in Figure 12 of Wright et al. (2010), as this removes most extragalactic sources.

Figure 6 shows an H-W2 vs. W1-W2 color-color plot. Note that some of the spectroscopically confirmed objects from WISE have H-W2 colors approaching 6.0 mag, and W1-W2 colors exceeding 4.0 mag. Other WISE candidates lacking spectroscopic follow-up have redder colors; one has extreme colors of H-W2 ≈ 8.5 and W1-W2 ≈ 4.3.

5. Conclusions

Three surveys – UKIDSS, CFBDS, and WISE – have made significant progress recently in uncovering colder brown dwarfs than those identified by the previous generation of surveys such as 2MASS, SDSS, and DENIS. Although we have yet to push significantly

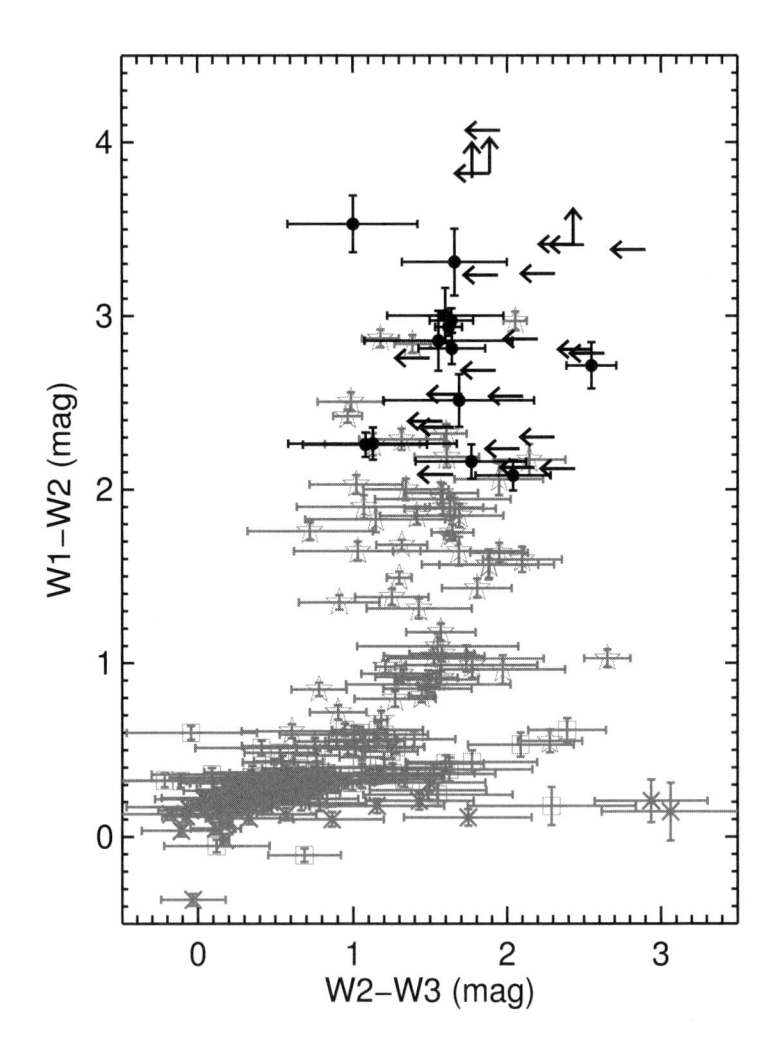

Figure 5. The W1-W2 vs. W2-W3 color plane. Symbols are the same as in the previous figure.

colder than $T_{eff} \approx 500K$ or to have positively identified an object we would call a bona fide Y dwarf, the outlook from WISE is very promising. Candidates with colors much more extreme than the current collection of >T8 dwarfs have already been identified, and follow-up has begun in earnest.

Acknowledgments. The author would like to thank the members of the extended WISE team for the years of work invested to make the mission a success. He would like to thank especially his indefatigable colleagues on the WISE brown dwarf team, without whose dedication the results presented in this paper would have been impossible: Mike Cushing, Peter Eisenhardt, Chris Gelino, Roger Griffith, Amy Mainzer,

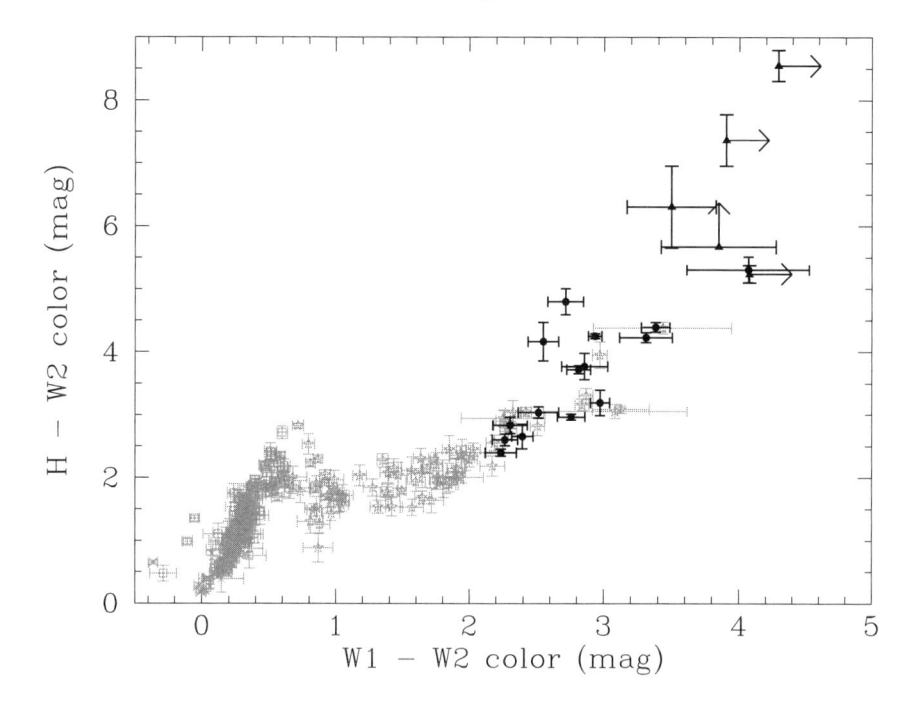

Figure 6. The H-W2 vs. W1-W2 color plane. Color coding is the same as in the previous two figures. In addition to the spectroscopically confirmed WISE discoveries (solid black dots), selected WISE candidates lacking spectra but having H-band follow-up are shown (solid black triangles) to demonstrate that candidates with even more extreme colors are being uncovered.

Mike Skrutskie, Maggie Thompson, and Ned Wright. This publication makes use of data products from the Wide-field Infrared Survey Explorer, which is a joint project of the University of California, Los Angeles, and the Jet Propulsion Laboratory/California Institute of Technology, funded by the National Aeronautics and Space Administration.

References

Apai, D., Pascucci, I., Bouwman, J., Natta, A., Henning, T., & Dullemond, C. P. 2005, Science, 310, 834
Burgasser, A. J. 2004, ApJS, 155, 191
Burgasser, A. J., Geballe, T. R., Leggett, S. K., Kirkpatrick, J. D., & Golimowski, D. A. 2006, ApJ, 637, 1067
Burningham, B., Pinfield, D. J., Leggett, S. K., Tamura, M., Lucas, P. W., Homeier, D., Day-Jones, A., Jones, H. R. A., Clarke, J. R. A., Ishii, M., Kuzuhara, M., Lodieu, N., Zapatero Osorio, M. R., Venemans, B. P., Mortlock, D. J., Barrado Y Navascués, D., Martin, E. L., & Magazzù, A. 2008, MNRAS, 391, 320
Burningham, B., et al. 2009, MNRAS, 395, 1237
— 2010, MNRAS, 406, 1885
Delorme, P., et al. 2008a, A&A, 482, 961
— 2008b, A&A, 484, 469
— 2010, A&A, 518, A39+

Eisenhardt, P. R. M., et al. 2010, AJ, 139, 2455
Epchtein, N., et al. 1997, The Messenger, 87, 27
Kirkpatrick, J. D. 2003, in Brown Dwarfs, edited by E. Martín, vol. 211 of IAU Symposium, 497
Mainzer, A., Cushing, M. C., Skrutskie, M., Gelino, C. R., Kirkpatrick, J. D., Jarrett, T., Masci, F., Marley, M., Saumon, D., Wright, E., Beaton, R., Dietrich, M., Eisenhardt, P., Garnavich, P., Kuhn, O., Leisawitz, D., Marsh, K., McLean, I., Padgett, D., & Rueff, K. 2010, ArXiv e-prints. 1011.2279
Marley, M. S., Seager, S., Saumon, D., Lodders, K., Ackerman, A. S., Freedman, R. S., & Fan, X. 2002, ApJ, 568, 335
Marois, C., Macintosh, B., Barman, T., Zuckerman, B., Song, I., Patience, J., Lafrenière, D., & Doyon, R. 2008, Science, 322, 1348
Metchev, S. A., Kirkpatrick, J. D., Berriman, G. B., & Looper, D. 2008, The Astrophysical Journal, 676, 1281. URL http://stacks.iop.org/0004-637X/676/i=2/a=1281
Reylé, C., Delorme, P., Willott, C. J., Albert, L., Delfosse, X., Forveille, T., Artigau, E., Malo, L., Hill, G. J., & Doyon, R. 2010, Astronomy and Astrophysics, 522, A112. URL http://www.aanda.org/10.1051/0004-6361/200913234
Saumon, D., & Marley, M. S. 2008, ApJ, 689, 1327
Skrutskie, M. F., Cutri, R. M., Stiening, R., Weinberg, M. D., Schneider, S., Carpenter, J. M., Beichman, C., Capps, R., Chester, T., Elias, J., Huchra, J., Liebert, J., Lonsdale, C., Monet, D. G., Price, S., Seitzer, P., Jarrett, T., Kirkpatrick, J. D., Gizis, J. E., Howard, E., Evans, T., Fowler, J., Fullmer, L., Hurt, R., Light, R., Kopan, E. L., Marsh, K. A., McCallon, H. L., Tam, R., Van Dyk, S., & Wheelock, S. 2006, AJ, 131, 1163
Warren, S. J., Mortlock, D. J., Leggett, S. K., Pinfield, D. J., Homeier, D., Dye, S., Jameson, R. F., Lodieu, N., Lucas, P. W., Adamson, A. J., Allard, F., Barrado Y Navascués, D., Casali, M., Chiu, K., Hambly, N. C., Hewett, P. C., Hirst, P., Irwin, M. J., Lawrence, A., Liu, M. C., Martín, E. L., Smart, R. L., Valdivielso, L., & Venemans, B. P. 2007, MNRAS, 381, 1400
Wright, E. L., Eisenhardt, P. R. M., Mainzer, A. K., Ressler, M. E., Cutri, R. M., Jarrett, T., Kirkpatrick, J. D., Padgett, D., McMillan, R. S., Skrutskie, M., Stanford, S. A., Cohen, M., Walker, R. G., Mather, J. C., Leisawitz, D., Gautier, T. N., McLean, I., Benford, D., Lonsdale, C. J., Blain, A., Mendez, B., Irace, W. R., Duval, V., Liu, F., Royer, D., Heinrichsen, I., Howard, J., Shannon, M., Kendall, M., Walsh, A. L., Larsen, M., Cardon, J. G., Schick, S., Schwalm, M., Abid, M., Fabinsky, B., Naes, L., & Tsai, C. 2010, AJ, 140, 1868
York, D. G., et al. 2000, AJ, 120, 1579

L Dwarf Kinematics

Sarah J. Schmidt,[1] Andrew A. West,[2] and Suzanne L. Hawley[1]

[1] *University of Washington Department of Astronomy, Box 351580, Seattle, WA 98195-1580*

[2] *Department of Astronomy, Boston University, CAS, 725 Commonwealth Ave, Boston, MA 02215*

Abstract. The L spectral class is composed of both stars and brown dwarfs. Because brown dwarfs cool as they age, it is not possible to assign a mass to an L dwarf based on its T_{eff}/spectral type. Due to this degeneracy between mass, age, and spectral type, it is especially important to determine the ages of L dwarfs. Indirect estimates of age, such as a relationship between age and $J - K_S$ color have been proposed, but a better calibration is needed. Kinematics have proven to be a useful age calibrator, and will likely be instrumental in future studies of L dwarf ages. We examine the differences between the kinematics of a large, magnitude limited sample of L dwarfs from the Sloan Digital Sky Survey, and a smaller, volume-limited sample of L dwarfs within 20pc. Both samples show a kinematically hot component, suggestive of an older population of stars. Additionally, L dwarfs with bluer $J - K_S$ colors have hotter kinematics than redder L dwarfs, indicating a relationship between $J - K_S$ color and age in both samples.

1. The Ages of Field L Dwarfs

The ages of L dwarfs are particularly important due to the spectral class's inclusion of both stars and brown dwarfs. While stars (defined by sustained hydrogen fusion in their cores) have a relatively stable luminosity and T_{eff} (and thus spectral type) as they age, brown dwarfs radiate their heat of formation and rapidly become cooler and dimmer, evolving through spectral types M, L, and T (e.g. Burrows et al. 2001).

Due to this degeneracy between spectral type, mass, and age, L dwarfs of a given spectral type cannot definitively be classified as stars or brown dwarfs. The inability to determine masses also makes it impossible to directly determine the mass function from the L dwarf luminosity function at and beyond the the star/brown dwarf transition (e.g., Allen et al. 2005; Deacon et al. 2008).

Ages have been successfully determined for L dwarfs that are members of open clusters (e.g., Gizis 2002; Bihain et al. 2006; Bannister & Jameson 2007). This method is limited to younger L dwarfs, as the oldest open cluster with known brown dwarf members is the Hyades (625 Myr; Perryman et al. 1998). Age calibration has been extended to young field dwarfs primarily through spectroscopic features sensitive to surface gravity (Kirkpatrick et al. 2008; Cruz et al. 2009). Young brown dwarfs are still contracting, which results in lower surface gravities for younger, more extended objects (Burrows et al. 2001). These gravity-sensitive features have been found to correlate with red $J - K_S$ color, with younger L dwarfs showing redder colors. A relation of

$J - K_S$ color and age has been calibrated using open clusters, but only extends to ~0.7 Gyr (Jameson et al. 2008).

The ages of most field L dwarfs are not known, as they do not have the advantage of association with open clusters, and their contraction essentially halts at 1 Gyr (Burrows et al. 2001) where gravity sensitive features are no longer useful. Recent work has suggested that a population of L dwarfs with anomalously blue $J - K_S$ colors could be associated with old ages (Burgasser et al. 2008) and preliminary kinematic analysis has confirmed the relation between color and age extends to the older field L dwarfs (Faherty et al. 2009; Schmidt et al. 2010), but the color-age relation has not been calibrated past 1 Gyr. In order to reliably determine ages for field L dwarfs, more comprehensive kinematic studies are needed.

2. L Dwarf Kinematics

Kinematics are a powerful tool for investigating the ages of stars (e.g., West et al. 2008) but the small number of L dwarfs with kinematic measurements have prohibited a detailed calibration of the age-kinematics relation. Initial work on L dwarf kinematics was limited to examining the estimated mean ages of the population of objects with measured velocities (Schmidt et al. 2007; Zapatero Osorio et al. 2007). However, in the past few years large surveys have recovered ~800 known L dwarfs, ~700 of which have proper motions (Schmidt et al. 2010; Faherty et al. 2009, and references therein), and ~400 that have radial velocities (Mohanty & Basri 2003; Bailer-Jones 2004; Zapatero Osorio et al. 2007; Blake et al. 2007, 2010; Schmidt et al. 2010; Seifahrt et al. 2010).

Faherty et al. (2009) measured and compiled proper motions for over 800 late-M, L, and T dwarfs. The mean ages of the ultracool dwarfs, derived from their velocity dispersions, were found to be consistent with the disk velocity dispersions. The size of the sample allowed an examination of the mean age of the sample, as well as a preliminary investigation of $J - K_S$ color. L dwarfs with $J - K_S$ color redder than the mean $J - K_S$ color for their spectral type were found to have a low mean velocity and velocity dispersion, while objects with bluer $J - K_S$ colors had a high mean velocity and wide velocity dispersion.

The Sloan Digital Sky Survey (SDSS) 7th data release (Abazajian et al. 2009) included spectra of ~500 L dwarfs, 306 of which had sufficient quality for radial velocity measurements and measurable proper motions from a cross-match of positions from SDSS and the Two-Micron All Sky Survey (Skrutskie et al. 2006). These 306 L dwarfs are the largest existing sample with measured three dimensional velocities (Schmidt et al. 2010). The size of the sample allowed the use of kinematic tools that had previously been used to examine in the velocities of stars, particularly M dwarfs (Reid et al. 1995; Bochanski et al. 2007)

Figure 1 shows a velocity distribution of 1000 fake stars generated to mimic the cold (90% of the population; $\sigma_W = 18$ km s^{-1}) and hot (10% of the population; $\sigma_W = 60$ km s^{-1}) kinematic components of the L dwarf W velocities from Schmidt et al. (2010). The histogram of the combined velocity distribution is nearly indistinguishable from the velocity distribution of the cold component, but the cumulative velocity distribution elucidates the difference between the two components in the combined distribution. The hot and cold components, modeled on gaussian distributions, show up as straight lines. When the two are combined, the distribution strongly devi-

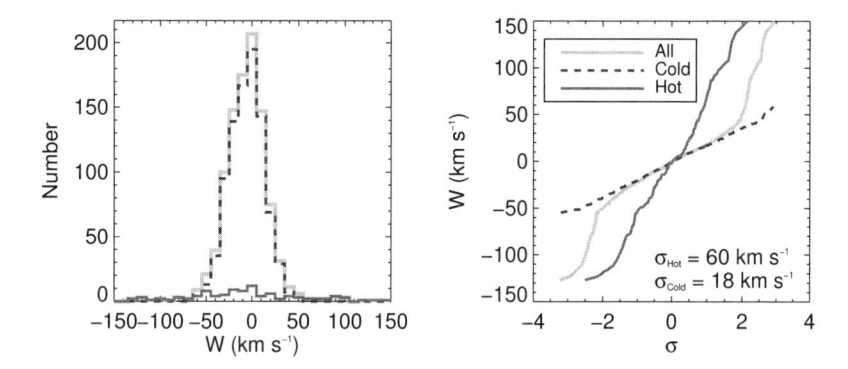

Figure 1. The W velocity histogram (left) and cumulative W velocity distribution (right) of 1000 data points generated to have the velocity dispersions and ratio of the cold (90% of the population; $\sigma_W = 18$ km s^{-1}) and hot (10% of the population; $\sigma_W = 60$ km s^{-1}) kinematic components of the SDSS L dwarfs from Schmidt et al. (2010). Even with a large sample of velocities, it is difficult to distinguish the two components in the histogram. When the data are illustrated in a cumulative distribution (right panel), the cold component (with a shallow slope, corresponding to a narrow distribution) and hot component (with a steep slope corresponding to a wide distribution) are much more easily distinguished.

ates from a straight line with the cold component dominating near the mean and the hot component more apparent at large standard deviations.

3. Hot and Cold Disk Components

While the SDSS sample has the benefit of a large number of L dwarfs, it is effectively a magnitude-limited sample, creating a bias towards earlier (intrinsically brighter) L dwarfs. To examine this bias, we can compare the SDSS L dwarfs to the sample of L dwarfs within 20pc, which is the same distance limit selected to calculate the L dwarf luminosity function (Cruz et al. 2007). All of the L dwarfs within 20pc have measured proper motions, but only half of them have measured radial velocities. We used these measurements to calculate three dimensional space motions for 59 nearby L dwarfs. The spectral type distributions of the SDSS and nearby samples of L dwarfs with measured UVW velocities are shown in Figure 2.

Schmidt et al. (2010) found that the SDSS L dwarfs show hotter (older) and colder (younger) kinematic components tracing a thin and thick disk. This can be seen in the W cumulative velocity distribution, shown in the left panel of Figure 3. The W cumulative velocity distribution for the 20pc sample is shown in the right panel of Figure 3. For each distribution, a dashed line shows a linear fit to the data closest to the mean ($-1 < \sigma < 1$). The data near the mean is dominated by the cold component, while the hot component is evidenced by the deviations from the fit at $\sigma < -1.5$ and $\sigma > 1.5$.

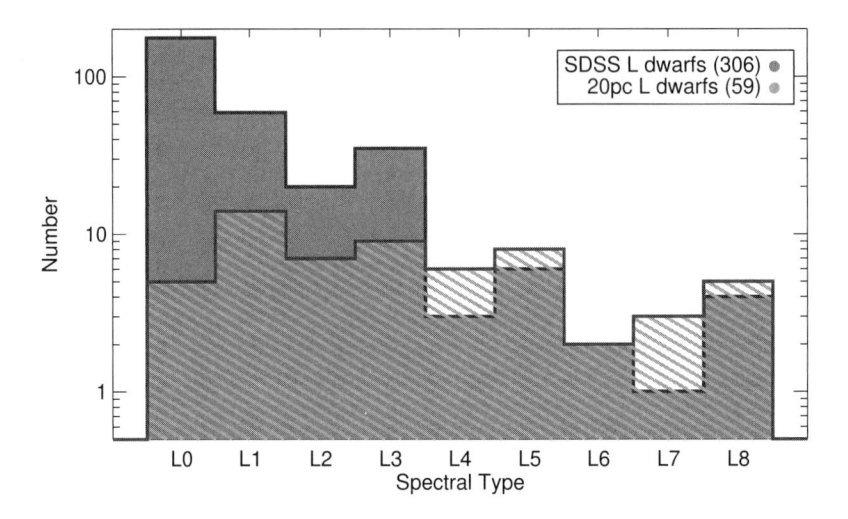

Figure 2. The spectral type distributions of the two samples of L dwarfs with measured UVW motions. While the SDSS sample (solid blue) has five times as many total L dwarfs, the 20pc sample (shaded green) has a larger number of L4-L8 dwarfs. This is due to the magnitude limit of the SDSS sample, which favors detection of brighter, earlier spectral type L dwarfs.

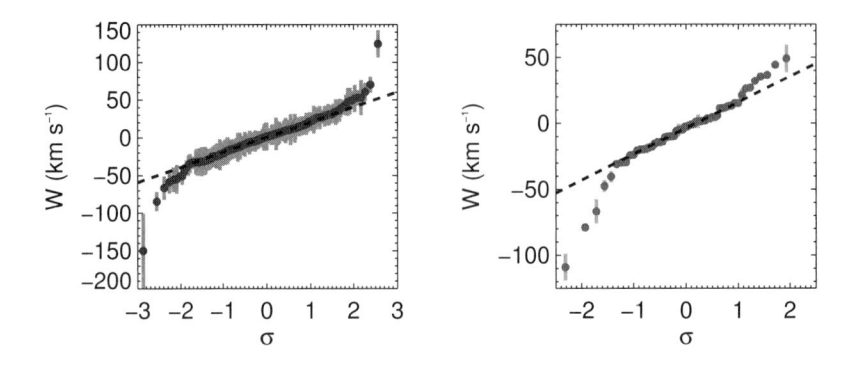

Figure 3. Cumulative velocity distributions of the W velocity data from SDSS (left panel) and the volume limited 20pc sample (right panel). The circles with error bars show the velocities, and the dashed lines show the best linear fit to the points between $-1 < \sigma < 1$, which is dominated by the cold kinematic component. The hot kinematic component can be seen in the deviations from the fit at $\sigma < -1.5$ and $\sigma > 1.5$.

4. $J - K_S$ Color and Age

We next used the kinematic results to examine the relation between $J-K_S$ color and age. In order to examine the entire L dwarf sample, we defined a spectral type independent

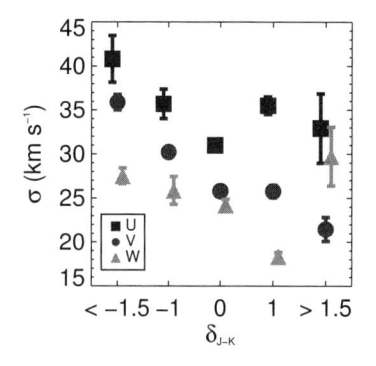

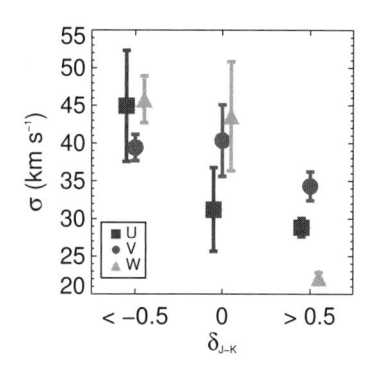

Figure 4. Velocity dispersion in U, V, and W as a function of δ_{J-K}. The velocity dispersions are based on a linear fit to the cumulative velocity dispersions of each of the bins in δ_{J-K} with uncertainties based on the uncertainties in the linear fit. In both the SDSS sample (left) and the volume limited 20pc sample (right), the velocity dispersions are larger for bluer $J - K_S$ colors and smaller for redder $J - K_S$ colors, which shows that bluer L dwarfs are older than redder L dwarfs. Additional data are needed to reduce the uncertainties in the 20pc sample and more conclusively investigate this trend.

color parameter, δ_{J-K}, which measures how many standard deviations an individual L dwarf's $J - K_S$ color is from the mean $J - K_S$ color of its spectral type (Schmidt et al. 2010). The two samples were then divided into bins of $\delta_{J-K} = 1$, resulting in ~20-150 objects in each bin. The U, V, and W velocity dispersions of each color bin were measured using a linear fit to the cumulative velocity distributions, measuring the cold (dominant) component.

Figure 4 shows the UVW velocity dispersions as a function of δ_{J-K} for the SDSS and 20pc L dwarf samples. The error bars shown are the uncertainties in the linear fit to each distribution (some error bars are smaller than the symbols). Generally, the bluer objects have larger velocity dispersions which are consistent with old ages, and the redder objects have smaller velocity dispersions. Not only are bluer objects kinematically older, but a statistically significant correlation between kinematic age and $J - K_S$ color was found for the SDSS sample. The results for the nearby sample are less conclusive due to the small sample size and therefore large uncertainty in the measurements. More data are needed to examine the relation between $J - K_S$ color and age over a larger range of spectral types than the early-L dwarfs that dominate the SDSS sample.

References

Abazajian, K., Adelman-McCarthy, J., Agueros, M., & Allam, S. 2009, ApJS, 182, 543
Allen, P. R., Koerner, D. W., Reid, I. N., & Trilling, D. E. 2005, ApJ, 625, 385
Bailer-Jones, C. A. L. 2004, A&A, 419, 703
Bannister, N. P., & Jameson, R. F. 2007, MNRAS, 378, L24
Bihain, G., Rebolo, R., Béjar, V. J. S., Caballero, J. A., Bailer-Jones, C. A. L., Mundt, R., Acosta-Pulido, J. A., & Manchado Torres, A. 2006, A&A, 458, 805
Blake, C. H., Charbonneau, D., & White, R. J. 2010, ApJ, 723, 684. 1008.3874

Blake, C. H., Charbonneau, D., White, R. J., Marley, M. S., & Saumon, D. 2007, ApJ, 666, 1198
Bochanski, J. J., Munn, J. A., Hawley, S. L., West, A. A., Covey, K. R., & Schneider, D. P. 2007, AJ, 134, 2418
Burgasser, A. J., Looper, D. L., Kirkpatrick, J. D., Cruz, K. L., & Swift, B. J. 2008, ApJ, 674, 451
Burrows, A., Hubbard, W. B., Lunine, J. I., & Liebert, J. 2001, Reviews of Modern Physics, 73, 719
Cruz, K. L., Kirkpatrick, J. D., & Burgasser, A. J. 2009, AJ, 137, 3345
Cruz, K. L., Reid, I. N., Kirkpatrick, J. D., Burgasser, A. J., Liebert, J., Solomon, A. R., Schmidt, S. J., Allen, P. R., Hawley, S. L., & Covey, K. R. 2007, AJ, 133, 439
Deacon, N. R., Nelemans, G., & Hambly, N. C. 2008, Astronomy and Astrophysics, 486, 283. URL http://www.aanda.org/10.1051/0004-6361:200809672
Faherty, J. K., Burgasser, A. J., Cruz, K. L., Shara, M. M., Walter, F. M., & Gelino, C. R. 2009, The Astronomical Journal, 137, 1. URL http://stacks.iop.org/1538-3881/137/i=1/a=1?key=crossref.0a38b8a7e1a1f71ed968384ee927da6b
Gizis, J. E. 2002, ApJ, 575, 484
Jameson, R. F., Lodieu, N., Casewell, S. L., Bannister, N. P., & Dobbie, P. D. 2008, MNRAS, 385, 1771
Kirkpatrick, J. D., et al. 2008, ApJ, 689, 1295
Mohanty, S., & Basri, G. 2003, ApJ, 583, 451. arXiv:astro-ph/0201455
Perryman, M. A. C., Brown, A. G. A., Lebreton, Y., Gomez, A., Turon, C., Cayrel de Strobel, G., Mermilliod, J. C., Robichon, N., Kovalevsky, J., & Crifo, F. 1998, A&A, 331, 81. arXiv:astro-ph/9707253
Reid, I. N., Hawley, S. L., & Gizis, J. E. 1995, AJ, 110, 1838
Schmidt, S. J., Cruz, K. L., Bongiorno, B. J., Liebert, J., & Reid, I. N. 2007, AJ, 133, 2258
Schmidt, S. J., West, A. A., Hawley, S. L., & Pineda, J. S. 2010, ArXiv e-prints. 1001.3402
Seifahrt, A., Reiners, A., Almaghrbi, K. A. M., & Basri, G. 2010, ArXiv e-prints. 1001.1780
Skrutskie, M. F., et al. 2006, AJ, 131, 1163
West, A. A., Hawley, S. L., Bochanski, J. J., Covey, K. R., Reid, I. N., Dhital, S., Hilton, E. J., & Masuda, M. 2008, AJ, 135, 785. 0712.1590
Zapatero Osorio, M. R., Martín, E. L., Béjar, V. J. S., Bouy, H., Deshpande, R., & Wainscoat, R. J. 2007, ApJ, 666, 1205

A Very Cool, Very Nearby Brown Dwarf Hiding in the Galactic Plane

P.W. Lucas,[1] C.G. Tinney,[2] Ben Burningham,[1] S.K. Leggett,[3] David J. Pinfield,[1] Richard Smart,[4] Hugh R.A. Jones,[1] Federico Marocco,[4] Robert J. Barber,[5] Sergei N. Yurchenko,[6] Jonathan Tennyson,[5] Miki Ishii,[7] Motohide Tamura,[8] Avril C. Day-Jones,[9] Andrew Adamson,[10,3] France Allard,[11] and Derek Homeier[12]

[1] *Centre for Astrophysics Research, University of Hertfordshire, College Lane, Hatfield AL10 9AB, UK*

[2] *Dept of Astrophysics, University of New South Wales, 2052, Australia*

[3] *Gemini Observatory, Northern Operations Centre, 670 N.A'ohoku Place, Hilo HI 96720, USA*

[4] *INAF/Osservatorio Astronomico di Torino, Strada Osservatorio 20, 10025 Pino Torinese, Italy*

[5] *Dept of Physics and Astronomy, University College London, London WC1E 6BT, UK*

[6] *Technische Universität Dresden, Institut für Physikalische Chemie und Elektrochemie, D-01062 Dresden, Germany*

[7] *Subaru Telescope, National Astronomical Observatory of Japan, 650 N.A'ohoku Place, Hilo, HI 96720, USA*

[8] *National Astronomical Observatory of Japan, 2-21-1 Osawa, Mitaka, Tokyo 181-8588, Japan*

[9] *Universidad de Chile, Camino el Observatorio, #1515, Las Condes, Santiago, Chile, Casilla 36-D, Chile*

[10] *Joint Astronomy Centre, 660 N.A'ohoku Place, University Park, Hilo, HI 96720, USA*

[11] *CRAL, Universite de Lyon, École Normale Supérieur, 46 allee d'Italie, 69364 Lyon Cedex 07, France*

[12] *Institut für Astrophysik Göttingen, Georg-August-Universität, Friedrich-Hund-Platz 1, D-37077 Göttingen, Germany*

Abstract. The UKIDSS Galactic Plane Survey (GPS) has provided near infrared data on several hundred million sources, most of which is now world public. We report the discovery of a very cool, isolated brown dwarf, UGPS 0722-05, which was identified as the sole candidate late T dwarf in the 6th Data Release via a simple SQL query, followed by inspection of a handful of images. The near-infrared spectrum UGPS 0722-05 displays deeper H_2O and CH_4 troughs than the coolest T dwarfs previously known, so we provisionally classify it as a T10 dwarf. The distance is measured by trigonometric parallax as $d=4.1^{+0.6}_{-0.5}$ pc, making it the closest known isolated brown dwarf. With the aid of *Spitzer*/IRAC we measure $H-[4.5] = 4.71$, which is redder than all previously known T dwarfs except the peculiar T7.5 dwarf SDSS J1416+13B, which is thought to be warmer and more luminous than UGPS 0722-05. We estimate that UGPS 0722-05 has $T_{eff}=520\pm40$ K. We place this discovery in the context of other recent discoveries from the UKIDSS Large Area Survey and note that there appears to be a deficit of late T dwarfs in the local field relative to predictions based on the IMF measured in young clusters. We comment on possible explanations for this.

1. Introduction

In recent years several brown dwarfs have been discovered in the local field that are cooler than 2MASS 0415-0935, the standard for the T8 spectral. In Table 1 we provide a list of these discoveries. Most of them were found in the UKIDSS surveys Lawrence et al. (2007), but two have come from the Canada-France Brown Dwarfs Survey. A large number of new discoveries are expected to be reported soon from the WISE survey (see article by Kirkpatrick in this volume) and from other new facilities such as VISTA and PanStarrs.

Table 1. Brown dwarfs cooler than the T8 standard 2MASS 0415-0935

Name	Spectral Type	Reference	Comment
ULAS 0034-0052	T9	1	Isolated
CFBDS 0059-0114	T9	2	Isolated
ULAS 1335+1130	T9	3	Isolated
ULAS 1238+0953	T8.5	3	Isolated
Wolf 940 B	T8.5	4	Companion to mature M dwarf
ULAS 1302+1308	T8.5	5	Isolated
CFBDSIR 1458+1013	T8.5±0.5	6	Isolated
Ross 458 C	T8/T8.5	7	Companion to young M dwarf binary
SDSS 1416+1348 B	T7.5p	8	Companion to blue L dwarf
2MASS 0939-2448	T8	9	Isolated
Photometric Candidates			
GJ 758 B	T8.5?	10,11	Likely companion to G8 dwarf
SDWFS 1433+3518	?	12	Very red and faint Spitzer source

References: (1) Warren et al. (2007); (2) Delorme et al. (2008); (3) Burningham et al. (2008); (4) Burningham et al. (2009); (5) Burningham et al. (2010a); (6) Delorme et al. (2010); (7) Goldman et al. (2010); (8) Burningham et al. (2010b); (9) Burgasser et al. (2006); (10) Thalmann et al. (2009); (10) Currie et al. (2010); (11) Eisenhardt et al. (2010)

Most of the UKIDSS T dwarfs have been discovered by the Large Area Survey (LAS), e.g. Burningham et al. (2010a), but have also been some discovered in the Galactic Cluster Survey (GCS) (Lodieu et al. 2009a) and some photometric candidates in the Deep Extraglactic Survey (Lodieu et al. 2009b). Little effort had been made to search the Galactic Plane Survey (Lucas et al. 2008) because source confusion exists in a significant fraction of the survey area and this makes colour-selected searches for rare objects quite difficult. For this reason very few brown dwarfs have previously been found in the Galactic plane, e.g. 2MASS 1126-5003 (Folkes et al. 2010) and DENIS 0817-6155 (Artigau et al. 2010). Both of these examples are at distances $d_¡10$ pc and the former has an unusual spectrum. This illustrates the value of searching in the Galactic plane despite the difficulties involved.

2. Candidate Selection

A search of the GPS 6th Data Release identified UGPS 0722-05 as the only candidate late-T or Y dwarf in the catalogue to satisfy the colour criteria $J-H<-0.2$ mag, $H-K<-0.1$ mag. We used several data quality restrictions to minimise the number of false candidates (see L08). The criteria were: *jmhPntErr*<0.3, *hmk_1PntErr*<0.3 (limiting the uncertainty in source colours); *jppErrbits*<256, *hppErrbits*<256, *k_1ppErrbits*<256 (removes sources with photometric quality warnings); *pstar*>0.9 (requires a point-source image profile); *jEll*<0.3, *hEll*<0.3, *k_1Ell*<0.3 (limits on ellipticity); $\sqrt{(hXi^2+hEta^2)}<0.3$, $\sqrt{(k\_1Xi^2+k\_1Eta^2)}<0.3$ (limits on coordinate shifts between passbands).

A further constraint was to limit the search to Galactic longitudes $l > 60°$ and $l < 358°$, which limited the search to 134 million sources. These restrictions were designed to select against blended stellar pairs with inaccurate photometry. Six candidates remained after this procedure. Of these, four were revealed as blended stellar pairs or defective data by inspection of the images and one (a candidate white dwarf) was ruled out by its detection in visible light in the USNO-B1.0 archive. An image of the remaining candidate, UGPS 0722-05, is shown in Fig. 1. The coordinates measured on 2 March 2010 were R.A.= $07^h22^m27.29^s$, Dec.=$-05^d40^m30.0^s$.

Additional late T dwarfs are likely to be found in the GPS as the survey continues and searches are made with less restrictive colour cuts. However the LAS and the GCS are more sensitive to such objects because they include Y band data. Searches based on the YJH filters probe a much larger volume for a given area on the sky (by a factor of 3 to 4) than JHK searches with the GPS.

3. Observations

In figure 2 (upper panel) we show a Gemini/NIRI spectrum of UGPS 0722-05 and the spectrum of a T9 dwarf, which appears fairly similar. The lower panel of figure 2, a ratio plot comparing UGPS 0722-05 with the average of three T9 dwarfs, shows that the broad molecular absorption troughs on either side of the flux peaks at 1.28 μm and 1.59 μm and on the long wavelength side of the 1.07 μm peak are between 10% and 30% deeper. A spectral type was determined by by measuring the depth of the water and methane absorption bands, using the indices that are commonly employed for late T dwarfs (Burningham et al. 2008). The values of these indices are smaller than those of T9 dwarfs by amounts similar to the differences between T8 and T9 dwarfs so the object is provisionally classified as a T10 dwarf. Details of the index values, and fuller details of multiaveband observations, are given in Lucas et al. (2010). There are signs of weak, narrow absorption features $\lambda=1.275$ μm and $\lambda=1.282$ μm, (not shown) which are unidentified at present.

UGPS 0722-05 has J mag = 16.52, which is brighter than any other T dwarf found in UKIDSS at present. Hence it offers opportunities for high resolution spectroscopy

Figure 1. UKIDSS J band discovery image (80″ square). The circles illustrate the proper motion. The centre circle marks the position at the time of the discovery image in 2006, while the circles to the left and right mark the positions at the time of the 2MASS image in 1998 and a UKIRT image from 2 March 2010 respectively.

to search for signs of ammonia, for example. The source is also bright enough for ground-based spectroscopy in the L′ band. With the aid of *Spitzer*/IRAC we measured H-[4.5] = 4.71. Gemini South/T-ReCS was used to measure the mid-infrared flux as N=10.28±0.24.

A preliminary trigonometric parallax has been measured as π=246±33 mas. This solution was calculated using 7 epochs of UKIRT Wide Field Camera data (including the original UKIDSS detection in 2006) and a weak 2MASS detection from 1998, which helped to constrain the proper motion.

4. Physical Properties of UGPS 0722-05 and Comparison with Other Very Cool Objects

UGPS 0722-05 has *(J-H)*=-0.37 and *(H-K)*=-0.18. These colours are similar to those of many T dwarfs with types T5 to T9. Y band data help to discriminate the latest spectral types: this object has *(Y-J)*=0.85, which is typical of T9 and T8.5 dwarfs and slightly bluer than T0-T8 dwarfs. The *i-z* and *z-Y* colours are also similar to those of T8.5 and T9 dwarfs (Lucas et al. 2010).

UGPS 0722-05 is redder in the H-[4.5] color than any other T dwarf except the T7.5p dwarf SDSS 1416+1348 B. The H-[4.5] colour is considered a good indicator of T_{eff} (e.g. Leggett et al. (2010b)) so this supports our inference from the near-infrared spectrum that UGPS 0722-05 is cooler than the three known T9 dwarfs, for which 4.0≤H-[4.5]≤4.6.

While the H-[4.5] colour is considered a good indicator of T_{eff}, it is also influenced by metallicity and gravity (Leggett et al. 2010b). Recent estimates of the distance to SDSS 1416+13B ($d \approx 8$ pc, e.g. Scholz (2010); Burgasser et al. (2010); Bowler et al. (2010)) indicate that it is more luminous than UGPS 0722-05 by a factor of ~2. Assuming that these estimates are not greatly in error, they indicate that the redder colour of

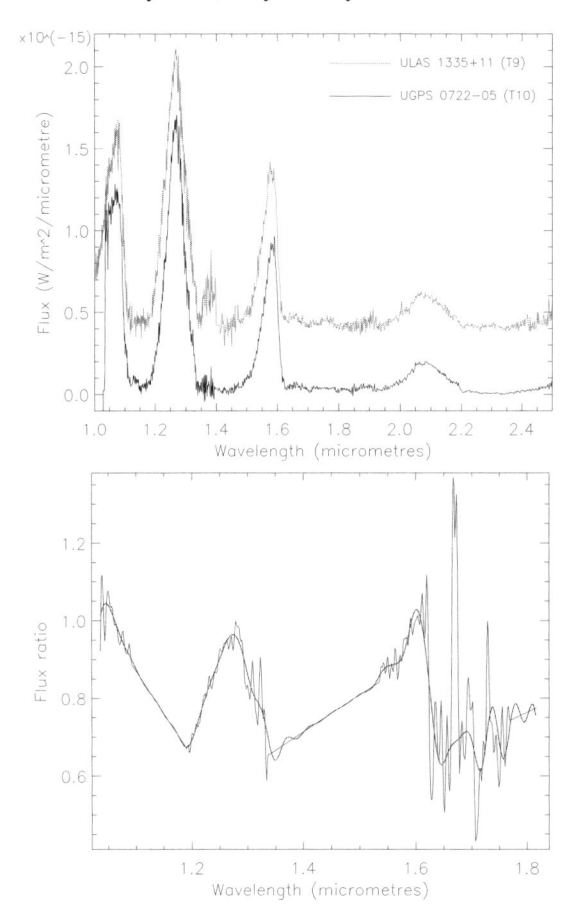

Figure 2. (upper panel) Gemini/NIRI spectrum of UGPS 0722-05, compared with that of a T9 dwarf. They appear quite similar but a ratio plot (lower panel) of the UGPS 0722-05 spectrum divided by the average of 3 T9 dwarfs shows that the broad molecular absorption troughs are signifcantly deeper in UGPS 0722-05.

SDSS 1416+1348 B is due to a combination of low metallicity and high gravity, which would also explain the extremely blue *(H-K)* colour (see e.g. Leggett et al. (2009)). As Burgasser et al. (2010) pointed out, this implies that the T_{eff} of SDSS 1416+1348 B is somewhat higher than the 500 K value that Burningham et al. (2010b) derived from the colours in the absence of a significant luminosity contraint. We therefore conclude that UGPS 0722-05 is the coolest brown dwarf known.

We note that the same logic can be applied to 2MASS 0939-2448, a T8 dwarf which resembles SDSS 1416+1348 B in that it has (i) a very red *H*-[4.5] colour for its spectral type, (ii) an exceptionally blue *(H-K)* colour and (iii) it's luminosity is relatively high, i.e. more consistent with its spectral type than its *H*-[4.5] colour. The simplest interpretation of properties of SDSS 1416+1348 B and 2MASS 0939-2448 is that their exceptional colours are due to low metallicity. Low metallicity T dwarfs are expected to be cooler than the norm for their spectral type but the high luminosities of these two sources indicate that they are not as cool as a simple H-[4.5] vs. T_{eff} relation would suggest. The alternative intepretation is that they are much cooler objects whose high luminosity is due to unresolved binarity (Burgasser et al. 2008). In this case, a low metallicity would still be required to explain their exceptionally blue *(H-K)* colors, so application of Occam's razor suggests that we should prefer the simpler interpretation.

The total luminosity of UGPS 0722-05 is L=9.2±3.1×10^{-7} $L_\odot$. Within these large error bars, this luminosity is comparable to that of T8.5 and T9 dwarfs (Marocco et al. 2010), (Leggett et al. 2010a). However, UGPS 0722-05 has fainter absolute magnitudes M_J and M_H.

Assuming an age in the range 0.2-10 Gyr, the evolutionary models of Saumon & Marley (2008) (hereafter SM08), allow a radius, R, between 0.085 $R_\odot$ (at 10 Gyr) and 0.12 $R_\odot$ (at 0.2 Gyr). Using the definition of T_{eff}: $L=4\pi R^2 \sigma T_{eff}^4$, we calculate T_{eff}=614±46 K for R=0.085 $R_\odot$ and T_{eff}=516±39 K for R=0.12 $R_\odot$.

Our preferred effective temperature is T_{eff}=520±40 K, which lies in the bottom half of the range allowed by the above calculation. Our reasoning is that UGPS 0722-05 has a later spectral type and a redder *H*-[4.5] colour than either the age benchmark T8.5 dwarf Wolf 940B (Leggett et al. 2010a), an object with a well constrained luminosity and radius, or the T9 dwarfs ULAS 0034-0052 and ULAS 1335+1130, which have well-measured luminosities. All three of these objects are thought to have $T_{eff} \leq$ 600 K. The location of the object in an *(H-K)* vs. *H*-[4.5] diagram or an *(H-K)* vs. M_H plot also indicate T_{eff} = 500 K, according to the SM08 models. At 480-560 K, the SM08 models indicate a mass in the range 5 to 15 M_{Jup}, $log(g)$ = 4.0 to 4.5 and age of 0.2 to 2.0 Gyr.

5. The Low Space Density of Late T Dwarfs in the Local Field

There is now strong evidence (Pinfield et al. 2008; Burningham et al. 2010a; Metchev et al. 2008; Reylé et al. 2010) that the space density of late T dwarfs in the local field with types T6 to T9 is lower than would be expected from measurements of the IMF in young clusters. When expressed as a power law α, where $N(M) \propto M^{-\alpha}$, the canonical value is quoted for young clusters at substellar masses is $\alpha \approx 0.6$. By contrast, the UKIDSS LAS data indicate that $-2 < \alpha < 0$.

The discovery of UGPS 0722-05 does not appreciably change the statistics. At a distance $d \approx 4$ pc it is the nearest brown dwarf primary to the sun. Adaptive optics imaging (Bouy et al. 2011) and a proper motion search of the SuperCosmos Science Archive find no sign of a companion. Only 4 brown dwarf primaries are known within 5 pc (Artigau et al. 2010; Tinney 1998; Costa et al. 2006), whereas there are 50 known stellar or substellar primaries in total within this volume.

There are several possibles explanations for this. One is that there is an error in the evolutionary models that are used to derive the masses and ages of late T dwarfs from their luminosity and effective temperature. This might be caused by an error in the cooling rate, due to errors in brown dwarf radii, or errors in the model atmospheres. The lack of good high temperature ammonia and methane line lists in present model atmospheres is a significant cause for concern. Fortunately, a high quality ammonia line list produced by the UCL group is now available at www.exomol.com (Yurchenko et al. 2010).

If this is the case then many brown dwarfs with T_{eff} < 500 K remain to be discovered in the local field. A related possibility is that the error lies in measurements of the cluster IMF. Uncertainties in very young clusters are known to be significant, e.g. because the ongoing effect of highly time-variable accretion in very young brown dwarfs are not included in the commonly used isochrones (Baraffe et al. 2009). More measurements of slightly older clusters with ages >10 Myr, such as IC4665 (deWit et al. 2006) and the Pleiades (Lodieu et al. 2007), would help to provide reassurance on this point.

Another possible explanation is that the IMF has changed over time, such that star formation regions now form more brown dwarfs than when the Galaxy was younger. This seems unlikely on the face of it because studies of the IMF in clusters in the Magellanic clouds (Liu et al. 2009) show no sign of a different IMF. However such studies do not extend much below a solar mass, so it is possible that brown dwarfs formed in fewer numbers when the Galactic disk was younger and had lower metallicity, e.g. due to the difficulty of cooling molecular cloud cores that contain less dust. A reason for considering this possibility is that if the models are correct then most of the UKIDSS T8-T9 brown dwarfs are young objects with very low masses (Leggett et al. 2010b) and so is UGPS 0722-05. While the number of such objects is still small (see Table 1) the paucity of old objects in the UKIDSS sample is starting to look significant.

Acknowledgments. The identification of the single good candidate late T dwarf found amongst several hundred million stars in the UKIDSS GPS is a tribute to the quality of UKIRT and the expertise of the staff at the Joint Astronomy Centre, the Cambridge Astronomical Survey Unit and the Wide Field Astronomy Unit at Edinburgh University. UKIRT is operated by the Joint Astronomy Centre on behalf of the Science and Technology Facilities Council of the UK. Gemini is is operated by the Association of Universities for Research in Astronomy, Inc., under a cooperative agreement with the NSF on behalf of the Gemini partnership, which consists of national scientific organisations in the USA, the UK, Canada, Chile, Australia, Brazil and Argentina (see www.gemini.edu). This research has made use of the USNOFS Image and Catalogue Archive operated by the United States Naval Observatory, Flagstaff Station.

References

Artigau, E., Radigan, J., Folkes, S., Jayawardhana, R., Kurtev, R., Lafrenière, R., Doyon, R., & Borissova, J. 2010, ApJ Letters, 718, 38
Baraffe, I., Chabrier, G., & Gallardo, J. 2009, ApJ, 702, L27
Bouy, H., Girard, J., Martn, E., Huélamo, N., & Lucas, P. 2011, A&A, 526, 55
Bowler, B. P., Liu, M. C., & Dupuy, T. J. 2010, ApJ, 710, 45. 0912.3796
Burgasser, A., Looper, D., & J.T., R. 2010, AJ, 139, 2448
Burgasser, A., Tinney, C., Cushing, M., Saumon, D., Marley, M., Bennett, C., & Kirkpatrick, J. 2008, ApJ Letters, 689, 53
Burgasser, A. J., Geballe, T. R., Leggett, S. K., Kirkpatrick, J. D., & Golimowski, D. A. 2006, ApJ, 637, 1067. arXiv:astro-ph/0510090
Burningham, B., et al. 2008, MNRAS, 391, 320
— 2009, MNRAS, 395, 1237
— 2010a, MNRAS, 406, 1885
— 2010b, MNRAS, 404, 1952
Costa, E., Méndez, R., Jao, W.-C., Henry, T., Subasavage, J., & Ianna, P. 2006, AJ, 132, 1234
Currie, T., Bailey, V., Fabrycky, D., Murray-Clay, R., Rodigas, T., & Hinz, P. 2010, ApJ Letters, 721, 177
Delorme, P., et al. 2008, A&A, 482, 961
— 2010, A&A, 518, 39

deWit, W., et al. 2006, A&A, 448, 189
Eisenhardt, P., Griffith, R., Stern, D., & Wright, E. 2010, ApJ, 139, 2455
Folkes, S., Pinfield, D., Kendall, T., & Jones, H. 2010, MNRAS Letters, 408, 56
Goldman, B., Marsat, S., Henning, T., Clemens, C., & Greiner, J. 2010, MNRAS, 405, 1140
Lawrence, A., et al. 2007, MNRAS, 379, 1599
Leggett, S., Saumon, D., Burningham, B., Cushing, M., Marley, M., & Pinfield, D. 2010a, ApJ, 720, 252
Leggett, S., et al. 2009, ApJ, 695, 1517
— 2010b, ApJ, 710, 1627
Liu, Q., de Grijs, R., Deng, L., Hu, Y., & Beaulieu, S. 2009, A&A, 503, 469
Lodieu, N., Burningham, B., Hambly, N., & Pinfield, D. 2009a, MNRAS, 397, 258
Lodieu, N., Dobbie, P., Deacon, N., Venemans, B., & Durant, M. 2009b, MNRAS, 395, 1631
Lodieu, N., Dobbie, P. D., Deacon, N. R., Hodgkin, S. T., Hambly, N. C., & Jameson, R. F. 2007, MNRAS, 380, 712
Lucas, P., et al. 2008, MNRAS, 391, 136
— 2010, MNRAS Letters, 408, 56
Marocco, F., et al. 2010, A&A, 524, 38
Metchev, S., Kirkpatrick, J., Berriman, G., & Looper, D. 2008, ApJ, 676, 1281
Pinfield, D., et al. 2008, MNRAS, 390, 304
Reylé, C., Delorme, P., Willott, C., Albert, L., Delfosse, X., Forveille, T., Artigau, E., Malo, L., Hill, G., & Doyon, R. 2010, A&A, 522, 112
Saumon, D., & Marley, M. S. 2008, ApJ, 689, 1327. 0808.2611
Scholz, R.-D. 2010, A&A Letters, 510, 8
Thalmann, C., et al. 2009, ApJ Letters, 707, 123
Tinney, C. G. 1998, MNRAS, 296, L42+
Warren, S., et al. 2007, MNRAS, 381, 1400
Yurchenko, S., Barber, R., & Tennyson, J. 2010, MNRAS (submitted) see arXiv:1011.1569, 1

Low–Mass Stars in the Sloan Digital Sky Survey: Galactic Structure, Kinematics, and the Luminosity Function

John J. Bochanski[1,2]

[1]*Astronomy & Astrophysics Dept., Pennsylvania State University, 525 Davey Laboratory, University Park, PA, 16802, USA, email: jjb29@psu.edu*

[2]*Kavli Institute for Astrophysics and Space Research, Massachusetts Institute of Technology, Building 37, 77 Massachusetts Avenue, Cambridge, MA 02139, USA*

Abstract. Modern sky surveys, such as the Sloan Digital Sky Survey and the Two–Micron All Sky Survey, have revolutionized the study of low–mass stars. With millions of photometric and spectroscopic observations, intrinsic stellar properties can be studied with unprecedented statistical significance. Low–mass stars dominate the local Milky Way and are ideal tracers of the Galactic potential and the thin and thick disks. Recent efforts, driven by SDSS observations, have sought to place the local low-mass stellar population in a broader Galactic context.

I highlight a recent measurement of the luminosity and mass functions of M dwarfs, using a new technique optimized for large surveys. Starting with SDSS photometry, the field luminosity function and local Galactic structure are measured simultaneously. The sample size used to estimate the LF is nearly three orders of magnitude larger than any previous study, offering a definitive measurement of this quantity. The observed LF is transformed into a mass function and compared to previous studies.

Ongoing investigations employing M dwarfs as tracers of Galactic kinematics are also discussed. SDSS spectroscopy has produced databases containing tens of thousands of low–mass stars, forming a powerful probe of the kinematic structure of the Milky Way. SDSS spectroscopic studies are complemented by large proper motion surveys, which have uncovered thousands of common proper motion binaries containing low–mass stars. Additionally, the SDSS spectroscopic data explore the intrinsic properties of M dwarfs, including metallicity and magnetic activity.

The highlighted projects demonstrate the advantages and problems with using large data sets and will pave the way for studies with next–generation surveys, such as PanSTARRS and LSST.

1. Introduction

Low–mass dwarfs ($0.08\ M_\odot < M < 0.8\ M_\odot$) are the dominant stellar component of the Milky Way, composing ~ 70% of all stars (Reid & Gizis 1997; Bochanski et al. 2010a) and nearly half the stellar mass of the Galaxy. However, despite their abundance, M dwarfs were studied in relatively small samples, due to their dim intrinsic brightnesses ($L \lesssim 0.05\ L_\odot$). However, the situation has been radically altered in the last decade, as deep surveys covering large areas of the sky were carried out. These projects, such as the Sloan Digital Sky Survey (SDSS, York et al. 2000) and the Two–Micron All Sky Survey (2MASS, Skrutskie et al. 2006), can trace their roots back to photographic surveys, epitomized by the National Geographic Society - Palomar Observatory Sky Survey (POSS-I, Minkowski & Abell 1963) and its successor, POSS-II (Reid et al. 1991). However, the photometric precision achieved by modern surveys distinguishes them from their photographic predecessors. For example, SDSS has imaged 1/4 of the sky to $r \sim 22$ and 2MASS imaged the entire sky to $J \sim 16.5$, with typical uncertainties of a few percent. The resulting databases contain accurate multi-band photometry of tens of millions of low–mass stars, enabling exciting new science. While there is a rich heritage of historical investigations, this article will focus on results derived from SDSS data.

A multitude of studies have focused on the intrinsic properties of low–mass stars using SDSS observations. The field luminosity function (LF) and corresponding mass function (MF) was measured using over 15 million stars (Bochanski et al. 2010a). The mass–radius relation of M dwarfs has been studied with eclipsing binary systems (Blake et al. 2008; Becker et al. 2008). Average photometric colors and spectroscopic features have also been quantified (Hawley et al. 2002; Bochanski et al. 2007b; Davenport et al. 2007; West et al. 2008, 2011). Multiple studies have attempted to estimate the absolute magnitude of low–mass stars in SDSS (Hawley et al. 2002; Jurić et al. 2008; Sesar et al. 2008; Bochanski 2008; Bilir et al. 2009; Bochanski et al. 2010b). However, due to the lack of precise trigonometric parallaxes for many of the M dwarfs in SDSS, absolute magnitudes and distances are derived by secondary means. Chromospheric activity, driven by magnetic dynamos within the stars, has been observationally traced with Hα emission measured in SDSS spectroscopy (West et al. 2004, 2008, 2011; Kruse et al. 2010). Flare rates have been measured for thousands of stars (Kowalski et al. 2009; Hilton et al. 2010b) and will be incorporated into predicting the flaring population observed by next–generation surveys (Hilton et al. 2010a).

While interesting objects in their own right, M dwarfs are also powerful tools for studying the Milky Way. Their ubiquity, combined with lifetimes much greater than a Hubble time (Laughlin et al. 1997), make them ideal tracers of Galactic structure and kinematics. Using photometry of over 15 million M dwarfs, Bochanski et al. (2010a) measured the scale heights and lengths of the thin and thick disks. A complementary study by Jurić et al. (2008) used higher mass stars to measure the stellar density profiles. The local gravitational potential has been probed using the kinematics of ~ 7000 M dwarfs along one line of sight with proper motions, SDSS spectroscopy and photometry (Bochanski et al. 2007a). This study has been expanded to the entire SDSS footprint, using a database of $\sim 25,000$ M dwarfs with velocities and distances (Pineda et al. 2010). Furthermore, Fuchs et al. (2009) employed proper motions and distances of ~ 2 million M dwarfs to estimate velocity distributions.

Below, I highlight several studies that have used SDSS observations to study both the intrinsic properties of low–mass stars *and* use them to study the Milky Way. In §2, the technical details of SDSS photometric and spectroscopic observations are briefly described. In §3, a new investigation measuring the M dwarf field LF and MF and Galactic structure parameters is detailed. Kinematics within the Milky Way are explored in §4. The importance of placing large, deep surveys of M dwarfs in a Galactic context is discussed in §5. Concluding remarks and avenues for future investigations are discussed in §6.

2. Observations

The Sloan Digital Sky Survey (SDSS, York et al. 2000) was a photometric and spectroscopic survey covering ~ 10,000 sq. deg. centered on the Northern Galactic Cap conducted with a 2.5m telescope (Gunn et al. 2006) at Apache Point Observatory (APO). Photometry was acquired in five filters (*ugriz*, Fukugita et al. 1996; Ivezić et al. 2007) down to a limiting magnitude r ~ 22 (Stoughton et al. 2002; Adelman-McCarthy et al. 2008). Calibration to a standard star network (Smith et al. 2002) was obtained through concurrent observations with the "Photometric Telescope" (PT; Hogg et al. 2001; Tucker et al. 2006). Absolute astrometric accuracy was estimated to better than 0.1″ (Pier et al. 2003), and has been used to measure proper motions with a precision of 3 mas yr^{-1} (Munn et al. 2004). When the Northern Galactic Cap was not visible at APO, a 300 sq. deg. stripe (Stripe 82) centered on zero declination was scanned repeatedly. Observations of Stripe 82 were used to empirically quantify photometric precision (Ivezić et al. 2007). Over 357 million unique photometric objects have been identified in the latest public data release (DR7, Abazajian et al. 2009). The photometric precision of SDSS is unrivaled for a survey of its size, with typical errors $\lesssim$ 0.02 mag (Ivezić et al. 2004, 2007).

When the conditions at APO were not photometric, SDSS operated in a spectroscopic mode. Twin fiber–fed spectrographs obtained medium–resolution ($R = 1,800$) optical spectra, covering 3800-9200 Å. Objects with SDSS photometry were targeted for spectroscopic followup by complex algorithms. The primary targets were galaxies (Strauss et al. 2002) and quasars (Richards et al. 2002). However, over ~ 70,000 M dwarfs have been identified within the SDSS spectroscopic database (West et al. 2011) from targeted and serendipitous (i.e., failed quasar) observations.

3. Luminosity Function and Galactic Structure

In the following section, the method and results of Bochanski et al. (2010a) are detailed. The Bochanski et al. (2010a) study produced two distinct yet intertwined results: it combined the investigation of an intrinsic stellar property (the MF and LF) with a measurement of Galactic structure and self–consistently described the local low–mass stellar population. The wide, deep, precise photometric coverage of SDSS enabled this type of analysis.

Previous investigations of the LF and MF have relied on one of two techniques: 1) nearby, volume–limited studies of trigonometric parallax stars (e.g. Reid & Gizis 1997); or 2) pencil–beam surveys of distant stars over a small solid angle (e.g. Zheng et al. 2001). In both cases, sample sizes were limited to a few thousand stars, prohibiting detailed statistical measurements. Using a sample drawn from SDSS, 2MASS and Guide Star Catalog photometry and supplemented with SDSS spectroscopy, Covey et al. (2008) performed the largest field low–mass LF and MF investigation prior to the Bochanski et al. (2010a) study. Covey et al. (2008) measured the LF with a sample covering 30 sq. deg. and containing ~ 30,000 low–mass stars. They also constrained their contamination rate to $\lesssim$ a few percent after obtaining spectra of every red point source in a 1 sq. deg. calibration region.

3.1. Sample Selection and Method

Please refer to Bochanski et al. (2010a) for a detailed description of sample selection. M dwarfs were selected from SDSS Data Release 6 (DR6; Adelman-McCarthy et al. 2008) footprint using $r-i$ and $i-z$ photometric colors. Noisy photometric observations were removed using flag cuts, yielding a final database containing over 15 million low–mass stars.

The Bochanski et al. (2010a) photometric sample comprised a data set three orders of magnitude larger (in number) than any previous LF study. Furthermore, it covered over 8,400 sq. deg., nearly 300 times larger than the sample analyzed by Covey et al. (2008). The large sky coverage of SDSS presented the main challenge in measuring the LF. Most previous studies either assumed a uniform density distribution (for nearby stars) or calculated a Galactic density profile, $\rho(r)$ along one line of sight. With millions of stars spread over nearly 1/4 of the sky, numerically integrating Galactic density profiles for each star was computationally prohibitive.

To address this issue, Bochanski et al. (2010a) employed the following technique for measuring the luminosity function. First, absolute magnitudes were assigned and distances to each star were computed using new $M_r, r-z$ and $M_r, r-i$ color–absolute magnitude relations (CMRs) estimated from nearby stars with $ugriz$ photometry and accurate trigonometric parallaxes. Next, a small range in absolute magnitude (0.5 mag) was selected and the stellar density was measured as a function of Galactic radius (R) and Galactic height (Z). This range in absolute magnitude was selected to provide high resolution of the LF, with a large number of stars ($\sim 10^6$) in each bin. Finally, a Galactic profile was fit to the R, Z density maps, solving for the shape of the thin and thick disks and the local stellar density. The LF was then constructed by combining the local density of each absolute magnitude slice. An example of one density map and the corresponding fit is shown in Figure 1.

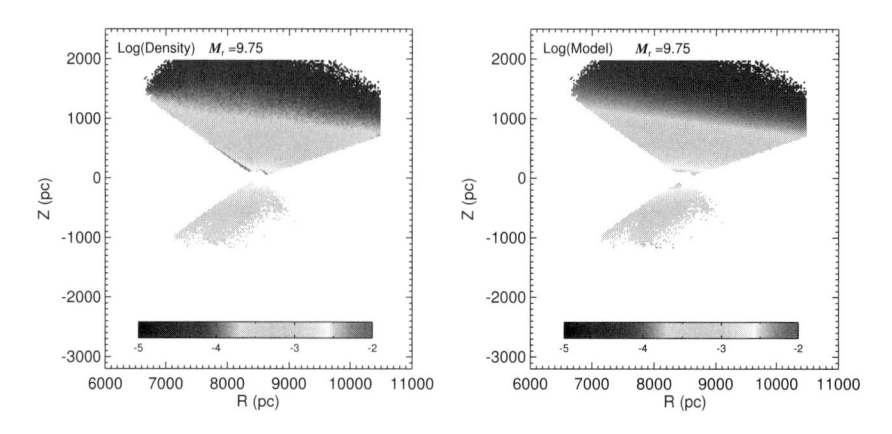

Figure 1. *Left Panel* - The stellar density distribution in R and Z for a 0.5 mag slice in M_r centered on $M_r = 9.75$. Over 1 million stars are contained within this slice. Note the smooth decrease in the vertical direction (due to the exponential scale heights of the Galactic thin and thick disks) and the signature of the Milky Way's scale length, manifested as a decrease in density with R. *Right Panel* - The corresponding two-disk model fit for the same M_r slice.

3.2. Systematic Challenges: Binarity, Metallicity and Absolute Magnitudes

The measured "raw" LF is subject to various systematic effects. Unresolved binarity, metallicity gradients within the Milky Way, interstellar extinction and Malmquist bias can all influence a star's estimated absolute magnitude and distance. Understanding these effects is important, since the photometric data set contains millions of stars and Poisson errors are no longer significant within a given LF bin. Rather, systematic errors in absolute magnitudes are the dominant source of uncertainty. In the Bochanski et al. (2010a) study, the uncertainty in each LF bin was calculated by repeating the entire analysis multiple times, using five different prescriptions for the CMR, which were selected to account for metallicity gradients, extinction and reddening.

The effects of unresolved binarity and Malmquist bias were quantified using Monte Carlo realizations of the Milky Way. Each model was populated with synthetic systems that were consistent with the observed Galactic structure and LF. Binaries were included within the simulations, using four different prescriptions for mass ratio distributions. The mock stellar catalog was analyzed with the same pipeline as the actual observations. The differences between the input and "observed" Galactic structure and LF were used to systematically correct the observed "raw" values from SDSS data. The single–star LF was also estimated during this step. The detailed analysis of Malmquist bias and unresolved binaries and their effect on the resultant LF are given in Bochanski et al. (2010a).

3.3. The SDSS Field LF and MF

The final adopted system and single–star M_r LFs are presented in Figure 2. The uncertainty in each LF bin was computed from the full spread due to differences among CMRs and binary prescriptions. The system LF (black circles) rises smoothly to $M_r \sim 11$, with a fall off to lower masses. The single–star LF (red circles) follows a similar track, but maintains a density increase of $\sim 2\times$ compared to the system LF at the same absolute magnitude. This implies that lower luminosity stars are easily hidden in binary systems, but isolated low–luminosity systems are not common.

Using the mass–magnitude relations of Delfosse et al. (2000), the field system and single–star LFs were converted to MFs, shown in Figure 3. The system MF exhibits similar behavior to the LF, peaking near 0.3 $M_\odot$. The single–star MF is compared to the local "eight parsec" MF of Reid & Gizis (1997) in Figure 3. The Reid & Gizis (1997) sample consisted of 558 stars spread over 3π steradians. There is broad agreement over most of the mass range probed, with the SDSS MF falling below the local MF at higher masses. The discrepancy is probably due to the assumed CMR, which may have overestimated the brightness of higher mass M dwarfs (see contribution by Hawley et al., this volume; Bochanski et al. 2010b). To first order, this would lead to overestimated distances and underestimated local stellar densities. Future investigations will rederive the LF and MF with updated CMRs. The SDSS MF was fit with two functions: a broken power law and lognormal. Both adequately represent the underlying data points, but a lognormal is slightly preferred due to the smaller number of free parameters.

The right panel of Figure 3 compares the SDSS system MF with the MF reported by Zheng et al. (2001). The Zheng et al. (2001) HST survey was a pencil beam survey of 1,400 M dwarfs over $\sim$ 1 sq. deg.. Their system MF agrees with most of the SDSS MF, especially at lower masses. The Zheng et al. (2001) power law fit agrees favorably with the low–mass component of the SDSS broken power law fit. At higher masses, there is a disagreement, which is likely due to CMR differences.

When searching for variations between MFs, it is crucial that MF *data* is compared, rather than best–fit functional forms. This point, first introduced in Covey et al. (2008) and reiterated by Bastian et al. (2010) is important. The comparison of *fits* and not *data* may have exaggerated the differences between the local census (i.e. Reid & Gizis 1997) and distant samples (i.e. Zheng et al. 2001), although other factors, such as unresolved binarity (e.g. Kroupa et al. 1991; Chabrier 2003) also contributed. Finally, the SDSS single-star MF is compared to selected seminal works in the field in Figure 4.

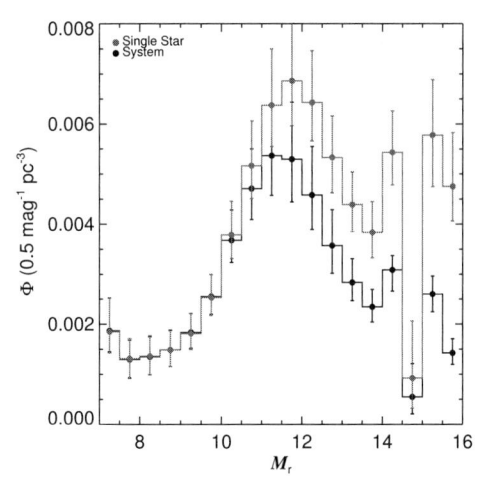

Figure 2. The single–star LF (red line) and system LF (black line) for SDSS M dwarfs. The LFs have been corrected for the effects of unresolved binarity and Malmquist bias.

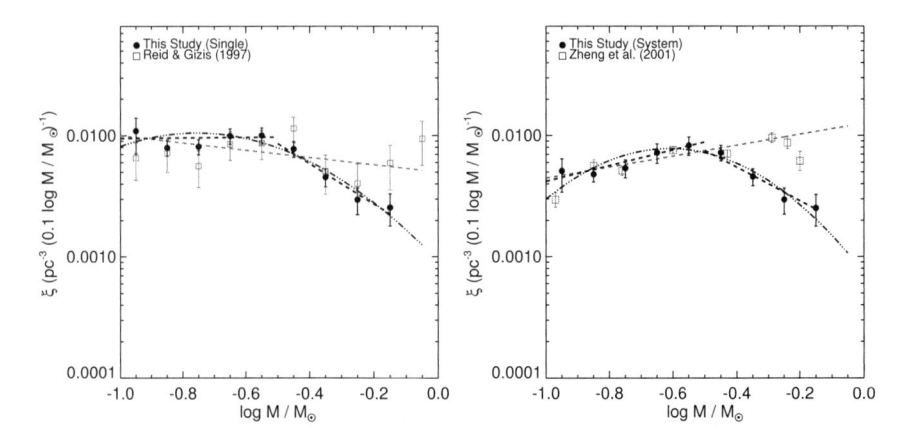

Figure 3. *Left Panel* - The SDSS single–star MF (filled circles) compared to the local census of Reid & Gizis (1997). The two samples show agreement over a large range in mass, suggesting the correction for unresolved binarity is valid. *Right Panel* - The SDSS system MF compared to the study of Zheng et al. (2001). The disagreement at large masses is most likely due to differences in the assumed CMRs.

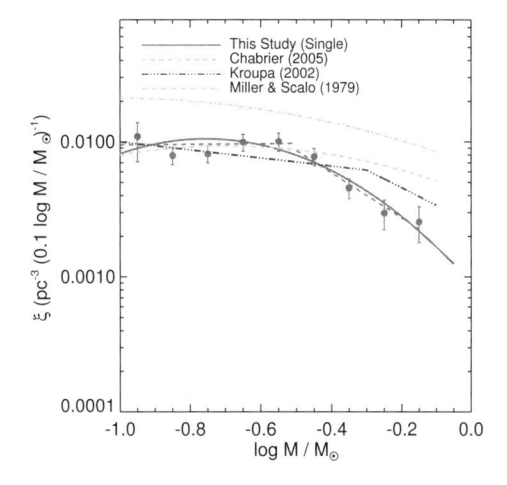

Figure 4. Shown is the MF data and best log-normal fit from Bochanski et al. (2010a, solid circles and red line), along with the analytic MF fits of Chabrier (2005, green dashed line), Kroupa (2002, dark blue dash–dot–dot–dot line) and Miller & Scalo (1979, light blue dash-dot line). I stress that comparing actual MF data is preferred to comparing analytic fits.

4. Kinematics

While photometric studies can be used to study the distribution of M dwarfs throughout the Galaxy and infer their fundamental properties, they offer a static glimpse of the population. The dynamics of stars through the Milky Way, as measured by radial velocities and proper motions, can be used to infer the underlying local mass distribution and gravitational potential. In particular, kinematic studies of M dwarfs have been useful for studying the properties of the thin and thick disks. Using a sample of ∼ 7,000 M dwarfs along one line of sight, Bochanski et al. (2007a) examined the UVW velocity distributions as a function of distance from the plane. Thin and thick disk components were identified with the use of probability plots (Lutz & Kelker 1973; Reid et al. 1995). In general, both the thin and thick disks dispersions increase with height above the Plane, with the W velocity dispersion being the smallest of the three directions. Low–mass stars follow the Galactic potential across the entire mass range. In other words, early and late–type M dwarfs, despite differing in mass by nearly a factor of ten, have similar velocity dispersions at the same height from the Galactic midplane. Similar trends with height were also quantified by Fuchs et al. (2009) using proper motions and distances of ∼ 2 million SDSS M dwarfs.

The analysis from Bochanski et al. (2007a) was expanded to the entire SDSS footprint by Pineda et al. (2010), using the largest spectroscopic sample of M dwarfs ever assembled (West et al. 2011). Pineda et al. (2010) determined mean velocities, dispersion and relative scalings between the thin and thick disk. They searched for velocity gradients radially in the disk, and observed a similar increase in dispersion away the Plane (see Figure 5). Pineda et al. (2010) kinematically estimated the local fraction of thick disk stars to be ∼ 5%, similar to the photometric result from Bochanski et al. (2010a). This study exemplifies the utility of low–mass stars as probes of the local Milky Way.

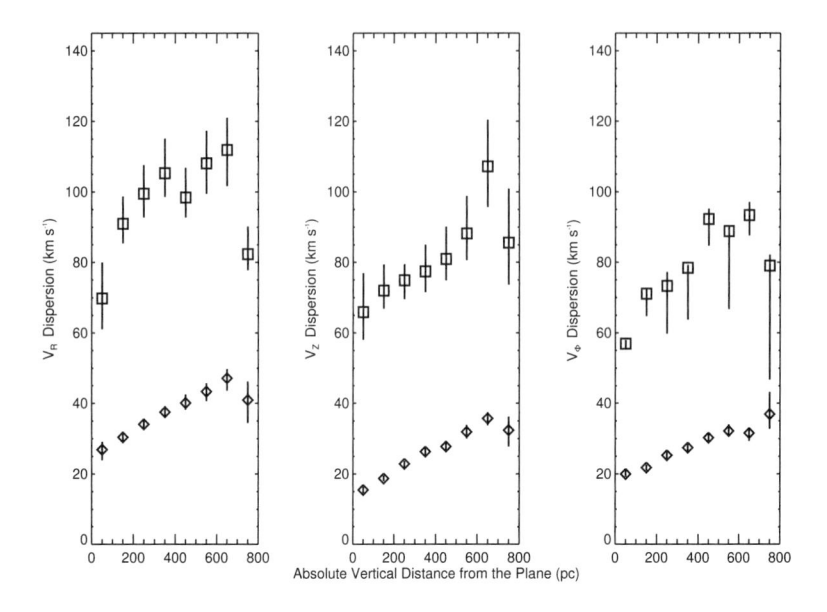

Figure 5. Thin disk (diamonds) and thick disk (open squares) velocity dispersions as a function of height in a Galactocentric coordinate system. Note the general increase with height above the Galactic plane. Figure from Pineda et al. (2010).

5. Low-Mass Stars in a Galactic Context

In addition to their utility as tracers of the local Milky Way, low–mass dwarfs have intrinsic properties, such as chromospheric activity (West et al. 2004; Bochanski et al. 2007a; Schmidt et al. 2007), that can be examined in a broader context. Additionally, metallicity may be studied using subdwarfs (Lépine et al. 2003), readily identified by their spectra, which show enhanced calcium hydride (CaH) absorption. Subdwarfs have been easily detected in large-scale surveys such as the SDSS (e.g., West et al. 2004; Lépine & Scholz 2008).

The study of M dwarf chromospheric activity in a Galactic context is described below. Similar investigations were severely limited before the advent of large spectroscopic databases, such as SDSS. It highlights the importance of studying M dwarfs in a broader context within the Milky Way, as their distributions within the Galaxy may be used as a valuable constraint on intrinsic properties.

5.1. Magnetic Activity: A Case Study

Low–mass stars harbor strong magnetic dynamos (Browning 2008), which in turn heat their chromospheres, producing Balmer line emission. For early-type stars (M0-M4), the dynamos are probably powered by stellar rotation, but the driving mechanism at later types is not well known. Ca H + K emission is also a tracer of activity in M dwarfs (Walkowicz & Hawley 2009), but is often not observationally accessible for these dim, red stars. Prior to SDSS, surveys that measured the fraction of active M dwarfs often consisted of < 1000 stars, with very few stars (< 100) in the faintest luminosity bins

(Hawley et al. 1996; Gizis et al. 2000). Since previous samples were flux-limited, they were heavily biased towards observing nearby, young stars.

The first major study of activity in M dwarfs with SDSS data was performed by West et al. (2004). This study observed many stars at distances 5-10 times larger than previous studies due to the multi-fiber capabilities of the SDSS spectrographs. The differences between the West et al. (2004) study and the past surveys were immediately obvious. First, some spectral types (M7 and M8) which were previously observed to have active fractions near 100% had lower active fractions, near 70%. The active fraction distributions for all spectral types declined with distance from the Galactic midplane with a spectral type dependent slope. West et al. (2004) suggested this was an age effect, as stars further from the Plane also exhibited larger velocity dispersions. Thus, older stars have undergone more dynamic interactions and also possess weaker magnetic fields. The M dwarf age-activity relation was calibrated by West et al. (2008), which determined the activity lifetime for M dwarfs as a function of spectral type. By using a simple model that fit both the velocity and active fraction distributions within the Galaxy, West et al. (2008) were able to constrain the ages of stars, a fundamental stellar property. It is worth noting that without large samples of M dwarfs, such as those available with SDSS, this type of analysis would not be possible. The West et al. (2004) study also highlights the potential biases that are encountered when limiting samples to the solar neighborhood. The Sun is very close to the midplane (15 pc; Cohen 1995), and nearby stars are predominately young (1-2 Gyrs; Nordström et al. 2004), biasing local samples.

5.2. Metallicity and Stellar Ages: The Next Frontiers

Current models of low-mass stars cannot accurately reproduce observable features. This has led to uncertainty in determining fundamental parameters, such as mass, age and luminosity. Thus, empirical relations have been developed to estimate these properties. Mass-magnitude relations determined from eclipsing binaries (Delfosse et al. 2000), have proven extremely useful, but other fundamental parameters, most notably age and metallicity have not been fully calibrated.

Recent advances in measuring metallicity are promising (e.g. Lépine et al. 2007; Johnson & Apps 2009; Rojas-Ayala et al. 2010, and see the review of the metallicity splinter session in this volume). In particular, Rojas-Ayala et al. (2010) have calibrated an infrared technique that delivers 0.15 dex precision from K band spectra. However, for the vast majority of M dwarfs with optical spectra, a similar level of precision has not been achieved. Further observations of M dwarfs with known metallicities (i.e., those in wide binaries with FGK companions (i.e., Dhital et al. 2010) or cluster members) need to be obtained before a suitable calibration can be determined. However, once precise metallicities can be measured for M dwarfs, they will be a powerful probe of the Milky Way's chemical history.

Age is notoriously difficult to estimate for field stars (of any mass). However, age-activity relations (i.e. West et al. 2008), suggest that observational tracers, such as chromospheric activity, can constrain the ages of M dwarfs to a few Gyr. Further work is needed to produce precise estimates of M dwarf ages. The next generation of synoptic surveys, such as LSST and PanSTARRS may be useful in this endeavor and provide a link between rotation periods and stellar age.

6. Conclusions

I have briefly detailed some of the investigations that examine both intrinsic M dwarf properties and Galactic structure and kinematics. The requisite data for such studies, namely precise, deep, multi-band photometry over large areas of the sky, will become more common in the future, as additional large surveys come online. Within the next decade, surveys such as PanSTARRS, LSST and GAIA (Kaiser et al. 2002; Ivezic et al. 2008; Perryman et al. 2001), will produce new data sets that are ideal for similar investigations.

While databases containing millions of low–mass stars will be the norm for future studies, it is important to note that systematic errors will be the dominant source of error in many areas. In SDSS, nearly all the observed M dwarfs do not have reliable trigonometric parallaxes, forcing the use of photometric parallax estimates. Reliably calibrating and testing these photometric parallaxes will only become more important as surveys push to larger distances. Other systematics, such as how metallicity affects the absolute magnitude of M dwarfs, will also be crucial to future investigations.

Finally, the importance of SDSS spectroscopy can not be overstated. The large spectroscopic samples of SDSS M dwarfs have enabled many novel investigations. Significant spectroscopic followup of the next generation of surveys should be a high priority.

Acknowledgments. I would like to thank the Cool Stars 16 SOC for the opportunity to present this work. I also thank Suzanne Hawley, Andrew West and Kevin Covey for their insight and many enlightening conversations over the years. I thank Sebastian Pineda for kindly providing a figure. I would also like to acknowledge the comments of Neill Reid, Ivan King, Niall Deacon and Gilles Chabrier, who all contributed to work presented in this article.

References

Abazajian, K. N., Adelman-McCarthy, J. K., Agüeros, M. A., Allam, S. S., Allende Prieto, C., An, D., Anderson, K. S. J., Anderson, S. F., Annis, J., Bahcall, N. A., & et al. 2009, ApJS, 182, 543. 0812.0649
Adelman-McCarthy, J. K., et al. 2008, ApJS, 175, 297
Bastian, N., Covey, K. R., & Meyer, M. R. 2010, ARA&A, 48, 339
Becker, A. C., et al. 2008, MNRAS, 386, 416
Bilir, S., Karaali, S., Ak, S., Coşkunoğlu, K. B., Yaz, E., & Cabrera-Lavers, A. 2009, MNRAS, 396, 1589
Blake, C. H., Torres, G., Bloom, J. S., & Gaudi, B. S. 2008, ApJ, 684, 635
Bochanski, J. J., Hawley, S. L., Covey, K. R., West, A. A., Reid, I. N., Golimowski, D. A., & Ivezić, Ž. 2010a, AJ, 139, 2679
— 2010b, AJ, 139, 2679
Bochanski, J. J., Munn, J. A., Hawley, S. L., West, A. A., Covey, K. R., & Schneider, D. P. 2007a, AJ, 134, 2418
Bochanski, J. J., West, A. A., Hawley, S. L., & Covey, K. R. 2007b, AJ, 133, 531
Bochanski, J. J., Jr. 2008, Ph.D. thesis, University of Washington
Browning, M. K. 2008, ApJ, 676, 1262
Chabrier, G. 2003, ApJ, 586, L133
— 2005, in The Initial Mass Function 50 Years Later, edited by E. Corbelli, F. Palla, & H. Zinnecker, vol. 327 of Astrophysics and Space Science Library, 41
Cohen, M. 1995, ApJ, 444, 874

Covey, K. R., Hawley, S. L., Bochanski, J. J., West, A. A., Reid, I. N., Golimowski, D. A., Davenport, J. R. A., Henry, T., Uomoto, A., & Holtzman, J. A. 2008, AJ, 136, 1778
Davenport, J. R. A., Bochanski, J. J., Covey, K. R., Hawley, S. L., West, A. A., & Schneider, D. P. 2007, AJ, 134, 2430
Delfosse, X., Forveille, T., Ségransan, D., Beuzit, J.-L., Udry, S., Perrier, C., & Mayor, M. 2000, A&A, 364, 217
Dhital, S., West, A. A., Stassun, K. G., & Bochanski, J. J. 2010, AJ, 139, 2566
Fuchs, B., et al. 2009, AJ, 137, 4149
Fukugita, M., Ichikawa, T., Gunn, J. E., Doi, M., Shimasaku, K., & Schneider, D. P. 1996, AJ, 111, 1748
Gizis, J. E., Monet, D. G., Reid, I. N., Kirkpatrick, J. D., Liebert, J., & Williams, R. J. 2000, AJ, 120, 1085
Gunn, J. E., et al. 2006, AJ, 131, 2332
Hawley, S. L., Gizis, J. E., & Reid, I. N. 1996, AJ, 112, 2799
Hawley, S. L., et al. 2002, AJ, 123, 3409
Hilton, E., et al. 2010a. AJ, in preparation
Hilton, E. J., West, A. A., Hawley, S. L., & Kowalski, A. F. 2010b, AJ, 140, 1402
Hogg, D. W., Finkbeiner, D. P., Schlegel, D. J., & Gunn, J. E. 2001, AJ, 122, 2129
Ivezić, Ž., et al. 2004, Astronomische Nachrichten, 325, 583
— 2007, AJ, 134, 973
Ivezic, Z., et al. 2008, ArXiv e-prints. 0805.2366
Johnson, J. A., & Apps, K. 2009, ApJ, 699, 933. 0904.3092
Jurić, M., et al. 2008, ApJ, 673, 864
Kaiser, N., et al. 2002, in Proceedings of the SPIE, Volume 4836, pp. 154-164 (2002)., edited by J. A. Tyson, & S. Wolff, vol. 4836, 154
Kowalski, A. F., Hawley, S. L., Hilton, E. J., Becker, A. C., West, A. A., Bochanski, J. J., & Sesar, B. 2009, AJ, 138, 633. 0906.2030
Kroupa, P. 2002, Science, 295, 82
Kroupa, P., Gilmore, G., & Tout, C. A. 1991, MNRAS, 251, 293
Kruse, E. A., Berger, E., Knapp, G. R., Laskar, T., Gunn, J. E., Loomis, C. P., Lupton, R. H., & Schlegel, D. J. 2010, ApJ, 722, 1352
Laughlin, G., Bodenheimer, P., & Adams, F. C. 1997, ApJ, 482, 420
Lépine, S., Rich, R. M., & Shara, M. M. 2003, AJ, 125, 1598
— 2007, ApJ, 669, 1235
Lépine, S., & Scholz, R.-D. 2008, ApJ, 681, L33
Lutz, T. E., & Kelker, D. H. 1973, PASP, 85, 573
Miller, G. E., & Scalo, J. M. 1979, ApJS, 41, 513
Minkowski, R. L., & Abell, G. O. 1963, The National Geographic Society-Palomar Observatory Sky Survey (Basic Astronomical Data: Stars and stellar systems, edited by K. A. Strand. Published by the University of Chicago Press, Chicago, IL USA, 1968, p.481), 481
Munn, J. A., et al. 2004, AJ, 127, 3034
Nordström, B., Mayor, M., Andersen, J., Holmberg, J., Pont, F., Jørgensen, B. R., Olsen, E. H., Udry, S., & Mowlavi, N. 2004, A&A, 418, 989
Perryman, M. A. C., de Boer, K. S., Gilmore, G., Høg, E., Lattanzi, M. G., Lindegren, L., Luri, X., Mignard, F., Pace, O., & de Zeeuw, P. T. 2001, A&A, 369, 339
Pier, J. R., Munn, J. A., Hindsley, R. B., Hennessy, G. S., Kent, S. M., Lupton, R. H., & Ivezić, Ž. 2003, AJ, 125, 1559
Pineda, J. S., et al. 2010. AJ, in preparation
Reid, I. N., & Gizis, J. E. 1997, AJ, 113, 2246
Reid, I. N., Hawley, S. L., & Gizis, J. E. 1995, AJ, 110, 1838
Reid, I. N., et al. 1991, PASP, 103, 661
Richards, G. T., et al. 2002, AJ, 123, 2945
Rojas-Ayala, B., Covey, K. R., Muirhead, P. S., & Lloyd, J. P. 2010, ApJ, 720, L113
Schmidt, S. J., Cruz, K. L., Bongiorno, B. J., Liebert, J., & Reid, I. N. 2007, AJ, 133, 2258

Sesar, B., Ivezić, Ž., & Jurić, M. 2008, ApJ, 689, 1244
Skrutskie, M. F., Cutri, R. M., Stiening, R., Weinberg, M. D., Schneider, S., Carpenter, J. M., Beichman, C., Capps, R., Chester, T., Elias, J., Huchra, J., Liebert, J., Lonsdale, C., Monet, D. G., Price, S., Seitzer, P., Jarrett, T., Kirkpatrick, J. D., Gizis, J. E., Howard, E., Evans, T., Fowler, J., Fullmer, L., Hurt, R., Light, R., Kopan, E. L., Marsh, K. A., McCallon, H. L., Tam, R., Van Dyk, S., & Wheelock, S. 2006, AJ, 131, 1163
Smith, J. A., et al. 2002, AJ, 123, 2121
Stoughton, C., et al. 2002, AJ, 123, 485
Strauss, M. A., et al. 2002, AJ, 124, 1810
Tucker, D. L., et al. 2006, Astronomische Nachrichten, 327, 821
Walkowicz, L. M., & Hawley, S. L. 2009, AJ, 137, 3297
West, A. A., Hawley, S. L., Bochanski, J. J., Covey, K. R., Reid, I. N., Dhital, S., Hilton, E. J., & Masuda, M. 2008, AJ, 135, 785. 0712.1590
West, A. A., et al. 2004, AJ, 128, 426
— 2011, AJ, in press
York, D. G., et al. 2000, AJ, 120, 1579
Zheng, Z., Flynn, C., Gould, A., Bahcall, J. N., & Salim, S. 2001, ApJ, 555, 393

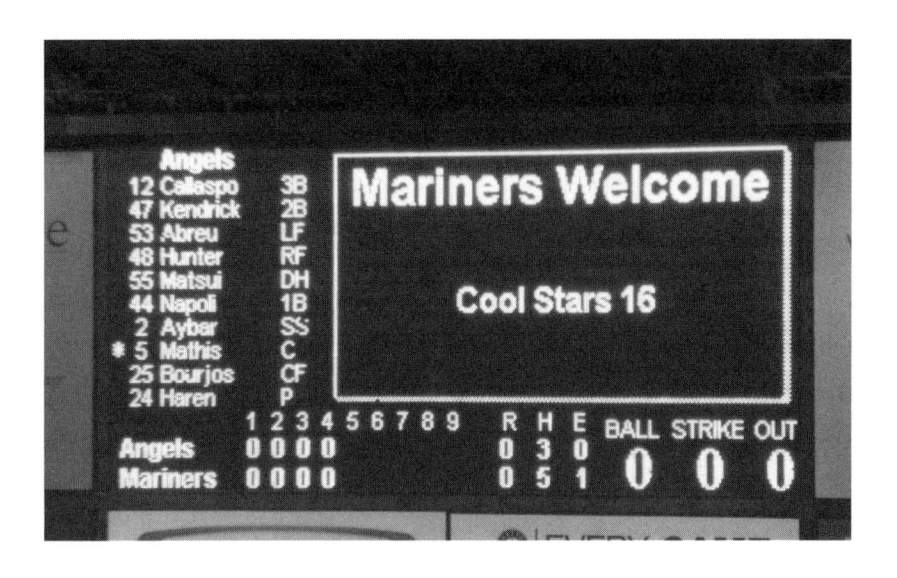

The Seattle Mariners Welcome Cool Stars 16

Looking for Systematic Variations in the Stellar Initial Mass Function

N. Bastian,[1] K. R. Covey,[2] and M. R. Meyer[3]

[1] *School of Physics, University of Exeter, Stocker Road, Exeter EX4 4QL, UK*

[2] *Department of Astronomy, Cornell University, Ithaca, NY 14853, USA*

[3] *Institute of Astronomy, ETH Zürich, Wolfgang-Pauli-Str. 27, 8093 Zürich, Switzerland*

Abstract. Few topics in astronomy initiate such vigorous discussion as whether or not the initial mass function (IMF) of stars is universal, or instead sensitive to the initial conditions of star formation. The distinction is of critical importance: the IMF influences most of the observable properties of stellar populations and galaxies, and detecting variations in the IMF could provide deep insights into the process by which stars form. In this contribution, we take a critical look at the case for IMF variations, with a view towards whether other explanations are sufficient given the evidence. Studies of the field, local young clusters and associations, and old globular clusters suggest that the vast majority were drawn from a "universal" IMF. Observations of resolved stellar populations and the integrated properties of most galaxies are also consistent with a "universal IMF", suggesting no gross variations in the IMF over much of cosmic time. Here we focus on 1) nearby star-forming regions, where individual stars can be resolved to give a complete view of the IMF, 2) star-burst environments, in particular super-star clusters which are some of the most extreme objects in the universe and 3) nearby stellar systems (e.g. globular clusters and dwarf spheroidal galaxies) that formed at high redshift and can be studied in extreme detail (i.e. near-field cosmology).

1. Introduction

In this contribution, we briefly summarise our recent review on stellar IMF variations (Bastian, Covey, & Meyer 2010). The IMF of the field has been studied extensively (e.g. Covey et al. (2008); Bochanski et al. (2010)), and we refer the reader to the contribution by Bochanski elsewhere in this volume for recent updates. In this contribution, we review the evidence for IMF variations in three distinct Galactic and extra-galactic environments. In § 2, we consider nearby star-forming regions and stellar clusters, where individual stars can be resolved and the IMF studied in exquisite detail. In § 3, we review measurements of the IMF in starburst galaxies, with particular attention to the super-star clusters they host, and which represent the most extreme environments in the Universe with respect to stellar densities and star-formation rate densities. In § 4 we consider IMF measurements in the context of "near-field cosmology", namely using local objects, such as globular clusters and dwarf spheroidal galaxies, to study the shape of the IMF at high redshift.

2. Nearby Star-Forming Regions and Open Clusters

Nearby open clusters and young star forming regions provide several advantages for IMF determinations: their proximity allow the IMF to be probed to the lowest masses via direct star counts; brown dwarfs are warm and bright at the youngest ages, easing their observational detection; their physical coherence provide a discrete sampling of the IMF as a function of metallicity and stellar density; and the short duration of their star formation events ameliorate ambiguities due to different potential star formation histories. For these reasons, open clusters and star forming regions have been frequent targets for local IMF measurements. It is beyond the scope of this contribution to review all such determinations, but an assortment of recent mass function measurements for star-forming associations, open and globular clusters is shown in Fig. 1 (from De Marchi et al. 2010). Here, we can see the characteristic shape of the IMF; a power-law distribution on the high mass side ($> 1 M_\odot$) and a lognormal (or multiple power laws) form below with a characteristic mass of $\sim 0.2 - 0.3 \, M_\odot$ below this. The vast majority of star-forming regions and open clusters appear to have stellar mass distributions that are qualitatively and quantitatively consistent with a Kroupa/Chabrier type IMF [1].

Most of the IMF variations reported in young regions can be ascribed to sampling effects, given the finite numbers of stars in each region, or to systematic effects, arising from differences in the analysis, such as the prescription adopted to correct for extinction or convert observational luminosities into model dependent stellar masses. Hence, it is imperative to systematically compare observations of different regions before claims of variations are made. Variations should be identified not solely on the basis of a comparison of the IMF index measured in two studies, but rather from a direct comparison of the underlying observational samples (i.e. color/luminosity distributions) and/or by analyzing the (model dependent) mass distribution in a more statistically robust way (i.e. through KS tests).

A handful of regions have been studied systematically in this way: of these, the prime example of a deviant IMF is that of the Taurus star-forming region. Comparing the IMF of Taurus, to those measured from the ONC, IC 348 and Chameleon I using the same methods, Luhman et al. (2009) find that Taurus has a significant excess of stars with mass $\sim 0.7 \, M_\odot$ compared to these other regions. Definitively identifying the root cause of this variation (e.g., a higher Jeans mass), and eliminating the possibility that this is a rare, statistical fluke, requires the discovery of other regions with similar physical properties and deviant IMFs, which we currently lack. To some extent, it is remarkable how well the low-mass end of the mass function agrees across different star-forming regions. Evolutionary tracks of low-mass PMS are highly uncertain, making the conversion from observed color/luminosity to mass (especially without spectroscopic typing) fraught with difficultly. Stellar rotation, accretion history and magnetic activity can all influence a star's position in the CMD, presenting serious, potentially insurmountable, challenges for using a (pre-main sequence) star's location in the CMD to accurately infer its age and/or mass (e.g. Jeffries 2007; Mayne & Naylor 2008; Baraffe et al. 2009). This re-iterates the need for authors to publish full observational catalogues for their mass function studies, enabling future tests for mass function varia-

---

[1] Dabringhausen et al. (2008) have shown that the log-normal parameterization of the IMF advanced by Chabrier (2003) is extremely similar to the two-part power-law defined by Kroupa et al. (1993); we therefore refer to all IMFs of this form as Kroupa/Chabrier-type IMFs.

tions to be based upon direct observable properties, rather than on second order inferred properties.

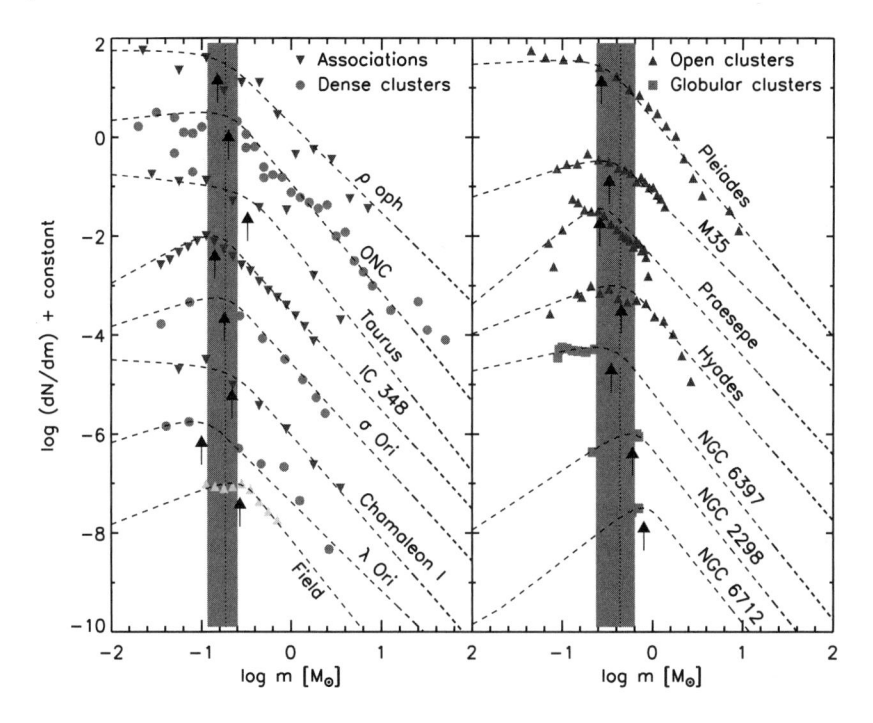

Figure 1. The derived present day mass function of a sample of young star-forming regions, open clusters spanning a large age range, and old globular clusters from the compilation of De Marchi et al. (2010). Additionally, we show the inferred field star IMF. The dashed lines represent "tapered power-law" fits to the data. The arrows show the characteristic mass of each fit (m_p), the dotted line indicates the mean characteristic mass of the clusters in each panel, and the shaded region shows the standard deviation of the characteristic masses in that panel (the field star IMF is not included in the calculation of the mean/standard deviation). The observations are consistent with a single underlying IMF, although the scatter at and below the stellar/sub-stellar boundary clearly calls for further study. The shift of the globular clusters characteristic mass to higher masses is expected from considerations of dynamical evolution. See Bastian, Covey, & Meyer (2010) for details.

3. Starburst Galaxies and Their Super-Star Clusters

3.1. The Technique and Application to SSCs

It is often stated that the IMF in starburst galaxies is significantly different than that seen locally. In particular, that the conditions prevailing in starbursts inhibit the formation of low-mass stars, resulting in a "top-heavy" IMF. It is difficult to place direct constraints on the IMF within starbursts due to the high star-formation rates, complicated extinction patterns (i.e. dust lanes) and possible presence of an AGN (not to mention their large distances making it impossible to resolve individual stars). One promising route, however, is through dynamical methods. This entails measuring the dynamical mass of the system, $M_{\rm dyn}$ (through Jeans modelling or simply adopting Virial equilibrium) and comparing that to what is expected from modelling of the stellar population, $M_{\rm pop}$. The latter depends on the age and metallicity of the stellar system and crucially on the adopted IMF. This technique relies on the fact that for "standard" IMFs the mass of the system is dominated by low-mass stars. This is demonstrated in Fig. 2. The light of these systems, however, is dominated by high-mass stars, so in essence this technique compares the mass-to-light ratio of a system to that expected for various IMFs.

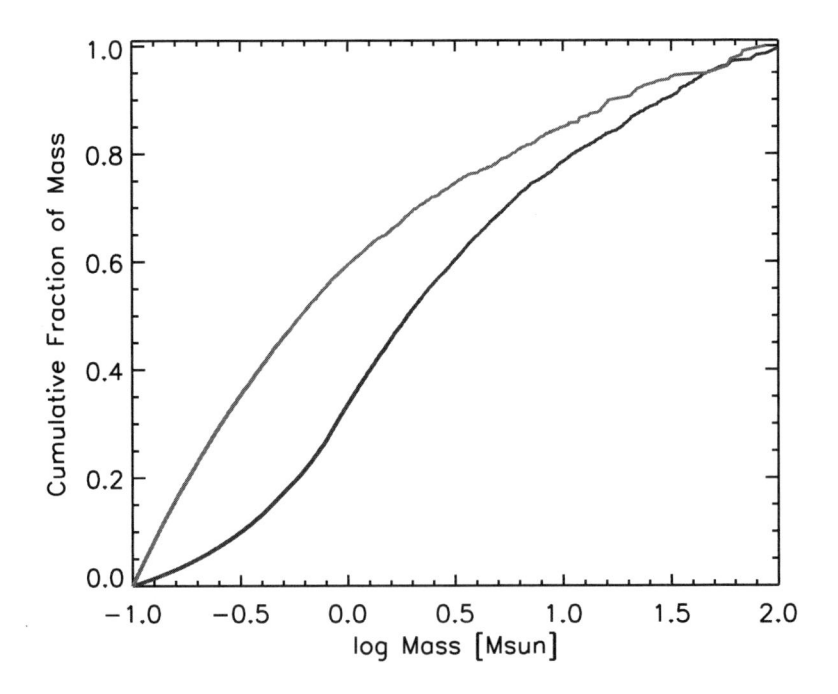

Figure 2. The cumulative fraction of mass in a stellar population with a Salpeter (1955) (red/top line) and Kroupa (2002) (blue/bottom line) type IMFs. Note that 50% of the mass of a (single age) stellar population is made up of stars less than 0.6 $M_\odot$ in the case of a Salpeter distribution or less than 2.0 $M_\odot$ in the case of a Kroupa type IMF. In either case, low mass stars dominate the mass of the stellar system (we have restricted the range considered from 0.1 to 100 $M_\odot$).

The simplest case is for stellar clusters, for which all the stars have the same age and metallicity. Within starburst (and quiescent) galaxies, there exist stellar clusters with high masses (in some cases up to ~ $10^8 M_\odot$ - Maraston et al. 2004) and ages between a < 10 to 1000s of Myr (see Portegies Zwart et al. 2010 for a recent review). These super-star clusters (SSCs) are thought to be young globular clusters, and their brightness allows their velocity dispersions to be obtained with current instrumentation. In Fig. 3, a selection of SSCs with estimated dynamical and population masses are shown as solid (red) squares (taken from Bastian et al. 2006). In the bottom panel, the horizontal lines show the expected ratio of $M_{\rm dyn}/M_{\rm pop}$ if the underlying stellar IMF has the Kroupa/Chabrier (dashed) or the Salpeter (dashed-dotted) form. Note that the points are consistent with a Kroupa/Chabrier form of the IMF, i.e. the same as that seen locally. While subtle variations are not ruled out by this method, drastic changes are clearly excluded.

SSCs are a particularly important benchmark to test for IMF variations due to their extreme nature. Their star-formation rates as well as their stellar surface and volume densities are orders of magnitudes higher than any galaxies in the nearby or distant universe. The fact that the IMFs of SSCs are similar to that seen in local diffuse star-forming regions argues strongly that the IMF does not change significantly as a function of the local environment. Hence, it would be strange indeed if trends were found linking the global environment (e.g. galaxy type or galaxy-wide star-formation rate) to IMF variations.

3.2. Early Type Galaxies

In a similar way to cluster investigations, one can carry out the same analysis for entire galaxies. This is complicated by the fact that galaxies are made up of multiple generations of stars (with different ages/metallicities) and extinction is often non-uniform. One way to circumvent this problem to some extent is to focus on Early Type Galaxies (ETGs) which are dominated by a single (old) stellar population. This has been done by e.g. Cappellari et al. (2006, 2009), who studied ETGs at low and high redshift (shown as empty circles and filled triangles respectively in Fig. 3). The authors conclude that both samples are consistent with a Chabrier/Kroupa type IMF and that there is no deficiency of low mass stars in these (presumably post-starburst) galaxies.

3.3. Direct Detection of Low-mass Stars in Integrated Spectra

With the advent of sensitive near-IR spectrographs on 8-10m class telescopes, it is now possible to detect the spectroscopic features of low-mass pre-main sequence (PMS) stars. The reason for this is that PMS are orders of magnitude brighter than their main sequence counterparts allowing for the detection of specific spectral features in the integrated light of young stellar clusters (Meyer & Greissl 2005)[2]. One can apply this technique directly to ongoing starbursts (and the stellar clusters within them) in order to detect the presence/absence of low-mass stars, hence directly constraining the IMF. A first test of this technique has been carried out on young clusters in the Antennae galaxies and low-mass PMS stars were unambiguously detected (Greissl et al. 2010). However, due to the relatively large slit-width ($\sim$ 90 pc) the spectra were contaminated by surrounding (unassociated) RGB stars, hence it was not possible to quantify the numbers of low-mass stars present and constrain the slope of the IMF. Future studies, concentrating on nearby systems, are a promising way to detect and quantify IMF variations in the low-mass regime in different extragalactic environments.

---

[2]A similar technique has recently been suggested by van Dokkum & Conroy (2010), to investigate the presence of old sub-solar mass stars in elliptical galaxies.

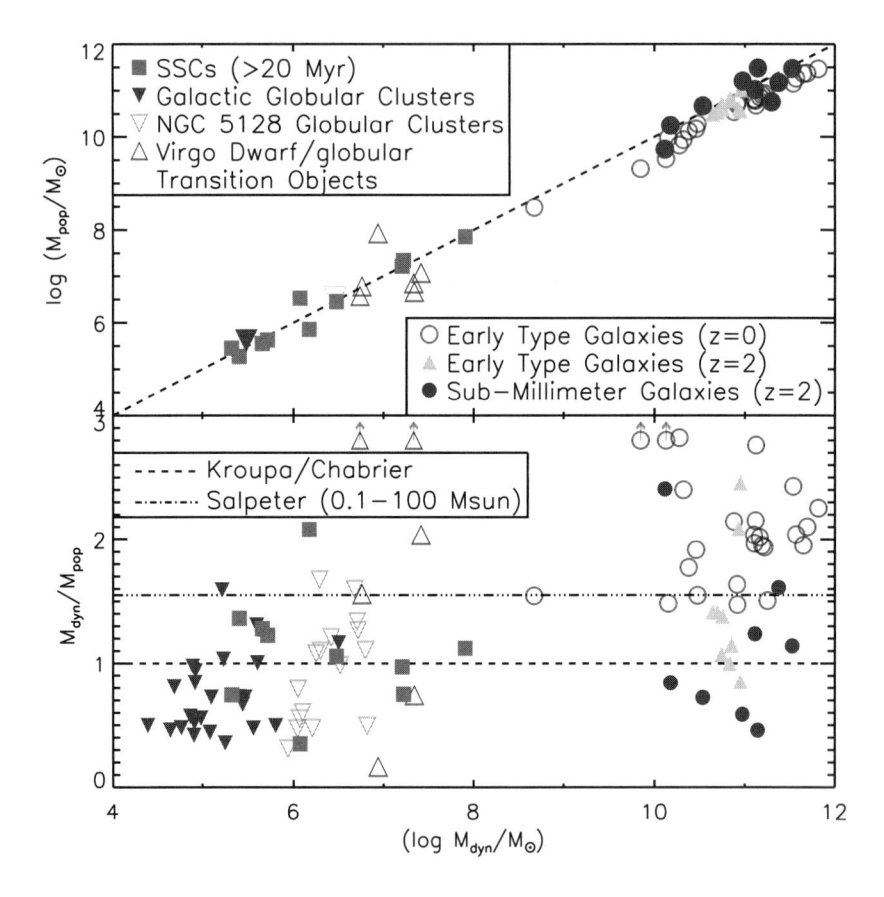

Figure 3. **Top panel:** The measured dynamical mass, M_{dyn} (or through Jeans dynamical modelling in the case of the Early Type Galaxies - see Cappellari et al. 2006) versus the stellar mass derived through modelling of their integrated light using simple/composite stellar population models which adopt an Kroupa type IMF. The solid red squares are super star clusters with ages greater than 20 Myr, the upside down blue filled triangle and open magenta upside down triangle represents the mean of 24 Galactic globular clusters and 16 globular clusters in NGC 5128, respectively, open circles show Early Type Galaxies in the local Universe and filled green triangles and blue circles show Early Type Galaxies and Sub-millimeter galaxies at high redshift. **Bottom panel:** The same as the top panel except now the ratio between M_{dyn} and M_{pop} is shown. If the underlying IMF was well described by a Kroupa-type distribution a ratio of 1 is expected in this representation (shown as a dashed line). If a Salpeter IMF (down to 0.1 $M_\odot$) is a good representation of the underlying IMF a value of 1.55 is expected (dash-dotted line). Note that the galaxy points are upper limits, as a fraction of the M_{dyn} measurement is expected to come from dark matter. See Bastian, Covey, & Meyer (2010) for details.

4. Near-Field Cosmology

While the most massive stars dominate the light that we see in star-forming galaxies at high redshift, it is the low mass stars which will remain for many Gyr, bearing the imprint of their IMF. Locally, there are abundant examples of the remnants of star formation from the early Universe ($z \geq 3$), namely globular clusters (GCs) and dwarf spheroidal galaxies.

4.1. Globular Clusters

Globular clusters (GCs) are thought to be, by and large, made up of a single population of stars with a common age and metallicity. GCs are some of the oldest objects in the Universe (Brodie & Strader 2006), although GC-like objects continue to form in the present day (see § 3.1). Due to the proximity of Galactic GCs, many studies of their mass functions have been carried out using deep HST and ground based imaging. Fitting their mass functions leads to characteristic masses (adopting the Chabrier (2003) form) of $M_c = 0.33$ (Paresce & De Marchi 2000). This characteristic mass is similar to, although slightly larger than, that found for young clusters/star-forming regions in the Milky Way disk ($M_c = 0.1 - 0.3$, see § 2). However, as pointed out by van Dokkum (2008), this characteristic mass is clearly lower than predicted by the results of van Dokkum (2008) and Davé (2008) (based on inferred IMF variations using galaxy scaling relations), which expect $M_c > 4 M_\odot$ for the globular cluster formation epoch ($z = 3 - 5$). Additionally, intermediate age (4.5 Gyr, $z_{\rm form} \sim 0.4$) clusters in the SMC also appear to have mass functions consistent with globular clusters and young clusters (Rochau et al. 2007).

The similarity between the IMF of young clusters and globular clusters is shown in Fig. 1. Fitting tapered power-laws to the young and old clusters results in a consistent picture where the stars in both young and old clusters formed from the same underlying IMF (De Marchi et al. 2010). The characteristic mass in older clusters does appear to be systematically larger than in young clusters and in the field. However, this is expected from a Hubble time of dynamical evolution, which systematically removes the lower mass members of clusters (e.g. Baumgardt et al. 2008; Kruijssen 2009).

4.2. Dwarf Spheroidal Galaixes

Dwarf spheroidal galaxies provide excellent laboratories in which to study the IMF while avoiding the dynamical evolution encountered by GCs. These galaxies have similar ages and star formation histories as GCs, however their stellar densities are orders of magnitude lower, so they experience correspondingly less dynamical evolution. Wyse et al. (2002) used HST data to construct a luminosity function (LF) of stars in Ursa Major, an old, local group dwarf spheroidal galaxy. To avoid the complication of transforming observed luminosities to stellar mass, the authors directly compare the Ursa Major LF to that of globular clusters of similar metallicity. They find that the LFs are indistinguishable down to $\sim 0.3 M_\odot$, and conclude that the IMF must be independent of density and the presence of dark matter into at least the sub-solar mass regime.

Of course the downside to such investigations is that we are limited to the low-mass regime, as higher mass stars have already ended their lives. However, the similarity in the low-mass regime and the characteristic mass already places severe restrictions on theories of IMF variations as a function of redshift, metallicity, stellar density, and the presence/absence of dark matter.

5. Future Work

Upcoming advances in ground/space based instrumentation as well as theoretical and computational power will lead to significantly tighter constraints on the form of the IMF and any potential variations as a function of environment that may exist. How exactly is IMF sampled? What is responsible for setting detailed shape? Are regions like Taurus truly different, or merely statistical flukes? Can "extreme" local environments (e.g. the Galactic center - Bartko et al. 2010) be used to test/rule-out competing star-formation theories? These are just some of the questions that will be addressed in detail over the next few years.

Realizing the promise of these technical advances, however, will require a similar advance in the statistical analysis of IMF measurements. As we enter this new era, we advocate a shift in the means used to characterize and search for variations in the IMFs of resolved stellar populations. Specifically, we recommend that future IMF studies publish their derived space densities, such that IMF variations can be tested by using a direct statistical comparison of two measured IMFs, such as with a KS test, rather than by comparing the parameters of the analytic fit adopted to characterize these increasingly rich datasets. If a functional form is fit to a IMF measurement, we suggest that statistical tools such as the F-test can provide quantitative guidance as to the most appropriate functional form to adopt, and that the uncertainties associated with the derived parameters be clearly reported. By providing a more statistically sound basis for IMF comparisons, we will be better poised to uncover IMF variations where they do exist, and quantify the limits on IMF variations imposed by measurements consistent with a "universal IMF".

References

Baraffe, I., Chabrier, G., & Gallardo, J. 2009, ApJ, 702, L27
Bartko, H., Martins, F., Trippe, S., Fritz, T. K., Genzel, R., Ott, T., Eisenhauer, F., Gillessen, S., Paumard, T., Alexander, T., Dodds-Eden, K., Gerhard, O., Levin, Y., Mascetti, L., Nayakshin, S., Perets, H. B., Perrin, G., Pfuhl, O., Reid, M. J., Rouan, D., Zilka, M., & Sternberg, A. 2010, ApJ, 708, 834
Bastian, N., Covey, K. R., & Meyer, M. R. 2010, ARA&A, 48, 339
Bastian, N., Saglia, R. P., Goudfrooij, P., Kissler-Patig, M., Maraston, C., Schweizer, F., & Zoccali, M. 2006, A&A, 448, 881
Baumgardt, H., De Marchi, G., & Kroupa, P. 2008, ApJ, 685, 247
Bochanski, J. J., Hawley, S. L., Covey, K. R., West, A. A., Reid, I. N., Golimowski, D. A., & Ivezić, Ž. 2010, AJ, 139, 2679
Brodie, J. P., & Strader, J. 2006, ARA&A, 44, 193
Cappellari, M., Bacon, R., Bureau, M., Damen, M. C., Davies, R. L., de Zeeuw, P. T., Emsellem, E., Falcón-Barroso, J., Krajnović, D., Kuntschner, H., McDermid, R. M., Peletier, R. F., Sarzi, M., van den Bosch, R. C. E., & van de Ven, G. 2006, MNRAS, 366, 1126
Cappellari, M., di Serego Alighieri, S., Cimatti, A., Daddi, E., Renzini, A., Kurk, J. D., Cassata, P., Dickinson, M., Franceschini, A., Mignoli, M., Pozzetti, L., Rodighiero, G., Rosati, P., & Zamorani, G. 2009, ApJ, 704, L34
Chabrier, G. 2003, PASP, 115, 763
Covey, K. R., Hawley, S. L., Bochanski, J. J., West, A. A., Reid, I. N., Golimowski, D. A., Davenport, J. R. A., Henry, T., Uomoto, A., & Holtzman, J. A. 2008, AJ, 136, 1778
Dabringhausen, J., Hilker, M., & Kroupa, P. 2008, MNRAS, 386, 864
Davé, R. 2008, MNRAS, 385, 147
De Marchi, G., Paresce, F., & Portegies Zwart, S. 2010, ApJ, 718, 105

Greissl, J., Meyer, M. R., Christopher, M. H., & Scoville, N. Z. 2010, ApJ, 710, 1746
Jeffries, R. D. 2007, MNRAS, 381, 1169
Kroupa, P. 2002, Science, 295, 82
Kroupa, P., Tout, C. A., & Gilmore, G. 1993, MNRAS, 262, 545
Kruijssen, J. M. D. 2009, A&A, 507, 1409
Luhman, K. L., Mamajek, E. E., Allen, P. R., & Cruz, K. L. 2009, ApJ, 703, 399
Maraston, C., Bastian, N., Saglia, R. P., Kissler-Patig, M., Schweizer, F., & Goudfrooij, P. 2004, A&A, 416, 467
Mayne, N. J., & Naylor, T. 2008, MNRAS, 386, 261. 0801.4085
Meyer, M. R., & Greissl, J. 2005, ApJ, 630, L177
Paresce, F., & De Marchi, G. 2000, ApJ, 534, 870
Portegies Zwart, S. F., McMillan, S. L. W., & Gieles, M. 2010, ARA&A, 48, 431
Rochau, B., Gouliermis, D. A., Brandner, W., Dolphin, A. E., & Henning, T. 2007, ApJ, 664, 322
Salpeter, E. E. 1955, ApJ, 121, 161
van Dokkum, P., & Conroy, C. 2010, ArXiv e-prints. 1009.5992
van Dokkum, P. G. 2008, ApJ, 674, 29
Wyse, R. F. G., Gilmore, G., Houdashelt, M. L., Feltzing, S., Hebb, L., Gallagher, J. S., III, & Smecker-Hane, T. A. 2002, New Astronomy, 7, 395

Implications of Radial Migration for Stellar Population Studies

Rok Roškar,[1,2] Victor P. Debattista,[3] Sarah R. Loebman,[1] Željko Ivezić,[1] and Thomas R. Quinn[1]

[1]*Astronomy Department, University of Washington, Box 351580 Seattle, WA, 98195, USA*

[2]*Institute for Theoretical Physics, University of Zürich, CH-8057 Zürich, Switzerland*

[3]*Jeremiah Horrocks Institute, University of Central Lancashire, Preston, PR1 2HE, UK*

Abstract. Recent theoretical work suggests that it may be common for stars in the disks of spiral galaxies to migrate radially across significant distances in the disk. Such migrations are a result of resonant scattering with spiral arms and move the guiding centers of the stars while preserving the circularities of their orbits. Migration can therefore efficiently mix stars in all parts of the Galactic disk. We are rapidly approaching an important confluence of theory and observation, where we may soon be able to uncover signatures of such processes in our own Milky Way. The resolution and robustness of the physical modeling in simulations has improved drastically, while observational datasets are increasing in depth and astrometric accuracy. Here, we discuss the results from our idealized N-body/SPH simulations of disk formation and evolution, emphasizing specifically the observational consequences of stellar migration on the solar neighborhood and the vertical structure of the disk. We demonstrate that radial mixing of stars is a crucial dynamical process that we must try to understand if we are to draw significant conclusions about our Galactic history from the properties of stars in our vicinity.

1. Introduction

Disks of spiral galaxies are kinematically cool, rotationally-supported, self-gravitating systems made up of stars and gas. Stars form in the thin gas layer and begin their lives on mostly circular orbits due to the efficiency with which gas is able to shed excess energy. The stellar orbits are subsequently heated through interactions with giant molecular clouds (GMCs), external perturbations from infalling substructure, and resonances with disk structure. Observations of stars in the solar neighborhood show that velocity dispersion is related to stellar age by a power-law (Holmberg et al. 2009). Given that the epicycle amplitude in the solar neighborhood is limited by $\Delta R \simeq \sqrt{2}\sigma_R/\kappa$ (equation 3.99 Binney & Tremaine 2008, hereafter BT08), $\sigma_R \sim 50$ km/s (Holmberg et al. 2009) for the oldest stars in the solar neighborhood, and $\kappa_0 = 37$ km/s/kpc (BT08), the largest orbital excursions for stars at the solar radius are $\lesssim 2$ kpc. Averaged over the entire sample, the amplitude of radial oscillations $\Delta R \simeq 1.3$ kpc, allowing stars to traverse across ~ 2 kpc during a single orbit.

Such radial orbital oscillations of stars naturally affect the distribution of stellar population properties in the solar neighborhood. If regions of the galaxy evolve approximately as closed-box systems, then one expects there to be a tight relationship between the age of a star and its metallicity, the so-called age-metallicity relation (AMR) (Twarog 1980). Radial oscillations of stars dilute such a relation, and have been considered in the past to explain the large scatter in the observed AMR (Edvardsson et al. 1993; Holmberg et al. 2009). However, if we naively assume a relatively steady metallicity gradient over the past few Gyr of ~ 0.06 dex kpc^{-1} (Daflon & Cunha 2004), the dispersion of the stellar metallicity in the solar neighborhood arising solely from such orbital excursions should be $\lesssim$ 0.1 dex. The observed dispersion in the AMR in the solar neighborhood is larger by at least a factor of 2 (Holmberg et al. 2009). Of course, the gradient could have been steeper in the past, but correcting for orbital eccentricities in the observed Geneva-Copenhagen sample does not explain away the large dispersion in the AMR (Nordström et al. 2004; Binney 2007).

The observationally-limited maximum epicycle amplitude of $\lesssim$ 2 kpc leads to a picture of galactic disks where stars must remain relatively close to their radii of origin, only increasing their epicyclic energies in response to perturbations in the disk. The need for more substantial radial mixing was first discussed by Wielen et al. (1996), who argued that the Sun could not have originated at its present radius based on its anomalously high metallicity with respect to the surrounding stars and the local ISM metallicity. Sellwood & Binney (2002, hereafter SB02) spearheaded the recent resurgence in the interest in radial mixing by showing that radial migrations of even larger magnitude than that postulated by Wielen et al. (1996) are not only possible but very likely in the presence of recurring transient spirals. SB02 demonstrated that the migration takes place due to scattering at the corotation resonance of the spiral, but the large changes in the stellar guiding centers are not accompanied by significant heating. The lack of heating is important because mostly circular stellar orbits are typically assumed to have had a relatively quiet history - SB02 showed that this assumption is not necessarily true. Subsequently, Lépine et al. (2003) explored the orbital evolution of stars under the influence of corotation resonance; Roškar et al. (2008b,a) studied the effects of radial migration in a full N-body simulation of disk formation; Schönrich & Binney (2009a,b) were the first to incorporate these ideas into a chemical evolution model of the Milky Way; Debattista et al. (2006) and Minchev & Famaey (2010) investigated the dependence of the mixing process on the presence of a central bar.

We examine this paradigm further by means of self-consistent, though idealized in its initial configuration, simulation of disk galaxy formation. The results of this simulation have previously been reported in Roškar et al. (2008b,a, R08 collectively hereafter) and Loebman et al. (2010). In this contribution, we specifically focus on the importance of radial migration for studies using local (solar neighborhood) stellar samples to infer something about our Galaxy's past evolution (such as West et al. 2008; Bochanski et al. 2010; see also Bochanski, this volume).

2. The Simulation

The details of our simulation method have been discussed previously in R08. Here, we recall the salient qualitative aspects. The basic picture is one of a hot, spherical gaseous halo embedded in a massive dark matter halo, such as one might expect to exist after the last major merger for a Milky Way (MW) type galaxy. Thus, our model is initialized with a spherical gas halo in hydrostatic equilibrium set by the potential well of a $10^{12} M_\odot$ dark matter (DM) halo. Both halos follow the same NFW profile (Navarro et al. 1997), but we impart a spin upon the gas corresponding to a cosmologically-motivated dimensionless spin parameter $\lambda = 0.039$ (Bullock et al. 2001). The two components are represented by 1 million particles each - this results in a mass resolution of $\sim 10^5 M_\odot$ for the gas and $\sim 10^6 M_\odot$ for the DM. We evolve the system with the parallel N-body + Smooth Particle Hydrodynamics code GASOLINE (Wadsley et al. 2004) for 10 Gyr. As the simulation proceeds, the gas is able to cool and collapse to the center of the potential well, forming a centrifugally-supported disk.

The crucial point here is that we do not insert by hand any properties of the disk. Instead, its final properties are a product of a complex interplay of various processes (gas cooling, stellar feedback, star formation, self-gravity). Our simulation code allows the gas to form stars once appropriate temperature and density are attained, thus forming a disk of stars self-consistently within our sub-grid star-formation framework. The stars in turn are modeled as evolving stellar populations, a fraction of which explode as supernovae (of type Ia and II), returning metals back to the ISM. We follow the yields of α elements and iron separately, allowing us to track gross abundance patterns.

Our model should therefore be regarded as a crude chemical evolution model coupled with self-consistent N-body dynamics. Since we do not set any properties of the disk a priori, our simulation is not a fit to the Milky Way. However, the simulation yields a disk that agrees qualitatively in many ways with the properties of our Galaxy. A critical concern is that the amount of structure that forms in the disk is unreasonable - should the amplitude and frequency of spirals be unusually high, this could result in unreasonable heating rates and unrealistic rates of migration. Importantly, the heating rates as derived from the age-velocity dispersion relationship in the model are even somewhat lower than the MW - power law fits give indices of 0.24, 0.21, 0.25, 0.37 for total, u, v, w dispersions respectively, compared with 0.40, 0.39, 0.40, and 0.53 for the same quantities derived for the solar neighborhood from the Geneva-Copenhagen survey (Holmberg et al. 2009). We speculate that the discrepancy arises from the presence of additional heating sources in the Galaxy, which are not captured by the simulation such as substructure (our disk evolves in isolation) and giant molecular clouds. In our model, the heating is primarily from spirals, and these numbers indicate that the amount of asymmetric structure is not unreasonable. Our simulated disk also yields a reasonably flat rotation curve, though it flattens at ~ 240 km/s compared to 220 km/s for the MW indicating that the model is slightly more massive. The scale length of the simulated disk is 3.2 kpc compared to 2.6 kpc derived for the MW from SDSS data (Jurić et al. 2008), and the gas fraction is $\sim 10\%$.

3. Radial Migration in Disks

On a Galactic scale, the solar neighborhood occupies a very small volume - the largest samples contain stars from a sphere only a few tens of parsecs in radius. It is therefore tempting to consider the grouping of stars around the Sun to be of a common origin, if not of direct relation through a common birth cluster (i.e. the "siblings" of the Sun Portegies Zwart 2009), then at least by virtue of having been born in the same part of the Galaxy. The latter assumption is made by virtually every chemical evolution model of the MW (e.g. Matteucci & Francois 1989; Carigi 1996; Boissier & Prantzos 1999; Chiappini et al. 2001).

Such assumptions, as we show below, oversimplify the dynamical evolution of stars in a galactic disk. Stellar orbits are profoundly affected by the spontaneous growth of spirals in a self-gravitating disk. The heating offered by such transient spirals has been extensively studied in the literature, and is required to explain the observed heating rates (BT08). However, transient spirals can also cause large changes in stellar orbital radii, while keeping the random energy of the orbit untouched, i.e. without heating (Sellwood & Binney 2002). This results in radial migration of stars, where individual radii in a MW-type disk can change by several kpc during the lifetime of the disk.

Figure 1 shows the probability density plot of formation radii as a function of final radii for the simulated disk at the end of the simulation. It is evident that the above assumption of stars found locally now being somehow related, is incorrect - whichever final radius one chooses, its stellar population is significantly contaminated by stars from other radii. The extent of this contamination is drastic - at 8 kpc, a $\sim 2\%$ chance exists that a star has formed at a radius as small as 2 kpc.

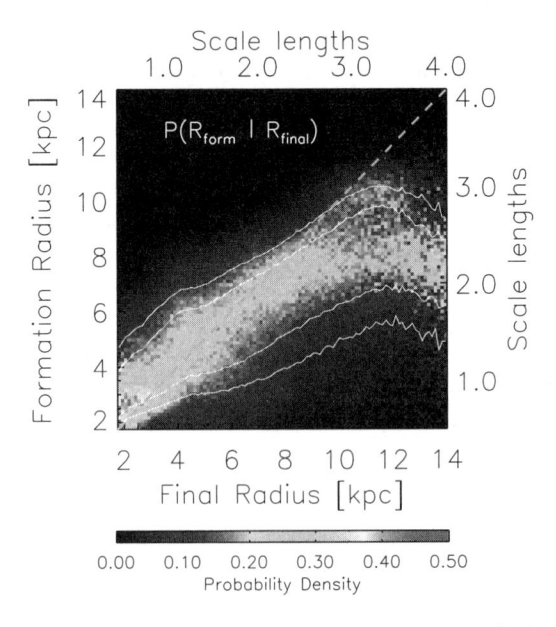

Figure 1. Probability density plot of formation radius given a final radius for stars in the simulation. The inner (outer) white contours enclose 75% (95%) of the mass. If radial migration was unimportant, the highest probabilities would concentrate along the dashed yellow 1:1 line.

For stellar population studies focusing on the solar neighborhood, the most important concern is how the narrow region around the Sun may be affected by such mixing. Due to the limited resolution of our simulation, we cannot isolate a single spherical region 50 pc in radius, but instead focus on the average properties of a "solar neighborhood" defined as radial annuli centered on 8 kpc with several different ΔR. Figure 2 shows the distributions of formation radii for these annuli. From the left panel, it is clear that the distribution is skewed to $R < 8$ kpc and that it is quite extended. A better measure of the contribution of stars born at smaller radii to the overall mix at the solar radius is given by the cumulative distribution function shown in the center panel.

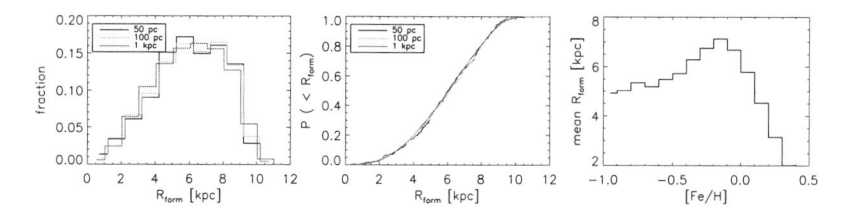

Figure 2. **Left:** Distribution of formation radii R_{form} for stars found in three different annuli centered on 8 kpc. The black, green and red lines show the distribution for annuli 50, 100, and 1000 pc wide respectively. **Center:** The cumulative distribution function of R_{form} for the same annuli. More than 50% of the stars come from $R < 6$ kpc. **Right:** Mean R_{form} as a function of metallicity. The ISM metallicity at 8 kpc is approximately Solar.

At least 50% of stars presently in the solar neighborhood have come from $R < 6$ kpc and $> 80\%$ of the particles currently in the solar neighborhood have come from radii smaller than 8 kpc.

In the right panel of Figure 2, we show the mean formation radii as a function of metallicity. Due to the fact that the disk grows from the inside-out, the majority of metal-rich stars therefore come from the interior of the disk. As a result, the high-metallicity end of the stellar distribution preferentially originates in the inner disk. Of course, because the local ISM metallicity is approximately solar, *all* stars with metallicities above solar must have come from another part of the galaxy, unless significant recent infall of pristine gas diluted the metals or the ISM is azimuthally very inhomogeneous.

Such distributions are of particular interest in light of the well-known planet-metallicity correlation for main-sequence FGK dwarfs hosting gas giants (Fischer & Valenti 2005). If the correlation is intrinsic (i.e. availability of planet-building material in the protoplanetary nebula) rather than a result of the pollution of a star's atmosphere by the accretion of one of the planets, the right panel of Figure 2 suggests that locally-found planet hosts should have spent considerable fractions of their lives in the interior of the Galaxy. Should the planet-metallicity correlation extend to hosts of lower-mass planets, the galactic history of such stars could have interesting consequences for the searches for signatures of extraterrestrial life (note that this correlation does not seem to exist for planets around giants, e.g. Pasquini et al. 2007; Ghezzi et al. 2010).

4. Vertical Evolution - Migrating into the Thick Disk

As stars move outward from the inner disk, they preserve their vertical energy, but the restoring force from the disk decreases due to the smaller surface density. Consequently, the amplitudes of their vertical oscillations also increase. The basic nature of this process is illustrated by Figure 3. At the midplane (top left), the stellar population mix is strongly influenced by local star formation so the majority of stars are locally born and relatively young. Note, however, that the distribution of R_{form} is very extended, as shown in Figure 2 (i.e. the projection of top left panel of Figure 3 along onto the y-axis). As we consider vertical slices at increasing heights from the midplane, the stellar populations become dominated by old stars that have come from the inner disk.

Fitting the vertical disk profile at 8 kpc with a double sech^2 profile yields two distinct components with scale heights of 381 pc and 913 pc, roughly consistent with the values found for the thin and thick disk of the MW (270 pc and 1200 pc Jurić et al. 2008).

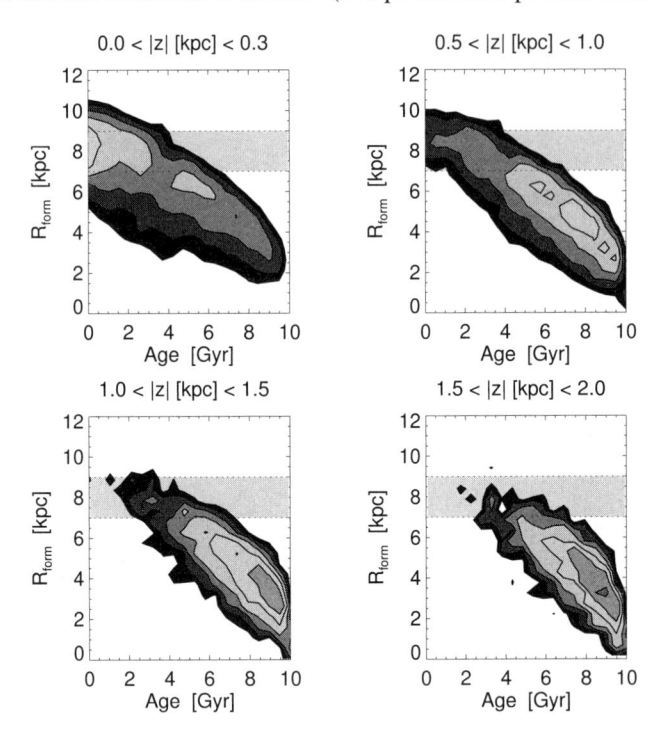

Figure 3. Distributions of formation radius vs. age for particles found in various slices above the midplane, in a 2 kpc annulus centered on 8 kpc (shown by the shaded region). Due to the high density of individual points, logarithmically-spaced contours are used as a proxy - colors indicate relative density, increasing from blue to red.

Because the disk grows and evolves, the birth environment of these migrated stars was quite different from their present surroundings. The metallicity gradient in the past was steeper (see Figure 2 of Roškar et al. 2008a), resulting in a large number of stars forming at low metallicity. The disk was also young and not yet polluted by supernova Ia metals so these old stars tend to be α-enhanced. As these stars are old, they have been heated through secular processes in the disk and therefore lag the local standard of rest. In short, the migrated population of old stars at the solar radius has all of the characteristics of the thick disk. See Schönrich & Binney (2009b) and Loebman et al. (2010) for more detailed comparisons between the properties of migrated stars and the observed MW thick disk.

In Figure 4 we show a part of this comparison. The left panel shows the Toomre diagram for the particles in the midplane at the solar radius. We select the thin and thick disk stars based on their kinematics, as is frequently done in the literature (e.g. Bensby et al. 2003, 2005), though because we do not know the kinematic properties of different components a priori we use a simple ad-hoc hard-cut dividing the thin and thick disk at

-70 km/s. Nonetheless, dividing the stars in this way gives us a rough idea of the interdependence between kinematics and chemistry. In the right panel of Figure 4, we show the α-element enrichment as a function of metallicity for the kinematically-identified thin and thick disk stars - the thick disk is enhanced, on average, in α elements, consistent with observational results for the MW thick disk (Bensby et al. 2005).

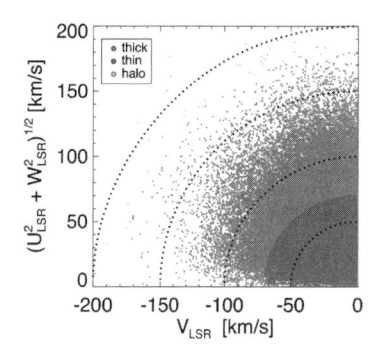

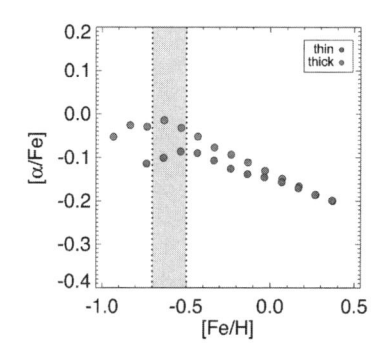

Figure 4. **Left:** Toomre diagram for stars in the midplane between 7-9 kpc. Color indicates a kinematic cut to define the thick and thin disks - stars with $V_{LSR} \lesssim -70$ km/s are assigned thin disk membership, while stars with $-150 \lesssim V_{LSR} < -70$ km/s are considered thick disk. **Right:** α-element enrichment as a function of metallicity of stars selected to be in the thick and thin disks based on kinematics.

Given the prevalence of sub-structure and streams found in the MW halo, presumably a consequence of the hierarchical merging process integral to the LCDM paradigm of galaxy formation, it is usually assumed that the thick disk is likewise a relic our Galaxy's cosmological history (see Wyse 2008 for a review). While we do not try to dispute the fact that our Galaxy is embedded in a tumultuous environment where disk-perturbing interactions are frequent, the fact remains that radial migration *will* contribute to the stellar populations found away from the disk plane. Radial migration must therefore be considered when assessing the cosmological significance of the stars found in the thick disk.

5. Conclusions

In galaxies with recurring spirals, stars do not remain near their birth radii. The guiding centers of their orbits can evolve drastically over the course of their lives. Conversely, no part of the disk in a spiral galaxy is free from contamination by stars that have come from other disk regions. Interpretations of stellar population properties found anywhere in the MW disk need to account for this fact. We have shown that in our models of MW-like disk formation, more than 50% of the stars have come from $R \lesssim 6$ kpc. The bias toward smaller formation radii is most pronounced at higher metallicities. Furthermore, we have argued that the migrated population also naturally forms a thick disk. This thick disk is in many ways reminiscent of the observed thick disk in the MW, yet its formation does not in any way depend on the galaxy's cosmological environment. Within the next decade, we can look forward to ground-breaking new surveys such as PanSTARRS, LSST and GAIA that will yield vast new datasets, detailing the structure of our Galaxy. Combined with detailed dynamical models, such data will allow us to finally discriminate between the various models and scenarios for the formation of our MW disk.

References

Bensby, T., Feltzing, S., & Lundström, I. 2003, A&A, 410, 527
Bensby, T., Feltzing, S., Lundström, I., & Ilyin, I. 2005, A&A, 433, 185
Binney, J. 2007, Dynamics of Disks (Island Universes - Structure and Evolution of Disk Galaxies), 67
Binney, J., & Tremaine, S. 2008, Galactic Dynamics: Second Edition (Galactic Dynamics: Second Edition, by James Binney and Scott Tremaine. ISBN 978-0-691-13026-2 (HB). Published by Princeton University Press, Princeton, NJ USA, 2008.)
Bochanski, J. J., Hawley, S. L., Covey, K. R., West, A. A., Reid, I. N., Golimowski, D. A., & Ivezić, Ž. 2010, AJ, 139, 2679
Boissier, S., & Prantzos, N. 1999, MNRAS, 307, 857
Bullock, J. S., Dekel, A., Kolatt, T. S., Kravtsov, A. V., Klypin, A. A., Porciani, C., & Primack, J. R. 2001, ApJ, 555, 240
Carigi, L. 1996, Revista Mexicana de Astronomia y Astrofisica, 32, 179
Chiappini, C., Matteucci, F., & Romano, D. 2001, ApJ, 554, 1044
Daflon, S., & Cunha, K. 2004, ApJ, 617, 1115
Debattista, V. P., Mayer, L., Carollo, C. M., Moore, B., Wadsley, J., & Quinn, T. 2006, ApJ, 645, 209
Edvardsson, B., Andersen, J., Gustafsson, B., Lambert, D. L., Nissen, P. E., & Tomkin, J. 1993, A&A, 275, 101
Fischer, D. A., & Valenti, J. 2005, ApJ, 622, 1102
Ghezzi, L., Cunha, K., Schuler, S. C., & Smith, V. V. 2010, ApJ, 725, 721
Holmberg, J., Nordström, B., & Andersen, J. 2009, A&A, 501, 941
Jurić, M., et al. 2008, ApJ, 673, 864
Lépine, J. R. D., Acharova, I. A., & Mishurov, Y. N. 2003, ApJ, 589, 210
Loebman, S. R., Roskar, R., Debattista, V. P., Ivezic, Z., Quinn, T. R., & Wadsley, J. 2010, ArXiv e-prints
Matteucci, F., & Francois, P. 1989, MNRAS, 239, 885
Minchev, I., & Famaey, B. 2010, ApJ, 722, 112
Navarro, J. F., Frenk, C. S., & White, S. D. M. 1997, ApJ, 490, 493
Nordström, B., Mayor, M., Andersen, J., Holmberg, J., Pont, F., Jørgensen, B. R., Olsen, E. H., Udry, S., & Mowlavi, N. 2004, A&A, 418, 989
Pasquini, L., Döllinger, M. P., Weiss, A., Girardi, L., Chavero, C., Hatzes, A. P., da Silva, L., & Setiawan, J. 2007, A&A, 473, 979
Portegies Zwart, S. F. 2009, ApJ, 696, L13
Roškar, R., Debattista, V. P., Quinn, T. R., Stinson, G. S., & Wadsley, J. 2008a, ApJ, 684, L79
Roškar, R., Debattista, V. P., Stinson, G. S., Quinn, T. R., Kaufmann, T., & Wadsley, J. 2008b, ApJ, 675, L65
Schönrich, R., & Binney, J. 2009a, MNRAS, 396, 203
— 2009b, MNRAS, 399, 1145
Sellwood, J. A., & Binney, J. J. 2002, MNRAS, 336, 785
Twarog, B. A. 1980, ApJ, 242, 242
Wadsley, J. W., Stadel, J., & Quinn, T. 2004, New Astronomy, 9, 137
West, A. A., Hawley, S. L., Bochanski, J. J., Covey, K. R., Reid, I. N., Dhital, S., Hilton, E. J., & Masuda, M. 2008, AJ, 135, 785
Wielen, R., Fuchs, B., & Dettbarn, C. 1996, A&A, 314, 438
Wyse, R. F. G. 2008, ArXiv e-prints

Searching for Ultra-cool Objects at the Limits of Large-scale Surveys

D. J. Pinfield,[1] K. Patel,[1] Z. Zhang,[1] J. Gomes,[1] B. Burningham,[1] A. C. Day-Jones,[2] and J. Jenkins[2],

1*Centre for Astrophysics Research, University of Hertfordshire, College Lane, Hatfield, AL10 9AB, UK*

2*Departamento de Astronomia, Universidad de Chile, Camino del Observatorio 1515, Santiago, Chile*

Abstract. We have made a search (to Y=19.6) of the UKIDSS Large Area Survey (LAS DR7) for objects detected only in the Y-band. We have identified and removed contamination due to solar system objects, dust specs in the WFCAM optical path, persistence in the WFCAM detectors, and other sources of spurious single source Y-detections in the UKIDSS LAS data-base. In addition to our automated selection procedure we have visually inspected the ~600 automatically selected candidates to provide an additional level of quality filtering. This has resulted in 55 good candidates that await follow-up observations to confirm their nature. Ultra-cool LAS Y-only objects would have blue Y-J colours combined with very red optical-NIR SEDs - characteristics shared by Jupiter, and suggested by an extrapolation of the Y-J colour trend seen for the latest T dwarfs currently known.

1. Introduction

Much effort in recent years has focused on colour-based searches for ultra-cool populations in large-scale survey combinations such as UKIDSS and SDSS (Burningham et al. (2010), Pinfield et al. (2008), Schmidt et al. (2010), Zhang et al. (2009)). Such work will continue to gain in strength as UKIDSS is completed and VISTA coverage grows. However, the most extreme types of object (e.g. lowest T_{eff}) are generally to be found at the noisy limits of these surveys, with intrinsically extreme colours often leading to detection in some bands but not in others. For example, late T dwarfs are often K-band non detections in the UKIDSS Large Area Survey (LAS), and in many cases are also H-band non detections (with good detections only in the Y- and J-bands). Allowing for this naturally pushes up levels of contamination as one relies on less and less information (i.e. fewer bands) to select candidate samples. Indeed, contamination becomes particularly pernicious in the case of single band detections, since solar system objects are also contributing - for surveys where multi-band imaging is obtained in a non-simultaneous way, solar system objects moving across the sky at many arcseconds per hour are a major sources of single-band false positives. This paper presents the results of an initial search for ultra-cool objects detected only in the Y-band, and summarises our methods for removing contamination.

2. Why do a Y-only Search?

The UKIDSS LAS loses sensitivity to the latest T dwarfs in the K and H bands due to their extremely blue J-H and J-K colours (e.g. Fig 1a). Observational trends (down to 500K, see Fig 1b) suggest it may also be the case that even cooler objects become blue in Y-J also. Although the NIR spectrum of Jupiter is a combination of emitted and reflected light, it is note-worthy that Jupiter does indeed have blue Y-J colour. The LAS is complete to Y=19.6, J=19.1 (about half a mag brighter than the 5-sigma detection limits), and thus becomes J-band incomplete (but Y-band complete to Y=19.6) in the colour range Y-J<0.5 (see Fig 1a). A search for Y-only detections down to Y=19.6 is thus much more sensitive to objects in this colour space than a search that requires J-band detection.

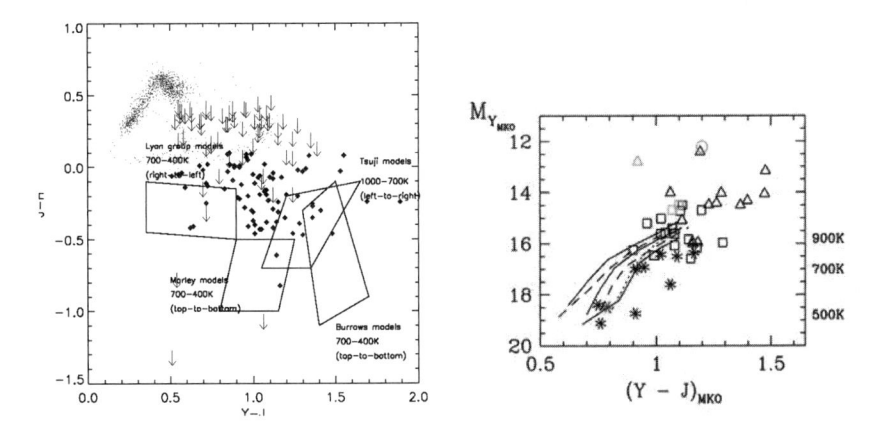

Figure 1. (a) Two colour diagram of T dwarfs from the LAS (from Burningham et al. (2010)). The LAS is Y-band complete (to Y=19.6) but becomes J-band incomplete in the region Y-J<0.5. (b) Cool dwarf colour magnitude diagram from Leggett et al. (2010). Circles are M, triangles are L, squares are T0-6.5, and asterisks are T7-9 dwarfs. There is a trend towards bluer Y-J colour for the latest objects.

3. Optical Information

To avoid contamination from stars, SDSS non detection is also important. For Y-J<0.5 stellar optical-NIR colours are well constrained, e.g. i-Y<2 (see Fig 2). The SDSS 5-sigma limit is i=22.3 and the completeness limit is i=21.3 (Abazajian et al. 2009), so we would expect the large majority of stars to be detected in SDSS-i down to Y=19.6 (i-Y<1.7).

4. The Y-only Sample

We have selected Y-detections with Y≤19.6 that are not detected in J H K or in SDSS, and are isolated (no neighbours within 8 arcsecs). Some source contamination may be expected to occur at the faint limit due to M dwarfs with i-Y∼2, but there are several other more pernicious sources of contamination that must be dealt with.

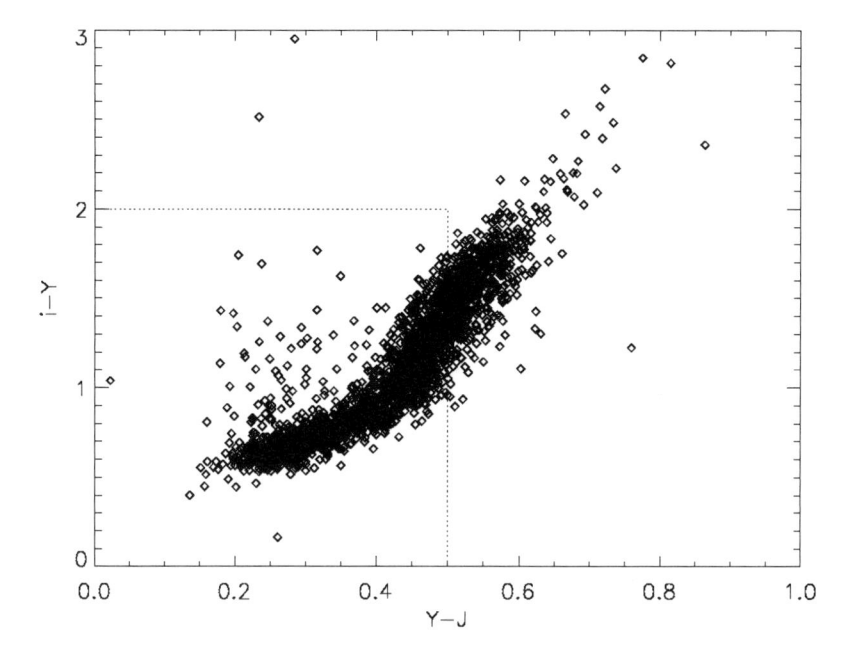

Figure 2. SDSS+LAS colours for a sample of typical stars. In the Y-J<0.5 range it can be seen that the vast majority of stars have i-Y<2.

5. Solar System Objects (SSOs)

We employ an ellipticity limit (e<0.1) to remove the rapid SSOs from our sample (see Fig 3). To identify slower moving SSOs we made a search for J-only LAS detections and cross-matched with our Y-only sample out to 30 arcseconds (requiring Y-J between -0.5 and +1.0 for SSOs). If the J-band image was measured within 26 mins of the Y-band image then we would expect to remove slower moving SSOs in this way. We thus limited our Y-only search to regions where Y(epoch)-J(epoch)<26 mins. This requirement is met by 60% of LAS DR7, and we identified 1378 likely SSOs in the way (see Fig 4).

6. Visual Inspection

After the automated removal of SSOs, we visually inspected 607 remaining sources. Other major sources of contamination include dust specks in the detector, and persistence from bright stars. Illustrative examples of such contamination are shown in Fig 5a-b and Fig 6a-b with explanation in the caption. Figures 7-8 show how SSOs can pass through our automated procedure, but be picked up by visual inspection. After visual inspection and identification of the different varieties of contamination, we rejected all but 55 Y-only candidates from our LAS DR7 search (see Fig 9).

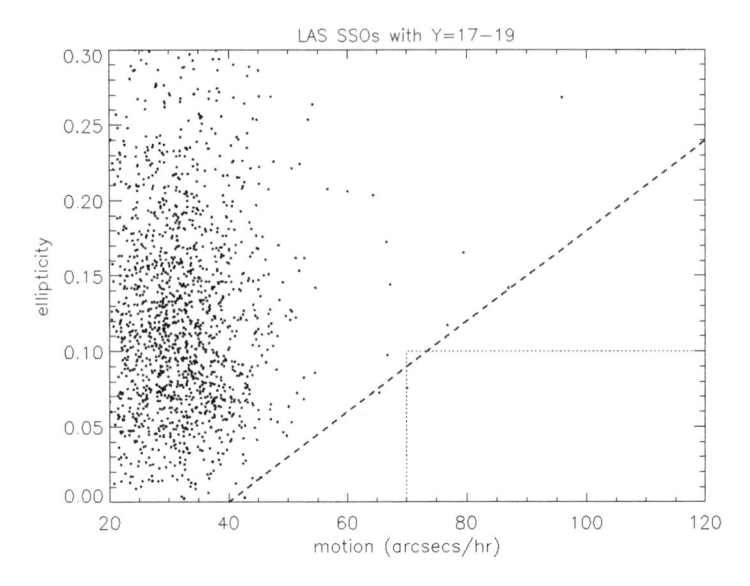

Figure 3. Ellipticity-motion plot for a sample of Y-only detections. The dashed line shows the increasing ellipticity due to the rapid motion of solar system objects (SSOs) across the sky (tens of arcseconds per hour). Our ellipticity<0.1 requirement should remove SSOs with motion >70 arcsecs/yr.

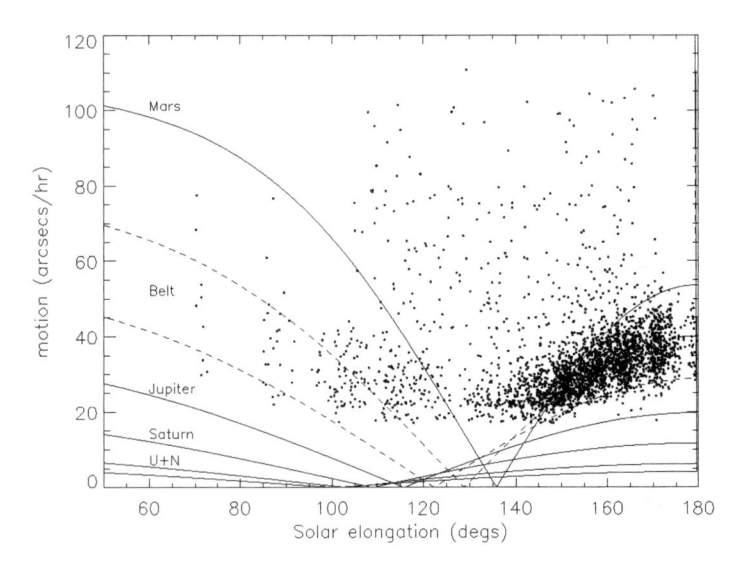

Figure 4. Motion against solar elongation plot for the SSOs identified in our Y-only sample. The lines indicate the motion of planets (solid) and members of the asteroid belt (dashed). It can be seen that the majority of SSOs are in or near the belt region.

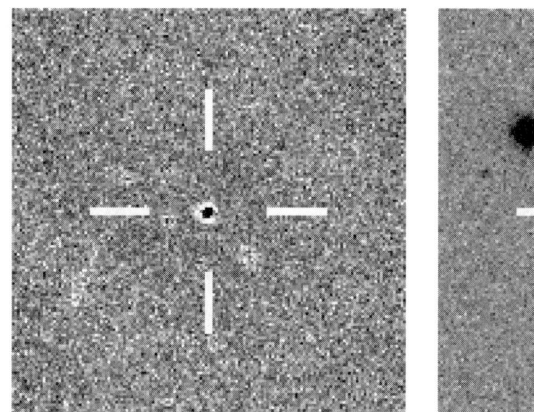

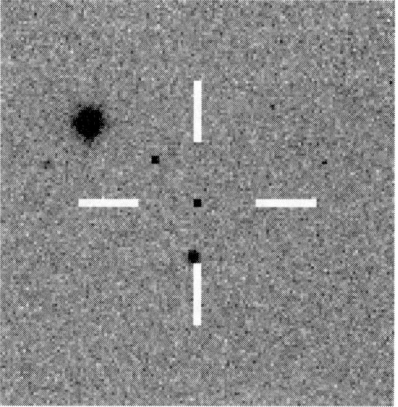

Figure 5. (a) Example of a dust speck. Optical effects can lead to a "fried egg-like" appearance. (b) Another example of a dust speck. The jitter movement between separate Y exposures (that are combined into the final LAS Y image) can lead to "Y-only pairs" of sources.

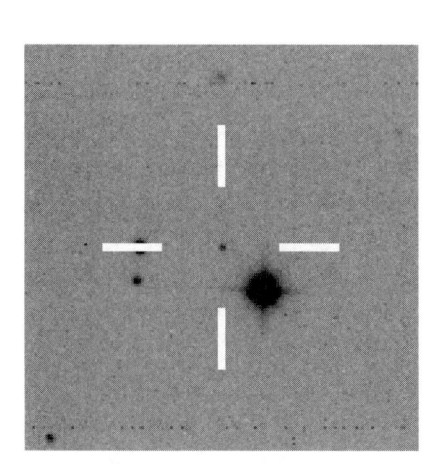

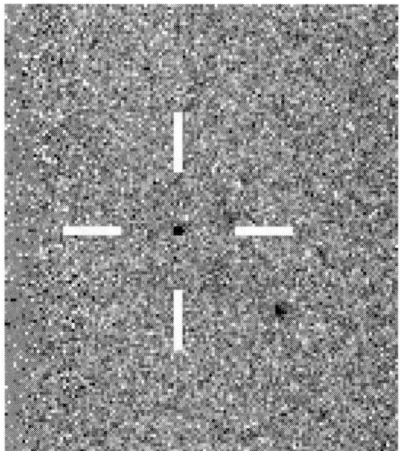

Figure 6. (a) An example of a "persistence" source. The jitter movement between separate Y exposures can lead to persistence on the array at the point where the bright star previously was. In the final LAS Y image this appears as a faint source offset from the bright star. (b) An example of a source <30 arcsecs away from the WFCAM detector edge. One cannot reliably rule out rapidly moving SSOs if the object is too close to the detector edge.

7. Following up the Final Sample

We are currently in the process of obtaining new epoch images of our Y-only sample. Additional Y-band images will quickly and simply identify dust specks which will not be present in the new images. Detected sources will also be followed up with additional photometry: z-band to rule out M dwarfs, and deeper JHK to reveal full range of colours. High proper motion objects will also be apparent from this follow-up and non moving objects may even turn out to be high-z quasars.

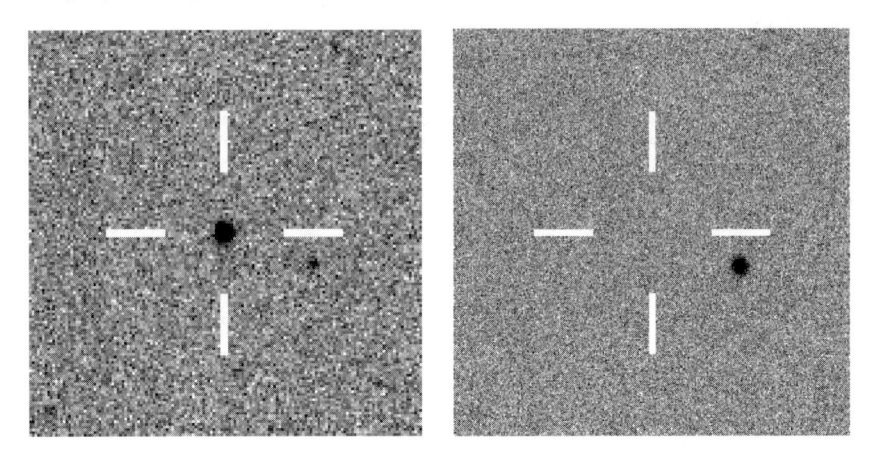

Figure 7. YJ images (see also Fig 8 for HK images) taken back-to-back showing the passage of an SSO. If the position of the SSO in the J-band image overlaps with another source, such objects can pass through our automated SSO identification, and needed to be identified through visual inspection.

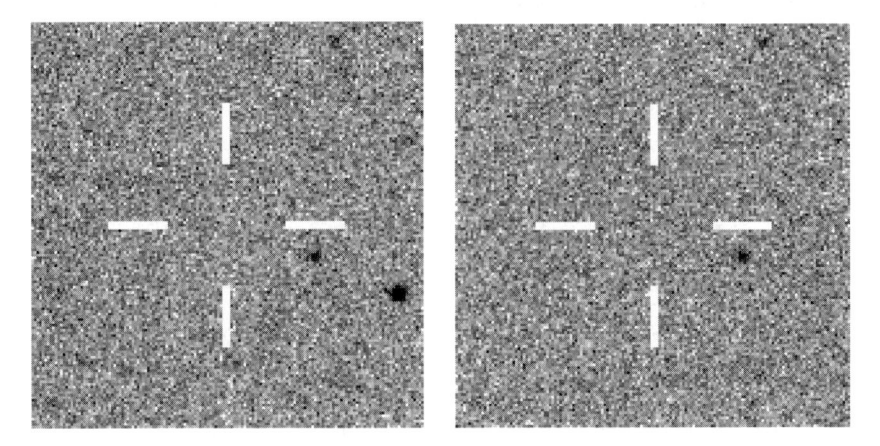

Figure 8. HK images showing the passage of an SSO (see Fig 7 for YJ images and caption).

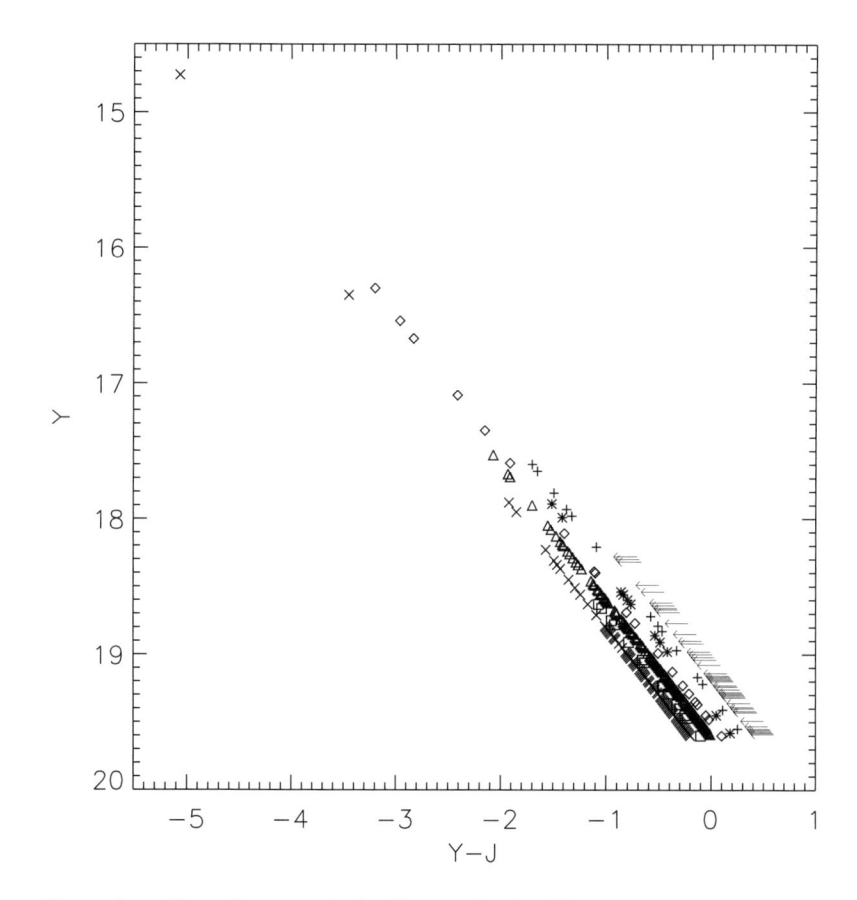

Figure 9. The colour-magnitude diagram for the 607 Y-only objects produced from our automated selection process. Our final sample of good Y-only sources are shown as arrows (indicating upper limits in Y-J), and range in brightness from Y=18.25-19.6.The different types of contamination are indicated with different symbols that have been offset blue-ward in colour by -0.2 to -0.7 for clarity: Plus symbols are dust speck "fried eggs", asterisks are persistence effects, open diamonds are SSOs, open triangles are dust speck "Y-only pairs", open squares are candidates for which a 2nd Y-image was available in UKIDSS (ruling out the candidate), and X symbols represent rejects due to several other issues (object <30 arcsecs from edge of array, source has a non-point-like profile, no useful SDSS coverage available, and a pin-cushion effect in J).

Acknowledgments. JG is supported by RoPACS, a Marie Curie Initial Training Network funded by the European Commission's Seventh Framework Programme.

References

Abazajian, K., Adelman-McCarthy, J., Agueros, M., & Allam, S. 2009, ApJS, 182, 543
Burningham, B., et al. 2010, MNRAS, 406, 1885

Leggett, S., Burningham, B., Saumon, D., & M.S., M. 2010, ApJ, 710, 1627
Pinfield, D. J., Burningham, B., Tamura, M., Leggett, S. K., Lodieu, N., Lucas, P. W., Mortlock, D. J., Warren, S. J., Homeier, D., Ishii, M., Deacon, N. R., McMahon, R. G., Hewett, P. C., Osori, M. R. Z., Martin, E. L., Jones, H. R. A., Venemans, B. P., Day-Jones, A. C., Dobbie, P. D., Folkes, S. L., Dye, S., Allard, F., Baraffe, I., Barrado Y Navascués, D., Casewell, S. L., Chiu, K., Chabrier, G., Clarke, F., Hodgkin, S. T., Magazzù, A., McCaughrean, M. J., Nakajima, T., Pavlenko, Y., & Tinney, C. G. 2008, Monthly Notices of the Royal Astronomical Society, 390, 304. URL http://blackwell-synergy.com/doi/abs/10.1111/j.1365-2966.2008.13729.x
Schmidt, S., West, A., Hawley, S., & Pineda, J. 2010, AJ, 139, 1808
Zhang, Z., R.S., P., Jones, H., & Pinfield, D. 2009, A&A, 497, 619

The Seattle skyline (Top) and CS16 participants chatting between sessions (Bottom)

Part VI

Splinter Sessions

Habitability of Planets Orbiting Cool Stars

Rory Barnes,[1,2] Victoria S. Meadows,[1,2] Shawn D. Domagal-Goldman,[2,3,4] René Heller,[5] Brian Jackson,[4,6] Mercedes López-Morales,[7,8] Angelle Tanner,[9] Natalia Gómez-Pérez,[8,10] and Thomas Ruedas[8]

[1]*Astronomy Department, University of Washington, Box 351580, Seattle, WA 98195, USA*

[2]*Virtual Planetary Lab, USA*

[3]*Planetary Sciences Division, NASA Headquarters, 300 E St. SW, Washington, DC, USA*

[4]*NASA Postdoctoral Program Fellow*

[5]*Astrophysikalisches Institut Potsdam (AIP), An der Sternwarte 16, 14482 Potsdam, Germany*

[6]*NASA Goddard Space Flight Center, Greenbelt, MD 20771, USA*

[7]*Institut de Ciències de L'Espai (CSIC-IEEC), Campus UAB, Fac. Ciències. Torre C5 parell 2, 08193 Bellaterra, Barcelona, Spain*

[8]*Carnegie Institution of Washington, Dept. of Terrestrial Magnetism, 5241 Broad Branch Road NW, Washington, DC, 20015, USA*

[9]*Department of Physics and Astronomy, Georgia State University, One Park Place, Atlanta, GA 30303, USA*

[10]*Departamento de Física, Universidad de los Andes, Cr 1E No 18A-10, Bogotá, Colombia.*

Abstract. Terrestrial planets are more likely to be detected if they orbit M dwarfs due to the favorable planet/star size and mass ratios. However, M dwarf habitable zones are significantly closer to the star than the one around our Sun, which leads to different requirements for planetary habitability and its detection. We review 1) the current limits to detection, 2) the role of M dwarf spectral energy distributions on atmospheric chemistry, 3) tidal effects, stressing that tidal locking is not synonymous with synchronous rotation, 4) the role of atmospheric mass loss and propose that some habitable worlds may be the volatile-rich, evaporated cores of giant planets, and 5) the role of planetary rotation and magnetic field generation, emphasizing that slow rotation does not preclude strong magnetic fields and their shielding of the surface from stellar activity. Finally we present preliminary findings of the NASA Astrobiology Institute's workshop "Revisiting the Habitable Zone." We assess the recently-announced planet Gl 581 g and find no obvious barriers to habitability. We conclude that no known phenomenon completely precludes the habitability of terrestrial planets orbiting cool stars.

1. Introduction

Initially dismissed as potential habitats, planets orbiting M dwarfs have lately seen renewed interest (Tarter et al. 2007; Scalo et al. 2007). With lower luminosities, their "habitable zones" (HZ), the range of orbits in which an Earth-like planet could support surface water (Kasting et al. 1993), are significantly closer-in than for solar-type stars. This proximity leads to new perils such as increased susceptibility to stellar activity and stronger tidal effects. The discovery of exoplanets spurred more detailed and careful evaluation of these phenomena, and many researchers now argue that they are not as dangerous for life as previously feared. On the contrary, these planets may be ideal laboratories to test models of geophysics, atmospheric dynamics, celestial mechanics, photochemistry, aeronomy, and ultimately habitability.

Without analogs in our Solar System or adequate remote sensing capabilities, the surface properties of M dwarf planets can only be considered theoretically. Nonetheless, progress has been made in several areas, such as modeling of planetary interiors (Sotin et al. 2007; O'Neill & Lenardic 2007), atmospheric mass loss (Yelle 2004; Segura et al. 2010), atmospheric dynamics (Joshi 2003; Heng & Vogt 2010), and tidal effects (Jackson et al. 2008; Heller et al. 2010). This chapter is a multidisciplinary study of the potential habitability of M star planets, but with an astrophysical bias. A full treatment would require far more space than permitted by this format. For a more comprehensive analysis of M dwarf planet habitability (and habitability in general), see Tarter et al. (2007), Scalo et al. (2007), and § 7.

This chapter is organized as follows. First we examine the current detection limits of terrestrial planets due to stellar variability. Second, we explore the different chemical reactions in planetary atmospheres due to different stellar spectral energy distributions. Next we examine tidal effects. Fourth, we consider the possible existence of "habitable evaporated cores" of giant planets. Fifth, we explore the magnetic fields of terrestrial planets. Finally, we report key interdisciplinary findings from a recent NASA Astrobiology Institute workshop titled "Revisiting the Habitable Zone."

2. Is M Dwarf Variability a Barrier to Detecting Earth-Mass Planets?

With > 500 extra-solar giant and super-Earth planets detected around nearby stars, we are now probing the dependence on stellar mass of planet formation and migration. At the same time, we are pushing the limits of radial velocity (RV) planet detection methods below "super-Earths" (planet mass $m_p \leq 10$ $M_\oplus$) and toward a bona fide Earth analog, e.g. Mayor (2009). M dwarfs play a critical role in both of these scientific pursuits. The diminutive masses of M dwarfs result in a higher sensitivity to lower-mass planets at a given RV precision, and many of the lightest exoplanets have been detected around M dwarfs with semi-amplitudes of ~ 1 m/s (Mayor 2009). New instruments have the potential to push the instrumental residuals down to 10 cm/s (Pepe & Lovis 2008), the amplitude of an Earth-mass planet in the habitable zone of a solar-type star. The amplitude of that same planet around an early-type M dwarf is 30-60 cm/s and > 1 m/s for late-M dwarfs. These limits imply that we currently have the instrumental precision necessary to detect terrestrial planets in the HZs of the nearest M dwarfs.

On the other hand, M dwarfs are hundreds of times fainter than G stars at optical wavelengths. As a result, there has been a surge in near-infrared RV survey efforts with high-resolution spectrographs such as NIRSPEC on Keck and CRIRES on the VLT.

Using telluric lines for wavelength calibration, these surveys are capable of reaching RV precisions of 20-100 m/s on M and L dwarfs depending on the rotation rate of the star (Blake et al. 2010). With additional calibration from either an ammonia or methane gas cell, these instruments can reach RV precisions of 5-10 m/s for mid to late M dwarfs (Bean et al. 2010a,b). While the amplitude of the RV signature is larger around these lighter mass stars, M dwarfs suffer from more flaring and starspot activity than their solar-type counterparts. These photospheric activity sources introduce perturbations, often called jitter, into both the photocenter of the star and the disk-averaged radial velocity. The degree of the perturbation depends on the rotation rate of the star, the flare occurrence rate, the starspot lifetime and the degree of starspot coverage. Makarov et al. (2009) used a simple starspot model to estimate that RV jitter for F, G and K stars can reach up to 69, 38 and 23 cm/s, respectively and, thus, inhibit our ability to detect Earth-mass planets in the HZ.

Stellar variability may be measured in several different ways. The ultra-precise photometry from the *Kepler* telescope shows that, in general, early-M dwarfs display less astrophysical jitter than their FGK counterparts (Basri et al. 2010; Ciardi et al. 2010). However, *Kepler* data include few mid-M dwarfs and no late-M dwarfs. Additionally, jitter levels may vary from star to star within the same spectral class suggesting we need to quantify the jitter for individual targets. In fact, in some instances the variability of the Hα line, a common proxy for stellar activity, is *anti-correlated* with the RV jitter (Zechmeister et al. 2009).

Starspots are another source of astrophysical noise, and many groups have developed basic (one continuous, circular spot) and complex (various spot lifetimes and temperatures, multiple spots at a range of latitudes) models. The expected contribution to RV jitter from starspots on M dwarfs was addressed by Reiners et al. (2010) using a model with a single spot and multiple wavelengths and stellar rotation rates. For example, a 2800 K star with a 2600 K spot and a 2 km/s rotation rate will have an RV jitter of 10 m/s in the optical and 2 m/s in the infrared. These levels go up to 14 and 10 m/s, respectively, for a rotation rate of 10 km/s. This jitter is larger than the 2 m/s precision needed to detect an Earth-mass planet in the habitable zone of an M6 star.

RV technology will soon reach the 10 cm/s precision necessary to detect Earth-mass planets in the HZs of nearby stars. However, we do not know the levels of intrinsic stellar jitter of stars in our local neighborhood – the majority of which are M dwarfs. Recent advances have demonstrated that broad assumptions based on spectral type are inadequate. Precise photometry and complex star spot modeling, as well as the forthcoming statistical results from the *Kepler, MOST* and *CoRoT* missions could provide valuable insight, but, in certain cases, jitter may prevent the detection of terrestrial planets.

3. Implications of M Dwarf Spectral Energy Distributions on Life Detectability

Most plans to search for and characterize life on extrasolar planets involve probing the atmospheric chemistry of those planets by analyzing their spectra to search for specific gaseous components which can only be produced by biological processes. For example, there have been calls to search for the simultaneous presence of methane (CH_4) and either molecular oxygen (O_2), or its photochemical by-product, ozone (O_3) (for a review see Des Marais et al. 2002). Others have suggested looking for life by searching for methyl-chloride (CH_3Cl) (Segura et al. 2005), or nitrous oxide (N_2O) (Sagan et al.

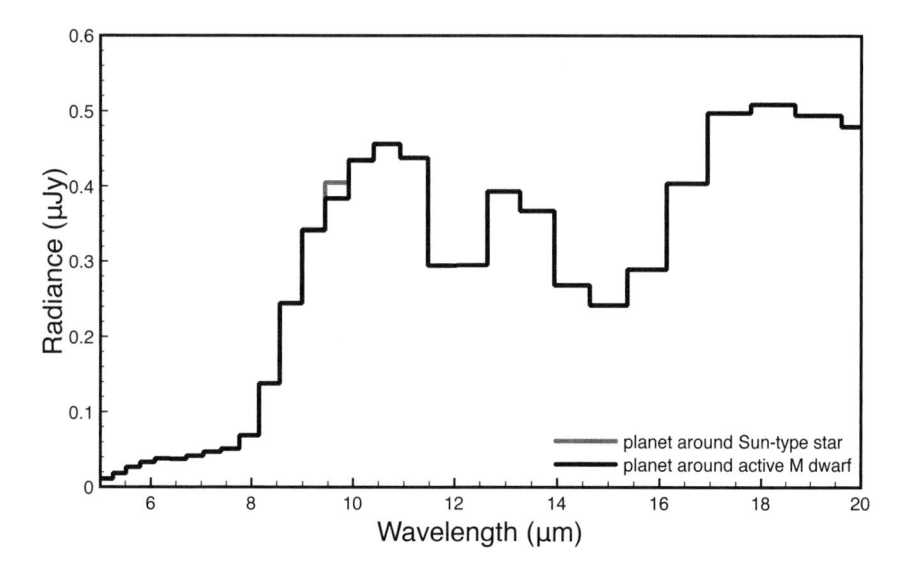

[Figure: Model spectra plot showing radiance (μJy) vs wavelength (μm) from 6 to 20 μm, comparing planet around Sun-type star and planet around active M dwarf.]

Figure 1. Model spectra of planets orbiting different stars. The black curve is the predicted reflection spectrum from a hypothetical planet orbiting AD Leo with about the resolution expected for *Terrestrial Planet Finder* missions. The red curve (see online version) shows the same, but with the planet orbiting the Sun. The difference, which is due to the abiotic build-up of O_3, is at the limits of detectability. Different planetary conditions and stellar parameters may make this discrepancy larger.

1993). In this section, we discuss how the stellar spectral energy distribution (SED) of M dwarfs can significantly impact our ability to identify biospheres.

Different SEDs profoundly affect atmospheric chemistry, as many reactions are the direct result of photolysis by UV photons, or indirectly the result of photolysis reactions that produce radicals that rapidly react with other species. Cool stars emit longer wavelength radiation, leading to a lower amount of energy in the UV region responsible for many photolytic reactions, and correspondingly slower photolysis rates. However, active M dwarfs such as AD Leonis (AD Leo) emit significant energy fluxes at wavelengths shortward of 200 nm, allowing a subset of photochemical reactions to proceed at a rate comparable to that around warmer stars. As a result, the photolysis reactions caused by these photons may proceed at Earth-like rates on planets around cool stars.

As an example we consider the effects of the SEDs of the Sun and AD Leo on a hypothetical planet orbiting in the HZ (see Segura et al. (2005)). The abundance of O_2 and O_3 in an atmosphere is a function of photolysis rates, and hence different SEDs could influence the abundance of these important biomarkers in a planetary atmosphere. On hypothetical planets around AD Leo, short-UV radiation from flares can lead to photolysis of O_2, H_2O, and CO_2, liberating O atoms. These O atoms can then react with O_2 to form O_3. However, because O_3 photolysis occurs at longer wavelengths then O_2 photolysis, and these longer wavelengths are relatively scarce in all M dwarf SEDs, the destruction of the O_3 will be slower. *This could lead to build-up of atmospheric O_3 from processes that are photochemical and not biological.*

The net impact of the influence of SEDs on a planetary spectrum can be seen in Fig. 1. The black line shows results from a coupling of our photochemistry model (Pavlov et al. 2001) with our line-by-line spectral model (Crisp 1997). This particular simulation is of a spectrum from an organic-rich planet without biological O_2 production in the HZ of AD Leo. The red line represents our simulations of the same planet in the habitable zone of the Sun. The offset between these two lines results from AD Leo's lack of emission of photons with wavelengths in the range 200–800 nm, which destroy O_3. This result demonstrates how ignoring the stellar context of a planetary environment could lead to a false positive for life. The SEDs of M stars likely lead to biosignatures of their inhabited planets that are qualitatively different than those expected from inhabited planets orbiting F, G, and K stars.

4. Tidal Constraints on Habitability

Terrestrial planets orbiting close to their host stars may be deformed by the gradient of the gravitational force across their diameters. The tidal bulge raised on the planet will generally not be aligned with the line between the gravitational centers of the two bodies as long as 1) the orbit is eccentric ($e \neq 0$), or 2) the rotational period is different from its orbital period ($P_p^{rot.} \neq P_p^{orb.}$), or 3) the spin has an obliquity with respect to the orbital plane ($\psi_p \neq 0$). Although gravity tries to align the bulge, friction within he body resists, resulting in "tidal heating." Conservation of energy and angular momentum forces the planet's semi-major axis a, e, $P_p^{rot.}$, and ψ_p to steadily change.

Initially tides drive $\psi_p \rightarrow 0$ ("tilt erosion"), and $P_p^{rot.}$ evolves towards the "equilibrium rotation" $P_p^{equ.}$. When $\psi_0 \approx 0$, the planet will not experience seasons, i.e. over the course of an orbit, the insolation distribution on the planet will not vary. If $P_p^{rot.} = P_p^{equ.}$, one hemisphere of the planet will permanently be irradiated by the star, while the night side will freeze, which may prevent global habitability (Joshi 2003). Moreover, "tidal heating" in the planet may cause global volcanism or rapid resurfacing as observed in the Solar System on Io.

We define the "tilt erosion time" t_{ero} to be the time tides require to decrease an initial Earth-like obliquity of $\psi_p = 23.5°$ to $5°$, which depends on the initial a, e, and $P_p^{rot.}$. In the left panel of Fig. 2, t_{ero} is projected on to the a-e plane, as calculated with the tidal model of Leconte et al. (2010). The tidal time lag of the planet τ_p, the interval between the passage of the perturber and the tidal bulge, is scaled by $\tau_p = 638\,s \times Q_\oplus/Q_p$ to fit the Earth's time lag $\tau_\oplus$ and dissipation value $Q_\oplus$ (Neron de Surgy & Laskar 1997). $Q_p = 100$ and an initial rotation period $P_p^{rot.} = 1\,d$ are assumed. An error estimate for Q_p of a factor 2 is indicated with dashed lines. The test planet has one Earth-mass and orbits a $0.25\,M_\odot$ star. The HZ of Barnes et al. (2008) is shaded in grey. Obviously, planets in the HZ experience $t_{ero} < 0.1\,Gyr$. For lower stellar masses, $t_{ero} \ll 0.1\,Gyr$ for terrestrial planets in the HZ (Heller et al. 2010). For terrestrial planets in highly eccentric orbits in the HZ, tilt erosion can occur within 10 Gyr for stellar masses as large as $1\,M_\odot$.

The tidal equilibrium rotation period $P_p^{equ.}$ is a function of both e and ψ_p (Hut 1981). As an example, the right panel of Fig. 2 shows $P_p^{equ.}$ for the Super-Earth Gl 581 d projected onto the e-ψ_p plane. Observations (Mayor 2009) provide $e = 0.38 \pm 0.09$ (grey line), while ψ_p is not known. At an age of $\gtrsim 2\,Gyr$ (Bonfils et al. 2005), an initial

Earth-like obliquity of the planet is already eroded (Heller et al. 2010). For $\psi_p \lesssim 40°$, then $P_p^{\mathrm{rot.}} \approx P_p^{\mathrm{orb.}}/2$.

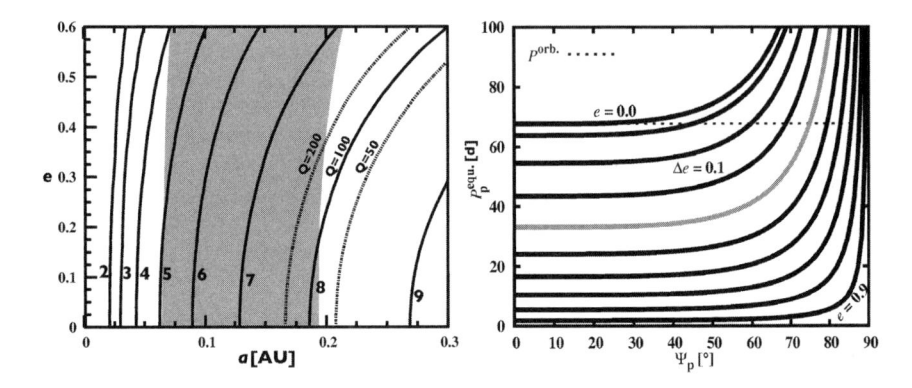

Figure 2. *Left:* Tilt erosion times in units of $\log(t_{\mathrm{ero}}/\mathrm{yr})$ for an Earth-mass planet orbiting a $0.25\,M_\odot$ star. The HZ is shaded in grey. *Right:* Equilibrium rotation period of Gl 581 d as a function of obliquity ψ_p for different values of e. The observed $e = 0.38 \pm 0.09$ is close to the grey line for $e = 0.4$. The orbital period, 67 days, is marked with a dashed line.

"Orbital shrinking" ($a \to 0$) may pull an adequately irradiated planet out of the HZ (Barnes et al. 2008). Thus, planets observed outside the HZ might have been habitable once in the past or become habitable in the future. Similarly planets currently in the HZ may have been inhospitable earlier. Terrestrial planets in the HZ of stars with masses $\leq 0.25\,M_\odot$ undergo significant tidal heating, potentially causing global volcanism (Jackson et al. 2008; Barnes et al. 2009; Heller et al. 2010), possibly rendering such planets uninhabitable. The consideration of tidal processes affects the concept of the habitable zone. Tilt erosion and equilibrium rotation need to be considered by atmospheric scientists, while orbital shrinking and tidal heating picture scenarios for geologists.

5. Evaporated/ing Cores of Gas Giants

Several hot Jupiters, e.g. HD 209458 b, show evidence of mass loss from their atmospheres (Vidal-Madjar et al. 2004). Many M dwarfs are far more active than the typical known planet-hosting stars. Hence, gas giants around these stars may also be losing significant mass. If these planets could be stripped of all their gas, a rocky/icy core could be left behind (Raymond et al. 2008). Furthermore, since *in situ* formation of large terrestrial planets (larger than Mars) appears challenging (Raymond et al. 2007), *detectable* rocky planets in the HZ of M stars may have followed just such an evolution. The core accretion model of planet formation (Pollack et al. 1996; Lissauer et al. 2009) posits that the cores of giant planets formed beyond the "snow line" (the region of a protoplanetary disk which is cold enough to permit the formation of water ice), and hence we may expect such cores to be volatile-rich. This section explores the possibility that terrestrial planets in the HZ of M dwarfs could be the remnant cores of ice or gas giants.

Atmospheric mass loss is most tightly coupled to the extreme ultraviolet flux, F_{XUV}, incident on a planet (Baraffe et al. 2004). On FGK stars, F_{XUV} drops over a timescale of Gyr (Ribas et al. 2005). For M dwarfs, stellar activity often produces XUV photons, but also decreases with time (West et al. 2008). Therefore, for a constant orbit, we expect the mass loss rate to decrease with time.

The complete removal of a gas giant's atmosphere would likely leave behind a core with a mass of perhaps several $M_\oplus$ (Baraffe et al. 2004; Raymond et al. 2008). On the other hand, some studies argue that the observations of HD 209458 b do not imply significant loss of mass (Ben-Jaffel 2007), and a theoretical study by Murray-Clay et al. (2009) suggested that complete evaporation of a gas giant's atmosphere is unlikely. These competing hypotheses may now be testable. With the detection capabilities of the *Kepler* and *CoRoT* missions, rocky planets arising from a variety of histories may be detected.

In addition to mass loss, tides will play an important and interrelated role in the evolution of an evaporating gas or ice giant. Jackson et al. (2010) showed that the coupling of mass loss and orbital evolution may have played a significant role in CoRoT-7 b's history. Several important feedbacks between mass loss and tidal evolution are possible. As a decreases, tides will accelerate orbital decay and mass loss. However, as mass decreases, orbits can decay more slowly. Orbital decay can occur on a timescale of Gyr, similar to the timescale for F_{XUV} for M dwarfs to diminish. Therefore, as tides pull a planet in, the mass loss increases due to proximity, but decreases due to less flux, but as mass is lost, tidal evolution slows.

Although Jackson et al. (2010) explored mass loss for CoRoT-7 b, the analogous problem for M dwarfs has yet to be tackled, in part because F_{XUV} is poorly constrained. In particular for M dwarfs, stellar activity does not evolve smoothly, as it is often punctuated by strong outbursts that may drive fast mass loss. Nonetheless, the example of CoRoT-7 b suggests that similar processes may be important for planets orbiting M dwarfs. As such planets are discovered, determination of their histories will be critical, as evaporated cores may not even be habitable.

6. Magnetic Shielding of Exo-Earths in the Habitable Zones of M Dwarfs

Many planets in the HZs of M dwarfs will be exposed to denser stellar winds and be tidally locked. As a consequence, the exoplanet community initially reached the consensus that the slow rotation of the planets will prevent the development of magnetic fields strong enough to shield surface life. In this section, we show that a planet in the HZ of an M dwarf, even if rotating very slowly, can have stronger magnetic shielding than previously thought.

Although planets in the HZs of M dwarfs do not necessarily rotate synchronously due to tides (see § 4, Correia et al. (2008) or Barnes et al. (2010)), tidal locking still leads to slow rotation, except for the very latest M dwarfs and/or large eccentricities. Planetary scientists have been working for decades on models to reproduce the magnetic moments M of planets and satellites in our Solar System. The best model generated to date, in the sense of reproducing the measured magnetic moments of most objects in the Solar System, is by Olson & Christensen (2006).

We applied that model to the case of hypothetical exo-Earths (with masses up to 12 $M_\oplus$), to determine the strengths of their magnetic fields. We assume their interiors are stratified in two separate layers, a mantle and a core. We also assume that the planets have thin atmosphere and ocean/crust layers which account only for 1% of the planetary radius. We use two different chemical compositions for each layer: a pure iron core and an iron alloy core, containing 10 mole % S, and for the mantle we use pure olivine and perovskite+ferropericlase compositions, which resemble, respectively, the upper and lower mantle compositions of the Earth.

We implemented all possible combinations of those model layers into the standard equations of planetary structure to obtain the density profile of the planets and from those density profiles computed the parameters needed to determine their magnetic moment, i.e. the radius of the core, the average bulk density of the convective zone, and its thickness.

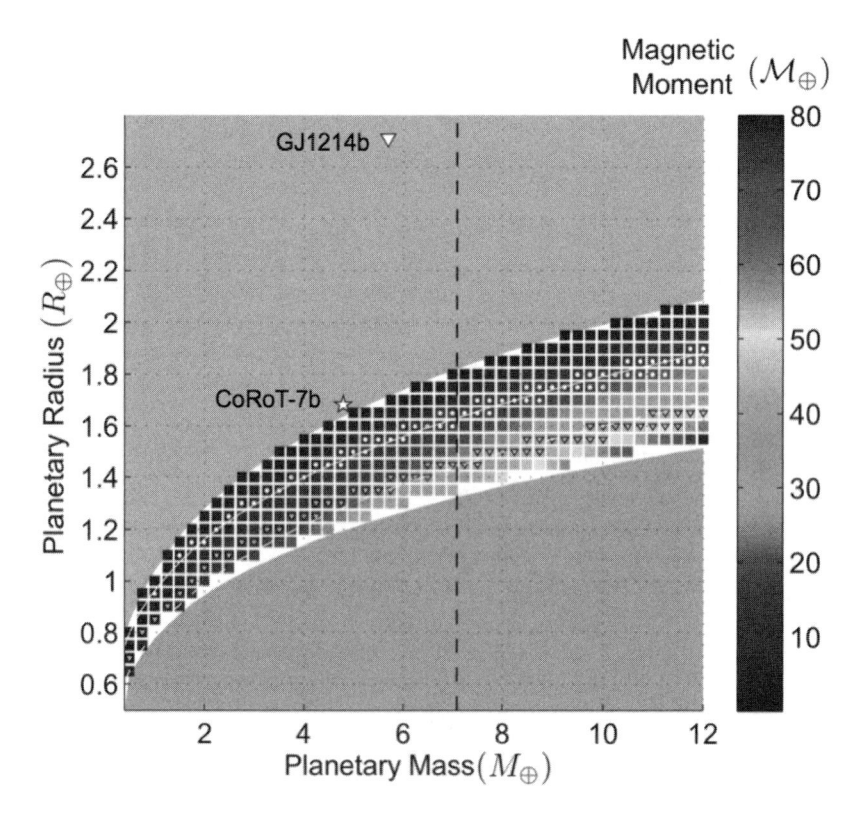

Figure 3. Magnetic moment model estimates for planets with masses up to 12 $M_\oplus$, a pure iron core, and perovskite+ferropericlase mantle compositions. The color scale (see online version) on the right corresponds to magnetic moment values between 0 and 80 times that of the Earth. The region below the colored points corresponds to planets made out of core materials denser than iron, while the region above corresponds to planets with radii too large, and therefore too low density, to have a core capable of generating a magnetic field. The triangle in the upper edge of the plot corresponds to GJ 1214 b, and the star symbol corresponds to CoRoT-7 b (Léger et al. 2009). For the core and mantle compositions in these models, neither of those two planets will have a magnetic field, but this result can change in the case of CoRoT-7 b for slightly different interior chemical compositions.

The magnitude of the magnetic moment of a planet varies with density and core-mantle composition. Pure iron core and perovskite+ferropericlase mantle models result in the smallest planetary core radius and therefore the smallest moment, but even in this case the value for Earth-sized planets like those in our simulations will still be at least $0.4 M_\oplus$, independent of the planets' rotation rates. All other models produce stronger dipoles. Fig. 3 summarizes our findings for a grid of simulated planetary mass and radius values of exo-Earths with masses up to 12 $M_\oplus$, assuming the mantle and core compositions that gave the weakest magnetic moment, as described above. Therefore, the values in the figure are lower limits to the expected magnetic moment strengths.

Planets with larger masses and smaller radii are more likely to have larger dipole moments and stronger magnetic fields at the surface, because denser planets tend to have larger cores. The main conclusions are: 1) *the magnetic moment of a planet does not depend on its rotation rate*; 2) the magnetic moment depends instead on its mass and size, its chemical composition, and the efficiency of convection in its interior; and 3) any terrestrial planet up to a few Earth masses in the HZ of an M-dwarf might have a strong enough magnetic field to shield its atmosphere and surface.

Notice, however, that these models do not account for changes in the thickness of the convective zone or the convective flux. Also, planets under extreme conditions, i.e. highly inhomogeneous heating or very strong stellar winds, will undoubtedly have their magnetic fields affected.

7. Revisiting the Habitable Zone: Summary of the NAI Workshop Discussions

To take stock of the current state of the field of planetary habitability, the NASA Astrobiology Institute's Habitability and Biosignatures Focus group organized a workshop in Seattle, WA on Aug 3–5, 2010. This workshop gathered together 38 scientists from the fields of biology, geology, atmospheric science, ecology and astronomy. The primary goal of the workshop was to identify, review and prioritize planetary and stellar characteristics that affect habitability, and to provide an interdisciplinary synthesis of these developments in our understanding of planetary habitability. As a secondary goal, the workshop initiated a discussion on the development of a multi-parameter means of assessing the likelihood of extrasolar planet habitability. In this section we will briefly describe workshop highlights in the major topic areas covered. The full report will be available in 2011 (Meadows et al., in prep.).

Life's Requirements: The recently-published NRC report on "The Limits of Organic Life in Planetary Systems"[1] finds that life requires a scaffolding element which can form covalent bonds with other elements that are relatively easily broken. Weaker, non-covalent electrostatic bonds are also crucial, as they are required to maintain the 3-D structure of proteins. As life's molecules require both covalent and non-covalent bonds to function, this strongly favors a polar solvent like water, and much of our discussion in this session was on scientific arguments for water as the most likely solvent for life. Energy is also required, but in the HZ of Kasting et al. (1993), stellar photons provide an essentially limitless supply. More speculatively, the boundaries of an HZ may be a function of the metabolic pathways utilized on a particular planet.

---

[1] http://www.nap.edu/catalog.php?record_id=11919

Stellar Radiative Effects: The star drives a planet's climate, but can also negatively impact habitability by subjecting the planet to high-energy radiation and particles. Modeling work on the effect of flares on Earth-like planets orbiting in the HZ of M dwarfs find that this latter effect may be mitigated if the planet has an existing ozone layer, and if flares are infrequent (Segura et al. 2010). UV radiation from the flare has little effect on O_3, and it is the chemistry associated with the proton flux from the flare that produces the most damage. However, DNA-damaging UVB flux reaching the surface at the peak of the flare was only 1.2 times Earth's level.

Planetary System Architecture: Discussion in this session emphasized the importance of planetary rotation rate, obliquity and eccentricity, characteristics that are often extremely difficult to constrain observationally. Tidal locking with synchronous rotation can also lead to a 0 or 180 degree obliquity (planetary pole perpendicular to the orbital plane) such that the planet's poles do not experience seasonal melting, see § 3. This state can lead to runaway glaciation as the poles freeze and planetary albedo steadily increases. High obliquity and moderate eccentricity can push out the outer edge of the HZ (Spiegel et al. 2010). Tidal heating can also be so strong as to render a planet in the HZ uninhabitable (Jackson et al. 2008; Barnes et al. 2009).

Forming Habitable Planets: Highlights of this session included a discussion of water and carbon worlds. Radioactive heating from supernova-sourced ^{26}Al can deplete water in planetesimals, leading to drier planets. Without ^{26}Al, Earth might have formed with 30–50 times as much water resulting in oceans 400 km deep. This "waterworld" might be uninhabitable however, as the overlying pressure could form an ice layer on the ocean floor, cutting off nutrient communication with the rocky interior. Planet formation and disk-chemistry modeling suggests that 1 in 3 planetary systems has a C/O ratio higher than about 0.8 and may form SiC as the principle planetary constituent Bond et al. (2010).

Planetary Characteristics: This session included an in depth discussion of planetary tectonic regimes. Internal energy is needed to drive convection for plates of a particular strength and planets may go through different tectonic regimes, including plate tectonics as we know it, but also episodic overturns and stagnant lid regimes. Even the Earth is likely to only have had smooth plate tectonics for the last billion years (Condie 1998). This could have ramifications for the carbonate-silicate cycle which may have buffered the Earth's climate for the past 4 Gyr (Walker et al. 1981). A Dune-like world, with only 1–10% of the Earth's water abundance, could both cool more efficiently at the inner HZ edge, and avoid an ice-albedo feedback at the outer edge, and hence may have a much broader HZ than predicted by Kasting et al. (1993).

Detecting Habitability: This session included discussions of detecting distant oceans. Robinson et al. (2010) have used a realistic 3-D spectral model of the Earth to show that photometric observations at specific extrasolar planetary phases could be used to discriminate between planets with and without significant bodies of liquid on their surfaces. Polarization signals also permit the detection of surface liquids, and the peak in polarization percentage may give clues to atmospheric thickness.

8. Conclusions

Less than one month after the Cool Stars XVI meeting, Vogt et al. (2010) reported the RV detection of a potentially rocky planet ($m_p \geq 3M_\oplus$) orbiting in the HZ of the M3 star Gl 581 in a nearly circular orbit. If confirmed, this planet is the first discovered near the middle of the HZ of a main sequence star, and, as expected, that star is an M dwarf. So how does this planet measure up in terms of potential habitability? Vogt et al. note that the host star is extremely quiescent, to the point that they cannot really detect any jitter, and hence stellar activity is not currently an issue for the planet. However, the star is at least a few Gyr old (Bonfils et al. 2005), hence we cannot exclude the possibility that fatal harm was done to a potential biosphere in the past. The planet is tidally locked, or in a spin-orbit resonance (Heller et al. 2010), but we do not know its rotation rate. The planet's mass is large enough that it can sustain tectonic activity for 10 Gyr, assuming it formed with a similar ratio of radiogenic isotopes as the Earth (tidal heating is minimal, even for $e = 0.2$). Therefore, from the RV data, we can not discern any major issue impeding habitability. Unfortunately, though, remote detection of its biosphere is not possible for the foreseeable future, as at 0.15 AU from its host star, reflection spectra will be unavailable from any currently planned space mission.

In spite of early skepticism, planets orbiting M dwarfs can be inhabited. Although many issues have been identified, such as tidal locking and atmospheric removal, more careful modeling has shown that these phenomena are not so deadly. While improvements in our knowledge via modeling will continue, transmission spectra of planets transiting M dwarfs will provide, at least for the next decade, our only observational means to directly assess the habitability of planets orbiting cool stars.

References

Baraffe, I., Selsis, F., Chabrier, G., Barman, T. S., Allard, F., Hauschildt, P. H., & Lammer, H. 2004, A&A, 419, L13. arXiv:astro-ph/0404101
Barnes, J. R., Barman, T. S., Jones, H. R. A., Leigh, C. J., Cameron, A. C., Barber, R. J., & Pinfield, D. J. 2008, MNRAS, 390, 1258
Barnes, R., Jackson, B., Greenberg, R., & Raymond, S. N. 2009, ApJ, 700, L30. 0906.1785
Barnes, R., Jackson, B., Greenberg, R., Raymond, S. N., & Heller, R. 2010, in Astronomical Society of the Pacific Conference Series, edited by V. Coudé Du Foresto, D. M. Gelino, & I. Ribas, vol. 430 of Astronomical Society of the Pacific Conference Series, 133. 0912.2095
Basri, G., Walkowicz, L. M., Batalha, N., Gilliland, R. L., Jenkins, J., Borucki, W. J., Koch, D., Caldwell, D., Dupree, A. K., Latham, D. W., Meibom, S., Howell, S., & Brown, T. 2010, ApJ, 713, L155. 1001.0414
Bean, J. L., Seifahrt, A., Hartman, H., Nilsson, H., Reiners, A., Dreizler, S., Henry, T. J., & Wiedemann, G. 2010a, ApJ, 711, L19
Bean, J. L., Seifahrt, A., Hartman, H., Nilsson, H., Wiedemann, G., Reiners, A., Dreizler, S., & Henry, T. J. 2010b, ApJ, 713, 410. 0911.3148
Ben-Jaffel, L. 2007, ApJ, 671, L61. 0711.1432
Blake, C. H., Charbonneau, D., & White, R. J. 2010, ApJ, 723, 684. 1008.3874
Bond, J. C., O'Brien, D. P., & Lauretta, D. S. 2010, ApJ, 715, 1050. 1004.0971
Bonfils, X., Delfosse, X., Udry, S., Santos, N. C., Forveille, T., & Ségransan, D. 2005, A&A, 442, 635
Ciardi, D. R., von Braun, K., Bryden, G., van Eyken, J., Howell, S. B., Kane, S. R., Plavchan, P., & Stauffer, J. R. 2010, ArXiv e-prints. 1009.1840
Condie, K. C. 1998, Earth and Planetary Science Letters, 163, 97
Correia, A. C. M., Levrard, B., & Laskar, J. 2008, A&A, 488, L63
Crisp, D. 1997, Geophys. Res. Lett., 24, 571
Des Marais, D. J., Harwit, M. O., Jucks, K. W., Kasting, J. F., Lin, D. N. C., Lunine, J. I., Schneider, J., Seager, S., Traub, W. A., & Woolf, N. J. 2002, Astrobiology, 2, 153
Heller, R., Barnes, R., & Leconte, J. 2010, A&A. (submitted)
Heng, K., & Vogt, S. S. 2010, ArXiv e-prints. 1010.4719

Hut, P. 1981, A&A, 99, 126
Jackson, B., Barnes, R., & Greenberg, R. 2008, MNRAS, 391, 237. 0808.2770
Jackson, B., Miller, N., Barnes, R., Raymond, S. N., Fortney, J. J., & Greenberg, R. 2010, MNRAS, 407, 910. 1005.2186
Joshi, M. 2003, Astrobiology, 3, 415
Kasting, J. F., Whitmire, D. P., & Reynolds, R. T. 1993, Icarus, 101, 108
Leconte, J., Chabrier, G., Baraffe, I., & Levrard, B. 2010, A&A, 516, A64+. 1004.0463
Léger et al. 2009, A&A, 506, 287
Lissauer, J. J., Hubickyj, O., D'Angelo, G., & Bodenheimer, P. 2009, Icarus, 199, 338. 0810.5186
Makarov, V. V., Beichman, C. A., Catanzarite, J. H., Fischer, D. A., Lebreton, J., Malbet, F., & Shao, M. 2009, ApJ, 707, L73. 0911.2008
Mayor, e. a., M. 2009, A&A, 507, 487
Murray-Clay, R. A., Chiang, E. I., & Murray, N. 2009, ApJ, 693, 23. 0811.0006
Neron de Surgy, O., & Laskar, J. 1997, A&A, 318, 975
Olson, P., & Christensen, U. R. 2006, Earth and Planetary Science Letters, 250, 561
O'Neill, C., & Lenardic, A. 2007, Geophys. Res. Lett., 34, 19204
Pavlov, A. A., Brown, L. L., & Kasting, J. F. 2001, J. Geophys. Res., 106, 23267
Pepe, F. A., & Lovis, C. 2008, Physica Scripta Volume T, 130, 014007
Pollack, J. B., Hubickyj, O., Bodenheimer, P., Lissauer, J. J., Podolak, M., & Greenzweig, Y. 1996, Icarus, 124, 62
Raymond, S. N., Barnes, R., & Mandell, A. M. 2008, MNRAS, 384, 663. 0711.2015
Raymond, S. N., Scalo, J., & Meadows, V. S. 2007, ApJ, 669, 606. 0707.1711
Reiners, A., Bean, J. L., Huber, K. F., Dreizler, S., Seifahrt, A., & Czesla, S. 2010, ApJ, 710, 432. 0909.0002
Ribas, I., Guinan, E. F., Güdel, M., & Audard, M. 2005, ApJ, 622, 680. arXiv:astro-ph/0412253
Robinson, T. D., Meadows, V. S., & Crisp, D. 2010, ApJ, 721, L67. 1008.3864
Sagan, C., Thompson, W. R., Carlson, R., Gurnett, D., & Hord, C. 1993, Nat, 365, 715
Scalo et al. 2007, Astrobiology, 7, 85
Segura, A., Kasting, J. F., Meadows, V., Cohen, M., Scalo, J., Crisp, D., Butler, R. A. H., & Tinetti, G. 2005, Astrobiology, 5, 706. arXiv:astro-ph/0510224
Segura, A., Walkowicz, L. M., Meadows, V., Kasting, J., & Hawley, S. 2010, Astrobiology, 10, 751. 1006.0022
Sotin, C., Grasset, O., & Mocquet, A. 2007, Icarus, 191, 337
Spiegel, D. S., Raymond, S. N., Dressing, C. D., Scharf, C. A., & Mitchell, J. L. 2010, ApJ, 721, 1308. 1002.4877
Tarter et al. 2007, Astrobiology, 7, 30
Vidal-Madjar, A., Désert, J., Lecavelier des Etangs, A., Hébrard, G., Ballester, G. E., Ehrenreich, D., Ferlet, R., McConnell, J. C., Mayor, M., & Parkinson, C. D. 2004, ApJ, 604, L69. arXiv:astro-ph/0401457
Vogt, S. S., Butler, R. P., Rivera, E. J., Haghighipour, N., Henry, G. W., & Williamson, M. H. 2010, ApJ, 723, 954
Walker, J. C. G., Hays, P. B., & Kasting, J. F. 1981, J. Geophys. Res., 86, 9776
West, A. A., Hawley, S. L., Bochanski, J. J., Covey, K. R., Reid, I. N., Dhital, S., Hilton, E. J., & Masuda, M. 2008, AJ, 135, 785. 0712.1590
Yelle, R. V. 2004, Icarus, 170, 167
Zechmeister, M., Kürster, M., & Endl, M. 2009, A&A, 505, 859. 0908.0944

Aspects of Multi-Dimensional Modelling of Substellar Atmospheres

Ch. Helling,[1] E. Pedretti,[1] S. Berdyugina,[2] A. A. Vidotto,[1] B. Beeck,[3,4] E. Baron,[5,6] A. P. Showman,[7] E. Agol,[8] and D. Homeier[4]

[1] *SUPA, School of Physics & Astronomy, University St Andrews, St Andrews, North Haugh, KY16 9SS, UK*

[2] *Kiepenheuer Institut für Sonnenphysik, Schöneckstr. 6, D-79104 Freiburg, Germany*

[3] *Max Planck Institute for Solar-System Research, Max-Planck-Straße 2, 37191 Katlenburg-Lindau, Germany*

[4] *Georg August University, Institute for Astrophysics, Friedrich-Hund-Platz 1, 37077 Göttingen, Germany*

[5] *Homer L. Dodge Department of Physics and Astronomy, University of Oklahoma, 440 W Brooks, Rm 100, Norman, OK 73019 USA*

[6] *Computational Research Division, Lawrence Berkeley National Laboratory, MS 50F-1650, 1 Cyclotron Rd, Berkeley, CA 94720 USA*

[7] *Department of Planetary Sciences, University of Arizona, Tucson, AZ, USA*

[8] *Dept. of Astronomy, Box 351580, University of Washington, Seattle, WA 98195*

Abstract. Theoretical arguments and observations suggest that the atmospheres of Brown Dwarfs and planets are very dynamic on chemical and on physical time scales. The modelling of such substellar atmospheres has, hence, been much more demanding than initially anticipated. This Splinter[1] has combined new developments in atmosphere modelling, with novel observational techniques, and new challenges arising from planetary and space weather observations.

1. Introduction

A rich molecular gas-phase chemistry coupled with cloud formation processes determines the atmosphere spectra of very low-mass, cool objects. Interferometry (E. Pedretti, Sect. 2) and polarimetry (S. Berdyugina, Sect. 3) can potentially provide more insight. However, present day interferometers are not capable of surface imaging Brown Dwarfs and planets due to financial constraints. Polarimetry, as a novel planet detection method, benefits from Rayleigh scattering on high-altitude sub-μm cloud particles. Such clouds were predicted to form by non-equilibrium processes several years ago

---

[1]http://star-www.st-and.ac.uk/~ch80/CS16/MultiDSplinter_CS16.html

(Woitke & Helling 2004). Wavelength dependent transit timing may reveal the interaction of the planetary exosphere with the stellar corona, and hence, limits may be set on planetary magnetic field strengths (A.A. Vidotto, Sect. 4). Radiative MHD simulations suggest that magnetic field driven convection significantly changes in fully convective objects compared to the Sun: M-dwarfs are suggested to exhibit darker magnetic structures (B. Beeck, Sect. 5). Studies of multi-dimensional radiative transfer emphasize that full solutions of physical problems are needed to access limits approximations (E. Baron, Sect. 6). The superrotation observed in planetary atmospheres is suggested to result from standing Rossby waves generated by the thermal forcing of the day-night temperature difference (A.P. Showman, Sect. 7). A search for transit-time variations at 8 μm reveals a difference between the transit and the secondary eclipse timing after subtracting stellar variability, and hence, confirms the superrotation on HD 189733b (E. Agol, Sect. 8). Results of multi-dimensional simulations are starting to be used as input for 1D model atmospheres for synthetic spectra production (D. Homeier, Sect. 9).

2. Combined Interferometry for Substellar Variability (Ettore Pedretti)

Optical and infrared long–baseline interferometry allows high–resolution imaging that is out of reach for the current large telescope facilities and for the planned 30m class telescopes. Examples of typical and future targets for long–baseline interferometry are stellar surfaces, planet–forming discs, active galactic nuclei and extrasolar planets. The main interferometric facilities in the northern hemisphere are the center for high angular resolution astronomy (CHARA) array and the Keck interferometer. CHARA is a visible and infrared interferometer composed of 6 one–metre telescopes on a 330m maximum baseline (see Pedtretti et al. 2009). The Keck interferometer is composed of two 10m telescopes on a 85m baseline and works mainly in the infrared. The main facility in the southern hemisphere is the very large telescope interferometer (VLTI), composed of four 8m telescopes and four 2m telescopes on a 200m maximum baseline. The Sidney university stellar interferometer (SUSI) in Australia has the longest available baseline in the world (640m) but so far has only used up to 80m baselines. Previous generation interferometers provided unique science by measuring the diameters of stars with two telescopes or by providing simple model dependent imaging combining up to 3 telescopes (Berger et al. 2001, Monnier et al. 2003, Pedretti et al. 2009) Model–independent imaging of complex objects was achieved quite recently at the CHARA array, that obtained the first image of a main-sequence star, Altair (Monnier et al. 2007). CHARA also imaged the most distant eclipsing system, the star β Lirae and witnessed the spectacular eclipse from the ϵ Aur system (Kloppenborg et al. 2010; Fig. 1). The VLTI imaged the young stellar object IRAS 13481-6124 (Kraus et al. 2010).

An interesting question is whether interferometry could resolve brown dwarfs and provide the same sort of high–resolution pictures offered to its larger stellar cousins. ϵ Indi B is the nearest brown dwarf (Scholz et al 2003). The distance is 3.6 pc, corresponding to an angular diameter of 0.3 milliarcseconds, and a magnitude at H band Mh = 11.3. ϵ Indi B is in the southern hemisphere, therefore it is only accessible by the VLTI and SUSI. The VLTI does not have long enough baselines, its maximum baseline being 200m. SUSI with its 640m baselines would achieve in the infrared, at H band a resolution of 0.5 milliarcseconds, therefore it could measure the diameter and effective temperature of ϵ Indi B if its bolometric flux was known. However SUSI has never used baselines longer than 80m and it is not sensitive enough to reach Mh = 11.3,

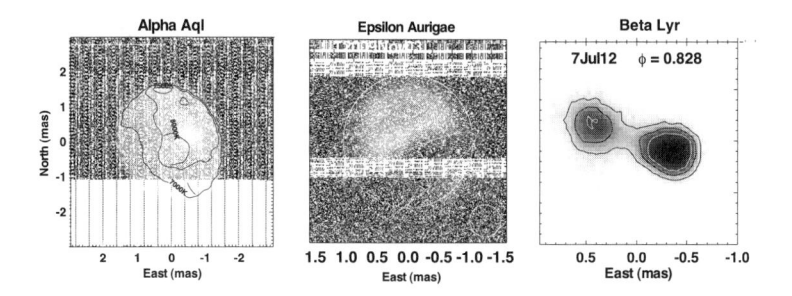

Figure 1. CHARA results. From left: Alpha Aql, first image of a main sequence star other than the Sun; the dust disc orbiting an unseen companion and eclipsing the star ϵ Aurigae; β Lyr, the closest interacting binary star ever imaged.

since it uses small 10cm siderostats. Resolved imaging of brown-dwarfs is out of reach for the present interferometric facilities. Brown–dwarfs are at least as challenging to image as Jupiter–size planets. A facility the size of the Atacama large millimetre array (ALMA) at infrared wavelengths would possibly achieve imaging of brown–dwarfs and Jupiter–size planets but it is unlikely that such facility will be financed in the short term. The remaining question is what could existing interferometers do in term of science, other than resolved imaging of the atmosphere of a brown–dwarf. The recent detection of brown dwarfs in binary systems within 5 AU from the main star from Corot (see CoRoT-3b, CoRot-6b and Super-WASP soon) opens up interesting possibilities. Many rejected planets in radial-velocity surveys may be brown dwarfs therefore there may be a large number available targets. Interferometry could potentially characterise brown dwarfs in binary and multiple systems very close to a brighter, more massive companion star through precision closure-phase measurement (Zhao 2009). Closure-phase nulling (Chelli et al 2009) a special case of precision closure–phase, where measurement are performed around nulls of visibility function and produce large change of closure-phase is potentially more sensitive. Interferometry could yield spectral and flux information about the brown dwarf and derive the mass of the brown–dwarf by measuring the inclination of the orbit in combination with radial velocity measurements.

3. Polarized Light in Stars and Planets (Svetlana Berdyugina)

Polarimetry is a powerful technique for revealing hidden structures in astrophysical objects, far beyond spatial resolution provided by direct imaging at any telescope. Polarization is a fundamental property of the light. It is its incredible sensitivity to asymmetries that empowers polarimetry and allows us to look inside unresolved structures. For instance, light can become polarized when it is scattered, or passes through magnetized matter, or is absorbed in an environment illuminated by anisotropic radiation.

Using molecular spectropolarimetry of cool stars enables us to obtain the first 3D view of starspots with the strongest field and the coldest plasma on the stellar surface. Such a phenomenon is common among stars possessing convective envelopes, where magnetic fields are believed to be generated (Berdyugina 2005). However, until recently magnetic fields have never been measured directly inside starspots. By selecting molecular lines which preferably form in cool starspots and at different heights in their atmosphere, such as TiO, CaH, and FeH, starspots and their internal structure are resolved (Berdyugina 2011).A new diagnostic technique of embedded stars and inner parts of protoplanetary disks based on radiative pumping of absorbers has been pioneered by (Kuhn et al. 2007). In the presence of anisotropic incident radiation, e.g. clumps of protoplanetary material illuminated by a star, the lower state magnetic sublevels of atoms or molecules in the intervening gas become unequally populated (even in the absence of magnetic fields). Such an 'optically pumped' gas will result in polarized line absorption along the line of sight. This provides novel insights into the structure and evolution of the innermost parts of circumstellar disks which are inaccessible to any other technique (Kuhn et al. 2011).

Detecting planetary atmospheres in polarized light provides a direct probe of exoplanets outside transits. The light scattered in the planetary atmosphere is linearly polarized and, when the planet revolves around the parent star, the polarization varies. Thus, the observed polarization variability exhibits the orbital period of the planet and reveals the inclination, eccentricity, and orientation of the orbit as well as the nature of scattering particles in the planetary atmosphere. HD189733b is a very hot Jupiter and the first exoplanet detected in polarized light (Berdyugina et al 2008). The observed polarization (Fig. 2) is caused by Rayleigh scattering, possibly on 20 nm $MgSiO_3$ dust condensates (Berdyugina 2011).

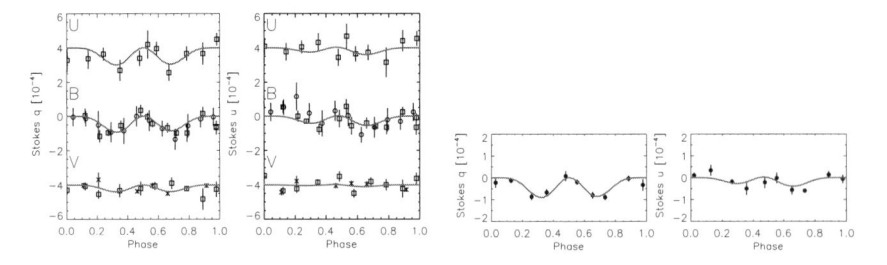

Figure 2. Polarimetric data (normalized Stokes q and u) for HD189733b in UBV bands. *Left:* squares – Berdyugina et al. (2011), open circles – binned B-band data from Berdyugina et al. (2008), crosses – broad-band filter data by Wiktorowicz (2009) centered at the V-band. The U and V data are shifted in vertical by $\pm 4 \cdot 10^{-4}$ for clarity. *Right:* All the U and B data from the years 2006-2008 binned together. The mean error of the binned data is $1.7 \cdot 10^{-5}$. Curves are the best-fit solutions for a model atmosphere with Rayleigh scattering on dust particles.

4. Stellar Influence on Planet Atmosphere is Shocking (Aline A. Vidotto)

WASP-12b is a transiting giant planet that was first identified in an optical photometric transit survey (Hebb et al. 2009). More recently, transit observations were also done in the near-UV (Fossati et al 2010a), revealing that while the time of the egress of the transit occurs almost simultaneously for both the optical and the near-UV observations, the ingress is first seen in the near-UV. This asymmetric behavior in the planet light curve has been explained by the presence of asymmetries in the planetary exosphere.

Motivated by this difference in transit durations, we proposed a model where the interaction of the stellar coronal plasma with the planet is able to modify the structure of the outer atmosphere of WASP-12b (Vidotto et al. 2010a; Paper1) WASP-12b is a giant planet with $M_p = 1.41\ M_J$ and $R_p = 1.79\ R_J$, where M_J and R_J are the mass and radius of Jupiter, respectively. It orbits its host star (a late-F star, with $M_* = 1.35\, M_\odot$, $R_* = 1.57\, R_\odot$) at an orbital radius of $R_{\rm orb} = 0.023$ AU $= 3.15\, R_*$, with an orbital period of $P_{\rm orb} = 1.09$ d. Due to such a close proximity to the star, the flux of coronal particles impacting on the planet comes mainly from the azimuthal direction, as the planet moves at a relatively high Keplerian orbital velocity of $u_K = (GM_*/R_{\rm orb})^{1/2} \sim 230$ km s^{-1} around the star. Therefore, stellar coronal material is compressed ahead of the planetary orbital motion, possibly forming a bow shock ahead of the planet. The condition for the formation of a bow shock is that the relative motion between the planet and the stellar corona is supersonic. Although we know the orbital radius of WASP-12b, we do not know if at this radius the stellar magnetic field is still capable of confining its hot coronal gas, or if this plasma escapes in a wind (see Paper 1). In the first case, where the coronal medium around the planet can be considered in hydrostatic equilibrium, the velocity of the particles that the planet 'sees' is supersonic if $\Delta u = |u_K - u_\varphi| > c_s$, where $u_\varphi = 2\pi R_{\rm orb}/P_*$ is the azimuthal velocity of the stellar corona, c_s is the sound speed, and P_* is the stellar period of rotation. From observations of the sky projected stellar rotation velocity, $P_* \gtrsim 17$ days (Fossati et al. 2010b). This implies that for a coronal temperature $T \lesssim (4-5) \times 10^6$ K, shock is formed around WASP-12b. Although stellar flares can raise the coronal plasma temperature above these values, it is unlikely that a corona would be hotter than this.

If we take the observationally derived stand-off distance from the shock to the center of the planet ($\sim 4.2\, R_p$, Lai et al. 2010) as approximately the extent of the planetary magnetosphere r_M, we showed that pressure balance between the coronal total pressure (i.e., ram, thermal, and magnetic pressures) and the planet total pressure requires that

$$B_c(R_{\rm orb}) \simeq B_p(r_M), \tag{1}$$

where $B_c(R_{\rm orb})$ is the magnetic field intensity of the star at $R_{\rm orb}$ and $B_p(r_M)$ is the magnetic field intensity of the planet at r_M. Note that we neglected the ram and thermal pressures in previous equation. Assuming that both the stellar and the planetary magnetic fields can be described as dipoles, from Eq. (1), we have

$$B_p = B_* \left(\frac{R_*/R_{\rm orb}}{R_p/r_M}\right)^3 = B_* \left(\frac{1/3.15}{1/4.2}\right)^3 \simeq 2.4 B_*, \tag{2}$$

where B_* and B_p are the magnetic field intensities at the stellar and planetary surfaces, respectively. Adopting the upper limit of the stellar magnetic field of 10 G suggested by Fossati et al. (2010b), our model predicts a maximum planetary magnetic field of about 24 G. It is likely that shock formation around close-in planets is a common feature of transiting systems. In fact, in a follow-up work Vidotto et al. (2010b), we showed that about 36 out of 92 known transiting systems (as of Sept/2010) would lie above a reasonable detection threshold. For these cases, the observation of bow-shocks may be a useful tool in setting limits on planetary magnetic field strengths.

5. MHD Simulations Reveal Crucial Differences Between Solar and Very-cool Star Magnetic Structures (Benjamin Beeck, Manfred Schüssler, Ansgar Reiners)

Cool main-sequence stars of spectral types F through L have a thick convective envelope or are fully convective. In many of such stars, magnetic fields of various strengths

Figure 3. Snapshots of the bolometric intensity $I_{\rm bol}(\mu = 1)$ (right panels) and vertical component $B_z(\tau_R \approx 1)$ of the magnetic field at the optical surface (left panels) for MHD models of (G2V, upper panels) and an M2 dwarf (lower panels).

have been detected. In the Sun, the surface magnetic field is observed to be highly structured owing to its interaction with the convective flows. In contrast to the Sun, the structure and properties of magnetic fields on cool stars are unknown. In the absence of spatially resolved observations, the effect of the magnetic structure on signatures of the magnetic field can be evaluated by numerical simulations of the magneto-convective processes. Using the MHD code MURaM, we carried out 3D radiative magnetohydrodynamic simulations of the convective and magnetic structure in the surface layers (uppermost part of the convection zone and photosphere) of main-sequence stars of spectral types F3 to M2. The code is a "box-in-the-star" code that solves the equations of (non-ideal) MHD in three spatial dimensions with constant gravitational acceleration. It includes compressibility, partial ionization, and non-grey radiative energy transport (for details see Vögler et al 2005). To fit the surface conditions for different stellar spectral types, gravity and effective temperature were adjusted and the opacity bins were recalculated. The size of the computational box and the spatial resolution were modified in order to cover the relevant length scales of the different convection patterns.

The model grid comprises six main-sequence stars of spectral types F3, G2, K0, K5, M0, and M2. The start models were run with $\mathbf{B} \equiv \mathbf{0}$ until a quasi-stationary state was reached. Then, a homogeneous vertical magnetic field with the field strength $B_0 = 20\,{\rm G}$, $100\,{\rm G}$, or $500\,{\rm G}$ was introduced.

The modelled magneto-convection shows significant differences between M-dwarfs and stars of earlier spectral types, e. g. the Sun. As illustrated by Fig. 3, the initially homogeneous magnetic flux is accumulated into very few structures of high field strength, the cause of which are stable downflows. While solar magnetic structures appear as bright features, the magnetic structures on M-dwarfs tend to be rather dark. In the case of the Sun, magnetic structures create a strong depression of the optical surface with hot side walls that can radiatively heat the interior of the magnetic structure. Owing to higher densities and shallower temperature gradient, this side wall heating is much less efficient for the magnetic structures in M-dwarf atmospheres. Since the magnetic field suppresses convective energy transport, the structures cool down.

These findings indicate that plage regions on M-stars might not show bright points but rather "pores" and small "star spots" of reduced intensity, which has a crucial impact on the interpretation of observational data such as M-dwarf spectra.

6. Generalized 3-D Radiative Transfer for Astrophysical Atmospheres (E. Baron)

PHOENIX is a generalized model atmosphere code which works in 1 or 3 spatial dimensions. The philosophy behind PHOENIX is that it should work in a wide range of astrophysical environments and that it should handle both static and moving flows in full relativity. PHOENIX is well calibrated on many astrophysical objects: Planets/BDs, Cool Stars, Hot Stars (βCMa, ϵCMa), α-Lyrae, Novae, and SNe (Iabc, IIP, IIb) (see Hauschildt & Baron 2010 an references therein). Much of the development work on PHOENIX has been devoted to handling radiative transfer in velocity flows. Velocities are important in many astrophysical objects: novae, supernovae, AGN, and γ-ray bursts. But of course velocities are also important in stars since the linewidth is determined by the convective velocity field as shown by Stein & Nordlund (2000).

There are two ways to deal with velocity fields: the Eulerian formulation and the co-moving formulation. Each has advantages and disadvantages. In the Eulerian formulation wavelengths are uncoupled, significantly reducing memory requirements; however, opacities are angle dependent, significantly increasing computational requirements. It is also extremely difficult to handle relativity in the Eulerian formulation. In the co-moving formulation one can include both special and general relativity exactly and opacities are isotropic, significantly reducing computational requirements; however, wavelengths are coupled, significantly increasing memory requirements. While the solution of the radiative transfer equation in the co-moving frame in 1-D has been understood for quite some time (Mihalas 1980); in 3-D it is much more complex and has been mostly approached via the cumbersome tetrad formalism (Morita & Kaneko 1984, 1986). A much simpler approach via affine parameters was developed in (Chen et al. 2007) and implemented in (Baron et al. 2009). The Eulerian Formulation is valid for velocities $v < 1000$ km s^{-1} and thus is of interest in stars with low velocities. The Eulerian formulation trades off the high memory requirement of the co-moving frame for the explicit coupling of angles and frequencies, that is, all momentum space variables are coupled. Thus, at each spatial point opacities must be calculated for each coupled wavelength direction point, leading to both a large amount of computation and storage. Nevertheless PHOENIX has been adapted to include the Eulerian formulation in 3-D (Seelmann et al. 2010).

In summary, PHOENIX 3-D solves the generalized atmosphere problem with both co-moving and Eulerian formulations in Cartesian, spherical, and cylindrical geometry. We still need to study which approach is computationally better, which may depend on the particular computer architecture, particularly with the advent of GPUs and very low memory per core exoscale computing. The next step is to go beyond test problems to production code. While full 3-D RT is too computationally complex for radiation hydrodynamics, some of the methods we have developed may be adapted to a more simplified approach. It is crucial to do full radiative transfer to determine abundances, perform detailed hydrodynamic model verification, and for other applications to observed data.

7. Multi-D Hydro-simulations of Substellar Atmospheres (Adam P.Showman)

Over 100 transiting hot Jupiters are now known, and observations from the Spitzer and Hubble Space Telescopes and groundbased facilities constrain the atmospheric composition and three-dimensional temperature structure of many such objects. Phase curves show that some hot Jupiters, such as HD 189733b, have modest (~200 K) day-night temperature variations (e.g., Knutson et al. 2007), while others have much larger day-night temperature differences. For the case of HD 189733b –the best-observed hot Jupiter– the Spitzer infrared light curves imply that the hottest region is not at the substellar point but rather is displaced 30 degrees of longitude to the east. This feature provides strong evidence of atmospheric circulation on these tidally locked planets.

The atmospheric dynamical regime of hot Jupiters differs from that of, for example, brown dwarfs. The atmospheric circulation on hot Jupiters is probably driven primarily by the $\sim 10^5 - 10^6$ W/m^2 net radiative heating on the dayside and cooling on the nightside; unlike the case of Jupiter or typical brown dwarfs, the absorbed stellar flux exceeds the convective fluxes in the planet's interior by 3-5 orders of magnitude. Evolution and structure models indicate that multi-Byr-old hot Jupiters have deep radiative zones extending from the top of the atmosphere to pressures of typically 100-1000 bars. Thus, the weather near the infrared photosphere on hot Jupiters occurs in a stably stratified radiative zone. Hot Jupiters are thought to be synchronously rotating with their 1-10-day orbital periods, implying that planetary rotation is less dominant in their dynamics than is the case for Jupiter or typical brown dwarfs.

A variety of 3D dynamical models of the atmospheric circulation on hot Jupiters have been published (e.g., Showman and Guillot 2002; Dobbs-Dixon & Lin 2008; Showman et al. 2008, 2009; Rauscher & Menou 2010; Thrastarson and Cho 2010). These models typically model the circulation of hot Jupiters on 2-4-day orbits assuming the interior is tidally locked. At the pressure of the infrared photosphere, the circulation in these models typically exhibits a banded structure, with 1-3 broad east-west jet streams whose speeds reach several km/sec. Day-night temperature differences are commonly hundreds of K at the photosphere. Real hot Jupiters probably exhibit a wide diversity of behaviors, which have yet to be thoroughly explored in circulation models.

Interestingly, the eastward offset of the hot region from the substellar longitude inferred from infrared light curves of HD 189733b – which provides our current best evidence for atmospheric circulation on hot Jupiters – was predicted five years before its discovery (Showman and Guillot 2002). In their model, the eastward offset results from advection by a robust, eastward flowing equatorial jet stream that dominates the circulation (Fig. 4). Subsequent models by several groups have generally confirmed the robustness of this feature (e.g., Showman et al. 2008, 2009; Dobbs-Dixon and Lin 2008; Rauscher and Menou 2010). However, to date, the mechanism for this so-called "equatorial superrotation" has not been identified. New, unpublished work by A.P. Showman and L.M. Polvani shows, however, that the superrotation results from standing Rossby waves generated by the day-night thermal forcing. The Rossby waves, which are planetary in scale, generate phase tilts such that equatorward-moving air

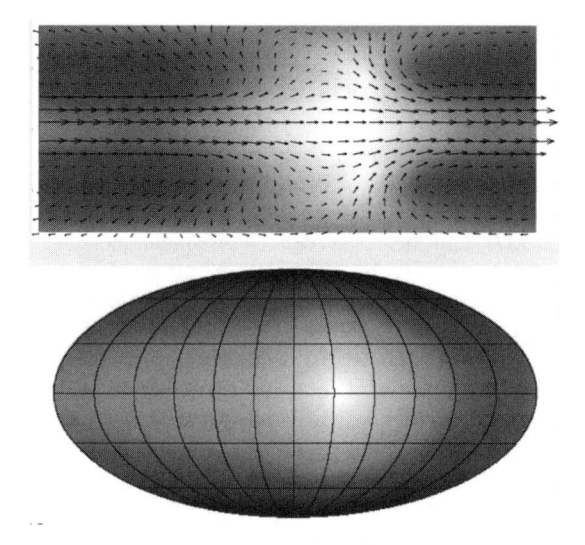

Figure 4. (Bottom): Longitudinal temperature structure of of hot Jupiter HD 189733b inferred from Spitzer 8-micron infrared light curve (Knutson et al. 2007), showing eastward offset of hottest region from substellar longitude. (Top) Temperature pattern (greyscale) and winds (arrows) from three-dimensional circulation model of a hot Jupiter by Showman and Guillot (2002). A common feature of such models is the eastward equatorial jet, which displaces the hottest regions to the east of the substellar point, as seen in the observations.

exhibits greater eddy angular momentum than poleward-moving air. This pumps eddy angular momentum from the midlatitudes to the equator and generates the equatorial superrotating jet. Showman and Polvani demonstrated the mechanism in an idealized, linear, analytic model, in simplified nonlinear models, and in full three-dimensional general circulation models. An implication is that the mechanism for producing the jet need not involve turbulent cascades or other eddy-eddy interactions, but rather results from a direct interaction between the standing, thermally generated quasi-linear eddies and the mean flow at the planetary scale.

8. Weather on a Hot Jupiter (Eric Agol)

Testing global models of weather on hot jupiters requires observations of their global properties. To date, the best means available for such a comparison is infrared observations of the phase variation of hot jupiters (Knutson et al. 2007). If the weather pattern changes on a timescale slower than the orbital time of the planet, then during the orbit the different faces of the planet will be observed, allowing a deconvolution of the longitudinal brightness of the planet (Cowan & Agol 2008). Since the planet cannot be resolved from the star, there are several possible ways such an analysis might go wrong: (1) stellar variability might swamp the planet phase variation; (2) planet variability might invalidate the steady-state assumption required for inversion; (3) planet-star interaction might cause stellar brightness variations on a similar timescale as the planet's orbit. In addition, instrumental effects can be present which may be stronger than the phase variation. So far the best target for this sort of observation is the exoplanet HD 189733b. It orbits a bright star, it is large in size compared to its host star, and it is hot enough and has a short enough period to enable observations of a reasonable duration. However, the host star is strongly variable in the optical, $\sim 1-2\%$, with a period of about 12 days, so the host star variability must be accounted for to properly measure the planet's infrared variability with the Spitzer Space Telescope. We measured the phase variation over slightly longer than half an orbital period at 8 μm with IRAC Channel 4 (Knutson et al. 2007), and then observed a subsequent six transits and six eclipses with the goals of determining the day-side variability and looking for transit-timing variations (Agol et al. 2008, Agol et al. 2010). We also obtained simultaneous ground-based monitoring in the optical which we used to correct for the stellar variability by extrapolating the optical stellar variation into the infrared (Winn & Henry 2008).

Based on these data, we found that the absolute flux of the system could be measured to <0.35 mmag after decorrelating with instrumental variations and stellar variability. We used this decorrelation to correct for the stellar variability, giving a more precise phase variation (Figure 5). The observed phase function is in good qualitative agreement with models of weather on this hot jupiter (e.g. Showman et al. 2009), albeit with an observed peak of the planet's flux that is closer to the secondary eclipse than predicted by the models. The location of this peak is primarily controlled by the ratio of the radiative timescale to the advection timescale, ϵ, so the data indicate that the models either overpredict the super-rotation speed of the equatorial jet, or they overpredict the cooling timescale at the 8 micron photosphere. We also find an offset in the secondary eclipse time; after correcting for light-travel time across the system, the offset can be accounted for by the asymmetric dayside flux caused by the super-rotating jet (this offset is due to the fact that we fit the secondary eclipse with a model in which the planet is uniform in surface brightness). The night side is about 64% of the brightness of the day side and the secondary eclipse depth variation has an RMS of < 2.7%, which is limited by the photometric precision of the data. These results are in good agreement with general circulation models for this planet which predict fluctuations of <1%.

9. Overshoot, Gravity Waves and Non-equilibrium Chemistry (Derek Homeier, France Allard, Bernd Freytag)

The PHOENIX BT-Settl models (Allard et al. 2010) combine a cloud formation and gas phase non-equilibrium chemistry model, using vertical diffusion profiles based on CO5BOLD RHD simulations as an input to our models, finding the mixing in the transition zone from carbon monoxide- to methane-dominated chemistry to be governed by gravity waves forming in the upper atmosphere (Freytag et al. 2010). We introduce updated reaction rates for the CO to CH_4 conversion from Visscher et al. (2010) and include departures from chemical equilibrium in the CO to CO_2 ratio to determine the molecular fractions of CH_4, CO and CO_2 for atmospheres spanning the range from L to T dwarfs. Synthetic spectra calculated for the resulting compositions reproduce the observed upmixing of carbon monoxide in brown dwarfs across the L/T transition. In addition we find carbon dioxide to appear in excess of its CE abundance in T dwarfs. The models produce an improved fit to the observed mid-infrared photometry of the coolest brown dwarfs (Burningham et al. 2010), and are confirmed by the identification of distinctive 4.2 μm CO_2 absorption features in several late T dwarf spectra by the AKARI satellite (Yamamura et al. 2010).

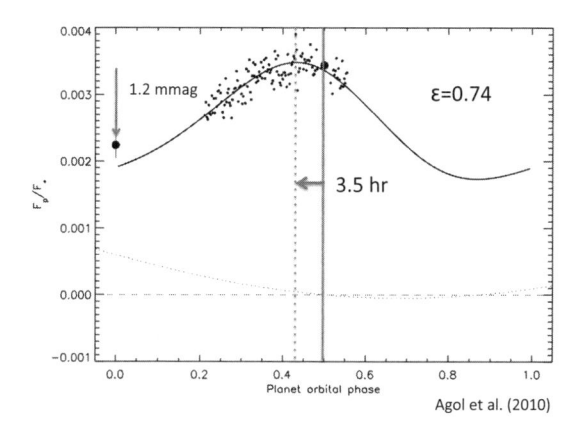

Figure 5. Measured phase-variation of HD 189733b (dots) at 8 μm after correction for stellar variability (dotted line). The peak of the phase function is offset 3.5 hours before secondary eclipse (orbital period: 53 hours). The night-side is 1.2 mmag fainter than the dayside, or about 64% of the flux. The phase variation may be fit by a toy model (solid line; Cowan & Agol 2010) in which the energy is advected in a super-rotating jet in which the ratio of the radiative to advection times $\epsilon = 0.74$.

Acknowledgments. EB was supported by NSF grant AST-0707704, US DOE Grant DE-FG02-07ER41517, and by program number HST-GO-12298.05-A which is supported by NASA through a grant from the Space Telescope Science Institute, which is operated by the Association of Universities for Research in Astronomy, Incorporated, under NASA contract NAS5-26555. This research used resources of the NERSC, which is supported by the Office of Science of the U.S. Department of Energy under Contract No. DE-AC02-05CH11231; and the Höchstleistungs Rechenzentrum Nord (HLRN). Support for EA was provided by NSF through CAREER Grant No. 0645416 and by NASA through an award issued by JPL/Caltech. APS acknowledges funding from NASA Origins. D. H. gratefully acknowledges support from a foreign travel grant by the DAAD. F. A. and B. F. acknowledge financial support from the ANR, and the PNPS of CNRS/INSU, France. Stellar and substellar atmosphere models for this study have been calculated at the GWDG.

References

Agol, E., Cowan, N. B., Bushong et al. 2009, vol. 253 of IAU Symposium, 209
Agol, E., Cowan, N. B., Knutson, H. A. et al. D. 2010, ApJ, 721, 1861
Allard, F., Homeier, D., & Freytag, B. 2010 *this vol.* 1011.5405
Baron, E., Hauschildt, P. H., & Chen, B. 2009, A&A, 498, 987
Berdyugina S. V., Berdyugin A. V., Fluri D. M., Piirola V. 2011, ApJ submitted
Berdyugina S. V., in Polarimetry of cool atmospheres: From the Sun to exoplanets, eds. J.R. Kuhn, 2011, (arXiv:1011.0751)
Berdyugina, S. V., Berdyugin, A. V., Fluri, D. M., Piirola, V. 2008, APJ 673, L83
Berdyugina, S. V. 2005, Liv. Rev. Solar Phys. 2, 8
Berger, J. P., Haguenauer, P., Kern, P. et al. 2001, A&A, 376, L31
Burningham, B., Leggett, S. K., Homeier et al. 2010, submitted to MNRAS.
Chelli, A., Duvert, G., Malbet, F., & Kern, P. 2009, A&A, 498, 321
Chen, B., Kantowski, R., Baron, E., Knop, S., & Hauschildt, P. 2007, MNRAS, 380, 104

Cowan, N. B., & Agol, E. 2008, ApJ, 678, L129
— 2010, arXiv:1011.0428
Dobbs-Dixon, I. & Lin, D.N.C. 2008. ApJ 673, 513.
Fossati, L., Haswell, C. A., Froning, C. S., & et al. 2010a, ApJ, 714, L222
Fossati, L., Bagnulo, S., Elmasli, A., & et al. 2010b, ApJ, 720, 872
Freytag, B., Allard, F., Ludwig, H., Homeier, D., & Steffen, M. 2010, A&A, 513
Hauschildt, P. H., & Baron, E. 2010, A&A, 509, A36
Hebb, L., Collier-Cameron, A., Loeillet, B., & et al. 2009, ApJ, 693, 1920
Henry, G. W., & Winn, J. N. 2008, AJ, 135, 68
Homeier, D., Freytag, B., & Allard, F. 2010 *this vol. suppl.*
Kloppenborg, B., Stencel, R., Monnier, J. D. et al. 2010, Nat, 464, 870
Knutson, H. A., Charbonneau, D., Allen, L. E. et al. 2007, Nat, 447, 183
Kraus, S., Hofmann, K., Menten, K. M. et al. 2010, Nat, 466, 339
Kuhn, J. R., Geiss, B., Harrington, D. M. 2011, in Solar Polarization 6, eds. Kuhn et al., (arXiv:1010.0705)
Kuhn, J. R., Berdyugina, S. V., Fluri, D. M., Harrington, D. M., Stenflo, J. O. 2007, ApJL 668, L63
Lai, D., Helling, Ch., & van den Heuvel, E. P. J. 2010, ApJ, 721, 923
Mihalas, D. 1980, ApJ, 237, 574
Monnier, J. D., Berger, J. P., Millan-Gabet, R. 2003, in Proc. SPIE, Interferometry for Optical Astronomy, W.A. Traub, editor, Vol. 4838, 1127–1138
Monnier, J. D., Zhao, M., Pedretti, E. et al. 2007, Science, 317, 342
Morita, K., & Kaneko, N. 1984, Ap&SS, 107, 333
— 1986, Ap&SS, 121, 105
Pedretti, E., Monnier, J. D., Brummelaar, T. T., & Thureau, N. D. 2009a, Nat, 53, 353
Pedretti, E., Monnier, J. D., Lacour, S. et al. 2009b, MNRAS, 397, 325
Rauscher, E. & Menou, K. 2010, ApJ 714, 1334
Seelmann, A., Hauschildt, P. H., & Baron, E. 2010, A&A, 522, A102
Scholz, R., McCaughrean, M. J., Lodieu, N., & Kuhlbrodt, B. 2003, A&A, 398, L29
Showman, A. P., Fortney, J. J., Lian, Y. et al. 2009, ApJ, 699, 564
Showman, A.P. et al. 2008, ApJ 682, 559
Showman, A.P., & Guillot T. 2002, A&A 385, 166.
Stein, R. F., & Nordlund, Å. 2000, Sol. Phys., 192, 91
Thrastarson, H.T. & Cho, J.Y-K. 2010 ApJ, 716, 144
Vidotto, A. A., Jardine, M., & Helling, Ch. 2010a, ApJ, 722, L168 (Paper 1)
— 2010b, MNRAS Letter, in press. arXiv:1011.3455
Visscher, C., Moses, J. I., & Saslow, S. A. 2010, Icarus, 209, 602. 1003.6077
Vögler, A., Shelyag, S., Schüssler, M. et al. 2005, A&A, 429, 335
Wiktorowicz, S. J. 2009, ApJ 696, 1116
Woitke, P. & Helling, Ch 2004, A&A 414, 335
Yamamura, I., Tsuji, T., & Tanabé, T. 2010, ApJ, 722, 682. 1008.3732
Zhao, M. 2009, PhD thesis, University of Michigan

The 16th Cambridge Workshop on Cool Stars, Stellar Systems and the Sun
ASP Conference Series, Vol. 448
Christopher M. Johns-Krull, Matthew K. Browning, and Andrew A. West, eds.
© 2011 Astronomical Society of the Pacific

Young Stars in the Time Domain: A CS16 Splinter Summary

Kevin R. Covey,[1,2] Peter Plavchan,[3] Fabienne Bastien,[4] Ettore Flaccomio,[5] Kevin Flaherty,[6] Stephen Marsden,[7] Maria Morales-Calderón,[8] James Muzerolle,[9] and Neal J. Turner[10]

[1]*Hubble Fellow; Cornell University, Department of Astronomy, 226 Space Sciences Building, Ithaca, NY 14853, USA.*

[2]*Visiting Researcher, Department of Astronomy, Boston University, 725 Commonwealth Ave, Boston, MA 02215, USA.*

[3]*NASA Exoplanet Science Institute, California Institute of Technology, MC 100-22, 770 S. Wilson Avenue, Pasadena, CA 91125, USA.*

[4]*Department of Physics and Astronomy, Vanderbilt University, Nashville, TN 37235*

[5]*INAF - Osservatorio Astronomico di Palermo, Piazza del Parlamento 1, 90134 Palermo, Italy.*

[6]*Steward Observatory, University of Arizona, Tucson, AZ 85721, USA.*

[7]*Center for Astronomy, School of Engineering and Physical Sciences, James Cook University, Townsville, 4811, Australia*

[8]*Spitzer Science Center, California Institute of Technology, Pasadena, CA 91125, USA.*

[9] *Space Telescope Science Institute, 3700 San Martin Dr., Baltimore, MD 21218, USA.*

[10]*Jet Propulsion Laboratory, California Institute of Technology, 4800 Oak Grove Drive, Pasadena, CA 91109, USA.*

Abstract. Variability is a defining characteristic of young stellar systems, and optical variability has been heavily studied to select and characterize the photospheric properties of young stars. In recent years, multi-epoch observations sampling a wider range of wavelengths and time-scales have revealed a wealth of time-variable phenomena at work during the star formation process. This splinter session was convened to summarize recent progress in providing improved coverage and understanding of time-variable processes in young stars and circumstellar disks. We begin by summarizing results from several multi-epoch *Spitzer* campaigns, which have demonstrated that many young stellar objects evidence significant mid-IR variability. While some of these variations can be attributed to processes in the stellar photosphere, others appear to trace short time-scale changes in the circumstellar disk which can be successfully modeled with axisymmetric or non-axisymmetric structures. We also review recent studies probing variability at shorter wavelengths that provide evidence for high frequency pulsations associated with accretion outbursts, correlated optical/X-ray variability in Classical T Tauri stars, and magnetic reversals in young solar analogs.

1. Introduction

The formation and early evolution of stars is inherently a time-domain process: the conversion of a dense molecular core into a zero-age main sequence star requires nearly every pertinent physical parameter (radius, density, temperature, v_{rot}, etc.) to change by multiple orders of magnitude over the relatively brief timescale of 10s of Myrs. These time-scales still dwarf that of a human lifetime[1], and one might *a priori* conclude that star formation is no more amenable to time-domain study than any other aspect of a stellar astrophysics.

Photometric variability has nonetheless been recognized for decades as a common trait of many young stars (e.g., Joy 1945). Historically, this variability has been best explored via optical photometry, most often sampling time-scales of days to (a few) years. These studies have provided detailed descriptions of the spot properties of optically revealed young stars (Vrba et al. 1988), a comprehensive inventory of stellar rotation in young clusters (see first the review by Herbst et al. 2007; Bouvier et al. 1995; Stassun et al. 1999; Rebull et al. 2004; Cieza & Baliber 2007; Irwin et al. 2008), and a quantitative statistical portrait of accretion and extinction-induced variability (Grankin et al. 2007). These monitoring programs have also identified a number of rare, astrophysically valuable systems: pre-main sequence eclipsing binaries (e.g., Cargile et al. 2008), disk occulting systems (ie, KH-15D: Hamilton et al. 2001), and stars undergoing massive accretion events (i.e., FU Ori, EX Lup, or V1647-like variables: Herbig 1977, 1989; Hartmann & Kenyon 1996; Reipurth & Aspin 2004; Lorenzetti et al. 2007).

Significant advances in observational time domain astronomy are uncovering new phenomena and systems for study. This expansion is due to many factors: a steady improvement in the size and sensitivity of optical arrays, enhancing the coverage and cadence possible for observations of optically revealed clusters; even greater advances in the capabilities of near-infrared arrays; and access to precise multi-epoch mid-infrared (mid-IR) photometry and spectroscopy from the *Spitzer Space Telescope*. These new observational capabilities have allowed variability studies to cover a greater number of targets at higher cadences, and characterize the variability properties of more deeply embedded, and presumably less evolutionarily advanced, sources. The signature of these advances is evident in our increased sensitivity to variability at the lowest masses within nearby star-forming regions (Cody & Hillenbrand 2010) as well as the increasing frequency with which we identify formerly rare young variables, including new eclipsing binaries (Hebb et al. 2010), new disk occulting systems (e.g., WL4: Plavchan et al. 2008), pulsating protostars (Morales-Calderón et al. 2009) and large-amplitude outbursts (Covey et al. 2010; Miller et al. 2010; Kóspál et al. 2010; Garatti et al. 2010). These observational advances have been accompanied by similar progress on the theoretical front, with increasingly detailed models of the physical processes underlying the observed variations (e.g., accretion variations, disk processes, etc.; Vorobyov & Basu 2010; Zhu et al. 2010; Baraffe & Chabrier 2010)

---

[1] or, perhaps more relevantly, the timescale of a PhD thesis

This splinter session was convened to review recent progress in characterizing, analyzing, and understanding variability in the youngest stars. The session incorporated presentations covering physical processes that develop over a range of time-scales, with observational signatures spanning a wide range of wavelengths. Following these presentations, the audience participated in a broader discussion of these new results and highlighted areas of particular promise for future observational or theoretical work. We provide a brief summary of each presentation below, and conclude with a recap of the open questions highlighted in the audience discussion.

2. Mid-infrared (Mid-IR) Variability

2.1. Maria Morales-Calderón: A Global View of Mid-IR Variability from the YSOVAR Orion Survey

The YSOVAR Orion program provides the first large-scale survey of the photometric variability of Young Stellar Objects (YSOs) in the mid-IR. In Fall 2009, *Spitzer*/IRAC observed a 0.9 sq. deg. area centered on the Trapezium cluster twice a day for 40 consecutive days, producing 3.6 and 4.5 μm high fidelity light curves (~3% typical photometric uncertainties) for over 2000 disked and diskless YSOs in Orion (as diagnosed with precursor 3.6-8.0 μm IRAC photometry). For brevity, we refer to the diskless stars as Weak T Tauri stars (WTTs); we refer to stars with disks, or which are otherwise heavily embedded in their natal cloud, as YSOs. For many of the stars, we also obtained complementary time-series photometry at optical (I_c) and/or near-infrared (JK_s) wavelengths. We find that 65% of the disked YSOs and 30% of the diskless WTTs are variable. The WTT variations are mostly spot-like, while the disked/embedded YSOs display a wider range of variability types: see Figure 1 for examples of the different types of variability evident in the YSOVAR light curves. Consistent with their greater propensity for spot-like variability, 65% of the variable WTTs are periodic, while periods are detected for only 16% of the YSOs. Those disked/embedded YSOs that are detected to be periodic, however, typically evidence larger amplitudes, as well as longer periods (see Figure 2).

YSOVAR monitoring has also identified several rare pre-main sequence variables. Among these are five new candidate PMS eclipsing binaries, including one that is fainter than all previously known ONC PMS eclipsing binaries. One of the most surprising and interesting classes of variables we find are characterized by short duration flux dips (less than one to a few days; ie, AA Tau analogs: Bouvier et al. 2003). We interpret these events as the star being extincted by either clouds of relatively higher opacity in the disk atmosphere, or geometric disk warps of relatively higher latitude which pass through the line of sight to the star. This hypothesis is consistent with both the duration and the wavelength dependence of the events (i.e., larger amplitudes at bluer wavelengths).

2.2. James Muzerolle: Spitzer Mid-Infrared Spectroscopic Variability in IC348

We initiated a multi-epoch study of the young cluster IC 348 with all instruments on board *Spitzer*. This region is an appealing target given its reasonably close distance and wide range of YSO evolutionary states, including a significant number of highly evolved and transitional disks. With prior GTO/legacy observations, our dataset includes a total of 10 maps of the entire cluster with MIPS, 7 maps with IRAC, and 4 to 7

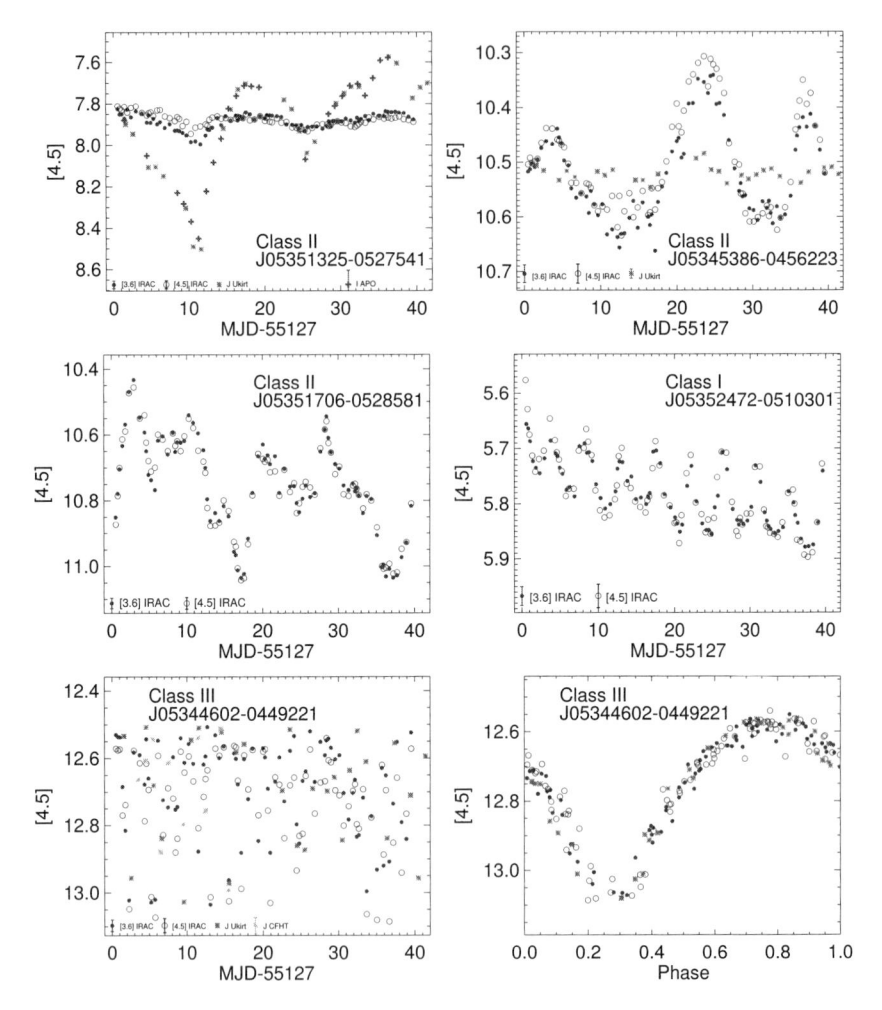

Figure 1. A sample of the range of variability types captured by the YSOVAR program. Most variations are relatively colorless (e.g., the bottom four panels), but some strong color variations are observed. Both red-dominant and blue-dominant color variations are detected (see top left and right panels, respectively), suggesting that different physical mechanisms (extinction, variable disk temperature & geometry, etc.) may be required to explain the full range of color-dependent variability. Aperiodic variability (which we typically interpret as disk/accretion driven) is the dominant mode of variability seen in the disked/embedded sources (e.g., middle-left panel), but we do see examples of periodic variability even in the most heavily embedded sources (e.g., middle-right panel). Unphased and phased light curves are shown for a particularly short period WTTs system (~0.27 d) in the bottom right and left panels, respectively. While we typically interpret periodic WTTs as exhibiting spot-induced variability, this system's short period and asymmetric light curve suggest it may be a close, tidally distorted binary.

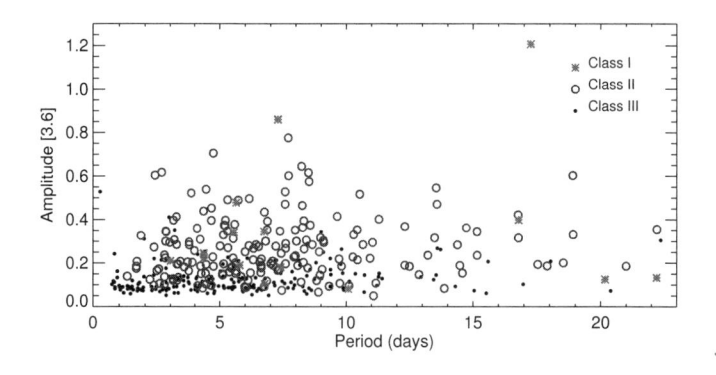

Figure 2. 3.6 µm amplitude vs. period for YSOs detected as periodic variables in YSOVAR's mid-IR monitoring of Orion. Both period and amplitude appear correlated with SED class: while periodic variability is most commonly detected for WTTs/Class III objects, those disked/embedded Class I-II YSOs that do demonstrate periodic variability have generally longer periods and larger amplitudes.

epochs of IRS spectroscopy for 14 individual cluster members (InfraRed Spectrograph; Houck et al. 2004). The measurements sample a range of cadences from days to years over the period 2004 - 2009. We find ubiquitous variability at all wavelengths: over half of the Class 0/I objects at 3.6 - 24 µm, nearly 70% of Class II objects at 3.6 - 8 µm, and 40% of Class II objects at 24 µm. Variability time-scales range from days to years at all wavelengths. We see apparent wavelength dependences with two general flavors: correlated variability over the 3.6 - 24 µm range, with or without color changes (most common in Class 0/I sources), and anti-correlated behavior with a pivot point typically around 6 - 10 microns (most common in Class II sources and transitional disks). From the IRS spectra, almost all of the variations are related to the continuum; we see measurable changes in the silicate features in only a few objects.

In most cases, the variations seen at longer wavelengths occur on time-scales too short to be representative of in-situ structural changes. For the Class II objects with anti-correlated wavelength behavior, the flux changes are likely connected to the innermost disk regions (puffed inner rim associated with dust sublimation, or dynamical truncation from close companions). Perturbations can cause a change in the inner disk height, which leads to direct changes in the short wavelength emission, and can cause shadowing of cooler material farther out in the disk, which leads to changes in the longer wavelength emission. The origin of the inner disk perturbations is not clear, but may be a result of changes in the accretion luminosity emitted by stellar accretion shocks, or inner disk warping created by tilted stellar magnetic fields or gravitational interactions with embedded companions.

2.3. Kevin Flaherty: LRLL 31, a Case Study of Mid-IR Variability in Transition Disks

In addition to providing a robust characterization of mid-IR photometric variability, the *Spitzer Space Telescope* has diagnosed mid-IR variability in specific young stars via multi-epoch spectroscopy with IRS. These observations identified the detailed wavelength dependence of mid-IR flux variations, pointing the way to new and unexpected results. One of the most interesting cases is that of LRLL 31, a G6 T Tauri star in IC 348: IRS spectra separated by one week revealed that the 5-8μm flux decreased to nearly photospheric levels while the 8-40μm flux increased by 60% (Muzerolle et al. 2009). What makes this star especially interesting is that it shows a deficit of flux around 10 μm compared to a traditional T Tauri star, indicative of a removal of small dust grains from the inner disk. This transition disk represents an evolutionary stage where the dust and gas are being removed from the circumstellar disk, and the fluctuations may be related to the process that is clearing the disk. Previous theories for explaining variability in young stellar objects (rotation of hot/cold spots across the stellar surface, changes in the accretion rate, extinction events) do not explain the wavelength dependence or the strength of the variability observed in LLRL 31. Fluctuations on daily to weekly time-scales are surprising, since thermal equilibrium arguments would indicate that mid-IR wavelengths probe regions of the disk from 0.5 to a few AU, where the dynamical timescale is closer to years. *In situ* changes in this region of the disk should not be possible on these time-scales, so LLRL 31's mid-IR variability must be rooted in the dynamics of the inner disk.

Motivated by these observations of LLRL 31, Flaherty & Muzerolle (2010) developed a model that explains the observed variability as arising from a warp in LLRL 31's disk, as opposed to the purely axisymmetric disks that had been previously assumed. Varying the height of a warp at the inner edge of the disk was able to explain the observed strength and wavelength dependence of the variability. As the warp grows it is more directly illuminated by the central star and heats up, emitting more flux at shorter wavelengths. The larger warp also shadows more of the outer disk, reducing the long-wavelength flux that originates far from the star. This model assumes the variability arises from structural changes in the disk warp: the precession of such a structure is unable to reproduce the variability.

While this model was successful at explaining much of LLRL 31's variability at $\lambda < 10\mu$m, it was still limited by the fact that the 5-40 μm IRS spectra only partially traces emission from the inner disk. Observations at shorter wavelengths, produced by emission from the hottest dust, are needed to better understand the properties of the material responsible for the mid-infrared variations. Further monitoring at 3.6 and 4.5 μm with Spitzer, and ground-based 0.8-5 μm spectra have revealed that the temperature of LLRL 31's inner disk remained constant while the emitting area changed by a factor of 10. Moreover, in 2009 there appeared to be a correlation between the accretion rate, as measured by the Paβ line, and the infrared flux, measured by the infrared photometry (Flaherty et al. submitted). This strong correlation suggests that the process that disturbs the dust must also disrupt the accretion flow. The most promising explanations are either a companion beyond 0.3 AU on an orbit misaligned with the disk (e.g. Fragner & Nelson 2009) or a dynamic interface between the stellar magnetic field and the disk (e.g., Goodson & Winglee 1999; Bouvier et al. 2007). Contemporaneous X-ray and Spitzer observations will be able to trace some of the interaction between the magnetic field and the disk, further probing this process. The wavelength dependence of the variability seen in LRLL 31 appears to be common among transition disks (Espaillat et al. submitted) and further observations of both the gas and dust in these objects could determine if the physical process occurring in LRLL 31 is common in these other objects.

2.4. Neal Turner: Models of Axisymmetric Disk Disturbances from Thermal Waves and Turbulent Eddies

T Tauri stars' diverse mid-IR variability likely arises from several different processes. Variations that correlate poorly with the accretion luminosity are most simply explained by changes in the way the starlight falls across the disk surface. Kevin Flaherty presents above some ways to produce non-axisymmetric disk disturbances. Among processes leading to approximately-axisymmetric disturbances, two explored here are thermal waves and turbulent eddies.

Thermal waves are driven by the stellar illumination. The front of each wave tilts into the starlight, receiving extra heating and expanding, while the back of each wave receives less heating and cools and contracts. The wave propagates inward, dying out when it enters the shadow of the disk's inner rim. The wavelength is about equal to the length of the shadow cast by each peak. Thermal waves operate at mass accretion rates below about 10^{-7} $M_\odot$/yr. For systems with larger accretion rates, accretion heating within the disk overwhelms the heating from external starlight, so that the waves no longer form or propagate. The wave periods are comparable to the heating and cooling time-scales, and are measured in years (Watanabe & Lin 2008).

But some T Tauri stars vary on time-scales of weeks, much shorter than the thermal wave period. The changes must arise near the disk inner edge, where the orbital period is about a week. On such short time-scales, the outer disk can be treated as static. A process producing changes in the disk photosphere height over times comparable to the orbital period is the same magneto-rotational turbulence that drives the accretion flow onto the star. The strength of the magnetic fields varies episodically, leading to vertical excursions of the disk photosphere by as much as 30% (Turner et al. 2010). The excursions are approximately axisymmetric, at least over the length scales modeled in shearing-box calculations. When higher, the rim intercepts more starlight. Monte Carlo radiative transfer calculations (Turner N. J. et al., in preparation) show raising the rim increases the area with temperatures around 1000 K, while shadowing the disk beyond so that the emission from cooler material is reduced. The resulting variations in the two warm-Spitzer bands are correlated, with amplitudes generally decreasing toward longer wavelengths (see Figure 3). Similar effects can be expected from the shadow cast by the outer edge of the gap opened in the disk by a giant planet, but the weaker frontal illumination would tend to decrease the amplitude.

3. Short Wavelength Variability

3.1. Fabienne Bastien: High frequency Variability of V1647 Orionis

Numerous young stars have been observed to undergo large amplitude photometric outbursts. While the observational record is still woefully incomplete, these events may play a significant role in the early evolution of many, possibly even most, young stars. These outbursts have traditionally been divided into two major classes: FU Ori-like large amplitude (> 4 mag), long duration (t> 10s years) outbursts, with spectra dominated by absorption features characteristic of cool (GKM) supergiants (Hartmann & Kenyon 1996), and EX Lup-like (often shortened to EXor) outbursts, which appear to be more frequent, of shorter duration, and characterized by heavily veiled spectra with strong emission and absorption components in accretion and outflow-sensitive features (Herbig 2007).

422   Covey et al.

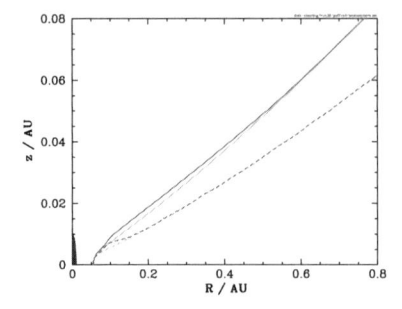

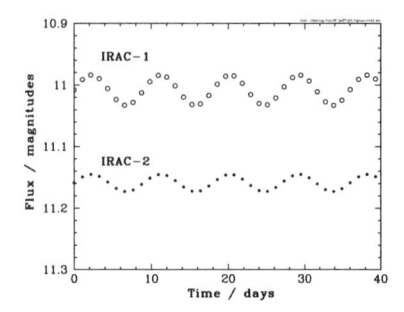

Figure 3. *(Left)* – A bump near the inner rim casts a long, cool shadow across the disk while itself being heated by the intercepted starlight. Curves show the surfaces of unit optical depth for light coming from the star (solid) and from directly overhead (dashed), for cases with (red) and without a bump (green). (Right) - Bumps whose height changes over time produce variability with amplitude greatest at shorter infrared wavelengths. Synthetic IRAC 3.6 and 4.5 μm light curves for a model disk with a magnetically inflated inner rim. (both panels, Turner et al., in prep).

In 2004, the star V1647 Orionis was observed to undergo a major accretion outburst. Numerous observations were obtained throughout the outburst, which persisted through 2006, but the source's photometric and spectroscopic properties resisted simple classification as a prototypical FU Ori-like or EX Lup-like outburst (e.g., Fedele et al. 2007). Archival photometry revealed that the 2003 event was not V1647 Ori's first outburst, and additional objects have been observed to undergo outbursts with similar spectral characteristics (e.g., Covey et al. 2010), suggesting that these outbursts may signify a new phenomena of import for a star's early evolution.

Our study of high cadence time-series photometry of the 2003-2004 and 2008-2009 eruptions of V1647 Orionis revealed interesting transient variability on timescales of a few hours: we detect a highly significant period of 0.13 days in our 2003 data that does not appear in our 2009 data. We attribute this period, which is inconsistent with the object's rotation period, to a short-term radial pulsational mode of the star excited by the sudden increase in its accretion rate at the time of our 2003 observations. This period, if also excited in V1647 Ori's latest outburst event, would have presumably rung down by the time we observed the star in 2009. More near infrared and optical observations with cadences of at least one image per hour over the course of several days during the rise, plateau and decline phases of such outbursts would allow us to better understand what short timescale phenomena occur during these events.

3.2. Ettore Flaccomio: Correlated Optical and X-ray Variability in Accreting Classical T Tauri Stars

The magnetospheric accretion model, in which material is funneled from the inner edge of the circumstellar disk to the stellar photosphere along magnetic field lines (e.g., Ostriker & Shu 1995; Bouvier et al. 2007), predicts that accreting material will be strongly shocked when it impacts the stellar surface. This extremely hot (T ~ 10,000 K; Herczeg & Hillenbrand 2008) shocked region is believed to produce significant continuum emission at optical, ultraviolet, and X-ray wavelengths (Günther et al. 2007). Clear differences have indeed been detected between the X-ray properties of accreting Classic T

Tauri stars (CTTs) and non-accreting WTTs: CTTs possess harder, more time-variable X-ray spectra than WTTs (Flaccomio et al. 2006), and detailed spectral studies suggest that accretion contributes a distinct, soft (2-3 MK) component to CTTs X-ray spectra (Güdel et al. 2007).

Many open questions remain, however, as to the exact mechanisms underlying the connection between accretion and X-ray emission. For instance, accreting systems are counter-intuitively seen to be significantly less X-ray luminous than predicted by standard magnetospheric accretion models (Johns-Krull 2007) as well as their WTTs counterparts (Preibisch et al. 2005), potentially due to the shocked emission being extincted by material in the overlying accretion columns (Gregory et al. 2007). Previous observations have also failed to identify a clear correlation between CTTs optical and X-ray emission. Accretion is expected to contribute a portion of the X-ray emission, originating in the photospheric accretion shock (as opposed to coronal X-rays; Stassun et al. 2006, 2007; Grosso et al. 2007).

In March 2008, Alencar et al. (2010) observed NGC 2264 with CoRoT for 23.5 days obtaining high-quality uninterrupted optical light-curves of its young stars. During the CoRoT pointing, two short Chandra observations were performed with a separation of 16 days, allowing us to study the correlation between optical and X-ray variability on this timescale, and thus the physical mechanism responsible for the variability (Flaccomio et al. 2010). The variabilities of CTTs in the optical and soft X-ray (0.5-1.5 keV) bands are correlated, while no correlation is apparent in the hard (1.5-8.0 keV) band. Also, no correlation in either band is present for WTTs. The correlation between soft X-ray and optical variability of CTTs can be naturally explained in terms of time-variable shading (absorption) from circumstellar material orbiting the star, in a scenario rather similar to the one invoked to explain the observed phenomenology in the CTT star AA Tau. The slope of the observed correlation implies (in the hypothesis of homogeneous shading) a significant dust depletion in the circumstellar material.

3.3. Marsden

One of the key aspects of the solar dynamo is the reversal of the Sun's global magnetic field every ~11 years. However, the dynamo mechanism for young rapidly-rotating solar-type stars is unclear and may well be fundamentally different to that of today's Sun (Donati et al. 2003). To learn more about the type of dynamo operating in young stars we have used the technique of Zeeman Doppler Imaging (ZDI; Semel 1989; Donati & Brown 1997) to map the magnetic topology of two young solar-type stars (HD 141943 and HR 1817) over several years, searching in particular for evidence of polarity reversals in the global magnetic field. Given the rapid differential rotation seen on such stars (Marsden et al. 010a) it is expected that they have significantly shorter magnetic cycles than the Sun, perhaps as short as a few years.

While HD 141943 shows evidence of changes in its magnetic topology it shows no evidence of a polarity reversal over the 3 years of observations (Marsden et al. 010a). This result is similar to another young solar-type star (HD 171488) which has been previously observed using similar techniques (Marsden et al. 2006; Jeffers & Donati 2008; Jeffers et al. 2010). In contrast, the preliminary results for HR 1817 indicate a reversal of the polarity of the star's radial magnetic field between 2008 and 2009, although a similar reversal is not seen in its azimuthal magnetic field (see Figure 4; Marsden et al., in prep.). Such behavior suggests a phase delay in polarity reversal between the radial and azimuthal fields.

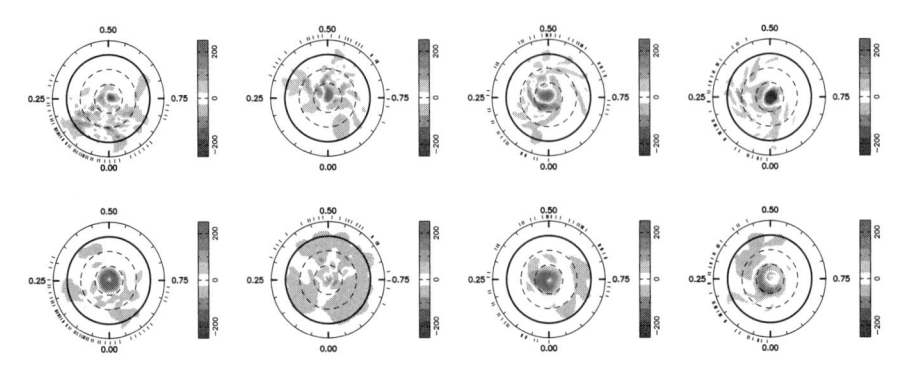

Figure 4. Preliminary radial (top) and azimuthal (bottom) magnetic field topologies for HR 1817 for the years (left to right), 2001, 2007, 2008 and 2009. The images are flattened polar projections extending down to -30° latitude. The bold line denotes the equator and the dashed lines are +30° and +60° latitude parallels. The radial ticks outside the plot indicate the phases at which the star was observed and the scale is in Gauss. (Marsden et al., in prep.)

In order to assess how young solar-type stars undergo magnetic polarity reversals, optical spectropolarimetric observations of a statistically useful number of such stars needs to be undertaken regularly over a number years. A determination of how young solar-type stars go through magnetic polarity reversals and whether they have regular or chaotic magnetic cycles should greatly help our understanding of the dynamo in young Suns.

4. Conclusions

As this splinter summary demonstrates, recent years have seen considerable progress in our ability to observe, characterize, and model time-variable processes in the formation and early evolution of stars, circumstellar disks and planets. These advances include the ability to study variability across a wider range of wavelengths and time-scales than previously possible, and to construct more detailed and computationally intensive models of star and disk processes.

Nonetheless, our current understanding of time-variable phenomena in early stellar evolution is significantly incomplete. In addition to the questions highlighted above, the audience identified several other outstanding challenges for studies of time-domain processes in the formation and early evolution of stars and planets. These include:

- Improving the mechanisms and capabilities for conducting multi-site and multi-wavelength studies with long time baselines.

- Achieving statistically robust constraints on variability (e.g., FU Ori outburst rate/duty cycle), rather than qualitative descriptions.

- Identifying a critical mass of individual, rare systems (i.e, KH 15D, McNeil's Nebula, etc.) such that we can infer universal lessons from their specific properties.

- Developing better observational and theoretical diagnostics to locate the source regions and determine the causes of variability. Two examples are detecting kinematic signatures of the underlying gas motions in the line emission from the star and disk, and linking the photometric variability of the inner disk to changes in the scattered light from the spatially-resolved outer disk.

- Understanding the implications for planet formation of time-variable structure in protoplanetary disks.

Meeting each of these challenges will require ingenuity, focused effort, and careful planning. The young stars presenting here today suggest that we will indeed be up to the task, but only time[2] will tell.

Acknowledgments. We thank the Cool Stars 16 SOC for organizing a productive and enjoyable meeting. We particularly thank the editors of this volume, whose enthusiasm and encouragement were critical to the "timely" preparation of this article.

References

Alencar, S. H. P., Teixeira, P. S., Guimarães, M. M., McGinnis, P. T., Gameiro, J. F., Bouvier, J., Aigrain, S., Flaccomio, E., & Favata, F. 2010, A&A, 519, A88+. 1005.4384
Baraffe, I., & Chabrier, G. 2010, A&A, 521, A44+. 1008.4288
Bouvier, J., Alencar, S. H. P., Harries, T. J., Johns-Krull, C. M., & Romanova, M. M. 2007, Protostars and Planets V, 479. arXiv:astro-ph/0603498
Bouvier, J., Covino, E., Kovo, O., Martin, E. L., Matthews, J. M., Terranegra, L., & Beck, S. C. 1995, A&A, 299, 89
Bouvier, J., Grankin, K. N., Alencar, S. H. P., Dougados, C., Fernández, M., Basri, G., Batalha, C., Guenther, E., Ibrahimov, M. A., Magakian, T. Y., Melnikov, S. Y., Petrov, P. P., Rud, M. V., & Zapatero Osorio, M. R. 2003, A&A, 409, 169. arXiv:astro-ph/0306551
Cargile, P. A., Stassun, K. G., & Mathieu, R. D. 2008, ApJ, 674, 329. 0709.3356
Cieza, L., & Baliber, N. 2007, ApJ, 671, 605. 0707.4509
Cody, A. M., & Hillenbrand, L. A. 2010, ApJS, 191, 389. 1011.3539
Covey, K. R., Hillenbrand, L. A., Miller, A. A., Poznanski, D., Cenko, S. B., Silverman, J. M., Bloom, J. S., Kasliwal, M. M., Fischer, W., Rayner, J., Rebull, L. M., Butler, N. R., Filippenko, A. V., Law, N. M., Ofek, E. O., Agueros, M., Dekany, R. G., Rahmer, G., Hale, D., Smith, R., Quimby, R. M., Nugent, P., Jacobsen, J., Zolkower, J., Velur, V., Walters, R., Henning, J., Bui, K., McKenna, D., Kulkarni, S. R., & Klein, C. 2010, ArXiv e-prints. 1011.2565
Donati, J., Cameron, A. C., Semel, M., Hussain, G. A. J., Petit, P., Carter, B. D., Marsden, S. C., Mengel, M., López Ariste, A., Jeffers, S. V., & Rees, D. E. 2003, MNRAS, 345, 1145
Donati, J.-F., & Brown, S. F. 1997, Astronomy and Astrophysics, 326, 1135
Fedele, D., van den Ancker, M. E., Petr-Gotzens, M. G., & Rafanelli, P. 2007, A&A, 472, 207. 0706.3281
Flaccomio, E., Micela, G., Favata, F., & Alencar, S. P. H. 2010, A&A, 516, L8+. 1006.0041
Flaccomio, E., Micela, G., & Sciortino, S. 2006, A&A, 455, 903. arXiv:astro-ph/0604243
Flaherty, K. M., & Muzerolle, J. 2010, ApJ, 719, 1733. 1007.1249
Fragner, M. M., & Nelson, R. P. 2009, A&A, 505, 873. 0912.1299
Garatti, A. C. o., Garcia Lopez, R., Scholz, A., Giannini, T., Eislöffel, J., Nisini, B., Massi, F., Antoniucci, S., & Ray, T. P. 2010, ArXiv e-prints. 1012.0281
Goodson, A. P., & Winglee, R. M. 1999, ApJ, 524, 159

---

[2](bad) Pun intended.

Grankin, K. N., Melnikov, S. Y., Bouvier, J., Herbst, W., & Shevchenko, V. S. 2007, A&A, 461, 183. arXiv:astro-ph/0611028
Gregory, S. G., Wood, K., & Jardine, M. 2007, MNRAS, 379, L35. 0704.2958
Grosso, N., Briggs, K. R., Güdel, M., Guieu, S., Franciosini, E., Palla, F., Dougados, C., Monin, J., Ménard, F., Bouvier, J., Audard, M., & Telleschi, A. 2007, A&A, 468, 391. arXiv: astro-ph/0608696
Güdel, M., Telleschi, A., Audard, M., Skinner, S. L., Briggs, K. R., Palla, F., & Dougados, C. 2007, A&A, 468, 515. arXiv:astro-ph/0609182
Günther, H. M., Schmitt, J. H. M. M., Robrade, J., & Liefke, C. 2007, A&A, 466, 1111. arXiv: astro-ph/0702579
Hamilton, C. M., Herbst, W., Shih, C., & Ferro, A. J. 2001, ApJ, 554, L201. arXiv:astro-ph/0105412
Hartmann, L., & Kenyon, S. J. 1996, ARA&A, 34, 207
Hebb, L., Stempels, H. C., Aigrain, S., Collier-Cameron, A., Hodgkin, S. T., Irwin, J. M., Maxted, P. F. L., Pollacco, D., Street, R. A., Wilson, D. M., & Stassun, K. G. 2010, A&A, 522, A37+. 1008.4312
Herbig, G. H. 1977, ApJ, 217, 693
— 1989, in European Southern Observatory Conference and Workshop Proceedings, edited by B. Reipurth, vol. 33 of European Southern Observatory Conference and Workshop Proceedings, 233
— 2007, AJ, 133, 2679
Herbst, W., Eislöffel, J., Mundt, R., & Scholz, A. 2007, Protostars and Planets V, 297. arXiv: astro-ph/0603673
Herczeg, G. J., & Hillenbrand, L. A. 2008, ApJ, 681, 594. 0801.3525
Houck, J. R., Roellig, T. L., Van Cleve, J., Forrest, W. J., Herter, T. L., Lawrence, C. R., Matthews, K., Reitsema, H. J., Soifer, B. T., Watson, D. M., Weedman, D., Huisjen, M., Troeltzsch, J. R., Barry, D. J., Bernard-Salas, J., Blacken, C., Brandl, B. R., Charmandaris, V., Devost, D., Gull, G. E., Hall, P., Henderson, C. P., Higdon, S. J. U., Pirger, B. E., Schoenwald, J., Sloan, G. C., Uchida, K. I., Appleton, P. N., Armus, L., Burgdorf, M. J., Fajardo-Acosta, S. B., Grillmair, C. J., Ingalls, J. G., Morris, P. W., & Teplitz, H. I. 2004, in Society of Photo-Optical Instrumentation Engineers (SPIE) Conference Series, edited by J. C. Mather, vol. 5487 of Presented at the Society of Photo-Optical Instrumentation Engineers (SPIE) Conference, 62
Irwin, J., Hodgkin, S., Aigrain, S., Bouvier, J., Hebb, L., & Moraux, E. 2008, MNRAS, 383, 1588. 0711.0329
Jeffers, S. V., & Donati, J.-F. 2008, Monthly Notices of the Royal Astronomical Society, 390, 635
Jeffers, S. V., Donati, J.-F., Alecian, E., & Marsden, S. C. 2010, Monthly Notices of the Royal Astronomical Society, submitted
Johns-Krull, C. M. 2007, ApJ, 664, 975. 0704.2923
Joy, A. H. 1945, ApJ, 102, 168
Kóspál, Á., Ábrahám, P., Acosta-Pulido, J. A., Arévalo Morales, M. J., Carnerero, M. I., Elek, E., Kun, M., Pál, A., & Szakáts, R. 2010, ArXiv e-prints. 1011.4009
Lorenzetti, D., Giannini, T., Larionov, V. M., Kopatskaya, E., Arkharov, A. A., De Luca, M., & Di Paola, A. 2007, ApJ, 665, 1182. 0707.0967
Marsden, S. C., Donati, J.-F., Semel, M., Petit, P., & Carter, B. D. 2006, Monthly Notices of the Royal Astronomical Society, 370, 468
Marsden, S. C., Jardine, M. M., Vélez, J. C. R., Alecian, E., Brown, C. J., Carter, B. D., Donati, J.-F., Dunstone, N., Semel, M., & Waite, I. A. 2010a, Monthly Notices of the Royal Astronomical Society, submitted
Miller, A. A., Hillenbrand, L. A., Covey, K. R., Poznanski, D., Silverman, J. M., Kleiser, I. K. W., Rojas-Ayala, B., Muirhead, P. S., Cenko, S. B., Bloom, J. S., Kasliwal, M. M., Filippenko, A. V., Law, N. M., Ofek, E. O., Dekany, R. G., Rahmer, G., Hale, D., Smith, R., Quimby, R. M., Nugent, P., Jacobsen, J., Zolkower, J., Velur, V., Walters, R., Henning, J., Bui, K., McKenna, D., Kulkarni, S. R., & Klein, C. R. 2010, ArXiv

e-prints. 1011.2063
Morales-Calderón, M., Stauffer, J. R., Rebull, L., Whitney, B. A., Barrado y Navascués, D., Ardila, D. R., Song, I., Brooke, T. Y., Hartmann, L., & Calvet, N. 2009, ApJ, 702, 1507. 0907.3360
Muzerolle, J., Flaherty, K., Balog, Z., Furlan, E., Smith, P. S., Allen, L., Calvet, N., D'Alessio, P., Megeath, S. T., Muench, A., Rieke, G. H., & Sherry, W. H. 2009, ApJ, 704, L15. 0909.5201
Ostriker, E. C., & Shu, F. H. 1995, ApJ, 447, 813
Plavchan, P., Gee, A. H., Stapelfeldt, K., & Becker, A. 2008, ApJ, 684, L37. 0807.4557
Preibisch, T., Kim, Y., Favata, F., Feigelson, E. D., Flaccomio, E., Getman, K., Micela, G., Sciortino, S., Stassun, K., Stelzer, B., & Zinnecker, H. 2005, ApJS, 160, 401. arXiv: astro-ph/0506526
Rebull, L. M., Wolff, S. C., & Strom, S. E. 2004, AJ, 127, 1029
Reipurth, B., & Aspin, C. 2004, ApJ, 606, L119. arXiv:astro-ph/0403667
Semel, M. 1989, Astronomy and Astrophysics, 225, 456
Stassun, K. G., Mathieu, R. D., Mazeh, T., & Vrba, F. J. 1999, AJ, 117, 2941
Stassun, K. G., van den Berg, M., & Feigelson, E. 2007, ApJ, 660, 704. arXiv:astro-ph/0701735
Stassun, K. G., van den Berg, M., Feigelson, E., & Flaccomio, E. 2006, ApJ, 649, 914. arXiv: astro-ph/0606079
Turner, N. J., Carballido, A., & Sano, T. 2010, ApJ, 708, 188. 0911.1533
Vorobyov, E. I., & Basu, S. 2010, ApJ, 719, 1896. 1007.2993
Vrba, F. J., Herbst, W., & Booth, J. F. 1988, AJ, 96, 1032
Watanabe, S., & Lin, D. N. C. 2008, ApJ, 672, 1183. 0709.1760
Zhu, Z., Hartmann, L., Gammie, C. F., Book, L. G., Simon, J. B., & Engelhard, E. 2010, ApJ, 713, 1134. 1003.1759

The 16th Cambridge Workshop on Cool Stars, Stellar Systems and the Sun
ASP Conference Series, Vol. 448
Christopher M. Johns-Krull, Matthew K. Browning, and Andrew A. West, eds.
© 2011 Astronomical Society of the Pacific

Ultracool Dwarf Science from Widefield Multi-Epoch Surveys

N. R. Deacon,[1] D. J. Pinfield,[2] P. W. Lucas,[2] Michael C. Liu,[1] M. S. Bessell,[3] B. Burningham,[2] M. C. Cushing,[4] A. C. Day-Jones,[5] S. Dhital,[6] N. M. Law,[7] A. K. Mainzer,[4] and Z. H. Zhang[2]

[1] *Institute for Astronomy, University of Hawai'i, 2680 Woodlawn Drive, Honolulu, Hawai'i, 96822-1839, USA*

[2] *Centre for Astrophysics Research, University of Hertfordshire, Hatfield, AL10 9AB, UK*

[3] *Research School of Astronomy and Astrophysics, Mount Stromlo Observatory, Cotter Rd, ACT 2611, Australia*

[4] *Jet Propulsion Laboratory, M/S 264-765, 4800 Oak Grove Drive, Pasadena, CA 91109, USA*

[5] *Universidad de Chile, Camino El Observatorio 1515, Las Condes, Santiago, Chile*

[6] *Department of Physics & Astronomy, Vanderbilt University, Nashville, TN 37235, USA*

[7] *Dunlap Institute for Astronomy and Astrophysics, University of Toronto, 50 St. George Street Room 101, Toronto, Ontario, M5S 3H4, Canada*

Abstract. Widefield surveys have always provided a rich hunting ground for the coolest stars and brown dwarfs. The single epoch surveys at the beginning of this century greatly expanded the parameter space for ultracool dwarfs. Here we outline the science possible from new multi-epoch surveys which add extra depth and open the time domain to study.

1. Introduction

Widefield multi-epoch surveys have been a crucial discovery tool for ultracool dwarfs. Photographic plate data was used by Luyten (1979) to identify nearby cool stars by their proper motion. As technology evolved these plates were digitised (Hambly et al. 2001, Monet et al. 2003) leading to more discoveries of cool nearby stars.

The study of brown dwarfs did not begin with widefield digital infrared and red optical surveys, but datasets such as 2MASS (Skrutskie et al. 2006) and SDSS (York et al. 2000) massively expanded the sample of brown dwarfs. Works such as Chiu et al. (2006), Reid et al. (2008) and Burgasser et al. (2004) provided large samples and objects from these surveys were used to define the new L and T spectral classes (Kirkpatrick et al. 1999, Burgasser et al. 2006).

Now a new generation of widefield surveys are expanding into relatively unexplored regions, identifying extremely cool objects, finding nearby brown dwarfs by their trigonometric parallax and opening the time domain to the search for variability. Here we outline the leading surveys currently in operation and those about to come online.

2. Current Surveys

2.1. The UKIRT Deep Infrared Sky Survey

Of the five UKIDSS surveys (Lawrence et al. 2007), three are contributing significantly to the study of substellar objects: the Galactic Clusters Survey (GCS), the Large Area Survey (LAS) and the Galactic Plane Survey (GPS). The first of these has is targeting 10 clusters, and totalling over 1400 sq degs to a depth of $K = 18.7$. Four of these will have two epochs to allow proper motion selection of members for robust IMF determination (Pleiades, Alpha Per, Praesepe and Hyades). Results already published for several clusters are painting a picture of the substellar IMF that is broadly consistent with previous studies. For an IMF of the form $\psi(M) \propto M^{-\alpha}$, a value of $\alpha = 0.6$ appears to be common (e.g. Lodieu et al. 2007; Lodieu et al. 2009).

The LAS search for field brown dwarfs has now resulted in the discovery of over 100 T dwarfs, making it the single largest contributor to the known sample of these objects, has now provided the most precise measurement of the space density of T6-T8 dwarfs to-date (Burningham et al. 2010), revealing a significant dearth these objects compared to what would be expected given the initial mass function observed in young clusters. The origin of this discrepancy is not clear, but possibilities include a variation in the birthrate of brown dwarfs over the lifetime of the Galaxy, a similarly variant substellar initial mass function or problems with the evolutionary models used to predict the space density. In addition to studies of the IMF, the LAS sample is producing additional results from searches for benchmark systems (e.g. Day-Jones et al. 2011; Zhang et al. 2010) and the search for halo T dwarfs (e.g. Murray et al. in prep.)

Despite the challenges associated with searching for T dwarfs in the Galactic plane, the GPS is also beginning to deliver very cold objects. The discovery of the extremely cool T10 dwarf UGPS J0722-0540 at a distance of just 4.1pc (Lucas et al. 2010), means that UKIDSS T dwarfs are now on the threshold of probing the sub-500K regime, where water clouds may start to be seen in the photosphere. Over the next few years the synergy of the UKIDSS data set with the WISE survey will likely reveal even cooler atmospheres, and possibly the first observation of water clouds beyond the Solar System.

2.2. VISTA Public Surveys

The Visible and Infrared Survey Telescope for Astronomy (VISTA) is a UK built 4 metre class, near-infrared survey telescope located at the ESO site on Cerro Paranal, Chile. VISTA's wide field camera has sixteen 2048x2048 Raytheon arrays with quantum efficiency >90%. It produces tiles with dimensions of 1.5x1.0 degrees, compiled from 6 dithered pawprints. The image scale is 0.34 arcsec per pixel. During its first 5 years of operation VISTA is mainly dedicated to performing six public surveys, which began taking data in late 2009. These surveys include: UltraVISTA, the smallest of the surveys covering only 0.73 square degrees to a depth of J=26.6. VIDEO (Vista

Deep Extragalactic observations), VMC (Vista Magellanic Cloud Survey), VVV (Vista Variables in the Via Lactea), a synoptic survey of the Galactic plane, VIKING (Vista Kilo-Degree Infrared Galaxy Survey) and the largest, VHS (Vista Hemisphere Survey), which will image the rest of the southern hemisphere.

The last two of these surveys will likely be the source of most brown dwarf discoveries from the VISTA surveys. With their large sky coverage in several filters, these surveys will probe deeper than any other near-infrared survey, for example VIKING alone will cover three times the volume of the UKIDSS Large Area Survey, providing data in the Z,Y,J,H and K_s filters. Both VIKING and VHS will have optical counterparts provided respectively from the ESO KIDS (Kilo degree survey) and DES (Dark Energy Survey), such that combining these surveys, and data from WISE, Skymapper and Pan-STARRS, will provide a powerful tool for the identification of brown dwarfs. There is the potential to increase the number of T dwarfs know by an order of magnitude. In addition this will allow the identification of many brown dwarfs that could be used as benchmark objects as well as objects that are cooler than those presently known. These discoveries will be able to test and help calibrate existing ultracool models and measurement of the mass function to new precision. However, a very efficient follow up strategy will be required to classify the large number of brown dwarf candidates, many of which will be challenging targets for ground-based spectroscopy.

2.3. The WISE All-Sky Survey

One of the two primary science objectives for the Wide-field Infrared Survey Explorer (WISE) is to find the coldest brown dwarfs, which represent the final link between the lowest mass stars and the giant planets in our own solar system. WISE is a NASA Medium-class Explorer mission designed to survey the entire sky in four infrared wavelengths, 3.4, 4.6, 12 and 22 microns (Wright et al. 2010; Liu et al. 2008; Mainzer et al. 2005). WISE consists of a 40 cm telescope that images all four bands simultaneously every 11 seconds. It covers nearly every part of the sky a minimum of eight times, ensuring high source reliability, with more coverage at the ecliptic poles. Astrometric errors are less than 0.5 arcsec with respect to 2MASS (Wright et al. 2010). The preliminary estimated SNR=5 point source sensitivity on the ecliptic is 0.08, 0.1, 0.8 and 5 mJy in the four bands (assuming eight exposures per band; Wright et al. 2010). Sensitivity improves away from the ecliptic due to denser coverage and lower zodiacal background. WISE's two shortest wavelength bands, centered at 3.4 and 4.6 m (W1 and W2, respectively), were specially designed to optimize sensitivity to the coolest types of brown dwarfs (Kirkpatrick et al. in prep). In particular, BDs cooler than ~1500 K exhibit strong absorption due to the nu-3 band of CH_4 centered at 3.3 microns, with the onset of methane absorption at this wavelength beginning at ~1700 K (Noll & Marley 2000). The as-measured sensitivities of 0.08 and 0.1 mJy in W1 and W2 allow WISE to detect a 300 K BD out to a distance of ~8 pc, according to models by Marley et al. (2002) and Saumon & Marley (2008). While the W1 and W2 bands are superficially similar to the Spitzer/IRAC bands 1 and 2 (which were also designed to isolate cool BDs; Fazio et al. 1998), the WISE W1 band is wider to improve discrimination between normal stars and ultra-cool BDs (Wright et al. 2010. This results in a systematically larger W1-W2 color compared to [3.6]-[4.5]. Currently, only about two dozen objects with spectral types later than T7 are known, and four objects are known to have type T9 or later (Warren et al. 2007; Delorme et al. 2008, Burningham et al. 2008, and Lucas et al. 2010). New spectral indices for typing these late-type T dwarfs have been developed as the CH4 absorption band depths may not continue to increase in the NIR with temperatures lower than those of T8/T9 dwarfs (Burningham et al. 2010), but spectral anchors have yet to be defined. We have recently reported the discovery of WISEPC J0458+64 (Mainzer et al. 2010) an object which is consistent with an extremely late T dwarf. The best-fitting model has an effective temperature of 600 K, log g=5.0, [Fe/H]=0, and evidence for the presence of vertical mixing in its atmosphere. As this remarkably cool object was found easily in some of the first WISE data, WISE is likely to find many more similar objects, as well as cooler ones, inevitably producing abundant candidates for the elusive Y class brown dwarfs. Scaling from the Spitzer sample of 4.5 micron selected ultra-cool BD candidates (Eisenhardt et al. 2010), we expect to find hundreds of new ultra-cool brown dwarfs with WISE. We compute that for most likely initial mass functions, WISE has a better than 50% chance of detecting a cool BD which may actually be closer to our Sun than Proxima Centauri (Wright et al. 2010).

2.4. Surveys with Pan-STARRS 1

The Pan-STARRS project (Panoramic Survey Telescope and Rapid Response System; Kaiser et al. 2002), led by the University of Hawaii's Institute for Astronomy, is developing a unique optical survey instrument consisting of four co-aligned wide-field telescopes based in the Hawaiian Island (http://panstarrs.ifa.hawaii.edu). As a pathfinder for the full system, the project has successfully completed a single telescope system, called Pan-STARRS 1 (PS1), on the summit of Haleakala on the Hawaiian island of Maui. PS1 is a full-scale version of one of the four Pan-STARRS telescopes, with a 1.8-m primary mirror which images a 7-degree^2 field-of-view on a 1.4-Gigapixel camera. The large étendue of PS1 allows for rapid surveying of the observable sky, obtaining ≈2500 sq. degs and 1.5 Tb of raw data each night.

The PS1 3.5-year science survey mission officially began in May 2010 and is carrying out a suite of pre-defined surveys, spanning the nearest asteroids to the cosmological horizon.

Of greatest interest to ultracool dwarf research is the Pan-STARRS 1 3π survey, the largest of the PS1 surveys, amounting to 56% of the observing time. The Pan-STARRS 1 3π survey will monitor three quarters of the sky (30,000 sq. degs north of declination $-30°$) in g, r, i, z and y with six epochs per filter spread over three years and two observations per epoch to detect solar system objects. The survey goes to depths roughly similar to SDSS in the bluest bands and is significantly more sensitive in the far-red, with the novel y-band filter (0.95-1.03 μm) providing the greatest gains for ultracool dwarf science. The 5σ single-epoch limits are $grizy$ = 23.2, 22.5, 22.2, 21.2, 19.8 mag (AB system).

The Pan-STARRS 1 3π cadence will allow objects to be selected based on their proper motions and parallaxes without a priori colour selection (Magnier et al. 2008). This will allow a complete survey of the ultracool dwarf population within 30 pc down to the mid/late T regime, as well as discovery of low-mass objects with extreme properties that might otherwise have gone undetected.

The resulting catalog unprecedented in its volume, sample size, and completeness will address several key areas, including: (1) the luminosity function of nearby ultracool objects with the first volume-limited sample of brown dwarfs; (2) rigorous tests of theoretical models of brown dwarf atmospheres and evolution using densely populated color-magnitude diagrams; and (3) exploring the physics of ultracool objects over a wide range of metallicities and gravities.

As of this writing, Pan-STARRS 1 is just completing its first pass of the sky. With these first epoch data, we have been carrying out a proper motion survey for nearby L and T dwarfs by combining them with 2MASS. The survey spans a ≈ 10 year baseline and is most sensitive to objects with proper motions from ≈ 0.2–$2.0''$/yr. So far this has resulted in the discovery of a number of interesting ultracool dwarfs, including bright T dwarfs not previously identified in 2MASS (Deacon et al., in prep.).

2.5. The Stromlo Southern Sky Survey

SkyMapper is a 1.3m telescope with a 5.7 sq degree field of view covered with 32 2Kx4K E2V CCDs that will carry out a 6-color, multi-epoch survey of the southern sky (including the galactic plane) - The Stromlo Southern Sky Survey. We aim to provide star and galaxy photometry to better than 3% global accuracy and astrometry to better than 50 mas. The sampling will be 4 hr, 1 day, 1 week, 1month and 1 yr, although not in all filters. It will take five years to complete the survey. Some time is reserved for non-survey work. The photometric system of u (like Stromgren), v (like DDO38), griz (SDSS) is designed to maximize precision in the derivation of stellar astrophysical quantities. We also have available an Halpha filter for limited use. We expect 6 epoch limiting magnitudes of 22.9, 22.9, 22.8, 22.8, 21.9, 21.2 in u,v,g,r,i,z, respectively. The data will be supplied to the community after science verification without a proprietary period. Young M dwarfs will be surveyed using Halpha, r and i. L and T dwarfs (high-z QSO contaminants) will be surveyed using i-z. VISTA photometry will also be cross-referenced and proper motions derived from our different epoch SkyMapper observations as well as through comparison with earlier epoch catalog positions. Follow-up spectroscopy will be carried out on the ANU 2.3m telescope and AAO 3.9m telescopes.

2.6. The Palomar Transient Factory

The Palomar Transient Factory (PTF; Law et al. 2009) is a new fully-automated, widefield survey conducting a systematic exploration of the optical transient sky. The transient survey is performed using a new 8.1 square degree camera installed on the 48 inch Samuel Oschin telescope at Palomar Observatory; colors and light curves for detected transients are obtained with the automated Palomar 60 inch telescope. With an exposure of 60 s the survey reaches a depth of m~21.3 and m~20.6 (5σ, median seeing). Cadences range between 10 minutes and 5 days, with a 3-day-cadence r-band supernova search making up the largest fraction of the survey. The survey covers 8000 square degrees, and fields are observed between tens and hundreds of times. PTF provides automatic, real-time transient classification and follow-up, as well as a database including every source detected in each frame. As of Nov 22 2010, PTF had discovered and spectroscopically classified 910 supernovae. The wide-area, many epoch dataset is being used for a wide variety of cool star science, including the activity / rotation / age relation, proper motion surveys, and a search for transiting planets around 100,000 M and L dwarfs.

Table 1. A summary of the surveys discussed in this paper.

Survey	Filters	Depth(Single Epoch)	Area (sq.deg.)	Epochs	Dates	Public Release
UKIDSS LAS	Y,J,H,K	Y=20	3700	2(3000sq.deg.)	2005-2012	2014
Pan-STARRS 1	g,r,i,z,y	y=19.8(AB, 5σ)	33000	6 per filter	2010-2013	2014
VISTA VHS	J,H, partial Z,Y,K	J=20.2(AB, 5σ)	19000	1	2010-2017	2011-
Stromlo SSS	u,v,g,r,i,z	z=20.6(AB, 5σ)	22000	6 per filter	2010-2015	After calibration
WISE	3,4, 4.6, 12, 22 μm	σ=0.1 mJy in 4.6μm	All Sky	1	2010	2011-2012
PTF	g,R	r=21	8000	300+	2009-	2012
LSST	u,g,r,i,z,y	z=23.3(AB)	20000	1000 total	2018-2028	2018-

3. Future Surveys

3.1. The Large Synoptic Survey Telescope

The Large Synoptic Sky Survey (LSST) will cover the 20000 square degrees of the Southern sky, with ~1000 total visits in the *ugrizy* bands over 10 years. The faintness limit will be $z \sim 23.3$ mags for each visit and $z \sim 26.2$ mags for the total survey (LSST Science Collaborations 2009). LSST's depth, all-sky coverage, and 1000 epochs provide an large and unique dataset for studying the ultracool dwarfs (UCDs; specifically of M, L, T, and the hypothetical Y spectral classes).

LSST will allow for a census of the ultracool dwarfs (UCDs) in the solar neighborhood, with M0, L6, and T9 dwarfs detectable to ~10 kpc, 200 pc, and 25 pc, respectively. In terms of absolute numbers, Galactic simulations predict>347,000 M, 35000 L, 2300 L, and ~ 18 Y dwarfs in the LSST fields (LSST Science Collaborations 2009). In addition, we will have accurate proper motions and parallaxes (see Figure 1). This will allow us not only to measure the mass function, luminosity function, and velocity distribution of UCDs but also complete the census of nearby moving groups and associations (also see chapter on Juvenile UCD session).

The Southern sky is very rich in large, nearby open clusters, like the Orion Nebula, NGC 3532, IC 2602/IC 2391, and Blanco 1. Detecting and characterizing brown dwarfs in these clusters, which have known ages, will allow us to constrain the theoretical brown dwarf cooling curves (e.g. Chabrier et al. (2000)). As most known brown dwarfs are in the field with very uncertain ages, they cannot be used to constrain the cooling curves. Similarly, LSST will detect a large of of UCD eclipsing binaries that will serve as benchmarks to constrain our stellar evolutionary models. Based on LSST data, 77 M and 2 L dwarfs EBs will be characterized, with perhaps ten times as many UCD EB candidates (J. Pepper et al. in prep.)

4. Combining Survey Data

As mentioned previously surveys often complement each other by providing information in different wavelengths or providing additional epochs for proper motion measurement. Works such as Burgasser et al. (2004) use infrared colours in combination with legacy optical data such as Monet et al. (2003) to select ultracool dwarfs while UKIDSS studies such as Burningham et al. (2010) make use of SDSS data. Examples of using widefield surveys in combination for proper motions include Sheppard & Cushing (2009) and Deacon et al. (2009). The new generation of surveys can also be used together and with legacy data increase their science yield.

4.1. Ultracool Subdwarf Binaries Discovered from Large Area Surveys

Ultracool subdwarfs (UCSDs; e.g., Burgasser et al. 2009) are metal- poor, halo counterpart of ultracool dwarfs. They have not been understood well both observationally and theoretically due to the limited number of known UCSDs, especially benchmarks (e.g., Pinfield et al. 2006). The number of known UCSDs is increasing benefit from current optical and NIR large area surveys (SDSS, York et al. 2000; 2MASS, Skrutskie et al. 2006; UKIDSS, Lawrence et al. 2007. This make it is possible to identified UCSDs in wide binary systems. I selected a sample of objects with proper motions large than 100 mas/yr from the SDSS and USNO-B proper motion catalog (Munn et al. 2004). Around

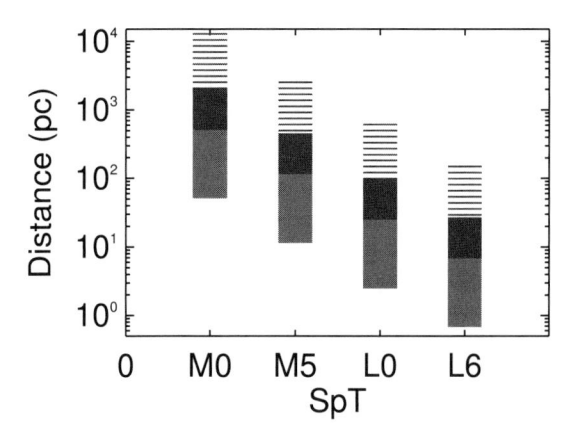

Figure 1. The distance at which stars of a given spectral type will be detected in LSST. The red (lighter grey in black and white) strip shows the distance for the brightness limit of $r = 16$ and $r = 21$ where proper motions greater than 1.5 mas yr^{-1} and parallaxes greater than 4.5 mas will be measured, with a 5=*sigma* significance. The solid blue (darker grey in black and white) shows the single epoch faintness limit, where proper motions greater than 7.5 masyr^{-1} and parallaxes greater than 22 mas will be measured. The hashed blue shows the limits of the entire 10 year survey, extending up to $r = 28$. The depth as well as the proper motion and parallax accuracies are unprecedented.

one thousand M subdwarfs are confirmed with SDSS spectra. Start with this subdwarf sample, I identified two extreme UCSDs (esdM6+esdK2, esdM6+esdM1) and six M subdwarfs (including an sdM1+WD system) in binary systems based on their common proper motions (Zhang et al. in preparation). The esdM6+esdK2 binary is the most exciting system with very high proper motions (444±3mas/yr). Distances for both components are estimated, and both consistent with 180±30 parsec. It is one of the widest ultracool systems (16740±2790AU), With a separation of 93″ at this distance. It's galactic velocity ($U = -325.3 km/s$; $V = -259.5 km/s$; $W = 8.4 km/s$) consistent with halo population. This is the first halo subdwarf benchmark in wide binary system. We have obtained optical spectra of both components and will do further analysis on them. This research indicated that the wide cool subdwarf binary fraction is similar to that of ultracool dwarfs ($\gtrsim 0.01$; e.g., Zhang et al. 2010; Faherty et al. 2010). It indicates that a large number of benchmarks are potentially identifiable, sampling the broad range in mass, age and metallicity need to adequately calibrate both atmospheric and evolutionary models.

5. Summary

Widefield surveys will continue to play a leading role in the discovery and study of ultracool dwarfs. From the widefield infrared surveys with UKIDSS, VISTA and WISE, to multi-epoch optical surveys with Skymapper, Pan-STARRS, PTF and in the future LSST, such surveys will identify the coolest objects, large samples for population studies and many interesting variable sources. A summary of the surveys discussed in this paper can be found in Figure 2.

Acknowledgments. The authors would like to thank the University of Washington for hosting Cool Stars 16, Local and Scientific Organising Committees and our SOC contact Andrew West.

References

Burgasser, A. J., Geballe, T. R., Leggett, S. K., Kirkpatrick, J. D., & Golimowski, D. A. 2006, ApJ, 637, 1067
Burgasser, A. J., McElwain, M. W., Kirkpatrick, J. D., Cruz, K. L., Tinney, C. G., & Reid, I. N. 2004, AJ, 127, 2856
Burgasser, A. J., Witte, S., Helling, C., Sanderson, R. E., Bochanski, J. J., & Hauschildt, P. H. 2009, ApJ, 697, 148
Burningham, B., Pinfield, D. J., Leggett, S. K., Tamura, M., Lucas, P. W., Homeier, D., Day-Jones, A., Jones, H. R. A., Clarke, J. R. A., Ishii, M., Kuzuhara, M., Lodieu, N., Zapatero Osorio, M. R., Venemans, B. P., Mortlock, D. J., Barrado Y Navascués, D., Martin, E. L., & Magazzù, A. 2008, MNRAS, 391, 320
Burningham, B., et al. 2010, MNRAS, 406, 1885
Chabrier, G., Baraffe, I., Allard, F., & Hauschildt, P. 2000, ApJ, 542, 464
Chiu, K., Fan, X., Leggett, S. K., Golimowski, D. A., Zheng, W., Geballe, T. R., Schneider, D. P., & Brinkmann, J. 2006, AJ, 131, 2722
Day-Jones, A. C., et al. 2011, MNRAS, 410, 705
Deacon, N. R., Hambly, N. C., King, R. R., & McCaughrean, M. J. 2009, MNRAS, 394, 857
Delorme, P., Delfosse, X., Albert, L., Artigau, E., Forveille, T., Reylé, C., Allard, F., Homeier, D., Robin, A. C., Willott, C. J., Liu, M. C., & Dupuy, T. J. 2008, A&A, 482, 961. `0802.4387`
Eisenhardt, P. R. M., et al. 2010, AJ, 139, 2455

Faherty, J. K., Burgasser, A. J., West, A. A., Bochanski, J. J., Cruz, K. L., Shara, M. M., & Walter, F. M. 2010, AJ, 139, 176
Fazio, G. G., et al. 1998, SPIE, 3354, 1024
Hambly, N., et al. 2001, MNRAS, 326, 1279
Kaiser, N., et al. 2002, in Survey and Other Telescope Technologies and Discoveries. Proceedings of the SPIE, Volume 4836, edited by S. Tyson, J. Anthony; Wolff, 154
Kirkpatrick, J. D., et al. 1999, ApJ, 519, 802
Law, N. M., Kulkarni, S. R., Dekany, R. G., Ofek, E. O., Quimby, R. M., Nugent, P. E., Surace, J., Grillmair, C. C., Bloom, J. S., Kasliwal, M. M., Bildsten, L., Brown, T., Cenko, S. B., Ciardi, D., Croner, E., Djorgovski, S. G., van Eyken, J., Filippenko, A. V., Fox, D. B., Gal-Yam, A., Hale, D., Hamam, N., Helou, G., Henning, J., Howell, D. A., Jacobsen, J., Laher, R., Mattingly, S., McKenna, D., Pickles, A., Poznanski, D., Rahmer, G., Rau, A., Rosing, W., Shara, M., Smith, R., Starr, D., Sullivan, M., Velur, V., Walters, R., & Zolkower, J. 2009, PASP, 121, 1395. 0906.5350
Lawrence, A., et al. 2007, MNRAS, 379, 1599
Liu, F., et al. 2008, SPIE, 7017, 16
Lodieu, N., Dobbie, P. D., Deacon, N. R., Hodgkin, S. T., Hambly, N. C., & Jameson, R. F. 2007, MNRAS, 380, 712
Lodieu, N., Zapatero Osorio, M. R., Rebolo, R., Martín, E. L., & Hambly, N. C. 2009, A&A, 505, 1115
LSST Science Collaborations, T. 2009. URL http://arxiv.org/abs/0912.0201
Lucas, P. W., et al. 2010, MNRAS, 408, L56
Luyten, W. J. 1979, LHS catalogue. A catalogue of stars with proper motions exceeding 0"5 annually (Minneapolis: University of Minnesota)
Magnier, E. A., Liu, M., Monet, D. G., & Chambers, K. C. 2008, IAUS, 248, 553
Mainzer, A., Cushing, M. C., Skrutskie, M., Gelino, C. R., Kirkpatrick, J. D., Jarrett, T., Masci, F., Marley, M., Saumon, D., Wright, E., Beaton, R., Dietrich, M., Eisenhardt, P., Garnavich, P., Kuhn, O., Leisawitz, D., Marsh, K., McLean, I., Padgett, D., & Rueff, K. 2010, ArXiv e-prints. 1011.2279
Mainzer, A. K., Eisenhardt, P., Wright, E. L., Liu, F.-C., Irace, W., Heinrichsen, I., Cutri, R., & Duval, V. 2005, SPIE, 5899, 262
Marley, M. S., Seager, S., Saumon, D., Lodders, K., Ackerman, A. S., Freedman, R. S., & Fan, X. 2002, ApJ, 568, 335
Monet, D. G., Levine, S. E., Canzian, B., Ables, H. D., Bird, A. R., Dahn, C. C., Guetter, H. H., Harris, H. C., Henden, A. A., Leggett, S. K., Levison, H. F., Luginbuhl, C. B., Martini, J., Monet, A. K. B., Munn, J. A., Pier, J. R., Rhodes, A. R., Riepe, B., Sell, S., Stone, R. C., Vrba, F. J., Walker, R. L., Westerhout, G., Brucato, R. J., Reid, I. N., Schoening, W., Hartley, M., Read, M. A., & Tritton, S. B. 2003, AJ, 125, 984. arXiv:astro-ph/0210694
Munn, J. A., et al. 2004, AJ, 127, 3034
Noll, K. S., & Marley, M. S. 2000, ApJ, 541, 75
Pinfield, D. J., Jones, H. R. a., Lucas, P. W., Kendall, T. R., Folkes, S. L., Day-Jones, a. C., Chappelle, R. J., & Steele, I. a. 2006, MNRAS, 368, 1281
Reid, I., Cruz, K. L., Kirkpatrick, J. D., Allen, P. R., Mungall, F., Liebert, J., Lowrance, P., & Sweet, A. 2008, AJ, 136, 1290
Saumon, D., & Marley, M. S. 2008, ApJ, 689, 1327
Sheppard, S. S., & Cushing, M. C. 2009, AJ, 137, 304
Skrutskie, M. F., Cutri, R. M., Stiening, R., Weinberg, M. D., Schneider, S., Carpenter, J. M., Beichman, C., Capps, R., Chester, T., Elias, J., Huchra, J., Liebert, J., Lonsdale, C., Monet, D. G., Price, S., Seitzer, P., Jarrett, T., Kirkpatrick, J. D., Gizis, J. E., Howard, E., Evans, T., Fowler, J., Fullmer, L., Hurt, R., Light, R., Kopan, E. L., Marsh, K. A., McCallon, H. L., Tam, R., Van Dyk, S., & Wheelock, S. 2006, AJ, 131, 1163
Warren, S. J., Mortlock, D. J., Leggett, S. K., Pinfield, D. J., Homeier, D., Dye, S., Jameson, R. F., Lodieu, N., Lucas, P. W., Adamson, A. J., Allard, F., Barrado Y Navascués, D., Casali, M., Chiu, K., Hambly, N. C., Hewett, P. C., Hirst, P., Irwin, M. J., Lawrence,

A., Liu, M. C., Martín, E. L., Smart, R. L., Valdivielso, L., & Venemans, B. P. 2007, MNRAS, 381, 1400

Wright, E. L., Eisenhardt, P. R. M., Mainzer, A. K., Ressler, M. E., Cutri, R. M., Jarrett, T., Kirkpatrick, J. D., Padgett, D., McMillan, R. S., Skrutskie, M., Stanford, S. A., Cohen, M., Walker, R. G., Mather, J. C., Leisawitz, D., Gautier, T. N., McLean, I., Benford, D., Lonsdale, C. J., Blain, A., Mendez, B., Irace, W. R., Duval, V., Liu, F., Royer, D., Heinrichsen, I., Howard, J., Shannon, M., Kendall, M., Walsh, A. L., Larsen, M., Cardon, J. G., Schick, S., Schwalm, M., Abid, M., Fabinsky, B., Naes, L., & Tsai, C. 2010, AJ, 140, 1868

York, D. G., et al. 2000, AJ, 120, 1579

Zhang, Z. H., et al. 2010, MNRAS, 1834, 1817

The 16th Cambridge Workshop on Cool Stars, Stellar Systems and the Sun
ASP Conference Series, Vol. 448
Christopher M. Johns-Krull, Matthew K. Browning, and Andrew A. West, eds.
© 2011 Astronomical Society of the Pacific

Splinter Session "Solar and Stellar Flares"

L. Fletcher,[1] H. Hudson,[1,2] G. Cauzzi,[3] K. V. Getman,[4] M. Giampapa,[5] S. L. Hawley,[6] P. Heinzel,[7] C. Johnstone,[8] A. F. Kowalski,[6] R. A. Osten,[9] and J. Pye[10]

[1] *School of Physics and Astronomy, SUPA, University of Glasgow, Glasgow G12 8QQ, UK (lyndsay@astro.gla.ac.uk)*

[2] *Space Sciences Laboratory, U. C. Berkeley, 7 Gauss Way, Berkeley, CA 94720, USA*

[3] *INAF - Osservatorio Astrofisico di Arcetri, Largo Enrico Fermi 5, 50125 Firenze, Italy*

[4] *Department of Astronomy & Astrophysics, 525 Davey Laboratory, Pennsylvania State University, University Park, PA 16802, USA*

[5] *National Solar Observatory, 950 North Cherry Ave, Tucson, AZ 85719, USA.*

[6] *Astronomy Department, University of Washington, Box 351580, Seattle, WA 98195, USA*

[7] *Astronomical Institute, Academy of Sciences of the Czech Republic, 25165 Ondřejov, Czech Republic*

[8] *School of Physics and Astronomy, University of St Andrews, St Andrews, Scotland KY16 9SS*

[9] *Space Telescope Science Institute, 3700 San Martin Drive, Baltimore, MD 21218, USA*

[10] *Department of Physics and Astronomy, University of Leicester, Leicester, LE1 7RH, UK*

Abstract. This summary reports on papers presented at the *Cool Stars–16* meeting in the splinter session "Solar and Stellar flares." Although many topics were discussed, the main themes were the commonality of interests, and of physics, between the solar and stellar flare communities, and the opportunities for important new observations in the near future.

1. Introduction

For several years, studies of solar flares and of stellar flares have – by necessity – evolved along diverging tracks. To be observable on an essentially unresolved stellar source, stellar flares must be bright, so that it is possible to make extremely high caliber spectroscopic observations, but without much spatial information. On the other hand, the solar observations now provide exquisitely detailed spatial information, so that the spatial evolution of a solar flare is typically very well characterized, but the difficulty of predicting its location means that good spectroscopic information, from a slit placed exactly over the flare, is rare – especially in the crucial "impulsive phase." This splinter meeting, which drew together participants from the solar and stellar flare communities (plus others) aimed to summarize the current state of knowledge in the field, and plot a possible way ahead. These notes do not cite the literature comprehensively, but instead mainly reflect current research work, and the topics discussed at the Splinter meeting.

2. An Overview of Solar Flares

Spatially-resolved flare observations in the modern era reveal that the energetically dominant optical and UV emission of a flare, occurring during the few minutes of the flare's impulsive phase, originates in two or more ribbons which spread across the chromosphere. The strongest optical emission is organized into very compact footpoints, smaller than a few thousand km on the Sun (see Figure 1) which evolve rapidly in position and intensity. This reflects the progression of the main flare energy release. Compared to chromospheric and transition-region ribbons, the optical footpoints have been very little studied in the recent era. White-light (WL) emission had previously been thought to be a big flare phenomenon, but has now been seen even in small C-class events (Hudson et al. 2006; Jess et al. 2008). WL footpoints have a very precise timing and spatial relationship to flare hard X-rays (Metcalf et al. 2003; Hudson et al. 2006; Fletcher et al. 2007; Watanabe et al. 2010). Hard X-rays are generated by electrons accelerated during a flare, which are a major energy-carrying component, and the peak of the flare energy spectrum, as far as we can pin it down, is in the optical-UV regime (Hudson et al. 2010). The implication is that energy carried by flare electrons is eventually radiated in the lower-atmospheric optical-UV, even though the energy has been built-up and stored in coronal magnetic structures. Studies of flare energetics in the two largest flares of the last cycle indicate that the energy contained in solar flare electrons and ions is of the order of 10^{30-32} ergs in each component, comparable to the kinetic energy of the coronal mass ejections associated with the events, and with the estimates of the free magnetic energy extrapolated from the photospheric boundary (Emslie et al. 2004, 2005). Non-thermal electrons are of core importance in solar flares, and the results from the *Ramaty High Energy Solar Spectroscopic Imager* (RHESSI; Lin & al. 2002) have added substantially to our understanding of this. RHESSI provides imaging spectroscopy from 3 keV up into the γ-ray range, with unprecedented dynamic range and spectral resolution. It thus uniquely complements the more recent fleet of solar spacecraft, none of which provide this key high-energy data; the latest addition, the *Solar Dynamics Observatory*, only reaches the EUV wavelength range but provides remarkably detailed and complete context information.

Interpreted as the chromospheric intersections of evolving magnetic (quasi-) separatrix surfaces,chromospheric flare ribbons allow the progression of field restructuring to be followed (Masson et al. 2009). Chromospheric flares have of course long been beautifully observed in Hα, but also now in other lines such as the 304 nm filter from the *Solar Dynamics Observatory* (SDO). These show abundant fine detail from 3D structures. Flare ribbons give insight into both energetics and magnetic restructuring, in the impulsive and the gradual phase. The fact that there is a small subset of the ribbons illuminated in optical and HXR indicates that there are special locations in the mag-

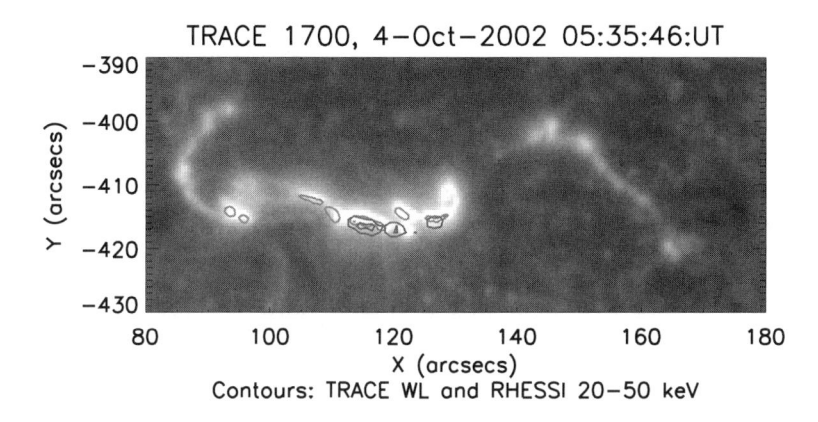

Figure 1. High spatial resolution image of a solar flare. The grayscale is the TRACE 1700Å channel, orange is the TRACE white light channel and green is RHESSI 25-50 keV footpoints. The field of view spans around $72,000 \times 29,000$ km^2.

netic field where very efficient electron acceleration takes place. It is not clear what the significance of these sites is; they may be associated with singular structures in the field called 'magnetic spines' (Des Jardins et al. 2009). Flare footpoints are also visible emitting in soft X-rays, at temperatures in the range 8-10 MK, with emission measures consistent with chromospheric densities and volumes (Mrozek & Tomczak 2004).

It is not clear whether the optical footpoint emission corresponds to heating of the photosphere, but during the impulsive phase there is most certainly evidence that the effects of a solar flare can reach this layer. The 'sunquakes' (Kosovichev & Zharkova 1998; Donea & Lindsey 2005), and the non-transient changes in the photospheric line-of-sight magnetic field (Wang 1992; Cameron & Sammis 1999; Sudol & Harvey 2005), though not visible in all flares, demonstrate the photosphere is mechanically perturbed, by a shock, by the Lorentz force, or by some other means.

In response to the impulsive-phase heating the chromosphere expands into the corona, seen spectroscopically in the extreme UV (Milligan & Dennis 2009), with a chromospheric downflow that approximately balances momentum (Zarro et al. 1988; Teriaca et al. 2006). The upflow speed is observed to vary with the (LTE) temperature of the emitting plasma, with higher temperature plasmas moving faster. Arcades of X-ray and EUV loops form, straddling the photospheric polarity inversion line between the chromospheric ribbons. As the ribbons spread away from this line, the loops become oriented more perpendicularly to it (Sakurai et al. 1992). This suggests that later phase energy release occurs in magnetic field which – in addition to weakening – may become less sheared. The loops cool from coronal temperatures, eventually becoming visible in Hα. The indication is that, at least during larger events, the temperature evolution is slower than would be suggested by radiative and conductive cooling (Harra-Murnion et al. 1998) indicating the presence of energy input even during the gradual phase.

Spatially-resolved information provides a clear picture of the flare evolution, and indicates also the physical processes at work. It is an embarrassment that there is no

444 Fletcher et al.

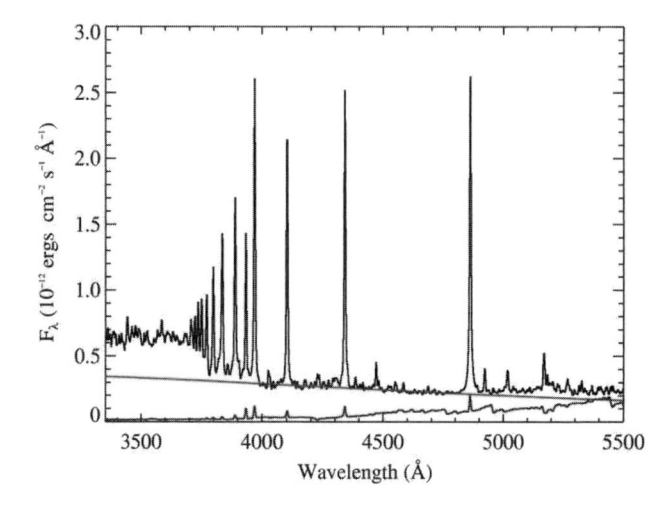

Figure 2. High spectral resolution observation of a megaflare on YZ CMi (Kowalski et al. 2010): red line, a blackbody fit longward of the Balmer jump; purple line, the preflare stellar spectrum.

optical or UV imaging flare spectroscopy from space, although the IRIS instrument will soon be available. There are also precious few ground-based spectra – extensive measurements have not been made since the 1980s and early 90s. The review in this splinter by Cauzzi highlighted some past and present work in ground-based spectroscopy of flares. Recently, this has predominantly been high resolution slit spectroscopy on a small number of spectral lines, e.g. the observations of Hα, Hβ, Ca H and K, Ca II λ 8542 with the refurbished instrument at Ondrejov (Kotrč 2007), and lower resolution multi-channel spectrograph work, in which images can be made simultaneously at several points across a spectral line, or a number of spectral lines (e.g. Radziszewski et al. 2007), including a flare presented which was observed in August 2010 with the IBIS instrument (Cavallini 2006). Observations over a broad spectral range have been very few and far between in recent times (Johns-Krull et al. 1997). The chromospheric lines are potentially rich in diagnostic information. For example, Metcalf et al. (1990a,b) have used Mg I lines to probe the temperature and density structure of the temperature minimum region. Both the Ca infrared triplet at $\lambda\lambda$ 8498, 8542, 8662 and the Balmer lines provide information on electron density (e.g. Švestka 1976; Ding et al. 2002), and potentially also diagnostics of non-thermal electron distributions (Zharkova & Kobylinskii 1993; Ding & Fang 2001; Kašparová & Heinzel 2002).

HMI on SDO is now providing imaging spectroscopy of the photospheric Fe I 6173 Å line, and SDO/EVE provides full EUV spectra of the Sun as a star. In the meanwhile, Figure 2 suggests the power of what may emerge from better spectroscopy: this is a stellar flare spectrum dominated by the Balmer series, but with white-light continuum (and photospheric lines) in the background (Kowalski et al. 2010). We do not have modern spectra for solar flares with quality approaching stellar ones.

3. An Overview of Stellar Flares

The Sun is a weak flare source. Its flaring energy output is tiny compared to what is seen on other stars (Schaefer et al. 2000). Even among the 'normal' single stars (leaving aside pre-main sequence stars), stellar flares can involve the release of 10^6 times more energy in total than solar flares, suggesting substantially higher magnetic fields and larger flaring volumes (the available energy is $\int (B^2 - B_{pot}^2)\, dV$, where B_{pot} is the potential field matching the same boundary conditions). Flares are observed in all types of stars with convective envelopes, including from M dwarfs, solar analogs, RS CVn stars, and also giants. Flare emission emerges in the optical and near-UV, as well as X-rays and radio but, as with the solar case, a large fraction of the radiated energy for dMe flares emerges in the optical part of the spectrum (Hawley & Pettersen 1991) during both the impulsive and gradual phases. In many cases, the stellar flares observed so far show a great deal of consistency with solar observations, i.e. most of the power is in the continuum, the ratios of radio to UV emission are similar (Osten et al. 2004), the impulsive phase can show similarly abrupt rises to the solar case (minutes) and the decay phases typically last for hours, with evidence for ongoing heating during the decay. Hard X-rays, above about 10 keV, have also been observed in stellar flares (though only in a few extremely large ones) and the ratio of hard and soft fluxes follow the same scaling relationship in solar and stellar flares (Isola et al. 2007), pointing towards the hard photons having a non-thermal spectrum. Solar flare HXR photons account for around 10^{-5} of the total flare energy; if the same is true for stellar flares it is no surprise that thus far good stellar flare hard X-ray spectroscopy has not been possible. The 'superflare' on II Peg (an "X4.4×10^6" event) reported by Osten et al. (2007) and observed with *Swift*/BAT is an exception. While in this event an unambiguously non-thermal spectral shape is not confirmed, it appears highly plausible. HXR spectroscopy of smaller events should become possible with the proposed Simbol-X mission. However, observations of gyrosynchrotron emission in stellar flares, which is in fact the principal diagnostic of non-thermal electrons in stellar flares, has long led us to expect that high energy, hard spectrum, non-thermal electrons are present (Güdel 2002); indeed (Güedel et al. 1996) found that stellar flares on UV Ceti are radio-overluminous with respect to their X-ray emission in comparison with solar events.

However, there are observed differences, for example in the shape of light curves. During the session an example was given by Pye et al. of an XMM Newton flare (on 2MASS 04072181-1210033) with a comparatively rapid decay, and the Chandra Orion Ultradeep Project (COUP) survey of pre main-sequence stars by Getman et al. (2008) reports many examples of slow-rise events with extended flat tops and very long decays, continuing for days. The soft X-ray rise timescale of a solar flare is characterised by the duration of the impulsive phase (usually a set of discrete energy input events) and the length of the coronal structure to be filled by evaporation – in other words by the configuration of the magnetic field in which the flare event takes place. These different Chandra characteristics imply a very much larger flaring structure: Getman et al. (2008) suggest they are enhanced analogues to solar long duration events.

There are also flares showing much more exotic X-ray light curves, with dips and multiple peaks, and it is conceivable that these could be used to deduce spatial information about stellar flares, which is at a premium. For example, the time variations in about a quarter of the SXR light curves in the COUP survey (Getman et al. 2008) can be explained by the rotational eclipsing of large emitting loops. Work by Johnstone et al. presented in this session identified events in this survey having a significant dip in the lightcurves, constituting about 30% of all of the flares in the survey. An estimate can be made of the expected number of eclipsing flares by sampling at random the distributions of flare loop lengths and durations (relative to rotation period of the star), estimated from SXR decays by Getman et al. (2008). This results in a much smaller estimate, i.e. that about 6% of all flares should show the effect of eclipses in their light curves (the larger, longer-lived ones). The implication is that eclipses are important even in flares of 'normal' size and duration'. The effects might be more subtle – small dips, a reduced peak emission measure compared to an uneclipsed flare, or a shorter apparent duration. How to disentangle the behavior due to eclipsing from the intrinsic flaring behavior is not clear. Some independent estimates of flare X-ray sizes may also be available from the technique of coronal seismology, such as is frequently used now in solar flare studies. For example, variations in the X-ray light curve of a flare loop on AT Mic were interpreted by Mitra-Kraev et al. (2005b) as damped oscillations of a standing magneto-acoustic wave, giving constraints on the loop length of 2.5×10^{10}cm and field strength of 100±50 G.

Given the spectral flux distribution it is also possible to estimate the size of the excited patch in the flare star chromosphere/photosphere. This has been done under the assumption that the optical flare corresponds to a patch of blackbody emission with an elevated temperature compared to the surrounding photosphere (Hawley & Fisher 1992; Hawley et al. 2003). Area coverage obtained in this way is typically small – on the order of 0.01% to 0.1% of the star's visible hemisphere (around a factor ten larger than the area coverage of optical patches in solar flares), though larger events are possible (e.g. 3% coverage in the Osten et al (2010) 'superflare'). The energy input implied by these areas and the luminous energy can be on the order of 10^{11}erg cm^{-2}s^{-2} (Hawley et al. 2003; Kowalski et al. 2010; Osten et al. 2010) corresponding to an intense solar flare. Evidence was also presented in this session for sub-structure in the spatial distribution of flare optical emission, namely the YZ CMi flare reported by Kowalski et al. (2010) analysis of which implies a Balmer continuum-emitting patch which is around 3-15 times larger than the blackbody patch. This may indicate the same behavior as is consistently observed in solar flares, namely relatively extended UV and Hα emitting ribbons compared to optical and HXR footpoints.

How invariable is the flare emission spectrum from star to star, considering all wavelengths? Mitra-Kraev et al. (2005a) find stellar flare UV and soft X-ray fluxes to correlate, and have comparable energies. On the other hand the first true bolometric observations of solar flares (Woods et al. 2004) suggest $L_X/L_{bol} \sim 0.01$, which would be quite different. A partial explanation may be that the solar soft X-ray photometry normally refers to GOES, which does not detect long-wavelength soft X-rays as well as typical stellar instruments. In contrast to this possible suggestion of a potential unification of flare spectra, Osten et al. (2005) find a surprising variety of flare detectability in disparate wavelength bands. Even with the same basic physics one would expect systematic differences across stellar variations in scale, abundances, and gravity, and so the global energetics of solar and stellar flares will be an interesting topic for the future.

4. Surveys and Catalogs

The Sun and flare stars present us with a wealth of events. On the Sun, we have catalogues of large numbers of flares, including collections of flares from individual active regions. These, and the spatial information available, permit us to generate sets of flares which have been selected for particular characteristics (e.g. flares without an eruption, flares with coronal sources, flares with only two footpoints.... etc). In principle it is possible also to identify sets of flares from an active region with very similar spatial characteristics – the so-called 'homologous flares'. Thus one can try and control the large number of variables which might affect the process of flare evolution, particle acceleration, etc. On the other hand, with the Sun we are restricted to events on a single class of object – a middle-sized, middle-aged, isolated, main-sequence star. The population of flare stars provides a richer range of flaring environments, as well as clues to how frequent flaring activity is in solar-type, and other types of stars. The first step in such analysis is the survey and cataloging step. The COUP survey mentioned above is one example as is the XMM-Newton Serendipitous Source catalogue presented in this session by Pye. This catalogue, obtained by analysis of the light curves of sources present in the field of XMM-Newton turned up around 130 flaring events, 40 of which came from previously unknown flare sources. Preliminary analysis by Pye et al. has shown that the serendipitous sources which flare more than once spend up to 10-20% of their time in a flaring state, only a factor two or so less than the 'known' flare stars among the XMM-Newton sample. Furthermore, a few percent of all stars for which an XMM-Newton light curve could be extracted show flaring activity, demonstrating that significant flaring is not so infrequent. As well as the light curves, some spectral information is available (Pye & Rosen, elsewhere in these Proceedings), represented for this analysis as the ratio of X-ray count rates in two XMM bands (0.2-1, 1-12 keV), giving a clue to the temperature evolution of the flare (assuming a thermal spectrum) – the kind of analysis which can also be carried out straightforwardly for solar flares based on their GOES X-ray light curves. Large and growing serendipitous catalogs also offer the opportunity to search for flaring activity in solar-type stars.

In the solar-flare arena, work has tended to proceed in two ways: catalog-based statistical analyses of moderate or large populations of flares, and detailed, often multi-wavelength analyses of spatial, spectral and temporal evolution of individual events. It has so far proved difficult to merge the two approaches, detailed studies proving so intricate and throwing up many different facets of flare behavior which would be very time-consuming to analyze in a statistically-significant sample. But the detailed studies provide general information which we believe to be of wide applicability. For example, imaging analysis of the energy input (i.e. the HXR evolution) of a small number of large, long-duration solar flares reveals without a doubt that they are due to a group, possibly a sequence, of events occurring at different locations throughout a spatially extended region, rather than one single loop or site undergoing repeated flaring activity (e.g. Grigis & Benz 2005). However, it must be said that the sheer richness of solar data is very tempting; there are relatively few statistical studies of simple properties of large populations compared to studies of detailed and complex evolution of individual flare events. It is clear that for comparisons with stellar populations, more effort has to be expended on the former type of solar flare analysis, informed also by stellar flare work.

5. Modelling

Solar and stellar flare modeling have many aspects in common, perhaps because stellar flare theorists and observers have often looked to theories for the better-observed solar flares for initial guidance. The beam-driven radiative hydrodynamic approach of Abbett & Hawley (1999) has been applied to solar flares and to dMe star flares as well (Allred et al. 2006), and remains the most sophisticated treatment of the impulsive phase and its consequences. Also the analysis of the cooling phase in terms of radiative and conductive losses calls on the same basic physics (and arrives at the same conclusion that there is often evidence for ongoing energisation in the late phase). Where solar modeling is required to be more detailed than stellar flare modeling is in the interpretation of the spatial structure of the flare, and in particular its relationship to the connectivity or topology of the solar magnetic field (Sweet 1958). Here a whole area of research has grown up which seeks to describe the three-dimensional magnetic topology of coronal structures rooted in the observed photospheric magnetic field, and to explain the evolution and the energy release in a flare as a consequence of flux transfer from one magnetic domain to another, facilitated by magnetic reconnection (e.g. Henoux & Somov 1987; Demoulin et al. 1992; Priest & Schrijver 1999). From such work we have learned the important of magnetic structures such as separators, separatrices and nulls. The topology of stellar magnetic fields is also being approached using extrapolations based on Zeeman and Doppler imaging, and particularly well-studied stars reveal topologies significantly more complex than a straightforward dipole (e.g. Donati 2001; Jardine et al. 2002; Phan-Bao et al. 2009; Morin et al. 2010). However, it should be cautioned that – based on the solar experience – the topology of the magnetic field on the scale of the whole star may have very little to say about the location and evolution of a flare. In the solar case it seems that we will have to understand the configuration and evolution of the non-potential magnetic field on the scale of less than ten thousand kilometers, and on the timescale of the impulsive phase, to be able to map out the locations of energy storage, the likely sites of a flare initiation or trigger, and the subsequent magnetic field restructuring.

In both stellar and solar flare cases, spectroscopic modeling tends to proceed along the same lines, again relying extensively on the radiation hydrodynamic simulations mentioned above. These complex simulations, however, are limited in the number of species and ionization stages that can be included, so other approaches include computing the state of the flare atmosphere under particular assumptions about energy input and redistribution, then using this as input for other spectral synthesis codes - such as the CHIANTI software (Dere et al. 1997) for the optically-thin regime (which at present makes the assumption of LTE), or the multi-level non-LTE radiative transfer approach (Rybicki & Hummer 1991, 1992, 1994) as implemented in the codes of Uitenbroek (2001) and Heinzel (1995) for the optically thick regime. In the solar case, at the present time we have rather little observational motivation for attempting to model the UV/optical lines and continuum, with the exception of a few, due to the paucity of observational data. But in the stellar case this is obviously of great value - not only the overall shape of the spectrum, but the detailed profiles of lines can be used to diagnose the flaring plasma. In the solar case, a small number of lines have been observed and modeled in some detail. These include Hα line, the profile and time evolution of which has been studied by many authors, e.g. Kašparová et al. (2009). Careful study of the line profiles offers the possibility of discriminatory tests for the nature of the chromospheric energy input, including the spectral index of the electron spectrum that excites the Hα radiation and its fine structuring in time. Other lines which have been modeled in some detail include Ca II K (Fang et al. 1993) and the Ca II λ8542 line (Ding 2003; Berlicki et al. 2008), i.e. those which have received observational attention.

Semi-empirical models of flaring atmospheres are much used, and certain of them have become standards (e.g. the 'F1' model of Machado et al. 1980). They are certainly useful, but are constructed by analyzing data from different times of different flares, so can only give some kind of averaged view of a flare chromosphere. They tend not to represent the impulsive phase very well, both because the input spectroscopic information is so limited and because the physics may be more complicated. Such models are in the single-fluid approximation with $T_e = T_p$, and therefore do not handle interesting parts of the plasma physics.

6. More Exotic Flaring Objects

The wide variety of astrophysical sources offers the possibility to study magnetically-driven energy storage and release in many situations, not just isolated stars. These include the magnetised star plus disk configuration of a young stellar object (Favata et al. 2005), and the binary configuration of RS CVns, in which the two components are interacting and possibly magnetically linked (Uchida & Sakurai 1985). The variety of configurations and activity levels may allow us to investigate questions like what determines the energy and evolution of a flare in a given magnetic configuration? Is it just the overall size or mean magnetic field, or are there other restrictions placed by the accessible topologies of the system? And what determines whether a flare results in heating primarily, or in particle acceleration? Is it only a function of the total energy, or some other parameter such as the mean density and temperature (i.e. collisionality) of the plasma in which the primary energy release takes place? Is the Sun unusual in converting stored magnetic energy efficiently into the KE of non-thermal electrons?

The pre-main sequence stars, which were the main target of the COUP survey mentioned above, represent "normal" stars in probably their most magnetically-active phase. Their X-ray luminosity can be $> 10^4$ times that of a solar flare's X-ray luminosity (Figure 8 in Getman et al. 2008) which is in turn about 10% of the total solar flare luminosity. As discussed by Getman in this session, flares from both disk bearing and diskless COUP stars tend to last 10-100 times longer than solar events, and also involve loops which are 100-1000 times as long as in the solar case - many times the radius of the star (length being determined from a cooling timescale analysis). One particular PMS object, DQ Tau, offers an unusual opportunity to study repeated flaring as it is in a binary with a period of just under 16 days and undergoes mm band flaring at periastron (Salter et al. 2010), attributed to interactions of the magnetospheres of the binary components - the typical loop lengths derived from the above-mentioned analyses are comparable with the periastron separation, mm and X-ray flares are causally related, and the energy released is comparable with what should be expected from reconnection in colliding magnetospheres (Getman et al. 2010 submitted).

More energetic yet are the magnetars - highly magnetised pulsars - which also show extreme flare outputs but represent physical conditions which differ so much from normal MS stars that it is not clear what can be learned from their study which is of direct relevance to interpreting the radiation signature of solar or stellar flares.

7. Proposals for the Future

At the end of this splinter session, a set of questions was proposed for future study, and some general comments made.

- When is a feature seen in a light curve to be classified as a flare?

- Do all major flares (solar and stellar) show an impulsive phase?
- What is the energy conversion process during the impulsive phase?
- What do the timescales in the energy-release phase represent?
- What is the relationship between microflaring and coronal heating?
- Are scaled-up versions of the solar mechanisms always appropriate for attempting to understand stellar flares?
- Can ideas of topology be usefully applied to stellar as well as solar flares?
- What is the role of accelerated particles in stellar flare energy budgets?

We still have a lot to learn about the theory of solar flares, so stellar flare physicists are cautioned to investigate models beyond the 'standard flare model'. This model, embodied in the 'Carmichael-Sturrock-Hirayama-Kopp-Pneuman' cartoon, is one of many many scenarios possible for flares (see e.g. http://solarmuri.ssl.berkeley.edu/~hhudson/cartoons/). The CSHKP cartoon may do a decent job at encapsulating the gradual phase of a flare but was never intended to describe the impulsive energy release. We do not know what triggers the solar flare instability, nor for certain how energy is transported through the flaring atmosphere (the long-standing 'number problem' and 'return current problem' with the electron beam model (e.g. Brown & Melrose 1977; Hoyng et al. 1976; Fletcher & Hudson 2008) are not solved satisfactorily). Proper kinetic modeling of the plasma processes, though crucial for some areas (e.g. coherent radio emission) is not nearly as widespread as it should be, and the treatment of chromospheric processes assumes that the plasma is a single Maxwellian fluid, despite the fact that collisionally-maintained LTE is unlikely in the extreme energy inputs received by small areas of the chromosphere.

Observationally, to understand better the solar processes, we advocate an emphasis on optical spectroscopy of the chromosphere. Relatively easily observed compared to hard X-rays, white-light (and UV-continuum) observations provide a view of the total energy of a flare complementary to that provided by HXR radiation. There is little effort to conduct studies of chromospheric flares from space, though there has been notable success in targeted programs using TRACE, and serendipitous observations with *Hinode*. It is to be expected that the *Solar Dynamics Observatory* will also reveal the detailed morphological evolution of ribbons, and possibly give some insight also into the structure of the bright ribbon sources. At its most basic, continuum spectroscopic information is required for a proper assessment of the total radiative intensity in flares (which has so far been obtained only in a small number of events), and to determine if there are white-light properties during solar flares not predicted by RHD models, as is the case for the hot, blackbody component of M dwarf flares. Of particular interest is the identification (Neidig 1989) of the Paschen jump (indicating the presence of the Paschen continuum from free-bound emission in the optical part of the spectrum) and flare-related changes to the Balmer and Paschen jumps in the flare spectrum, indicating variations in ionization fraction. Enhanced continuum shortwards of the Paschen jump may also suggest increased H^- opacity, and photospheric excitation. This basic measurement has been made in very few flares (see e.g. Neidig 1983; Mauas et al. 1990), and usually not for the strongest impulsive-phase optical flare kernels, but has major implications for the energy transport model (Neidig 1989; Fletcher & Hudson 2008).

Acknowledgments. LF would like to thank the CS16 Program Committee for selecting this Splinter Session to run, and all the participants for a lively and stimulating discussion. She would also like to acknowledge financial support from the European Community & Research Training Network project SOLAIRE(MTRN-CT-2006-035484), from UK STFC Rolling Grant ST/F002637/1, and from the International Space Sciences Institute, Bern whose support of the 'Chromospheric Flares' team led directly to this Splinter Session. P. Heinzel acknowledges travel support by the grant No. P209/10/1680 of the Grant Agency of the Czech Republic.

References

Abbett, W. P., & Hawley, S. L. 1999, ApJ, 521, 906
Allred, J. C., Hawley, S. L., Abbett, W. P., & Carlsson, M. 2006, ApJ, 644, 484. arXiv: astro-ph/0603195
Berlicki, A., Heinzel, P., Schmieder, B., & Li, H. 2008, A&A, 490, 315
Brown, J. C., & Melrose, D. B. 1977, Solar Phys., 52, 117
Cameron, R., & Sammis, I. 1999, ApJ, 525, L61
Cavallini, F. 2006, Solar Phys., 236, 415
Demoulin, P., Henoux, J. C., & Mandrini, C. H. 1992, Solar Phys., 139, 105
Dere, K. P., Landi, E., Mason, H. E., Monsignori Fossi, B. C., & Young, P. R. 1997, A&AS, 125, 149
Des Jardins, A., Canfield, R., Longcope, D., Fordyce, C., & Waitukaitis, S. 2009, ApJ, 693, 1628
Ding, M. D. 2003, Journal of Korean Astronomical Society, 36, 49
Ding, M. D., & Fang, C. 2001, MNRAS, 326, 943. arXiv:astro-ph/0105097
Ding, M. D., Liu, Y., & Chen, P. F. 2002, Solar Phys., 207, 125
Donati, J. 2001, in Astrotomography, Indirect Imaging Methods in Observational Astronomy, edited by H. M. J. Boffin, D. Steeghs, & J. Cuypers, vol. 573 of Lecture Notes in Physics, Berlin Springer Verlag, 207
Donea, A.-C., & Lindsey, C. 2005, ApJ, 630, 1168
Emslie, A. G., Dennis, B. R., Holman, G. D., & Hudson, H. S. 2005, Journal of Geophysical Research (Space Physics), 110, 11103
Emslie, A. G., Kucharek, H., Dennis, B. R., & al. 2004, Journal of Geophysical Research (Space Physics), 109, 10104
Fang, C., Henoux, J. C., & Gan, W. Q. 1993, A&A, 274, 917
Favata, F., Flaccomio, E., Reale, F., & al. 2005, ApJS, 160, 469. arXiv:astro-ph/0506134
Fletcher, L., Hannah, I. G., Hudson, H. S., & Metcalf, T. R. 2007, ApJ, 656, 1187
Fletcher, L., & Hudson, H. S. 2008, ApJ, 675, 1645. 0712.3452
Getman, K. V., Feigelson, E. D., Broos, P. S., Micela, G., & Garmire, G. P. 2008, ApJ, 688, 418. 0807.3005
Grigis, P. C., & Benz, A. O. 2005, ApJ, 625, L143. arXiv:astro-ph/0504436
Güdel, M. 2002, ARA&A, 40, 217. arXiv:astro-ph/0206436
Güedel, M., Benz, A. O., Schmitt, J. H. M. M., & Skinner, S. L. 1996, ApJ, 471, 1002
Harra-Murnion, L. K., Schmieder, B., van Driel-Gesztelyi, L., & al. 1998, A&A, 337, 911
Hawley, S. L., Allred, J. C., Johns-Krull, C. M., & al. 2003, ApJ, 597, 535
Hawley, S. L., & Fisher, G. H. 1992, ApJS, 78, 565
Hawley, S. L., & Pettersen, B. R. 1991, ApJ, 378, 725
Heinzel, P. 1995, A&A, 299, 563
Henoux, J. C., & Somov, B. V. 1987, A&A, 185, 306
Hoyng, P., van Beek, H. F., & Brown, J. C. 1976, Solar Phys., 48, 197
Hudson, H. S., Fletcher, L., & Krucker, S. 2010, MmSAIt, 81, 637. 1001.1005
Hudson, H. S., Wolfson, C. J., & Metcalf, T. R. 2006, Solar Phys., 234, 79
Isola, C., Favata, F., Micela, G., & Hudson, H. S. 2007, A&A, 472, 261. 0707.2322

Jardine, M., Collier Cameron, A., & Donati, J. 2002, MNRAS, 333, 339. arXiv:astro-ph/0205132
Jess, D. B., Mathioudakis, M., Crockett, P. J., & Keenan, F. P. 2008, ApJ, 688, L119. 0810.1443
Johns-Krull, C. M., Hawley, S. L., Basri, G., & Valenti, J. A. 1997, ApJS, 112, 221
Kašparová, J., & Heinzel, P. 2002, A&A, 382, 688
Kašparová, J., Varady, M., Heinzel, P., Karlický, M., & Moravec, Z. 2009, A&A, 499, 923. 0904.2084
Kosovichev, A. G., & Zharkova, V. V. 1998, Nat, 393, 317
Kotrč, P. 2007, in The Physics of Chromospheric Plasmas, edited by P. Heinzel, I. Dorotovič, & R. J. Rutten, vol. 368 of Astronomical Society of the Pacific Conference Series, 559
Kowalski, A. F., Hawley, S. L., Holtzman, J. A., Wisniewski, J. P., & Hilton, E. J. 2010, ApJ, 714, L98. 1003.3057
Lin, R. P., & al. 2002, Solar Phys., 210, 3
Machado, M. E., Avrett, E. H., Vernazza, J. E., & Noyes, R. W. 1980, ApJ, 242, 336
Masson, S., Pariat, E., Aulanier, G., & Schrijver, C. J. 2009, ApJ, 700, 559
Mauas, P. J. D., Machado, M. E., & Avrett, E. H. 1990, ApJ, 360, 715
Metcalf, T. R., Alexander, D., Hudson, H. S., & Longcope, D. W. 2003, ApJ, 595, 483
Metcalf, T. R., Canfield, R. C., Avrett, E. H., & Metcalf, F. T. 1990a, ApJ, 350, 463
Metcalf, T. R., Canfield, R. C., & Saba, J. L. R. 1990b, ApJ, 365, 391
Milligan, R. O., & Dennis, B. R. 2009, ApJ, 699, 968. 0905.1669
Mitra-Kraev, U., Harra, L. K., Güdel, M., & al. 2005a, A&A, 431, 679. arXiv:astro-ph/0410592
Mitra-Kraev, U., Harra, L. K., Williams, D. R., & Kraev, E. 2005b, A&A, 436, 1041. arXiv:astro-ph/0503384
Morin, J., Donati, J., Petit, P., Delfosse, X., Forveille, T., & Jardine, M. M. 2010, MNRAS, 407, 2269. 1005.5552
Mrozek, T., & Tomczak, M. 2004, A&A, 415, 377
Neidig, D. F. 1983, Solar Phys., 85, 285
— 1989, Solar Phys., 121, 261
Osten, R. A., Brown, A., Ayres, T. R., & al. 2004, ApJS, 153, 317. arXiv:astro-ph/0402613
Osten, R. A., Drake, S., Tueller, J., Cummings, J., Perri, M., Moretti, A., & Covino, S. 2007, ApJ, 654, 1052. arXiv:astro-ph/0609205
Osten, R. A., Godet, O., Drake, S., & al. 2010, ApJ, 721, 785. 1007.5300
Osten, R. A., Hawley, S. L., Allred, J. C., Johns-Krull, C. M., & Roark, C. 2005, ApJ, 621, 398. arXiv:astro-ph/0411236
Phan-Bao, N., Lim, J., Donati, J., Johns-Krull, C. M., & Martín, E. L. 2009, ApJ, 704, 1721. 0909.2355
Priest, E. R., & Schrijver, C. J. 1999, Solar Phys., 190, 1
Radziszewski, K., Rudawy, P., & Phillips, K. J. H. 2007, A&A, 461, 303
Rybicki, G. B., & Hummer, D. G. 1991, A&A, 245, 171
— 1992, A&A, 262, 209
— 1994, A&A, 290, 553. arXiv:astro-ph/9404019
Sakurai, T., Shibata, K., Ichimoto, K., Tsuneta, S., & Acton, L. W. 1992, PASJ, 44, L123
Salter, D. M., Kóspál, Á., Getman, K. V., & al. 2010, A&A, 521, A32+. 1008.0981
Schaefer, B. E., King, J. R., & Deliyannis, C. P. 2000, ApJ, 529, 1026. arXiv:astro-ph/9909188
Sudol, J. J., & Harvey, J. W. 2005, ApJ, 635, 647
Sweet, P. A. 1958, in Electromagnetic Phenomena in Cosmical Physics, edited by B. Lehnert, vol. 6 of IAU Symposium, 123
Teriaca, L., Falchi, A., Falciani, R., Cauzzi, G., & Maltagliati, L. 2006, A&A, 455, 1123. 0903.0232
Švestka, Z. 1976, Solar Flares (Springer-Verlag Berlin Heidelberg)
Uchida, Y., & Sakurai, T. 1985, in Unstable Current Systems and Plasma Instabilities in Astrophysics, edited by M. R. Kundu & G. D. Holman, vol. 107 of IAU Symposium, 281

Uitenbroek, H. 2001, ApJ, 557, 389
Wang, H. 1992, Solar Phys., 140, 85
Watanabe, K., Krucker, S., Hudson, H., Shimizu, T., Masuda, S., & Ichimoto, K. 2010, ApJ, 715, 651. 1004.4259
Woods, T. N., Eparvier, F. G., Fontenla, J., & al. 2004, Geophys. Res. Lett, 31, 10802
Zarro, D. M., Canfield, R. C., Metcalf, T. R., & Strong, K. T. 1988, ApJ, 324, 582
Zharkova, V. V., & Kobylinskii, V. A. 1993, Solar Phys., 143, 259

The 16th Cambridge Workshop on Cool Stars, Stellar Systems and the Sun
ASP Conference Series, Vol. 448
Christopher M. Johns-Krull, Matthew K. Browning, and Andrew A. West, eds.
© 2011 Astronomical Society of the Pacific

The Radio–X-ray Relation in Cool Stars: Are We Headed Toward a Divorce?

Jan Forbrich,[1] Scott J. Wolk,[1] Manuel Güdel,[2] Arnold Benz,[3] Rachel Osten,[4] Jeffrey L. Linsky,[5] Margaret McLean,[1] Laurent Loinard,[6] and Edo Berger[1]

[1]*Harvard-Smithsonian CfA, 60 Garden St, Cambridge, MA 02138, USA*

[2]*University of Vienna, Department of Astronomy, Türkenschanzstr. 17, A-1180 Vienna, Austria*

[3]*Institute of Astronomy, ETH Zurich, 8093 Zurich, Switzerland*

[4]*Space Telescope Science Institute, 3700 San Martin Drive, Baltimore, MD 21218, USA*

[5]*JILA, University of Colorado and NIST, Boulder, CO 80309-0440, USA*

[6]*Centro de Radioastronomía y Astrofísica, Universidad Nacional Autónoma de México, Apartado Postal 3-72, 58090, Morelia, Michoacán, Mexico*

Abstract. This splinter session was devoted to reviewing our current knowledge of correlated X-ray and radio emission from cool stars in order to prepare for new large radio observatories such as the EVLA. A key interest was to discuss why the X-ray and radio luminosities of some cool stars are in clear breach of a correlation that holds for other active stars, the so-called Güdel-Benz relation. This article summarizes the contributions whereas the actual presentations can be accessed on the splinter website[1].

1. Radio Emission and X-rays as Diagnostics of Coronal Energy Release

X-rays and radio emission are excellent diagnostic probes to study energy release in magnetized stellar coronae. Solar observations have been key to deciphering the plethora of phenomena seen in these wavelength ranges. In brief, X-rays trace the presence of dense, hot (million-degree) plasma trapped in closed coronal magnetic fields, heated by processes that are still not fully understood. Radio observations, in contrast, probe both thermal atmospheric components from the chromosphere to the corona, and populations of non-thermal, accelerated electrons, typically residing in low-density, open or closed coronal magnetic fields.

A solar-stellar analogy is, however, complicated by phenomenology in *magnetically active stars* that is, at first sight, not present in the Sun. X-ray emission becomes much stronger toward more active stars, most likely as a result of increased surface coverage with active regions; however, the average, "characteristic" temperature of the corona increases along with the coronal luminosity (e.g., Schrijver et al. 1984; Güdel

---

[1]http://cxc.harvard.edu/cs16xrayradio/

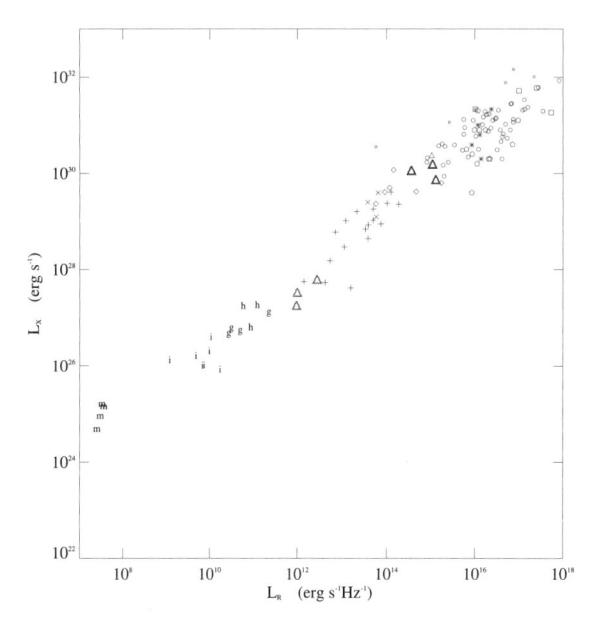

Figure 1. Radio vs. X-ray correlation. Different symbols in the upper right part of the figure refer to different types of active stars. The letters in the lower left show the loci of solar flares (luminosities averaged over the duration of the X-ray flare; see Benz & Güdel 1994 for details). The red and blue triangles show examples of stellar flares from M dwarfs (Güdel et al. 1996) and RS CVn binaries (Osten et al. 2004).

et al. 1997), a trend that requires additional physical explanation. At radio wavelengths, magnetically active stars also show a different face. Solar radio emission in the 1-10 GHz range is dominated by bremsstrahlung from various chromospheric/transition region levels and by optically thick gyroresonance emission from coronal layers above magnetic active regions. In contrast, observed radio brightness temperatures and radio spectra from active stars indicate gyrosynchrotron radiation from electrons with much higher energies than typically present in the solar atmosphere. The Sun occasionally features gyrosynchrotron emission, often accompanied by a variety of coherent radiation types (see Sect. 2), but such radiation is confined to episodes of flaring.

Magnetically active stars reveal further radio properties lacking any clear solar analogy such as extremely large coronal structures with size scales of order a stellar radius and more (e.g., Benz et al. 1998; Mutel et al. 1998; Peterson et al. 2010). In contrast, X-ray coronae tend to be rather compact even in extremely active stars (e.g., Walter et al. 1983; Ottmann et al. 1993); this is a consequence of the X-ray brightness scaling with the square of the electron density combined with relatively small pressure scale heights. Radio and X-ray sources are therefore not necessarily co-spatial and probe rather different plasmas or particle populations in stellar atmospheres, different atmospheric layers and structures, and perhaps even different energy sources. There should be little reason to expect that the two types of radiation are correlated in stars.

It therefore came as a surprise when such a correlation was uncovered for the steady radio and X-ray luminosities of RS CVn close binary stars. Drake et al. (1989)

found that the soft X-ray and radio (6 cm) luminosities are correlated over a few orders of magnitude albeit somewhat different from a linear trend ($L_R \propto L_X^{1.37\pm0.13}$). They suggested a self-consistent scheme in which the radio emission originates from the tail of the Maxwellian electron distribution of a very hot (>50 MK) plasma through the gyrosynchrotron process; such a plasma component is suggested from X-ray observations. This model is elegant as it links X-rays and radio emission by suggesting a common source. However, serious difficulties remain. Gyrosynchrotron spectra from a thermal plasma reveal a steep decline toward higher frequencies not observed in any magnetically active star; an acceptable spectral fit requires an extraordinary setup of the coronal magnetic field, such as a field strength decreasing with radius as r^{-1} (Chiuderi Drago & Franciosini 1993; Beasley & Güdel 2000). Non-thermal (power-law) electron distributions, in contrast, readily produce shallow spectra as observed, especially if the "aging" of an injected electron distribution, leading to spectral modifications due to synchrotron and collisional losses, is taken into account (Chiuderi Drago & Franciosini 1993).

A closer inspection of stars more akin to the Sun is in order. To that end, Güdel et al. (1993) studied X-ray and radio luminosities for M dwarfs, followed by other spectral types including G dwarfs (e.g., Benz & Güdel 1994; Güdel et al. 1995). Again, active stars of all late-type spectral classes followed a similar trend, best described by a proportionality, $L_X/L_R \approx 10^{15.5\pm0.5}$ Hz (in the following referred to as the GB relation). Combining these samples with samples of RS CVn binaries, Algol binaries, FK Com-type stars and also pre-main sequence weak-lined T Tauri stars, a coherent trend is found over 5-6 orders of magnitude in L_R and L_X (Figure 1). It is important to note that the L_X/L_R ratio is by no means universal. It has been demonstrated exclusively for *magnetically active stars* but does clearly not apply to inactive stars like the Sun; such stars keep appreciable levels of quasi-steady soft X-ray emission but are *not* sources of continuous radio emission of the gyrosynchrotron type. In fact, present-day radio observatories still cannot systematically detect nearby cool stars except for extremely active examples.

Active stars stand out by two properties mentioned above - their extremely hot plasma seen in X-rays, and their non-thermal electron populations evidenced by their radio emission. Let us assume that the energy initially contained in the accelerated electrons eventually heats the coronal plasma. If the corona releases energy at a rate $\dot{E}$ by injecting accelerated electrons at an energy-dependent rate $\dot{n}_{in}(\epsilon)$, then

$$\dot{E} = \frac{1}{a}\int_{\epsilon_0}^{\infty}\dot{n}_{in}(\epsilon)\epsilon d\epsilon = \frac{1}{b}L_X \quad (1)$$

where a is the fraction of the total energy that is channeled into particle acceleration and b is the fraction of the total energy that is eventually radiated as soft X-rays. Eq. 1 assumes an equilibrium between energy injection and energy loss. After introducing radiation processes into Eq. 1, one finds (Güdel & Benz 1993)

$$L_R = 3 \times 10^{-22} B^{2.48} \frac{a}{b}\tau_0(\alpha+1)L_X \quad (2)$$

i.e., a proportionality if several parameters on the right-hand side take characteristic, constant values, in particular B and the ratio a/b (α is the power-law index for the energy dependence of the electron lifetime). Conversely, comparing Eq. 2 with observations, one finds (e.g., for $a/b \approx 1$) the time scale τ_0 for electron trapping (i.e., the lifetime of the population), concluding that the radiation must decay on time scales of minutes to hours in most cases. This necessitates frequent or quasi-continuous replenishment of the corona by high-energy electrons.

This mechanism is what the "standard model" for a solar flare would predict. The standard solar flare model, the chromospheric evaporation scenario, posits that electrons initially accelerated in reconnecting magnetic fields propagate to chromospheric layers where they heat and ablate material which escapes into closed magnetic loops and cools by X-ray radiation. The best observational evidence for this model is the "Neupert Effect", stating that the time derivative of the flare X-ray light curve resembles the radio (or hard X-ray or U-band) light curve, $dL_X/dt \propto L_R$.

This prediction follows from assuming that L_X roughly scales with the thermal energy content in the hot plasma accumulated from the high-energy electrons, while radio emission scales with the number of such electrons present at any given time. The Neupert Effect is frequently observed in solar flares (e.g., Dennis & Zarro 1993), but has also frequently been seen in stellar flares, both extremely large events and the smallest yet discerned in stellar X-rays (e.g., Güdel et al. 1996, 2002; Osten et al. 2004).

We need one further ingredient, relating flares to the observed quiescent emission. During the past decade, a number of studies have shown that the occurrence rate of stellar flares in X-rays is distributed as a power law in radiated energy (a concept familiar to solar physics), $dN/dE \propto E^{-\alpha}$, with $\alpha \gtrsim 2$ (e.g., Audard et al. 2000; Kashyap et al. 2002). In that case, and assuming that the power law continues toward smaller energies, the energy integration, $\int_{E_0}^{\infty} E(dN/dE)dE$ diverges for $E_0 \to 0$, i.e., a lower cut-off is required. More relevant here, the entire apparently steady emission level could be explained by the large number of small flares that superpose to a quasi-steady emission level while not recognizable individually in light curves.

Assembling the above pieces, we then suggest to solve the $L_X - L_R$ puzzle as follows: Radio and X-ray emission correlate in magnetically active stars because the radiation we perceive as "quiescent" emission is made up of contributions from numerous small flares; each of these flares heats plasma by transforming kinetic energy from accelerated electrons; a portion of the latter is evident from their radio emission, while the heated plasma is observed by its X-ray emission. The L_X/L_R ratio therefore reflects the energy loss ratio of individual flares. As a check, we consider whether *solar and stellar flares* reveal radiative output ratios similar to those of the "quiescent" radiation. Average X-ray and radio luminosities for a range of solar flares (with specified total flare durations) as well as a sample of stellar flares are overplotted in Fig. 1. Indeed, the solar flares continue the trend seen in magnetically active stars (Benz & Güdel 1994) and the stellar flares show luminosity ratios in perfect agreement with the trend for quiescent emission. These observations support a picture in which flares are at the origin of coronal heating, of the steady radiation in magnetically active stars, and consequently of the L_X/L_R correlation.

2. The Sun

Correlation of Solar and Stellar X-rays with Gyrosynchrotron Radio Emission.
X-ray emission from stellar and solar flares and coronae is produced by bound-bound transitions and by bremsstrahlung of rapidly moving electrons deflected on ions. The emission comes in two flavors depending on the energy distribution of the electrons: thermal or non-thermal. Thermal X-rays range from less than 0.1 keV to more than 10 keV, resulting from temperatures in the range between 10^{6-8} K. Non-thermal X-rays are

emitted by energetic electrons, accelerated by plasma processes. Their bremsstrahlung emission is observed in solar flares from about 10 keV up to 100 MeV. It is approximately a power law in the photon energy spectrum, often with some breaks caused by a change in power-law index. The observed spectrum indicates a power law also in the energy distribution of electrons.

Gyrosynchrotron radio emission is produced by individual particles, usually mildly relativistic (> 100 keV) non-thermal electrons. The emission is caused by their spiraling motion in magnetic fields (e.g., Dulk & Marsh 1982). Each high-energy electron radiates both bremsstrahlung and gyrosynchrotron emission. Although different parameters enter the emissivity (most notably the magnetic field in the gyrosynchrotron case), it is not surprising that non-thermal X-rays and gyrosynchrotron emission correlate in solar flares. Kosugi et al. (1988) find a linear correlation between non-thermal X-ray and radio peak fluxes with deviations of less than a half order of magnitude. It is more surprising that in both stellar quiescent and flaring coronae, the *thermal* (soft) X-ray luminosity also correlates with gyrosynchrotron radio emission (see above).

The correlation of non-thermal gyrosynchrotron radiation and thermal X-ray emission fits the standard flare scenario of flares: A significant fraction of the flare energy is released in the form of non-thermal electrons, which precipitate to a dense medium and heat it to the point of thermal X-ray emission. The scenario has not been confirmed in solar flares at a quantitative level. It is surprising, nevertheless, that over the large range and widely different objects the correlation remains within half an order of magnitude (Krucker & Benz 2000). In particular, the magnetic field and electron life time are expected to change. Some deviations are noted: RS CVn binaries, Algols, and BY Dra binaries tend to be radio-rich (Güdel & Benz 1993), nanoflares are radio-poor (Krucker & Benz 2000). These differences may be the result of different parameters in Eq. (2).

The excellent correlation of the radio/X-ray relation has several consequences:
1) The identical relation for flares and active star quiescent X-ray emission strongly suggests that their X-ray emitting corona is flare heated.
2) The radio/X-ray relation has allowed the discovery of radio emission of K, G, and F stars. Selecting bright X-ray emitters among these stellar types, they were easily detected for the first time in radio emission (Güdel et al. 1994; Güdel 2002).
3) A large deviation from the relation can be used to identify radio emission originating by a mechanism *other than gyrosynchrotron* (e.g., Benz 2001).

Correlation of Solar Soft X-rays with Coherent Radio Emission. Cosmic radio emission originates in two ways: coherent and incoherent. Gyrosynchrotron and thermal radiation are incoherent and result from the emission of individual particles. Conversely, coherent emission is produced by a group of particles emitting in phase. The phasing may be caused by a wave in the plasma driven by an instability. Such instabilities occur in plasmas where electrons have non-thermal velocity distributions, such as a beam component, a loss-cone or strong electric current (A. Benz 2002). In wave terminology, the plasma wave is transformed into radio waves (Melrose 1980). Excitation and transformation are highly non-linear. Thus the correlation with X-rays is weak or absent. Characteristics of coherent radio emission are a narrowband spectrum ($\Delta \nu / \nu \ll 1$), high polarization ($\gtrsim 40\%$), and extremely high brightness temperature ($\gtrsim 10^{10}$ K). An example of various coherent emissions is shown in Fig 2. At frequencies above 3000 MHz, gyrosynchrotron radiation appears as diffuse broadband structures increasing to higher frequencies beyond the instrumental limit at 4000 MHz.

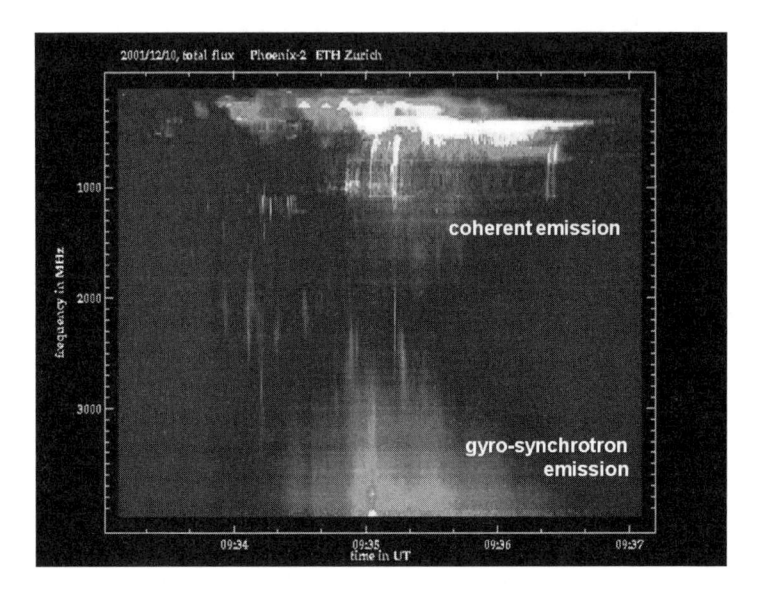

Figure 2. Spectrogram of solar flare radio emission in meter and decimeter waves observed with Phoenix-2 at ETH Zurich. The frequency increases downward. The example shows a clear separation of incoherent gyrosynchrotron emission at high frequencies and various coherent emissions at low frequencies.

Coherent radio emissions from the Sun have been reported at all wavelengths longer than 3 cm. At meter wavelengths, they are classified as the well-known Type I to V of radio bursts. At shorter wavelength, the metric types gradually disappear except Type III bursts which have been identified at up to 8.5 GHz. Other types, classified as DCIM, dominate at decimeter wavelengths. They consist of broadband pulsations, patches of continuum or narrowband spikes. Figure 5 in Benz (2009) displays the highest fluxes reported. Gyrosynchrotron emission (marked therein as IVμ) dominates up to 10 cm, followed by DCIM emission possibly directly associated with flare energy release, and Type II emission caused by coronal shock waves dominating above 100 cm. We note that a flux density of 10^6 solar flux units (sfu) at 1 AU corresponds to 100 μJy at 50 pc.

Solar coherent radio emissions were extensively compared to X-ray observations by the RHESSI and GOES satellites in a survey by Benz et al. (2005). The survey finds that 17% of the X-ray flares larger than GOES class C5 are not associated with coherent radio emission between 0.1 and 4 GHz. A later study, however, showed that all of the "radio-quiet" flares occurred either near the limb or showed emission below 0.1 GHz (Benz et al. 2007). Thus all flares appear to be associated by coherent radio emission.

Figure 2 shows that gyrosynchrotron emission may be associated with coherent radio emission but not correlate in detail. Correlation has been investigated not with gyrosynchrotron emission, but non-thermal X-rays which can be taken as a proxy. Some cases with good temporal correlation have been reported by Dabrowski & Benz (2009). However, recent imaging observations show that even correlating coherent radio emission does not originate from the location of coronal X-ray emission where electron ac-

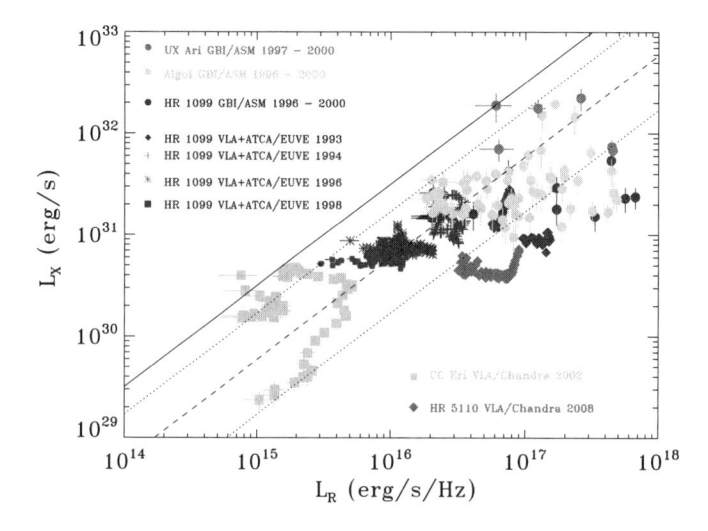

Figure 3. X-ray vs. radio luminosity for different simultaneous multi-wavelength observations of 5 different active stars: two Algol systems (Algol, HR 5110), two RS CVn binary systems (HR 1099 and UX Ari), and one BY Dra binary system (CC Eri). The dotted lines indicate the $L_X/L_R = \kappa \times 10^{15.5 \pm 0.5}$ Hz (κ=0.17) relationship in Benz & Güdel (1994). The dashed line gives the average L_X/L_R =5.9×10^{14} Hz from these data, and the solid line shows the GB relation with κ=1, $L_X/L_R = 10^{15.5}$ Hz.

celeration is expected (Benz, Battaglia & Vilmer, *in prep.*). The current interpretation of coherent radio emission still associates some of the DCIM emissions with particle acceleration, but not necessarily with the major acceleration site. The new imaging observations therefore require a flare model that has a much larger volume than the coronal X-ray sources currently observable.

Coherent radio emissions from solar flares do not correlate with thermal X-rays in detail except for the 'big flare syndrome' (larger flares being generally more luminous at all wavelengths). Benz et al. (2006) find that the total energy emitted in decimeter radio waves scales on the average with the peak soft X-ray flux as $E_{\text{radio}} = 1.92 \times 10^8 F_{\text{SXR}}^{1.42}$ [erg] where E_{radio} is in units of 10^{20} erg and F_{SXR} is the GOES flux from 1.8 to 12.4 keV in units of W/m^2. The correlation coefficient is 0.75, with a 95% significance range between 0.34 and 0.92. The scatter is more than an order of magnitude.

In most active stars and planets where the plasma frequency, ω_p, exceeds the electron gyrofrequency Ω_e, coherent radio emission is dominated by gyromagnetic maser emission at the first harmonic $\nu = \Omega_e/2\pi = 2.80 \times 10^6 B$ [Hz]. Thus coherent radio emission, although not a measure for flare importance, may be used to measure the magnetic field strength in stellar and planetary atmospheres.

3. RS CVn, Algol, and BY Dra Systems

Active binaries consisting of RS CVn, BY Dra, and Algol systems all lie at the upper right corner of the L_X vs. L_R diagram (Fig. 1), exhibiting the extremes of magnetic activity. However, these objects are also known to be highly time-variable sources. This gives pause to whether relationships connecting time-averaged behaviors hold when strictly simultaneous observations are used.

Based on simultaneous data summarized in the caption, Figure 3 shows the radio-X-ray correlation for five objects in comparison to the GB relation. Radio and X-ray variability are apparent, but most of the time these hyperactive stars do display L_X and L_R values which are consistent with the GB relation, albeit with a fair amount of scatter. Note, however that the range of variability differs between the two. For HR 1099 alone, L_R spans a range of 230 while only spanning a factor of 6 in L_X.

The conventional explanation for the nearly linear relationship between the two luminosities argues that there is a common energy reservoir out of which both plasma heating and particle acceleration occur, and these processes occur at a roughly fixed ratio between different classes of active stars. In the current situation, however, we can see that while this may be true in a time-averaged sense, the examples of uncorrelated radio and X-ray flaring which produce the extreme values of L_X/L_R have different ratios of particle acceleration and plasma heating. This is also true for flares which exhibit the Neupert effect (the green light curve of CC Eri contains one such example), where the instantaneous L_X/L_R ratio varies by a factor of 20 during the initial stages of the flare, spanning preflare conditions to the time when the flare-associated particle acceleration is occurring (radio luminosity peak) to the maximum flare-associated plasma heating (peak of the X-ray luminosity).

4. Young Stellar Objects

High-energy processes in Young Stellar Objects (YSOs) as observed in both the radio and the X-ray wavelength regime have been known for some time (e.g., Feigelson & Montmerle 1999). Very early evolutionary stages of YSOs, class I protostars, emit strong X-ray emission and nonthermal radio emission, presumably both due in large part to magnetically induced flares. However, it is still unclear in which stage YSOs become radio-active. Since most YSOs are relatively weak radio continuum sources, an observational difficulty has been to unambiguously find genuine nonthermal radio sources, e.g., by their polarization. YSOs can also produce free-free thermal radio emission in ionized material, for example at the base of outflows or jets (for a review of YSOs in the context of stellar radio astronomy, see Güdel et al. 2002).

Over the last decade, YSO X-ray variability research has been put on a much improved statistical basis, particularly for short timescales. However, centimetric radio variability on similar scales has only been studied toward a few sources (e.g., Forbrich et al. 2008). The correlation of variability in both wavelength ranges as well as the overall correlation of time-averaged radio and X-ray luminosities, as could be expected if the GB relation applies, remains unclear. While necessary to eliminate the effects of variability, there have only been a few simultaneous radio and X-ray observations.

Feigelson et al. (1994) obtained the first simultaneous radio and X-ray observations of a T Tauri star, targeting V773 Tau. They found uncorrelated variability, suggesting that the two emission mechanisms are decoupled. Following the discovery of the first class I protostar with nonthermal radio emission (Feigelson et al. 1998), the question of when protostars begin to be radio-active became more acute. Gagné et al. (2004) carried out a simultaneous radio and X-ray observing campaign of ρ Oph, a region rich in YSOs of all classes. They detected several T Tauri stars in both wavelength ranges. The first simultaneous X-ray and radio detections of class I protostars were reported by Forbrich et al. (2007) (see also Forbrich et al. 2006), targeting the *Coronet* cluster in CrA. With observations of the LkHα101 cluster, Osten & Wolk (2009) for the first time used simultaneous X-ray and multi-frequency radio observations to show that some sources show an inverse correlation between radio flux and spectral index. Most recently, observations of IC 348 and NGC 1333 have been carried out, but only few YSOS are detected in both bands (Forbrich, Osten, & Wolk, *submitted*).

All these simultaneous measurements agree on the location of YSOs on an L_X vs. L_R plot. Both luminosities are among the highest observed toward stars. YSOs do not seem to fall directly on the GB relation, but they are shifted toward higher radio luminosities for a given X-ray luminosity. Still, there are only a few YSOs that seem to be clearly off the GB relation and their numbers are too low to provide firm conclusions. To date observations have been limited by the radio sensitivity, but the Expanded Very Large Array will soon lead to far more radio detections among YSOs. Higher signal-to-noise ratios will also allow us to better distinguish non-thermal (e.g., gyrosynchrotron) and thermal radio sources. Interesting insights are also coming from VLBI radio observations (e.g., Dzib et al. 2010).

5. The X-ray-Radio Disconnect in Ultracool Dwarfs

Prior to the first detection of radio emission from an ultracool dwarf (Berger et al. 2001), measuring radio emission from very low mass stars and brown dwarfs seemed improbable. Given the tight correlation between radio and X-ray emission in higher-mass, coronally-active stars, initial X-ray results predicted radio emission well below detectable limits. The detection of a radio flare from LP944-20 required a severe violation of the GB relation, with $L_{\nu,\mathrm{rad}}/L_X \gtrsim 10^{-11.5}$ Hz^{-1}. Subsequent radio detections of ultracool dwarfs (Berger 2002) have yielded similar results, indicating a general violation of the correlation in this regime.

With a large sample of simultaneous observations, Berger et al. (2010) conclusively demonstrated this correlation no longer holds for objects beyond spectral type M6. For M dwarfs in the range M0-M6, the ratio of radio to X-ray emission is $L_{\nu,\mathrm{rad}}/L_X \approx 10^{-15.5}$. This ratio steadily increases for cooler objects. In the range M7-M8 it is around 10^{-14} and beyond M9 it is greater than 10^{-12}. It should be noted that there are a few objects which have been detected in the X-rays without corresponding radio emission, and hence may not violate the correlation. However, with the exception of a marginal detection from the L dwarf Kelu-1, these are all earlier objects in the range M7-M9 and most do not have deep radio luminosity limits and may yet be detected.

To investigate the nature of the breakdown of X-ray/radio correlation, it useful to examine the trends of the X-ray and radio emission in ultracool dwarfs separately. Berger et al. (2010) showed a sharp decline in X-ray activity for objects with spectral types beyond M7. In contrast, they also found indications for an increase in the ratio activity ($L_{\mathrm{rad}}/L_{\mathrm{bol}}$) for those objects cooler than M7. It is clear, then, that the breakdown is the due to the decline in X-ray luminosity combined with the sustained strength of the radio emission. But why do particle acceleration and plasma heating no longer correlate? As demonstrated by the radio emission, neither magnetic field dissipation nor particle acceleration appear to be affected by the increasingly neutral atmospheres of ultracool dwarfs. The total radio luminosities remain roughly constant even in these cooler objects, indicating the fraction of magnetic energy that goes into accelerating the electrons responsible for the radio-emission does not change, nor does the efficiency of field dissipation.

The difference appears to lie in the efficiency of the plasma heating, which is responsible for the hot X-ray producing gas. If the radio-emitting electrons are directly responsible for plasma heating in higher mass stars, the enhanced trapping of these electrons could account for the breakdown in the correlation. It could also be produced by a change in the geometry of the radio-emitting regions, should they evolve to smaller sizes in lower-mass objects and hence have less of an effect on the large-scale coronal heating. However, this is unlikely given that rotationally stable, quiescent radio emission and periodic Hα emission has been detected from several ultracool dwarfs, indicative of large magnetic filling factors.

Since radio emission requires only a small population of relativistic electrons, *a decline in the bulk coronal density could suppress the X-ray heating without affecting the radio emission*. The decreased X-ray emission, if caused by a loss in coronal density, could be attributed to coronal stripping. There are hints of super-saturation in the X-ray-rotation relation in the fastest rotators among the ultracool dwarfs (Berger et al. 2008). For objects later than M7, there is decline in L_X/L_{bol} for objects with decreasing rotation periods. The median value of L_X/L_{bol} is $\approx 10^{-4}$ for $P > 0.3$ d, while for $P < 0.3$ d it is $L_X/L_{bol} \approx 10^{-5}$. The combination of rapid rotation and the shrinking co-rotation radius in lower mass stars may lead to the decrease in X-ray emission among these objects. In comparision, there are indications that the fastest rotators among ultracool dwarfs are more likely to be detected in the radio and are the most severe violaters of the X-ray-radio correlation (Berger et al. 2008).

In contrast to hotter stars which primarily produce radio emission through gyrosynchrotron radiation, several ultracool dwarfs have been observed with short-duration, highly polarized, narrow band, bursts which can be attributed to a coherent emission process, such as the electron cyclotron maser (ECM) instability (Benz 2001; Hallinan et al. 2007, 2008; Berger et al. 2009). Although this process cannot account for all radio emission detected from ultracool dwarfs, this change in emission mechanism may be indicative of a change in the properties of the relativistic electron population and its impact on large-scale coronal heating.

6. Conclusions

In short, we are not headed toward a divorce. Instead, the X-ray and radio luminosities of cool stars appear to be in an "open relationship" where a lot depends on the type of radio emission that is present. Until now, observational data have not always been sensitive enough to unambiguously identify dominant emission mechanisms, for example in the case of YSOs. Clearly, the L_X/L_R relation does not apply to all stars. As mentioned above, inactive stars violate this relation as they do not show non-thermal radio (gyrosynchrotron) emission. The "non-flaring" Sun is an example. Either, there are additional heating mechanisms at work in these stars that do not involve high-energy electrons, or the energy transformation process is more efficient in heating the plasma to sufficiently high temperatures so as to become visible in X-rays. The other important class of stars violating the relation are brown dwarfs, but coherent radio radiation mechanisms may matter here. Similarly, a number of protostellar objects do not follow the standard trend although they are magnetospheric radio and X-ray sources. Here, however, the measurement of either of the luminosities is difficult. Radio gyrosynchrotron emission may be attenuated by overlying ionized winds (e.g., the jets easily detected as thermal radio sources); X-ray emission could be partially attenuated by neutral gas masses, such as neutral winds, accretion streams, or molecular outflows. If only part of the coronal emission is (fully) attenuated, e.g., by accretion streams, then an assessment of the intrinsic luminosities becomes impossible. Further observations, particularly deeper radio observations with new instruments such as EVLA, as well as large X-ray surveys with *Chandra* and XMM-*Newton* will help shed light on the limits of the radio/X-ray correlation in cool stars.

Appendix: What is the Connection Between Nonthermal Radio Emission and Thermal X-ray Emission?

The lack of consensus among the theoreticians concerning which physical models best explain heating and particle acceleration and the GB correlation suggest to Jeff Linsky that there must be a simple approach to understanding the relevant physics. Given that flares occur when there are rapid changes in the structure of complex magnetic fields, Maxwell's equations require the presence of electric fields and currents. In a plasma the acceleration force of the electric field $F_{accel} = eE$ is opposed by a collisional drag force $F_{drag} \sim (n_e/T)(v_{th}/v)^2$ when the speed v exceeds the thermal speed $v_{th} = (kT/m)^{1/2}$. Those electrons travelling at greater than a critical speed $v_c \sim (n_e/E)^{1/2}$ set by $F_{accel} > F_{drag}$, will be accelerated to high speeds, a process called "runaway". If the electric field exceeds the Dreicer field, $E > E_D = 4\pi e^3 \ln\Lambda/kT$ (Holman 1985), then all electrons will runaway, but this rarely occurs because large E fields produce turbulence and anomalous resistivity (Norman & Smith 1978).

When $E < E_D$ only those electrons in the tail of the Maxwell-Boltzmann velocity distribution, $f(v)$, with $v > v_c$ are accelerated. In this "sub-Dreicer regime", the ratio of nonthermal to thermal electrons is $N_{nonth}/N_{th} = \int_{v_c}^{\infty} f(v)dv / \int_0^{\infty} f(v)dv$. When $E \ll E_D$, $N_{nonth}/N_{th} \approx 0.5 e^{-0.5[(E_D/E)^{1/2} - E/E_D]^2}$ (Norman & Smith 1978). In addition to accelerating electrons, electric fields can heat the plasma by several mechanisms, the simplest being the Joule heating rate $iE = i^2 r$, where i is the current and the resistance r can be anomalous due to current-driven instabilities. Holman (1985) developed the theory of heating and acceleration by electric fields in the context of solar flares. For the example of $T = 10^7$ K, $B = 300$ G, and $EM = 10^{45}$ cm^{-3}, he finds that the time scales for Joule heating and acceleration of 10^{32} electrons are both about 30 seconds and the energy that goes into particle acceleration is always less that the heating rate. In his Case I, the assumption of $E = 0.03 E_D$ corresponds to $v_c = 5.8 v_{th}$, electron acceleration to maximum energy $W_{max} = 10$ MeV, and $N_{nonth}/N_{th} = 3.4 \times 10^{-8}$. In his Case II, doubling the electric field strength to $E = 0.06 E_D$ results in $v_c = 4.2 v_{th}$, $W_{max} = 100$ keV, and $N_{nonth}/N_{th} = 1.5 \times 10^{-4}$. Thus the N_{nonth}/N_{th} ratio and the maximum energy of the nonthermal electrons both depend very strongly on E/E_D, but in the opposite sense.

Since gyrosynchrotron radiation is broad band with $\Delta \nu \approx 1 \times 10^{10}$ Hz, the ratio of radio to X-ray emission ratio is $R = L_R \Delta \nu / L_X = 10^{-5.5 \pm 0.5}$, much less than unity as predicted by Holman (1985). Since the gyrosynchrotron radio emission rate per relativistic electron is proportional to its energy squared, the main contribution to $L_R \Delta \nu$ will be from the highest energy electrons. Free-free X-ray emission from $T = 10^7$ K electrons is proportional to $N_{th}^2 V$. Thus R is proportional to $(N_{nonth}/N_{th}) W_{max}^2 V_J / N_{th} V$, where V is the volume of the thermal gas from which the nonthermal electrons have been swept up and V_J is the volume of the nonthermal electrons in the current channel. For Holman's Case I, $V_J/V = 10^{-4}$ and $R \propto 3.4 \times 10^{-8}(10^7 \text{eV})^2 10^{-4}/10^9 \text{cm}^{-3} = 3.4 \times 10^{-7}$. For Case II, $V_J/V = 1/200$ and $R \propto 1.5 \times 10^{-4}(10^5 \text{eV})^2(1/200)/10^{11} \text{cm}^{-3} = 0.75 \times 10^{-7}$. The similar ratios for the two theoretical cases, despite the very different N_{nonth}/N_{th} values, show that DC electric fields can produce both runaway electrons in the sub-Dreicer regime and Joule heating consistent with a constant L_R/L_X relation.

References

A. Benz (ed.) 2002, Plasma Astrophysics, second edition, vol. 279 of Astrophysics and Space Science Library
Audard, M., Güdel, M., Drake, J. J., & Kashyap, V. L. 2000, ApJ, 541, 396
Beasley, A. J., & Güdel, M. 2000, ApJ, 529, 961
Benz, A. O. 2001, Nat, 410, 310
— 2009, in SpringerMaterials - The Landolt-Börnstein Database. DOI: 10.1007/978-3-540-88055-4_13, edited by J. Trümper (Heidelberg: Springer-Verlag)
Benz, A. O., Brajša, R., & Magdalenić, J. 2007, Solar Phys., 240, 263
Benz, A. O., Conway, J., & Gudel, M. 1998, A&A, 331, 596
Benz, A. O., Grigis, P. C., Csillaghy, A., & Saint-Hilaire, P. 2005, Solar Phys., 226, 121
Benz, A. O., & Güdel, M. 1994, A&A, 285, 621
Benz, A. O., Perret, H., Saint-Hilaire, P., & Zlobec, P. 2006, Advances in Space Research, 38, 951
Berger, E. 2002, ApJ, 572, 503
Berger, E., Ball, S., Becker, K. M., Clarke, M., Frail, D. A., Fukuda, T. A., Hoffman, I. M., Mellon, R., Momjian, E., Murphy, N. W., Teng, S. H., Woodruff, T., Zauderer, B. A., & Zavala, R. T. 2001, Nat, 410, 338
Berger, E., Basri, G., Fleming, T. A., Giampapa, M. S., Gizis, J. E., Liebert, J., Martín, E., Phan-Bao, N., & Rutledge, R. E. 2010, ApJ, 709, 332
Berger, E., Basri, G., Gizis, J. E., Giampapa, M. S., Rutledge, R. E., Liebert, J., Martín, E., Fleming, T. A., Johns-Krull, C. M., Phan-Bao, N., & Sherry, W. H. 2008, ApJ, 676, 1307
Berger, E., Rutledge, R. E., Phan-Bao, N., Basri, G., Giampapa, M. S., Gizis, J. E., Liebert, J., Martín, E., & Fleming, T. A. 2009, ApJ, 695, 310
Chiuderi Drago, F., & Franciosini, E. 1993, ApJ, 410, 301
Dabrowski, B. P., & Benz, A. O. 2009, A&A, 504, 565
Dennis, B. R., & Zarro, D. M. 1993, Solar Phys., 146, 177
Drake, S. A., Simon, T., & Linsky, J. L. 1989, ApJS, 71, 905
Dulk, G. A., & Marsh, K. A. 1982, ApJ, 259, 350
Dzib, S., Loinard, L., Mioduszewski, A. J., Boden, A. F., Rodríguez, L. F., & Torres, R. M. 2010, ApJ, 718, 610
Feigelson, E. D., Carkner, L., & Wilking, B. A. 1998, ApJ, 494, L215+
Feigelson, E. D., & Montmerle, T. 1999, ARA&A, 37, 363
Feigelson, E. D., Welty, A. D., Imhoff, C., Hall, J. C., Etzel, P. B., Phillips, R. B., & Lonsdale, C. J. 1994, ApJ, 432, 373
Forbrich, J., Menten, K. M., & Reid, M. J. 2008, A&A, 477, 267
Forbrich, J., Preibisch, T., & Menten, K. M. 2006, A&A, 446, 155
Forbrich, J., Preibisch, T., Menten, K. M., Neuhäuser, R., Walter, F. M., Tamura, M., Matsunaga, N., Kusakabe, N., Nakajima, Y., Brandeker, A., Fornasier, S., Posselt, B., Tachihara, K., & Broeg, C. 2007, A&A, 464, 1003
Gagné, M., Skinner, S. L., & Daniel, K. J. 2004, ApJ, 613, 393
Güdel, M. 2002, ARA&A, 40, 217
Güdel, M., Audard, M., Skinner, S. L., & Horvath, M. I. 2002, ApJ, 580, L73
Güdel, M., & Benz, A. O. 1993, ApJ, 405, L63
Güdel, M., Benz, A. O., Schmitt, J. H. M. M., & Skinner, S. L. 1996, ApJ, 471, 1002

Güdel, M., Guinan, E. F., & Skinner, S. L. 1997, ApJ, 483, 947
Güdel, M., Schmitt, J. H. M. M., & Benz, A. O. 1994, Science, 265, 933
Güdel, M., Schmitt, J. H. M. M., & Benz, A. O. 1995, A&A, 302, 775
Güdel, M., Schmitt, J. H. M. M., Bookbinder, J. A., & Fleming, T. A. 1993, ApJ, 415, 236
Hallinan, G., Antonova, A., Doyle, J. G., Bourke, S., Lane, C., & Golden, A. 2008, ApJ, 684, 644
Hallinan, G., Bourke, S., Lane, C., Antonova, A., Zavala, R. T., Brisken, W. F., Boyle, R. P., Vrba, F. J., Doyle, J. G., & Golden, A. 2007, ApJ, 663, L25
Holman, G. D. 1985, ApJ, 293, 584
Kashyap, V. L., Drake, J. J., Güdel, M., & Audard, M. 2002, ApJ, 580, 1118
Kosugi, T., Dennis, B. R., & Kai, K. 1988, ApJ, 324, 1118
Krucker, S., & Benz, A. O. 2000, Solar Phys., 191, 341
Melrose, D. B. 1980, Plasma astrophysics: Nonthermal processes in diffuse magnetized plasmas. Volume 2 - Astrophysical applications
Mutel, R. L., Molnar, L. A., Waltman, E. B., & Ghigo, F. D. 1998, ApJ, 507, 371
Norman, C. A., & Smith, R. A. 1978, A&A, 68, 145
Osten, R. A., Brown, A., Ayres, T. R., Drake, S. A., Franciosini, E., Pallavicini, R., Tagliaferri, G., Stewart, R. T., Skinner, S. L., & Linsky, J. L. 2004, ApJS, 153, 317
Osten, R. A., & Wolk, S. J. 2009, ApJ, 691, 1128
Ottmann, R., Schmitt, J. H. M. M., & Kuerster, M. 1993, ApJ, 413, 710
Peterson, W. M., Mutel, R. L., Güdel, M., & Goss, W. M. 2010, Nat, 463, 207
Schrijver, C. J., et al. 1984, A&A, 138, 258
Walter, F. M., Gibson, D. M., & Basri, G. S. 1983, ApJ, 267, 665

Cool Stars 16 Splinter Sessions

Planet Formation Around M-dwarf Stars: From Young Disks to Planets

I. Pascucci,[1] G. Laughlin,[2] B. S. Gaudi,[3] G. Kennedy,[4] K. Luhman,[5] S. Mohanty,[6] J. Birkby,[4] B. Ercolano,[4] P. Plavchan,[7] and A. Skemer[8]

[1] *Space Telescope Science Institute, Baltimore, USA*

[2] *University of California, Santa Cruz, USA*

[3] *The Ohio State University, USA*

[4] *Institute of Astronomy, Cambridge, UK*

[5] *The Pennsylvania State University, USA*

[6] *Imperial College London, UK*

[7] *NASA Exoplanet Science Institute, USA*

[8] *University of Arizona, Steward Observatory, USA*

Abstract. Cool M dwarfs outnumber sun-like G stars by ten to one in the solar neighborhood. Due to their proximity, small size, and low mass, M-dwarf stars are becoming attractive targets for exoplanet searches via almost all current search methods. But what planetary systems can form around M dwarfs? Following up on the Cool Stars 16 Splinter Session "Planet Formation Around M Dwarfs", we summarize here our knowledge of protoplanetary disks around cool stars, how they disperse, what planetary systems might form and can be detected with current and future instruments.

1. Disk Observations

1.1. The Properties of Disks Around M Dwarfs

There is growing observational evidence that young (1-2 Myr) very low-mass stars (M dwarfs) and brown dwarfs undergo a T Tauri phase: they are surrounded by optically thick gas-rich dust disks, they accrete disk gas, and have jets/outflows like their more massive counterparts. **Subhanjoy Mohanty** from the Imperial College, London, presented an extensive overview of the properties of these young very low-mass stars and brown dwarfs. He reviewed the disk accretion and the jet diagnostics detected around young cool stars with special emphasis on the broad (~100 km/s) Hα emission lines (e.g., Mohanty et al. 2005), and the broad and blueshifted forbidden [OI] line at 6300 Å (e.g., Whelan et al. 2009). He pointed out that accretion and infall are detected even in ~10 Myr-old very low-mass stars and brown dwarfs, similarly to what is found for sun-like stars (Mohanty et al. 2003; Herczeg et al. 2009; Looper et al. 2010). There also appears to be a strong correlation between the mass of the central star and the amount of gas accreted onto it ($\dot{M} \propto M_\star^2$, Muzerolle et al. 2003; Natta et al. 2006; Herczeg

& Hillenbrand 2008), a correlation which challenges our theoretical understanding of viscous disk evolution. Infrared emission in excess to the stellar photosphere is often detected around young very low-mass stars and brown dwarfs pointing to the presence of circumstellar disks within which planets might form. The disk frequency of ~50% is found to be similar to that around young sun-like stars indicating that the raw material for planet formation is often available regardless of the star mass (e.g., Luhman et al. 2007).

In spite of similar diagnostics for the T Tauri phase, differences are emerging in the properties of disks as a function of stellar mass. Dust disks around very low-mass stars and brown dwarfs appear to have smaller scale heights than disks around sun-like stars of similar age (Szűcs et al. 2010). This could result from a real difference in the disk structure, due for instance to faster grain growth and hence decoupling of the dust from the gas followed by settling, or from an opacity effect caused by infrared observations probing closer to the disk midplane in disks around cool stars (Kessler-Silacci et al. 2007). Supporting the first scenario, dust grains in the disk atmosphere traced via prominent silicate emission features at 10 and 20 μm appear to be more evolved around brown dwarfs than around sun-like stars (Apai et al. 2005; Kessler-Silacci et al. 2006; Pascucci et al. 2009 and Sect. 1.2).

The outer disk radius and disk mass are among the most important parameters to assess the likelihood of forming planets around very low-mass stars and brown dwarfs (see Sect. 2.2). Mohanty presented preliminary results from a large JCMT/SCUBA-2 survey which, combined with previous measurements, suggest that: i) M_{disk}/M_{star} at a given age (e.g., Taurus) is consistent with being constant from brown dwarfs to sun-like stars, at ~1%, but ii) the same ratio appears to significantly decrease with increasing age (e.g., by TWA). The latter finding is consistent with a substantial increase in grain size/settling with age in brown dwarf disks, as noted previously for sun-like stars. Disk radii are notoriously difficult to determine. Ercolano et al. (2009b) recently proposed to use the ratio of two forbidden [CI] lines, one in the near- and the other in the far-infrared, as a diagnostic for outer disk radii. The first near-infrared [CI] line detection is now available toward TWA30B (Looper et al. 2010). Far-infrared observations of [CI] lines at 370 and 609μm will be possible with the Herschel Space Observatory.

One system of special interest is the brown dwarf 2MASS1207 in the TW Hya association, which is surrounded by a dust disk and has a planetary mass companion (2MASS1207b) at a projected separation of ~50 AU. 2MASS1207b is a peculiar object in that it is under-luminous by two orders of magnitudes than expected at all bands from I to L. **Andy Skemer** from the University of Arizona summarized what scenarios have been proposed to explain its under-luminosity and evaluated their likelihood. Based on modeling of the full spectral energy distribution (SED), and absence of photometric variability, he argued that gray extinction by a nearly edge-on disk is highly unlikely. He pointed to clouds of micron-sized dust grains in the atmosphere of 2MASS1207b as a possible sink of luminosity. Unfortunately, current atmospheric models do not treat self-consistently the effect of dust grains in planets' atmospheres, hence a quantitative comparison between theory and observations is yet not feasible.

1.2. Disk Evolution Around Sun-like Stars and M Dwarfs

As summarized in the previous section, very low-mass stars and brown dwarfs undergo a T Tauri phase that in some aspects resembles that around sun-like stars. In this part of the CS16 session, **Kevin Luhman** from The Pennsylvania State University addressed the question of how disks evolve and disperse with emphasis on similarities and differences between sun-like stars and M dwarfs/brown dwarfs. He showed that the frequency of infrared excess emission appears to decrease less steeply with age as the mass of the central object decreases, although this result is not yet conclusive because few disk fractions have been measured for low-mass stars older than 3 Myr (Carpenter et al. 2006; Luhman 2009). If this trend is confirmed the implication is that optically thick dust disks, providing the raw material to form planets, persist longer around very low-mass stars and brown dwarfs than around sun-like stars.

Detailed studies of SED shapes demonstrate that very low-mass stars and brown dwarfs undergo the same disk clearing phases as their higher mass counterparts. There is evidence for transitional disks (reduced mid- but large far-infrared emission caused by a gap or hole) around brown dwarfs as well as for settled disks (reduced emission at all infrared wavelengths), see Fig. 1.2 and Muzerolle et al. (2006). The fraction of disks with transitional SEDs is about ~15% in disks around sun-likes stars as well as M dwarfs in young (a few Myr-old) star-forming regions. The paucity of transition disks in both star samples point to a rapid disk clearing, on a timescale of ~0.1 Myr (Luhman et al. 2010).

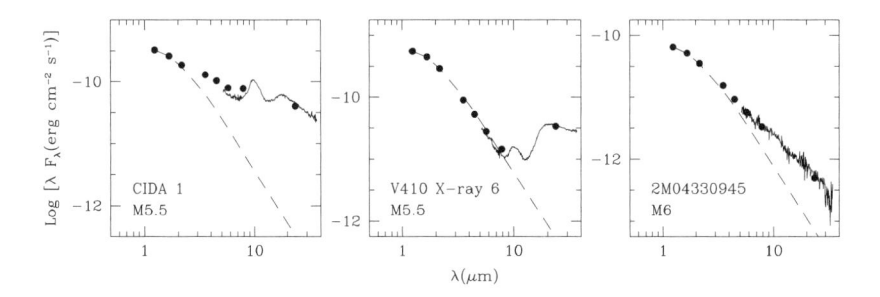

Figure 1. Sample SEDs around very low-mass stars and brown dwarfs. Left panel: SED of a primordial optically thick dust disk. Middle panel: SED of a transitional disk. Right panel: SED of an evolved/settled disk.

In addition to the dispersal of primordial material, brown dwarf disks also present the same signs of dust processing as disks around sun-like stars. Grain growth as well as crystalline grains have been detected in brown dwarf disks (Apai et al. 2005; Furlan et al. 2005) with some evidence for more efficient processing than in disks around sun-like stars belonging to the same star-forming region (Scholz et al. 2007; Morrow et al. 2008; Riaz et al. 2009). There is also some hint that grain growth may be affected by the presence of sub-stellar companions (Adame et al. submitted).

Finally, thanks to the sensitivity of new infrared and UV instrumentation, we are starting to have detections of gaseous species in disks around very low-mass stars and brown dwarfs (Pascucci et al. 2009; France et al. 2010). Differences in the column densities of the detected species might result from the different stellar radiation field impinging onto the disk surface (Pascucci et al. 2009). It will be interesting to determine how much the chemistry of the disk midplane is impacted by this different radiation field and what are the effects on the composition of forming planets.

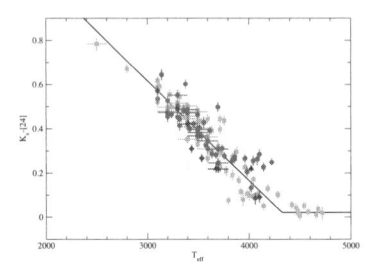

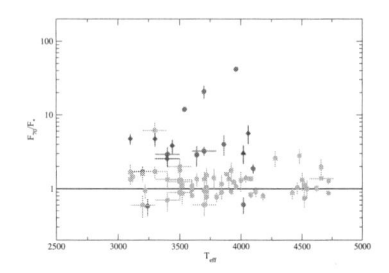

Figure 2. Spitzer 24 and 70 micron excess plots for the following samples: the control sample of nearby, older M dwarfs (yellow), Xray active M dwarfs (red), rapid rotators (blue) and IRAS and sub-mm excess candidates (purple). These plots show that only Xray active M dwarfs tend to have a higher frequency of small warm 24μm excesses, while both Xray active and rapid-rotator M dwarfs tend to have a higher disk frequency at 70μm when compared to nearby, older M dwarfs.

Moving into older/nearby M dwarfs, **Peter Plavchan** from the NASA Exoplanet Science Institute reported on his latest results from a search of debris dust disks, second generation dust disks, around M-dwarf stars. He showed that there is a decreasing rate of debris disks detections with spectral type going from A stars to FGK stars down to the M dwarfs (Plavchan et al. 2009). Because at young ages (<200 Myr) the frequency of debris disks around M dwarfs is similar to that around old sun-like stars, there could be an efficient mechanism to remove debris dust around M-dwarf stars. Plavchan et al. (2009) proposed stellar wind drag as the most likely mechanism. They are currently testing this scenario via a large survey of debris dust disks with Spitzer around Xray bright M dwarfs likely to be young and not having large radial stellar winds. Preliminary results confirm that Xray bright M dwarfs have detectable debris disks with a frequency that is very similar to that of debris around sun-like stars (Plavchan et al. in prep. and Fig. 1.2). Because debris dust is very likely linked to the formation of planetesimals, these results indicate that planetesimal formation occurs as often around M-dwarf stars as around sun-like stars.

2. Theory

After discussing the observational properties of young and old disks around M dwarfs, we turned to the theory of disk dispersal and planet formation with the goal of understanding what type of planetary systems can form around M-dwarf stars.

2.1. Theory of Disk Dispersal Around M Dwarfs

Grant Kennedy from the Institute of Astronomy, Cambridge, reviewed in detail the physical mechanisms dispersing protoplanetary disks and how they might act to remove the primordial gas and dust around M dwarfs. Viscous accretion of gas onto the central star, accompanied by outward spreading, is certainly a major player in disk evolution. As summarized in Sect. 1.1 young M dwarfs and even brown dwarfs experience disk mass accretion in the first few Myr of their lives. Although ubiquitous, viscous evolution alone cannot reproduce the relatively short disk lifetimes of 1-10 Myr inferred from observations, as noted e.g. by Clarke et al. (2001). Photoevaporation driven by the central star has been proposed as the next ubiquitous and most relevant disk dispersal mechanism. In brief, radiation from the central star (X-ray, EUV and FUV photons) heat the gas in the upper layers of the disk, depositing enough energy to leave the gas unbound, thus establishing a photoevaporative flow (Hollenbach et al. 2000; Dullemond et al. 2007 for reviews). When the accretion rate drops below the wind loss rate, photoevaporation limits the supply of gas to the inner disk, which drains onto the star on the local viscous timescale— of order 10^5 years (see Fig. 2.1). With the inner disk gone, the remaining outer disk is rapidly dispersed by direct stellar radiation. Though the original model of Clarke et al. invoked EUV photons as the main cause of photoevaporation (see also Alexander et al. 2006 for following developments), recent work suggests that FUV and Xray radiation, which launch winds from denser disk regions, are more important (Ercolano et al. 2008, 2009a; Gorti et al. 2009). The first strong observational evidence for photoevaporation driven by the central star has been recently presented by Pascucci & Sterzik (2009) for the disk around the ~10 Myr-old star TW Hya.

Gorti & Hollenbach (2009) modeled the dispersal of protoplanetary disks around stars of different masses and different radiation fields. They found that the disk lifetime substantially decreases for stars more massive than $7 M_\odot$, due to their high UV fields, but does not vary much for stellar masses in the range 0.3-3 $M_\odot$, hence from M dwarfs up to a few times the mass of the Sun. This is likely because lower mass accretion rates and lower photoevaporative loss rates in M dwarfs are accompanied by lower disk masses. On this topic, **Barbara Ercolano** from the Institute of Astronomy, Cambridge, presented a very interesting color-color diagram to observationally distinguish disks around M dwarfs that are cleared inside-out (as expected from photoevaporation) and those that have been subject to uniform draining, due for instance to viscous evolution. Comparison of this theoretical diagram, obtained by extensive radiative transfer modelling, to available photometric observations of young stellar objects in nearby star forming regions, strongly supports a dispersal mechanism that operates from the inside out over a rapid transition timescale (Ercolano et al. 2010).

External photoevaporation induced by high-energy photons from OB stars in a cluster might be another important dispersal mechanism especially for disks around low-mass M dwarfs which have weak gravity fields. Adams et al. (2004) calculated that disks around M dwarfs are evaporated and shrink to disk radii ≤15 AU on short timescales ≤10 Myr under moderate FUV fields which are present in small stellar groups and clusters. Disks around sun-like stars are more durable and would require a much more intense FUV field (10 times higher) to evaporate and shrink on a similarly short (~10 Myr) timescale.

In summary, disk evolution and dispersal around isolated M dwarfs is expected to proceed as for sun-like stars. This similarity means that at least the early stages of planet formation are also likely to proceed in a similar way (as suggested in Sect. 1.2). However, as discussed below, disk dispersal is a crucial ingredient in planet formation models and both viscous evolution and (internal and external) photoevaporation should be included to properly evaluate the timescale and likelihood of forming planets.

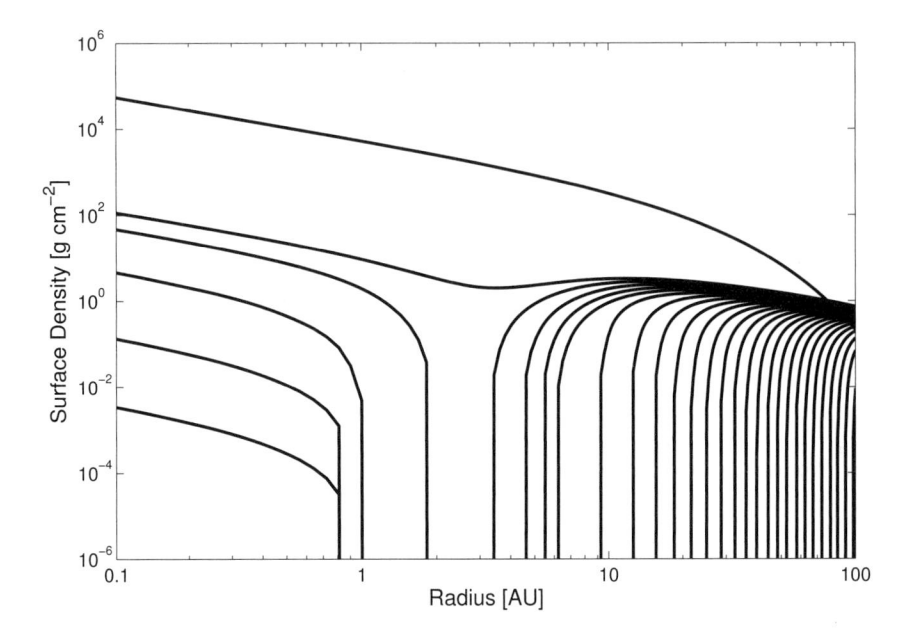

Figure 3. An example of disk dispersal due to X-ray photoevaporation and viscous evolution. The first (highest) line shows the initial surface density profile, the next shows the profile at 75% of the disk lifetime. Remaining lines show the surface density at 1% steps in disk lifetime. Figure courtesy of James Owen, see also Owen et al. (2010).

2.2. Planet Formation Around M Dwarfs

Greg Laughlin from the University of Santa Cruz provided a comprehensive review of the theory of planet formation around M dwarfs. First, he pointed out the main characteristics of planets detected around M dwarfs (see also Sect. 3): a clear paucity of close-in giant planets (thought exceptions exist, see the case of Gliese 876 by Rivera et al. 2010), mostly low eccentricities for the detected planets, and a possible correlation between the metallicity of the M-dwarfs and the detection of giant planets around them (Johnson & Apps 2009a). This last relation extends up to the more massive sun-like stars (Fischer & Valenti 2005) and points to giant planet formation via core accretion as the most likely formation scenario.

According to the core-accretion paradigm giant planet formation can be broken up into 3 phases. In the first phase, the growing planet consists mostly of solid material. The planet experiences runaway growth until the feeding zone is depleted. Solid accretion occurs much faster than gas accretion in this phase. During the second phase, both solid and gas accretion rates are small and nearly independent of time. Finally, in the last stage runaway gas accretion occurs (starting when the solid and gas masses are roughly equal). The baseline model by Pollack et al. (1996) predicted a long ($\sim$ 10 Myr) timescale to form a planet like Jupiter at 5 AU. This timescale is clearly too long when compared to the observed disk lifetimes (Sect. 1.2) hence several ideas have been proposed to shorten the longest of the phases, the core mass accretion followed by gas accretion.

Laughlin et al. (2004) have extended the core-accretion theory of planet formation around M dwarfs. They showed that giant planet formation is highly suppressed around a 0.4 solar mass star in comparison to sun-like stars due to the lower disk surface densities beyond the snow line and longer timescales both for the formation of planetesimals as well as for their accumulation. Hence, the zeroth order predictions from the core-accretion theory seem to be correct: higher metallicity – more planets (lower metallicity – fewer planets); higher stellar mass – more planets (lower stellar mass – fewer planets). The case of the planetary system around Gliese 581 (at least four planets comprising super-Earths and hot Neptunes) suggests that the planet formation process responsible for the majority of the planets in the galaxy occurs robustly in M-star systems. The formation of super-Earths has not been investigated in detail. Simulations from Raymond et al. (2007) showed unambiguously that if the protoplanetary disks associated with the lowest mass stars are scaled down versions of the Sun's protoplanetary disk, then the resulting terrestrial planets rarely exceed the mass of Mars, are dry, and inhospitable. Montgomery & Laughlin (2009) investigated a more optimistic scenario in which the disk mass is largely independent of the stellar mass and showed that habitable planets can form easily even around a 0.1 $M_\odot$ star. Observations of disk masses are an essential input to the theory as discussed in Sect. 1.1.

3. Planets Around M Dwarfs

We started this session with the contributed talk by **Jayne Birkby** from the Institute of Astronomy, Cambridge. She presented the first results from the WFCAM Transit Survey carried out with the 3.8m UKIRT at infrared wavelengths. The main goal of the survey is to place meaningful observational constraints on the occurrence of rocky and giant planets around low-mass stars using time-series photometric observations of a large ($\sim 6,000$) sample of M dwarfs. The survey is half-way to completion, with extensive scrutiny and follow-up already performed on approximately one third of the M-dwarf sample. So far, the survey shows no hot Jupiter detections nor the more commonly expected hot-Neptune detections. However, with only a third of the sample thoroughly explored at this time, it is too early to speculate on the planet fraction for the M-dwarf sample. The team also reported the detection of a number of M-dwarf eclipsing binary systems with masses $< 0.6 M_\odot$ which can be used to further test the mass-radius relation, and they have found a very low mass ($\sim 0.2 + 0.1 M_\odot$) eclipsing binary, which is near the theorized limiting mass for formation via disk fragmentation.

We concluded our CS16 splinter session with the review talk by **Scott Gaudi** from The Ohio State University on the demographics of planets around M dwarfs. Although the faintness of M-dwarf stars makes it hard to detect planets around them with almost any technique, their smaller mass and size present some advantages. In the case of the radial velocity (RV) technique, a giant planet around a M dwarf will produce a larger radial velocity signal than the same planet around a sun-like star. However, precision RV surveys in the optical are limited to bright early M dwarfs (e.g., Mayor et al. 2009, Johnson et al. 2010, Vogt et al. 2010). Infrared RV searches are just starting and are promising especially for detecting planets around the very low-mass M dwarfs (e.g., Bean et al. 2010) . In the case of transits, depths of transits due to planets orbiting M dwarfs are larger ($\sim 10\%$ for Jupiter-size planets), but the duty cycles and transit probabilities are lower for fixed period. On the other hand, for habitable planets, the transit depths, probabilities, and duty cycles are dramatically larger for low-mass stars, thus favoring the discovery of habitable planets around M dwarfs (Gould et al. 2003). The main challenge with discovering transiting planets around M dwarfs is their intrinsically low-luminosity. Surveys for transiting planets around bright M dwarfs such as MEarth (Nutzman & Charbonneau 2008; Charbonneau et al. 2009) must contend with the paucity and sparsity of targets (i.e., there are only 2,000 M-dwarfs with $V \lesssim 12$ over the entire sky), whereas deep, pencil-beam surveys (see the contribution from J. Kirby) must contend with the difficulties with RV follow-up for the faint candidates. Microlensing has the advantage of being insensitive to host star luminosity and thus sensitive to planets throughout the Galaxy. Furthermore, since M dwarfs are the most common stars in the Galaxy, the microlensing host stars are likely to be M-dwarf stars. Although the faintness of the host star presents challenges for characterization of the detected systems, often the host star and planet masses can also be measured with followup observations to within $\sim 10 - 20\%$ (Bennett et al. 2007). Microlensing is most sensitive to planets with projected separations near the Einstein ring of the host star, which is $\sim 3.5 (M/M_\odot)^{1/2}$ and thus a factor of ~ 3 times the snow line for typical parameters (Gould et al. 2010). Thus microlensing discoveries are complementary to those of close-in planets detected via RV or transit searches.

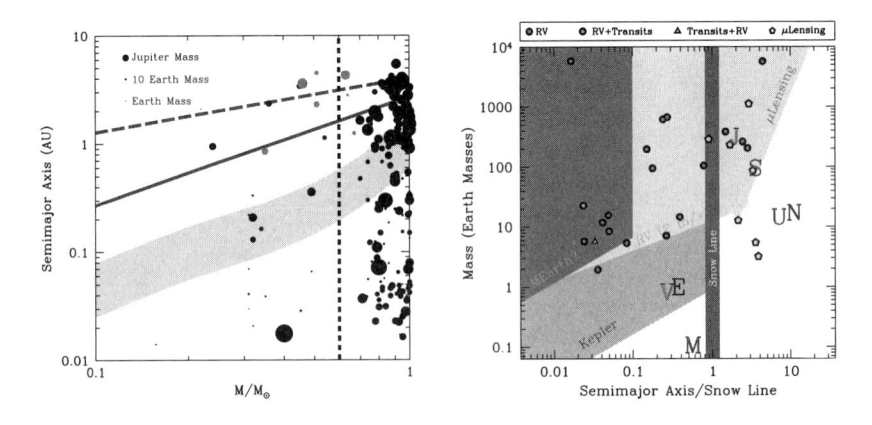

Figure 4. Left Panel: Semimajor axis versus host star mass for planets discovered by radial velocity or transits (black), and planets discovered by microlensing (red). The size of the points are proportional to the mass of the planets to the 1/3 power. The vertical dashed line shows the approximate mass limit for M dwarfs. The cyan shaded region shows the "conservative" habitable zone according to Selsis et al. (2007). The blue solid line is the location of the snow line, assuming a scaling of $a_{SL} = 2.7$ AU$(M/M_\odot)$. The dashed purple line is the typical Einstein ring radius for microlens host stars. Right Panel: Planet mass versus semimajor axis in units of the snow line for the 27 planets orbiting stars with $M < 0.6\ M_\odot$ known at the time of the talk. Also shown are the location of the solar system planets, as well as approximate regions of sensitivity for various transit, radial velocity, and microlensing surveys.

At the time of the talk 27 planets were known orbiting M-dwarf stars: 19 of them were discovered via the RV technique, 1 via transit, and 7 via microlensing. One system identified via microlensing is particularly interesting since it appears to be a scaled-down version of the Solar System (Gaudi et al. 2008; Bennett et al. 2010). Several of the planets discovered so far present extreme planet/star mass ratios, examples are HD41004b (Zucker et al. 2003) and the GJ876 system (Marcy et al. 2001), which may be challenging to explain via the core-accretion theory (Laughlin et al. 2004). RV searches point to a paucity of close-in (distances < a few AU) giant planets around M dwarfs (Johnson et al. 2007) and to a possible correlation between the presence of giant planets and the metallicity of the central star (Johnson & Apps 2009b; Schlaufman & Laughlin 2010). On the contrary, microlensing shows that giant planets are relatively common around M dwarfs at radial distances of a few times the snow line (Gould et al. 2010). These results together might indicate that giant planets form in M-dwarf disks but do not migrate inward. Additional possible trends discussed in the literature include the more likely presence of low-mass planets around low-mass stars (Bonfils et al. 2007) and a higher fraction of low-mass planets in multiple systems (Mayor et al. 2009).

4. Summary

Because of their proximity and numbers M dwarf-stars are attractive targets to search for Earth-like planets with near-future instruments. Surveys to detect giant- and down to Super-Earth-size planets are well under way and have already identified interesting planetary systems around M dwarfs. While some of them seem to be scaled down versions of our Solar System, others are very different in planet/star mass ratio and challenge our current understanding of how planets form from the circumstellar gas and dust around young M dwarfs. It is clear that a complete understanding of planet formation does require inputs both from the demographics of planets as well as from the properties and evolution of the protoplanetary disks within which planets form. The evolution of protoplanetary disk masses as a function of central star mass is one of most important input parameters to planet formation theories. In addition, properly accounting for the mechanisms dispersing protoplanetary disks is necessary to understand both when and in what environment planets form as well as whether they can migrate after their formation.

References

Adams, F. C., Hollenbach, D., Laughlin, G., & Gorti, U. 2004, ApJ, 611, 360
Alexander, R. D., Clarke, C. J., & Pringle, J. E. 2006, MNRAS, 369, 216
Apai, D., Pascucci, I., Bouwman, J., Natta, A., Henning, T., & Dullemond, C. P. 2005, Science, 310, 834
Bean, J. L., Seifahrt, A., Hartman, H., Nilsson, H., Wiedemann, G., Reiners, A., Dreizler, S., & Henry, T. J. 2010, ApJ, 713, 410
Bennett, D. P., Anderson, J., & Gaudi, B. S. 2007, ApJ, 660, 781
Bennett, D. P., Rhie, S. H., Nikolaev, S., & et al. 2010, ApJ, 713, 837
Bonfils, X., Mayor, M., Delfosse, X., Forveille, T., Gillon, M., Perrier, C., Udry, S., Bouchy, F., Lovis, C., Pepe, F., Queloz, D., Santos, N. C., & Bertaux, J. 2007, A&A, 474, 293. 0704.0270
Carpenter, J. M., Mamajek, E. E., Hillenbrand, L. A., & Meyer, M. R. 2006, ApJ, 651, L49
Charbonneau, D., et al. 2009, Nat, 462, 891
Clarke, C. J., Gendrin, A., & Sotomayor, M. 2001, MNRAS, 328, 485
Dullemond, C. P., Hollenbach, D., Kamp, I., & D'Alessio, P. 2007, Protostars and Planets V, 555

Ercolano, B., Clarke, C. J., & Drake, J. J. 2009a, ApJ, 699, 1639
Ercolano, B., Clarke, C. J., & Hall, A. C. 2010, MNRAS, 1505
Ercolano, B., Drake, J. J., & Clarke, C. J. 2009b, A&A, 496, 725
Ercolano, B., Drake, J. J., Raymond, J. C., & Clarke, C. C. 2008, ApJ, 688, 398
Fischer, D. A., & Valenti, J. 2005, ApJ, 622, 1102
France, K., Linsky, J. L., Brown, A., Froning, C. S., & Béland, S. 2010, ApJ, 715, 596
Furlan, E., et al. 2005, ApJ, 621, L129
Gaudi, B. S., Bennett, D. P., Udalski, A., & et al. 2008, Science, 319, 927
Gorti, U., Dullemond, C. P., & Hollenbach, D. 2009, ApJ, 705, 1237
Gorti, U., & Hollenbach, D. 2009, ApJ, 690, 1539
Gould, A., Dong, S., Gaudi, B. S., & et al. 2010, ApJ, 720, 1073
Gould, A., Pepper, J., & DePoy, D. L. 2003, ApJ, 594, 533
Herczeg, G. J., Cruz, K. L., & Hillenbrand, L. A. 2009, ApJ, 696, 1589
Herczeg, G. J., & Hillenbrand, L. A. 2008, ApJ, 681, 594
Hollenbach, D. J., Yorke, H. W., & Johnstone, D. 2000, Protostars and Planets IV, 401
Johnson, J. A., Aller, K. M., Howard, A. W., & Crepp, J. R. 2010, PASP, 122, 905
Johnson, J. A., & Apps, K. 2009a, ApJ, 699, 933. 0904.3092
— 2009b, ApJ, 699, 933. 0904.3092
Johnson, J. A., Butler, R. P., Marcy, G. W., Fischer, D. A., Vogt, S. S., Wright, J. T., & Peek, K. M. G. 2007, ApJ, 670, 833
Kessler-Silacci, J., et al. 2006, ApJ, 639, 275
Kessler-Silacci, J. E., Dullemond, C. P., Augereau, J., Merín, B., Geers, V. C., van Dishoeck, E. F., Evans, N. J., II, Blake, G. A., & Brown, J. 2007, ApJ, 659, 680
Laughlin, G., Bodenheimer, P., & Adams, F. C. 2004, ApJ, 612, L73. arXiv:astro-ph/0407309
Looper, D. L., et al. 2010, ApJ, 714, 45
Luhman, K. L. 2009, in American Institute of Physics Conference Series, edited by E. Stempels, vol. 1094 of American Institute of Physics Conference Series, 55
Luhman, K. L., Allen, P. R., Espaillat, C., Hartmann, L., & Calvet, N. 2010, ApJS, 186, 111
Luhman, K. L., Joergens, V., Lada, C., Muzerolle, J., Pascucci, I., & White, R. 2007, Protostars and Planets V, 443
Marcy, G. W., Butler, R. P., Fischer, D., Vogt, S. S., Lissauer, J. J., & Rivera, E. J. 2001, ApJ, 556, 296
Mayor, M., et al. 2009, A&A, 507, 487
Mohanty, S., Jayawardhana, R., & Barrado y Navascués, D. 2003, ApJ, 593, L109
Mohanty, S., Jayawardhana, R., & Basri, G. 2005, ApJ, 626, 498
Montgomery, R., & Laughlin, G. 2009, Icarus, 202, 1
Morrow, A. L., et al. 2008, ApJ, 676, L143
Muzerolle, J., Hillenbrand, L., Calvet, N., Briceño, C., & Hartmann, L. 2003, ApJ, 592, 266
Muzerolle, J., et al. 2006, ApJ, 643, 1003
Natta, A., Testi, L., & Randich, S. 2006, A&A, 452, 245
Nutzman, P., & Charbonneau, D. 2008, PASP, 120, 317
Owen, J. E., Ercolano, B., Clarke, C. J., & Alexander, R. D. 2010, MNRAS, 401, 1415
Pascucci, I., Apai, D., Luhman, K., Henning, T., Bouwman, J., Meyer, M. R., Lahuis, F., & Natta, A. 2009, ApJ, 696, 143
Pascucci, I., & Sterzik, M. 2009, ApJ, 702, 724
Plavchan, P., Werner, M. W., Chen, C. H., Stapelfeldt, K. R., Su, K. Y. L., Stauffer, J. R., & Song, I. 2009, ApJ, 698, 1068
Pollack, J. B., Hubickyj, O., Bodenheimer, P., Lissauer, J. J., Podolak, M., & Greenzweig, Y. 1996, Icarus, 124, 62
Raymond, S. N., Scalo, J., & Meadows, V. S. 2007, ApJ, 669, 606
Riaz, B., Lodieu, N., & Gizis, J. E. 2009, ApJ, 705, 1173
Rivera, E. J., Laughlin, G., Butler, R. P., Vogt, S. S., Haghighipour, N., & Meschiari, S. 2010, ApJ, 719, 890
Schlaufman, K. C., & Laughlin, G. 2010, A&A, 519, A105+. 1006.2850

Scholz, A., Jayawardhana, R., Wood, K., Meeus, G., Stelzer, B., Walker, C., & O'Sullivan, M. 2007, ApJ, 660, 1517
Szűcs, L., Apai, D., Pascucci, I., & Dullemond, C. P. 2010, ApJ, 720, 1668
Vogt, S. S., Butler, R. P., Rivera, E. J., Haghighipour, N., Henry, G. W., & Williamson, M. H. 2010, ApJ, 723, 954
Whelan, E. T., Ray, T. P., Podio, L., Bacciotti, F., & Randich, S. 2009, ApJ, 706, 1054
Zucker, S., Mazeh, T., Santos, N. C., Udry, S., & Mayor, M. 2003, A&A, 404, 775

Juvenile Ultracool Dwarfs

Emily L. Rice,[1] Jacqueline K. Faherty,[1,2] Kelle Cruz,[1,3] Travis Barman,[4] Dagny Looper,[5] Lison Malo,[6] Eric E. Mamajek,[7] Stanimir Metchev,[2] and Evgenya L. Shkolnik[8]

[1]*Department of Astrophysics, American Museum of Natural History, 79th Street and Central Park West, New York, NY 10024, USA*

[2]*Department of Physics & Astronomy, State University of New York, Stony Brook, Stony Brook, NY 11794, USA*

[3]*Department of Physics & Astronomy, Hunter College, 695 Park Avenue, New York, NY 10065, USA*

[4]*Lowell Observatory, 1400 West Mars Hill Road, Flagstaff, AZ 86001, USA*

[5]*Institute for Astronomy, University of Hawaii at Manoa, 2680 Woodlawn Drive, Honolulu, Hawaii 96822, USA*

[6]*Département de Physique and Observatoire du Mont-Mégantic, Université de Montréal (Québec), H3C 3J7, Canada*

[7]*Department of Physics & Astronomy, University of Rochester, 500 Wilson Boulevard, Rochester, NY 14627, USA*

[8]*Department of Terrestrial Magnetism, Carnegie Institute of Washington, 5241 Broad Branch Road, NW, Washington, DC 20015, USA*

Abstract. Juvenile ultracool dwarfs are late spectral type objects (later than ~M6) with ages between 10 Myr and several 100 Myr. Their age-related properties lie intermediate between very low mass objects in nearby star-forming regions (ages 1–5 Myr) and field stars and brown dwarfs that are members of the disk population (ages 1–5 Gyr). Kinematic associations of nearby young stars with ages from ~10–100 Myr provide sources for juvenile ultracool dwarfs. The lowest mass confirmed members of these groups are late-M dwarfs. Several apparently young L dwarfs and a few T dwarfs are known, but they have not been kinematically associated with any groups. Normalizing the field IMF to the high mass population of these groups suggests that more low mass (mainly late-M and possibly L dwarf) members have yet to be found. The lowest mass members of these groups, along with low mass companions to known young stars, provide benchmark objects with which spectroscopic age indicators for juvenile ultracool dwarfs can be calibrated and evaluated. In this proceeding, we summarize currently used methods for identifying juvenile ultracool dwarfs and discuss the appropriateness and reliability of the most commonly used age indicators.

1. Introduction

Juvenile ultracool dwarfs are very low mass stellar and substellar objects. *Juvenile* refers to objects with intermediate ages (10 Myr < age $\lesssim$ 600 Myr). They lack substantial ongoing accretion and primordial circumstellar material, but they still exhibit some signatures of youth that are not seen in typical field objects. *Ultracool* dwarfs have a spectral type of ~M6 or later. Juvenile ultracool dwarfs are typically identified by the combination of a late spectral type with one or more youth indicators: activity signatures, low gravity spectral features, membership in a young cluster or nearby moving group, and/or companionship to a known young star. A small but significant population of these objects are currently known, including such benchmark objects as 2MASS J1207−39, a member of the ~10 Myr TW Hydrae moving group and the host of a planetary-mass companion.

The properties of juvenile ultracool dwarfs play a role in many aspects of star formation and stellar evolution. A complete census of the low mass population of young, nearby moving groups is essential for understanding how the initial mass function varies across stellar environments. Characterization of the physical and circumstellar properties of juvenile ultracool dwarfs is crucial for complete understanding of any evolutionary phenomenon with a mass or age dependence, for example: planet formation, disk dissipation, angular momentum evolution, companion frequency, and chromospheric activity. Benchmark juvenile ultracool dwarfs (i.e. objects with well-characterized kinematic and physical properties) will essentially provide calibration data for evolutionary models. Finally, juvenile ultracool dwarfs provide excellent targets for exoplanet searches because they are nearby and young, thus potentially hosting self-luminous giant planets that provide a favorable contrast ratio and angular separation for direct imaging instruments (Beichman et al. 2010; Kataria & Simon 2010).

The specific questions about juvenile ultracool dwarfs addressed in the splinter session were:

1. What is the most efficient and accurate method for identifying juvenile ultracool dwarfs and associating them with young nearby moving groups?

2. What properties/features are reliable age indicators for late-M, L, and T spectral types?

3. How do juvenile ultracool dwarfs fit in with our current understanding of star formation, e.g., mass function, number density, multiplicity, disk fraction, etc.?

The first question is addressed in Section 2 and the second in Section 3. Question 3 is not explicitly discussed in this proceeding. As a result of the splinter session it became clear that a more complete answer to 1 and 2 will further our understanding of point 3. Section 4 discusses important caveats for identifying young moving groups, evaluating membership, and using membership as an age indicator.

2. Finding Low Mass Members

The concept of moving groups emerged in the late 19th century when Proctor (1869) and Huggins (1871) noted that five of the A stars in the Ursa Major constellation were moving toward a common convergence point. Since that time, kinematic and activity-based studies have uncovered several other co-moving associations (e.g., Figure 1, Eggen 1965, 1958, Zuckerman & Song 2004). The most studied of these to date include TW Hydrae, Tucana-Horologium, β Pictoris, AB Doradus, and η Chamaeleontis, which are all nearby ($\lesssim$100 pc) and span ages from ~10–100 Myr (see Zuckerman & Song 2004, Torres et al. 2008). Moving groups are older and more dispersed than star-forming regions with members widely spread-out on the sky. However, their proximity also makes them convenient laboratories for studying juvenile ultracool dwarfs because more distant ultracool dwarfs are too faint for detailed observations.

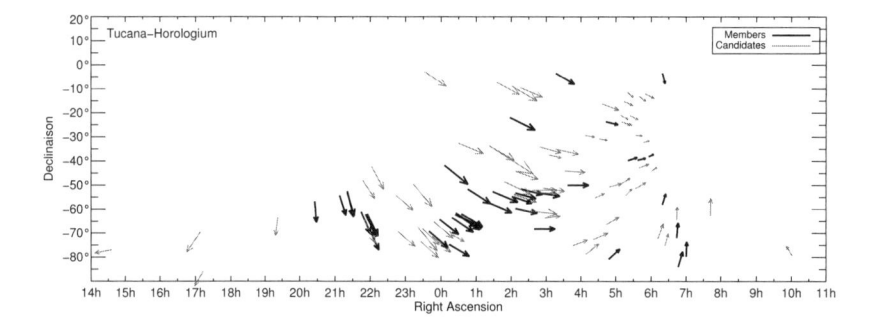

Figure 1. Position on the sky and vector of proper motion for the known Tucana-Horologium members (black arrows) listed in Zuckerman & Song (2004) and new candidates (red arrows) from Malo et al. (in prep.). The size of the arrows is proportional to the proper motion amplitude. While distributed over a large fraction of the celestial sphere, all members follow a coherent and distinctive movement. A color version of this figure is available in the online edition.

Observational studies to discover new low-mass members are motivated by the apparent lack of M dwarfs in moving groups relative to the field initial mass function (Torres et al. 2006, 2008). However identifying and confirming low mass members can be difficult as these associations are sparse and widely dispersed on the sky. Age-indicative characteristics such as strong X-ray and Hα activity, lithium absorption, and low surface gravity have been used as criteria for establishing youth among field objects (e.g. McGovern et al. 2004, Kirkpatrick et al. 2008, West et al. 2008, Cruz et al. 2009). Once proper motion, radial velocity, and distance are known, complete UVW space motion and XYZ positions can be used to robustly establish membership (e.g., Figure 2). However, parallaxes are time-consuming measurements rarely available for ultracool field objects; therefore, kinematic membership is often established without independent distance measurements. The high-resolution spectroscopy required to measure radial velocities and unambiguous youth indicators (Hα 10% width, lithium absorption, alkali line equivalent widths, etc., see Section 3.1) are also time consuming for ultracool dwarfs. Caveats about evaluating the membership of objects with incomplete kinematic and spectral characterization are discussed in Section 4.

Proper motion is available through numerous astrometric catalogs (e.g. USNO, LSPM-N, Hipparcos, Tycho, UCAC, etc.), but radial velocity measurements require high-resolution spectroscopy. Therefore, a number of studies have combined proper motion with near-IR or optical colors to search color-magnitude diagrams for new low

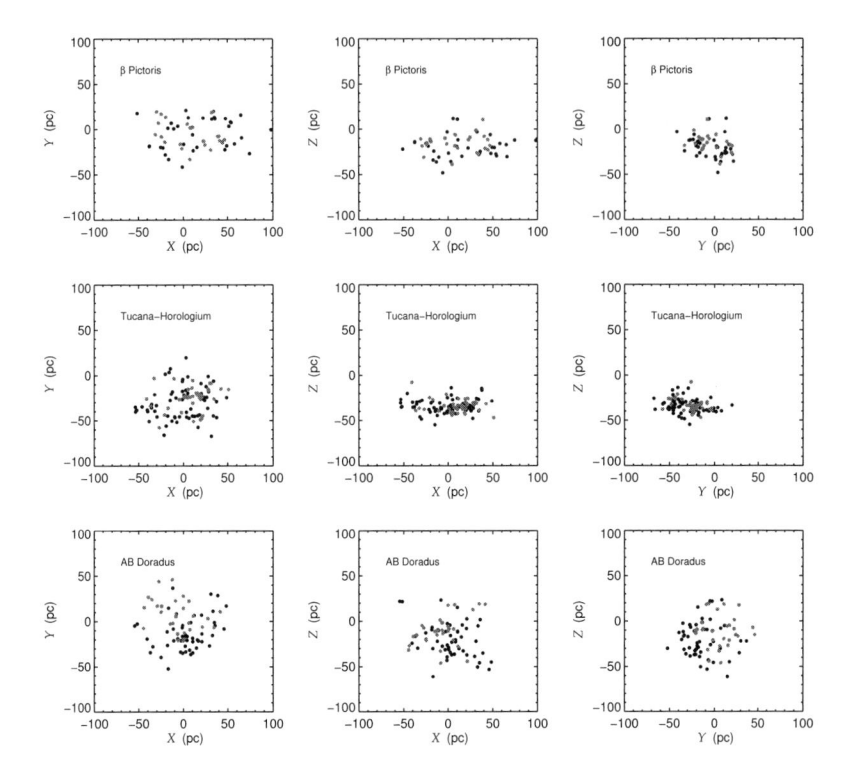

Figure 2. Galactic position (XYZ) of known members (black) and new candidates (red) for three nearby young moving groups (Malo et al. in prep.). A color version of this figure is available in the online edition.

mass members (e.g., Montes et al. 2001, Gizis 2002, Ribas 2003, Bannister & Jameson 2005, Clarke et al. 2010). Follow-up spectroscopic observations to measure radial velocities and confirm kinematic association are only performed for high probability candidate members. In this manner, Lépine & Simon (2009) and Schlieder et al. (2010) identify new members of β Pictoris and AB Doradus, Rice et al. (2010b) identify the lowest mass free-floating member of β Pictoris, and Malo et al. (in prep.) identify candidate members in Tucana-Horologium and employ a Bayesian model to evaluate the membership probability and the most probable distance based on measured properties (Figures 1 and 2). TW Hydrae, with an age estimate as young as 8 Myr, is on the younger end of "juvenile", and some members still show evidence of accretion, although there is no associated molecular cloud (Tachihara et al. 2009), and some members have debris disks. Looper et al. (in prep.) identify new low mass members of the TW Hydrae association (Figure 3) including TWA 30A and B, a low-mass co-moving system exhibiting signatures of an accretion disk and jet (Looper et al. 2010a,b).

Juvenile ultracool dwarfs that are confirmed members of young groups are particularly important because their age is constrained by higher mass stars and properties of the group as a whole. Therefore their observed activity and spectroscopic properties can be used to calibrate models and constrain the ages of individual objects that lack a kinematic association.

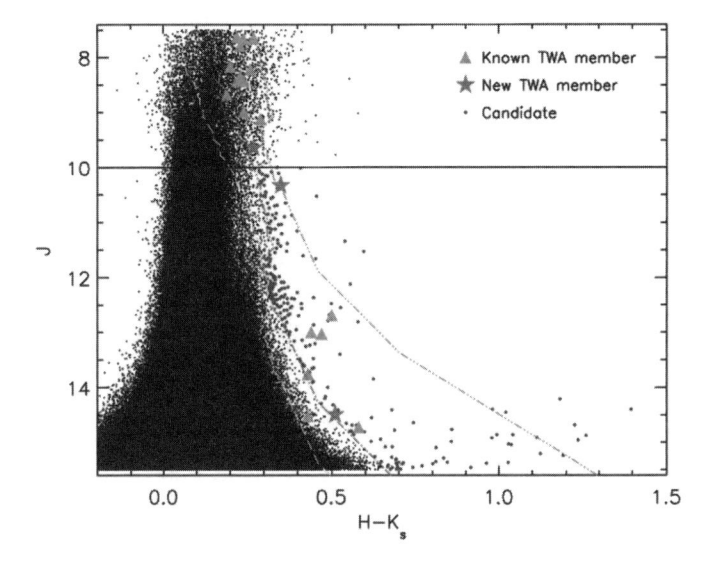

Figure 3. Near-infrared color-magnitude diagram for known (green triangles), new (red stars), and candidate (blue filled circles) TW Hydrae members with isochronal tracks of Baraffe & et al. (1998) combined with Chabrier & et al. (2000) of 10 Myr at 10 pc (rightmost/orange dash-dotted line), 60 pc (middle/red dash-dotted line) and 100 pc (leftmost/yellow dashed line). Small black dots show the >800,000 targets after spatial and J-band magnitude selection. Figure from Looper et al. 2011, in preparation. A color version of this figure is available in the online edition.

3. Evaluating Spectral Age Indicators

3.1. M dwarfs

There are several established age indicators for early-to mid-M dwarfs that can be applied to objects that are not (yet) kinematically associated with young moving groups. Figure 4 summarizes upper limits on age as a function of mass provided by four diagnostic properties: UV and X-ray emission, low surface gravity, lithium depletion, and accretion (as indicated by Hα emission).

X-ray and UV emission are related to magnetic activity, which can provide an upper age limit for early M dwarfs because magnetic activity is expected to decrease with age as angular momentum is dissipated over time (Preibisch & Feigelson 2005). However, for later spectral types ($\geq$M4) the activity lifetime is several Gyr, which is likely a consequence of the objects being fully convective and having a different mechanism for generating magnetic fields (West et al. 2008). Nevertheless, activity evidenced by UV and/or X-ray emission has been successfully used to identify candidate members of nearby young moving groups (e.g., Shkolnik et al. 2009; Schlieder et al. 2010). The use of UV emission as an age diagnostic is less established than X-ray emission, but the

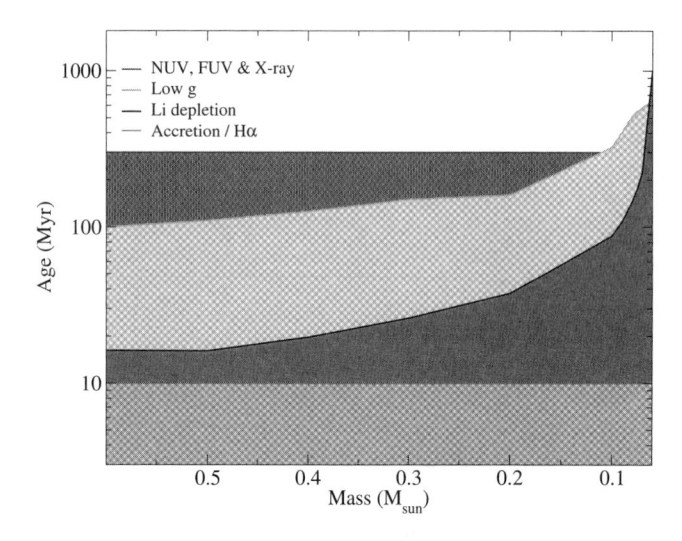

Figure 4. A summary of age diagnostics used in Shkolnik et al. (2010). Each technique provides an upper limit. The limits set by low gravity are from evolutionary models of Baraffe & et al. (1998) and lithium depletion from models of Chabrier et al. (1996). Barrado y Navascués & Martín (2003) set an upper limit of 10 Myr for a star still undergoing accretion. A color version of this figure is available in the online edition.

sensitivity and sky coverage of the GALEX satellite compared to X-ray missions like *Chandra* and *ROSAT* enable promising early results (Shkolnik et al. 2010; Rodriguez et al. 2010). Shkolnik et al. (2010) discovered two new mid-M dwarf members of TW Hydrae, TWA 31 and 32, using GALEX NUV and FUV emission.

Lithium abundance in low mass objects is a strong function of age, but the constraint it provides varies with mass. Lithium depletion models provide an age diagnostic that can be applied to individual objects via spectroscopic detection of lithium as well as to entire clusters via the determination of the lithium depletion boundary (e.g., Mentuch et al. 2008; Yee & Jensen 2010, and references therein). A core temperature of 2.5×10^6 K is required to burn lithium; therefore, objects with M < 0.06 $M_\odot$ will never deplete their lithium. Thus for the lowest mass objects, lithium becomes a diagnostic of mass rather than age. Even for stars with M > 0.06 $M_\odot$, the age determined by comparing measured lithium abundances to lithium depletion models is often inconsistent with the age determined from the H-R diagram (Figure 5). A possible explanation for this discrepancy is found in Baraffe & Chabrier (2010), who show that episodic accretion can temporarily increase core temperatures enough to burn lithium more efficiently. This results in prematurely depleted lithium compared to models that do not incorporate the effects of episodic accretion. However, the discrepancy between lithium age and H-R diagram age (Figure 5) shows some mass dependence (later spectral types are typically more depleted in lithium than their H-R diagram ages imply), suggesting that there might still be a mass-dependent systematic uncertainty in the models (E. Jensen, priv. comm., 2010).

Several spectral features of ultracool dwarfs are gravity sensitive, including alkali lines (e.g., Na, K, e.g., Gorlova et al. 2003), metal hydride bands (e.g., CaH, CrH,

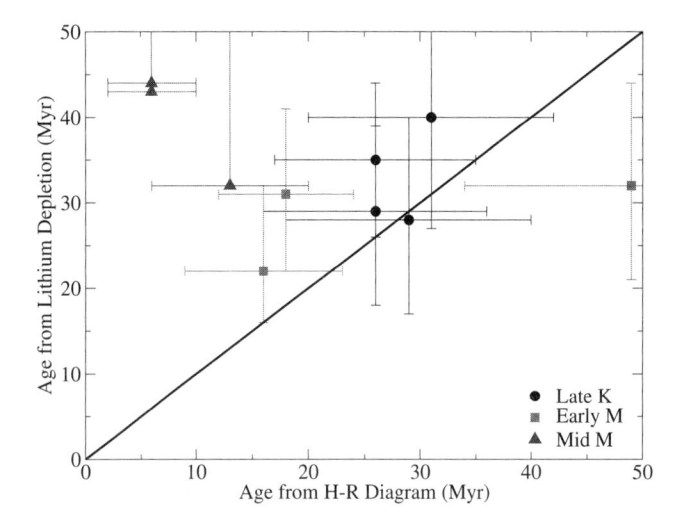

Figure 5. Data from Yee & Jensen (2010) comparing the ages of 10 late-K to mid-M dwarfs derived from lithium depletion model versus those derived from H-R diagram (Baraffe & et al. 1998). For most objects the age from the lithium depletion models are substantially higher than the age inferred from the H-R diagram, lending support to the theory that increased frequency of episodic accretion can increase the efficiency of lithium burning, resulting in young objects having depleted their lithium earlier than would be expected from lithium depletion models that do not take episodic accretion history into account (Baraffe & Chabrier 2010). A color version of this figure is available in the online edition.

FeH, e.g., Shkolnik et al. 2009), and metal oxide bands (e.g., VO, TiO, Kirkpatrick et al. 2008). CaH in particular is used as a gravity indicator in M dwarfs, but weak CaH bands can be a result of high metallicity as well as low surface gravity. Many gravity-sensitive features are sensitive to temperature and/or metallicity so they must be interpreted with caution. Gravity-sensitive spectral features are discussed in more detail for L and T dwarfs below.

The strictest age constraint is obtained by detecting Hα emission produced by ongoing accretion, providing an upper limit of 10 Myr (Barrado y Navascués & Martín 2003). Hα emission can be reliably attributed to ongoing accretion (as opposed to chromospheric activity) if the width of the emission at 10% the maximum strength is ≥ 200 km s^{-1} (White & Basri 2003). Weak, narrow Hα emission will persist for billions of years in most M dwarfs as a result of chromospheric activity.

The spectroscopic and activity-based age indicators described above are very useful, but in and of themselves they are not failsafe methods of inferring the age of individual very low mass stars. The interpretation of many age indicators also depends on temperature, metallicity, and possibly more ambiguous properties like accretion history. Therefore it is necessary to approach age indicators with caution and to realistically assess the degeneracies and systematic uncertainties inherent in inferring the age of a very low mass star via spectral age indicators.

488   Rice et al.

3.2. L and T dwarfs

Age indicators are even more ambiguous and uncertain for L and T dwarfs, but important advances have been made in the past several years. Estimating the ages of substellar objects is complicated by their long cooling time and lack of a main sequence, which provides age-independent constraints on mass, luminosity, and effective temperature for hydrogen-burning stars. Brown dwarfs with $M < 0.06\ M_\odot$ will never reach temperatures high enough to burn lithium so the detection of lithium constrains mass instead of age for these objects. Furthermore, their cool, complex atmospheres include significant opacity from molecules and dust, inhomogeneous cloud structure, and non-equilibrium chemistry, further muddling the interpretation of their spectra and any potentially gravity-sensitive features.

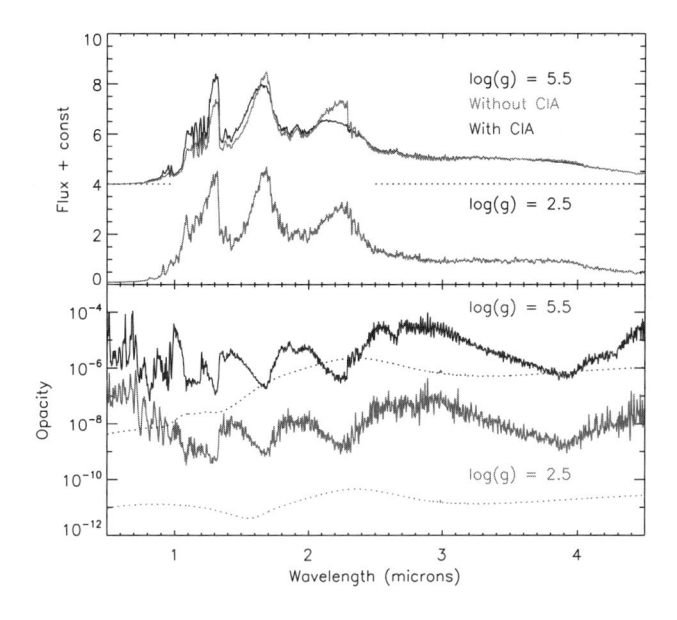

Figure 6. Top: synthetic spectra calculated with the PHOENIX model atmosphere code at two surface gravities, including and removing CIA from H_2 for the higher gravity. Bottom: opacity of H_2 CIA (dotted) and H_2O absorption (solid) at two surface gravities (Barman et al., in prep., after Borysow et al. 1997). A color version of this figure is available in the online edition.

Substellar objects cool and shrink over their entire lifetimes, and gravity is the parameter that changes most with time (e.g., Baraffe & et al. 2003). Gravity and effective temperature uniquely determine age and mass, unlike luminosity which is degenerate with mass and age. There are several spectral features that are gravity-sensitive, but they are also typically dependent on temperature and/or metallicity, if not higher order parameters like dust, clouds, and chemistry. Nonetheless several gravity-sensitive spectral features are routinely used to identify young, low mass objects. The broadest feature is a peaked H-band spectral morphology, first observed by Lucas et al. (2001) in spectra of substellar objects in Orion and later by Luhman et al. (2005) for objects in IC 348 and by Allers et al. (2007) for objects in Chamaeleon II and Ophiuchus. Figure 6 shows the underlying physical explanation for the peaked *H*-band morphology. The feature is prominent for low surface gravity objects because the H$_2$O opacity dominates over collisionally-induced absorption (CIA) from H$_2$. For higher gravity objects, the opacity of the H$_2$ CIA is larger than the opacity from H$_2$O in the *H*- and *K*-bands, effectively flattening the peaks.

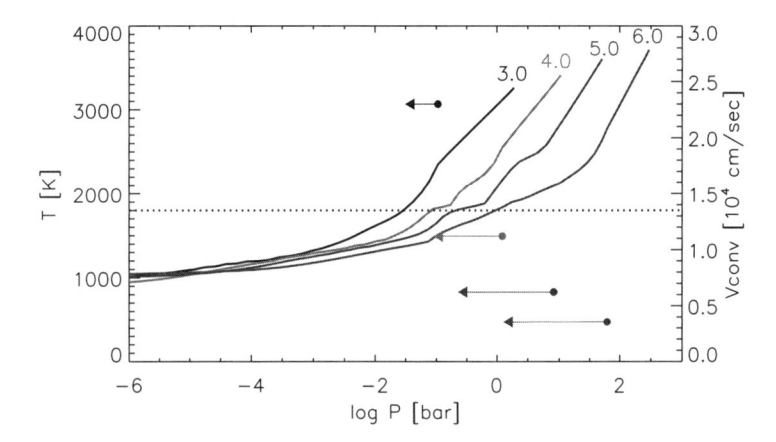

Figure 7. Temperature-pressure atmosphere profiles (solid lines, use left y-axis) for T$_{eff}$=1800 K and four values of log(g): 3 (black), 4 (red) , 5 (blue), 6 (green), calculated with the PHOENIX code. The photosphere for each atmosphere is approximately where the solid line intersects the dotted line. The filled circles locate the radiative-convective boundary in pressure (x-axis), and the right vertical axis shows the maximum velocity just below this boundary in the convection zone. The length of the arrow shows the relative proximity of the radiative-convective boundary to the photosphere in pressure space. For lower surface gravities, the separation is smaller and the maximum convective velocity is higher. Thus, more efficient vertical mixing is expected at lower surface gravities, perhaps resulting in stronger non-equilibrium chemistry and thicker clouds (Barman et al., in prep.). A color version of this figure is available in the online edition.

At moderate spectral resolutions ($R \gtrsim 1000$), the strengths of alkali lines like Na I and K I have been shown to be sensitive to surface gravity, but they are also sensitive to temperature, resulting in a strong degeneracy that is evident at high resolution (Zapatero Osorio et al. 2004; Rice et al. 2010a). Molecular features like CrH and VO have also been shown to be gravity-sensitive (McGovern et al. 2004; Kirkpatrick et al. 2008; Cruz et al. 2009).

Two further properties of L and T dwarfs possibly related to youth are: red near-infrared colors and underluminosity. Red colors are expected to be linked to dust-enhanced atmospheres resulting from low surface gravity. Many unusually red (for their spectral type) objects found in the field show multiple signatures of youth (e.g., Cruz et al. 2009). There are several objects with red colors lacking youth signatures (Kirkpatrick et al. 2010), and high metallicity could also producer redder spectra (Burrows et al. 2006; Looper et al. 2008). While overluminosity on a color-magnitude diagram is a hallmark for youth in low mass stars (Luhman & et al. 2007), young L and T dwarfs appear *under*luminous (Metchev & Hillenbrand 2006). Moreover, from a parallax survey of eight low surface gravity L dwarfs, Faherty et al. (in prep.) determine that these objects are ~1 magnitude underluminous on a brown dwarf near-IR H-R diagram.

Disentangling low gravity and other spectroscopic youth indicators becomes even more problematic at and beyond the L-T transition. It is becoming apparent that low gravity objects with effective temperatures comparable to known T dwarfs (e.g., the young planetary-mass object 2MASS J1207−39b) have L dwarf spectral types because they lack CH_4 absorption. However, the atmosphere is probably lacking CH_4 not because the atmosphere is too hot but because low gravity strengthens the effects of vertical mixing (Figure 7). This issue is particularly important for wide, self-luminous extrasolar planets for which low resolution near-infrared spectra can now be obtained, like HR 8799b (Bowler et al. 2010).

4. Caveats for Young Groups and their Members

Because of the difficulty in assigning ages to isolated field objects (E. E. Mamajek, D. R. Soderblom, & R. F. G. Wyse 2009; Soderblom 2010), young stellar groups play an important role in studies of age-dependent phenomena. Groups are observed (or assumed) to be approximately coeval, and group membership is usually used as a *primary* indicator of age. However, membership must be very carefully evaluated when used as an age indicator.

Assigning membership to an object and adopting the group age should be done cautiously and with as much corroborating evidence as possible. Other youth indicators such as rotation, activity, lithium, low gravity, full three-dimensional kinematics (radial velocity and parallax in addition to proper motion) and common proper motion with another member should also be considered.

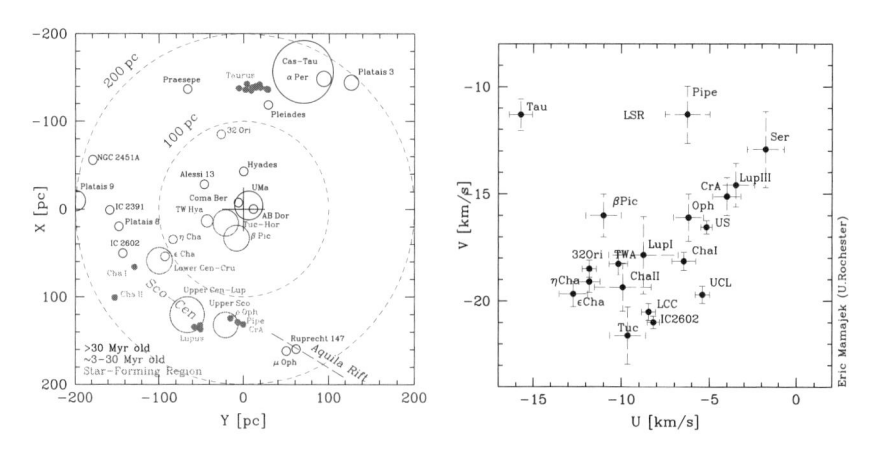

Figure 8. Left: XY distributions of star-forming regions and young groups within 200 pc of the Sun. Right: UV distributions of star-forming regions and young groups. The close proximity of many groups in XYZ and UV requires that membership be evaluated carefully, particularly when kinematic and distance measurements are incomplete (Mamajek 2011, submitted). A color version of this figure is available in the online edition.

In particular, caution is urged when assigning membership with just proper motions because stellar groups of different ages can have similar velocities. As shown in Figure 8, the UV distributions of the nearest, youngest groups are tightly clustered. However, the obvious nuclei of these groups all have velocity dispersions of only ~1 km s^{-1}, independent of density. In order to reliably assign membership, complete and accurate UVW velocities and XYZ coordinates are needed.

Furthermore, similar velocities are not sufficient to warrant the definition of a new group. Stars with consistent space motions could be a "supercluster" kinematic stream and not related to formation at all. Superclusters are now known to be dynamical streams in the Milky Way galaxy as a result of spiral density waves, and the common motion of constituent stars does not imply a common age (e.g., Famaey et al. 2005). Certain proposed young moving groups are unphysical – that is, do not share a common or age – because of large scatters in their H-R diagrams, radial velocities, distances, and/or peculiar velocities. An upcoming study of the revised Hipparcos astrometry for young stellar groups within 100 pc by Mamajek 2011 (submitted) shows that some candidate groups appear to be unphysical: Chereul 2, Chereul 3, Latyshev 2, and Polaris (Chereul et al. 1999; Latyshev 1977; Turner 2004).

Acknowledgments. The authors would like to thank the organizers of the Cool Stars 16 meeting, particularly Suzanne Hawley, the head of the SOC, for providing the opportunity to have this splinter session; Adam Burgasser, the SOC liaison, for helping us organize it; and the participants for engaging in a productive discussion.

References

Allers, K. N., et al. 2007, ApJ, 657, 511
Bannister, N. P., & Jameson, R. F. 2005, Astronomische Nachrichten, 326, 1020
Baraffe, I., & Chabrier, G. 2010, A&A, 521, A44+
Baraffe, I., & et al. 1998, A&A, 337, 403
— 2003, A&A, 402, 701
Barrado y Navascués, D., & Martín, E. L. 2003, AJ, 126, 2997. arXiv:astro-ph/0309284
Beichman, C. A., et al. 2010, PASP, 122, 162
Borysow, A., Jorgensen, U. G., & Zheng, C. 1997, A&A, 324, 185
Bowler, B. P., Liu, M. C., Dupuy, T. J., & Cushing, M. C. 2010, ApJ, 723, 850
Burrows, A., Sudarsky, D., & Hubeny, I. 2006, ApJ, 640, 1063. arXiv:astro-ph/0509066
Chabrier, G., Baraffe, I., & Plez, B. 1996, ApJ, 459, L91+
Chabrier, G., & et al. 2000, ApJ, 542, 464
Chereul, E., Crézé, M., & Bienaymé, O. 1999, A&AS, 135, 5
Clarke, J. R. A., et al. 2010, MNRAS, 402, 575
Cruz, K. L., Kirkpatrick, J. D., & Burgasser, A. J. 2009, AJ, 137, 3345
E. E. Mamajek, D. R. Soderblom, & R. F. G. Wyse (ed.) 2009, The Ages of Stars, vol. 258 of IAU Symposium
Eggen, O. J. 1958, MNRAS, 118, 65
— 1965, in Galactic Structure, edited by A. Blaauw & M. Schmidt, 111
Famaey, B., Jorissen, A., Luri, X., Mayor, M., Udry, S., Dejonghe, H., & Turon, C. 2005, A&A, 430, 165
Gizis, J. E. 2002, ApJ, 575, 484
Gorlova, N. I., Meyer, M. R., Rieke, G. H., & Liebert, J. 2003, ApJ, 593, 1074
Huggins, W. 1871, Royal Society of London Proceedings Series I, 20, 379
Kataria, T., & Simon, M. 2010, AJ, 140, 206
Kirkpatrick, J. D., Looper, D. L., Burgasser, A. J., Schurr, S. D., Cutri, R. M., Cushing, M. C., Cruz, K. L., Sweet, A. C., Knapp, G. R., Barman, T. S., Bochanski, J. J., Roellig, T. L.,

McLean, I. S., McGovern, M. R., & Rice, E. L. 2010, ApJS, 190, 100. 1008.3591
Kirkpatrick, J. D., et al. 2008, ApJ, 689, 1295
Latyshev, I. N. 1977, Astronomicheskij Tsirkulyar, 969, 7
Lépine, S., & Simon, M. 2009, AJ, 137, 3632
Looper, D. L., Bochanski, J. J., Burgasser, A. J., Mohanty, S., Mamajek, E. E., Faherty, J. K., West, A. A., & Pitts, M. A. 2010a, AJ, 140, 1486
Looper, D. L., et al. 2008, ApJ, 686, 528
— 2010b, ApJ, 714, 45
Lucas, P. W., Roche, P. F., Allard, F., & Hauschildt, P. H. 2001, MNRAS, 326, 695
Luhman, K. L., & et al. 2007, ApJ, 654, 570
Luhman, K. L., Lada, E. A., Muench, A. A., & Elston, R. J. 2005, ApJ, 618, 810
McGovern, M. R., Kirkpatrick, J. D., McLean, I. S., Burgasser, A. J., Prato, L., & Lowrance, P. J. 2004, ApJ, 600, 1020
Mentuch, E., Brandeker, A., van Kerkwijk, M. H., Jayawardhana, R., & Hauschildt, P. H. 2008, ApJ, 689, 1127
Metchev, S. A., & Hillenbrand, L. A. 2006, ApJ, 651, 1166
Montes, D., López-Santiago, J., Gálvez, M. C., Fernández-Figueroa, M. J., De Castro, E., & Cornide, M. 2001, MNRAS, 328, 45
Preibisch, T., & Feigelson, E. D. 2005, ApJS, 160, 390. arXiv:astro-ph/0506052
Proctor, R. A. 1869, Royal Society of London Proceedings Series I, 18, 169
Ribas, I. 2003, A&A, 400, 297
Rice, E. L., Barman, T., Mclean, I. S., Prato, L., & Kirkpatrick, J. D. 2010a, ApJS, 186, 63
Rice, E. L., Faherty, J. K., & Cruz, K. L. 2010b, ApJ, 715, L165
Rodriguez, D. R., Bessell, M. S., Zuckerman, B., & Kastner, J. H. 2010, ArXiv e-prints, astro-ph/1010.2493
Schlieder, J. E., Lépine, S., & Simon, M. 2010, AJ, 140, 119
Shkolnik, E., Liu, M. C., & Reid, I. N. 2009, ApJ, 699, 649
Shkolnik, E. L., Liu, M. C., Reid, I. N., Dupuy, T., & Weinberger, A. 2010, ArXiv e-prints, astro-ph/1011.2708
Soderblom, D. R. 2010, ARA&A, 48, 581
Tachihara, K., Neuhäuser, R., & Fukui, Y. 2009, PASJ, 61, 585
Torres, C. A. O., Quast, G. R., da Silva, L., de La Reza, R., Melo, C. H. F., & Sterzik, M. 2006, A&A, 460, 695
Torres, C. A. O., Quast, G. R., Melo, C. H. F., & Sterzik, M. F. 2008, Young Nearby Loose Associations, 757
Turner, D. G. 2004, in Bulletin of the American Astronomical Society, vol. 36 of Bulletin of the American Astronomical Society, 744
West, A. A., Hawley, S. L., Bochanski, J. J., Covey, K. R., Reid, I. N., Dhital, S., Hilton, E. J., & Masuda, M. 2008, AJ, 135, 785. 0712.1590
White, R. J., & Basri, G. 2003, ApJ, 582, 1109
Yee, J. C., & Jensen, E. L. N. 2010, ApJ, 711, 303
Zapatero Osorio, M. R., Lane, B. F., Pavlenko, Y., Martín, E. L., Britton, M., & Kulkarni, S. R. 2004, ApJ, 615, 958
Zuckerman, B., & Song, I. 2004, ARA&A, 42, 685

Frontiers in X-ray Astronomy - CS16 Splinter Session

Jan Robrade, P. Christian Schneider, and Katja Poppenhaeger

Hamburger Sternwarte, Gojenbergsweg 112, 21029 Hamburg, Germany
e-mail: jrobrade@hs.uni-hamburg.de

Abstract. The "Frontiers in X-ray astronomy" splinter session presented recent discoveries and new topics of the field of stellar X-ray astronomy. The focus was on three topics that go beyond X-ray emission from solar-type dynamo action. We covered the fields of X-ray emission at the edges of the "stellar magnetic activity strip," magnetic Ap stars and brown dwarfs, further high-energy phenomena in young stars and jets and multi-wavelength campaigns and finally the interaction of stars and planets. These science themes were chosen to highlight new X-ray phenomena and debated topics that are of interest for a broader community including non X-ray scientists.

1. Introduction

With now ten years having passed since the launch of the new generation X-ray telescopes *Chandra* and *XMM-Newton*, X-ray astronomy still continuously reveals surprising new discoveries and deepened insights in the world and life of stars. Its full diversity goes well beyond what can be discussed in a 'Cool Stars' splinter session, thus we focus on a few selected topics, specifically:

a. The hottest and coolest magnetically active stars as well as brown dwarfs
Fast rotating stars with a thin outer convective layer and fully convective stellar and substellar objects with very cool atmospheres are not simply scaled cases of solar-like activity. A wealth of new X-ray data has been gathered in the regime of borderline X-ray activity with implications reaching from magnetic field generation in the stellar interior up to the energy release in the outer atmospheric layers. In addition to cool stars results from X-ray observation of magnetic Ap stars as well as young and old brown dwarfs are discussed.

b. X-ray emission and jets from young stars
Deep X-ray imaging and spectroscopic studies have e.g. revealed the existence of stellar X-ray jets and allowed the study of outflow and accretion phenomena to an unprecedented detail. These are important for star-formation scenarios and link stars to the field of jet formation and X-ray generation in collimated outflows. Further, results from multi-wavelength observations of young stellar outbursting objects are presented.

c. X-rays, Stars & Planets
An up to date topic with a range of new science themes like star-planet interaction or planet evaporation. High energy emission influencing the evolution of planetary systems and the interaction stars with their planets, especially close-in hot ones, is a new and fast evolving field where more comprehensive data has become available.

2. The Hottest and Coolest Magnetically Active Stars

Speaker: J. Robrade, B. Stelzer

Jan Robrade presented results from a deep *XMM-Newton* observations of the A7 star Altair that obtained the first high-resolution X-ray spectrum of a late A-type star. Altair is detected as a rather faint, but long-term stable X-ray source with an X-ray luminosity of $L_X = 1.4 \times 10^{27}$ erg/s (Robrade & Schmitt 2009), confirming that X-ray emission is generated by magnetic activity in the regime of the 'hottest cool stars' up to spectral types mid/late-A ($T_{\text{eff}} \approx 8000$ K). Its X-ray activity level of $\log L_X/L_{\text{bol}} = -7.4$ is extremely low and the coronal plasma is rather cool with dominant temperatures of 1–4 MK. Coronal abundances show a solar-like FIP effect and Ne/O ratio; overall Altair's X-ray properties resemble those of the inactive Sun, despite very different underlying stars. A detailed study of the X-ray data from Altair calls for a modification of the standard picture, where magnetic activity is produced in a thin outer convective layer. Instead it indicates, that localized coronae exist in the region of the equatorial bulge of these fast rotating, non-spherical stars that exhibit strong gravity darkening (Monnier et al. 2007). The study of the OVII f/i ratio, that is sensitive to the stellar UV-field, shows that coronal structures on the stellar surface are located above the cooler parts of Altair's photosphere, i.e. equatorial or low-latitude regions. This finding is further supported by the presence of rotational modulation of the X-ray brightness and matches results obtained for its chromosphere and transition region. The X-ray detection of the A5 star HR 8799 with *Chandra* at $\log L_X/L_{\text{bol}} = -6.2$ is only apparently contradicting this picture (Robrade & Schmitt 2010). Actually, its full spectral type is kA5 hF0 mA5 v λ Boo and the A5 classification is based on metal lines, whereas from hydrogen lines a spectral type F0 was derived, indicating $T_{\text{eff}} \approx 7450$ K. Thus, from the X-ray point of view, the F0 classification is the 'right' one and stars like Altair are still the hottest stars that show significant magnetic activity and X-ray emission.

Also intermediate mass, main sequence stars may be intrinsic X-ray sources. Prime candidates are the so-called Ap/Bp stars, i.e. chemically peculiar stars that have large scale magnetic fields of up to many kG. Their X-ray emission is interpreted in the framework of the magnetically confined wind-shock (MCWS) model, originally developed for the A0p star IQ Aur (Babel & Montmerle 1997). Advanced variants of this model were applied to a variety of magnetic hot stars and it became the standard model to explain phenomena like X-ray overluminosity, hard spectral components, flares or rotational modulation. First results obtained from new X-ray observations of the prototypical A0p stars IQ Aur and α^2CVn clearly show, that the X-ray properties of these stars differ extremely. While IQ Aur is clearly detected with an quasi-quiescent X-ray luminosity of $\log L_X = 29.6$ erg/s, α^2CVn remained undetected at an upper limit of $\log L_X \lesssim 26$ erg/s. Additionally IQ Aur exhibited a strong flare as shown in the left panel of Fig. 1, likely the first X-ray flare ever observed from an A0p star. (Robrade & Schmitt in prep.) Possible interpretations in the framework of magnetically channeled stellar winds were discussed.

Further, JR presented results from X-ray observations of nearby, very low-mass (VLM) stars. The solar neighborhood population is a mixture of stellar and substellar objects at a various ages, mainly overlapping in spectral classification at late-M. The presence of lithium, modeling of evolutionary tracks or exploitation of binarity are common methods to separate the populations, but especially substellar objects that are close to the stellar mass-threshold and at a specific age are hard to distinguish from real stars. X-ray emission is again an important diagnostic of studying magnetic activity in

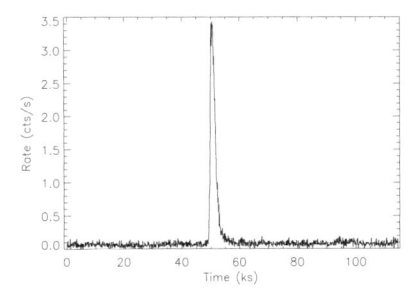

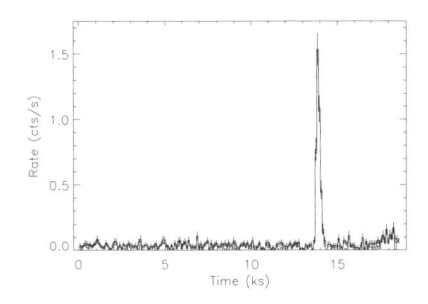

Figure 1. X-ray light curves of two very different stars. *Left:* the A0p star IQ Aur, where X-ray emission is thought to originate from magnetically confined wind-shocks. *Right:* the M8.5 dwarf SCR 1845-6357, a very low mass star with an presumed age of a few Gyr. Figure courtesy of J. Robrade.

fully convective stars with a virtually neutral photosphere. An increasing number of stars beyond spectral type M7 was detected over the last years in X-rays, specifically also in their quasi-quiescent state, allowing to study their X-ray activity outside strong flares. These observations show that stars down to the end of the main sequence with $M < 0.1 M_\odot$ and $T_{\rm eff} < 2500$ K, can be highly active X-ray emitter at $\log L_X/L_{\rm bol} = -3 \ldots -4$. While the highest activity level of VLM stars are comparable to those of more massive stars and similarly the average plasma temperature increases with activity, they also differ in certain aspects. The typical coronal temperatures of VLM stars are significantly lower compared to e.g. early M dwarfs of the same activity level. Further, strong X-ray flares are produced by VLM stars (see e.g. Fig. 1, right) but intermediate events are lacking, pointing to a different flare energy distribution. There are also indications that strong magnetic activity of VLM stars may persist longer than for solar-like stars or brown dwarfs. The highly active ($\log L_X/L_{\rm bol} = -3.8$) M8.5 star SCR 1845-6357 (Robrade et al. 2010) has an estimated age of a few Gyr, while the ~ 500 Myr old M9 brown dwarf LP 944-20 remained undetected ($\log L_X/L_{\rm bol} < -6.3$) in quasi-quiescence (Martín & Bouy 2002). These trends are deduced from a rather limited sample and merit further investigation; nevertheless the overall X-ray properties of very low-mass stars require that an efficient dynamo mechanism is operating in fully convective stars.

Beate Stelzer presented results from X-ray observations of brown dwarfs. In addition to stars, also brown dwarfs are able to generate X-ray emission and the first X-ray detection of a brown dwarf in Cha I was already made with ROSAT (Neuhäuser & Comeron 1998). Studies of star forming regions like Orion (Preibisch et al. 2005) or Taurus (Grosso et al. 2007) revealed that brown dwarfs are especially active at very young age up to a few Myr, when they are still rather bright and hot. However, star forming regions are typically quite far away and the number of X-ray detected young brown dwarfs is still small and it is debated if they are as active, expressed by $\log L_X/L_{\rm bol}$, as stars or not. Some very young brown dwarfs also show signs of accretion as deduced e.g. from Hα profiles, analog to the more massive T Tauri stars. However, in contrast to T Tauri stars the activity level of accreting and non-accreting brown dwarfs was found to be apparently quite similar. A remarkable individual object is the optically overluminous, young, accreting brown dwarf FU Tau A, that was recently detected in X-rays (Stelzer et al. 2010). It shows an X-ray spectrum with a strong soft component that is overall similar to the accretion dominated X-ray spectrum of TW Hya (see Fig. 2). FU Tau is a young brown dwarf binary and while FU Tau A (M7) is X-ray detected at $\log L_X = 29.7$ erg/s, the secondary FU Tau B (M9) remained undetected at an two magnitudes lower upper limit. A possible explanation would be a very steep decline of X-ray luminosity towards later spectral type.

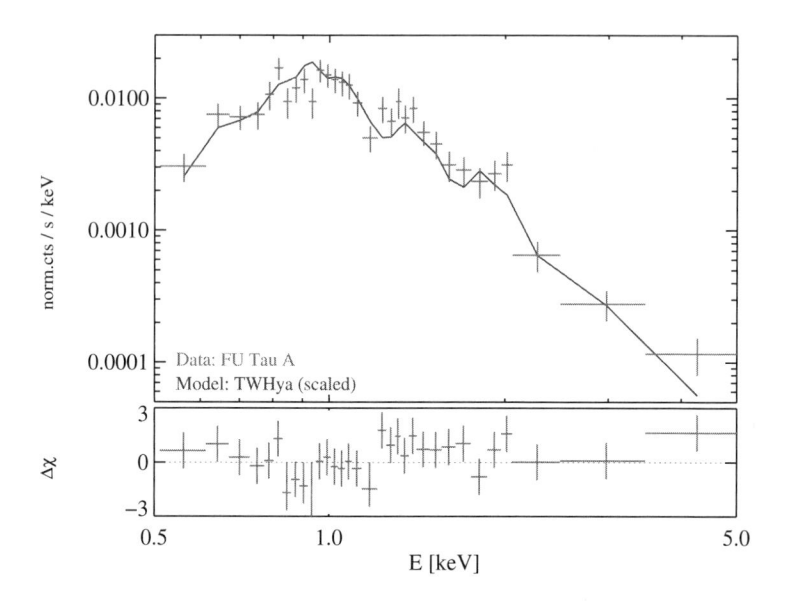

Figure 2. The X-ray spectrum of FU Tau A; overlayed is a scaled model of the X-ray spectrum from the CTTS TW Hya. Figure courtesy of B. Stelzer.

For brown dwarfs there is in general an interdependency between effective temperature, age and mass and outside well known stellar associations only in a few cases reasonable age estimates can be derived either by studying binaries or utilizing partial lithium depletion. It is instructive to compare only hotter or only high-mass objects and study the influence of the other parameter. These studies show that X-ray activity of brown dwarfs fades rather quickly with age and/or T_{eff} (Stelzer et al. 2006). Some higher-mass brown dwarfs exhibited X-ray emission at ages of up to a few hundred Myr, i.e. where they still reside around spectral type late M. Usually these objects were detected during larger flares, e.g. Gl 569 Bab (Stelzer 2004), LP 944-20 (Rutledge et al. 2000). Possible X-ray detections of older brown dwarfs in quiescence, including one early L dwarf binary (Kelu 1, Audard et al. 2007) are extremely rare and the few X-ray photons were often seen as flare afterglow. While the exact breakdown of X-ray emission in brown dwarfs is at the moment indeterminable, substellar objects older than about 0.5 Gyr are virtually X-ray dark with typical upper limits being at around $\log L_X \lesssim 24 \ldots 25$ erg/s.

Abandoning the often unknown mass as separating parameter between brown dwarfs and VLM stars, all objects at spectral type M7 and beyond are often denoted as ultracool dwarfs. Ultracool dwarfs roughly separate into two groups in X-rays. The first group shows X-ray flares, but usually no radio emission (see Fig. 3). On the other

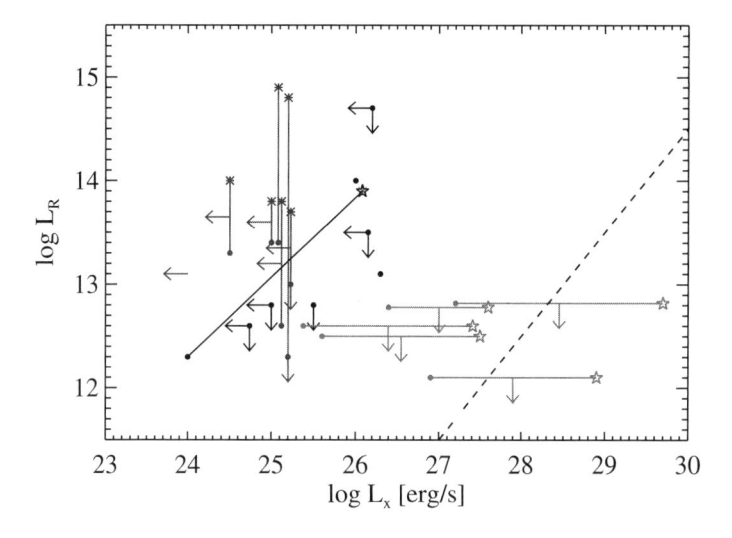

Figure 3. X-ray vs. Radio luminosity for radio bright ultracool dwarfs (blue) and X-ray flaring ultracool dwarfs (red); star symbols and asterisks denote flaring emission. Figure courtesy of B. Stelzer.

hand, many other ultracool dwarfs are strong radio emitters (see e.g. Berger 2002), but have no or extremely weak X-ray emission. They significantly violate the Guedel-Benz relation - indicated as dashed line in Fig. 3 - describing the correlation between X-ray and radio emission in magnetically active stars. These latter objects are typically fast rotators, arising the question of a general activity/rotation relation in ultracool dwarfs. Further, they often show a rotation modulated radio flux or radio bursts. In contrast to coronal sources, electron cyclotron maser emission generated at the magnetic poles is thought to be responsible for the high radio flux (Hallinan et al. 2006).

3. X-ray Emission and Jets from Young Stars

Speaker: J. Kastner, P.C. Schneider

During the evolution towards the main sequence, young stars or protostars are actively accreting matter from their disk and drive spectacular outflows/jets. X-ray signatures of both processes allow a detailed study of the physical mechanisms involved and pose tight constraints on the star formation process.

Joel Kastner presented a results from an ongoing Target of Opportunity (ToO) program for FUor/EXor-like outbursts. Young stellar objects sometimes undergo dramatic changes in their accretion rate accompanied by sudden increases of their luminosity. These events are loosely termed FUors or EXors, depending mainly on the duration of these events. EXors, a hybrid of the prototypical EX Lup and FUor coined by G. Herbig, show chaotic variability on monthly or yearly time scale, while FUor outbursts last decades. It is currently not clear, which mechanism triggers these accretion bursts. Only about 15 FUor prototypes/candidates and approximately the same number of EXor candidates are known to date. One of these is McNeil's Star (V1647 Ori), which went into outburst in late 2003 and returned to its pre-outburst brightness level in early 2006. The short duration of the outburst qualifies V1647 Ori rather as an EXor than a FUor and a second outburst already began early in 2008. The X-ray emission of V1647 Ori follows roughly the trend of the optical/IR emission; modulated by strong short-term variations (Kastner et al. 2004, 2006). The X-ray/optical correlation is unlikely caused by a chance coincidence with the usual X-ray flares, but points to some kind of magnetic star-disk interaction. The 2008 outburst of V1647 Ori triggered ToO observations with *Chandra* and *Suzaku*, revealing a striking similarity with the 2003 outburst, when comparable rapid, short-time flux variations and spectral properties were observed. The *Suzaku* spectrum clearly shows a hard spectral component and strong Fe Kα emission (see Fig. 4 and Hamaguchi et al. 2010). Line-of-sight obscuration of the irradiating plasma might explain the absence of associated X-ray flares, which are generally observed for the Sun's Kα emission. JK also presented *XMM-Newton* observations of EX Lup, the prototype of the EXor objects, which shows an X-ray spectrum consisting of a hot, strongly absorbed component and a soft, less absorbed component. Based on the absence of optical forbidden line emission usually tracing protostellar jets and the near pole-on viewing geometry, the soft X-ray emission can be interpreted as arising from accretion shocks, while the hot component, on the other hand, might be caused by a "smothered" corona, absorbed by the accretion streams (Grosso et al. 2010). Thus, hard X-ray emission in FUors and EXors can be generated during accretion bursts, but cannot be explained by accretion shocks and might support a magnetic origin of the accretion bursts. However, other FUor or EXor objects often do not show a strong correlation between their defining optical appearance and their X-ray behavior.

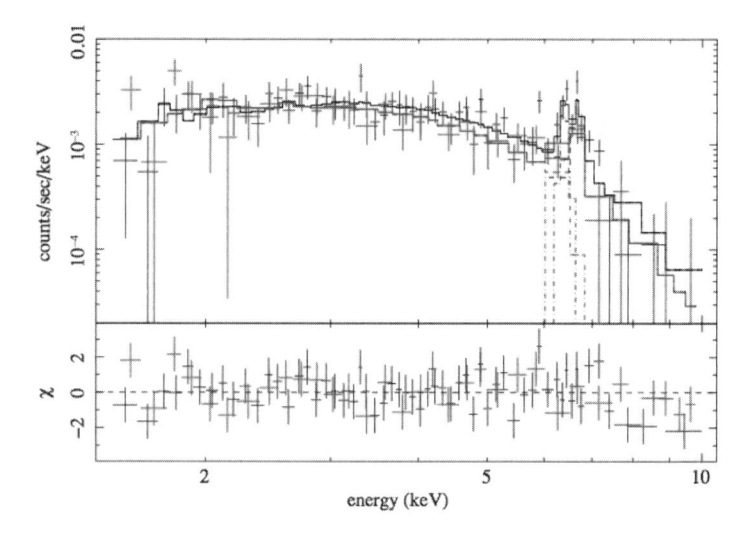

Figure 4. Suzaku X-ray spectrum of the EXor V1647 Sgr showing the Fe Kα emission line at 6.4 keV. Figure courtesy of J. Kastner (from Hamaguchi et al. 2010).

The second topic JK presented, "X-ray-irradiated gaseous disks orbiting nearby pre-MS stars" concerns the feedback of X-ray emission on the evolution of circumstellar material (Lepp & Dalgarno 1996). Only four, rather old ($\gtrsim$ 10 Myr), objects within 100 pc are known, which still possess molecular disks. Among them, the well studied TW Hya and V4046 Sgr share important properties. However, V4046 Sgr is a binary system and possesses a circumbinary disk, which extends out to at least 370 AU and which is probably coplanar with the orbital plane of the two YSOs. The first results from a large XMM campaign on V4046 Sgr were presented (Montmerle et al., in prep.). X-ray and optical light curves and the time resolved evolution of the density sensitive Ne IX f/i ratio were shown. The X-ray emission measure distribution of V4046 Sgr is comparable to that of TW Hya, but with a larger emission measure at higher temperatures. JK concluded this part by comparing various molecular tracers in gaseous circumstellar disks around TW Hya and V4046 Sgr. Pointing out similarities between them, shows that X-rays likely modify the disk chemistry. Further steps will involve the detailed investigation of the disks around other nearby classical T Tauri, e.g., MP Mus, which also still possess a molecular disk as shown in Fig. 5 (Kastner et al. 2010).

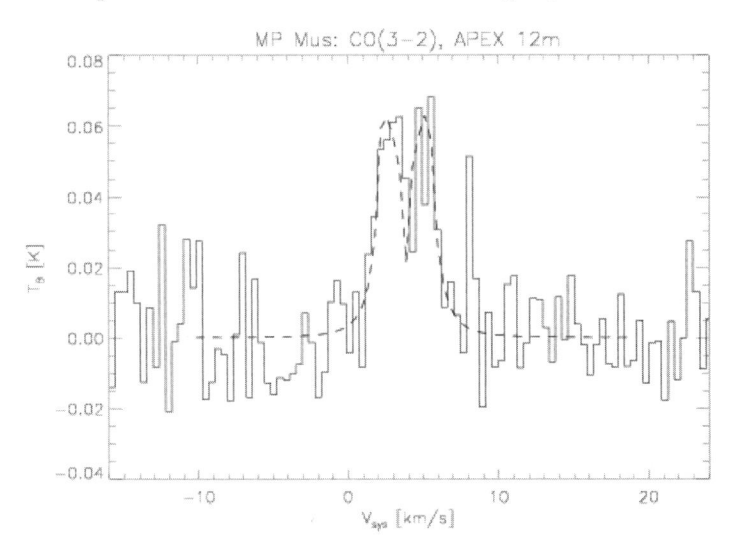

Figure 5.   CO(3-2) emission around MP Mus indicating disk-like emission. Figure courtesy of J. Kastner (Kastner et al. 2010).

Christian Schneider presented the results of a new observation of the X-ray jet of L1551 IRS 5, a deeply embedded protostellar binary system. Outflow activity of forming protostars is common, but is usually observed in relatively low temperature forbidden emission lines ($T \approx 10^4$ K). The X-ray emission of this protostellar jet was the second detection of X-ray emission from a protostellar jet (Favata et al. 2002; Bally et al. 2003) and points to the existence of a million degree plama within the jet. The L1551 IRS 5 jet, often termed HH 154, is particularly fruitful for X-ray studies due to (a) the strong absorption of the central binary system allowing to trace the jet close to the driving sources without contaminating stellar X-ray emission and (b) due to its proximity to the Sun ($d = 140$ pc). In combination, these properties make high resolution *Chandra* X-ray imaging sensitive to spatial, i.e., morphological changes on time scales of a few years (Favata et al. 2006). The new X-ray observation (Schneider et al., in prep.), however, does not show new X-ray emitting knots, as tentatively identified in the second *Chandra* observation, nor the proper-motion of individual knots (Fig. 6).

In particular, the inner X-ray emission complex still persists. This pattern contrasts the proper-motion observed for the optical part of the jet at greater distances from the driving sources. Also, the total X-ray luminosity, albeit nominally slightly decreasing, is compatible with a constant luminosity. Performing a spatially resolved spectroscopic study, CS showed that the X-ray emitting plasma is likely cooler farther from the driving sources and that this trend remained constant over almost a decade. Comparing these properties with other X-ray emitting jets, notably with the X-ray jet of DG Tau, an explanation involving some kind of standing shock responsible for the heating of the jet to X-ray emitting temperatures was put forward, as it would naturally explain the constant appearance of the X-ray emission.

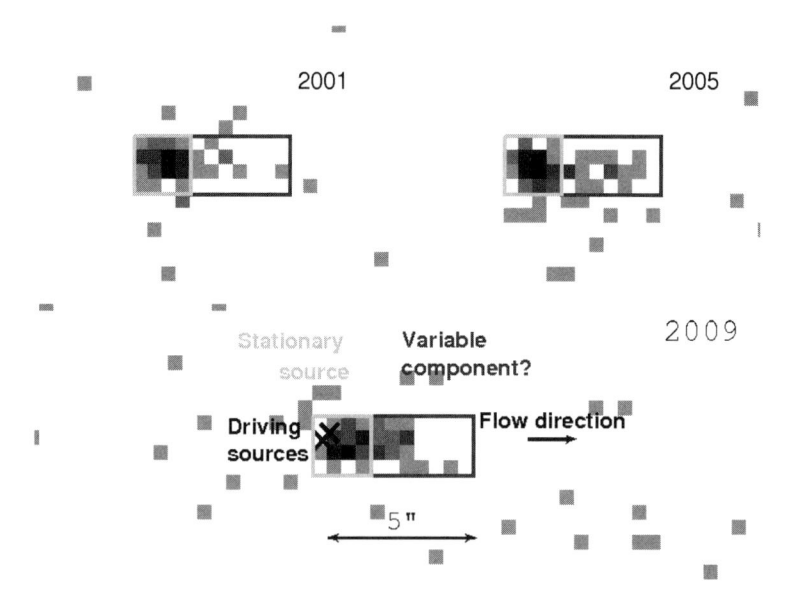

Figure 6. X-ray emission of the L1551 IRS 5 jet in the 0.5 to 3.0 keV band observed by *Chandra*, important regions are labeled. Figure courtesy of P.C. Schneider.

4. X-rays, Stars and Planets

Speaker: K.Poppenhaeger, S. Wolk, J. Sanz-Forcada

The interplay between X-ray emission of stars and the presence of planets has been debated lively during the last years. Open questions discussed here are the possibility of Star-Planet Interactions (SPI), where planets are thought to influence their host star's activity level by magnetic or tidal interaction, as well as evaporation of planetary atmospheres which is driven by high-energy irradiation.

Katja Poppenhaeger presented results from an X-ray analysis of all planet-hosting stars[1] within 30 pc distance from the Sun (Poppenhaeger et al. 2010b). They tested this complete sample for correlations between planetary parameters (mass M_{pl} and semi-major axis a_{pl}) and stellar X-ray properties (L_X and L_X/L_{bol}). The only significant correlation detected was between L_X and $M_{pl} \times a_{pl}^{-1}$ (Fig. 7, left panel). The radial velocity method, with which almost all planets in the sample were detected, produces a selection effect, since stellar activity masks the RV signal; this trend is similar to the one detected in the data set. To check if there is an *additional* trend present on top of this bias, a linear regression of $\log L_X$ versus $\log(M_{pl} \times a_{pl}^{-1})$ was conducted for two subsamples, heavy close-in planets and small far-out planets. The trend in the far-out sample should be dominated by the RV selection effect, since SPI effects are expected only for small distances. The close-in sample should show the RV trend plus a potential trend from SPI. However, the two trends overlap well within errors; no *additional* activity trend which might be induced by SPI is detectable (Fig. 7, right panel).

Additionally, the υ Andromedae system, an F8V star with a Hot Jupiter and two other known planets, was investigated for signatures of Star-Planet Interactions in the chromosphere and the corona (Poppenhaeger et al. 2010a). While there is periodic low-level variability detected ($P = 9.5$ d) for the chromospheric activity indicator, Ca II K line emission, it does not match the planetary orbital period of 4.6 d, but rather the stellar rotation period which is loosely determined to be between 8.5 – 12 d (Wright et al. 2004; Henry et al. 2000). Also the coronal emission of υ And shows no signs of SPI.

Scott Wolk presented new X-ray data on the planet-hosting star HD 189733 (Pillitteri et al. 2010). The star hosts a Hot Jupiter in an 2.2 d orbit and was observed with *XMM-Newton* during transit in 2007 and during secondary eclipse in 2009. The transit observations were inconspicuous, showing only low-level stellar variability. The 2009 observations covered a stellar flare ca. 3 ks after the end of the secondary eclipse; the timing of this event might be due to interactions between star and planet. Also, a softening of the stellar X-ray spectrum was observed during the secondary eclipse. Magnetohydrodynamic simulations were conducted, showing that the magnetic interaction between the star and the planet can enhance the density and the magnetic field in a region between them. There were hints for a blue-shift of oxygen lines in spectra recorded with *XMM-Newton's* Reflection Grating Spectrometer (RGS), which might be due to a flow of material from the planet to the star.

Also, the HD 189733 system was found to be possibly older than previously thought. HD 189733A has a stellar companion of spectral type M4 at 216 AU distance, which was not detected in the two X-ray pointings. This hints towards an age of the system of ~ 2 Gyr.

Jorge Sanz-Forcada presented results on the atmospheric evaporation of planets due to high-energy irradiation by their host stars (Sanz-Forcada et al. 2010). X-ray and EUV irradiation is thought to be the main driver for planetary evaporation; such evaporation may have been observed for the planet HD 209458b (Vidal-Madjar et al. 2003). Sanz-Forcada calculated the planetary mass loss rate to depend in approximation linearly on the received XUV flux and the reciprocal planetary density, following Erkaev

---

[1]to access this data, go to tinyurl.com/spixray

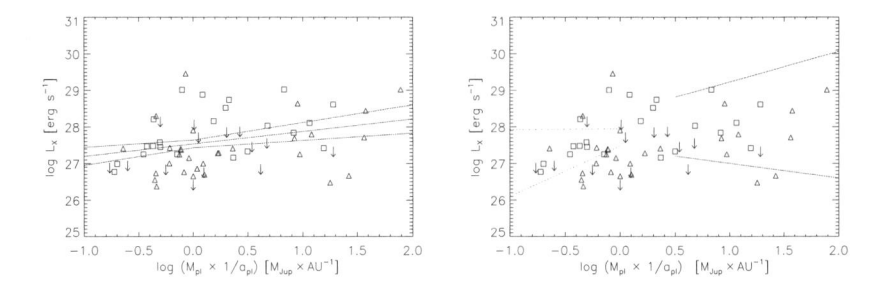

Figure 7. *Left:* X-ray luminosity versus planetary mass and inverse semi-major axis for the investigated stellar sample; *XMM-Newton* detections plotted as triangles, ROSAT detections as squares. *Right:* separate regression analyzes performed for the two subsamples. Both trends overlap well within errors, indicative of no SPI-related trend in this sample. Figure courtesy of K. Poppenhaeger.

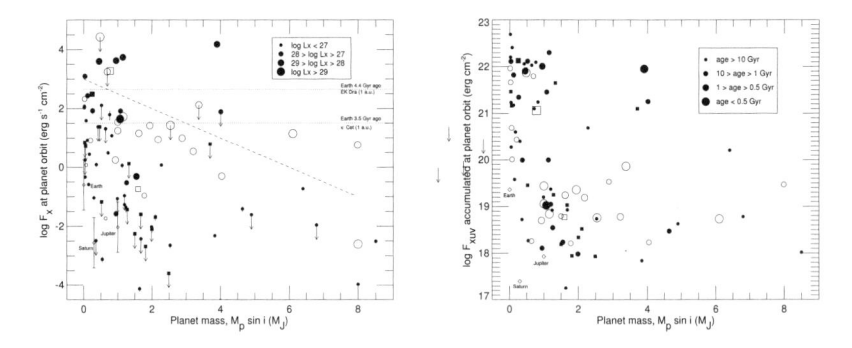

Figure 8. *Left:* Distribution of planetary masses ($M_p \sin i$) with X-ray (1–100 Å) flux at the planet orbit. Filled symbols (squares for subgiants, circles for dwarfs) are *XMM-Newton* and *Chandra* data. Arrows indicate upper limits. Open symbols are ROSAT data without error bars. Diamonds represent Jupiter, Saturn, and the Earth. The dashed line marks the "erosion line" that apparently separates a phase of strong erosion from weaker erosion. Dotted lines indicate the X-ray flux of the younger Sun at 1 AU *Right:* Distribution of planetary masses ($M_p \sin i$) with the XUV flux (1–920 Å) accumulated at the planet orbit since an age of 20 Myrs to the present day. Symbols as in left panel. Figure courtesy of J. Sanz-Forcada

et al. (2007), Lammer et al. (2003) and Baraffe et al. (2004). Comparing the X-ray flux at the planetary orbit to the projected planetary masses (Fig. 8, left panel) yielded that there are very few high-mass planets at high irradiation levels. The X-ray irradiation levels were calculated backwards in time by using relationships between $\log L_X$ and stellar age (see for example Ribas et al. 2005). Using this and calculating synthetic EUV spectra from the measured X-ray spectra, the accumulated XUV irradiation of planets was computed. Fig. 8, right panel, shows that only three planets with masses higher than 1.5 M_{jup} have survived an irradiating flux of more than 10^{19} erg cm^{-2}, while these exceptions are somewhat special cases due to low or not well determined age or stellar companions. Measurements of planetary densities, which are only available for a few targets up to now, will enable more detailed studies of evaporation scenarios.

Acknowledgments. We thank all the contributing speakers for their talks, the organizers for setting up the meeting and all participants for joining us at the X-ray splinter session. J.R acknowledges support from DLR under grant 50QR0803, P.C.S. and K.P. under grant 50OR0703.

References

Audard, M., Osten, R. A., Brown, A., Briggs, K. R., Güdel, M., Hodges-Kluck, E., & Gizis, J. E. 2007, A&A, 471, L63. 0707.1882
Babel, J., & Montmerle, T. 1997, A&A, 323, 121
Bally, J., Feigelson, E., & Reipurth, B. 2003, ApJ, 584, 843
Baraffe, I., Selsis, F., Chabrier, G., Barman, T. S., Allard, F., Hauschildt, P. H., & Lammer, H. 2004, A&A, 419, L13. arXiv:astro-ph/0404101
Berger, E. 2002, ApJ, 572, 503. arXiv:astro-ph/0111317
Erkaev, N. V., Kulikov, Y. N., Lammer, H., Selsis, F., Langmayr, D., Jaritz, G. F., & Biernat, H. K. 2007, A&A, 472, 329. arXiv:astro-ph/0612729
Favata, F., Bonito, R., Micela, G., Fridlund, M., Orlando, S., Sciortino, S., & Peres, G. 2006, A&A, 450, L17. arXiv:astro-ph/0603186
Favata, F., Fridlund, C. V. M., Micela, G., Sciortino, S., & Kaas, A. A. 2002, A&A, 386, 204. arXiv:astro-ph/0110112
Grosso, N., Briggs, K. R., Güdel, M., Guieu, S., Franciosini, E., Palla, F., Dougados, C., Monin, J., Ménard, F., Bouvier, J., Audard, M., & Telleschi, A. 2007, A&A, 468, 391. arXiv:astro-ph/0608696
Grosso, N., Hamaguchi, K., Kastner, J. H., Richmond, M. W., & Weintraub, D. A. 2010, A&A, 522, A56+. 1007.2838
Hallinan, G., Antonova, A., Doyle, J. G., Bourke, S., Brisken, W. F., & Golden, A. 2006, ApJ, 653, 690. arXiv:astro-ph/0608556
Hamaguchi, K., Grosso, N., Kastner, J. H., Weintraub, D. A., & Richmond, M. 2010, ApJ, 714, L16
Henry, G. W., Baliunas, S. L., Donahue, R. A., Fekel, F. C., & Soon, W. 2000, ApJ, 531, 415
Kastner, J. H., Hily-Blant, P., Sacco, G. G., Forveille, T., & Zuckerman, B. 2010, ApJ, 723, L248. 1010.1174
Kastner, J. H., Richmond, M., Grosso, N., Weintraub, D. A., Simon, T., Frank, A., Hamaguchi, K., Ozawa, H., & Henden, A. 2004, Nat, 430, 429. arXiv:astro-ph/0408332
Kastner, J. H., Richmond, M., Grosso, N., Weintraub, D. A., Simon, T., Henden, A., Hamaguchi, K., Frank, A., & Ozawa, H. 2006, ApJ, 648, L43. arXiv:astro-ph/0607653
Lammer, H., Selsis, F., Ribas, I., Guinan, E. F., Bauer, S. J., & Weiss, W. W. 2003, ApJ, 598, L121
Lepp, S., & Dalgarno, A. 1996, A&A, 306, L21
Martín, E. L., & Bouy, H. 2002, New Astronomy, 7, 595. arXiv:astro-ph/0210360
Monnier, J. D., Zhao, M., Pedretti, E., Thureau, N., Ireland, M., Muirhead, P., Berger, J.-P., Millan-Gabet, R., Van Belle, G., ten Brummelaar, T., McAlister, H., Ridgway, S., Turner, N., Sturmann, L., Sturmann, J., & Berger, D. 2007, Science, 317, 342. arXiv:0706.0867
Neuhäuser, R., & Comeron, F. 1998, Science, 282, 83
Pillitteri, I., Wolk, S. J., Cohen, O., Kashyap, V., Knutson, H., Lisse, C. M., & Henry, G. W. 2010, ApJ, 722, 1216. 1008.3566
Poppenhaeger, K., Lenz, L. F., Reiners, A., & Schmitt, J. H. M. M. 2010a, ArXiv e-prints. 1010.5632
Poppenhaeger, K., Robrade, J., & Schmitt, J. H. M. M. 2010b, A&A, 515, A98+. 1003.5802

Preibisch, T., McCaughrean, M. J., Grosso, N., Feigelson, E. D., Flaccomio, E., Getman, K., Hillenbrand, L. A., Meeus, G., Micela, G., Sciortino, S., & Stelzer, B. 2005, ApJS, 160, 582. arXiv:astro-ph/0506049

Ribas, I., Guinan, E. F., Güdel, M., & Audard, M. 2005, ApJ, 622, 680. arXiv:astro-ph/0412253

Robrade, J., Poppenhaeger, K., & Schmitt, J. H. M. M. 2010, A&A, 513, A12+. 1002.2389

Robrade, J., & Schmitt, J. H. M. M. 2009, A&A, 497, 511. 0903.0966

— 2010, A&A, 516, A38+. 1004.1318

Rutledge, R. E., Basri, G., Martín, E. L., & Bildsten, L. 2000, ApJ, 538, L141. arXiv:astro-ph/0005559

Sanz-Forcada, J., Ribas, I., Micela, G., Pollock, A. M. T., García-Álvarez, D., Solano, E., & Eiroa, C. 2010, A&A, 511, L8+. 1002.1875

Stelzer, B. 2004, ApJ, 615, L153. arXiv:astro-ph/0409617

Stelzer, B., Micela, G., Flaccomio, E., Neuhäuser, R., & Jayawardhana, R. 2006, A&A, 448, 293. arXiv:astro-ph/0511168

Stelzer, B., Scholz, A., Argiroffi, C., & Micela, G. 2010, MNRAS, 408, 1095. 1006.2717

Vidal-Madjar, A., Lecavelier des Etangs, A., Désert, J., Ballester, G. E., Ferlet, R., Hébrard, G., & Mayor, M. 2003, Nat, 422, 143

Wright, J. T., Marcy, G. W., Butler, R. P., & Vogt, S. S. 2004, ApJS, 152, 261. arXiv:astro-ph/0402582

The 16th Cambridge Workshop on Cool Stars, Stellar Systems and the Sun
ASP Conference Series, Vol. 448
Christopher M. Johns-Krull, Matthew K. Browning, and Andrew A. West, eds.
© 2011 Astronomical Society of the Pacific

The M4 Transition: Toward a Comprehensive Understanding of the Transition into the Fully Convective Regime

Keivan G. Stassun,[1] Leslie Hebb,[1] Kevin Covey,[2] Andrew A. West,[3] Jonathan Irwin,[4] Richard Jackson,[5] Moira Jardine,[6] Julien Morin,[7] Dermott Mullan,[8] and I. Neill Reid[9]

[1]*Department of Physics & Astronomy, Vanderbilt University, VU Station B 1807, Nashville, TN 37235, USA*

[2]*Department of Astronomy, Cornell University, 226 Space Sciences Building, Ithaca, NY 14853, USA*

[3]*Department of Astronomy, Boston University, 725 Commonwealth Ave, Boston, MA 02215, USA*

[4]*Harvard-Smithsonian Center for Astrophysics, 60 Garden St., Cambridge, MA 02138, USA*

[5]*Astrophysics Group, Keele University, Keele, Staffordshire ST5 5BG, UK*

[6]*School of Physics and Astronomy, University of St Andrews, St Andrews, Scotland KY16 9SS, UK*

[7]*Dublin Institute for Advanced Studies, 31 Fitzwilliam Place, Dublin 2, Ireland*

[8]*Department of Physics and Astronomy, University of Delaware, Newark, DE 19716, USA*

[9]*Space Telescope Science Institute, Baltimore, MD 21218, USA*

Abstract. The difference in stellar structure above and below spectral type ~M4 is expected to be a very important one, connected directly or indirectly to a variety of observational phenomena in cool stars—such as rotation, activity, magnetic field generation and topology, timescales for evolution of these, and even the basic mass-radius relationship. In this Cool Stars XVI Splinter Session, we aimed to use the M4 transition as an opportunity for discussion about the interiors of low-mass stars and the mechanisms which determine their fundamental properties. By the conclusion of the session, several key points were elucidated. Although M dwarfs exhibit significant changes across the fully convective boundary, this "M4 transition" is not observationally sharp or discrete. Instead, the properties of M dwarfs (radius, effective temperature, rotation, activity lifetime, magnetic field strength and topology) show smooth changes across M3–M6 spectral types. In addition, a wide range of stellar masses share similar spectral types around the fully convective transition. There appears to be a second transition at M6–M8 spectral types, below which there exists a clear dichotomy of magnetic field topologies. Finally, we used the information and ideas presented in the session to construct a framework for how the structure of an M dwarf star, born with specific mass and chemical composition, responds to the presence of its magnetic field, itself driven by a feedback process that links the star's rotation, interior structure, and field topology.

1. Introduction

The interior structure of a star is a function of its mass. Above a threshold mass of ~0.3 $M_\odot$, the interior is expected to resemble that of the Sun, having a radiative zone and a convective envelope. In contrast, below the threshold mass the convection zone extends all the way to the core.

Despite not having a strong radiative/convective interface (like that found in the Sun), mid- to late-type M dwarfs (spectral type of M4 and greater) are observed to have strong magnetic fields and subsequent surface heating that results in observed magnetic activity (traced by strong line or continuum emission). Recent theoretical modeling of stars with convective envelopes that extend to the core (and a lack of a strong radiative/convective interface) were able to produce strong, long lived, large scale magnetic fields (Browning 2008). Observational results also suggest that in mid- to late-type M dwarfs, the fraction of the magnetic field in large scale components is larger than in early type M dwarfs (Donati et al. 2008; Reiners & Basri 2008).

These are just some examples of the phenomena and questions that play out around the M4 spectral type. The purpose of this Cool Stars XVI Splinter Session was to use this specific transition point—the M4 transition—as an opportunity for focused discussion and "out of the box" collective thinking about the interiors of low-mass stars, what we know, what we don't know, what we wish we understood better. We intended this to be a rich, fun session to bring together observational evidence from multiple directions with the goal of disentangling the effects of multiple physical phenomena and leading toward a clearer set of guiding questions for future work.

2. The M4 Transition

Neill Reid began by reminding us of some of the basic properties of M dwarfs and of the basic predictions of theoretical stellar evolution models for M dwarf interior structure. Most theoretical stellar evolution models predict the transition into the fully convective regime to occur at a stellar mass of ~0.3 $M_\odot$, but differ in detail. For example, Chabrier & Baraffe (1997) predict the boundary to occur at $M \approx 0.35$ $M_\odot$ (spectral type M2), whereas Dorman et al. (1989) place the fully convective boundary at $M \approx 0.25$ $M_\odot$ (spectral type M4). For the fiducial properties of M2 stars, refer to the benchmark systems Gliese 22A (Henry & McCarthy 1993) and Gliese 411 (Lane et al. 2001), both with empirical masses of ≈ 0.36 $M_\odot$ and the latter with an interferometrically measured radius. For the fiducial properties of M4 stars, refer to the benchmark eclipsing binary CM Dra, comprising two M4.5 stars with masses of 0.22 $M_\odot$ (e.g. Morales et al. 2009). Table 1 gives a summary of the physical properties of spectral type M2 and M4.5 stars.

Table 1. Basic physical properties of exemplar M2 and M4.5 dwarfs.

Name	Mass ($M_\odot$)	Radius ($R_\odot$)	$T_{\rm eff}$	M_V	$V - K$
GJ 22 A	0.352 ± 0.036	...	3380 ± 150	10.99 ± 0.06	4.55
GJ 411	0.403 ± 0.020	0.393 ± 0.008	3828 ± 100	8.08 ± 0.09	4.25
CM Dra A	0.2310 ± 0.0009	0.2534 ± 0.0019	3130 ± 70	12.78 ± 0.05	5.10
CM Dra B	0.2141 ± 0.0010	0.2396 ± 0.0015	3120 ± 70	12.91 ± 0.05	5.10

2.1. Basic Stellar Properties

Neill Reid discussed the observed properties of M dwarfs across the fully convective boundary. It is useful to recall up front that certain fundamental stellar properties do not evince any obvious manifestation of the changes in stellar structure that accompany the transition into the fully convective regime. For example, the mass-luminosity relation is smooth through the M spectral types. Similarly, Figure 1a shows the radius-luminosity relation for a large sample of M dwarfs (Ribas 2006; Demory et al. 2009): The observed radius-luminosity relation is smooth through the M spectral types, as predicted by stellar evolution models. Evidently, energy generation and output in these stars does not much know or care about changes in structure or in energy transport that may be occurring across the fully convective boundary.

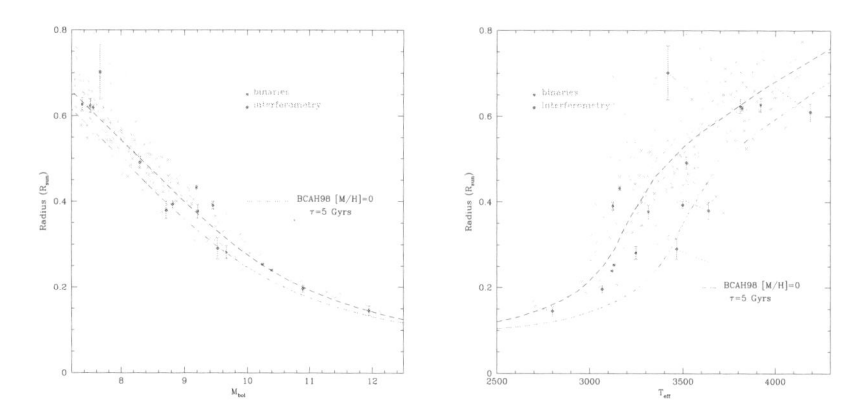

Figure 1. *Left:* Radius-luminosity relation for M dwarfs. Crosses are nearby field M dwarfs with reliable parallaxes, and temperatures derived from $V - I$ colors. The relationship is smooth across the M spectral types, as predicted by stellar evolution models (dashed lines). *Right:* Radius-T_{eff} relation. As highlighted Clemens et al. (1998), there is a marked decrease in radius near spectral type M4. Note, however, that this change in the radius-T_{eff} relationship is not infinitely sharp, but rather is a finite transition from spectral types M3 to M6.

However, other basic stellar properties do manifest clear changes across a mass of ~ 0.3 $M_\odot$. Figure 1b shows the radius-temperature relation for the same stars in Figure 1a. Unlike the featureless radius-luminosity relation, here we see a clear change in slope—with the stellar radii decreasing rapidly with decreasing effective temperature—centered around $T_{eff} \sim 3150$ K (spectral type M4). The change in stellar radius here is a factor of ~ 2. In other words, there is a relatively large range of stellar mass and radius over a relatively small range of spectral types. Indeed, stars with a wide range of masses around the fully convective transition are predicted to have similar spectral types (cf. Fig. 4 in Chabrier & Baraffe 2000), a consequence of the flatness of the mass-temperature relation caused by H_2 formation. There is a strong morphological similarity between the observed and predicted relations, but the two are offset by ~ 250K, with the observations cooler than the models (compare pink and black curves in Fig. 1).

Notably, the change in the radius-temperature relation near M4 is not an abrupt step function; rather, it is a *finite transition* between $T_{eff} \approx 3300$ K and $T_{eff} \approx 2850$ K (i.e. spectral types M3 to M6). At later spectral types, the stellar radii asymptotically approach the value of ≈0.1 $R_\odot$ dictated by electron degneracy pressure.

As Leslie Hebb discussed, recent direct radius measurements of M dwarfs in eclipsing binaries show the stellar radii to be inflated relative to the radii predicted by theoretical stellar evolution models (e.g. Ribas 2006). This has been attributed to the effects of stellar magnetic fields and activity (e.g. Morales et al. 2008). The magnitude of this effect does change across the fully convective boundary (Figure 2), but as above it does not appear to change in an abrupt fashion.

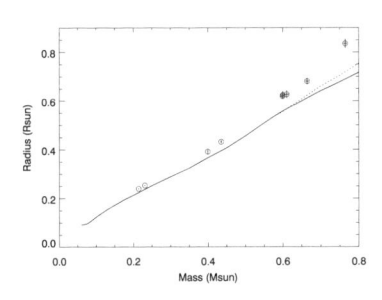

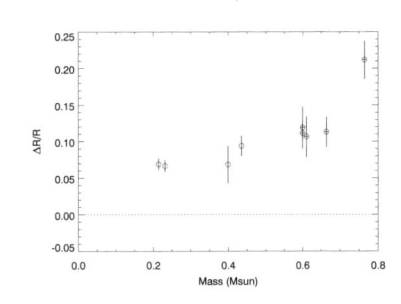

Figure 2. *Left:* Mass-radius relation for eclipsing binary systems with masses and radii measured to better than 3%. *Right:* Fractional increase in the radii over the theoretically predicted radii (Baraffe et al. 1998). The enlarged radii are thought to be caused by the affects of magnetic fields. Early type M dwarfs show a larger increase in radii compared to late-type, fully convective M dwarfs, consistent with predictions of theoretical models (Mullan & MacDonald 2001; Chabrier et al. 2007).

2.2. Stellar Activity and Stellar Rotation: Effects of Stellar Age

As Jonathan Irwin and Richard Jackson discussed, recent observational results have confirmed that the M4 transition plays an important part in shaping the rotation and magnetic activity evolution of M dwarfs. Large samples of M dwarfs show an increase in activity fractions around spectral types of ~M4 (West et al. 2008).

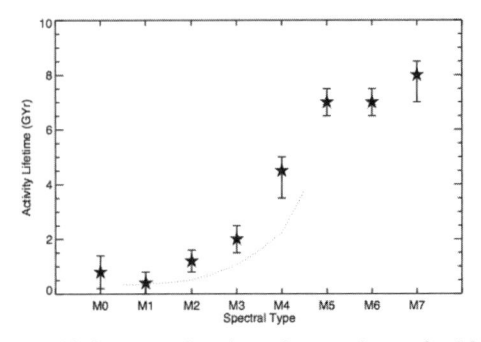

Figure 3. Activity lifetime as a function of spectral type for M dwarfs (West et al. 2008). The lifetime over which an M dwarf stays active is a strong function of spectral type, from 1–2 Gyr for M3 dwarfs to 7–8 Gyr for M6 dwarfs. Evidently, activity lifetime does respond to the change in internal structure across the fully convective boundary, but the transition is a finite one between M3 and M6.

This result has been shown to be dependent on the ages of the stars. When looking at the activity fractions as a function of the M dwarfs' location above the Galactic plane, and using a simple model of dynamical heating which scatters M dwarfs to larger Galactic scale heights over time, West et al. (2008) determine the amount of time that M dwarfs of different spectral types remain chromospherically active. They find that the "activity lifetime" changes dramatically between spectral types M3 and M6 (from 1–2 Gyr for M3 dwarfs to 7–8 Gyr for M6 dwarfs; Fig.3).

This important transition was bolstered by the rotation analysis of Reiners & Basri (2008), who found that the average rotation velocities of a nearby sample of M dwarfs increases dramatically and monotonically as one looks at later spectral types. The average $v \sin i$ is <3 km/s at M2, increases to ~10 km/s at M6, and continues increasing to ~30 km/s at L4 (Figure 4). This almost certainly points to a lengthening of the timescale for stellar spindown as the stellar structure changes from solar-type (i.e. radiative core + convective envelope) to fully convective. However, the transition is not a step function, thus again pointing to the importance of evolution over time.

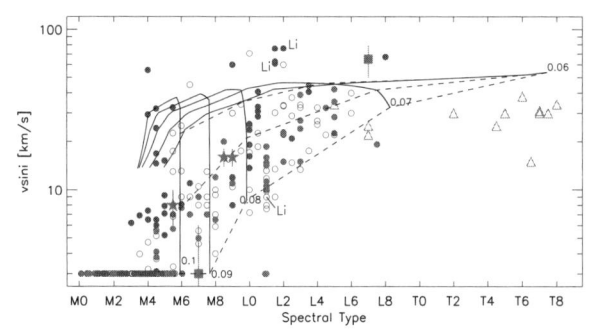

Figure 4. Rotation ($v \sin i$) of late-type stars (Reiners & Basri 2008). Starting at spectral type ~M3, mean $v \sin i$ increases monotonically, implying that the spindown timescale increases monotonically across the fully convective boundary.

The importance of time evolution is seen most clearly in rotation-period distributions in populations of different ages. Fig. 5 shows the evolution of stellar rotation periods from ~1 Myr to $\gtrsim$10 Gyr. The situation among the pre–main-sequence populations is complex, likely due to the effects of multiple competing phenomena (Bouvier 2009). However, at ages of ~100 Myr and older, the stars show a clear pattern with two dominant sequences of rapid and slow rotators; the slow-rotator sequence becomes increasingly dominant with increasing stellar age. These are the so-called 'C' and 'I' sequences of the Barnes "gyrochronology" paradigm (Barnes 2003, 2007).

Importantly, it is also clear from this figure that at all ages there is a spread of stellar rotation periods near the fully convective boundary; age alone does not determine whether a given M4 star will be on the 'C' sequence or on the 'I' sequence. Thus, while all stars appear to follow a similar pattern of evolution in stellar rotation, presumably connected to evolution of their interior structure and magnetic field properties, there is evidently star-to-star scatter in these properties at any given age.

2.3. Magnetic fields: Topology and Effects on Stellar Properties

As discussed by Moira Jardine and Julien Morin, M dwarfs also show clear topological changes in their surface magnetic fields as a function of spectral type. Figure 6a shows

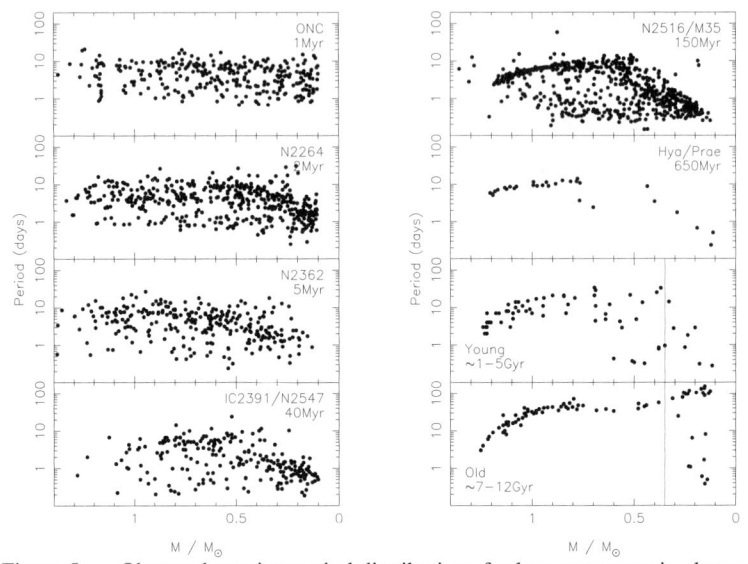

Figure 5. Observed rotation period distributions for low-mass stars in clusters of varying ages. Among clusters with ages >100 Myr, two sequences of rapid and slow rotators are observed, with the slow-rotator branch becoming increasingly dominant with time. At the fully convective boundary (vertical line), the stars do not transition to the slow-rotator sequence instantaneously, but rather on a finite timescale.

a simple cartoon illustration summarizing the basic change that is observed: from globally weak but highly structured (i.e. multipolar and non-axisymmetric) fields at early spectral types, to globally strong and well ordered (i.e. dipolar) fields at later spectral types. In other words, the energy in the dipole component, relative to higher-order components, is higher in the lower-mass stars. Figure 6b depicts these changes in reference to three specific low-mass stars (Gl 494, EV Lac, EQ Peg b) whose magnetic field topologies have been modeled in detail (Jardine et al. 2002) using Zeeman Doppler Imaging (ZDI) maps as boundary conditions (Donati et al. 2008; Morin et al. 2008).

This is shown more quantitatively in Figure 7, which depicts the field topology and strength as a function of stellar mass, derived for a large number of low mass stars using the ZDI technique. Interestingly, we see a hint of two transitions here. The first occurs near the fully convective boundary, where as already mentioned the fields transition from a globally weak, highly structured topology, to a globally strong, dipolar topology. A second transition appears at spectral type M6–M8, such that below M~ 0.15 $M_\odot$, stars with similar stellar masses and rotational properties can have very different magnetic topologies. This apparent M6–M8 transition could be reflecting the point at which the stellar atmospheres become completely neutral (e.g. Mohanty & Basri 2003), though it could again reflect an age effect as witnessed in the bifurcated rotational properties of M dwarfs (see Fig. 5).

Finally, it is important to bear in mind that the ZDI measurements, because they rely upon observations of polarized light, are probing field components that represent only a few percent of the total magnetic energy. There is likely a large amount of magnetic flux in small-scale field components (which do not contribute much net polarized light) in addition to the larger scale components that the ZDI technique is mapping.

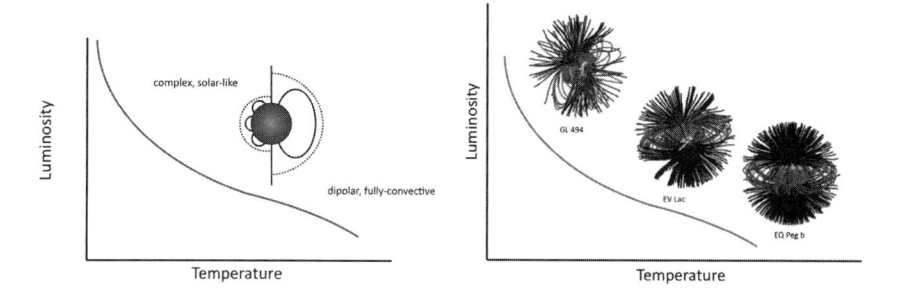

Figure 6. *Left:* Cartoon representing observed differences in surface magnetic field topology across the fully convective regime. *Right:* Reconstructed surface field topologies for three low-mass stars are shown (top to bottom: Gl 494, EV Lac, EQ Peg b). The field changes from a strong, highly structured, multi-polar configuration at early M types, to a weaker, more well organized dipole at the late M types.

Dermott Mullan discussed recent modeling efforts to account for the observed inflated radii of M dwarfs compared to the predictions of most (non-magnetic) theoretical stellar evolution models. Chabrier and collaborators have suggested that strong *surface* fields can inhibit surface convection and also create cool starspots, both of which act to decrease the surface temperature and then—because the stellar luminosity generated in the core is unaffected—requires an increase in the stellar radius.

Mullan also discussed alternate models in which a strong field threads the entire star. These models can successfully reproduce for example the observed temperature reversal in the brown-dwarf eclipsing binary 2M0535–05 (Figure 8), which at an age of only ~1 Myr has a spectral type of M6. These models do predict a change in the stellar luminosity as a result of the very strong (~MG) fields in the core (Figure 9).

3. Putting it all Together: A Conceptual Framework

Figure 10 is an initial attempt to conceptually summarize the inter-related phenomena discussed during our Session. This is not intended to be inclusive of all mechanisms or observable phenomena possibly relevant to M dwarfs, and it is by necessity a simplified representation of what is likely to be a more complex set of inter-relationships.

The figure attempts to represent the complex set of causal relationships—stellar structure, magnetic field topology, rotation, magnetic field strength—that mediate the fundamental relationships between basic stellar properties (e.g., how mass and composition ultimately translate into radius and effective temperature). As depicted in the figure, we imagine that, absent any magnetic field effects, there would be a single, fundamental mass–radius–temperature relationship (modulo metallicity). This relationship is, however, modified by the presence of a magnetic field that manifests itself on the surface through various activity tracers (e.g. chromospheric Hα emission). This magnetic field affects the observed surface properties of the star (effective temperature and radius) via surface effects (i.e. star spots; Chabrier et al. (2007)) and/or via a strong field that inhibits convection throughout the stellar interior (Mullan & MacDonald 2001).

The mechanism that drives the magnetic field's action (and perhaps its generation) is likely complex, but almost certainly involves at its heart a rotational dynamo. We

512 Stassun et al.

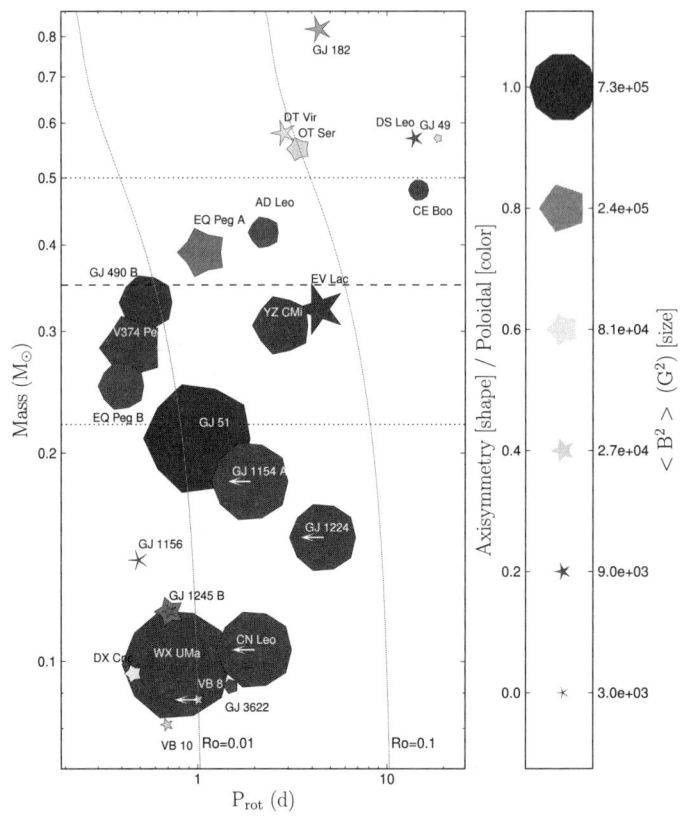

Figure 7. Properties of the large-scale magnetic topologies of a sample of M dwarfs inferred from spectropolarimetric observations and Zeeman Doppler Imaging (ZDI) (Donati et al. 2008; Morin et al. 2008; Phan-Bao et al. 2009) as a function of rotation period and mass. Larger symbols indicate stronger fields, symbol shapes depict the degree of axisymmetry of the reconstructed magnetic field (from decagons for purely axisymmetric to sharp stars for purely non axisymmetric), and colours the field configuration (from blue for purely toroidal to red for purely poloidal). Leftward white arrows indicate stars for which only an upper limit on P_{rot} is derived and the field properties are extrapolated. For VB8 the same symbol as VB10 is arbitrarily used, no magnetic field was detected from spectropolarimetric observations. For GJ 1245 B symbols corresponding to 2007 and 2008 data are superimposed in order to emphasize the variability of this object. Solid lines represent contours of constant Rossby number $R_0 = 0.1$ (saturation threshold) and 0.01. The theoretical full-convection limit (M~ 0.35 $M_\odot$, Chabrier & Baraffe (1997)) is plotted as a horizontal dashed line, and the approximate limits of the three stellar groups identified are represented as horizontal dotted lines.

envision that the star's rotation directly affects the strength of the magnetic field (more rapid rotation → stronger field) for stars below the saturation threshold. In general, a stronger field will drive a more rapid evolution of the star's angular momentum, leading to a decline of the star's rotation on a certain spindown timescale, and this spindown in turn feeds back into the strength of the field, dialing it down over time.

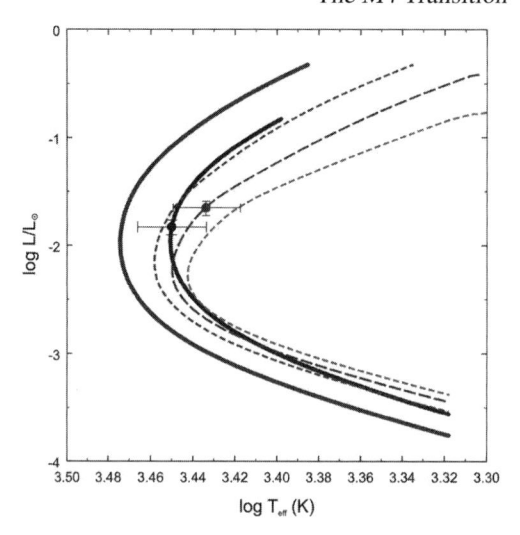

Figure 8. The reversal of effective temperature with mass for the brown-dwarf eclipsing binary 2M0535–05 (Stassun et al. 2006, 2007) can be explained by recent models that include the effects of strong magnetic fields threading through the entire interior of a low-mass star (MacDonald & Mullan 2009). Solid curves show the non-magnetic models for the primary (more massive) brown dwarf (red) and the secondary brown dwarf (blue), dashed lines show the magnetic models.

Importantly, the magnetic field acts according to an intricate feedback process such that the magnetic field impacts the stellar structure (i.e. convection zone depth) and rotation rate of the star which in turn affects the (re)generation of the field. The nature of the dynamo itself depends on the interior structure (i.e., convection zone depth). As the star approaches a more fully convective state (that is, as the size of the radiative zone recedes), there must be a transition from a solar-type ($\alpha - \omega$) dominated dynamo to a purely turbulent (α^2) dynamo. The spindown timescale is probably intimately coupled to the field topology, but exactly how the two are connected is not obvious. For example, later type M dwarfs have longer spindown timescales, yet these same stars have magnetic field topologies which posses more open field lines (i.e. dipolar fields), which would seem to enable them to more effectively drive mass and angular momentum loss. Clearly, we do not yet understand entirely how magnetic topology affects the observed rotational evolution of these stars.

It is moreover not clear whether the dynamo-driven strength of the field can feed back into the depth of the convection zone and/or the field topology. For example, it is thought that a very strong field can cause an otherwise fully convective star to generate a radiative core, either directly through the inhibition of convection at the center of the star (e.g. Mullan & MacDonald 2001) and/or indirectly through altering the boundary conditions at the stellar surface (e.g. Chabrier et al. 2007). Similarly, in some models very rapid stellar rotation can cause the opening of surface field lines (e.g. Jardine et al. 2010). Thus, the inter-relationships between rotation, field strength, interior structure, and surface field topology are likely complex, non-linear, and in any event dependent on the age of the star through its rotational history.

The rotation-period distributions of low-mass stars as a function of stellar age show that rotation is not simply a function of time alone. Indeed, current thinking identifies

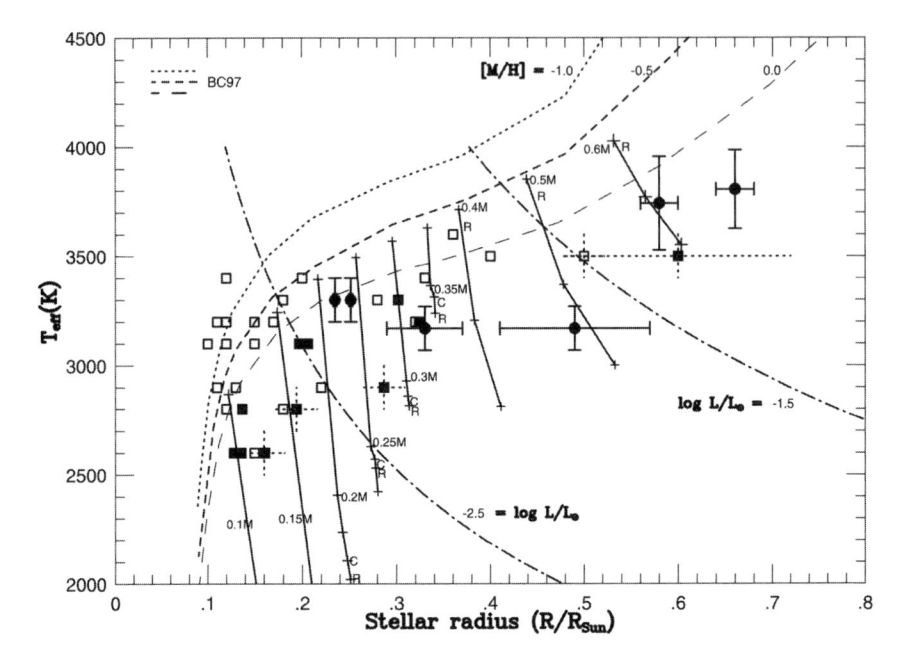

Figure 9. Models of Mullan & MacDonald (2001) showing the change in stellar radius, temperature, and luminosity, resulting from strong internal magnetic fields.

at least two dominant sequences of rotational properties in clusters (e.g., the 'C' and 'I' sequences of Barnes), such that a star of a given mass and age is likely to have one of two possible rotation periods. Perhaps these distinct rotational sequences reflect stars in different states of the complex feedback mechanisms between field generation, dynamo type, and field topology explored above. Indeed, the ZDI maps of field topologies appear to bear out this picture of multiple types of field configurations intermixed among the M3–M6 spectral types (Figure 7). Given this, and in light of the time-dependence of these effects, it is perhaps not surprising that the "M4 transition" across the fully convective boundary is not a sharp transition, but rather "smeared" from approximately M3 to M6. For example, the two eclipsing binaries CM Dra and CU Cnc (see Fig. 1), with masses and radii that differ by a factor of ~2 but with nearly identical effective temperatures (spectral types ≈M4), may be exemplars of this smearing effect.

4. Open Questions and Future Work

We concluded with an open discussion leading to a set of questions to help guide ongoing work by the community interested in the "M4 transition" and related phenomena.

1. What field strengths are necessary and reasonable in the interiors of stars to potentially explain the observed mass-radius relationship?

2. Abundance measurements needed! (e.g. [O/Fe]) The mass-radius relationship is actually a mass-radius-abundance relationship. FGK+M binaries are a promising approach, tying M dwarf abundances to FGK primary stars.

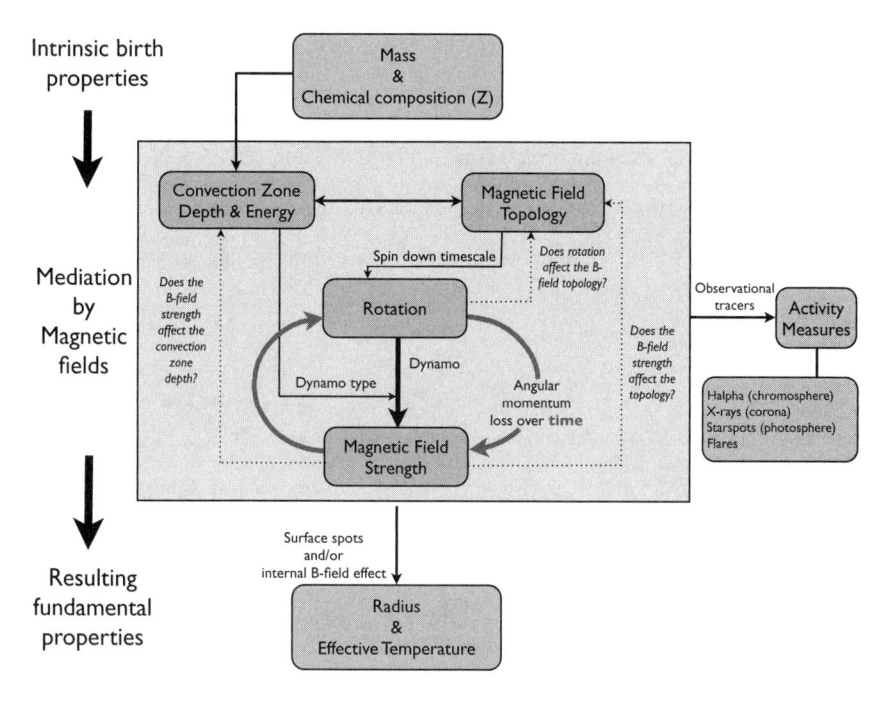

Figure 10. Representation of the process by which magnetic fields influence the interior structure and observed fundamental properties of M dwarf stars. M dwarfs are defined intrinsically by mass and chemical composition which determine the initial depth of the convection zone. This in turn affects the magnetic field topology and strength through the dynamo type (α–Ω or α^2). The rotation rate, which is moderated by angular momentum loss through stellar winds determined largely by the field topology, in turn affects the field strength for stars in the non-saturated regime via the dynamo process. X-rays, chromospheric emission, starspots, and flaring all trace the magnetic activity. Over time, stars spin down and the magnetic field weakens, causing the entire system to evolve. It remains unclear how much feedback there is in the system. Does the strength of the magnetic field affect the topology or the convection zone depth, and how much does the rotation rate influence the field topology? Taken together, the interlinked magnetic fields, internal structure, and rotation properties affect the resulting fundamental parameters: radius and temperature. However, the detailed mechanism by which the latter occurs is not yet fully understood.

3. Continue to identify low-mass eclipsing binaries. Long-period systems are needed to disentangle effects of magnetic activity on stellar radii and temperatures.

4. Examine selection effects in samples, especially with regard to activity. Ages of stellar samples matters.

5. What is the mass of the fully convective boundary? How are the rotations and magnetic activity of stars affected across the fully convetive boundary? Does it depend on age?y

6. Other observational clues: Period gap in CVs (2–3 hr periods) → kink in mass-radius relationship?

7. Independent radius measurements: interferometry, $P_{rot} + v\sin i$ (gives $R\sin i$)
8. Bulk of field is in very small-scale structures → How to fully characterize the fields? Constraints from many techniques/tracers are needed to fully characterize magnetic fields: Zeeman broadening, spectropolarimetry, radio, X-rays, etc.
9. "M6–M8 transition" in field topologies, caused by neutral atmospheres? Age?
10. How does the fully- to partly-convective evolution on the pre–main-sequence affect the early magnetic and rotational history of low-mass stars?

References

Baraffe, I., Chabrier, G., Allard, F., & Hauschildt, P. H. 1998, A&A, 337, 403
Barnes, S. A. 2003, ApJ, 586, 464. arXiv:astro-ph/0303631
— 2007, ApJ, 669, 1167. 0704.3068
Bouvier, J. 2009, in EAS Publications Series, edited by C. Neiner & J.-P. Zahn, vol. 39 of EAS Publications Series, 199. 0810.4850
Browning, M. K. 2008, ApJ, 676, 1262. 0712.1603
Chabrier, G., & Baraffe, I. 1997, A&A, 327, 1039
— 2000, ARA&A, 38, 337
Chabrier, G., Gallardo, J., & Baraffe, I. 2007, A&A, 472, L17
Clemens, J. C., Reid, I. N., Gizis, J. E., & O'Brien, M. S. 1998, ApJ, 496, 352. arXiv:astro-ph/9710304
Demory, B., Ségransan, D., Forveille, T., Queloz, D., Beuzit, J., Delfosse, X., di Folco, E., Kervella, P., Le Bouquin, J., Perrier, C., Benisty, M., Duvert, G., Hofmann, K., Lopez, B., & Petrov, R. 2009, A&A, 505, 205
Donati, J., Morin, J., Petit, P., Delfosse, X., Forveille, T., Aurière, M., Cabanac, R., Dintrans, B., Fares, R., Gastine, T., Jardine, M. M., Lignières, F., Paletou, F., Velez, J. C. R., & Théado, S. 2008, MNRAS, 390, 545. 0809.0269
Dorman, B., Nelson, L. A., & Chau, W. Y. 1989, ApJ, 342, 1003
Henry, T. J., & McCarthy, D. W., Jr. 1993, AJ, 106, 773
Jardine, M., Donati, J., Arzoumanian, D., & de Vidotto, A. 2010, ArXiv e-prints. 1008.4885
Jardine, M., Wood, K., Collier Cameron, A., Donati, J., & Mackay, D. H. 2002, MNRAS, 336, 1364. arXiv:astro-ph/0207522
Lane, B. F., Boden, A. F., & Kulkarni, S. R. 2001, ApJ, 551, L81
MacDonald, J., & Mullan, D. J. 2009, ApJ, 700, 387
Mohanty, S., & Basri, G. 2003, in The Future of Cool-Star Astrophysics: 12th Cambridge Workshop on Cool Stars, Stellar Systems, and the Sun, edited by A. Brown, G. M. Harper, & T. R. Ayres, vol. 12, 683
Morales, J. C., Ribas, I., & Jordi, C. 2008, A&A, 478, 507
Morales, J. C., et al. 2009, ApJ, 691, 1400. 0810.1541
Morin, J., Donati, J., Petit, P., Delfosse, X., Forveille, T., Albert, L., Aurière, M., Cabanac, R., Dintrans, B., Fares, R., Gastine, T., Jardine, M. M., Lignières, F., Paletou, F., Ramirez Velez, J. C., & Théado, S. 2008, MNRAS, 390, 567. 0808.1423
Mullan, D. J., & MacDonald, J. 2001, ApJ, 559, 353
Phan-Bao, N., Lim, J., Donati, J., Johns-Krull, C. M., & Martín, E. L. 2009, ApJ, 704, 1721
Reiners, A., & Basri, G. 2008, ApJ, 684, 1390. 0805.1059
Ribas, I. 2006, Ap&SS, 304, 89. arXiv:astro-ph/0511431
Stassun, K. G., Mathieu, R. D., & Valenti, J. A. 2006, Nat, 440, 311
— 2007, ApJ, 664, 1154. 0704.3106
West, A. A., Hawley, S. L., Bochanski, J. J., Covey, K. R., Reid, I. N., Dhital, S., Hilton, E. J., & Masuda, M. 2008, AJ, 135, 785. 0712.1590

The 16th Cambridge Workshop on Cool Stars, Stellar Systems and the Sun
ASP Conference Series, Vol. 448
Christopher M. Johns-Krull, Matthew K. Browning, and Andrew A. West, eds.
©2011 Astronomical Society of the Pacific

Fundamental Stellar Properties from Optical Interferometry

Gerard T. van Belle,[1] Jason Aufdenberg,[2] Tabetha Boyajian,[3,9] Graham Harper,[4] Christian Hummel,[1] Ettore Pedretti,[5] Ellyn Baines,[6] Russel White,[3] Vikram Ravi,[7] and Steve Ridgway[8]

[1]*European Southern Observatory, Karl-Schwarzschild-Str. 2, 85748 Garching, Germany*

[2]*Embry-Riddle Aeronautical University, 600 S. Clyde Morris Blvd. Daytona Beach, Florida 32114, USA*

[3]*Department of Physics and Astronomy, Georgia State University, P.O. Box 4106, Atlanta, GA 30302-4106, USA*

[4]*Astrophysics Research Group, School of Physics, Trinity College, Dublin 2, Ireland*

[5]*School of Physics and Astronomy, University of St Andrews, North Haugh, St Andrews, Fife, KY16 9SS, UK*

[6]*Remote Sensing Division, Naval Research Laboratory, 4555 Overlook Avenue SW, Washington, DC 20375, USA*

[7]*University of California, Berkeley Space Sciences Lab 7 Gauss Way Berkeley, CA 94720-7450, USA*

[8]*Kitt Peak National Observatory National Optical Astronomy Observatories P.O. Box 26732 Tucson, AZ 85726-6732, USA*

[9]*Hubble Fellow*

Abstract. High-resolution observations by visible and near-infrared interferometers of both single stars and binaries have made significant contributions to the foundations that underpin many aspects of our knowledge of stellar structure and evolution for cool stars. The CS16 splinter on this topic reviewed contributions of optical interferometry to date, examined highlights of current research, and identified areas for contributions with new observational constraints in the near future.

1. Introduction

Observations of the fundamental parameters of stars - their masses, radii and effective temperatures - are crucial components in understanding stellar formation and evolution. For many types of stars, obtaining measurements with sufficient precision to constrain theoretical models can be challenging. A discussion of the contributions of optical interferometry to this field is timely given the technological maturation and recent scientific contributions of modern facilities such as VLTI, the Keck Interferometer, and the

CHARA Array. Furthermore, a promising road lies ahead, as the 2-telescope measurements used in most of the work described below, are now being expanded by multi-way multi-channel combiners at the VLTI and CHARA which provide significant additional constraints by adding closure phase, differential phase, and in a few cases, imaging.

Areas where optical interferometry is making compelling contributions towards parameterizing the fundamental properties of cool stars include: **Radius** - The sub-milliarcsecond resolutions of modern optical interferometers allow for direct measurement of the linear sizes of nearby cool stars of all luminosity classes. **Effective temperature** - Angular sizes, in combination with measured spectral energy distributions, provide direct quantification of this macroscopic quantity. **Mass** - Dynamical masses from orbits determined with optical interferometers and radial velocity measurements are the highest-precision determinations of this most important fundamental parameter. Additionally, asteroseismological measurements of single stars provide direct measures of stellar mean density; coupled with interferometric radii, mass determinations are possible. **Distance** - Parallactic measurements of stellar distances from optical interferometers calibrate stellar luminosities. **Temperature Structure** - Limb-darkening measurements from optical interferometers probe the vertical temperature structure of a stellar atmosphere, providing constraints on model atmosphere structures in general, and models for convective flux transport in particular.

Specific scientific questions that were considered during the splinter session were: (1) What have been the constraints on fundamental parameters for cool stars provided by optical interferometry? What are the limits on those constraints? (2) Where are the most attractive areas for guiding development of cool stars astrophysical models with observations from optical interferometers? (3) Are there specific areas of cool star evolution that are particularly well suited for studies by optical interferometry? (4) What are the observing opportunities available today for cool star research? (5) What are the prospects for future developments in optical interferometry that can significantly advance our knowledge of cool stars?

Answers to these questions are of considerable interest to the Cool Stars 16 audience. Furthermore, the most recent event formally addressing fundamental stellar parameters was over a decade ago[1], predating the current generation of operational facilities, and did not specifically address cool stars.

2. Angular Sizes

With available mass estimates, temperature and size predictions of stellar evolutionary models as a function of age can be compared to values obtained with interferometry. Historically, this had been limited to eclipsing binary systems (Andersen 1991). Long-baseline optical/infrared interferometry has changed this dramatically by providing sufficient resolution to measure the angular sizes of dozens of nearby stars. Angular size measurements with errors under 1% are now possible. The limb-darkening corrections, normally modeled, are now subject to direct interferometric verification (e.g. Lacour et al. 2008).

---

[1]IAU 189 in Sydney, Australia, "Fundamental Stellar Properties: The Interaction Between Observation and Theory" (T. R. Bedding, A. J. Booth, & J. Davis 1997)

Measurements of single M-dwarf radii show that they are potentially 10-15% larger than currently predicted by models (Berger et al. 2008) with a suggestion that the discrepancy increases with elevated metallicity, although not all studies are in agreement on this point (Demory et al. 2009). Examples of ultra-precise radii and temperatures have been measured for coeval binary stars (Kervella et al. 2008), metal poor population II stars (Boyajian et al. 2008), giant stars in the Hyades (Boyajian et al. 2009), pulsating Mira variables (Thompson et al. 2002; van Belle et al. 2002), exoplanet host stars (van Belle & von Braun 2009) and calibration of the giant and supergiant effective temperature scales (van Belle et al. 1999, 2009, respectively). In the precise measurements of radii, interferometry has also revealed very low levels of circumstellar emission around main sequence stars (Ciardi et al. 2001; Absil et al. 2008; Akeson et al. 2009) that are too faint and close to the star to be otherwise detected.

Furthermore, interferometry is showing how limited the very concept of effective temperature can be. Rapidly rotating stars have significant temperature gradients on their surfaces, but even more leisurely stars like Procyon cannot be fit precisely with a single $T_{\rm EFF}$ model. For Procyon, multi-wavelength angular diameters from Mark III and VLTI showed that 1-D mixing length convection doesn't correctly predict stellar limb darkening (Aufdenberg et al. 2005). Recent 3-D models of the solar atmosphere (Asplund et al. 2009) have significantly revised the solar abundance; such 3-D models for more distant stars can be tested with interferometric limb darkening measurements in ways that spectroscopy alone cannot.

2.1. Observational Results of Diameters, Temperatures - Tabetha Boyajian

Studies targeting the discovery of previously unknown nearby stars, such as RECONS (Henry et al. 2006; Subasavage et al. 2008), have uncovered a wealth of new objects to be examined with optical interferometers - particularly cool main sequence stars. Angular diameters measured by interferometry can be combined with distances (well known for nearby objects) to produce linear radii; with bolometric fluxes to directly produce effective temperatures ($T_{\rm EFF} \propto (F_{\rm BOL}/\theta^2)^{1/4}$). There are well over 500 objects for which angular sizes have been measured (Richichi et al. 2005), roughly a four-fold increase over just a decade ago (Davis 1997), with many of the new measurements being produced for main sequence and not evolved objects (Boyajian 2009).

For stars similar to the Sun, comparison of the stellar effective temperature measured directly versus semi-empirically (Allende Prieto & Lambert 1999; Holmberg et al. 2007; Takeda 2007) show the temperature values are lower (by $\sim 1.5 - 4\%$), and radii values are higher (by $\sim 4 - 10\%$) then those semi-empirical approaches (Boyajian 2009). These offsets are such that the resulting luminosities, regardless of method, are in agreement with each other. Additionally, Boyajian (2009) point out that no correlation is seen when investigating the deviation in temperature and radii with respect to color index or metallicity.

Ages and masses of these nearby, single, stars measured with interferometry can be found by fitting Y^2 isochrones (Yi et al. 2001; Kim et al. 2002; Demarque et al. 2004) to the directly measured temperatures and luminosities, and these results compare very well with results from eclipsing binaries in Andersen (1991) (see Figure 2.1). Alternatively, the masses of this same sample of stars are derived by combining spectroscopic surface gravity measurements log g with the measured interferometric radii (see Figure 2.1). However, it is apparent here that the spectroscopic log g measurements tend to be overestimated, leading to over-predicted masses (and younger ages), and is thought to be a direct consequence of the spectroscopic temperature being over-estimated.

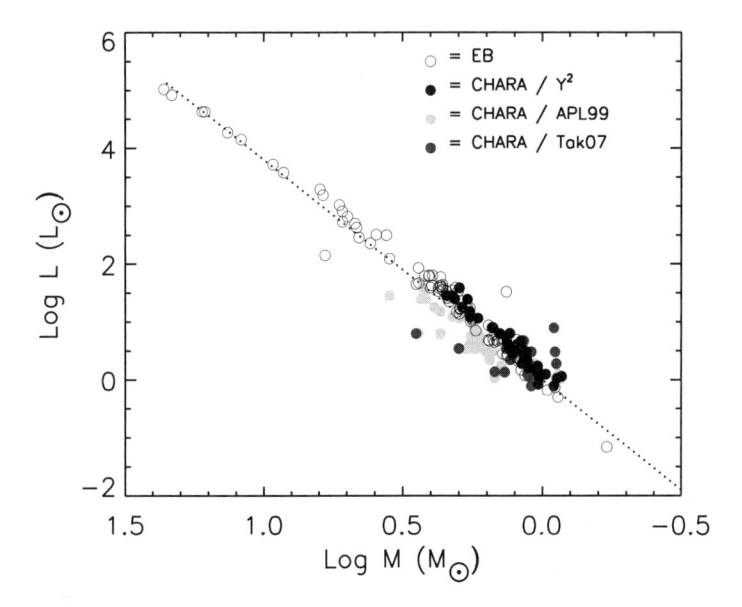

Figure 1. Mass-Luminosity plot of non-evolved eclipsing binary (EB) stars in Andersen (1991) (open circles). Also shown are data from a large sample of main sequence solar-type stars measured with interferometry Boyajian (2009) where masses are found from fitting the directly determined temperatures and luminosities to the Yonsei-Yale (Y^2) isochrones. Masses for these same data are solved for when applying the interferometrically measured radii with spectroscopic log g values from Allende Prieto & Lambert (1999) and Takeda (2007), shown as green and blue filled points, respectively.

A solid foundation to build empirical relations to stellar temperature and radius are beginning to emerge with the sensitivity of current interferometers. Temperature scales for giants and supergiants are explored in van Belle et al. (1999, 2009) and it is shown that we have now reached accuracy of ~ 2.5% level, only to be limited by the distances to these objects. Empirical relations to the temperatures and radii of main sequence stars are reviewed from compiling the data available in van Belle & von Braun (2009) and Boyajian (2009). Additionally we introduce new results from a large interferometric survey of late-type dwarfs at the CHARA Array (Boyajian et al., in preparation) and exoplanet host stars (von Braun et al., in preparation), which double the number of published values to-date for these types of stars. Statistically, there is no difference between the luminosity class of an object and the T_{EFF} versus $(V - K)$ or T_{EFF} versus spectral type relations until the main sequuence stars are introduced, showing much steeper slopes, and somewhat of a dis-continuity at spectral type ~ M1 ($V - K \sim 4$). These empirical solution for a muliti-parameter T_{EFF}:Color:Metallicity relation based on these data has now finally reached the 1% level.

We conclude this discussion with a look into the of the disagreement with stellar radii as predicted by models compared to observations. We include the (currently unpublished) work mentioned above from (Boyajian et al., in preparation) and (von Braun et al., in preparation) along with all other interferometric observations of K-M type dwarfs presented in (Lane et al. 2001; Ségransan et al. 2003; Boyajian et al. 2008; Kervella et al. 2008; Demory et al. 2009; van Belle & von Braun 2009). Figure 2.1 shows the mass-radius relationship for binary and single K and M dwarfs. We see that for both single and binary stars, models begin to underpredict radii by an average of $\sim 10\%$ (N.B. the bottom panel restricts the data plotted to radii measurements better than 5%). We also notice a difference in mixing lengths for stars $> 0.6 M_\odot$, where the single stars are modeled better with larger mixing lengths.

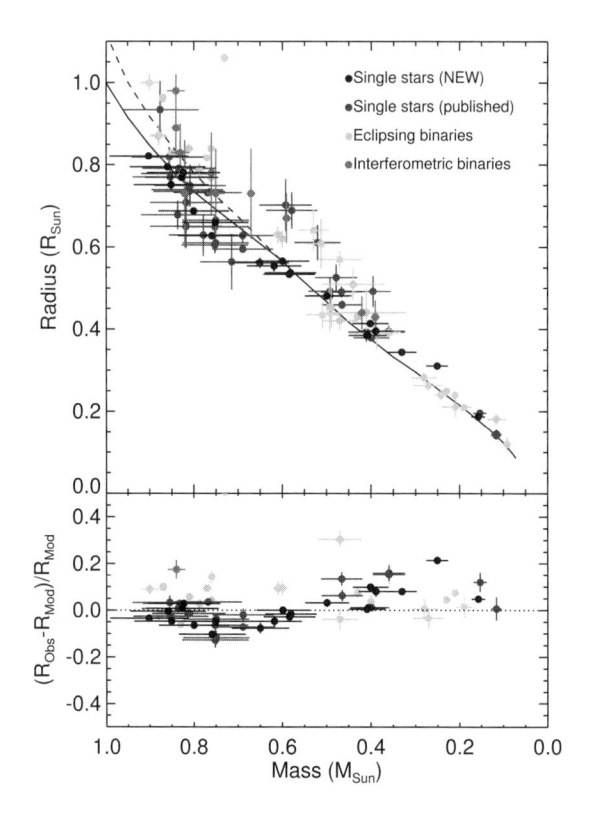

Figure 2. *TOP* The Mass-Radius relationship for K and M stars. The solid black line is a 5 Gyr isochrone from the BCAH98 models (Baraffe et al. 1998). For stars with mass > 0.6 M$_\odot$, the dashed line indicates $L_{mix} = H_p$ and the solid line indicates $L_{mix} = 1.9 H_p$. *BOTTOM* Deviation in radius versus mass for stars with radii measurements better than 5%. Masses for single stars are derived from the K-band mass-luminosity relation from Delfosse et al. (2000), and assume a 10% error.

2.2. Diameters and Modeling - Jason Aufdenberg

Historically speaking, the characterizations of stellar limb darkening have only been done for a single object - our Sun. The Sun is the only star for which angular resolution has been sufficiently fine to examine limb darkening laws predicted by stellar atmosphere theory. However, using our Sun as a guide is an imperfect approach, since this object does not represent the full range of possibilities for stellar atmospheric structure.

Recently interferometric observations of α Cen B have been compared against the predictions of stellar limb darkening obtained from both 1D ATLAS atmospheric models and 3D radiative hydrodynamic (RHD) simulations Bigot et al. (2006). Significantly, since convection is fundamentally a three dimensional process, the 1D models have an inherent weakness in proper treatment of this phenomenon. It is no surprise that there are deviations in the 1D and 3D models, with the latter showing less limb darkening - resulting in diameters that are 0.1% smaller in the near-IR, and up to 1.5% smaller in the optical.

Further observational evidence that we are departing the era where 1D models are sufficient is seen in the data on Procyon presented in Aufdenberg et al. (2005). Comparison of models to the measured stellar diameters of Procyon at both the optical and near-IR wavelengths shows a good match to 3D model predictions as well as 1D models which include an overshooting correction. Furthermore, the nature of 3D models allow for a range of temperature structures, consistent with granulation structure (like that which is known to appear on the solar surface) and its multi-spectral nature, required to match Procyon's spectral energy distribution. This also appears to be the case for 3D models of late-type giants where 3-D models show significant flux deviations (~5-20%) in the visible and shortwards, along with notable deviations in the near-infrared (~5%) (Kučinskas et al. 2009) relative to 1D models. Precision angular diameter measurements are needed in the optical, for example on α Cen B (K1 V), where the 3D convection models now applied to our Sun can be further tested.

Early closure phase imaging experiments have shown that giant and supergiant stars have mottled, inhomogenous surfaces (Tuthill et al. 1997). Recent comparisons of interferometric data against 1D models have show increasing differences with increasing spatial frequency data (Haubois et al. 2009), indicative of poor treatment of convection. In contrast, 3D RHD models show much better agreement when confronted by similar data sets (Chiavassa et al. 2009, 2010). Extending these 3D models from gray opacities to multi-wavelength radiative transfer is a necessary next step. The current use of gray opacities means the thermal gradient is likely too shallow, and neglect of radiation pressure means the predicted atmosphere is probably too compact. Multi-wavelength angular diameters of objects like α Ori in the near- and mid-IR (Perrin et al. 2004) indicate a strong dependence on diameter with wavelength, a dependence only exacerbated at higher spectral resolution (Wittkowski et al. 2004, 2006).

3. Orbits and Masses - Christian Hummel

Arguably the most fundamental parameter for a star is its mass, as this sets the timescale for evolution and determines the star's ultimate fate. The high resolution capabilities of optical interferometry have the potential to greatly increase the number and types of stars for which we have dynamical mass estimates, and greatly improve overall mass estimates. The traditional technique of using eclipsing binaries to provide these mea-

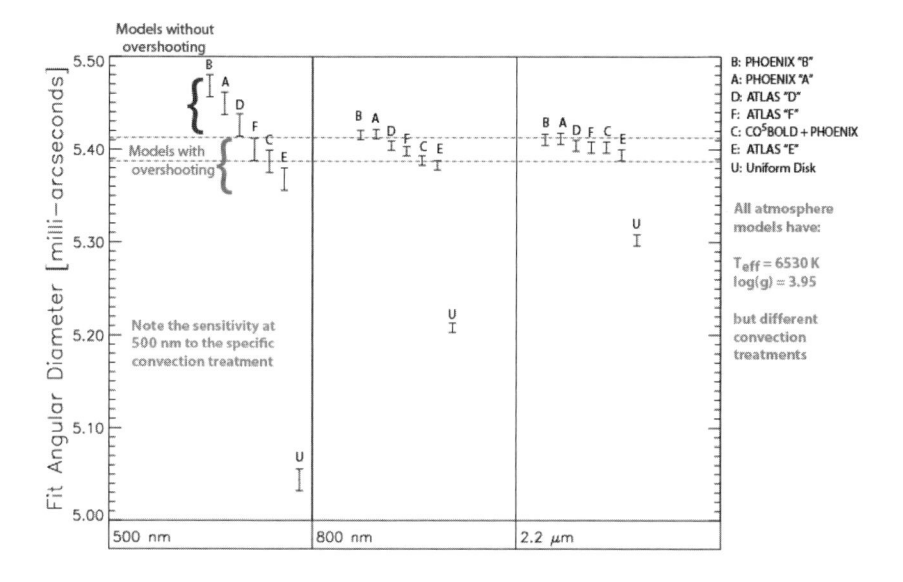

Figure 3. Comparison of the best-fit angular diameters of Procyon at 500 nm, 800nm and 2.2 microns for seven atmosphere models with different convection treatments. Based on a figure from Aufdenberg et al. (2005)

surements limits studies to biased samples where a history of mass exchange is more likely - high angular resolution is needed to have the freedom to observe "typical" binaries

Although an early example of interferometric observations of a binary can be found as early as 40 years ago (Hanbury Brown et al. 1970), routine modern observations began roughly two decades later with small-aperture interferometers in the visual and NIR going after easy targets to show the power of milli-arcsec resolution, with a limiting magnitude about V=5 (e.g. Armstrong et al. 1992). After many successful orbit determinations on systems that were 'forgiving' from the standpoint of brightness and/or orbital parameters, the emphasis is now being put on observing astrophysically more challenging and interesting targets. Simply put, interferometry of these objects is now a mature technique offered in service mode at Keck and VLTI.

Dynamical masses of young binaries from interferometry (Boden et al. 2005a; Schaefer et al. 2008) are an area where only interferometry allows the study of non-eclipsing systems: mass estimates at this early age are especially important because of the poorly understood input physics (e.g. convection, opacities) of pre-main sequence stars. With precise distance estimates (some from FGS interferometry on board HST, e.g. Benedict et al. 2006) relative orbits will provide dynamical masses of stars in rarer evolutionary states, such as those transitioning to the giant phases (Boden et al. 2005b) and traversing the Hertzsprung Gap (Boden et al. 2006). Additionally, only just recently have the prospects of high contrast imaging via non-redundant aperture masking on

524   van Belle et al.

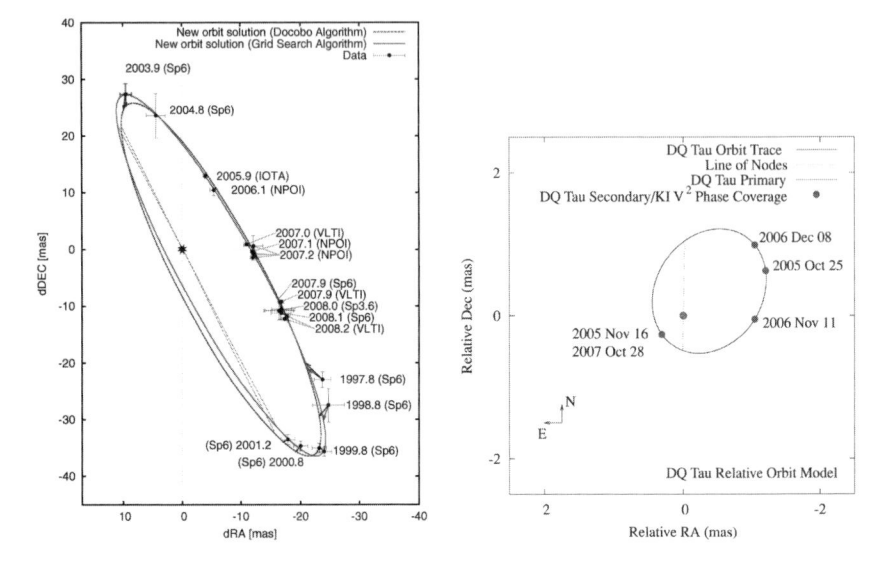

Figure 4. Observations of binaries being done with modern interferometers: (Left) the orbit of θ^1 Ori C, from Figure 8 from Kraus et al. (2009), using data from IOTA/NPOI/VLTI; (right) the orbit of DQ Tau from Figure 3 of Boden et al. (2009) with KI data.

large-diameter telescopes been realized. This work can provide dynamical masses for sub-stellar objects (Ireland et al. 2008), a mass range where evolutionary models are very poorly constrained due to the age/temperature/mass degeneracies. As an example of the current state of the art, the results found in Kraus et al. (2009) and Boden et al. (2009) are seen in Figure 3.

References on binary star work that are useful include the general review of optical interferometry in astronomy by Monnier (2003) and the recent examination of accurate masses and radii of normal stars byTorres et al. (2010). Online, the web sites for "Optical Long Baseline Interferometry News"[2] and "Database of Publications in Stellar Interferometry"[3] are both useful.

4. Recent Observational Results and Advances

4.1. CHARA MIRC - Ettore Pedretti

The Georgia State University (GSU) Center for High Angular Resolution Astronomy (CHARA) Array is a six-element optical interferometer (ten Brummelaar et al. 2005) that is beginning to produce true dilute aperture images. With the Michigan Infrared Combiner (MIRC) instrument (Monnier et al. 2006), initial 4-way imaging has begun to produce remarkable results. It is further interesting to note that the VLBA has a

---

[2] http://olbin.jpl.nasa.gov/index.html

[3] http://apps.jmmc.fr/bibdb/

physical size of roughly 10,000km (Hawaii to Puerto Rico), while its optical analog in the CHARA Array spans only 330m - and yet it has comparable angular resolution, given the shorter wavelength of operation. The capability to directly image of the surfaces of hotter stars such as Altair (Monnier et al. 2007) and α Oph & α Cep (Zhao et al. 2009) is now being directed towards cooler objects such as the cloud occulting the ϵ Aur (Kloppenborg et al. 2010) or ζ And. The latter object is of particular interest given availability of Doppler imaging data for comparison (Kővári et al. 2007). An upgrade to full 6-telescope operation will further enhance the 'snapshot' capabilities of the system (Monnier et al. 2010).

4.2. The 'A-List' - Ellyn Baines

In addition to their many other peculiarities, the ages of A-type stars are poorly known. While these objects are on the main sequence, they evolve in T_{EFF} and in L/R, which affords an opportunity to establish their ages through interferometric means. By obtaining their angular sizes, and deriving values for R and T_{EFF}, observational data can be compared to models on an H-R diagram, which will then indicate their ages and masses. Such investigations are particularly timely given the recent discovery of potential planetary objects orbiting HR8799 (Marois et al. 2008, 2010).

For this particular star, its estimated age is 30-160Myr - based on a evidence that only poorly constrains that parameter: the object's galactic space motion, its placement on a color-magnitude diagram, and 'typical' ages for λ Boo or γ Dor stars (Marois et al. 2008, and references therein). The current age estimate of ~60Myr for HR8799 is presented with considerable uncertainty.

Marois et al. (2008) indicate that an age of > 300Myr would be necessary for all companion objects to be brown dwarfs. An initial investigation of this object using the CHARA Array has preliminary results that could be consistent with this object being well in excess of this age; this result is consistent with the asteroseismic finding of Moya et al. (2010).

4.3. Young Star Sizes - Russel White

Using temperature, photometry, and distance measurements of young stars in the nearest and youngest moving groups (β Pictoris and AB Dor), we predict the stellar radii and angular sizes of known members to identify if any can be spatially resolved with the long-baseline near-infrared interferometers operating in the northern and southern hemispheres. The motivation is to potentially constrain pre-main sequence evolutionary models from direct radius measurements. In the northern hemisphere ($\delta > -20^o$), 3 stars have sizes large enough to be spatially resolved ($\theta \gtrsim 0.4$) with the CHARA Array, which has a baseline of 331-m; this subsample includes the low mass M2 dwarf GJ 393 which is near the fully convective boundary. All 3 have now been successfully spatially resolved with a precision of of a few percent; the analysis of these results is still maturing. In the southern hemisphere ($\delta < +20^o$), 9 stars have sizes larger than this angular diameter, including the high-profile low mass debris disk star AU Mic. However, the current longest baselines of the VLTI are not able to resolve any of these systems; opening new longer-baseline stations may alleviate this problem.

4.4. Betelgeuse with the ISI, 2006-2009 - Vikram Ravi

Closure phase observations of Betelgeuse (α Ori) at 11 μm over the four year interval 2006-2009 have been able to probe surface features and the size of this object as it evolves in time (Wishnow et al. 2010). During this interval, contributions of individual spots to the overall surface appear with a measure of time variability, possibly consistent with evolution of convection cells on the surface. The effective surface temperature of the star appears to drift slightly in time as well, from 2750 ± 100K in 2006 to 2350 ± 250K in 2008, back up to 2650 ± 75 in 2009. This temperature variation is found to be directly correlated with the size of the star. Ongoing operations of the ISI for 2010-2011 include a upgrade for very high spectral resolution, which can simultaneously measure spectral lines and continuum.

5. The 'Outsider Perspective' on Where Interferometry Can Contribute - Graham Harper

There are a number of areas where even a small number of 'magic bullet' interferometric observations can make significant contributions. (This list is intended to be instructive rather than comprehensive.)

#1 Break the photospheric models. For the current stock of photospheric models a certain number of assumptions are inherent in their construction - assumptions which do not always hold good. For example, outer boundary conditions, assumptions of hydrostatic equilibrium, dimensionality of model (e.g. 1D versus 3D), all can lead to weaknesses in the models - weakness which can be exposed by probing those stellar candidates that are most sensitive to such assumptions. The best current candidates are bright giants & supergiants, which have lower surface gravity and less convective cells; later spectral types are also better in that the influence of molecules and neutral opacity goes up. The best observational techniques to employ here are measurements of wavelength-dependent limb darkening, taking data beyond the first null of the visibility curve.

#2 Even more accurate stellar θ_{LD}. Specifically, measures where $\Delta\theta/\theta \leq 1$-2%. There are a number applications: first spectroscopic absolute flux calibrations are reaching the ~5% level, and stellar chromospheric surface fluxes can be measured at this level. These reflect the radiative cooling that matches the chromospheric heating, whose origin is a source of active research. Since $F_\oplus \propto \theta^2 F_\star$, and $\Delta F_\star/F_\star \sim 2\Delta\theta/\theta + \Delta F_\oplus/F_\oplus$, the desire is for 1-2% errors on θ to match the flux errors. Second, when considering ALMA observations, errors in radio brightness temperature (T_{Br}) are related to angular size errors in a way that is directly analogous to our first case, e.g. $\Delta T_{Br}/T_{Br} \sim 2\Delta\theta/\theta + \Delta F/F$. However, when considering ALMA maps where the errors in mean electron temperature, T_e, are directly proportional to T_{Br}, a 2% error in T_e in the mid-chromosphere can lead to errors in electron density of $\Delta n_e/n_e \sim 100\%$.

#3 Calibrate surface brightness relations. In particular, for late-type low amplitude variability K, M and C-stars. An important question is, what is limiting the precision of existing surface brightness relations? For the foreseeable future there are always going to be stars that do not have measured angular diameters, so it is important to understand the limitations of such indirect techniques.

#4 Geometry and gravity darkening. The rotationally oblate star Altair exhibits unusual geometry and gravity darkening due to its rapid rotation (van Belle et al. 2001; Ohishi et al. 2004; Monnier et al. 2007). Similarly, gravitational darkening may be present in eclipsing binary ζ Aurigae systems distorted near periastron (Guinan & McCook 1979; Eaton et al. 2008). Resolving out the orbital geometry *and* gravity darkening of such systems would be instructive for stellar structure studies.

In summary, there are a number of things that are desirable to know. Specifically, and non-exhaustively: (1) How significant are existing differences in limb-darkening fits? (2) For what spectral-types and luminosity classes do MOL-spheres exist. What is their distribution in the HR diagram? (3) Can we resolve the geometry/gravity darkening in convective low gravity stars? (4) What limits the precision of surface brightness relations?

Of particular value for addressing such questions raised by non-interferometrists, in a timely fashion, would be some mechanism for requesting θ_{LD} for "special interest" stars. Perhaps a simple web-based system could be established where astronomers could post their needs and potential collaborations identified.

6. Opportunities for Observers - Steve Ridgway

Currently there are a number of facilities world-wide for which optical interferometric observations are possible. There are two facility-class installations (Keck Interferomeer, VLTI) with open calls, and a third instrument with some open time (CHARA). Additionally, there are three that are primarily accessible through collaborations with members (CHARA, NPOI, SUSI). Additionally, a substantial body of unpublished archival data from the now-closed Palomar Testbed Interferometer is available online.

7. Conclusion

Optical interferometry is a challenging technique that rewards those who take up that challenge with observational data unobtainable via any other approach. In particular, contributions of the technique in the area of fundamental parameters of cool stars are substantial, highlighting areas of concern and interest for theorists. Furthermore, the facilities that provide access to this technique have matured substantially over the last 10 years, making it more accessible to all manners of astronomer.

Acknowledgments. The splinter organizers would like to thank the splinter organizing committee (Rachel Akeson, Jason Aufdenberg, Andrew Boden, Tabetha Boyajian, Michelle Creech-Eakman, Christian Hummel, Stephen Ridgway, Peter Tuthill, GvB) for their contributions to the development of the splinter program, and to the organizers of CS16 for providing the splinter opportunity.

References

Absil, O., et al. 2008, A&A, 487, 1041. 0806.4936
Akeson, R. L., et al. 2009, ApJ, 691, 1896. 0810.3701
Allende Prieto, C., & Lambert, D. L. 1999, A&A, 352, 555. arXiv:astro-ph/9911002
Andersen, J. 1991, A&A Rev., 3, 91
Armstrong, J. T., et al. 1992, AJ, 104, 2217
Asplund, M., Grevesse, N., Sauval, A. J., & Scott, P. 2009, ARA&A, 47, 481. 0909.0948
Aufdenberg, J. P., Ludwig, H., & Kervella, P. 2005, ApJ, 633, 424. arXiv:astro-ph/0507336
Baraffe, I., Chabrier, G., Allard, F., & Hauschildt, P. H. 1998, A&A, 337, 403. arXiv:astro-ph/9805009
Benedict, G. F., et al. 2006, AJ, 132, 2206. arXiv:astro-ph/0610247
Berger, D. H., et al. 2008, in 14th Cambridge Workshop on Cool Stars, Stellar Systems, and the Sun, edited by G. van Belle, vol. 384 of Astronomical Society of the Pacific Conference Series, 226

Bigot, L., Kervella, P., Thévenin, F., & Ségransan, D. 2006, A&A, 446, 635
Boden, A. F., Akeson, R. L., Sargent, A. I., Carpenter, J. M., Ciardi, D. R., Bary, J. S., & Skrutskie, M. F. 2009, ApJ, 696, L111. 0903.2521
Boden, A. F., Torres, G., & Hummel, C. A. 2005a, ApJ, 627, 464. arXiv:astro-ph/0502250
Boden, A. F., Torres, G., & Latham, D. W. 2006, ApJ, 644, 1193. arXiv:astro-ph/0601515
Boden, A. F., et al. 2005b, ApJ, 635, 442. arXiv:astro-ph/0508331
Boyajian, T. S. 2009, Ph.D. thesis, Georgia State University
Boyajian, T. S., et al. 2008, ApJ, 683, 424. 0804.2719
— 2009, ApJ, 691, 1243. 0810.2238
Chiavassa, A., Haubois, X., Young, J. S., Plez, B., Josselin, E., Perrin, G., & Freytag, B. 2010, A&A, 515, A12+. 1003.1407
Chiavassa, A., Plez, B., Josselin, E., & Freytag, B. 2009, A&A, 506, 1351. 0907.1860
Ciardi, D. R., van Belle, G. T., Akeson, R. L., Thompson, R. R., Lada, E. A., & Howell, S. B. 2001, ApJ, 559, 1147. arXiv:astro-ph/0105561
Davis, J. 1997, in IAU Symposium, edited by T. R. Bedding, A. J. Booth, & J. Davis, vol. 189 of IAU Symposium, 31
Delfosse, X., Forveille, T., Ségransan, D., Beuzit, J., Udry, S., Perrier, C., & Mayor, M. 2000, A&A, 364, 217. arXiv:astro-ph/0010586
Demarque, P., Woo, J., Kim, Y., & Yi, S. K. 2004, ApJS, 155, 667
Demory, B., et al. 2009, A&A, 505, 205. 0906.0602
Eaton, J. A., Henry, G. W., & Odell, A. P. 2008, ApJ, 679, 1490. 0802.2238
Guinan, E. F., & McCook, G. P. 1979, PASP, 91, 343
Hanbury Brown, R., Davis, J., Herbison-Evans, D., & Allen, L. R. 1970, MNRAS, 148, 103
Haubois, X., et al. 2009, A&A, 508, 923. 0910.4167
Henry, T. J., Jao, W., Subasavage, J. P., Beaulieu, T. D., Ianna, P. A., Costa, E., & Méndez, R. A. 2006, AJ, 132, 2360. arXiv:astro-ph/0608230
Holmberg, J., Nordström, B., & Andersen, J. 2007, A&A, 475, 519. 0707.1891
Ireland, M. J., Kraus, A., Martinache, F., Lloyd, J. P., & Tuthill, P. G. 2008, ApJ, 678, 463. 0801.1525
Kervella, P., et al. 2008, A&A, 488, 667. 0806.4049
Kővári, Z., Bartus, J., Strassmeier, K. G., Oláh, K., Weber, M., Rice, J. B., & Washuettl, A. 2007, A&A, 463, 1071
Kim, Y., Demarque, P., Yi, S. K., & Alexander, D. R. 2002, ApJS, 143, 499. arXiv:astro-ph/0208175
Kloppenborg, B., et al. 2010, Nat, 464, 870. 1004.2464
Kraus, S., et al. 2009, A&A, 497, 195. 0902.0365
Kučinskas, A., Ludwig, H., Caffau, E., & Steffen, M. 2009, MmSAI, 80, 723. 0910.3412
Lacour, S., et al. 2008, A&A, 485, 561. 0804.0192
Lane, B. F., Boden, A. F., & Kulkarni, S. R. 2001, ApJ, 551, L81
Marois, C., Macintosh, B., Barman, T., Zuckerman, B., Song, I., Patience, J., Lafrenière, D., & Doyon, R. 2008, Science, 322, 1348. 0811.2606
Marois, C., Zuckerman, B., Konopacky, Q. M., Macintosh, B., & Barman, T. 2010, ArXiv e-prints. 1011.4918
Monnier, J. D. 2003, Reports on Progress in Physics, 66, 789. arXiv:astro-ph/0307036
Monnier, J. D., et al. 2006, in Society of Photo-Optical Instrumentation Engineers (SPIE) Conference Series, vol. 6268 of Society of Photo-Optical Instrumentation Engineers (SPIE) Conference Series
— 2007, Science, 317, 342. 0706.0867
— 2010, in Society of Photo-Optical Instrumentation Engineers (SPIE) Conference Series, vol. 7734 of Society of Photo-Optical Instrumentation Engineers (SPIE) Conference Series
Moya, A., Amado, P. J., Barrado, D., García Hernández, A., Aberasturi, M., Montesinos, B., & Aceituno, F. 2010, MNRAS, 405, L81. 1003.5796
Ohishi, N., Nordgren, T. E., & Hutter, D. J. 2004, ApJ, 612, 463. arXiv:astro-ph/0405301
Perrin, G., Ridgway, S. T., Coudé du Foresto, V., Mennesson, B., Traub, W. A., & Lacasse, M. G. 2004, A&A, 418, 675. arXiv:astro-ph/0402099

Richichi, A., Percheron, I., & Khristoforova, M. 2005, A&A, 431, 773
Schaefer, G. H., Simon, M., Prato, L., & Barman, T. 2008, AJ, 135, 1659. 0802.1692
Ségransan, D., Kervella, P., Forveille, T., & Queloz, D. 2003, A&A, 397, L5. arXiv: astro-ph/0211647
Subasavage, J. P., Henry, T. J., Bergeron, P., Dufour, P., & Hambly, N. C. 2008, AJ, 136, 899. 0805.2515
T. R. Bedding, A. J. Booth, & J. Davis (ed.) 1997, Fundamental stellar properties: the interaction between observation and theory. Proceedings. 189th Symposium of the International Astronomical Union, Sydney (Australia), 13 - 17 Jan 1997., vol. 189 of IAU Symposium
Takeda, Y. 2007, PASJ, 59, 335
ten Brummelaar, T. A., et al. 2005, ApJ, 628, 453. arXiv:astro-ph/0504082
Thompson, R. R., Creech-Eakman, M. J., & van Belle, G. T. 2002, ApJ, 577, 447
Torres, G., Andersen, J., & Giménez, A. 2010, A&A Rev., 18, 67. 0908.2624
Tuthill, P. G., Haniff, C. A., & Baldwin, J. E. 1997, MNRAS, 285, 529
van Belle, G. T., Ciardi, D. R., Thompson, R. R., Akeson, R. L., & Lada, E. A. 2001, ApJ, 559, 1155
van Belle, G. T., Creech-Eakman, M. J., & Hart, A. 2009, MNRAS, 394, 1925. 0811.4239
van Belle, G. T., Thompson, R. R., & Creech-Eakman, M. J. 2002, AJ, 124, 1706. arXiv: astro-ph/0210167
van Belle, G. T., & von Braun, K. 2009, ApJ, 694, 1085. 0901.1206
van Belle, G. T., et al. 1999, AJ, 117, 521
Wishnow, E. H., Mallard, W., Ravi, V., Lockwood, S., Fitelson, W., Wertheimer, D., & Townes, C. H. 2010, in Society of Photo-Optical Instrumentation Engineers (SPIE) Conference Series, vol. 7734 of Society of Photo-Optical Instrumentation Engineers (SPIE) Conference Series
Wittkowski, M., Aufdenberg, J. P., Driebe, T., Roccatagliata, V., Szeifert, T., & Wolff, B. 2006, A&A, 460, 855. arXiv:astro-ph/0610150
Wittkowski, M., Aufdenberg, J. P., & Kervella, P. 2004, A&A, 413, 711. arXiv:astro-ph/0310128
Yi, S., Demarque, P., Kim, Y., Lee, Y., Ree, C. H., Lejeune, T., & Barnes, S. 2001, ApJS, 136, 417. arXiv:astro-ph/0104292
Zhao, M., Monnier, J. D., Pedretti, E., Thureau, N., Mérand, A., ten Brummelaar, T., McAlister, H., Ridgway, S. T., Turner, N., Sturmann, J., Sturmann, L., Goldfinger, P. J., & Farrington, C. 2009, ApJ, 701, 209. 0906.2241

Determining the Metallicity of Low-Mass Stars and Brown Dwarfs: Tools for Probing Fundamental Stellar Astrophysics, Tracing Chemical Evolution of the Milky Way and Identifying the Hosts of Extrasolar Planets

Andrew A. West,[1,10] John J. Bochanski,[2,3] Brendan P. Bowler,[4] Aaron Dotter,[5] John A. Johnson,[6] Sebastian Lépine,[7] Bárbara Rojas-Ayala,[8] and Andreas Schweitzer[9]

[1] *Department of Astronomy, Boston University, 725 Commonwealth Avenue, Boston, MA 02215, USA, email: aawest@bu.edu*

[2] *Astronomy & Astrophysics Dept., Pennsylvania State University, 525 Davey Lab, University Park, PA, 16802, USA, email: jjb29@psu.edu*

[3] *Kavli Institute for Astrophysics and Space Research, Massachusetts Institute of Technology, Building 37, 77 Massachusetts Avenue, Cambridge, MA 02139, USA*

[4] *Institute for Astronomy, University of Hawai'i; 2680 Woodlawn Drive, Honolulu, HI 96822, USA, email: bpbowler@ifa.hawaii.edu*

[5] *Space Telescope Science Institute, 3700 San Martin Dr., Baltimore, MD, 21218, USA, email: aaron.dotter@gmail.com*

[6] *Department of Astrophysics, California Institute of Technology, MC 249-17, Pasadena, CA 91125, USA, email: johnjohn@astro.caltech.edu*

[7] *Department of Astrophysics, Division of Physical Sciences, American Museum of Natural History, Central Park West at 79th Street, New York, NY 10024, USA, email: lepine@amnh.org*

[8] *Department of Astronomy, Cornell University, 610 Space Sciences Building, Ithaca, NY 14853, USA, email: babs@astro.cornell.edu*

[9] *Hamburger Sternwarte, University of Hamburg, Gojenbergsweg 112, D-21029, Hamburg, Germany, email: Andreas.Schweitzer@hs.uni-hamburg.de*

[10] *Visiting Investigator, Department of Terrestrial Magnetism, Carnegie Institute of Washington, 5241 Broad Branch Road, NW, Washington, DC 20015, USA*

Abstract. We present a brief overview of a splinter session on determining the metallicity of low–mass dwarfs that was organized as part of the Cool Stars 16 conference. We review contemporary spectroscopic and photometric techniques for estimating metallicity in low–mass dwarfs and discuss the importance of measuring accurate metallicities for studies of Galactic and chemical evolution using subdwarfs, creating metallicity benchmarks for brown dwarfs, and searching for extrasolar planets that are orbiting around low–mass dwarfs. In addition, we present the current understanding of the effects of metallicity on stellar evolution and atmosphere models and discuss some of the limitations that are important to consider when comparing theoretical models to data.

1. Introduction

Low–mass dwarfs are the most numerous stellar constituents of the Milky Way and have main sequence lifetimes that exceed the current age of the Universe (at least for those that are not brown dwarfs). They therefore form an important laboratory for probing the structure and evolution of the Milky Way's disks. Because of their ubiquity, cool dwarfs may represent the largest population of stars with orbiting planets, especially low–mass planets in their respective habitable zones, which are considerably closer for cool dwarf systems. In addition, the diminutive sizes of these stars makes the detection of transiting planets easier than for higher mass stars (for any given planetary radius). Previous results have demonstrated that planets are more likely to be found orbiting metal-rich stars (e.g. Fischer & Valenti 2005). There were preliminary indications that the M dwarfs with known planets had sub-solar metallicities (Bonfils et al. 2005; Bean et al. 2006), in stark contrast to their high-mass counterparts. However, recent results have shown that the M dwarfs with attending planets appear to be metal-rich (see Section 2; Johnson & Apps 2009). With low–mass dwarfs becoming important sites for planet hunting (e.g. MEarth; Irwin et al. 2009; Endl et al. 2003; Johnson et al. 2007), the observational efficiency of these searches could be vastly increased with prior knowledge of stellar metallicity.

Because the ages of low–mass dwarfs span the lifetime of the Milky Way, they can provide important insight into the history and evolution of the Galaxy. Recent improvements in kinematic modeling and magnetic activity analysis have provided enhanced statistical age estimates for populations of low–mass dwarfs (West et al. 2006, 2008). Coupled with metallicity information, these ages can provide valuable insight into the chemical evolution history of the Milky Way disks. Without large samples of low–mass dwarfs, the utility of the statistically derived ages is limited. Fortunately, the advent of large surveys such as SDSS and 2MASS has produced photometric samples of low–mass dwarfs that number in the tens of millions (Bochanski et al. 2010) and spectroscopic samples that contain more than 70,000 M dwarfs (West et al. 2008; Kruse et al. 2010; West et al. 2011) and almost 500 L dwarfs (Schmidt et al. 2010b). In addition, these large catalogs of low–mass dwarfs have identified significant samples of metal–poor subdwarfs. The detailed metallicities of these objects, coupled with their kinematic distributions, establish important constraints on the structure and composition of the Milky Way halo.

Historically, the metallicity of low–mass dwarfs has been an elusive fundamental property due to the complex atmospheres of M, L and T dwarfs that have restricted the accuracy of detailed model atmospheres. Over the past several years, new observational techniques as well as independent theoretical advancements in atmospheric models have produced results that appear to link the metallicity of low–mass dwarfs to both their photometric and spectroscopic properties (e.g., Bean et al. 2006; Bonfils et al. 2005; Woolf & Wallerstein 2006; Johnson & Apps 2009; Hauschildt & Baron 2010; Rojas-Ayala et al. 2010). While these relations and the resulting metallicities provide fundamental measurements for stellar astrophysics, they also play a crucial role in studies of Galactic evolution and the environments that host extrasolar planets.

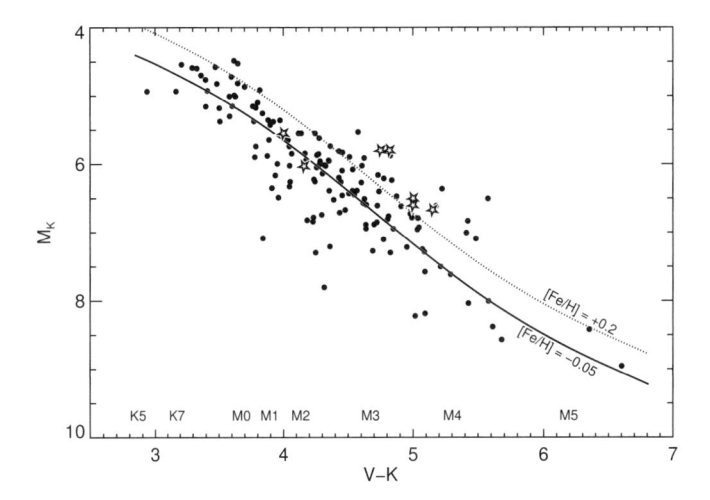

Figure 1. The low–mass target stars of the California Planet Survey in the M_K vs $V - K$ plane (filled circles), and the M dwarfs known to harbor one or more gas giant planets (five-point stars). The isometallicity contours for [Fe/H] = 0 (solid) and [Fe/H] = +0.2 (dotted line) are based on the broad-band photometric metallicity calibration of Johnson & Apps (2009).

2. Calibrating M Dwarf Metallicity Using Photometry

Several previous studies have estimated M dwarf metallicities using wide binary pairs that consist of both an M dwarf and a higher mass star (e.g., Bean et al. 2006; Bonfils et al. 2005; Woolf & Wallerstein 2006). Because binaries are assumed to be both coeval and have the same metallicity, the composition of the higher mass star (which can be accurately derived from comparison to theoretical models) can be applied to the companion M dwarf. Some of these studies have used optical and infrared spectroscopy to tie spectroscopic features to a metallicity scale (e.g., Bean et al. 2006; Woolf et al. 2009; Rojas-Ayala et al. 2010). Although the spectroscopic method has shown great promise in deriving M dwarf metallicities (see Section 3), spectroscopy is considerably more time consuming than photometry and may not be easy to obtain for large samples of stars.

Bonfils et al. (2005) used M dwarfs in wide binaries to derive a relation between the absolute K-band magnitude and the $V - K$ color (higher metallicity M dwarfs are slightly brighter at a given color). Given the large number of M dwarfs for which there exist photometric observations, this relation may prove exceedingly useful. However, there were 2 problems with resulting analyses: 1) using the Bonfils et al. (2005) relation, planet hosting M dwarfs appeared to be metal poor compared to their FGK star counterparts; and 2) and the relation yielded a mean metallicity of M dwarfs in the solar neighborhood that was almost 0.1 dex below the mean [Fe/H] of higher mass stars. These discrepancies were resolved by Johnson & Apps (2009), who discovered a systematic uncertainty in the photometry used by Bonfils et al. (2005). Johnson & Apps

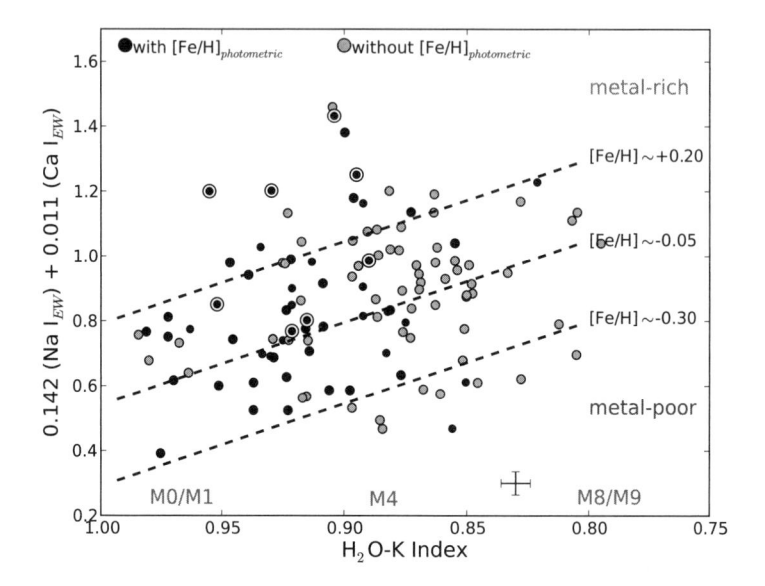

Figure 2. A linear combination of the EWs of the Ca I and Na I features versus the H$_2$O-K index for northern 8 pc M-dwarfs. The black dots represent M dwarfs with photometric metallicities and the yellow dots represent M dwarfs with only near-infrared (NIR) spectroscopic metallicities. The big black dots (with circles) represent the M dwarf planet hosts. Typical errors in EWs and H$_2$O-K index are represented by the error bars. The dashed lines in the top panel are iso-metallicity contours for [Fe/H] values of -0.30, -0.05 and +0.20, calculated from the NIR [Fe/H] calibration. The NIR [Fe/H] calibration allows to cover a larger sample of cooler and distant M dwarfs (yellow dots)

(2009) used corrected photometry to re-derive a relation between the metallicity, M_K and $V - K$ color of M dwarfs (see also Schlaufman & Laughlin 2010).

Figure 1 shows the low–mass target stars of the California Planet Survey in the M_K vs $V - K$ plane (filled circles), and the M dwarfs known to harbor one or more gas giant planets (five-point stars). The isometallicity contours for [Fe/H] = 0 (solid) and [Fe/H] = +0.2 (dotted line) are based on the broad-band photometric metallicity calibration of Johnson & Apps (2009). The distribution of stars illustrates the tendency of planet-hosting M dwarfs to be metal-rich compared to stars in the Solar Neighborhood. The planet-metallicity relationship therefore holds for M dwarfs as well as Sun-like FGK stars.

3. Calibrating M Dwarf Metallicity Using Infrared Spectroscopy

Most of the attempts to estimate the overall metal content of M dwarfs have been performed at visible wavelengths (e.g., Gizis 1997; Bonfils et al. 2005; Johnson & Apps 2009). Since M dwarfs are optically faint, this limited past analyses to early-type M dwarfs and few specific nearby stars, which are bright and have accurate parallaxes. To avoid this limitation, Rojas-Ayala et al. (2010) developed a near–infrared (NIR)

[Fe/H] spectroscopic calibration using strong absorption features in the K-band spectra of M dwarfs. Rojas-Ayala et al. (2010) adopted a similar approach to that of Bonfils et al. (2005) and Johnson & Apps (2009), assuming that binary systems share the same metallicity since both components formed from the same original molecular cloud. Seventeen FGK+M binary systems in the SPOCS catalog (Valenti & Fischer 2005) were used as metallicity calibrators. The NIR [Fe/H] calibration uses the Equivalent Widths (EWs) of the Na I doublet and the Ca I triplet, and a water absorption index (H_2O, Covey et al. 2010) to differentiate between metal-rich and metal-poor M dwarfs ($\sigma \sim 0.15$ dex). The results obtained with the NIR spectroscopic [Fe/H] calibration are in agreement with the results obtained with the photometric calibration by Johnson & Apps (2009). The eight M dwarf planet hosts analyzed by Rojas-Ayala et al. (2010) have metallicities higher than -0.05 dex, with the Jovian planets hosts being more metal-rich that their Neptune analogs. This corroborates the Johnson & Apps (2009) conclusion that planets are found preferentially around metal-rich stars, like in their Sun-like counterparts.

As a moderate resolution K-band spectrum can be efficiently obtained for most M dwarfs with current spectrographs (e.g. TripleSpec, FIRE), the NIR [Fe/H] calibration allows observations of cooler and distant M dwarfs (Figure 2). Thus, this technique will enable the identification of likely planet hosts at lower masses than is possible with optical [Fe/H] techniques. However, the NIR [Fe/H] calibration is currently limited to M dwarf spectral types earlier than $\sim$ M7 and [Fe/H] > -0.7, due to the lack of FGK dwarf/late-type M dwarf wide binary systems with measured spectroscopic metallicities, and subdwarfs with $\lambda/\Delta\lambda \approx 3000$ NIR spectra, to be used as calibrators.

4. Spectral Features of Low-Metallicity Brown Dwarfs

Brown dwarfs are expected to have a similar metallicity distribution to the stellar components of our Galaxy, but reliably determining the chemical compositions of individual brown dwarfs is a difficult task. Atmospheric models remain largely untested at non-solar metallicities, and there are no known benchmark brown dwarf companions to stars with significantly super- or sub-solar chemical compositions ([Fe/H] $\gtrsim$ +0.3 or [Fe/H] $\lesssim$ –0.3). The latest-type ultracool subdwarf companion known is the d/sdM9 benchmark HD 114762B ([Fe/H]= -0.7); atmospheric models do a reasonably good job of reproducing the medium-resolution ($\lambda/\Delta\lambda \sim 3800$) near-infrared spectral features of this object, but fits to the low resolution ($\lambda/\Delta\lambda \sim 150$) near-infrared spectrum are unreliable (Bowler et al. 2009).

Although atmospheric models are not yet grounded by brown dwarfs with known metallicities, trends in the models have provided qualitative indications of deviations from solar metallicity for a growing number of L and T dwarfs with peculiar spectra. The optical spectra of peculiar L dwarfs are marked most notably by enhanced metal-hydride and metal-oxide bands compared to normal L dwarfs of the same spectral type (Figure 3). These variations are likely caused by subsolar metallicities and possibly suppressed condensate formation. A reduced metallicity also increases collision-induced absorption by H_2 (CIA H_2), resulting in bluer NIR colors for a given optical spectral type. Cloud properties also influence the NIR colors of L dwarfs and there is no clear way to distinguish clouds from a mild deviation from solar metallicity from NIR colors or spectra alone (Burgasser et al. 2008). Among the $\sim$20 known objects that make up this class of "blue L dwarfs" (Kirkpatrick et al. 2010), which is distinct from L subdwarfs, two benchmark blue L dwarfs provide important clues about the

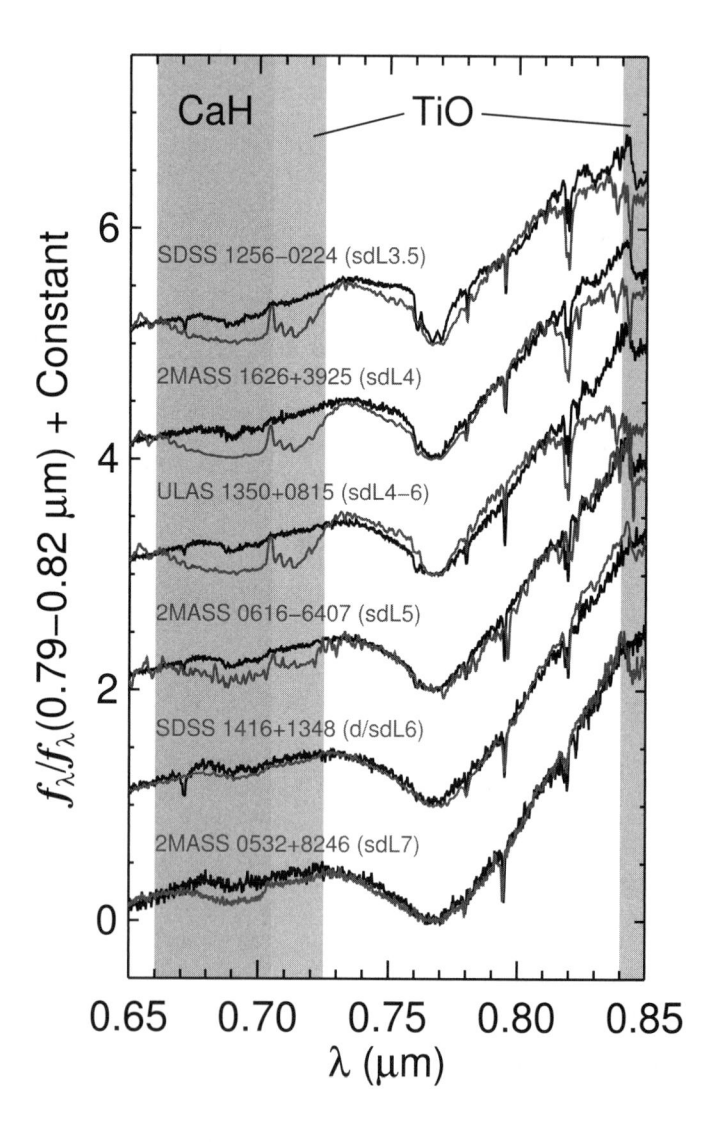

Figure 3. Optical spectra of L subdwarfs (red). The most notable differences in metal-poor L dwarfs compared to ordinary field objects are a stronger CaH absorption band at 6800 Å and stronger TiO absorption bands at 7100 Å and 8400 Å. From top to bottom the optical spectra originate from Burgasser et al. (2009), Burgasser et al. (2007), Lodieu et al. (2010), Cushing et al. (2009), Bowler et al. (2010), and Burgasser et al. (2003). Comparison spectra (black) are from Kirkpatrick et al. (1999); from top to bottom they are 2MASS 1146+2230 (L3), 2MASS 1155+2307 (L4), DENIS-P J1228.2−1547 (L5), DENIS-P J1228.2−1547 (L5), 2MASS 0850+1057 (L6), and DENIS-P J0205.4−1159 (L7). The spectra are normalized between 7900 Å and 8200 Å and are offset by a constant.

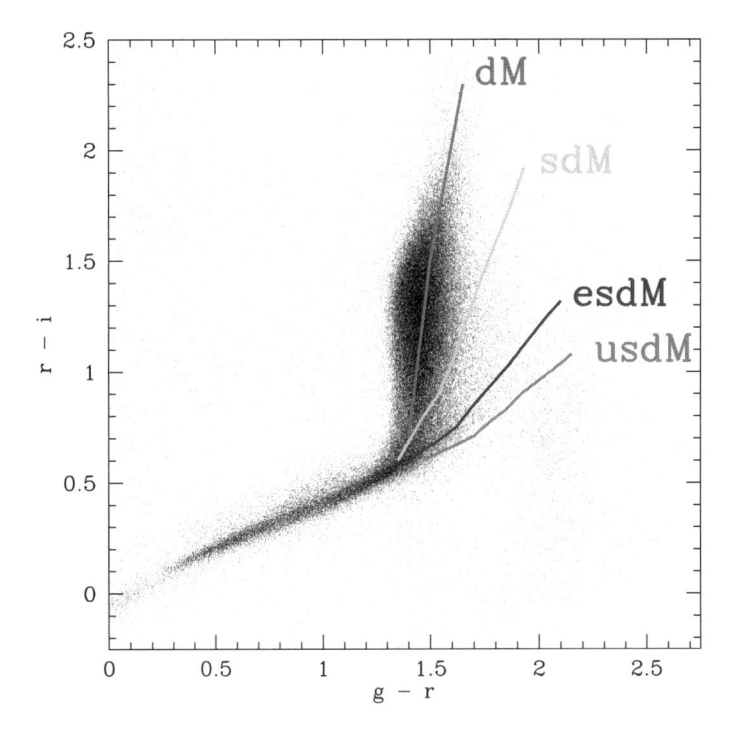

Figure 4. The $r-i$ vs. $g-r$ color-color digram for stars in the SDSS spectroscopic catalog. The thick colored lines show the mean loci for the 4 metallicity classes of M dwarfs (dM:red; sdM:green; esdM:blue; and usdM :purple).

nature of the spectral peculiarities. The blue L dwarf 2MASS J17114559+4028578 (L4.5) orbits a solar-metallicity star (Radigan et al. 2008) and the blue L dwarf SDSS J141624.08+134826.7 (d/sdL6, Bowler et al. 2010; Schmidt et al. 2010a) has a peculiar T7.5 companion with spectral features indicative of being mildly metal-poor (Burningham et al. 2010; Burgasser et al. 2010); this is the first evidence that blue L dwarfs may span a range of metallicities. For T dwarfs, gravity and metallicity both affect the K-band flux by influencing CIA H_2 (e.g., Liu et al. 2007). Metallicity (and to a lesser extent gravity) also affects the Y-band flux, offering a way to distinguish between these parameters (e.g., Leggett et al. 2007). Ongoing sensitive all-sky surveys like WISE and Pan-STARRS are expected to greatly increase the census of non-solar metallicity isolated and benchmark L and T dwarfs, enabling rigorous testing of atmospheric models and an empirical calibration of spectral classification schemes.

5. The Colors and sub-Classes of Subdwarfs in SDSS

Low–mass stars with very low metallicities, typical of the Galactic thick disk and halo population, have a spectral energy distribution that is significantly different from the more metal-rich disk stars. The reason lies in the reduced absorption from metal oxide

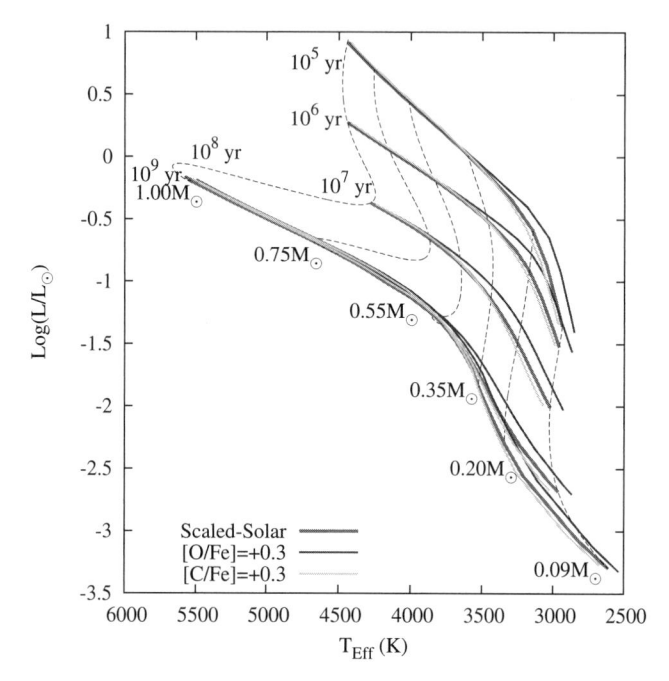

Figure 5. Luminosity vs. temperature diagram for a series of isochrones with masses between 0.09 and 1 $M_\odot$ and ages between 10^5 and 10^9 years with [Fe/H]=0. The red lines have Solar C and O abundances, while the blue and green lines are enhanced in O and C respectively.

bands, in particular TiO. M dwarfs are classified in four so-called "metallicity classes" based on the relative strengths of their TiO bands: from the metal-rich dwarf M dwarfs (dM), to subdwarfs (sdM), extreme subdwarfs (esdM), and the very metal-poor ultra-subdwarfs (usdM). The classification follows the system of Gizis (1997) recently upgraded by Lépine et al. (2007). The sequence usdM→esdM→sdM→dM is believed to form a sequence of increasing metallicity (Gizis & Reid 1997; Woolf et al. 2009), with [Fe/H]≈-0.5 for sdM, [Fe/H]≈-1.0 for esdM, and [Fe/H]$\lesssim$-1.5 for usdM, although the metallicity calibration remain relatively uncertain to this date.

A recent search of the Sloan Digital Sky Survey (SDSS) spectroscopic database (Lépine et al. in preparation) has produced over 7,600 M subdwarfs. Their color distribution reveals significant differences with the metallicity class. Figure 4 shows the $r - i$ color as a function of $g - r$. The dots show a typical distribution for nearby field stars, displaying the well-known "elbow" with a strong inflection point at $g - r = 1.4$, $r - i = 0.6$. The thick colored lines show the mean loci for the dM (red), sdM (green), esdM (blue), and usdM (purple). There is a clear segregation as a function of the metallicity class, which reflects the strong effect that the TiO bands have on the spectral energy distribution. Ultrasubdwarfs simply extend the linear relationship between $g - r$ and $r - i$, as one would expect from a blackbody. As the metallicity increases, the "elbow" becomes increasingly pronounced. This happens because the r-band gets increasingly depressed in the more metal rich stars, as the TiO opacity increases. This strong dependence of color on metallicity opens the possibility of estimating metallicities in low–mass stars based on broadband photometry alone. As it turns out, even dM show a significant scatter in $g - r$ which could be entirely explained by differences in metallicity. Should this be confirmed, this would provide a formidable tool for quick and easy metallicity estimates of low–mass stars. A proper calibration of the $g - r$ and $r - i$ color terms as a function of metallicity should be a priority.

6. Metallicity and Stellar Evolution Models

"Metallicity" loosely describes the heavy element content of a star or stellar population. Metallicity and [Fe/H] are often used interchangeably, with the implicit assumption that the other heavy elements scale with Fe as they do in the Sun. If they don't, then it's important to understand how changing a given element alters the spectrum, hence the opacity, hence the effective temperature scale of the star (Dotter et al. 2007).

After H and He, the two most abundant elements in the sun by mass or number fraction are C and O (e.g., Asplund et al. 2009). When C or O is enhanced relative to solar at fixed [Fe/H] the most dramatic effect appears in the molecular opacities. Figure 5 shows a series of isochrones with masses between 0.09 and 1 $M_\odot$ and ages between 10^5 and 10^9 years with [Fe/H]=0. As Figure 5 indicates, on the one hand, enhancing C actually makes the lowest mass stars hotter while, on the other hand, enhancing O makes them cooler. This behavior can be understood in terms of the contribution of water molecules to the opacity. When analyzing the physical properties of low mass stars with effective temperatures below about 4,000K it is important to consider that non-solar abundance ratios can skew the results.

7. Metallicity and Atmosphere Models

One of the primary tools for measuring metallicities of low–mass objects is the comparison of data to synthetic spectra (created using atmospheric models). Calculating such models is common practice (e.g., Sordo et al. 2010) using modern atmosphere codes (Hauschildt & Baron 1999). For low–mass objects in general, it is crucial to account for dust formation in the atmosphere. This dust formation needs to be treated as a microphysical growth and destruction process (e.g. Helling et al. 2008). Furthermore, molecules, both as an opacity source and as material that affects the equation of state, need to be accounted for with accurate input data such as formation constants or related quantities and line lists or equivalent opacity data. The situation is further complicated when calculating models specifically for low metallicity objects, such as the models of Witte et al. (2009). The decreasing metal content does not change or even simplify any of the main physical processes, but, in contrast, adds another dimension of parameter space. In particular, dust keeps forming in significant amounts down to metallicities of about [Fe/H]=-4.0 (Witte et al. 2009).

Recently, it has become possible to apply synthetic spectra to observations of L subdwarfs and to attempt to measure metallicities (Burgasser et al. 2009). However, the quality of the fits and the derived metallicities still vary (Witte et al. 2010) depending on the quality of the implemented physics (see also Fig. 6). However, it is worth noting that the derived metallicities of the sdL class do not need to be the same of the sdM class as derived by e.g. Gizis (1997) or Schweitzer (1999). Measuring metallicities directly

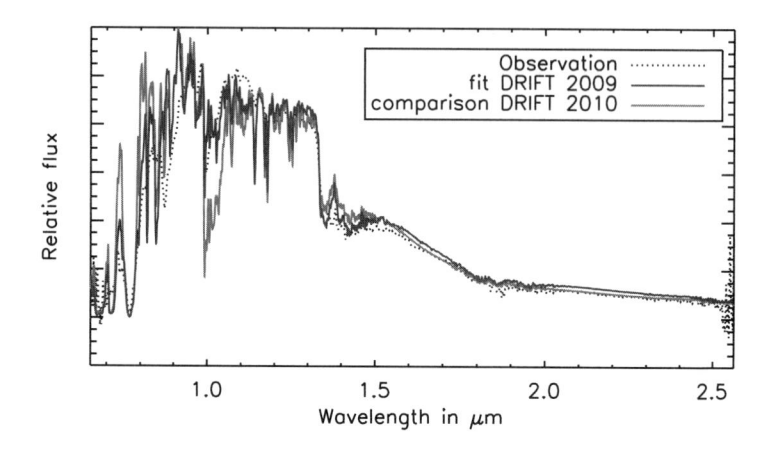

Figure 6. The sdL4 dwarf 2MASS1626+3925 (black, dotted; Burgasser 2004) and a fit with the DRIFT 2009 models (blue, solid; Witte et al. 2009). The comparison with the DRIFT 2010 models (red, solid; Witte et al. 2010) is a comparison with the same model parameters showing the differences in model details. Both models have $T_{\text{eff}} = 2100$K, $\log(g)=5.0$ and a metallicity [Fe/H]= -1.5.

at a resolution of 0.1 dex or higher has not been attempted yet since the molecular background lines add too much uncertainty.

8. Conclusions

During the first half of the last century, spectroscopic observations and radiative transfer theory began to unlock the composition of stars. Determining the metallicity of stars has been very important to a wide range of astronomical investigations, from planetary to cosmological scales. Yet, despite the progress made for most of the main sequence, measuring the metallicity of the Galaxy's most populous members, M dwarfs, remains a daunting task.

At the start of this century, astronomers are beginning to unlock the metal content of these stars. However, there is much work to be done by both observers and theorists. Observationally, there are promising new results suggesting that IR observations may be important for estimating metallicities. This is strengthened by the relative agreement between models and spectra in this regime. However, these methods need further testing (with M dwarf binaries or clusters). In the optical bandpass, the relative metallicity classes described in Section 5 display a clear separation in photometric colors. This will be crucial for estimating the metal content of these stars in the next generation of surveys, which will be largely photometric. Yet, these classes have not been rigorously tied to an absolute metallicity scale. Once this occurs, the chemical composition of M dwarfs will be a powerful tool for studying the Galaxy and identifying the most likely exoplanet hosts. Identifying new benchmarks, for both M dwarfs and brown dwarfs, will be crucial in calibrating optical observations.

On the theoretical front, new line lists, opacity calculations and the inclusion of dust grains have resulted in better agreement with observations. The effects of carbon and oxygen abundance differences can be modeled and explain observations of the lower main sequence of globular clusters. As computational power and techniques advance, these models will grow in sophistication and should offer a more realistic picture of the important physics within these stars.

Unlocking the metallicity of M dwarfs will profoundly benefit the astronomical community in a variety of ways, such as identifying exoplanet hosts and studying chemical evolution. The work presented here represents the first steps in solving this problem.

Acknowledgments. The authors would like to thank the Cool Stars 16 SOC for the opportunity to hold this productive splinter session. BPB gratefully acknowledges the Cool Stars 16 Accommodation Stipend Award, funded by the NASA Astrobiology Institute.

References

Asplund, M., Grevesse, N., Sauval, A. J., & Scott, P. 2009, ARA&A, 47, 481. 0909.0948
Bean, J. L., Benedict, G. F., & Endl, M. 2006, ApJ, 653, L65. arXiv:astro-ph/0611060
Bochanski, J. J., Hawley, S. L., Covey, K. R., West, A. A., Reid, I. N., Golimowski, D. A., & Ivezić, Ž. 2010, AJ, 139, 2679. 1004.4002
Bonfils, X., Delfosse, X., Udry, S., Santos, N. C., Forveille, T., & Ségransan, D. 2005, A&A, 442, 635. arXiv:astro-ph/0503260
Bowler, B. P., Liu, M. C., & Cushing, M. C. 2009, ApJ, 706, 1114. 0910.1604
Bowler, B. P., Liu, M. C., & Dupuy, T. J. 2010, ApJ, 710, 45. 0912.3796
Burgasser, A. J. 2004, ApJ, 614, L73. arXiv:astro-ph/0409179
Burgasser, A. J., Cruz, K. L., Cushing, M., Gelino, C. R., Looper, D. L., Faherty, J. K., Kirkpatrick, J. D., & Reid, I. N. 2010, ApJ, 710, 1142. 0912.3808
Burgasser, A. J., Cruz, K. L., & Kirkpatrick, J. D. 2007, ApJ, 657, 494. arXiv:astro-ph/0610096
Burgasser, A. J., Kirkpatrick, J. D., Burrows, A., Liebert, J., Reid, I. N., Gizis, J. E., McGovern, M. R., Prato, L., & McLean, I. S. 2003, ApJ, 592, 1186. arXiv:astro-ph/0304174
Burgasser, A. J., Looper, D. L., Kirkpatrick, J. D., Cruz, K. L., & Swift, B. J. 2008, ApJ, 674, 451. 0710.1123
Burgasser, A. J., Witte, S., Helling, C., Sanderson, R. E., Bochanski, J. J., & Hauschildt, P. H. 2009, ApJ, 697, 148. 0903.1567
Burningham, B., Leggett, S. K., Lucas, P. W., Pinfield, D. J., Smart, R. L., Day-Jones, A. C., Jones, H. R. A., Murray, D., Nickson, E., Tamura, M., Zhang, Z., Lodieu, N., Tinney, C. G., & Zapatero Osorio, M. R. 2010, MNRAS, 404, 1952. 1001.4393
Covey, K. R., Lada, C. J., Román-Zúñiga, C., Muench, A. A., Forbrich, J., & Ascenso, J. 2010, ApJ, 722, 971. 1007.2192
Cushing, M. C., Looper, D., Burgasser, A. J., Kirkpatrick, J. D., Faherty, J., Cruz, K. L., Sweet, A., & Sanderson, R. E. 2009, ApJ, 696, 986. 0902.1059
Dotter, A., Chaboyer, B., Ferguson, J. W., Lee, H., Worthey, G., Jevremović, D., & Baron, E. 2007, ApJ, 666, 403. 0706.0808
Endl, M., Cochran, W. D., Tull, R. G., & MacQueen, P. J. 2003, AJ, 126, 3099. arXiv:astro-ph/0308477
Fischer, D. A., & Valenti, J. 2005, ApJ, 622, 1102
Gizis, J., & Reid, I. 1997, PASP, 109, 1233. arXiv:astro-ph/9708244
Gizis, J. E. 1997, AJ, 113, 806. arXiv:astro-ph/9611222
Hauschildt, P. H., & Baron, E. 1999, Journal of Computational and Applied Mathematics, 109, 41. arXiv:astro-ph/9808182

— 2010, A&A, 509, A36+. 0911.3285
Helling, C., Woitke, P., & Thi, W. 2008, A&A, 485, 547. 0803.4315
Irwin, J., Charbonneau, D., Berta, Z. K., Quinn, S. N., Latham, D. W., Torres, G., Blake, C. H., Burke, C. J., Esquerdo, G. A., Fürész, G., Mink, D. J., Nutzman, P., Szentgyorgyi, A. H., Calkins, M. L., Falco, E. E., Bloom, J. S., & Starr, D. L. 2009, ApJ, 701, 1436. 0906.4365
Johnson, J. A., & Apps, K. 2009, ApJ, 699, 933. 0904.3092
Johnson, J. A., Butler, R. P., Marcy, G. W., Fischer, D. A., Vogt, S. S., Wright, J. T., & Peek, K. M. G. 2007, ApJ, 670, 833. 0707.2409
Kirkpatrick, J. D., Looper, D. L., Burgasser, A. J., Schurr, S. D., Cutri, R. M., Cushing, M. C., Cruz, K. L., Sweet, A. C., Knapp, G. R., Barman, T. S., Bochanski, J. J., Roellig, T. L., McLean, I. S., McGovern, M. R., & Rice, E. L. 2010, ApJS, 190, 100. 1008.3591
Kirkpatrick, J. D., Reid, I. N., Liebert, J., Cutri, R. M., Nelson, B., Beichman, C. A., Dahn, C. C., Monet, D. G., Gizis, J. E., & Skrutskie, M. F. 1999, ApJ, 519, 802
Kruse, E. A., Berger, E., Knapp, G. R., Laskar, T., Gunn, J. E., Loomis, C. P., Lupton, R. H., & Schlegel, D. J. 2010, ApJ, 722, 1352. 0911.2712
Leggett, S. K., Marley, M. S., Freedman, R., Saumon, D., Liu, M. C., Geballe, T. R., Golimowski, D. A., & Stephens, D. C. 2007, ApJ, 667, 537. 0705.2602
Lépine, S., Rich, R. M., & Shara, M. M. 2007, ApJ, 669, 1235. 0707.2993
Liu, M. C., Leggett, S. K., & Chiu, K. 2007, ApJ, 660, 1507. arXiv:astro-ph/0701111
Lodieu, N., Zapatero Osorio, M. R., Martín, E. L., Solano, E., & Aberasturi, M. 2010, ApJ, 708, L107. 0912.3364
Radigan, J., Lafrenière, D., Jayawardhana, R., & Doyon, R. 2008, ApJ, 689, 471. 0808.1575
Rojas-Ayala, B., Covey, K. R., Muirhead, P. S., & Lloyd, J. P. 2010, ApJ, 720, L113
Schlaufman, K. C., & Laughlin, G. 2010, A&A, 519, A105+. 1006.2850
Schmidt, S. J., West, A. A., Burgasser, A. J., Bochanski, J. J., & Hawley, S. L. 2010a, AJ, 139, 1045. 0912.3565
Schmidt, S. J., West, A. A., Hawley, S. L., & Pineda, J. S. 2010b, AJ, 139, 1808. 1001.3402
Schweitzer, A. 1999, Ph.D. thesis, PhD Thesis, Landessternwarte Heidelberg/Königstuhl (1999).
Sordo, R., Vallenari, A., Tantalo, R., Allard, F., Blomme, R., Bouret, J., Brott, I., Fremat, Y., Martayan, C., Damerdji, Y., Edvardsson, B., Josselin, E., Plez, B., Kochukhov, O., Kontizas, M., Munari, U., Saguner, T., Zorec, J., Schweitzer, A., & Tsalmantza, P. 2010, Ap&SS, 328, 331
Valenti, J. A., & Fischer, D. A. 2005, ApJS, 159, 141
West, A. A., Bochanski, J. J., Hawley, S. L., Cruz, K. L., Covey, K. R., Silvestri, N. M., Reid, I. N., & Liebert, J. 2006, AJ, 132, 2507. arXiv:astro-ph/0609001
West, A. A., Hawley, S. L., Bochanski, J. J., Covey, K. R., Reid, I. N., Dhital, S., Hilton, E. J., & Masuda, M. 2008, AJ, 135, 785. 0712.1590
West, A. A., et al. 2011, AJ, in press
Witte, S., Helling, C., & Hauschildt, P. H. 2009, A&A, 506, 1367. 0908.3597
Witte, S., et al. 2010, A&A, submitted
Woolf, V. M., Lépine, S., & Wallerstein, G. 2009, PASP, 121, 117
Woolf, V. M., & Wallerstein, G. 2006, PASP, 118, 218. arXiv:astro-ph/0510148

Waiting for the boat to the Cool Stars 16 banquet at Tillicum Village

Author Index

Aarnio, A. N., 43
Adamson, A., 339
Agüeros, M. A., 269
Agol, E., 403
Allard, F., 91, 339
Allers, K., 31
Artigau, É, 187
Aufdenberg, J., 517

Baines, E., 517
Barber, R. J., 339
Barman, T., 481
Barman, T. S., 147
Barnes, R., 391
Baron, E., 403
Basri, G. S., 177
Bastian, N., 361
Bastien, F., 415
Beeck, B., 403
Beichman, C. A., 53
Benz, A., 455
Berdyugina, S., 403
Berger, E., 455
Bessell, M. S., 61, 131, 429
Biller, B., 31
Birkby, J., 469
Bochanski, J. J., 347, 531
Bowler, B. P., 531
Boyajian, T., 517
Boyle, R. P., 219
Brickhouse, N. S., 23
Brown, B. P., 277
Browning, M. K., 277
Brun, A. S., 277
Burningham, B., 339, 379, 429
Butler, R. F., 219

Canovas, H., 15
Cantiello, M., 155
Catalán, S., 139

Cauzzi, G., 441
Chen, W., 285
Close, L. M., 31
Coughlin, J. L., 121
Covey, K., 505
Covey, K. R., 269, 361, 415
Crockett, C. J., 53
Cruz, K., 481
Cummings, J., 293
Cushing, M. C., 429

Day-Jones, A. C., 339, 379, 429
Deacon, N. R., 429
Debattista, V. P., 371
Delgado Mena, E., 81
Dhital, S., 429
Domínguez Cerdeña, C., 81
Domagal-Goldman, S. D., 391
Dotter, A., 531
Drake, S., 293
Dupuy, T., 31
Dupuy, T. J., 111

Ercolano, B., 469

Faherty, J. K., 481
Fernandes, J., 81
Fitelson, W., 207
Flaccomio, E., 415
Flaherty, K., 415
Fletcher, L., 441
Forbrich, J., 455
France, K., 69
Freytag, B., 91

Güdel, M., 455
Garcés, A., 139
Gaudi, B. S., 469
Getman, K. V., 441
Ghez, A. M., 147

Giampapa, M., 441
Gilliland, R. L., 167
Godet, O., 293
Golden, A., 219
Golenetskii, S., 293
Gomes, J., 379
Gómez-Pérez, N., 391

Hallinan, G., 147, 219
Harding, L. K., 219
Harper, G., 517
Harrison, T. E., 121
Hartigan, P., 53
Hawley, S. L., 197, 333, 441
Hebb, L., 505
Heinzel, P., 441
Heller, R., 391
Helling, C., 403
Hilton, E. J., 197
Hoffman, D. I., 121
Holtzman, J., 197
Homeier, D., 91, 339, 403
Hudson, H., 441
Hughes, W. J., 43
Hummel, C., 517

Ireland, M. J., 111
Irwin, J., 505
Ishii, M., 339
Israelian, G., 81
Ivezić, Ž., 371

Jackson, B., 391
Jackson, R., 505
Jaffe, D. T., 53
Jardine, M., 505
Jayawardhana, R., 187
Jeffers, S. V., 15, 245
Jenkins, J., 379
Johns-Krull, C., 285
Johns-Krull, C. M., 53, 245
Johnson, J. A., 531
Johnstone, C., 441
Jones, H. R. A., 339
Jordi, C., 99

Kürster, M., 313
Keller, C. U., 15, 245
Kennedy, G., 469
Kirkpatrick, J. D., 323

Kochukhov, O., 245
Konopacky, Q. M., 147
Kowalski, A. F., 197, 441
Kraus, A. L., 269
Krimm, H., 293

Lépine, S., 531
López-Morales, M., 121
Lafrenière, D., 187
Langer, N., 155
Laughlin, G., 469
Law, N. M., 269, 429
Lawson, W. A., 61
Leggett, S. K., 339
Lemonias, J. J., 269
Linsky, J. L., 69, 455
Liu, M., 31
Liu, M. C., 111, 429
Lockwood, S., 207
Loebman, S. R., 371
Loinard, L., 455
Looper, D., 481
Lucas, P. W., 339, 429
Luhman, K., 469
López-Morales, M., 391

Macintosh, B. A., 147
Mahmud, N., 53
Mainzer, A. K., 429
Makaganiuk, V., 245
Mallard, W., 207
Malo, L., 481
Mamajek, E. E., 481
Marocco, F., 339
Marsden, S., 415
Matt, S. P., 43
Mayor, M., 81
McGregor, S. L., 43
McLean, M., 455
Meadows, V. S., 391
Melandri, A., 293
Metchev, S., 481
Meyer, M. R., 361
Miesch, M. S., 277
Min, M., 15
Mistry, H., 207
Mohanty, S., 469
Morales, J. C., 99
Morales-Calderón, M., 415

Morin, J., 505
Mullan, D., 505
Murphy, S. J., 61
Muzzerolle, J., 415

Neilson, H. R., 155

Oates, S. R., 293
Osten, R., 455
Osten, R. A., 293, 441

Page, M. J., 293
Pal'shin, V., 293
Pandey, J. C., 301
Pascucci, I., 469
Patel, K., 379
Pedretti, E., 403, 517
Pinfield, D. J., 339, 379, 429
Piskunov, N., 245
Plavchan, P., 415, 469
Poppenhaeger, K., 493
Prato, L., 53
PTF Collaboration, 269
Pye, J., 293, 441

Quinn, T. R., 371

Radigan, J., 187
Randich, S., 81
Ravi, V., 207, 517
Reale, F., 293
Rebolo, R., 81
Rebull, L. M., 5
Reid, I. N., 505
Reiners, A., 255, 313
Ribas, I., 99, 139
Rice, E. L., 147, 481
Ridgway, S., 517
Roškar, R., 371
Robrade, J., 493
Rodenhuis, M., 15, 245
Rojas-Ayala, B., 531
Ruedas, T., 391

Santos, N. C., 81
Schmidt, S. J., 333
Schneider, P. C., 493
Schrijver, C. J., 231
Schweitzer, A., 531
Seemann, U., 313

Seifahrt, A., 313
Sheehan, B., 219
Shkolnik, E. L., 481
Showman, A. P., 403
Singh, K. P., 301
Skemer, A., 469
Smart, R., 339
Snik, F., 245
Sousa, S. G., 81
Stassun, K. G., 43, 505
Stempels, H. C., 245

Tamura, M., 339
Tanner, A., 391
Tennyson, J., 339
Tinney, C. G., 339
Toomre, J., 277
Townes, C., 207
Tueller, J., 293
Turner, N., 415

Udry, S., 81
Ule, N., 121

Valenti, J. A., 245
van Belle, G. T., 517
Vidotto, A. A., 403

Walkowicz, L. M., 177
Werthimer, D., 207
West, A. A., 333, 505, 531
White, R., 517
White, R. J., 147
Wishnow, E., 207
Wolk, S. J., 455

Yang, H., 69
Yurchenko, S. N., 339

Zhang, Z., 379
Zhang, Z. H., 429

Mt. Rainier (Top) and Tillicum Village totem pole (Bottom)

ASTRONOMICAL SOCIETY OF THE PACIFIC

THE ASTRONOMICAL SOCIETY OF THE PACIFIC is an international, nonprofit, scientific, and educational organization. Some 120 years ago, on a chilly February evening in San Francisco, astronomers from Lick Observatory and members of the Pacific Coast Amateur Photographic Association—fresh from viewing the New Year's Day total solar eclipse of 1889 a little to the north of the city—met to share pictures and experiences. Edward Holden, Lick's first director, complimented the amateurs on their service to science and proposed to continue the good fellowship through the founding of a Society "to advance the Science of Astronomy, and to diffuse information concerning it." The Astronomical Society of the Pacific (ASP) was born.

The ASP's purpose is to increase the understanding and appreciation of astronomy by engaging scientists, educators, enthusiasts, and the public to advance science and science literacy. The ASP has become the largest general astronomy society in the world, with members from over 70 nations.

The ASP's professional astronomer members are a key component of the Society. Their desire to share with the public the rich rewards of their work permits the ASP to act as a bridge, explaining the mysteries of the universe. For these members, the ASP publishes the Publications of the Astronomical Society of the Pacific (PASP), a well-respected monthly scientific journal. In 1988, Dr. Harold McNamara, the PASP editor at the time, founded the ASP Conference Series at Brigham Young University. The ASP Conference Series shares recent developments in astronomy and astrophysics with the professional astronomy community.

To learn how to join the ASP or to make a donation, please visit http://www.astrosociety.org.

ASTRONOMICAL SOCIETY OF THE PACIFIC
MONOGRAPH SERIES

Published by the Astronomical Society of the Pacific

The ASP Monograph series was established in 1995 to publish select reference titles. For electronic versions of ASP Monographs, please see
http://www.aspmonographs.org.

INFRARED ATLAS OF THE ARCTURUS SPECTRUM, 0.9-5.3μm
eds. Kenneth Hinkle, Lloyd Wallace, and William Livingston (1995)
ISBN: 1-886733-04-X, e-book ISBN: 978-1-58381-687-5

**VISIBLE AND NEAR INFRARED ATLAS
OF THE ARCTURUS SPECTRUM 3727-9300Å**
eds. Kenneth Hinkle, Lloyd Wallace, Jeff Valenti, and Dianne Harmer (2000)
ISBN: 1-58381-037-4, e-book ISBN: 978-1-58381-688-2

ULTRAVIOLET ATLAS OF THE ARCTURUS SPECTRUM 1150-3800Å
eds. Kenneth Hinkle, Lloyd Wallace, Jeff Valenti, and Thomas Ayres (2005)
ISBN: 1-58381-204-0, e-book ISBN: 978-1-58381-689-9

**HANDBOOK OF STAR FORMING REGIONS: VOLUME I
THE NORTHERN SKY**
ed. Bo Reipurth (2008)
ISBN: 978-1-58381-670-7, e-book ISBN: 978-1-58381-677-6

**HANDBOOK OF STAR FORMING REGIONS: VOLUME II
THE SOUTHERN SKY**
ed. Bo Reipurth (2008)
ISBN: 978-1-58381-671-4, e-book ISBN: 978-1-58381-678-3

A complete list and electronic versions of ASPCS volumes may be found at
http://www.aspbooks.org.

All book orders or inquiries concerning the ASP Conference Series, ASP Monographs, or International Astronomical Union Volumes published by the ASP should be directed to:

Astronomical Society of the Pacific
390 Ashton Avenue
San Francisco, CA 94112-1722 USA
Phone: 800-335-2624 (within the USA)
Phone: 415-337-2126
Fax: 415-337-5205
Email: service@astrosociety.org

For a complete list of ASP publications, please visit
http://www.astrosociety.org.